Chemistry in Context

A Project of the American Chemical Society

Fourth Edition

Chemistry in Context

Applying Chemistry to Society

Fourth Edition

Conrad L. Stanitski
University of Central Arkansas

Lucy Pryde Eubanks
Clemson University

Catherine H. Middlecamp
University of Wisconsin-Madison

Norbert J. Pienta
University of Iowa

A Project of the American Chemical Society

Boston Burr Ridge, IL Dubuque, IA Madison, WI New York San Francisco St. Louis
Bangkok Bogotá Caracas Kuala Lumpur Lisbon London Madrid Mexico City
Milan Montreal New Delhi Santiago Seoul Singapore Sydney Taipei Toronto

McGraw-Hill Higher Education

*A Division of The **McGraw-Hill** Companies*

CHEMISTRY IN CONTEXT: APPLYING CHEMISTRY TO SOCIETY
FOURTH EDITION

Published by McGraw-Hill, a business unit of The McGraw-Hill Companies, Inc., 1221Avenue of the Americas, New York, NY 10020.

This book is printed on recycled, acid-free paper containing 10% postconsumer waste.

International 1 2 3 4 5 6 7 8 9 0 QPD/QPD 0 9 8 7 6 5 4 3 2
Domestic 1 2 3 4 5 6 7 8 9 0 QPD/QPD 0 9 8 7 6 5 4 3 2

ISBN 0–07–241015–9
ISBN 0–07–115128–1 (ISE)

Publisher: *Kent A. Peterson*
Senior developmental editor: *Shirley R. Oberbroeckling*
Marketing manager: *Thomas D. Timp*
Senior project manager: *Vicki Krug*
Lead production supervisor: *Sandy Ludovissy*
Senior media project manager: *Stacy A. Patch*
Senior media technology producer: *Phillip Meek*
Design manager: *Stuart Paterson*
Cover designer: *Emily E. Feyen*
Interior designer: *Lloyd Lemna Design*
Cover image: *Corbis Images*
Lead photo research coordinator: *Carrie K. Burger*
Photo research: *Toni Michaels/PhotoFind LLC*
Supplement producer: *Brenda A. Ernzen*
Compositor: ***TECH****BOOKS*
Typeface: *10/12 Times Roman*
Printer: *Quebecor World Dubuque, IA*

The credits section for this book begins on page 543 and is considered an extension of the copyright page.

Library of Congress Cataloging-in-Publication Data

Chemistry in context : applying chemistry to society. — 4th ed. / Conrad L. Stanitski . . . [et al.].
p. cm.
Includes index.
ISBN 0–07–241015–9 (acid-free paper)
1. Biochemistry. 2. Environmental chemistry. 3. Geochemistry. I. Stanitski, Conrad L.

QD415 .C482 2003
540—dc21

2002067087
CIP

INTERNATIONAL EDITION ISBN 0–07–115128–1

www.mhhe.com

Brief Contents

Contents

Chapter 5

The Water We Drink 197

Chapter 6

Neutralizing the Threat of Acid Rain 243

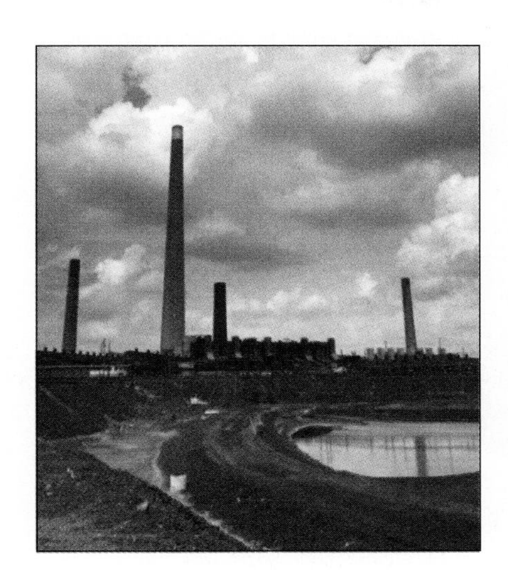

Chapter 7

The Fires of Nuclear Fission 283

Chapter 8

Energy from Electron Transfer 327

Chapter 9

The World of Plastics and Polymers 361

Chapter 10

Manipulating Molecules and Designing Drugs 393

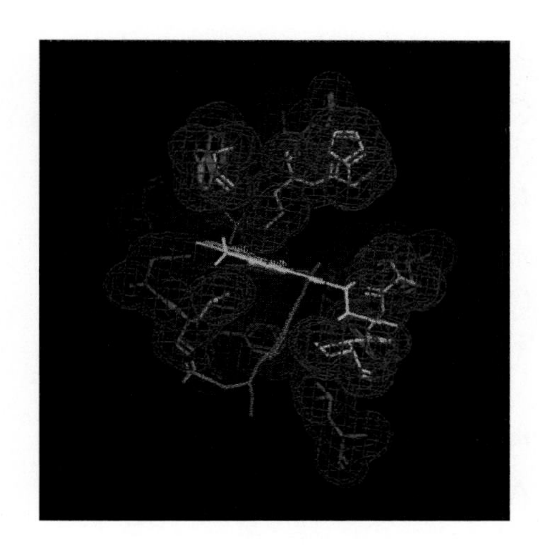

Preface

Following in the tradition of its first three editions, the goal of *Chemistry in Context,* fourth edition, is to establish chemical principles on a need-to-know basis within a contextual framework of significant social, political, economic, and ethical issues. We believe that by using this approach, students not majoring in a science develop critical thinking ability, the chemical knowledge and competence to better assess risks and benefits, and the skills that lead them to be able to make informed and reasonable decisions about technology-based issues. The word "context" derives from the Latin word meaning "to weave." Thus, the spider web motif on the cover, used for the first three editions, continues with this edition because a web exemplifies the complex connections between chemistry and society.

Chemistry in Context is not a traditional chemistry book for non-science majors. In this book, chemistry is woven into the web of life. The chapter titles of *Chemistry in Context* reflect today's technological issues and the chemistry principles imbedded within them. Global warming, alternate fuels, nutrition, and genetic engineering are examples of such issues. To understand and respond thoughtfully in an informed manner to these vitally important issues, students must know the chemical principles that underlie the socio-technological issues. This book presents those principles as needed, in a manner intended to better prepare students to be well-informed citizens.

Organization

The basic organization and premise remain the same as in previous editions. The focal point of each chapter is a real-world societal issue with significant chemical context. The first six chapters are core chapters in which basic chemical principles are introduced and expanded upon on a need-to-know basis. These six chapters provide a coherent strand of issues focusing on a single theme—the environment. Within them, a foundation of necessary chemical concepts is developed from which other chemical principles are derived in subsequent chapters. Chapters 7 and 8 consider alternate (non-fossil fuel) energy sources—nuclear power, fuel cells, batteries, and photovoltaics. The emphases in the remaining chapters are carbon-based issues and chemical principles related to polymers, drugs, nutrition, and genetic engineering. Thus, one-third of the text has an organic/biochemistry flavor. These latter chapters provide students with the opportunity to focus on additional interests beyond the core topics, as time permits. Most users teach seven to nine chapters in a typical one-semester course.

All content has been thoroughly updated. Tables, figures, and data are as up to date as possible using a printed format. Icons within the chapters direct the student to the Internet to get the latest information from the Web for answering questions and evaluating the information obtained from the Web.

Chapter 8 has been refocused to concentrate on electron transfers in chemical reactions, leading to an expansion of the coverage of fuel cells, including their latest designs, modern batteries, and photovoltaics.

Pedagogy

This text abounds in helpful pedagogy for students. The Chapter Overview, Conclusion, Chapter Summary, and Marginal Notes are study tools for the student. The Chapter Overview and Conclusion draw together the major themes at the beginning and end of each chapter. The Chapter Summary calls attention to the most important skills and applications developed in the chapter. Marginal Notes are used to succinctly summarize and emphasize key points or to link to sections in other chapters.

The in-chapter features—**Your Turn, Consider This,** and **Sceptical Chymist**—are activities to practice skills, to raise issues for thought, and to use critical thinking in extending and applying chemical principles. **Your Turn** activities provide opportunities for students to practice a skill or calculation that has been illustrated in the text. **Consider This** questions are decision-making activities requiring risk-benefit analysis, consideration of opposing viewpoints, speculation on the consequences of a particular action, or formulation and defense of a personal position. The **Sceptical Chymist** activities require analytical skills in response to various statements and assertions made in the popular media. These activities take their title, and its peculiar spelling, from an influential book written in 1661 by Robert Boyle, an important scientist and an early investigator of the properties of air.

Icons for Green Chemistry, the Web Exercises, and Figures Alives link special content in the text and the extension exercises on the *Chemistry in Context* Online Learning Center. Green chemistry is integrated within the text. The icon pulls the student attention to examples in which green chemical principles are applied. A web icon identifies each Consider This or Sceptical Chymist activity in which students should go to the Chemistry in Context Online Learning Center for further exploration. For the Figures Alive! interactives, the icon is adjacent to the text figure featured in the interactive. The icon signals the student to go to the Online Learning Center and learn by doing the interactive activities associated with the figure and its extensions.

Problem Solving

There are two main locations of problem-solving activities in *Chemistry in Context.*

The in-chapter problems are the Your Turn, Consider This, and Sceptical Chymist activities, as described earlier.

End-of-chapter problems are divided into three categories. *Emphasizing Essentials* are questions to practice and sharpen chemistry skills developed in the chapter. *Concentrating on Concepts* questions focus on chemical concepts and their relationships to the socio-technological topics under discussion. Questions in the *Exploring Extensions* category present a challenge to go beyond the textbook material by providing an opportunity to extend and integrate skills, concepts, and communication. The latter two categories of questions also incorporate the use of the Internet as a source of data and opinions. The questions with blue numbers are answered in Appendix 5.

Media

- *Web Exercises*—Continued from the third edition, the Web exercises use the Internet to answer various questions posed in many of the Consider This and Sceptical Chymist activities, as well as in chapter-end questions. Web-based exercises allow students to apply information to their own lives, use real-time data, get up-to-date information, and evaluate controversies using the Web. The Web presents students with the opportunity and responsibility to critically evaluate Web information among web sites of widely differing quality and validity. Many of the Web-based activities are linked to the *Chemistry in Context* web site (www.mhhe.com/physsci/chemistry/cic/).
- *Supplemental Readings*—These are listed on the *Chemistry in Context* web site rather than in the text so that updates can be provided in a timely way.
- **NEW!** *Figures Alive!*—These interactive materials in multimedia format are tied to a figure in each chapter. Each Figures Alive! interactive allows students to better understand the concepts in the figure by using interactive web exercises to develop this knowledge. The activities are based upon the same categories as the chapter-end problems—Emphasizing Essentials, Concentrating on Concepts, and Exploring Extensions.
- **NEW!** *Quiz Questions*—This set of questions allows students to test their knowledge of the material pre-

sented. Questions based on skills and concepts presented in the chapter give students a quick assessment of their knowledge and what they need to study further.

As with previous editions, a very detailed *Instructor's Resource Guide* (IRG) is available, compiled by Marcia Gillette. Unlike its predecessors, which were printed, the fourth edition's IRG is available only on the *Chemistry in Context* web site. For those whose course includes a laboratory component, a *Laboratory Manual,* compiled and edited by Wilmer Stratton, is available. The experiments use microscale equipment (wellplates and Beral-type pipets) and common materials. Cooperative/collaborative laboratory experiments are included.

It is always a pleasure to bring a new textbook or new edition to fruition. But the work is not done by just one individual. It is a team effort, one comprised of the work of many talented individuals. We have been fortunate to have the continuing, unstinting support of Sylvia Ware, Director of the ACS Division of Education and International Activities, who helped to create the first edition of *Chemistry in Context*. We also recognize the able assistance of Dr. Jerry Bell and Dr. Marta Gmurczyk of that ACS Education Division office. The McGraw-Hill team has been superb in all aspects of the project. Kent Peterson (Publisher) leads this outstanding team of Shirley Oberbroeckling (Senior Developmental Editor, Chemistry), Vicki Krug (Senior Project Manager), Phil Meek (Senior Media Producer), and Stacy Patch (Senior Media Project Manager), a team that does its job with the enviable combination of high quality and good humor.

The fourth edition is the product of a collaborative effort among writing team members—Catherine Middlecamp, Lucy Pryde Eubanks, Norbert Pienta, and Conrad Stanitski. This is the maiden voyage in this realm for Norbert Pienta, a new co-author and colleague. We have benefited from his diverse expertise.

We are very excited by the new features of this fourth edition, which exemplify how we continue to "press the envelope" to bring chemistry in creative, appropriate ways to non-science majors, while being honest to the science. We look forward to your comments.

Conrad Stanitski
Senior Author and Editor-in-Chief
conrads@mail.uca.edu
March 2002

Instructor Resources

Instructor's Resource Guide

The *Instructors Resource Guide,* edited by Marcia Gillette (Indiana University—Kokomo), can be found on the Online Learning Center under Instructor Resources. Included:

- A chemical topic matrix provides listing of chemical principles commonly covered.
- Course syllabi give some indication about the scope, pace, and scheduling of the course.
- Topical Essays give a variety of background material and pragmatic suggestions for teaching strategies and student development goals.
- Answers are given for suggested responses to many of the open-ended questions in the Consider This activities and the solutions to the in-chapter and chapter-end exercises and questions.
- Provides the instructors guide for the laboratory experiments.

Online Learning Center (www.mhhe.com/cic)

The Online Learning Center (OLC) is a comprehensive, book-specific web site offering excellent tools for both the instructor and the student. Instructors can create an interactive course with the integration of this site, and a secured Instructor Center stores their essential course materials to save prep time before class. This Instructor Center offers the Instructors Resource Guide. The Student Center offers Web Exercises, Figures Alive! interactives, and quiz questions. The Online Learning Center content has been created for use in PageOut, WebCT, and Blackboard course management systems.

Digital Content Manager

The Digital Content Manager is a multimedia collection of visual resources allowing instructors to utilize artwork from the text in multiple formats to create customized classroom presentations, visually based tests and quizzes, dynamic course web site content, and/or attractive printed support materials. The Digital Content Manager is a cross-platform CD containing an image library, photo library, and a table library.

Transparency Set

The transparency set contains selected four-color illustrations from the text reproduced on acetate for overhead projection.

Course Management Systems

PageOut is specifically designed to help you with your individual course needs. PageOut assists you in integrating your syllabus with *Chemistry in Context* and state-of-the-art new media tools. At the heart of PageOut you find integrated multimedia and a full-scale Online Learning Center.

The content from the Online Learning Center is available in WebCT and Blackboard on request to your sales representative.

Student Resources

Laboratory Manual

The laboratory manual, compiled and edited by Wilmer Stratton, includes experiments using microscale equipment and common materials. The experiments have been chosen and designed to reflect and amplify the contents of *Chemistry in Context*. Chemical information about the world around us can be obtained with simple chemical equipment and procedures.

Online Learning Center

The Online Learning Center (OLC) (www.mhhe.com/cic) is a comprehensive, exclusive web site that provides the student access to the Web-related activities in selected Consider This and Sceptical Chymist questions marked by the icon and the end-of-chapter questions marked by icon.

New to the Online Learning Center is the Figures Alive! interactives (marked by an icon near the figure in the text) that lead students through the discovery of various layers of knowledge inherent in the figure.

The web site also includes quizzing and other study tools for the students.

Acknowledgements

We would like to thank these individuals, whose comments were of great help to us in preparing this revision.

Paula Abel
McNeese State University
Susmita Acharya
Cardinal Stritch University
Bobby Adams
College of Alameda
Ibrahim Al-Ansari
University of Qatar
Monica Ali
Oxford College
Shawn Allin
Lamar University-Beaumont
David Anderson
University of Missouri
Karen Anderson
Madison Area Technical College
Steven Anderson
University of Wisconsin-Whitewater
Ray Baechler
Russell Sage College
Felicia Corsaro Barbieri
Gwynedd-Mercy College
John Bauman
University of Missouri-Columbia
Robert Bauman, Jr.
Amarillo College
Ronald Baumgarten
University of Illinois-Chicago
Elisabeth Bell-Loncella
University of Pittsburgh-Johnstown
David Bergbreiter
Texas A&M University-College Station
Wolfgang Bertsch
University of Alabama-Tuscaloosa
Rebecca Bilek
Muskingum College
Leah Blau
Stern College for Women
Pam Brown
New York City Technical College
Richard Bryant
Southwest Oklahoma State University-Weatherford
Paul Buckley
Hartwick College
Lisa Buller
University of Wisconsin-Platteville
Roy Burlington
Central Michigan University
Gertrude Busdiecker
Lansing Community College
Kabuika Butamina
Antioch College
Deborah Carey
Marywood University
Donald Cass
College of the Atlantic
Ralph Christensen
Southern Utah University
Jim Collier
Truckee Meadows Community College
A. W. (Wally) Cordes
University of Arkansas-Fayetteville
Philip Crawford
Southeast Missouri State University

Roger Crawford
University of Akron
Scott Davis
Mansfield University
Salim M. Diab
University of St. Francis-Joliet
Anthony Dribben
Mississippi College
William Dunn
University of Illinois-Chicago
Evelyn Erenrich
Rutgers University-New Brunswick
Jack Espinal
Park College
Ted Fickel
University of Judaism
K. Thomas Finley
State University College-Brockport
Douglas Flournoy
Indian Hills Community College-Ottumwa
Paula Getzin
Kean University
Marcia Gillette
Indiana University-Kokomo
Steve Goldberg
Adelphi University
Albert Gotch
Benedictine College-Atchison
Daniel Gregory
St. Cloud State University
Laura Hall
Washington State Community College
Mildred (Midge) Hall
Clark State Community College
Mary Handley
James Madison University
Milton Hanson
Augustana College
Alton Hassell
Baylor University
LeRoy Haynes
College of Wooster
Jonathan Heath
Horry Georgetown Technical College
C. E. Heltzel
Transylvania University
Bruce Heyen
Tabor College
Carol A. Higginbotham
Central Oregon Community College
Kathleen House
Illinois Wesleyan University
Judith M. Iriarte-Gross
Middle Tennessee State University
Warren Johnson
University of Wisconsin-Green Bay
Cindy Kepler
Bloomsburg University of Pennsylvania
Kevin Kolack
Cooper Union
Martha Kurtz
Central Washington University-Ellensburg
Brian Lamp
Truman State University
Raima Larter
Indiana University/Purdue University-Fort Wayne
Carol Lasko
Humboldt State University
Bernard Liburd
Guilford College
Robley Light
Florida State University-Tallahassee
William Loffredo
East Stroudsburg University
Brian Love
East Carolina University
Dahong Lu
Fitchburg State College
Kenneth Maloney
Baton Rouge Community College
Joseph Maloy
Seton Hall University
Stanley Manahan
University of Missouri-Columbia
Paul Marshall
University of North Texas
Albert Martin
Moravian College
Kenneth Marx
University of Massachusetts-Lowell
Eugene Mash Jr.
University of Arizona-Tucson
Mark Masthay
Murray State University
Martha McBride
Norwich University
Julie T. Millard
Colby College
Ray Miller
York College
Carl Minnier
Essex Community College
Susan Morgan
Southern Illinois University-Edwardsville

David Reed Myers
Simons Rock College
Alexander Nazarenko
Southern Illinois University-Carbondale
Raphael M. Ottenbrite
Virginia Commonwealth University
Charlotte Ovechka
University of St. Thomas
Linda Pallack
Washington-Jefferson College
Gus Palenik
University of Florida
Yasmin Patell
Kansas State University
Julie Peller
Indiana University-Northwest Gary
Richard Peterson
Northern State University
Shawn Phillips
Vanderbilt University
Robert Pike
College of William and Mary
Neil Potter
Susquehanna University
Jill Rawlings
Auburn University-Montgomery
Margaret Rempe
Seattle University
David Robertson
University of Missouri-Columbia
Kay Rowberg
Purdue University-Calumet-Hammond
Lynette Rushton
South Puget Sound Community College
Richard Scamehorn
Ripon College
Russ Selzer
Western Connecticut State University
Susan Shadle
Boise State University
Ike Shibley
Pennsylvenia State University-Berks-Lehigh Valley
Phil Silberman
Scottsdale Community College
JoElla Siuda
Illinois Institute of Art
Wayne Smith
Colby College
William Smith
Grand Valley State University
Steven M. Socol
McHenry County College
Kathryn Springsteen
Colby-Sawyer College
Wayne Stalick
George Mason University
Gail Steehler
Roanoke College
Karen Stevens
Whitworth College
James Streator
Manchester College
Margaret Suerth
Illinois Valley Community College
Keith Symcox
University of Tulsa
Erach Talaty
Wichita State University
Agnes Tenney
University of Portland
Suresh Tewani
Long Island University-Brooklyn
Mark Thomson
Xavier University
Gary Trammell
University of Illinois-Springfield
Tod Treat
Parkland College
Eric Trump
Emporia State University
Janet Truttmann
Pomona College
Michael Vaksman
University of Wisconsin-Superior
John B. Vincent
University of Alabama-Tuscaloosa
William Voige
James Madison University
Margorie Welch
Southwestern Iowa Community College
Bruce Wilcox
Bloomsburg University of Pennsylvania
Donald Williams
Hope College
Gary Wood
Valdosta State University
Robert Zumwalt
Westminster College
Lisa Zuraw
The Citadel

Guided Tour

Real-Life Applications

The material is presented to students in a manner that demonstrates how chemistry actually impacts their lives. The focal point of each chapter is a real-word societal issue with significant chemical context.

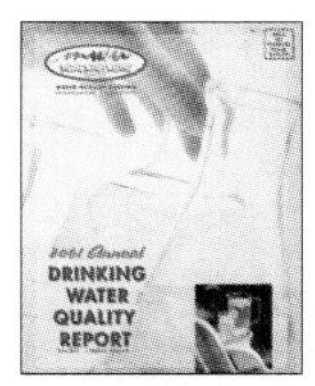

A "Right-to-Know" Report about a community's water supply.

So begins an article entitled "On Tap or Bottled, Pursuing Purer Water" that appeared in the personal health column of the *New York Times.* Similar questions about water quality may be raised about your water supply, even about the water supply for your college or university. You may even have received a Right-To-Know Report, also called a Consumer Confidence Report. The Safe Drinking Water Act as amended in 1996 mandates that such reports be delivered once a year to consumers of water from community water systems, public or private. As a person studying chemistry, you are in a good position to understand the meaning of the measurements being reported and the standards of quality that must be met. You might be asked to help friends or family understand the measurements and conclusions reached in a report they might receive about their water quality.

Such reports raise a number of questions that we address in this chapter. What is a regulated chemical and what are its maximum levels allowed in drinking water? How do such chemicals get into drinking water? Should we be concerned about any of them? Who establishes the rules? What are the major provisions of the Safe Drinking Water Act of 1996? What options are there for alternative sources of drinking water? In particular, why is bottled water cited as the most likely alternative? Are all water filtration systems equally effective?

Water is, indeed, a special compound. In spite of its commonness, this colorless liquid called water is amazing stuff. The noted anthropologist and essayist Loren Eiseley speaks poetically of the wonders of it: "If there is magic on this planet, it is contained in water . . . Its substance reaches everywhere; it touches the past and prepares the future; it moves under the poles and wanders thinly in the heights of the air." Arguably, water is the most important chemical compound on the face of the Earth. In fact, it covers about 70% of that face, giving the planet the lovely blue color in the famous "blue marble" photos taken from outer space by astronauts. Water is essential to all living species; without it humans would die within a week. Our bodies are approximately 60% water, blood is at least 50% water, and a human brain is an astonishing 77% water! Water is so important to life as we know it that speculation about life elsewhere in the universe hinges first and foremost on the availability of water. Water refreshes us, dominates weather systems, provides for many types of recreation on and in it, and even gives us aesthetic and relaxing pleasures.

• See the opening of Chapter 1 for a "Blue Marble" photo.

CHAPTER OVERVIEW

Although we generally take water for granted, it is a remarkable chemical compound with unique properties that account for its essential life-supporting role. In this chapter, we will consider water from the perspective of those who drink it. First a question of aesthetics: What makes a glass of water pleasing to the eye and to the palate? There is more to water, however, than can be seen or tasted. Unseen impurities in water, depending on their identities and amounts, can impart a crisp, fresh taste or produce an unpleasant illness. And so, we next look at water as a solvent and at some of the things that may be dissolved in drinking water. How much of a substance dissolves in water makes concentration an important part of the story. The concentrations of substances dissolved in water can be expressed in several ways, including descriptions of the extremely low concentrations. To better comprehend aqueous solutions and why some sub-

In-Chapter Features

Your Turn, Consider This, and *Sceptical Chymist* are activities to practice skills, to raise issues for thought, and to use critical thinking in extending and applying chemical principles. *Your Turn* activities provide opportunities for students to practice a skill or calculation that has been illustrated in the text. *Consider This* questions are decision-making activities requiring risk-benefit analysis, consideration of opposing viewpoints, speculation on the consequences of a particular action, or formulation and defense of a personal position. The *Sceptical Chymist* activities require analytical skills in response to various statements and assertions made in the popular media.

If more high-energy photons reach the surface of the Earth from the Sun, the potential for significant biological damage increases. All evidence shows that the average stratospheric ozone concentration has dropped significantly in the last 20 to 30 years. This phenomenon has been documented in many regions of the world, using data gathered from high-flying aircraft, ground-based systems, and satellites. Later in this chapter we will explore the special reasons why the observed percent ozone depletion is the greatest in the Antarctic. However, stratospheric ozone depletion means that the ability of the atmosphere to screen out UV radiation with wavelengths below 320 nm has decreased. Although this has happened to varying extents in different regions, living things are now exposed to greater intensities of potentially damaging radiation. Scientists have made calculations predicting that a given percent decrease in stratospheric ozone will increase the flux of biologically damaging UV radiation by twice that percentage. For example, a 6% decrease in stratospheric ozone could mean a 12% rise in skin cancer, especially the more easily treated form, non-melanoma skin cancers such as basal cell and squamous cell cancers. This condition is considerably more common among Caucasians than among those with more heavily pigmented skin. People of African and Indian origin are better equipped to withstand the high levels of UV radiation in the intense sunlight that strikes the Earth near the equator.

There is good evidence linking the incidence of non-melanoma skin cancers, the intensity of UV radiation, and latitude. For example, the disease generally becomes more prevalent as one moves farther south in the Northern Hemisphere (Figure 2.11).

2.17 Consider This: Geography of Skin Cancer

Many generations of immigrants have come to the United States. There are fair-skinned Northern Europeans, for example, who have settled in the area around San Antonio, TX; there are many other equally fair-skinned immigrants who have settled in the area around Seattle. Based on Figure 2.11, compare their relative risks of developing skin cancer. Identify several other factors that may affect the risk for any individual in these two populations.

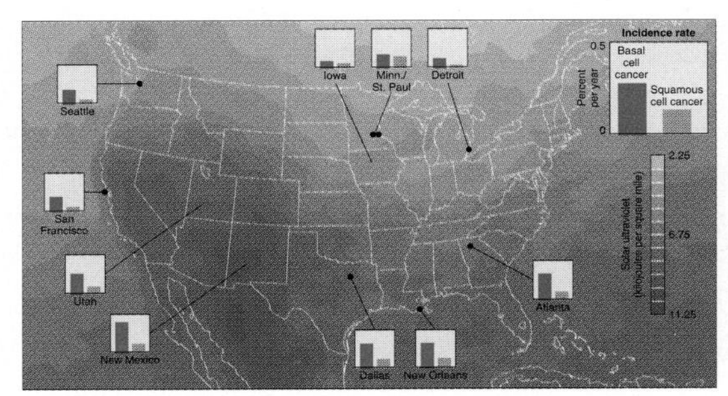

Figure 2.11
Skin cancer risks in selected U.S. locations.

Problem Solving

There are two main locations of problem-solving activities in *Chemistry in Context.* In-chapter problem solving is located in the *Your Turn, Consider This,* and *Sceptical Chymist* activities. End-of-chapter questions are divided into three categories—Emphasizing Essentials, Concentrating on Concepts, and Exploring Extensions.

tribute to an increase in the acidity of rain. Offer a possible explanation for this observation.

b. In Chapter 2, stratospheric ice crystals in the Antarctic were involved in the cycle leading to the destruction of ozone. Is this effect related to the observations in part **a**? Why or why not?

38. a. Several strategies to reduce SO_2 emissions are described in the text. The most effective ones in the last 10 years have been coal switching and stack gas scrubbing. Prepare a list of the advantages and disadvantages associated with each of these methods.

b. Explain why coal cleaning has not been an effective strategy.

39. Discuss the validity of the statement, "Photochemical smog is a local problem, acid rain is a regional one, and the enhanced greenhouse effect is a global one." Describe the chemistry behind each of these air quality problems and explain why the problems affect different geographical areas.

Exploring Extensions

40. The text makes this statement. "By trying to use a technological fix for one problem, we inadvertently create another."

a. Explain how the problems associated with acid rain fit this statement.

b. Pick another example from any of the issues explored in Chapters 1–5. Briefly explain how your choice fits the statement as well.

41. Here are two substances that both contain OH in their chemical formulas. Explain why you cannot write an ... o equation 6.3 for either one.

... Consult a solubility table.

... Consider bond energy and the nature ...

... you listed ions present in aqueous so... bases, and common salts. In question ...lecular substances to the list. To quan...

... molar concentration of all molecular ...ies in a 1.0 M solution of NaOH.

... molar concentration of all molecular ...ies in a 1.0 M solution of HCl.

...f the Electric Power Research Institute, ...tation in a workshop to establish re...nd criteria on factors that govern pre...ry, made this statement.

...ol strategy is hit upon is successful in ... in half, an evil conspiracy of chemists ... pH of precipitation to increase by 0.3."

...ymist in attendance at this worksho... ...espond to this statement? Explain t... ...esponse. *Hint:* See Appendix 3.

...es that energy must be added to get N... and O_2 to react to form NO. A Sceptical Chymist wants to check this assertion and determine how much energy is required. Show the Sceptical Chymist how this can be done. *Hint:* Draw the Lewis structures for the reactants and products and then check Table 4.1 for bond energies.

45. The text describes a green chemistry solution to reducing NO emissions for glass manufacturers.

a. Identify the strategy.

b. Use the Web to research what other industries might use this green chemistry strategy. Write a report to summarize your findings.

46. Many things have been suggested to help reduce acid rain, and some examples of what an individual can do are given here. For each item, explain the connection between what you would or wouldn't do and the generation of acid rain:

a. Hang your laundry to dry it.

b. Walk, ride a bicycle, or take public transportation to work.

c. Run the dishwasher and washing machine only with full loads.

d. Add additional insulation on hot water heaters and hot water pipes.

e. Buy locally produced food and other items.

47. How do researchers determine whether the negative effects of acid deposition on aquatic life are a direct consequence of low pH or the result of Al^{3+} released from rocks and soil? Find at least one article that gives the details of such a study. In your own words write a summary of the experimental plan and its results.

48. One way to compare the acid neutralizing capacity (ANC) of different substances is to calculate the mass of the substance required to neutralize one mole of hydrogen ion, H^+.

a. Write a balanced equation for the reaction of $NaHCO_3$ with H^+, and use it to calculate the ANC for $NaHCO_3$.

b. Determine the cost to neutralize one mole of H^+ with $NaHCO_3$ if $NaHCO_3$ costs \$9.50 per kilogram.

49. Why are developing countries likely to emit an increasingly higher percentage of the global amount of SO_2? Pick a nation, research its current emissions of SO_2 and calculate its percentage of global emissions. Speculate on whether increasing emissions are likely to continue in the future and offer an explanation for your prediction.

50. Like diesel trucks, sport utility vehicles (SUVs) emit more than their share of pollutants. Do SUVs emit NO_x, SO_2, or both? What are the current proposals to clean up their emissions? Use the resources of the ...

4.31 Consider This: The Price of Gasoline

Oil is a valuable resource, even beyond its use for home heating and gasoline. If we continue to use our petroleum supply to extract gasoline from it, we may lose our starting materials to make other petroleum-based products, such as many pharmaceuticals and plastics. Up to now, voluntary conservation of gasoline has not been effective. The government could force more conservation by rationing gasoline or by heavily taxing it. If our government increased the price of gasoline to \$4.00 per gallon (a price typical of that in western Europe and Japan), sales of gasoline likely would drop. This price would have serious consequences for the American work force because the price of a gallon of gasoline would be raised compared to the current minimum hourly wage.

Suppose a bill has been introduced in Congress to raise the price of gasoline to \$4.00 per gallon. Draft a letter to a friend at another college, either supporting or protesting the bill. Include the reasons for your position and your opinion on what should be done with this new revenue if the bill is passed.

CONCLUSION

To a considerable extent, choice ultimately influences what technology can do to conserve energy. As individuals and as a society, we must decide what sacrifices we are willing to make in speed, comfort, and convenience for the sake of our dwindling fuel supplies and the good of the planet. The costs might include higher taxes, more expensive gasoline and electricity, fewer and slower cars, warmer buildings in summer and cooler ones in winter, perhaps even drastically redesigned homes and cities. During the 1970s a series of energy crises occurred because of a dramatic rise in the cost of imported crude oil, principally from the Middle East. Although we have broadened our sources of imported crude oil to others than solely the Middle East, our reliance on imported fossil fuels remains high, regardless of their sources. This ongoing dependence keeps alive the specter of whether supply and demand factors for crude oil could precipitate another energy crisis, perhaps on a global basis. One thing seems to be clear. The best time to examine our options, our priorities, and our will is before we face another full-blown energy crisis. Quite obviously energy, chemistry, and society are closely intertwined. This chapter is an attempt to untangle them.

Chapter Summary

Having studied this chapter, you should be able to:

- Distinguish between energy and heat, and be able to convert among energy units (joules, kilocalories, Calories)
- Describe the factors related to the United States' dependency on fossil fuels for energy (4.2)
- Apply the terms exothermic, endothermic, and activation energy to chemical systems (4.3–4.5)
- Interpret chemical equations and basic thermodynamic relations to calculate heats of reaction, particularly heats of combustion (4.4)
- Use bond energies to describe the energy content of materials (4.4)
- Evaluate the risks and benefits associated with petroleum, coal, and natural gas as fossil fuel energy sources (4.7–4.8)
- Relate energy use to atmospheric pollution and global warming (4.7, 4.8)
- Understand the physical and chemical principles associated with petroleum refining (4.8–4.9)
- Describe "octane rating" and how refining, leaded gasoline, ethanol, and MTBE relate to it (4.9)
- Discuss approaches to alternative (supplemental) automobile fuels (4.9–4.11)
- Describe why reformulated and oxygenated gasolines are used (4.10)
- Relate the energy potentially available from a process with the efficiency of that process (4.13)
- Use entropy as a concept to explain the second law of thermodynamics (4.15)
- Take an informed stand on what energy conservation measures are likely to produce the greatest energy savings (4.16)
- With confidence, examine news articles on energy crises and energy conservation measures to interpret the accuracy of such reports (4.16)

Ethanol: H–C(H)(H)–C(H)(H)–Ö–H   Ethylene glycol: H–Ö–C(H)(H)–C(H)(H)–Ö–H

Figure 5.16
Structures of ethanol and ethylene glycol.

Ethanol

Figure 5.17
Hydrogen bonding of ethanol with water.
— covalent bond
– – – hydrogen bond.

water and ethanol to have a great affinity for each other, a conclusion consistent with the fact that they form solutions in all proportions. Ethylene glycol is also an alcohol with two —OH groups available for hydrogen bonding with H_2O. Therefore, ethylene glycol is highly water soluble, a necessary property for an antifreeze ingredient.

5.20 Your Turn

Sketch a diagram to show hydrogen bonding between ethylene glycol and water.

Finally, we consider sucrose, the compound that introduced this section. Examination of its structure (Figure 5.15) discloses that the sucrose molecule contains eight —OH groups and three additional oxygen atoms that can also participate in hydrogen bonding. This accounts for the high solubility of sugar in water.

5.21 Consider This: Three-Dimensional Representations of Molecules

Three-dimensional representations of molecules can be viewed on the Web using CHIME, a free plug-in that you can download and install. Three-dimensional representations of ethanol, ethylene glycol, and sucrose are there. Use these molecular representations to identify the places in each compound where hydrogen bonding occurs. Has your mental picture of these molecules changed after seeing these 3-D representations? Explain.

On the other hand, molecular compounds that differ in composition and molecular structure do not attract each other strongly. It has often been observed that "oil and water don't mix." They don't mix because they are structurally very different; unlike compounds don't like each other, which is just the reverse of "like dissolves like." Water is a highly polar compound, whereas oil consists of nonpolar hydrocarbon compounds. When placed in contact, they remain apart in separate layers (Figure 5.18). Even if shaken vigorously, the oil and water return to their own layers. But oily, nonpolar compounds generally dissolve readily in hydrocarbons or chlorinated hydrocarbons. For this reason, the latter compounds have often been used in dry cleaning solvents.

The tendency of nonpolar compounds to mix with other nonpolar substances affects how fish and animals store certain highly toxic substances such as PCBs (polychlorinatedbiphenyls) or the pesticide DDT. PCB and DDT molecules are nonpolar, and so when fish absorb them from water, the molecules are stored in body fat (which is also nonpolar) rather than in the blood (which is a highly polar water solution).

Solvents used to dry clean clothes are usually chlorinated compounds such as tetrachloroethylene, C_2Cl_4, also known as "perc" (perchlorinated ethylene), which is a human carcinogen. These materials also have serious environmental consequences. Dr. Joe DeSimone of the University of North Carolina-Chapel Hill has discovered a substitute for chlorinated compounds by synthesizing cleaning detergents that work in liquid carbon dioxide. The key to the process are the detergents, whose molecules are designed so that one end of the molecule is soluble in nonpolar substances like grease and oil stains, while the other end dissolves in the liquid CO_2. The new process recycles carbon dioxide produced as a waste product from industrial processes. Replacing large volumes of "perc" by using recycled CO_2 reduces the negative impact of "perc" on the workplace and the environment. The breakthrough process is paving the way for designing replacements for conventional halogenated solvents currently used

Oil
Water

Figure 5.18
Oil and water do not dissolve in each other.

Pedagogy

Chapter Overview, Conclusion, Chapter Summary, and Marginal Notes are study tools for the students. The Chapter Overview and Conclusion draw together the major themes at the beginning and end of each chapter. The Chapter Summary calls attention to the most important skills and applications developed in the chapter. Marginal Notes are used to succinctly summarize and emphasize key points or to link to sections in other chapters.

Green Chemistry

Green chemistry principles are introduced and used repeatedly to demonstrate creative contemporary chemical approaches that minimize and prevent generation of hazardous substances. An icon identifies areas within the text where green chemistry principles are discussed.

Protecting the Ozone Layer **75**

in Antarctica first observed it in 1985, they thought their instruments were malfunctioning. The area covered by ozone levels less than 220 DU, defined as the "ozone hole," was larger in early September 2000, than in any previous year. Total ozone destruction in the hole occurred from an altitude of 15 km to 20 km, consistent with measurements taken in recent years. The minimum total ozone value of 98 DU was not the lowest ever recorded. In early October 1993, the total ozone there dropped to 86 DU, the lowest recorded anywhere in the world in 36 years of measurement, less than 30% of the 1956 value. Keep in mind that there has always been a seasonal variation in ozone concentration over the South Pole, with a minimum in late September or early October, the Antarctic spring. What is unprecedented is the dramatic decrease in this minimum that has been observed over the 40 years.

- The reasons for the Antarctic ozone hole are discussed in Section 2.14.

More importantly, the stratospheric ozone concentrations are lower than those predicted using the simple Chapman cycle mechanism. That by itself is neither cause for alarm nor proof that the lower stratospheric ozone concentrations are the consequence of human intervention. In fact, many of the factors influencing the stratospheric ozone layer are natural in origin. We know that the processes establishing the steady-state concentration of stratospheric ozone are more complicated than originally believed.

For one thing, the natural concentration of stratospheric ozone is not uniform over all parts of the globe. On average, the total O_3 concentration increases the closer one gets to either pole (with the exception of the seasonal "hole" over the Antarctic). The formation of ozone via steps 1 and 2 of the Chapman cycle (p. 71) is triggered when an O_2 molecule absorbs a photon of ultraviolet light. Therefore, ozone production increases with the intensity of the radiation striking the stratosphere, an intensity that is not constant. Intensity varies with the seasons, reaching its maximum (in the Northern Hemisphere) in March and its minimum in October (just the reverse of the Southern Hemisphere). Consequently, stratospheric ozone concentrations also follow this seasonal pattern. In addition, the amount of radiation emitted by the Sun changes over an 11- to 12-year cycle related to sunspot activity. This variation also influences O_3 concentrations, but only by 1–2%. The winds blowing through the stratosphere cause other variations in ozone concentrations, some on a seasonal basis and others over a 28-month cycle. To further complicate matters, seemingly random fluctuations often occur. Finally, it is well established that certain gases from both natural and human-made sources are also responsible for the destruction of stratospheric ozone.

Dramatic TOMS images, such as the one that opens this chapter, are color coded to show stratospheric ozone concentrations in Dobson Units. The violet and purple regions are those where the greatest destruction of O_3 occurred, that is, lowest ozone concentrations. Since the early to mid-1990s, the size of the depleted ozone region annually equals nearly the total area of the North American continent, in some cases exceeding it. The next Consider This gives you the opportunity to examine a series of these images, each taken annually in October.

2.21 Consider This: Purple Octobers

NASA satellites provide stratospheric ozone data over time that can be tabulated in a number of ways, including global images, Antarctica ozone minima, and size of the ozone hole. All three are provided at the *Chemistry in Context* web site.

a. Using the web site, look first at the global images centered on Antarctica. Describe what is happening with the passage of time.
b. Now look at the graphs that show the minimum ozone levels and the size of the region affected. What information does each plot give you?
c. Use the information from all three views to write a description of what is meant by the term "ozone hole." In your statement, include references to the region of the globe, area affected, amount of ozone, and time.

The problem of ozone depletion is not limited to Antarctica. As another example, consider Figure 2.17. Scientists have been gathering data in Arosa, Switzerland, for more than 70 years, including satellite measurements started in the early 1980s. This figure shows that ozone levels in this Northern Hemisphere location, although lower than

Web-based Activities

Web-based activities are built into the in-chapter and chapter-end questions. Students can explore a variety of issues, including evaluating the accuracy of various web sites. The Web icon marks the activities and questions.

Online Learning Center

The *Chemistry in Context* Online Learning Center is a vital component of *Chemistry in Context* and contains an abundance of information for the student and the instructor. Students use Web exercises and Figures Alive! interactives that are integrated into the text. New to this edition is the addition of quiz questions for student self-assessment.

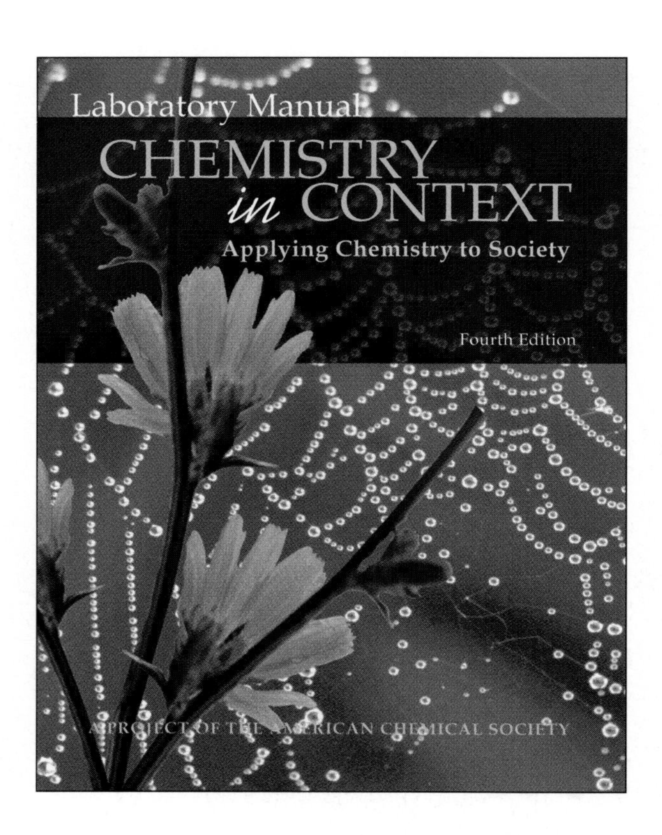

Instructor's Resource Manual and the Laboratory Manual

The new editions of the detailed Instructor's Resource Guide (IRG) and the Laboratory Manual are closely coordinated with the text. The IRG is available only on the Web.

Instructor Media

- Online Learning Center (OLC) is a secure, book-specific web site. The OLC is the doorway to a library of resources for instructors. The instructor will find the Instructor's Resource Guide and additional laboratory experiments in the Instructor Center.
- The Digital Content Manager CD-ROM is an instructor tool containing figures, photos, and tables from the text to use in PowerPoint presentations.
- Course Management Systems—PageOut, WebCT, and Blackboard—are available.

Digital Content Manager

Chemistry in Context
Applying Chemistry to Society, 4/e
American Chemical Society

CD 1 of 1

SYSTEM REQUIREMENTS

Adobe Acrobat Reader 4.0 or higher.
Windows: Windows NT, Windows 2000, Windows ME, Windows XP, Pentium 233 MHz or better, 128 MB RAM, 800 x 600 x 16 bit color, 32 bit preferred, Sound card and speakers, 8x CD-ROM drive, Mouse.
Macintosh: Power Mac OS X (Classic 9) or above, 200 MHz or faster recommended. 128 MB RAM, 800 x 600 x 16 bit color, 32 bit preferred, 8x CD-ROM drive, Mouse.

Quickstart Instructions:
If the CD doesn't launch automatically:

Windows:
1. Click the Windows **Start** button.
2. Choose **Run.**
3. Type the letter of your CD drive followed by **:\Digital-Content-Manager.exe.**
4. Click **OK.**

Macintosh:
1. Double click on the file named **DCM** on the CD.

For further information, see the readme, faq, and links_to_installers files located in the "How to use the CD" folder.

MADE WITH macromedia

Mc Graw Hill

ISBN 0-07-282902-8

1

The Air We Breathe

The "blue marble," our Earth, as seen from outer space.

"The first day or so, we all pointed to our countries. The third or fourth day, we were pointing to our continents. By the fifth day, we were aware of only one Earth."

Prince Sultan Bin Salmon Al-Saud, Saudi Arabian astronaut

Figure 1.1
The Great Lakes (Lake Superior upper left); image taken by the SeaWiFS spacecraft.

Individually and collectively, we take the air we breathe for granted. Yet our atmosphere is a fragile, thin veil of essential gases interspersed with various pollutants in differing amounts. It surrounds the third planet from the Sun, helping to make habitable the place we call home. The striking words of astronaut James Erwin compel us to consider the awesome spectacle of our home planet: "Finally it shrank to the size of a marble, the most beautiful marble anyone can imagine." Only a few women and men have actually observed what James Erwin saw in May 1969, but most of us have seen the spectacular photographs of the Earth taken from outer space. From that vantage point, our planet looks magnificent—a blue and white ball compounded of water, earth, air, and fire. It is where thousands upon thousands of species of plants and animals live, all interrelated in a global community. More than six billion of us belong to one particular species with special responsibilities for the protection of our beautiful "blue marble."

As we move in from outer space, a closer aerial view of Earth from a satellite reveals more detail about the "blue marble." On closer examination we see mountains, forests, deserts, prairies, glaciers, jungles, lakes, and rivers (Figure 1.1). In some cases, these natural features are boundaries that, for better or worse, separate our kind into national communities. The landforms visible in computer-enhanced photographs remind us of the great geological diversity of our planet and the many biological species that have adapted to these varied environments.

Figure 1.2
A crowd at a festival in Los Angeles.

At ground level we arrive at the communities that we know best, those closest to home—the cities, towns, ranches, and farms where we live, study, play, work, and sleep. Our families, friends, and neighbors are here (Figure 1.2). The people, customs, habits, and laws that create and constitute these regional environments shape each one of us.

As individuals, we simultaneously inhabit these concentric communities. Our personal lives are imbedded not just in our immediate surroundings, but also the countries we live in, and the entire globe. Changes in any of these environments affect us and we, in turn, have obligations at each level of community, from personal to global. This book is about some of those responsibilities and the ways in which a knowledge of chemistry can help us meet them with intelligence, understanding, and wisdom.

CHAPTER OVERVIEW

To be an informed (and healthy) citizen, you should know about the air you breathe. What is in air and the quality of it are essential for your existence, so are those things that might endanger it. Because air is a complex mixture of substances, knowing about it requires familiarity with certain fundamental chemical facts and concepts. Therefore, this chapter begins by considering the chemical composition of air, its major components and minor constituents including pollutants, and how their presence is expressed. Although pollutants are present, there is good news about the dramatic improvement in U.S. air quality over the past 30 years. The matter of air quality raises a theme that will frequently recur in this text—the difficult challenge of risk assessment.

In addition, you will learn about the structure of the atmosphere, and encounter the ways in which chemists categorize matter into elements and compounds and the symbols and formulas used to represent these substances. A consideration of atoms and molecules then provides a submicroscopic view of matter that helps further our comprehension. Elements and compounds, atoms and molecules undergo an amazing range of transformations. Such chemical reactions are at the very heart of chemistry, and we will focus on a few of the more important reactions that occur in the atmosphere. These, in turn, will provide an opportunity to explore the powerful shorthand of the chemist—chemical equations. Thus prepared, we will examine reasons for the significant enhancement in the quality of air in the U.S. over the past 30 years and consider the sources of the most important air pollutants in the United States and abroad. Also included is a brief discussion of indoor air quality. The chapter ends by reexamining, at the molecular level, the breath that initiated it.

1.1 Take a Breath

We begin by asking you to do something you do automatically and unconsciously thousands of times each day—take a breath. You certainly do not need textbook authors to tell you to breathe! A doctor may have encouraged your first breath with a well-placed slap, but from then on nature took over. Childish threats to hold your breath forever probably did not worry your parents. They knew that within a minute or two you would involuntarily gasp a lungful of that invisible stuff we call air. Indeed, you could not survive more than 5–10 minutes without a fresh supply of air.

1.1 Consider This: Take a Breath

One breath does not use much air, but what total volume of air do you exhale in a typical day? One approach to this question is simply to guess, but an educated guess is more reliable than a wild one. By designing and executing a simple experiment, you can come up with a reasonably accurate answer. You will need to determine how much air you exhale in a single breath and how many breaths you take per minute. Once you establish this information, determine how much air you exhale in a day (24 hours). Describe the experiment you performed, the data you obtained, and any factors that you can identify that affect the accuracy of your answer.

1.1 Consider This addresses how much air you breathe, but not the equally important topic of *what* you breathe, and whether it might be harmful. For a commentary on air quality we turn to this statement from a Shakespearean play about a troubled young

Table 1.1

Percent Changes in Air Quality 1980–1999*

Pollutant	Percent Decrease
Carbon monoxide	57
Nitrogen oxides	25
Ozone	20
Sulfur dioxide	50
PM	**
Lead	94

Source: EPA Annual Report "Latest Findings on National Air Quality: 1999 Status and Trends."

*These are overall percent changes. Changes for a particular urban area may differ.

**1990–1999 PM_{10} standards were first set in 1990.

- At ground level, ozone is an air pollutant. At high altitudes, ozone is beneficial, as you will find out in Chapter 2.
- Particulate matter (PM) is discussed in detail in Section 1.4.

man, Hamlet: "This most excellent canopy, the air, look you, this brave o'erhanging firmament, this majestical roof fretted with golden fire, why, it appears no other thing to me but a foul and pestilent congregation of vapors." To be sure, the speaker had a lot on his mind, especially the allegation that his uncle killed Hamlet's father before marrying his mother. Hamlet spends the rest of the play trying to decide what to do about his dysfunctional family, and no further reference to air pollution is made—except perhaps in the observation that "Something is rotten in the state of Denmark."

Actually, the air is probably worse in Los Angeles, Mexico City, or Bangkok than in Elsinore (Hamlet's home) or nearby Copenhagen. But, wherever you live, there is a good chance that the lungful of air you just inhaled contains some substances that, depending on their amount, could be harmful to your health. The health threat can be so serious that laws are passed in an effort to limit pollution by curtailing some of the ways we normally do things. Unfortunately, it is nearly impossible to completely avoid polluted air or to remove all pollutants from it. But governmental actions or proposed actions, chemical research, and the natural regenerative properties of the atmosphere have improved air quality in many parts of the world, including the United States, over the past two decades. In this country, the passage of the Clean Air Act in 1970 set into motion a federal mandate to reduce air pollution and improve air quality, and established national air quality standards. The dramatic results are shown in Table 1.1, where the percent decreases in six major pollutants for the period 1980–1999 are given. These six pollutants serve as the criteria for air quality. Four of the six pollutants are atmospheric gases—carbon monoxide, nitrogen oxides, ozone, and sulfur dioxide. The other two pollutants—particulate matter (PM) and lead—are solids. Later in the chapter (Section 1.11) we revisit the improvement of air quality.

1.2 Consider This: A Visit to the EPA

The U.S. Environmental Protection Agency (EPA) maintains an extensive web site providing resources for both scientists and the general public. Go to the EPA Air and Radiation web site, which contains a number of "consumer-friendly" documents on air quality. Explore it for a few minutes and provide this information about a document you found at the web site.

Title of EPA document
URL of document
Summary of contents
Last updated (if provided)
Notable features (good or bad)
Something interesting you learned

Get directions from your instructor if you are to share the information you found with your classmates, and how to do so.

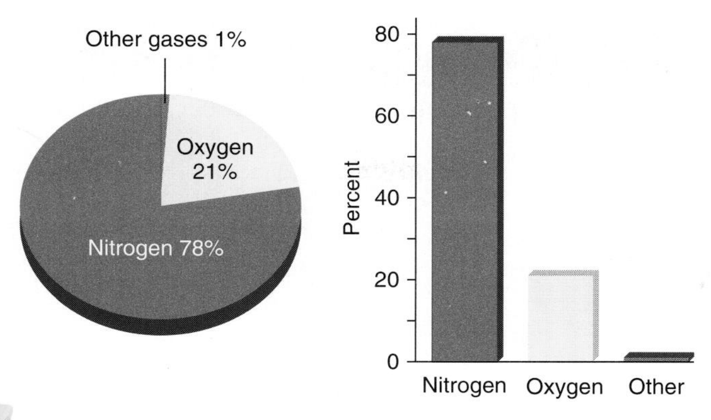

Figure 1.3
The composition of dry air, by volume.

- You can think about these percentages by considering that if you had 100 quarts of air, you would have
  - 78 quarts of nitrogen
  - 21 quarts of oxygen
  - 1 quart of other gases

 **Figures Alive!**

1.2 What's in a Breath? The Composition of Air

The air we breathe is a mixture of several substances. For the moment we will focus on only five: oxygen, nitrogen, argon, carbon dioxide, and water. The first four normally exist as gases. Although we usually think of water as a liquid, it can also be a gas—in which case we often call it "water vapor" to distinguish between the two physical states. The concentration of water vapor in air varies widely; it can be close to 0% in very dry desert air or 5–6% in a tropical rain forest. Because of this variability, reference tables typically list the composition of dry air. The normal composition of dry air is 78% nitrogen, 21% oxygen, and 1% other gases by volume. Percent means "parts per hundred (pph)." In this case, the parts are particles. Thus, 21% oxygen means 21 particles of oxygen per 100 air particles = 21 particles oxygen/100 particles air.

Figure 1.3 displays the composition of air in the form of a pie chart and a bar graph. Both of these are important, widely used methods for displaying numerical information and we will use each at various places in this text. The pie chart emphasizes the fractions of the total, whereas the bar graph emphasizes the relative sizes of each. Regardless of how we present the data, notice that 99% of the atmosphere is made up of only two substances, nitrogen and oxygen.

The 21% oxygen is what is most immediately essential for sustaining life. Oxygen is absorbed into our blood via the lungs and reacts with the foods we eat to release the energy needed for all life processes within our bodies (Chapter 11). Life on Earth bears the stamp of oxygen. Indeed, it is difficult to conceive of life on any planet without this remarkable chemical. Oxygen is also a participant in burning, in rusting, and in other corrosion reactions. Because it is a constituent of water and many rocks, oxygen is the most abundant element in the Earth's crust and the human body. Given this broad distribution and high reactivity, it is somewhat surprising that oxygen was not isolated as a pure substance until the 1770s. But once oxygen was isolated and purified, it proved to be of great significance in establishing the principles of the young science of chemistry.

- Table 11.1 gives the elemental composition of the human body.

Nitrogen is the most abundant substance in the air and constitutes over three-fourths of the air we inhale. However, it is much less reactive than oxygen, and it is exhaled from our lungs unchanged (Table 1.2). Although nitrogen is essential for life and is a part of all living things, most plants and animals obtain the nitrogen they require from other sources, not directly from the atmosphere.

The remaining 1% of air is mostly argon, a substance so unreactive that it is said to be "chemically inert." This inertness is recognized in the name *argon*, which means "lazy." Because argon refuses to form a chemical combination with anything (even itself), it does not normally make its presence known. Hence, it was not discovered until 1894.

Table 1.2

Typical Composition of Inhaled and Exhaled Air

Substance	Inhaled Air (%)	Exhaled Air (%)
Nitrogen	78.0	75.0
Oxygen	21.0	16.0
Argon	0.9	0.9
Carbon dioxide	0.04	4.0
Water	0.0	4.0

1.3 Consider This: Increasing the Oxygen in the Atmosphere

Humans are accustomed to living in a world with an atmosphere containing 21% oxygen. In such a world, a match from a paper matchbook burns completely in less than a minute and a log with a diameter of 3 inches is consumed in about 20 minutes in a fireplace fire. You have already determined how many times a minute you exhale air. Burning, rusting, and most metabolic processes in humans, plants, and animals depend on oxygen. How would life on Earth be different if the oxygen content in the atmosphere were increased to 40%, or higher? List at least four effects such a change would have on life as we know it.

- 1 liter = 1 L = 1.06 quart
- Conversion factors relating units are found in Appendix 1.

It is important to remember that the percentages we have been using to describe the composition of the atmosphere are based on volume. Thus, a mixture of 78 liters (L) of nitrogen, 21 L oxygen, and 1 L argon would yield 100 L of a mixture of gases very closely approximating the composition of dry air—78% nitrogen, 21% oxygen, and 1% argon. But because the volume of a gas sample increases with increasing temperature and decreases with increasing pressure, all gas volumes must be measured at the same temperature and pressure.

An alternative way to represent composition is in terms of the relative amounts of the particles of the various components present in the mixture. Using volumes or relative amounts of particles are equivalent ways of expressing composition because equal volumes of gases at the same temperature and pressure contain equal numbers of particles. In a total of 100 particles of air, 78 are nitrogen, 21 are oxygen, and 1 is argon. Consequently, 78% nitrogen means 78 particles of nitrogen per 100 air particles = 78 particles nitrogen/100 particles air. We will soon be more specific about the identity of those particles.

- The relation still applies when volume units other than liter are used:

  78 quarts nitrogen and 21 quarts oxygen and 1 quart argon

  78 gallons nitrogen and 21 gallons oxygen and 1 gallon argon

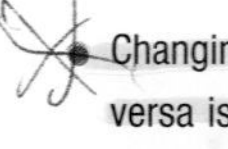

- Changing from percent to ppm or vice versa is a matter of moving the decimal point four places:

  move decimal point four places to the **right**

  parts per hundred (%) ⟶ / ⟵ **parts per million (ppm)**

  move decimal point four places to the **left**

- One part per million corresponds to:
  - one second in nearly 12 days
  - one step in a 568-mile journey
  - one penny out of $10,000
  - a pinch of salt on 20 pounds of potato chips

Some atmospheric components are present at less than parts per hundred, such as carbon dioxide, a very important constituent of the atmosphere. Its concentration of 0.0365% can be written as a fraction or ratio: 0.0365 parts carbon dioxide/100 parts air. Because it is difficult to envision less than one part, in the case of carbon dioxide 0.0365 parts, we scale up the measurement from parts per hundred to **parts per million** (abbreviated as **ppm**); 0.0365 parts per hundred equals 365 parts per million (365 ppm). It is fairly simple to do this conversion. We set up a pair of ratios, each representing the concentration, and set the ratios equal to each other.

$$\frac{0.0365 \text{ particles carbon dioxide}}{100 \text{ particles air}} = \frac{\text{number of carbon dioxide particles}}{1{,}000{,}000 \text{ air particles}}$$

The next step is to solve the equation for the number of carbon dioxide particles for every million air particles.

$$(100)\ (\text{number of carbon dioxide particles}) = (0.0365)(1{,}000{,}000)$$

$$\text{number of carbon dioxide particles} = \frac{0.0365 \times 1{,}000{,}000}{100} = (0.0365 \times 10{,}000) = 365$$

This means that a sample consisting of 1,000,000 particles of air will contain 365 particles of carbon dioxide. Hence, the carbon dioxide concentration is 365 ppm, which is equivalent to 0.0365%.

1.4 Your Turn

a. The EPA permissible limit for carbon monoxide is 9 ppm. Express this concentration as a percentage.
b. Exhaled air is typically 75% nitrogen gas. Express this concentration in parts per million.

Answers
a. 0.0009% **b.** 750,000 ppm

We human beings and our fellow members of the animal kingdom add carbon dioxide to the atmosphere every time we exhale. Table 1.2 indicates the difference in composition between inhaled dry air and exhaled air. Clearly some changes have taken place that use up oxygen and give off both carbon dioxide and water. Not surprisingly, chemistry is involved. In the biological process of metabolism, oxygen reacts with foods to yield carbon dioxide and water. However, most of the water in exhaled air is simply the result of evaporation from the moist surfaces within the lungs. Note that even exhaled air still contains 16% oxygen. Some people mistakenly think that in respiration most of the oxygen is replaced with carbon dioxide. But if this were true, mouth-to-mouth resuscitation would not work.

- You will learn a good deal more about carbon dioxide in Chapter 3.

1.3 What Else Is in a Breath? Minor Components

Our noses tell us that the air is obviously different in a pine forest, a bakery, an Italian restaurant, a locker room, or a barnyard. Even blindfolded, we can *smell* where we are. Pine needles, fresh bread, garlic, sweat, and manure all have distinctive odors. These odors are carried by matter. Hence, air must contain trace quantities of substances not included among the five substances listed in Table 1.1. The major components of air are odorless (at least to our noses), but many other airborne substances have pronounced odors. In fact, the human nose is an extremely sensitive odor detector. In some cases, only a minute trace of a substance is needed to trigger the olfactory receptors. Thus, tiny amounts of substances can have a powerful effect on our noses—as well as on our emotions.

Some of the trace substances that we can smell are pleasant and quite harmless; others can be very dangerous depending on their concentrations and the length of our exposure to them. Our noses warn us to avoid certain places because of the odors. But some of the most dangerous air pollutants have no odor at all, and others are dangerous at concentrations low enough that they cannot be detected by smell. As a result, it is often necessary to rely on specialized scientific equipment to measure the presence and concentrations of such substances in the air. It is rather ironic that the gases that cause serious air pollution are present in relatively small amounts, generally in the ppm to ppb range. Yet even at such low concentrations, they can do significant harm if humans and materials are exposed to them long enough.

In this chapter we will concentrate on four gases that contribute to air pollution at the surface of the Earth. One of these gases, carbon monoxide, is odorless; the other three—ozone, sulfur dioxide, and nitrogen oxides—have distinctly harsh smells. With sufficient exposure to them, each of these substances is hazardous to health, even at concentrations well below 1 ppm. Together they represent the most serious air pollutants at the Earth's surface. But as seen in Table 1.1, there is good news, because the concentrations of these four gases have decreased significantly over the last 20 years (Section 1.11).

Table 1.3 lists concentrations for the four major gaseous air pollutants as measured over 12 cities across the United States during 1996 and 2000. The bold numbers in the top row of the table are the air quality standards established by the U.S. EPA. These values, based on scientific studies, are the maximum concentrations considered to be safe for the general population. Notice that the air quality standards are quite different in terms of length of exposure:

- 9 ppm per eight hours for carbon monoxide
- 0.12 ppm per one hour for ozone

Table 1.3

Major Gaseous Air Pollutants for Selected Cities in the United States, 1996 and 2000

City	Carbon Monoxide* 9 ppm		Ozone** 0.12 ppm#		Sulfur Oxides*** 0.030 ppm		Nitrogen Oxides**** 0.053 ppm	
	1996	2000	1996	2000	1996	2000	1996	2000
Atlanta	4	2.5	0.14	0.158	0.022	0.018	0.027	0.023
Boston	5	2.3	0.11	0.089	0.037	0.029	0.031	0.029
Chicago	5	2.1	0.13	0.095	0.032	0.019	0.032	0.032
Detroit	6	4.5	0.11	0.088	0.079	0.042	0.021	0.024
Houston	15	3.8	0.20	0.17	0.011	0.010	0.045	0.016
Indianapolis	3	2.8	0.12	0.10	0.041	0.015	0.018	0.017
Los Angeles	7	5.0	0.18	0.12	0.046	0.007	0.023	0.040
New Orleans	4	4.0	0.11	0.11	0.035	—	0.018	0.019
New York City	6	4.6	0.12	0.11	0.055	0.025	0.042	0.026
Pittsburgh	4	2.4	0.11	0.11	0.070	0.037	0.030	0.022
San Francisco	5	2.8	0.10	0.055	0.007	0.007	0.022	0.020
Saint Louis	6	2.2	0.13	0.11	0.102	0.021	0.025	0.026

Source: EPA AIRDATA (2000) www.epa.gov/air/data/monvals.html.

*Second highest 8-hr avg

**Second highest 1-hr avg

***Second highest 24-hr avg

****Yearly avg

#Changed to 0.08 ppm for an 8-hr avg

- Even less for the sulfur oxides and nitrogen oxides based on a second highest 24-hour and yearly averages

According to these numbers, ozone is about 100 times more hazardous to breathe than carbon monoxide, and sulfur oxides are four times as hazardous as ozone. Although the concentrations are all expressed as parts per million, the concentrations of sulfur oxides and nitrogen oxides are sufficiently low that they also could conveniently be reported in parts per billion, ppb (sulfur oxides 0.030 ppm = 30 ppb; nitrogen oxides 0.053 ppm = 53 ppb). You must also be aware that the measured values reported in Table 1.3 are *not* the same sorts of averages for all four gases. For carbon monoxide and ozone, the data represent high values measured over relatively brief periods (hours). Sulfur oxides and nitrogen oxides are reported as second highest 24-hour and yearly averages.

- To convert ppm to ppb, move the decimal point three places to the right. For example, 0.0075 ppm equals 7.5 ppb; 6 ppm = 6000 ppb.

As you can see from the table, there has been a significant decrease in the concentrations of most of the four pollutants between 1996 and 2000. The measured values of nitrogen oxides are generally well below the permissible limit, although the concentrations of nitrogen oxides in a number of these cities have not decreased from 1996 to 2000. Also notice that there is a good deal of variability in the pollution levels in these cities.

1.5 Your Turn

Using data from Table 1.3, explain the statement "Ozone is about 100 times more hazardous to breathe than carbon monoxide, and sulfur oxides are four times as hazardous as ozone."

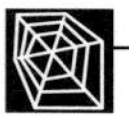

1.6 Consider This: Gaseous Pollutant Levels for Selected Cities

In Table 1.3, the top line that appears in boldface type lists the EPA accepted limits for different pollutants. Use both the data you gain from the table and your knowledge of chemistry to answer these questions.

a. List the cities that had the greatest decrease from 1996 to 2000 in
(1) carbon monoxide (2) sulfur oxides (3) nitrogen oxides

b. Which cities exceeded or came close to exceeding the ozone accepted limit in 2000?
c. By examining the data provided, which city had the lowest overall pollution in 1996? In 2000?
d. Use the Web to find out the concentrations of the major pollutants listed in Table 1.3 for the metropolitan area nearest you (if it is not listed in Table 1.3). Which, if any, of the pollutants there exceeds the EPA's accepted limit?

We now turn to consider briefly the behavior and damaging effects of these four pollutants. The first substance, carbon monoxide, enters the bloodstream and disrupts the delivery of oxygen throughout the body. In extreme cases (as in long exposure in a confined space to auto exhaust or emissions from a faulty furnace), it can lead to death. The health threat from carbon monoxide is especially serious for individuals suffering from cardiovascular disease, but healthy individuals are also affected. Visual perception, manual dexterity, and learning ability can all suffer.

Ozone is a special form of oxygen. It has a characteristic sharp odor that is frequently detected around photocopiers, electric motors, transformers, and welding torches. Unlike normal oxygen, ozone is very toxic. It affects the respiratory system, and even very low concentrations will reduce lung function in normal, healthy people during periods of exercise. Symptoms include chest pain, coughing, sneezing, and pulmonary congestion. At the Earth's surface, ozone is definitely a bad actor, but you will see in Chapter 2 that it plays an essential role at high altitudes (10–30 miles).

Sulfur oxides and nitrogen oxides are respiratory irritants that can affect breathing and lower resistance to respiratory infections. People most susceptible include the elderly, young children, and individuals with emphysema or asthma. A particularly severe example of these effects was the London fog of December 1952. The fog lasted five days and led to approximately 4000 deaths. The oxides of sulfur and nitrogen also contribute to acid precipitation—a subject explored in considerable depth in Chapter 6.

1.7 Consider This: Ozone Across the Country

Data on atmospheric ozone can be found at the *Chemistry in Context* web site where AIRNOW ground-level ozone maps are provided. Use the maps to estimate the ozone levels for one of the cities listed in Table 1.3. First select the appropriate state or region and then see what you can find about its ozone values by using the color-coded data provided. Notice that the data are reported in several different ways. Compare what you find on the Web with the values listed in Table 1.3. What factors might contribute to any differences you observe?

1.4 Taking and Assessing Risks

Air quality provides an opportunity for our first look at the subject of risk, an important topic and one to which we will return repeatedly throughout this text. Indeed, it is an issue that is central to life itself, because everything we do carries a certain level of risk, although these levels vary greatly. We are often presented with warnings about certain activities that are believed to carry high risk. For example, the law requires cigarette packages to carry the message "WARNING: Smoking cigarettes may be dangerous to your health." Still other practices have been declared illegal because the level of risk is judged unacceptable to society. On the other hand, there are many other activities that carry no warning, presumably because the degree of risk is quite low, the risk is obvious or unavoidable, or the benefits of the activity far outweigh the risk.

One feature of such warnings is a characteristic of risk itself. The warnings do not say that a specific individual *will* be affected by a particular activity. They only indicate the statistical probability or chance that an individual will be affected. For example, if the odds of dying from an accident while traveling 300 miles in a car are one in a million, this means that, on average, one person out of every million people traveling 300 miles by car would be killed in an accident. Such predictions are not simply guesses, but are the result of evaluating scientific data and making predictions in an organized manner about the probabilities of an occurrence. These studies are referred to as **risk assessment.**

For air pollutants, the assessment of risk requires knowing two factors: the **toxicity**—the intrinsic hazard of a substance, and the **exposure**—the amount of the substance encountered. Exposure is the easier factor to evaluate; it depends simply on the concentration of the substance in the air, the length of time a person is exposed, and the amount of air inhaled into the lungs in a given time. As you already know, the latter depends on lung size and breathing rate.

- One microgram is one one-millionth of a gram, 1 μg = 0.000001 g; 1 cubic meter is 1000 liters, 1 m^3 = 1000 L, which is about 250 gallons.

Concentrations of pollutants in air are usually expressed either as parts per million (ppm) or as micrograms per cubic meter ($\mu g/m^3$). Consider the risk related to carbon monoxide, one of the four major gaseous air pollutants. Billions of tons of carbon monoxide are spread throughout the atmosphere. Yet, by itself, this prodigious amount does not tell the true story of risk regarding this substance. Because carbon monoxide is not evenly distributed atmospherically, its concentration varies widely by location. In some places the concentration is so low, far less than the air quality standards, that it is not really considered a pollutant. In others, such as in an operating car's tailpipe emissions, the carbon monoxide concentration is sufficiently high to create a hazardous situation. Therefore, to assess our risk to this possible pollutant, we need to consider our exposure to it as well as its potential toxicity.

Depending on our size and other factors, each of us breathes 10,000 to 20,000 L of air daily. If we were to breathe 10,000 L (10 m^3) of air of moderate quality that contains carbon monoxide at a concentration of 10,000 μg carbon monoxide/m^3 of air, we would be exposed to 100,000 μg of carbon monoxide in a day:

- 100,000 μg = 0.100 g, which is equal to the mass of two postage stamps.

$$10 \text{ m}^3 \text{ air} \times \frac{10{,}000 \ \mu\text{g carbon monoxide}}{\text{m}^3 \text{ air}} = 100{,}000 \ \mu\text{g carbon monoxide in a day}$$

- This text does not make extensive use of calculations, but it is important for you to be able to do and understand calculations of the type shown here.

Someone who breathes twice as much of this air in a day obviously would be exposed to twice as much carbon monoxide (200,000 μg).

For large values such as 100,000, as in 100,000 μg carbon monoxide per day, we will use **scientific notation** to avoid turning the text into strings of zeros. This particular number, 100,000, is written in scientific notation as 1×10^5. The easy way to make this conversion is to simply count the number of places to the right of the initial 1. There are five of them, and 5 becomes the exponent of 10. The number 1 is then multiplied by 10^5 to obtain the number of micrograms of carbon monoxide per day, 1×10^5. Consider the number of molecules in a typical breath, a much larger number, where the usefulness of scientific notation is more readily apparent. There are more than 20,000,000,000,000,000,000,000 of them, a number large enough to take your breath away! In scientific notation this particular number is written as 2×10^{22}. Why 20,000,000,000,000,000,000,000 equals 2×10^{22} takes a bit more explaining. Remember that 10^{22} means 10 multiplied by itself 22 times. This is simply another instance in the following series:

$$1 \times 10^1 = 10$$

$$1 \times 10^2 = 10 \times 10 = 100$$

$$1 \times 10^3 = 10 \times 10 \times 10 = 1000$$

Note that 10^1 is 1 followed by 1 zero, that is, 10; 10^2 is 1 followed by 2 zeros; and 10^3 is 1 followed by 3 zeros. Continuing this pattern, 10^{22} is 1 followed by 22 zeros. Therefore, 2×10^{22} must equal 2 followed by 22 zeros or 20,000,000,000,000,000,000,000. If exponents and scientific notation are new to you, help with them is available in Appendix 2.

The 1×10^5 μg carbon monoxide/day might seem like a large quantity of carbon monoxide. But keep in mind that this quantity, when inhaled over a 24-hour period, is not toxic because it is less than the amount determined as hazardous. For carbon monoxide to be considered hazardous over 24 hours, we would need to be exposed to about 5×10^5 to 9×10^5 μg of it in air, nearly 5 to 10 times the amount found in air of moderate quality, and about 100 times more than in good quality air. The important point here is that when evaluating exposure with toxicity, it is necessary to compare the exposure level to the minimum amount required for the pollutant to become a public health risk or to be present at toxic levels. For example, 1×10^5 μg carbon monoxide/day is the exposure level and 5×10^5 to 9×10^5 μg is the minimum toxic threshold. Even though a pollutant may be present, it is not considered hazardous unless it exceeds the toxic threshold, the amount that causes harmful effects.

- Specific chemicals can be beneficial or harmful, depending on their location and concentration.

The toxicity of any substance to humans is more difficult to know with accuracy, in part because it is considered unethical to do controlled experiments with human subjects. This leaves scientists with three choices: human population studies, animal studies, and bacterial studies. Population studies involve collecting data on affected groups of people. For example, a researcher may determine what percentage of people who smoke one pack of cigarettes per day get lung cancer. Such studies are necessarily limited and may require many years of observation to obtain results that are statistically significant and reflect accurately the long-term risk. For this reason, animal studies have been a widely used substitute for those involving human subjects. Animals are given controlled doses of the substance being tested and observed for harmful effects. Aside from questions of animal rights, the problem here is that we do not know with certainty whether specific animal species respond the same as humans. There is a growing awareness among scientists that animal studies must be interpreted with great caution. A more recent area of toxicity measurements relies on studies with bacteria. An important advantage of bacteria is that they grow and reproduce very rapidly, allowing many studies to be done quickly and inexpensively.

Even if data are available to calculate the risks from a given pollutant, we still have to ask what level of risk is acceptable and for what groups of people. Various government agencies are charged with establishing safe limits of exposure for the major air pollutants. Table 1.4 gives current outdoor air quality standards established by the U.S.

Table 1.4

National Ambient Air Quality Standards, 1999

Pollutant	Limit, ppm	Limit, $\mu g/m^3$ (concentration approximately equivalent to ppm)
Carbon monoxide		
8-hr average	9	10,000
1-hr average	35	40,000
Nitrogen dioxide		
Annual arithmetic mean	0.053	100
Ozone		
1-hr average	0.12	235
8-hr average#	0.08	157
Lead		
Quarterly average	—	1.5
Particulates		
$\geq$ 10 μm (PM_{10}), 24-hr average*	—	150
$\geq$ 10 μm (PM_{10}), annual arithmetic mean*	—	50
$\leq$ 2.5 μm ($PM_{2.5}$), 24-hr average*#	—	65
$\leq$ 2.5 μm ($PM_{2.5}$), annual arithmetic mean*#	—	15
Sulfur dioxide		
Annual arithmetic mean	0.03	80
24-hr average	0.14	365
3-hr average	0.50	1300

*Refers to airborne particles that are greater or equal to 10 μm in diameter (PM_{10}) and equal to or smaller than 2.5 μm in diameter ($PM_{2.5}$).

#Included for information only. A 1999 Federal Court ruling blocked implementation of these standards, which EPA proposed in 1997. The EPA has asked the Supreme Court to reconsider that decision, which it has agreed to do.

EPA for the pollutants discussed in this chapter. Some states, including California and Oregon, have their own, stricter, standards.

A recent study reported in *Circulation*, a journal of the American Heart Association, has implicated $PM_{2.5}$ particles in urban air with an increased risk of heart attacks. In the study, the risk for heart attack in Boston peaked both two hours and 24 hours after patients were exposed to increased levels of the particles.

1.8 Your Turn

In a day, a person breathed in 15 m^3 of air that contained 1050 μg of SO_2. Did this air exceed the National Ambient Air Quality Standards (Table 1.4, annual arithmetic mean) for SO_2? Show calculations to support your answer.

An important factor in dealing with risks is not only the actual risk, but people's perception of a particular risk. For example, the majority of people in a survey were willing to accept a one-in-ten-thousand (1 in 1×10^4) to one-in-a-hundred thousand (1 in 1×10^5) risk of cancer from eating peanut butter that contains a naturally occurring carcinogen (cancer-causing material). Yet, these same people found it unacceptable to use a synthetic material with a one-in-a-million (1 in 1×10^6) risk factor, a case whose cancer risk is 10 to 100 times *less* than from eating peanut butter. The perception was that something (peanut butter) made from a natural material had to be less risky than to use a synthetic material, a common misconception.

1.9 Consider This: Communicating Air Pollution Levels

Newspapers often provide a color-coded "Pollution Index" to give the public a qualitative indication of the level of air pollution expected in the region. Typically, the index is based on computer-based models predicting anticipated levels of ground-level ozone. The pollution index color codes are green—low levels, with no cautionary action required; yellow—moderate levels and often leads to warnings that outdoor activity should be limited, particularly for children and adults with respiratory problems; red levels—ozone that is likely to exceed federal standards and outdoor activity should be avoided.

a. What are the advantages to using a color-coded qualitative index rather than simply reporting the predicted parts per million of ozone?
b. A "red alert" is forecast for tomorrow. Prepare a list of at least five actions that you could take that would help reduce air pollution tomorrow, particularly if everyone were to follow your suggestions.
c. Check the weather page of your local newspaper or the newspaper's or a television station's web site to see if an air pollution index is included. You may also want to check an appropriate state or regional web site for this information. Describe the pollution indices you have located and comment on their effectiveness in communicating with the public.

1.10 Consider This: Risk Analysis

"The general public is uncomfortable with uncertainties. Too often we think in terms of absolutes and demand that scientists and decision makers be held accountable for their risk decisions."*

Do you agree or disagree with these statements? Support your opinion with reasonable arguments, giving a specific example from your personal experience in considering a risk of importance to you.

**Chemical Risk: A Primer.* Department of Government Relations and Science Policy, American Chemical Society, 1996, p. 11.

1.5 The Atmosphere: Our Blanket of Air

The most familiar kinds of air pollution and their influences on air quality occur in the **troposphere,** *the part of the atmosphere that lies directly at the surface of the Earth.* Figure 1.4 provides the names of regions of the atmosphere and some reference points in relation to altitude. As one rises in the troposphere, the temperature decreases until it reaches about −40°C (also −40°F). That temperature roughly marks the beginning of

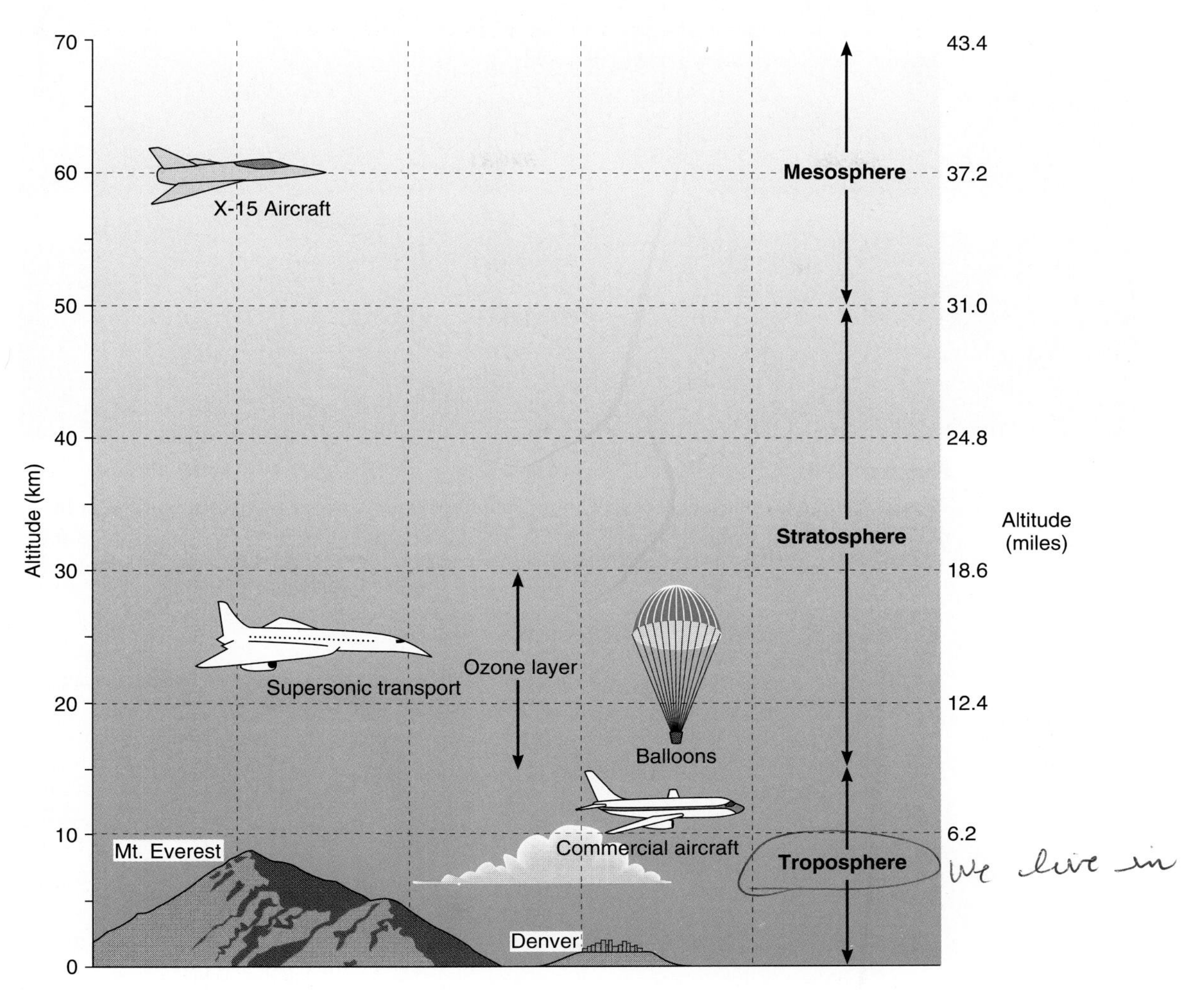

Figure 1.4
The regions of the atmosphere at various altitudes.

the **stratosphere,** which includes the ozone layer, the subject of Chapter 2. The temperature of the stratosphere increases from about −40°C at 20 km to 0°C (32°F) at 50 km. Above that altitude, the temperature of the atmosphere again begins to decrease on passing through the **mesosphere.** The issues we will study in the first three chapters of this book will take us to these various regions, which differ in atmospheric properties and phenomena (Figure 1.5). Bear in mind that no sharp physical boundaries separate these layers. The atmosphere is a continuum with gradually changing composition, concentrations, pressure, and temperature. In fact, temperature changes account for the organization of the atmosphere.

The relative concentrations of the major components of the atmosphere are nearly constant at all altitudes. In other words, the concentration of oxygen remains about 21% and that of nitrogen is 78%. However, you might know from reading about it, from the experience of hiking in high mountains, or flying in a jet plane that the air gets "thinner" with increasing altitude. As you climb up into the blanket of air, there is less of it—fewer particles in a given volume. Moreover, as you move up through the atmosphere, the mass of air above you decreases. Therefore, the **atmospheric pressure,** the force with which the atmosphere presses down on a given area, decreases with increasing altitude as shown in Figure 1.6. Atmospheric pressure is measured with a device

Figure 1.5

The Earth's fragile atmosphere is seen in this photo of a sunrise over the West Indies taken by the crew of the Space Shuttle Discovery (July 1995). The thin blue band is the stratosphere and the red-orange band is the troposphere. A cloud layer (white) is seen in the upper troposphere.

- As altitude ↑, atmospheric pressure ↓.

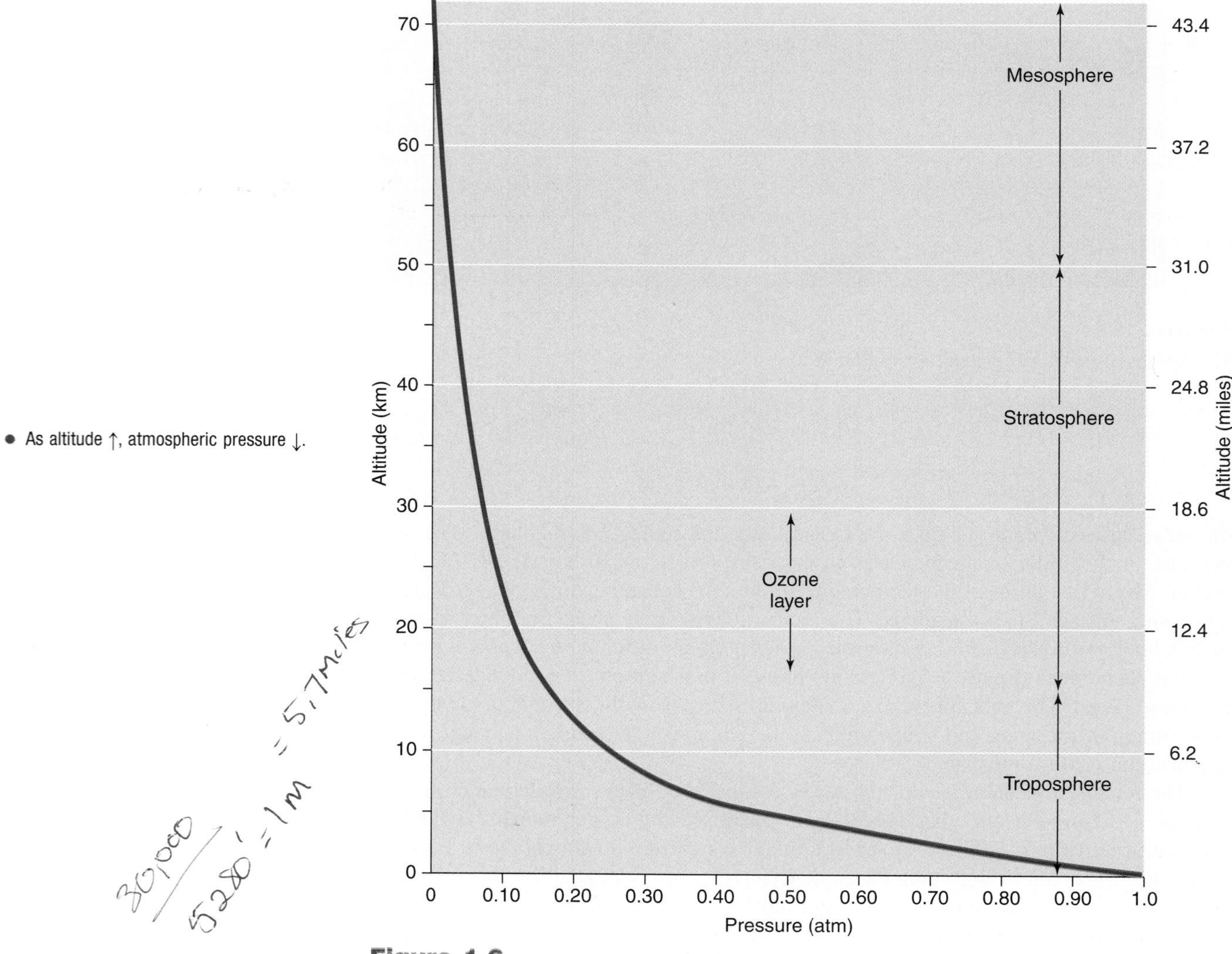

Figure 1.6

The pressure of the atmosphere changes with altitude. An increase in altitude is accompanied by a drop in air pressure.

called a barometer. At sea level, such as in Boston or Los Angeles, the barometric pressure is 14.7 pounds per square inch. A pressure of this magnitude is defined as 1 atmosphere (1 atm). In Denver, the "mile high city," the pressure is 12.0 pounds per square inch—about 0.8 atmosphere. You will note from Figure 1.6 that the plot of pressure versus altitude is not a straight line. Above about 20 kilometers (km) the pressure drops very sharply with increasing altitude. In this region, the pressure decreases by about 50% for every 5-kilometer increase in altitude (1 km = 0.62 mi, 5 km = 3.1 mi). At higher altitudes, the pressure decreases more gradually. Somewhere above 100 km, the atmosphere simply fades into the almost perfect vacuum of outer space.

1.11 Consider This: The Pressure Is on You

Use the Web to determine the atmospheric pressure at: (a) your locale; (b) Albuquerque, NM; (c) another U.S. city; (d) Mexico City, Mexico.

1.6 Classifying Matter: Mixtures, Elements, and Compounds

In describing the atmosphere and air quality, we referred to a number of substances and used a bit of chemical terminology. Therefore, before proceeding further, some clarification is probably necessary. For one thing, we need some understanding of the way chemists describe the composition of different types of matter. Chemists classify matter as mixtures or pure substances; the pure substances are elements or compounds (Figure 1.7). A breath of air is a **mixture**—a physical combination of two or more substances that may be present in variable amounts. Pure air and polluted air differ in composition, and we have seen that exhaled air differs from inhaled air. As composition varies, so do many of the properties of mixtures. Much of the matter we encounter in everyday life is in the form of mixtures. The fuels we burn, the foods we eat, the beverages we drink, our bodies themselves are all complex mixtures of individual substances. By appropriate experiments, these individual substances can be isolated and shown to have reproducible, characteristic properties and fixed composition.

- Matter can be classified as a mixture or a pure substance. Elements and compounds are pure substances; mixtures are not.

The two most plentiful components of air are nitrogen and oxygen. These are examples of chemical **elements**—substances that cannot be broken down into simpler stuff by any ***chemical*** means. There are over 100 of these fundamental building blocks and all common forms of matter are composed of them. About 90 elements occur naturally on planet Earth and, as far as we know, in the universe. The remainder have been created from other elements through artificially induced nuclear reactions. (Plutonium is probably the best known of the artificially produced elements, although it does occur in very low concentrations in nature.) In some cases, the total amount of a newly created

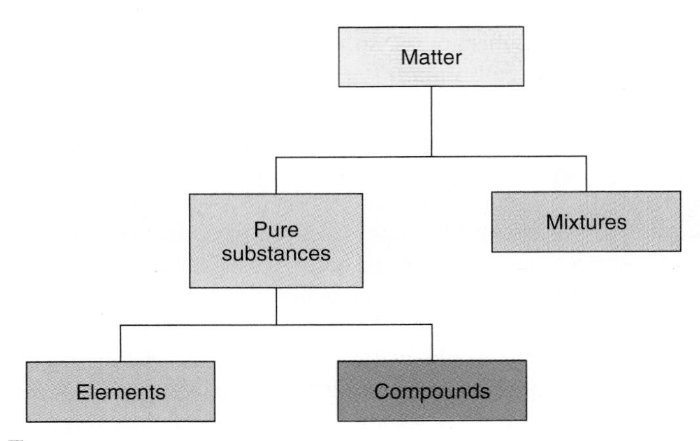

Figure 1.7
Classification of matter.

form of matter is so small that there is some uncertainty (perhaps even some controversy) over just how many elements have been identified. As this book went to press, there are 112 known (confirmed) elements. It should be noted that in 1999 the Joint Institute for Nuclear Research at Dubna, Russia, announced the formation of element 114 created by bombarding plutonium atoms with high-energy calcium atoms. In that same year, the Lawrence Berkeley National Laboratory at Berkeley, California reported the synthesis of two new elements—element 118 formed by colliding lead atoms and high-energy krypton atoms, and the formation of element 116 by the nuclear decay of element 118. These discoveries by both groups have not yet been confirmed by other researchers, a vital step in establishing unequivocally the presence of a new element. In fact, in July 2001, the Berkeley group retracted its claim of the element 118 synthesis. The retraction was based on a re-examination of the original experimental synthesis data by the Berkeley group and other experts in the field. On re-examination, the experimental data did not support the conclusion that element 118 was synthesized nor element 116, which was reported to have formed from 118. The retraction of the seeming synthesis of element 118 is an example of the self-correcting mechanism by which scientific discoveries are evaluated and changed if circumstances warrant a change, including rejection of the original discovery. Elements 113, 115, and 117 had not yet been reported when this book went into production.

1.12 Sceptical Chymist: Braving the Elements

a. Use the Web to determine the latest information about whether elements 116 and 118 have been synthesized, and whether the syntheses have been confirmed.

b. What information has been found about how the faulty conclusion was reached that element 118 had been synthesized?

c. Have elements 113, 115, or 117 been synthesized? If any of them have, write a report about the methods used to synthesize and detect the new element.

An alphabetical list of the known and named elements and their **symbols** appears in the inside back cover of the text. The symbols have been established by international agreement and are used throughout the world. The origins of some of the symbols are quite obvious to those who speak English. For example, oxygen is O, nitrogen is N, carbon is C, and sulfur is S. Most symbols consist of two letters, again based on the name of the element: Ni for nickel, Cl for chlorine, Ca for calcium, and so on. Other symbols appear to have little relationship to their English names. Thus, Fe is iron, Pb is lead, Au stands for gold, Ag is silver, Sn stands for tin, Cu is copper, and Hg is mercury. All these metals were known to the ancients and hence were given Latin names long ago. The symbols reflect those Latin names, for example, *ferrum* for iron, *plumbum* for lead, and *hydrargyrum* for mercury.

Elements have been named for properties, planets, places, and people. Hydrogen (H) means "water former," a name that reflects the fact that this flammable gas burns in oxygen to form water. Neptunium (Np) and plutonium (Pu) were named after the two most recently discovered members of our solar system. Berkeley and California are honored in berkelium (Bk) and californium (Cf). And Albert Einstein, Dmitri Mendeleev, and more recently Lise Meitner (codiscoverer of nuclear fission) have attained elementary immortality—einsteinium (Es), mendelevium (Md), and meitnerium (Mt).

It is particularly appropriate that Mendeleev should have his own element, because the most common way of arranging the elements reflects the periodic system developed by this 19th century Russian chemist. Figure 1.8 is the **periodic table.** We will explain the significance of the numbers and the elements order in Chapter 2. For the moment, it is sufficient to note that about the time of the American Civil War, Mendeleev arranged the 66 then-known elements so that the elements with similar chemical and physical properties fell in families or groups designated by vertical columns. Thus, the members of Group 1A include lithium (Li), sodium (Na), potassium (K), and three other very reactive metals. Similarly, Group 7A consists of very reactive nonmetals, including fluorine (F), chlorine (Cl), bromine (Br), and iodine (I). Nitrogen and oxygen, the two most common elements in the atmosphere, are side by side in Groups 5A and 6A. Some helpful generalizations

24 — Atomic number
Cr
52.00 — Atomic mass

1 1A	2 2A	3 3B	4 4B	5 5B	6 6B	7 7B	8	9 8B	10	11 1B	12 2B	13 3A	14 4A	15 5A	16 6A	17 7A	18 8A
1 **H** 1.008																	2 **He** 4.003
3 **Li** 6.941	4 **Be** 9.012											5 **B** 10.81	6 **C** 12.01	7 **N** 14.01	8 **O** 16.00	9 **F** 19.00	10 **Ne** 20.18
11 **Na** 22.99	12 **Mg** 24.31											13 **Al** 26.98	14 **Si** 28.09	15 **P** 30.97	16 **S** 32.07	17 **Cl** 35.45	18 **Ar** 39.95
19 **K** 39.10	20 **Ca** 40.08	21 **Sc** 44.96	22 **Ti** 47.88	23 **V** 50.94	24 **Cr** 52.00	25 **Mn** 54.94	26 **Fe** 55.85	27 **Co** 58.93	28 **Ni** 58.69	29 **Cu** 63.55	30 **Zn** 65.39	31 **Ga** 69.72	32 **Ge** 72.61	33 **As** 74.92	34 **Se** 78.96	35 **Br** 79.90	36 **Kr** 83.80
37 **Rb** 85.47	38 **Sr** 87.62	39 **Y** 88.91	40 **Zr** 91.22	41 **Nb** 92.91	42 **Mo** 95.94	43 **Tc** (98)	44 **Ru** 101.1	45 **Rh** 102.9	46 **Pd** 106.4	47 **Ag** 107.9	48 **Cd** 112.4	49 **In** 114.8	50 **Sn** 118.7	51 **Sb** 121.8	52 **Te** 127.6	53 **I** 126.9	54 **Xe** 131.3
55 **Cs** 132.9	56 **Ba** 137.3	57 **La** 138.9	72 **Hf** 178.5	73 **Ta** 180.9	74 **W** 183.9	75 **Re** 186.2	76 **Os** 190.2	77 **Ir** 192.2	78 **Pt** 195.1	79 **Au** 197.0	80 **Hg** 200.6	81 **Tl** 204.4	82 **Pb** 207.2	83 **Bi** 209.0	84 **Po** (210)	85 **At** (210)	86 **Rn** (222)
87 **Fr** (223)	88 **Ra** (226)	89 **Ac** (227)	104 **Rf** (261)	105 **Db** (262)	106 **Sg** (266)	107 **Bh** (264)	108 **Hs** (269)	109 **Mt** (268)	110	111	112	(113)	(114)	(115)	(116)	(117)	(118)

58 **Ce** 140.1	59 **Pr** 140.9	60 **Nd** 144.2	61 **Pm** (145)	62 **Sm** 150.4	63 **Eu** 152.0	64 **Gd** 157.3	65 **Tb** 158.9	66 **Dy** 162.5	67 **Ho** 164.9	68 **Er** 167.3	69 **Tm** 168.9	70 **Yb** 173.0	71 **Lu** 175.0
90 **Th** 232.0	91 **Pa** 231.0	92 **U** 238.0	93 **Np** (237)	94 **Pu** (244)	95 **Am** (243)	96 **Cm** (247)	97 **Bk** (247)	98 **Cf** (251)	99 **Es** (252)	100 **Fm** (257)	101 **Md** (258)	102 **No** (259)	103 **Lr** (262)

Metals

Metalloids

Nonmetals

The 1–18 group designation has been recommended by the International Union of Pure and Applied Chemistry (IUPAC) but is not yet in wide use. In this text we use the standard U.S. notation for group numbers (1A–8A and 1B–8B). No names have been assigned for elements 110–112. Elements 113–118 have not yet been synthesized.

Source: Raymond Chang, *General Chemistry: The Essential Concepts*, Third Edition, Copyright 2003 The McGraw-Hill Companies, New York, NY.

Figure 1.8
The periodic table of the elements.

about elements come from the table; the vast majority of elements are solids, some are gases, and only two—bromine and mercury—are liquids at room temperature and pressure. Note also that most elements are metals; far fewer of them are nonmetals. The Group 8A elements, such as neon and radon, are known as the *noble gases.* Some of them, such as helium, neon, and argon do not combine chemically with any elements.

The periodic table is a very handy database, an amazingly useful way of organizing the stuff of the universe, and we will return to it throughout the text. The periodically repeating properties can be beautifully explained by a knowledge of atomic structure. Moreover, the table continues to grow as new elements are made.

- Radioactivity is an important topic in Chapter 7.

By tradition, the discoverers of elements have been given the right to name them. Things get a bit complicated with recently discovered elements, especially when only a few atoms of the new element exist, and then for only a fraction of a second before they radioactively decompose. This is the case with the elements numbered 104 to 109. These new forms of matter were variously created in particle accelerators or "atom smashers" in the United States, Germany, and Russia. In some cases, more than one research group claimed priority of discovery. The International Union of Pure and Applied Chemistry (IUPAC) is the body that formally approves the names of elements, but its preliminary recommendations for elements 104–109 were controversial. After more than a decade of debate, IUPAC made a final decision in 1997 regarding these elements. Their proper symbols appear in Figure 1.8 as well as names in the table of elements in the inside back cover of the text.

- There is no residual uncombined oxygen or carbon in carbon dioxide. In CO_2, the two elements are chemically combined and are no longer in their elemental forms.

In addition to mixtures and elements, there are also chemical **compounds,** the third class of matter. A compound is a pure substance made up of two or more elements in a fixed, characteristic chemical composition and combination. Although there are only about 100 elements, over 20 million chemical compounds have been isolated, identified, and characterized. Among these are some of the most familiar naturally occurring substances, including water, salt, and sugar. But most of the known compounds do not exist in nature; they did not exist before they were synthesized by chemists. The motivation for making new compounds is almost as varied as the compounds themselves—to make synthetic fibers and plastics, to find drugs to cure AIDS or cancer, or just for the creativity and intellectual fun of it. In later chapters we will look at some examples of how chemists synthesize new compounds (Chapters 9 and 10).

1.13 Your Turn

Classify each of these as an element, compound, or mixture.

a. water **b.** nickel **c.** U.S. nickel coin
d. diamond **e.** sulfur dioxide **f.** lemonade

Answers

a. compound **b.** element **c.** mixture

1.14 Consider This: Adopt an Element

Periodic tables are available on the Web that list the properties of elements, their date of discovery, their naturally occurring isotopes, and much more. Thus, the Web can give you quick access to information that it might take you hours to find using reference books.

Use a search engine to bring up a list of periodic tables. Go to one of the periodic tables to find out more about an element of your choice. You probably will obtain more complete information if you select an element with atomic number of 94 (plutonium) or less.

Find out what year your element was discovered; whether it occurs naturally as a solid, liquid, or gas; its appearance; where it is found; and any two other facts, such as toxicity, cost, uses, etc.

Following the directions given by your instructor, get together with other students in your class to answer questions such as: Are most elements gases, solids, or liquids? Which elements were discovered first? Last? Are most elements found "free" in nature, that is, not combined with any other element? Do the elements chosen combine with other elements to form compounds?

Two important compounds in the atmosphere have already been mentioned: carbon dioxide and water. As its name implies, carbon dioxide is made up of chemically combined carbon and oxygen in a fixed composition. All pure samples of carbon dioxide contain 27% carbon and 73% oxygen by **weight** (or **mass**). Thus, a 100 gram (g) sample of carbon dioxide will always consist of 27 g of carbon and 73 g of oxygen, chemically combined to form this particular compound. These values never vary, no matter what the source of the carbon dioxide. This is simply one example of the fact that every compound exhibits a constant characteristic chemical composition. This is reflected in the fact that although carbon monoxide is also a compound of carbon and oxygen, the combination results in carbon monoxide with 43% carbon and 57% oxygen, by weight. Thus, 100 g of carbon monoxide contain 43 g of carbon and 57 g of oxygen, a much different composition than that of carbon dioxide, but not surprising because carbon monoxide and carbon dioxide are two different compounds.

- 1 gram = 0.00220 pound = 0.0352 ounce. A gram is approximately the mass of a peanut or a small paper clip.

Moreover, the composition of a compound is constant, as are its physical properties (such as boiling point) and its chemical reactivity. Consider water, a compound consisting of 11 g of hydrogen and 89 g of oxygen, (11% hydrogen and 89% oxygen by weight). At room temperature, water is a colorless, tasteless liquid. It boils at 100°C and it freezes at 0°C. It is an excellent solvent and participates in many chemical reactions. Like any compound, water can be broken down into its constituent elements and so 100 grams of water will decompose to yield 11 grams of hydrogen and 89 grams of oxygen.

1.7 Atoms and Molecules

The definitions of elements and compounds given earlier are valid, in spite of the fact that no assumptions were made about the physical structure of matter. But modern insights into the organization of matter help us better understand matter, the "stuff" of the universe. It is now well established that elements are made up of **atoms.** An atom is the smallest unit of an element that can exist as a stable, independent entity. The word *atom* comes from the Greek for "uncuttable." Today we know that atoms consist of smaller particles, and that atoms can be "split" by high-energy processes. However, atoms remain indivisible to chemical or mechanical means. Atoms are extremely small—many billions of times smaller than anything we can detect directly with our senses. Because of this small size, there must be huge numbers of atoms in any sample of matter that we can see or touch or weigh by conventional means. But, as Figure 1.9 reveals, scanning tunneling microscopy has recently made the invisible visible.

- Section 2.2 includes information about atomic structure and Section 7.2 discusses atom "splitting."

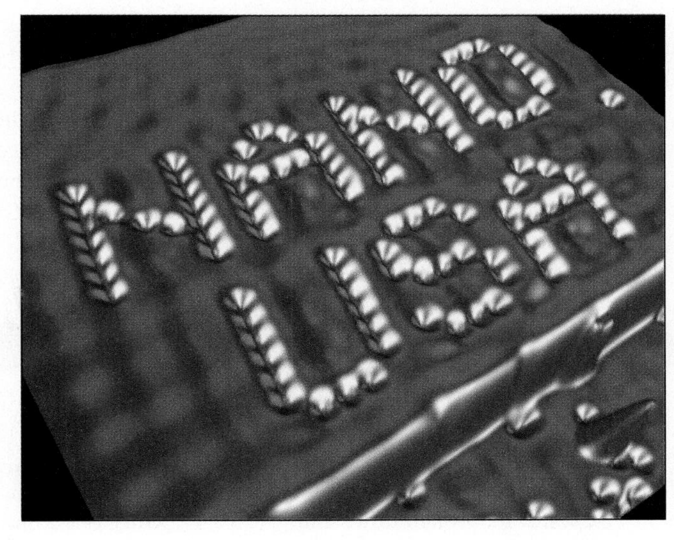

Figure 1.9

Using a scanning tunneling microscope, scientists at the IBM Almaden Research Center lined up 112 carbon monoxide molecules on a copper surface to spell Nano USA. Nanotechnology refers to work at the atomic and molecular (nanometer) scale: 1 nanometer (nm) = 1×10^{-9} m. Each letter is 4 nm high by 3 nm wide. At this size, about 250 million nanoletters could fit on a cross section of a human hair, corresponding to 300 300-page books.

The existence of atoms provides a means of refining our earlier definitions of elements and compounds. Each element has a different kind of atom, but within a sample of any given element all the atoms are chemically the same. By contrast, compounds are made up of the atoms of two or more elements. For example, the compound carbon dioxide has a ratio of one carbon atom for every two oxygen atoms and is symbolized as CO_2. In a similar manner, the formula for water, H_2O, indicates two hydrogen atoms for each oxygen atom. CO_2 and H_2O are examples of **chemical formulas,** which are symbolic representations of the elementary composition of chemical compounds. Note that, when an atom is used once in a formula, such as oxygen in water, it carries no subscript of "1."

- The number of atoms in a single drop of water is huge—about 5×10^{21} atoms. This is about a trillion times greater than the approximately 6 billion people on Earth, enough to give each person a trillion atoms from that water drop.

Carbon dioxide, water, and millions of other compounds exist as **molecules.** A molecule is a combination of a fixed number of atoms held together in a certain geometric arrangement. Thus, one molecule of carbon dioxide consists of one carbon atom bonded to two oxygen atoms. Like the compound itself, the molecule is represented by the formula CO_2. Similarly, a molecule of water, H_2O, contains exactly two hydrogen atoms and one oxygen atom, never any other combination. The constant ratio of atoms that characterizes each compound explains why compounds always have fixed composition.

- Information about molecular structure is found in Chapters 2 and 3.

Elements can exist either as molecules or as single atoms. Thus, the nitrogen and oxygen of the atmosphere are made up of N_2 and O_2 *diatomic molecules* (two atoms per molecule), whereas argon consists of individual Ar *atoms.*

1.15 Your Turn

Identify the elements in each of these compounds.

a. sulfur dioxide, SO_2 **b.** carbon tetrachloride, CCl_4
c. hydrogen peroxide, H_2O_2 **d.** sucrose, $C_{12}H_{22}O_{11}$

Answers

a. sulfur, oxygen **b.** carbon, chlorine

Table 1.5 describes elements, compounds, and mixtures in two ways: the behavior we can observe experimentally and the theory or model scientists use to explain what is happening at the incredibly small atomic level. Both are correct, and they complement each other. We can now apply these concepts to the atmosphere. Air is a mixture, which means that its composition can vary with time and place. Some of its components, such as nitrogen, oxygen, and argon, are individual elements; others, notably carbon dioxide and water, are compounds. All of the compounds and some of the elements are present as molecules (for example, CO_2 and O_2), but some elements exist as uncombined atoms (for example, Ar). Furthermore, we can give a more precise definition of the concentrations of gases in air. Recall that earlier we referred to "particles" of nitrogen and "particles" of air. These particles are the smallest units in which the constituents of air normally exist in their stable form—N_2 molecules, O_2 molecules, Ar atoms, CO_2 molecules,

Table 1.5

Classification of Matter

	Observable Properties	Atomic Theory
Element	Cannot be broken down into simpler substances	Only one kind of atom
Compound	Fixed composition, but capable of being broken down into elements	Two or more atoms in fixed combination
Mixture	Variable composition of elements and/or compounds	Variable assortment of atoms and/or molecules

and so on. When we say that air is 78% nitrogen, we mean that in every 100 particles of air, 78 of them will be N_2 molecules. Similarly, 100 particles of air will include 21 O_2 molecules and 1 Ar atom. A concentration of 365 ppm carbon dioxide means that there will be 365 CO_2 molecules in 1×10^6 molecules and atoms of air.

1.16 Sceptical Chymist: The Chemistry of Lawn Care

News reports and advertisements should be viewed with a critical eye—for scientific accuracy, bias, timeliness, and other criteria. Consider the following statement in a lawn care service advertisement that identified the fertilizers it uses as "a balanced blend of nitrogen, phosphorus, and potassium. They have an organic nature, made up of carbon molecules. These fertilizers are biodegradable and turn into water." Comment on the chemical correctness of this information. What changes would you suggest for the advertisement?

1.8 Formulas and Names: The Vocabulary of Chemistry

If elementary symbols are the alphabet of chemistry, then chemical formulas are the words. And the language of chemistry, like any other language, has rules of spelling and syntax. The symbols of the elements must be combined in ways that correctly correspond to the composition of the compounds in question. Although this system of chemical symbolism and nomenclature is logical, precise, and extremely useful (at least to chemists), it does present somewhat of a barrier to others studying the discipline. This, of course, is true of any specialized vocabulary, which often sounds like jargon to the uninitiated, whether in chemistry, accounting, music, sports, or any other area. This book does not seek to make its readers expert in all aspects of chemical nomenclature; however, some familiarity with the rules of writing formulas and naming simple chemical compounds will be helpful. In this chapter we will consider only the simplest compounds, those consisting of two elements.

The chemical formula of a compound reveals the elements that make it up (by chemical symbols) and the atom ratio of those elements (by the subscripts). The name usually conveys similar information. For example, the name "magnesium oxide" indicates a compound consisting of magnesium (Mg) and oxygen (O). Because the magnesium and oxygen combine in a one-to-one atomic ratio, the formula of magnesium oxide is MgO. Similarly, "sodium chloride" indicates a compound composed of the elements sodium (Na) and chlorine (Cl) with a formula of NaCl. The rule for naming such two-element compounds is simple: The name of the more metallic element comes first, followed by the name of the less metallic element, modified to end in "ide." So let's try to name the compound composed of potassium (K) and iodine (I). First we need to determine which of these two elements is more metallic, and for this we turn to the periodic table. As noted from the table, an important generalization is that the metallic elements are on the left side of the periodic table and the nonmetallic elements on the right. From this it follows that potassium must be more metallic than iodine. Applying the rule, potassium is first and the compound is named potassium iod*ide*. The formula turns out to be KI.

1.17 Your Turn

Name the compounds that contain each pair of elements.

a. bromine and magnesium **b.** oxygen and barium
c. hydrogen and chlorine **d.** sodium and sulfur

Answers

b. barium oxide **c.** hydrogen chloride

1.18 Your Turn

Name the compounds that have these formulas.

a. $ZnCl_2$ **b.** Al_2O_3
c. CaS **d.** Li_3N

Answers

a. zinc chloride **b.** aluminum oxide

1.19 Your Turn

Go to the *Chemistry in Context* web site and find the "naming compounds" practice exercises there. Complete the exercises given there as assigned by your instructor.

- The reason why calcium chloride has the formula $CaCl_2$ is discussed in Chapter 5.

Unfortunately, writing chemical formulas is a bit more complicated than we just suggested. Not all compounds exhibit one-to-one atomic ratios such as KI and MgO. For example, in calcium chloride there are two atoms of chlorine for each atom of calcium and the formula is therefore $CaCl_2$. Someone familiar with the periodic table and atomic structure should be able to predict the atomic ratio and the formula of just about any two-element compound and to name it. But we have not yet provided you with enough information to develop that skill. At present it is probably sufficient if you can name a two-element compound when presented with its formula. Thus, the formula H_2S represents a compound called hydrogen sulfide, a gas with the unmistakable smell of rotten eggs.

One of the complications associated with the rules we have been applying is that the name of the compound does not always unambiguously reveal its formula. A way of eliminating the ambiguity is to use prefixes that indicate the number of atoms of an element specified by the formula. A good example is one of the atmospheric compounds we have already introduced—carbon dioxide. *Di-* means "two" and thus the name carbon *di*oxide implies that each molecule of the compound includes two oxygen atoms. The corresponding formula, CO_2, indicates the two oxygen atoms with a subscript 2 on the symbol O. The use of the prefixes listed in Table 1.6 makes it possible to distinguish between two or more compounds consisting of the same elements, but in different atomic ratios. Thus, carbon *mon*oxide also consists of carbon and oxygen, but the prefix *mon-* or *mono-* reveals that only one oxygen atom is in each molecule of this compound. It follows that the formula of carbon monoxide is CO. Using the same logic and the same set of prefixes, the chemical name for SO_3 is sulfur trioxide. Note that in most compounds in which a formula contains only one atom of an element, the *mono-* prefix is omitted. Carbon monoxide is an exception; the *mono-* prefix is used to avoid possible confusion with carbon dioxide.

1.20 Your Turn

What information does each formula convey about these substances? Each is an important atmospheric component.

a. NO_2 **b.** SO_2 **c.** N_2O_4 **d.** O_3

Answer

a. A molecule of the compound represented by the formula NO_2 consists of one atom of the element nitrogen combined with two atoms of the element oxygen.

1.21 Your Turn

Use appropriate prefixes to name the compounds whose formulas are given in 1.20 Your Turn.

Answers

a. nitrogen dioxide **c.** dinitrogen tetraoxide

Table 1.6

Prefixes Used in Naming Compounds

Prefix	Meaning	Prefix	Meaning
Mono-	One	Hexa-	Six
Di- or bi-	Two	Hepta-	Seven
Tri-	Three	Octa-	Eight
Tetra-	Four	Nona-	Nine
Penta-	Five	Deca-	Ten

1.9 Chemical Change: Reactions and Equations; Oxygen's Role in Burning

The first pollutant listed in Table 1.3 is carbon monoxide, CO, whereas all air, polluted or unpolluted, contains carbon dioxide, CO_2. Carbon monoxide and carbon dioxide can both arise from the same source: combustion. **Combustion** (or burning) is the rapid combination of oxygen with another material. When carbon or carbon-containing compounds burn in air, oxygen combines with the carbon to form CO_2 and/or CO. Similarly, combustion reactions produce water (H_2O) and sulfur dioxide (SO_2) by the burning of hydrogen and sulfur, respectively.

Combustion is a major type of **chemical reaction** or a **chemical change.** A chemical reaction is a process whereby substances described as **reactants** are transformed into different substances called **products.** The process can be represented by an expression called a **chemical equation.** Chemical equations are the sentences in the language of chemistry. They are made up of elementary symbols (corresponding to letters), which are often combined in the formulas of compounds (the "words" of chemistry). Like a sentence, a chemical equation conveys information, in this case about the chemical change taking place. But a chemical equation must also obey some of the same constraints that apply to a mathematical equation.

At its most fundamental level, a chemical equation is very simple indeed. It is a qualitative description of the reaction:

$$\text{Reactant(s)} \longrightarrow \text{Product(s)}$$

By convention, the reactants are always written on the left and the products on the right. The arrow represents a chemical transformation and is read as "is converted to" or "yields." Thus, reactants are converted to products in the sense that the reaction creates products whose properties are different from those of the starting materials, the reactants.

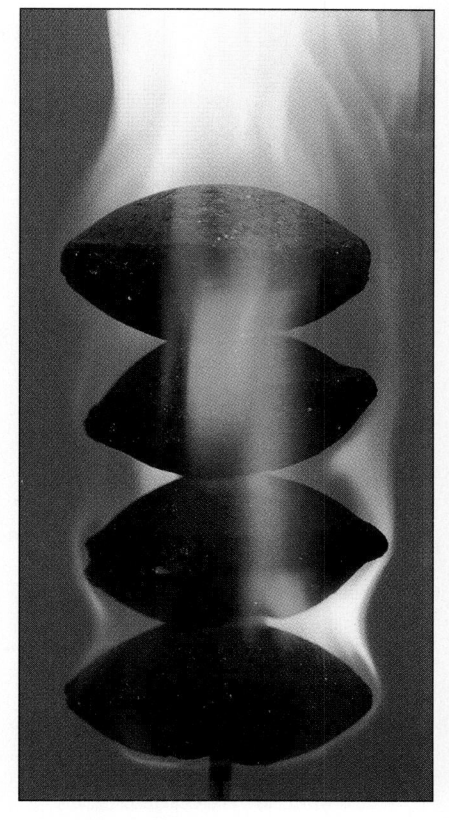

Figure 1.10
The burning of charcoal in air.

The combustion of carbon to produce carbon dioxide (for example, the burning of charcoal in air, Figure 1.10) can be represented in several ways. One way is by a "word equation":

$$\text{carbon} + \text{oxygen} \longrightarrow \text{carbon dioxide}$$

It is much more common to use chemical symbols and formulas for the elements and compounds involved:

$$C + O_2 \longrightarrow CO_2 \qquad [1.1]$$

This compact symbolic statement conveys a good deal of information to a chemist. A translation of equation 1.1 into words might read something like this: "One atom of the element carbon reacts with one molecule of the element oxygen (consisting of two oxygen atoms joined to each other) to yield one molecule of carbon dioxide, a compound consisting of one carbon atom linked to two oxygen atoms."

- Remember that oxygen molecules are diatomic when oxygen is not combined with any other element.

If we use a black sphere to represent a carbon atom and a red sphere to represent an oxygen atom, the rearrangement of atoms by this reaction looks something like this:

Likewise, the burning of sulfur to produce the air pollutant sulfur dioxide can be represented by a chemical equation

$$S + O_2 \longrightarrow SO_2 \qquad [1.2]$$

or symbolically with spheres, where a yellow sphere represents a sulfur atom and red spheres represent oxygen atoms.

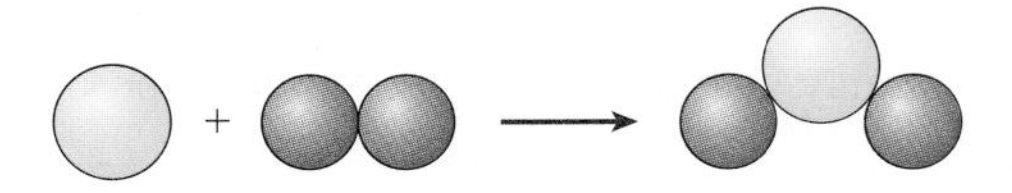

It is possible to pack more information into an equation by specifying the physical states of the reactants and products. A solid is designated by (*s*) following the symbol or formula; a liquid is designated by (*l*); and a gas is indicated by (*g*). Because carbon and sulfur are solids, and oxygen, carbon dioxide, and sulfur dioxide are gases at ordinary temperatures and pressures, equations 1.1 and 1.2 become:

$$C(s) + O_2(g) \longrightarrow CO_2(g)$$

$$S(s) + O_2(g) \longrightarrow SO_2(g)$$

In this text we will designate the physical states of the substances participating in a reaction when that information is particularly important, but in most cases we will omit it for simplicity.

You will note that equation 1.1 has some of the characteristics of a mathematical equation—in this case, the number and kinds of atoms on the left equal those on the right. One carbon atom and two oxygen atoms are on the left side of the arrow and one carbon atom and two oxygen atoms are on the right. This is the test of a correctly balanced equation: The number and kinds (which element) of the atoms on the reactant side of the arrow must equal the number and identity of atoms on the product side. Atoms are neither created nor destroyed in a chemical reaction, and the elements present do not change when converted from reactants to products. This relationship is called the law of **conservation of matter and mass:** In a chemical reaction, matter and mass are conserved. The mass of the reactants consumed equals the mass of the products formed. The total mass does not change, because no matter is created or destroyed.

- Equation 1.2 is balanced because an equal number of sulfur atoms are in the reactants and products and the numbers of oxygen atoms in the reactants and products are also equal.

Atoms are rearranged during a chemical reaction. That is what a chemical change is all about: The atoms in the products are in a different arrangement than they were as reactants. Therefore, there is no requirement that the number of molecules must be the same on both sides of the arrow. In fact, the number of molecules changes during many reactions. In equation 1.1, one atom of carbon plus one molecule of oxygen yields one molecule of carbon dioxide. This looks suspiciously like $1 + 1 = 1$. This is not a cause for alarm; a chemical equation is not exactly the same as a mathematical equation. Remember, a chemical equation represents a transformation, not a simple equality. In a correctly balanced chemical equation, some things must be equal, others need not be. Table 1.7 summarizes these peculiarities.

- Consider this analogy in terms of the reorganization of reactants to form products in a chemical reaction. The building materials used to construct an apartment building (reactants) can be disassembled and rearranged to build three houses and a garage (products).

Equation 1.1 describes the combustion of pure carbon in an ample supply of oxygen. However, if the oxygen supply is limited, the product is carbon monoxide, CO, not carbon dioxide. Like any reaction, this one can be expressed in equation form. First, we write the symbols and formulas of the reactants and products:

$$C + O_2 \longrightarrow CO \text{ (unbalanced equation)}$$

A quick count of atoms on either side of the arrow reveals that the expression does not balance. There are two oxygen atoms on the left and one on the right. We cannot bal-

Table 1.7

Characteristics of Chemical Equations

Always Conserved

Number of atoms in reactants = Number of atoms in products
Identity of atoms in reactants = Identity of atoms in products
Total mass of reactants = Total mass of products

Not Necessarily Conserved

Number of molecules of reactants may or may not equal number of molecules of products
Volume of reactants may or may not equal volume of products

ance the equation by simply adding an additional oxygen atom to the product side. Once we write the *correct* symbols and formulas for the reactants and products, we cannot change them, including their subscripts. To do so would imply a different reaction. All we can do is to use whole-number coefficients in front of the various symbols and formulas in the equation. In simple cases like this, the coefficients can be found quite easily by inspection or simple trial and error. If we place a 2 to the left of the symbol CO, it signifies two molecules of carbon monoxide. This corresponds to a total of two carbon atoms and two oxygen atoms. Because there are also two oxygen atoms on the left side of the arrow, the oxygen atoms have been balanced.

$$C + O_2 \longrightarrow 2\ CO \text{ (equation still not balanced)}$$

But now the carbon atoms are out of balance. There are two on the right but only one on the left. Fortunately, this is easily corrected by placing a 2 in front of the C.

$$2\ C + O_2 \longrightarrow 2\ CO \text{ (balanced equation)} \qquad [1.3]$$

- A subscript follows a chemical symbol, as in O_2 or CO_2; a coefficient precedes a symbol or a formula, for example 2 C or 2 CO.

The balanced equation can also be represented by colored spheres—black for carbon and red for oxygen atoms.

Note that the balanced equation includes quantitative and qualitative information. It tells us qualitatively what atoms are present, and quantitatively how many carbon atoms and how many oxygen molecules react to form carbon monoxide. It is evident from comparing equations 1.1 and 1.3 that, relatively speaking, more oxygen is required to form CO_2 from carbon than is needed to form CO.

The same kind of reasoning can be applied to balance the equation for the formation of another air pollutant, nitrogen monoxide (commonly called nitric oxide), created by the reaction of nitrogen with oxygen. We begin by writing down the unbalanced equation, giving the correct symbols and formulas of the reactants and products.

$$N_2 + O_2 \longrightarrow NO \text{ (unbalanced equation)}$$

- Remember that nitrogen as well as oxygen molecules are diatomic when they are not combined with any other element.

We notice that the equation is not balanced because there are two oxygen atoms on the left, and only one on the right. The same is true for nitrogen atoms. Placing a two to the left of the formula NO supplies two nitrogen *and* two oxygen atoms to that side, the same number that are to the left of the arrow; the equation is now balanced.

$$N_2 + O_2 \longrightarrow 2\ NO$$

- Nitrogen atoms are represented by blue spheres.

1.22 Your Turn

Balance each of these equations and then draw a representation of each equation using spheres.

a. $H_2 + O_2 \longrightarrow H_2O$
b. $N_2 + O_2 \longrightarrow NO_2$

Answer

a. Balanced equation: $2\ H_2 + O_2 \longrightarrow 2\ H_2O$

1.23 Your Turn

Go to the *Chemistry in Context* web site and find the "balancing equations" practice exercises there. Complete the exercises given there as assigned by your instructor.

1.24 Consider This: Advice from Grandma

A grandmother offered this advice to rid the garden of pesky caterpillars. "Hammer some iron nails about a foot up from the base of your trees, spacing them every four to five inches." According to this grandmother, the iron nails convert the tree sap (a sugary substance containing carbon, hydrogen, and oxygen atoms) into ammonia (NH_3), a substance the caterpillars cannot stand. Comment on grandma's advice from a chemical standpoint.

1.10 Fire and Fuel: Air Quality and Burning Hydrocarbons

• Chapter 4 contains many examples of the burning of fuels.

We saw in the previous section that the combustion of some elements—carbon, nitrogen, and sulfur—produces air pollutants. The combustion of fuels can also produce carbon monoxide and carbon dioxide. Many fuels, including those obtained from petroleum, are **hydrocarbons,** compounds of hydrogen and carbon. The simplest of these is methane, CH_4, the primary component of natural gas. When a hydrocarbon burns completely, all the carbon combines with oxygen to form carbon dioxide and all the hydrogen combines with oxygen to form water. We can use this reaction to again illustrate the process of balancing equations. First, we write formulas that qualitatively represent the complete combustion of methane—reactants are colored **magenta**; products are colored **cyan.**

$$CH_4 + O_2 \longrightarrow CO_2 + H_2O$$

The expression is already balanced with respect to carbon; there is one C, indicating one carbon atom, on each side of the arrow. But the equation is not balanced with respect to hydrogen and oxygen. It is easier to start with hydrogen because that element is present in only one substance on each side of the arrow: CH_4 on the left and H_2O on the right. Oxygen, on the other hand, winds up in both CO_2 and H_2O. Currently four hydrogen atoms are on the left of the expression (in CH_4) and two hydrogen atoms are on the right (in H_2O). To bring the hydrogen atoms into balance, we place the number 2 in front of H_2O.

$$CH_4 + O_2 \longrightarrow CO_2 + 2\ H_2O$$

The coefficient 2 is multiplied through the substance to the right of it, in this case water. Thus, it signifies 4 H atoms and 2 O atoms. Because a CO_2 molecule contains 2 O atoms, there are now a total of 4 O atoms on the right of the equation and 2 O atoms on the left. We equalize the number of O atoms by placing a 2 before O_2.

$$CH_4 + 2\ O_2 \longrightarrow CO_2 + 2\ H_2O \qquad [1.4]$$

That should balance the equation, but it is always a good idea to check. The nice thing about writing balanced equations is that you can always tell if you are correct by counting and comparing atoms on either side of the arrow.

Carbon: Left: 1 CH_4 molecule × 1 C atom/CH_4 molecule = 1 C atom

Right: 1 CO_2 molecule × 1 C atom/CO_2 molecule = 1 C atom

Hydrogen: Left: 1 CH_4 molecule × 4 H atoms/CH_4 molecule = 4 H atoms

Right: 2 H_2O molecules × 2 H atoms/H_2O molecule = 4 H atoms

Oxygen: Left: 2 O_2 molecules × 2 O atoms/O_2 molecule = 4 O atoms

Right: 1 CO_2 molecule × 2 O atoms/CO_2 molecule = 2 O atoms

+ 2 H_2O molecules × 1 O atom/H_2O molecule = 2 O atoms

This tabulation confirms that the equation is indeed balanced.

One of the most widely used hydrocarbon fuels in automobiles is gasoline, a mixture of dozens of different compounds. One of the compounds is octane, C_8H_{18}. If a sufficient supply of oxygen is delivered to the auto engine when the octane burns, only carbon dioxide and water are formed.

$$2\ C_8H_{18} + 25\ O_2 \longrightarrow 16\ CO_2 + 18\ H_2O \qquad [1.5]$$

In practice, however, not all of the carbon is converted to carbon dioxide. The amount of oxygen present and amount of time available for reaction (before the materials are ejected in the exhaust) are insufficient for the reaction represented by equation 1.5 to occur entirely. Instead, some CO is formed. An extreme situation is represented by equation 1.6, in which all of the carbon in the octane is converted to carbon monoxide.

$$2\,C_8H_{18} + 17\,O_2 \longrightarrow 16\,CO + 18\,H_2O \qquad [1.6]$$

Note that the coefficient of O_2 in equation 1.5 is 25, whereas the corresponding coefficient in equation 1.6 is 17. This indicates that less oxygen is used in the latter reaction.

What really happens in a car's engine is a combination of the two reactions. Most of the carbon released in automobile exhaust is in the form of CO_2, although there is also some CO. The relative amounts of these two gases indicate how efficiently the car burns the fuel, which is evidence of how well tuned the engine is. States that monitor auto emissions check for this by sampling exhaust emissions using a probe that detects CO. The measured CO concentrations are compared to established standards (1.20% in the state of Minnesota). A car whose CO emissions exceed the standard must be serviced so that it complies (Figure 1.11).

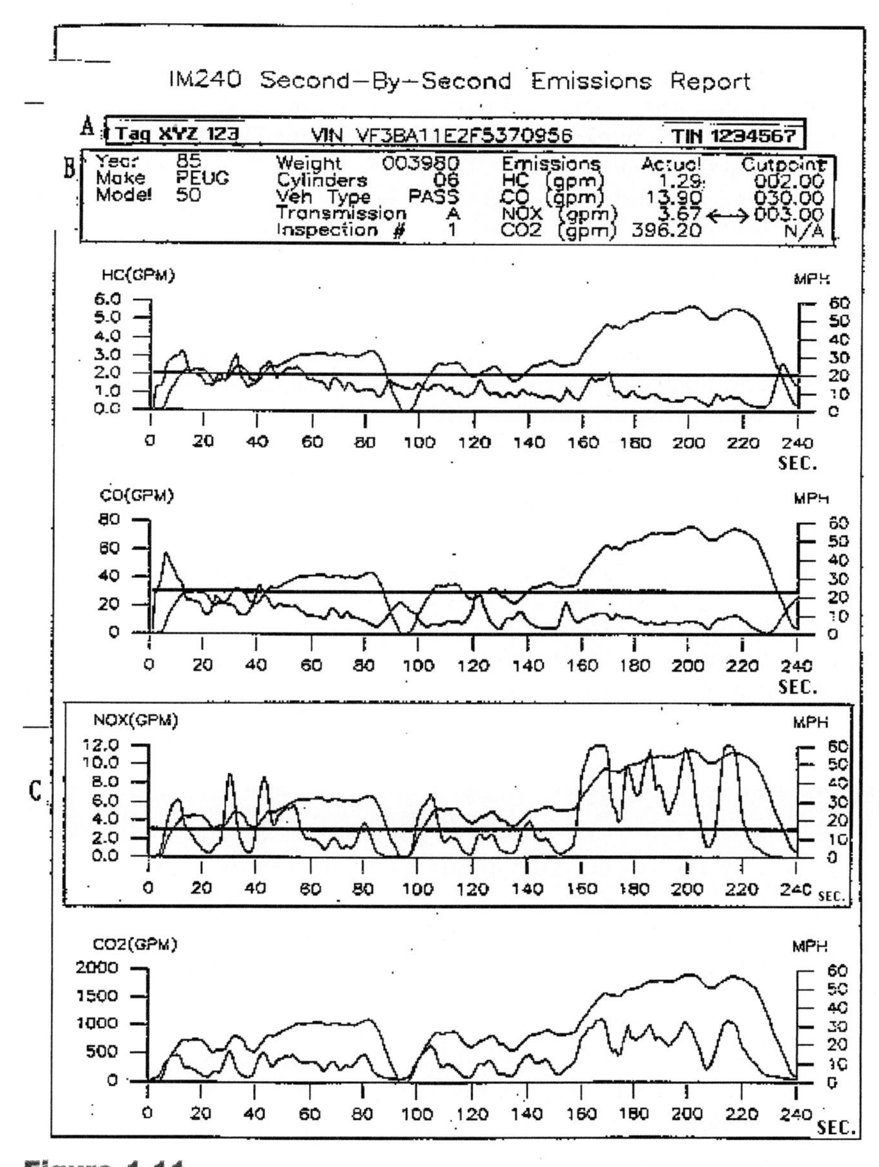

Figure 1.11
An auto emission report.

- GPM is the abbreviation for grams per mile.
- HC is the abbreviation for hydrocarbon.

1.25 Your Turn

a. "Bottled gas" or "liquid petroleum gas" (LPG) is mostly propane, C_3H_8. Balance this equation that represents the burning of propane.

$$C_3H_8 + O_2 \longrightarrow CO_2 + H_2O$$

b. Cigarette lighters burn butane, C_4H_{10}. Write a balanced equation for this combustion reaction.

Answer

a. $C_3H_8 + 5\ O_2 \longrightarrow 3\ CO_2 + 4\ H_2O$

1.26 Your Turn

Demonstrate that equations 1.5 and 1.6 are balanced by counting the number of atoms of each element on either side of the arrow.

Answer

In equation 1.5 there are 16 C, 36 H, and 50 O on each side of the equation.

1.11 Air Quality: Some Good News

Bad news always seems to get more attention than good news. Newspaper headlines and newscasts dwell on atmospheric pollutants, not the components of the atmosphere that are essential for life. And because pollutants such as carbon monoxide and sulfur dioxide are compounds generated by chemical processes, chemistry often gets blamed for pollution. Of course, oxygen, nitrogen, and carbon dioxide are also chemicals, and we would not be here without them.

Obviously, it is essential that science and society be concerned with hazardous components in the air we breathe. Chemists help monitor the concentrations of these pollutants and work to reduce them. As noted in Section 1.1, air quality in the U.S. has been improving over the past two decades. The Air Quality Index (AQI) has been developed by the EPA for reporting ozone levels and the concentrations of other common air pollutants. Formerly known as the Pollutants Standards Index, the AQI is developed from data measurements recorded by monitoring systems at more than a thousand locations throughout the country. The measurements are then converted to the AQI, whose values range from zero to 500, although levels above 300 are rare in the U.S. The higher the AQI number is for a pollutant, the lower the air quality and the greater the danger from the presence of that pollutant. In general, an AQI value of 100 usually corresponds to the national ambient air quality standard (NAAQS) for the pollutant (Table 1.4). The AQI ranges and the descriptors of air quality for various levels of tropospheric ozone are given in Table 1.8.

1.27 Consider This: Ozone Over Your Head

Data on tropospheric ozone in the form of AIRNOW ozone maps can be found at the *Chemistry in Context* web site. Use the maps to estimate the tropospheric ozone levels for a city in your locale. First, select the

Table 1.8

Air Quality Index and Air Quality Descriptors for Ozone

Ozone Concentration, ppm (8-hr average unless noted)	Air Quality Index Values	Air Quality Descriptor
0.00 to 0.064	0 to 50	Good
0.065 to 0.084	51 to 100	Moderate
0.085 to 0.104	101 to 150	Unhealthy for sensitive groups
0.105 to 0.124	151 to 200	Unhealthy
0.125 (8 hr) to 0.404 (1 hr)	201 to 300	Very unhealthy

Table 1.9

Metro Areas: Changes in Number of Days With AQI Greater Than 100

City	Days with AQI >100 1991	1998
Atlanta	23	43
Boston	13	7
Chicago	22	7
Detroit	28	17
Houston	37	38
Indianapolis	12	19
Los Angeles	169	56
New Orleans	2	7
New York City	49	17
Pittsburgh	21	39
San Francisco	0	0
Saint Louis	32	23

appropriate state or region, and then see what you can find about its ozone values by using the color-coded data provided. Notice that the data are reported in several different ways. Compare what you find on the Web with the values listed in Table 1.2. What factors might contribute to any differences you observe?

Another piece of evidence that air quality is improving is seen in the decrease in the number of days that the air quality index exceeded 100 in major metropolitan areas between 1991 and 1998. Table 1.9 lists 12 cities, the same cities as given in Table 1.3, which listed major pollutant levels in those cities. From Table 1.9, we see that the number of days that the air quality index exceeded 100 declined significantly in half of the 12 cities, including the country's three largest metro areas—New York City (nearly 300%), Los Angeles (more than 300%), and Chicago (more than 300%). Increases occurred in Atlanta, Pittsburgh, and New Orleans; slight increases took place in Houston and Indianapolis.

Earlier, in Table 1.1, we saw the percent decreases in six major air pollutants. Figure 1.12 indicates such changes for those six pollutants as percent changes from 1980 to 1999, and from 1990 to 1999.

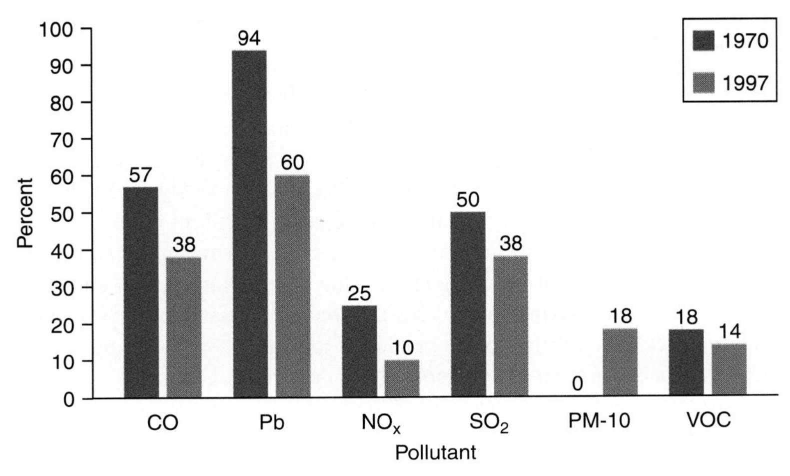

Figure 1.12

Percent changes in the average amounts of air pollutants in the United States, 1980–1999.

Data from *National Air Quality and Emissions Trends Report,* 1999, United States Environmental Protection Agency.

• VOC means volatile organic compound; PM-10 means particulate matter ⩾10 μm in diameter.

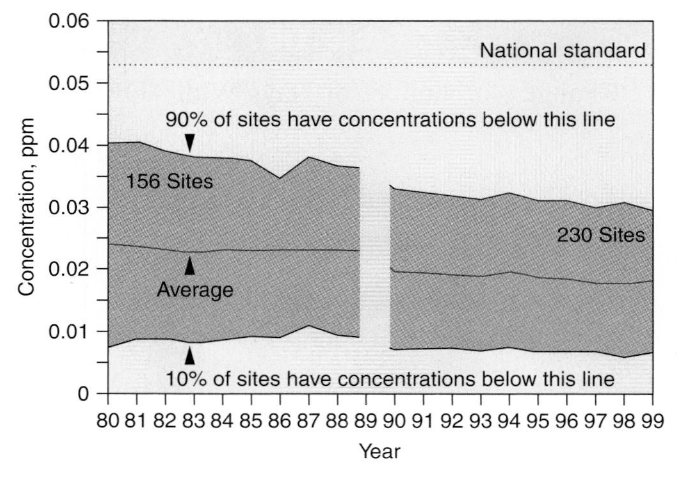

Figure 1.13

Trends in Annual NO_2 Average Concentration. The data were taken from 156 monitoring sites (1980–1989) and 230 sites (1990–1999).

Source of data: *National Air Quality and Emissions Trends Report,* 1999, United States Environmental Protection Agency.

Looking at the graph in Figure 1.12 it is easy to see that the concentrations of all these six pollutants have decreased in the United States, some markedly. The most dramatic changes from 1980 to 1999 have been in lead (94%) and carbon monoxide (57%). In 1999, atmospheric carbon monoxide concentration was the lowest it has been in 20 years. The 94% decrease in lead is principally due to the phasing out of leaded gasoline. Since 1980, the national nitrogen dioxide, NO_2, average concentration in the atmosphere has decreased 25%. There has been a steady decline in atmospheric NO_2 concentration since 1989, except for 1994 (Figure 1.13). Note that all the monitoring sites were below the national standard (0.053 ppm).

1.28 Consider This: Slowing Down

The data in Figure 1.13 indicate a slowing in the rate of decrease in the percentages by which all six pollutants changed from 1990 to 1999 compared with the changes from 1980 to 1999. Speculate why the changes were less during the more recent period than for the longer period.

• This is the Green Chemistry icon.

An obvious way to reduce pollution is not to have the pollutants form in the first place. Over the past decade, an important initiative known as "green chemistry," the use of chemistry to prevent pollution, has taken place. **Green chemistry** is the designing of chemical products and processes that reduce or eliminate the use and/or generation of hazardous substances. Begun under the EPA's Design for the Environment Program, green chemistry reduces pollution through fundamental chemical breakthroughs in designing and redesigning processes that make chemical products, with an eye toward making them environmentally friendly, that is "benign by design." In this regard, Dr. Barry Trost, a Stanford University chemist, advocates an "atom economy" approach to the synthesis of commercial chemical products such as pharmaceuticals, plastics, or pesticides. Such syntheses would be designed so that all reactant atoms end up as desired products, not as wasteful by-products. This approach will save money, as well as materials; undesired products would not be produced as waste, which requires disposal.

Dr. Darryl Busch, a recent president of the American Chemical Society, speaking at a conference on green chemistry said "The principles of green chemistry represent the pillars that have the potential to hold up our sustainable future. . . . A commitment to green chemistry will allow our economy to progress without creating new messes for the next generation to worry about." Dr. Lynn R. Goldman, an EPA administrator, says "Green chemistry is preventative medicine for the environment." Innovative "green" chemical methods already have made an impact on a wide variety of chemical manufacturing processes by decreasing or eliminating the use or creation of toxic substances. For example, the use of green chemical principles has led to cheaper, less wasteful, and less toxic production of ibuprofen, pesticides, new materials for disposable diapers and contact lenses, new dry cleaning methods, and recyclable silicon wafers for integrated circuits. The research chemists and chemical engineers who developed these and other green chemistry approaches have received the Presidential Green Chemistry Challenge Award. Begun in 1995, it is the only Presidential-level award recognizing chemists and the chemical industry for their innovations for a less polluted world; its theme is "Chemistry is not the problem, it's the solution." Since 1995, the products and processes developed by winners of the Presidential Green Chemistry Challenge Award have eliminated more than 102 billion pounds of hazardous chemicals and saved more than 3 billion gallons of water and 26 million barrels of oil. At various places throughout this book, we will discuss applications of green chemistry. They are designated by the Green Chemistry icon.

To better understand how pollutant levels can be reduced, we need to know some chemistry about the pollutants and their sources.

1.12 Air Pollutants: Sources

By now you should recognize that the atmospheric concentration of air pollutants such as carbon monoxide, the oxides of sulfur and nitrogen, and ozone is far less than that of nitrogen and oxygen, the major atmospheric components. But by their presence, these minor components can compromise the quality of the air we breathe, even at parts per million or parts per billion levels. Much of the carbon monoxide, sulfur dioxide, nitrogen oxides, and ozone in air result from modern society's demands for energy. Most of the energy is used in generating electricity and in transportation, for which gasoline powers millions of cars and trucks. The burning of coal is the major U.S. source of electric power and of SO_2. Coal is mostly carbon and hydrogen, and thus the major products of its combustion are carbon dioxide and water. But coal is a complex mixture of variable composition, not merely carbon and hydrogen. Most coals contain rock-like minerals and 1–3% sulfur. When coal is burned, the sulfur is converted into gaseous sulfur dioxide, and the minerals are converted into fine ash particles. If they are not removed, the particles and the sulfur dioxide gas go up the smokestack.

- You will learn in Chapter 3 that the CO_2 released by combustion is believed to contribute to global warming.

Sulfur dioxide can react with more oxygen to form sulfur trioxide, SO_3.

$$2\ SO_2 + O_2 \longrightarrow 2\ SO_3 \qquad [1.7]$$

This reaction is normally quite slow, but it is much faster in the presence of small ash particles. The ash particles also aid another process. If there is high humidity in the air, they promote the conversion of water vapor into an aerosol of tiny water droplets, which we call fog. An **aerosol** is a form of liquid in which the droplets are so small that they stay suspended in the air rather than settle out. Once sulfur trioxide is formed, it dissolves readily in the water droplets to form an aerosol of sulfuric acid, H_2SO_4.

$$H_2O + SO_3 \longrightarrow H_2SO_4 \qquad [1.8]$$

- The shapes of molecules are important, as we will discuss in Chapter 3.

When inhaled, the sulfuric acid aerosol droplets are small enough to be trapped in the lung tissue where they cause severe damage. Moreover, the sulfur oxides and sulfuric acid are also major contributors to acid precipitation, the topic of an entire chapter in this book (Chapter 6).

As noted in Figure 1.12, atmospheric levels of SO_2 have been decreasing slowly as a result of the Clean Air Act of 1970, which mandated reductions in power plant emissions. More stringent regulations were established in the Clean Air Act Amendments of 1990. Progress will not come cheaply. More information about the strategies and technologies available to reduce atmospheric SO_2 and their economic and political costs is included in Chapter 6.

The vast majority of cars are powered by internal combustion engines that run on gasoline. Because gasoline contains only very small quantities of sulfur, the automobile is not a significant source of sulfur dioxide. However, the ubiquitous motor car, as well as light trucks and SUVs, contribute to atmospheric concentrations of carbon monoxide, lead, nitrogen oxides, ozone, and a number of other unhealthful substances. The problem is particularly acute in America because the United States has more automobiles per capita than any other nation. There are more than 220 million vehicles, greater than one for every two Americans. In some cities, such as Denver, Houston, and Los Angeles, 90% of the working population commutes to work by car—often with only one person per vehicle.

The combustion reaction of octane, a component of gasoline, has already been discussed. From an energy and environmental standpoint, the ideal combustion products would be CO_2 and H_2O (equation 1.5). But a modern high-performance automobile, capable of operating at high speeds and with fast acceleration, is a source of carbon monoxide and some incompletely burned fragments of gasoline molecules called *volatile organic compounds* (VOCs). This incomplete combustion is caused by either insufficient oxygen or insufficient time in the engine cylinders for all the hydrocarbons to be burned to carbon dioxide and water. The problem of automobile carbon monoxide emissions has been partially solved by the installation of catalytic converters, devices that accelerate the conversion of CO in the exhaust stream to CO_2. The dramatic reduction in CO emissions shown in Figure 1.12 has occurred even though the number of cars has doubled in the past 25 years. The CO decrease is due to several factors: better engine design; computerized sensors that better adjust the fuel/oxygen mixture; and most importantly, all new cars since the mid-1970s have catalytic converters (Figure 1.14).

There is even more good news. For more than 50 years, a compound named tetraethyl lead was added to gasoline to make it burn more smoothly by eliminating premature ex-

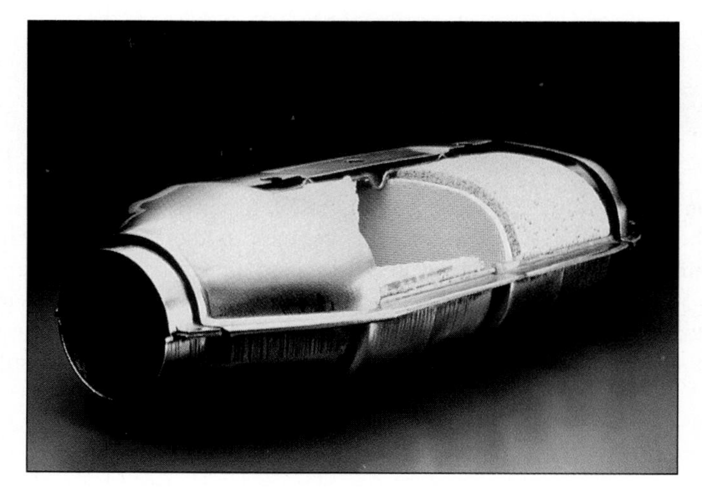

Figure 1.14

Catalytic Converter. A cutaway view of an automobile catalytic converter. Semiprecious metals such as platinum and rhodium are coated on the surface of ceramic beads. The metals catalyze the combustion of CO to CO_2. Other catalysts in the converter accelerate the conversion of nitrogen oxides to N_2 and O_2.

plosion or "knocking." It worked beautifully, but unfortunately the lead was released through the tailpipe to the atmosphere and ultimately into water supplies. Lead is a highly toxic element—a cumulative poison that can cause a wide variety of neurological problems, especially if ingested by children. Moreover, lead can also destroy the effectiveness of catalytic converters. With the advent of catalytic converters it became necessary to formulate gasoline without tetraethyl lead. Since 1976, all new cars and trucks sold in the United States have been designed to use only unleaded fuel. The result has been a dramatic decrease in lead emissions—from more than 219,000 tons in 1970 to 5,000 tons in 1997 (see Figure 1.12). Leaded fuel was banned by law in the United States in 1997.

The U.S. has had less success curbing the emission of nitrogen oxides. Cars are also a major source of these noxious gases. Nitrogen and oxygen are always present wherever there is air. And, when this mixture is subjected to high temperatures, as in an internal combustion engine, the following reaction occurs.

$$N_2 + O_2 \longrightarrow 2\ NO \qquad [1.9]$$

Subsequent reactions with atmospheric oxygen can generate other oxides of nitrogen, including nitrogen dioxide, NO_2. These oxides can be highly toxic and contribute to acid rain, the subject of Chapter 6.

The quantity of nitrogen oxides emitted into the atmosphere has decreased 21% since 1980. Given the significantly increased number of vehicles and miles driven, the decrease is impressive. The decline, in part, is due to the emission controls mandated starting in 1970. The Clean Air Act of 1970 set tailpipe emission standards, to go into effect in 1975, and these were subsequently revised. These are the current standards.

CO	3.4 g/mi
NOx	0.4 g/mi
NMHCs*	0.25 g/mi

*Non-methane hydrocarbons.

New NMHC standards began with 1996 models. If the EPA finds that the1999 data indicate the need to lower the standards, the standards will be cut in half, starting with the 2004 model year. EPA has been collecting and studying the 1999 data to make that assessment.

Despite early claims from the auto industry that it would be impossible, or too costly, to meet the standards, the industry has, in fact, achieved these goals by using improved catalytic converters, engine designs, and gasoline formulations.

1.29 Consider This: Electric Cars

Many people believe that the only true solution to the pollution caused by gasoline-powered cars is to promote widespread development and use of electric cars. Such cars are no longer just a hope for the future, but are currently available in some areas. What are the criteria that you would use when deciding whether to buy an electric car?

- Electric cars and other alternatives to gasoline-powered vehicles are discussed in Chapter 8.

1.13 Air Quality at Home and Abroad

Anyone who reads the newspapers or listens to broadcast news analysis knows that the control of automobile emissions is not a matter of technology alone. Economics and politics are also part of this complex web. For example, the U.S. automobile industry did not act to reduce emissions until forced to do so by federal legislation in 1970. There were no significant changes in the standards during the 1980s, when the national administration advocated less governmental regulation of private industry. The Clean Air Act signed by former-President George Bush in November 1990 was the first major new clean air legislation in 20 years, but it did not come about easily. There was a great deal of political infighting as Congress and the administration struggled to find acceptable compromises. The controversy continues as members of Congress seek a balance

- Clean air legislation is revisited in Section 6.14.

between government regulations that protect the environment and initiatives by commercial and industrial interests that could lessen environmental quality.

Air pollution is primarily an urban problem, and more than 50% of all Americans live in cities with populations over 500,000. Many of these cities fail to meet the national air quality standards, at least during periodic pollution alerts, and some at other times as well. In spite of recent improvement in air quality, we still have difficulties, especially with nitrogen oxides and ozone. Children and adults with chronic respiratory problems or heart disease are most at risk from exposure to these pollutants, and the risk is increased by vigorous physical activity. Furthermore, there is evidence that the present air quality standards provide little margin of safety in protecting public health. The American Lung Association estimates that $50 billion in health benefits could be realized annually in the United States if air quality standards were met throughout the country.

Increasingly, U.S. industries have been held accountable for their waste emissions. Annually since 1987, the EPA has published the Toxics Release Inventory (TRI), a national directive requiring companies to make available to the public data on the amounts of certain chemicals they have released into the air, water, and land. TRI is part of the Community Right To Know Law, and a kind of snapshot for a given time of the state of pollution from a list of pollutants as reported by U.S. industries. Former Vice-President Al Gore praised the program noting: "It has spurred innovation to help business work smarter and cleaner and become more profitable."

The law has had a significant impact; toxic emissions from all industries declined by nearly 50% since the law took effect. The EPA data indicate that the U.S. chemical industry has decreased its emissions 53%, the largest reduction by any manufacturing industry, in the years since the law began. In 1999, the releases into the environment continued to decline, down 19% since 1995. Figure 1.15 illustrates the nature of the environmental releases by the chemical industry; 52% of the total chemical emissions were released into the air. Just seven compounds make up nearly 65% of the emissions from the chemical industry, an indication of how directly these substances are associated with synthesizing the compounds produced by that industry. The compounds are ammonia (NH_3), methanol (CH_3OH), carbon disulfide (CS_2), hydrochloric acid (HCl), and three compounds important in plastics production—ethylene (C_2H_4), propylene (C_3H_6), and toluene (C_7H_8).

The EPA has the legal power to fine companies and agencies that violate pollution standards. During 2000, the EPA assessed fines of nearly 225 million dollars against violators who, in addition, spent 2.6 billion dollars to decrease their emissions, clean up their polluting, and improve their monitoring of releases to the environment.

TRI data are available for each state. The following exercise gives you the opportunity to check on the toxic emissions in your state or locale and the progress made in reducing those emissions.

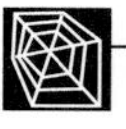

1.30 Consider This: TRI and You

Go to the EPA web site and find the Toxic Release Inventory (TRI) for your state and for your locale.

a. Compare the current levels of toxic emissions with what they were one or two years ago.

b. Which emissions (if any) have decreased and which have remained the same or increased? Determine, if you can, a reason for the changes.

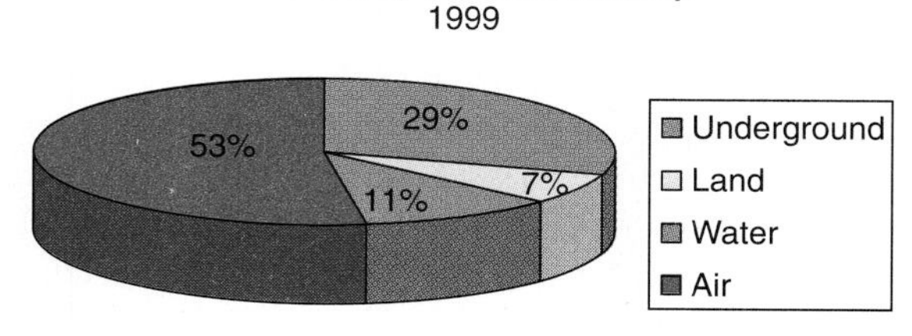

Figure 1.15

Environmental releases from the chemical industry.

Source of data: United States Environmental Protection Agency.

We face difficult political and economic choices. Are we willing to spend the money that would be needed to really clean up the air we breathe? What would happen if regulations were dropped or relaxed, as some have proposed? Would the supposed gains by the American economy compensate for the hidden costs? In considering the risks with the benefits, a tightening of regulations could mean a boon to health and a significant reduction in health care costs associated primarily with respiratory conditions. The improved air quality that we have enjoyed could be short-lived. Our environment is a fragile system that without mandated emission controls could quickly revert to a status of severe air pollution such as what existed just three decades ago in this country.

International comparisons indicate that air pollution does not respect nations (Table 1.10). It has the potential to be a major problem anywhere that has the defining

Table 1.10

International Air Quality

A. World Heath Organization (WHO) Air Quality Guidelines (1993–98)

Pollutant	Guideline, $\mu g/m^3$
SO_2	5–400
Particulates	Standard being revised
Lead	0.01–2
CO	500–7000
NO_2	10–150
O_3	10–100

B. Air Quality In Megacities Around the World (1992 Data). Data from the Atmospheric Research and Information Centre, Manchester Metropolitan University, Manchester, England.

City	SO_2	Particulate Matter	Lead	CO	NO_2	O_3
Bangkok, Thailand	Low	Serious	Moderate to Heavy	Low	Low	Low
Beijing, China	Serious	Serious	Low	—	Low	Moderate to Heavy
Cairo, Egypt	—	Serious	Serious	Moderate to Heavy	—	—
Jakarta, Indonesia	Low	Serious	Moderate to Heavy	Moderate to Heavy	Low	Moderate to Heavy
London, England	Low	Low	Low	Moderate to Heavy	Low	Low
Mexico City, Mexico	Serious	Serious	Moderate to Heavy	Serious	Moderate to Heavy	Serious
Moscow, Russia	—	Moderate to Heavy	Low	Moderate to Heavy	Moderate to Heavy	—
Rio de Janeiro, Brazil	Moderate to Heavy	Moderate to Heavy	Low	Low	—	—
Sao Paulo, Brazil	Low	Moderate to Heavy	Low	Moderate to Heavy	Moderate to Heavy	Serious
Tokyo, Japan	Low	—	Low	Low	Serious	

Low pollution—normally meets WHO guidelines (on occasion may exceed guidelines short-term).

Moderate to heavy pollution—WHO guidelines exceeded up to twice as much (short-term guidelines regularly exceeded at certain locations).

Serious pollution problem—WHO guidelines exceeded at greater than twice as much.

—Data not available

characteristics of modern industrial development—electrical power generation and many automobiles. Air pollution problems in the United States pale by comparison with those in other parts of the world. Many countries have few or no controls on pollutant emissions. In some cases this is because of the political system. Regulatory laws have been passed, but enforcement of them has been lax, often because of the enormous pressures to put economic development before environmental quality. Germany has 10 times as many automobiles per square mile as the United States, yet it has fewer emission controls on cars (or on electric power plants). As a result, parts of Germany face much more serious air pollution problems than we have in this country.

The situation in Eastern Europe is especially bad because heavy industrialization has occurred without the constraints of pollution controls. Portions of the region have become nearly uninhabitable. A very low grade of coal, called "brown coal," is widely used; large lead smelters release huge amounts of lead into the atmosphere. In China, 28 northern cities have SO_2 and particulate concentrations that are three to eight times higher than the guideline limits set by the World Health Organization (WHO). And the 20 million residents of Mexico City breathe ozone levels more than 50% above WHO guidelines for most of the year.

Several steps have been taken by the Mexican government to improve air quality in Mexico City. Catalytic converters are required on new cars; regulations have been placed on industries, and unleaded and cleaner-burning fuels have been introduced. In addition, a certified analysis of exhaust gases from automobiles is mandated each six months. Analyses classify cars into three groups, identified by a windshield tag. Category Zero includes those cars manufactured since 1992 and those with exhaust within legal limits for pollutants. Such cars can be driven daily. Category One cars have a higher, but not excessive, level of exhaust pollutants. These cars can be driven six days a week with the "off" days distributed evenly across the week. Category Two cars produce high levels of exhaust pollutants; driving these cars is limited during an "air quality alert" to either day during the first two days of such an alert. No cars in this category can be driven from the third day onward until the air quality alert is lifted.

1.31 Your Turn

Based on the data in Table 1.10,

a. Which city has the worst overall air quality?
b. Of the 10 cities, which one has the best air quality?
c. In which cities do automobiles contribute significantly to air pollution? Give a rationale for your answer.

1.32 Consider This: Growing Interest in Air Pollution

Air pollution has not occurred overnight. It has been a growing problem since at least the time of the Industrial Revolution. Why have we as a nation and a world community become so concerned with it lately? Through discussion and/or library and web research, identify at least four factors that have combined to make air pollution an important issue at present.

1.14 Breathing Lessons—Indoor Air

The 1.1 Consider This exercise had you "take a breath." That one breath adds up to between 1×10^4 to 2×10^4 L (10–20 m^3) of air each of us breathes daily. As we have learned from previous sections, higher air quality standards have decreased the allowable concentrations of various air pollutants by controlling emissions from automobiles and industries. But air quality depends on where we are. Ironically, such standards have been established for outdoor air, but not for indoor air. Yet, most of us sleep, work, study, and play indoors, spending up to 95% of our time in our dorm rooms, classrooms, offices, and residences. Consequently, we should be concerned not just with

Table 1.11

Indoor Air Pollutants and Their Sources

Phases of Matter	Source	Pollutant
Solid/particulate	Floor tile	Asbestos
	Pets	Pet dander, dust
	Plants	Molds, mildew, bacteria, viruses
Liquid/gas	Carpet	Styrene
	Cigarette smoke	Formaldehyde, carbon monoxide
	Clothes	Dry-cleaning fluid, moth balls
	Electric arcing	Ozone
	Faulty furnace or space heater	Carbon monoxide
	Furniture	Formaldehyde
	Glues and solvents	Acetone, toluene
	Paint and paint thinners	Methanol, methylene chloride
	Soil and rocks under house	Radon

outdoor air quality, but that of indoor air as well. With the exception of sulfur dioxide, the average indoor concentrations of the pollutants exceed their average outdoor concentrations.

Indoor air is a complex mixture; typically nearly 1000 substances are detectable in it at the ppb level or higher. However, indoor air sources are limited to either the outside air that enters buildings, or air that comes from within buildings. Tobacco smoke, cooking by-products, substances emitted from rugs, furniture, construction materials, and office products are some of the many materials that can degrade indoor air quality. Table 1.11 lists some of the sources of indoor air pollutants.

How quickly air pollutants build up indoors depends on the rates at which outdoor air moves inside and indoor air moves out. The buildup of pollutants also depends on how rapidly they are generated indoors. An insufficient exchange of outside air can cause the concentration of indoor air pollutants to build up to troublesome levels. Consider the risk-benefit trade-off in which buildings constructed within the past two decades have been more airtight to minimize drafts and increase energy efficiency. Although enhanced energy efficiency has been achieved, it has been at the cost of decreasing the flow of outside air into the building to replace the indoor air. When this reduction in air exchange occurs, it allows the concentration of indoor air pollutants to increase. Therefore, initially what was a benefit (better energy efficiency) can turn into an increased risk (increased pollutants concentration). Construction of some large office buildings has been so highly energy efficient that little exchange of outside air occurs within them. In some of these cases, the reduced air exchange has allowed indoor air pollutants to reach levels hazardous to the health of some individuals, creating a condition known as "sick building syndrome."

Radon, a colorless, odorless, tasteless, but radioactive gas, is a possible indoor air pollutant, depending on your location. Radon is generated naturally by the nuclear decomposition of uranium in the rocks and soil on which buildings rest. In some regions of the country, sufficient radon gas is present to seep into buildings through cracks in the foundations. Extended inhalation of radon gas can result in lung cancer. But the exposure levels of radon required to create health problems is controversial. Home radon testing kits are commercially available (Figure 1.16).

- Radioactivity is discussed in detail in Chapter 7.

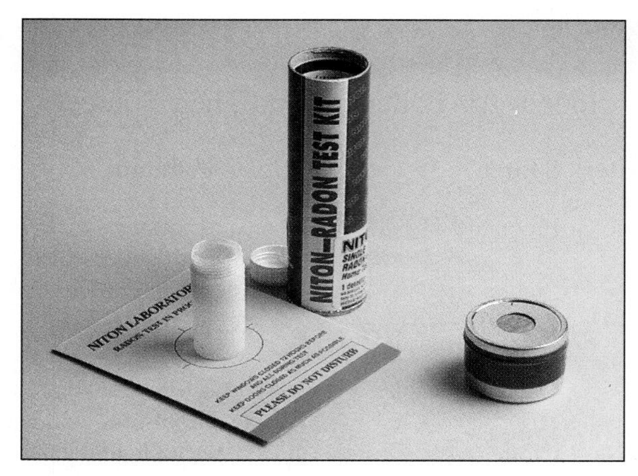

Figure 1.16
A home radon test kit.

1.33 Consider This: Rating Radon

As a public service, local and national agencies provide documents on the Web about radon. Search the Web to bring up a list about radon. In addition to searching just for "radon," you might want to add the terms "detection," "air quality," and/or "EPA."

a. Find two web sites about radon provided by government agencies. For each list the title, the source, and the URL.
b. How can you measure the radon levels in your home? Search the Web for a company that sells radon test kits. Describe the kit, including its price. If you don't find anything, switch to another search engine.
c. Is information from commercial sources about radon any different in its objectivity from that provided by agencies as a public service? If so, discuss the differences and reasons for them.

Whether we breathe indoor or outdoor air, we inhale (and exhale) a truly prodigious number of molecules and atoms during a lifetime. On a molecular and atomic level, these particles have some fascinating properties, ones that we consider next.

1.15 Back to the Breath—at the Molecular Level

The maximum concentrations of pollutants specified in Table 1.2 seem very small—and they are. Nine CO molecules out of one million particles of the mixture called air is a tiny fraction. But, as we will soon calculate, at this CO concentration a breath of air contains a staggering number of carbon monoxide molecules. This apparent contradiction is a consequence of the minuscule size of molecules and the immense numbers of them. Recall 1.1 Consider This: Take A Breath. If you are an average-sized adult in good physical condition, the total capacity of your lungs is between five and six liters. But you do not exchange this volume of air each time you take a breath. Even now as you are reading this, you are inhaling (and exhaling) only about 500 milliters (0.500 L) of air, approximately a half quart. Determining the number of molecules in this volume of air is no easy task, but it can be done. As a result of experiments (as well as theories), we know that a typical breath of 500 milliliters contains about 2×10^{22} particles – molecules such as N_2, O_2, CO, and individual atoms like Ar.

Using the number of particles in a breath, we can calculate the number of CO molecules in the breath you just inhaled. We will assume the breath contained 2×10^{22} molecules and that the CO concentration in the air was the national ambient air quality standard of 9 ppm. This means that out of every million (1×10^6) molecules of air, nine will be CO molecules. To compute the number of CO molecules in the breath we multiply the total number of air molecules by the fraction of them that are carbon monoxide molecules.

$$\text{number of CO molecules} = 2 \times 10^{22} \text{ air particles} \times \frac{9 \text{ CO molecules}}{1 \times 10^{6} \text{ air particles}}$$

$$= \frac{18 \times 10^{22}}{1 \times 10^{6}} \text{ CO molecules}$$

$$= \frac{18}{1} \times \frac{10^{22}}{10^{6}} \text{ CO molecules}$$

Note that in writing out this problem, we retain the labels on the numbers. This is a reminder of the physical entities involved, but it also provides a guide for setting up the problem correctly. The labels "air particles" cancel each other, and we are left with what we want—CO molecules.

However, we need to divide 10^{22} by 10^6 to convert the answer into a simpler number. To *divide* powers of 10, you simply *subtract* the exponents. In this case,

$$\frac{10^{22}}{10^{6}} = 10^{(22-6)} = 10^{16}$$

So, this means that there are $\frac{18}{1} \times \frac{10^{22}}{10^{6}}$ CO molecules $= 18 \times 10^{16}$ CO molecules in the breath.

Carbon monoxide monitors are commercially available to be used in residences, as well as offices and business sites.

The preceding answer is mathematically correct, but in scientific notation it is customary to have only one digit to the left of the decimal point. Here we have two: 18. Therefore, our last step will be to rewrite 18×10^{16} as 1.8×10^{17}. We can make this conversion because $18 = 1.8 \times 10$, which is the same as 1.8×10^1. We *add* exponents to *multiply* powers of 10. Thus, $18 \times 10^{16} = (1.8 \times 10^1) \times 10^{16}$ CO molecules, which equals 1.8×10^{17} CO molecules in that last breath you inhaled. (If all of this use of exponents is coming at you a little too fast, please consult Appendix 2.)

It may sound surprising, but it would be more accurate to round off the answer and report it as 2×10^{17} CO molecules. Certainly 1.8×10^{17} looks more accurate, but the data that went into our calculation were not very exact. The breath contains *about* 2×10^{22} molecules, but it might be 1.6×10^{22}, 2.3×10^{22}, or some other number. The jargon is that 2×10^{22} expresses a physically based property to "one **significant figure**." Only one digit, the initial 2, is used, and so there is only one significant figure in 2×10^{22}. That means that the number of molecules in the breath is closer to 2×10^{22} than to 1×10^{22} or to 3×10^{22}, but we cannot say with certainty much beyond that. Similarly, unless the analytical data are very good, the concentration of carbon monoxide is also known to only one significant figure, 9 ppm. The product 2×9 equals 18. That is certainly correct mathematically, but our question about CO is based on physical data. The answer, 1.8×10^{17} CO molecules, includes two significant figures, the 1 and the 8. It implies a level of knowledge that is not justified. The accuracy of a calculation is limited by the *least accurate* piece of data that goes into it. In this case, both the concentration of CO and the number of air particles in the breath were each known to only one significant figure (9 and 2, respectively); two significant figures in the answer are unjustified. The rule is that you cannot improve the accuracy of experimental measurements by ordinary mathematical manipulations like multiplying and dividing. Therefore, the answer must also contain only one significant figure; hence 2×10^{17}.

1.34 Your Turn

The local news has just reported that today's ground-level ozone readings are right at the acceptable standard, 0.12 ppm. How many molecules of ozone, O_3, are in each breath of this air, assuming there are 2×10^{22} molecules of air in each breath?

Answer

$$2 \times 10^{22} \cancel{\text{air particles}} \times \frac{0.12\ O_3 \text{ molecules}}{10^6 \cancel{\text{air particles}}} = 2 \times 10^{15}\ O_3 \text{ molecules in each breath}$$

You may well question the significance of all of this talk about significant figures, but it is very important in interpreting numbers associated with physical quantities. It has been observed that "figures don't lie, but liars can figure." Numbers often lend an air of authenticity to newspaper or television stories, so popular press accounts are full of numbers. Some are meaningful and some are not, and the informed citizen must be able to discriminate between the two types. For example, the assertion that the concentration of carbon dioxide in the atmosphere is 365.5537 ppm should be taken with a rather large grain of sodium chloride (salt). The estimate of 365 ppm (three significant figures) is reasonable; the previous assertion with seven significant figures is not valid.

There are other ways in which numbers sometimes introduce ambiguity. You have just encountered some conflicting information. The concentration of CO in air is very small, 9 parts per million. Nevertheless, the number of CO molecules in a breath is still large, about 2×10^{17}. Both statements are true. The consequence of these numbers is that it is *impossible* to completely remove pollutant molecules from the air. "Zero pollutants" is an unattainable goal; using the most sophisticated detection methods you still could not even determine whether it had been achieved. At present, our most sensitive methods of chemical analysis are capable of detecting one target molecule out of a trillion. One part per trillion corresponds to: moving six inches in the 93 million-mile trip to the sun; a single second in 320 centuries; or a pinch of salt in 10,000 tons of potato chips. A chemical could be undetectable at this level, and yet a breath might still include 2×10^{10} molecules of the substance.

- Absence of evidence is not the same as evidence of absence. The substance may be present, but in undetectable amounts.

1.35 Your Turn

To help you comprehend the magnitude of the 2×10^{17} CO molecules in just one of your breaths, assume that they were equally distributed among the 6 billion (6×10^9) human inhabitants of the Earth. Calculate each person's share of the 2×10^{17} CO molecules you just inhaled.

Hint: You are trying to distribute the huge number of molecules in a breath among all the human inhabitants of the Earth. Each person's share can be found by dividing the total number of CO molecules by the total number of humans:

$$\text{Each person's share is } \frac{2 \times 10^{17} \text{ CO molecules}}{6 \times 10^9 \text{ people}}$$

Now see if you can demonstrate that each person's share is 3×10^7 or 30,000,000 molecules of CO per person (to one significant figure).

A breath of air typically contains molecules of hundreds—perhaps thousands—of different compounds, most in minuscule concentrations (Figure 1.17). For almost all these substances, it is impossible to say whether the origin is natural or artificial. Indeed, many trace components, including the oxides of sulfur and nitrogen, come from both natural sources and those related to human activity. And, as with all chemicals, "natural" is not necessarily good and "human-made" is not necessarily bad. As you read in Section 1.4, what matters is exposure, toxicity, and the assessment of risk.

In addition to being extremely small, the particles in your breath possess other remarkable characteristics. In the first place, they are in constant motion. At room temperature and pressure, a nitrogen molecule travels at about 1000 feet per second and experiences approximately 400 billion collisions with other molecules in that time interval. Nevertheless, relatively speaking, the molecules are quite far apart. The actual volume of the extremely tiny molecules making up the air is only about 1/1000th of the total volume of the gas. If the particles in your 1-liter breath were all squeezed together, their volume would be about 1 milliliter (1 mL)—about one-third of a teaspoon. Sometimes people mistakenly think that air is empty space. It's 99.9% empty space, but the matter that is in it is literally a matter of life and death!

pph = 1 yd out of length of one field (100 yards)

ppthousand = 1 yd out of length of ten fields

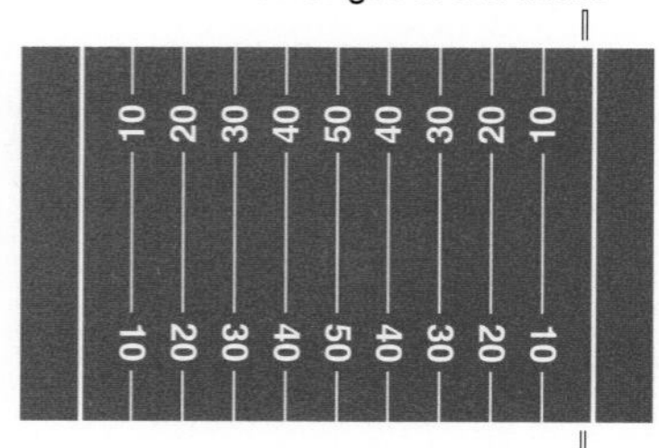

ppm = 1 yd out of length of 10^4 fields

ppb = 1 yd out of length of 10^7 fields

Parts per hundred, parts per million (ppm), and parts per billion (ppb).

Moreover, it is matter that we continuously exchange with other living things. The carbon dioxide we exhale is used by plants to make the food we eat, and the oxygen that plants release is essential for our existence. Our lives are linked together by the elusive medium of air. With every breath we exchange millions of molecules with each other. As you read this, your lungs contain 4×10^{19} molecules that have been previously breathed by other human beings, and 6×10^{8} molecules that have been breathed by some *particular* person— say Julius Caesar, Marie Curie, or Martin Luther King, Jr. Pick your favorite hero or heroine—your body almost certainly contains atoms that were once in his or her body. In fact, the odds are very good that right now your lungs contain one molecule that was in Caesar's *last* breath. The consequences are breathtaking!

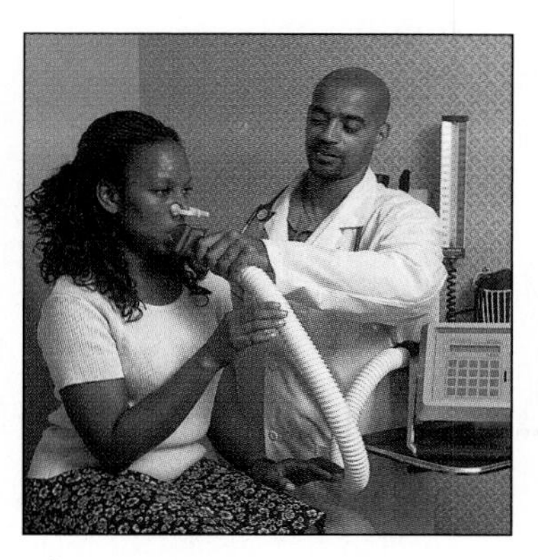

Figure 1.17
A spirometer is an instrument used for measuring an individual's breathing capacity.

1.36 The Skeptical Chymist: Caesar's Last Breath

We just claimed that your lungs currently contain one molecule that was in Caesar's last breath. That assertion is based on some assumptions and a calculation. We are not asking you to reproduce the calculation, but rather to identify some of the assumptions and arguments that we might have used. Are they reasonable?

Hint: Here is a start. The calculation assumes that all of the molecules in Caesar's last breath have been uniformly distributed throughout the atmosphere.

In starting this chapter with "taking a breath," we began a pattern for a kind of activity that we will ask you to reflect on in several places in this text. The next "take a . . . " activity will come up in Chapter 5 where you will be asked to "take a drink" of water. Several of the themes that arose in Chapter 1 regarding air will also appear in that later chapter dealing with water—the quality of a natural resource essential to life, its sources, its special properties, and the risks associated with its quality or lack of it.

CONCLUSION

The air we breathe has a personal and immediate effect on our health. Our very existence depends on having a large supply of relatively clean, unpolluted air with its essentials for life—the elements, oxygen and nitrogen, and two compounds, water and carbon dioxide. But air can be polluted with potentially toxic substances such as carbon monoxide, ozone, sulfur oxides, and nitrogen oxides. This is true especially in the urban environments of our large cities, the very places where the majority of Americans live. The major pollutants are, for the most part, relatively simple chemical substances. Carbon monoxide and the oxides of sulfur and nitrogen are compounds that exist as molecules made from atoms of their constituent elements. These compounds are formed by chemical reactions, often as unavoidable consequences of our dependence on coal for energy production in power plants and gasoline in internal combustion engines. Over the past 30 years, governmental regulations, industrial participation, modern technology, and green chemistry have resulted in large reductions in many pollutants. But it is not possible to reduce pollutant concentrations to zero because of the minuscule size of atoms and molecules and their immense numbers. Rather we must ask what the risk is from a given level of pollutant and then what level of risk is acceptable for various population groups.

The oxygen-laden air we breathe, whether indoors or out, is, of course, very close to the surface of the Earth. But the Earth's atmosphere extends upward for considerable distance and contains other substances that are also essential for life on this planet. In the next two chapters, we consider two of these substances and how they are changing, perhaps as a result of human activities.

Chapter Summary

Having studied this chapter, you should be able to:
(The numbers that follow indicate the sections in which the topics are introduced and explained):

- Recognize the composition of air and reasons for local and regional variations of it (1.1–1.3);
- Understand factors behind air quality and the chief components of air pollution (1.3, 1.11, 1.12);
- Evaluate conditions significant in risk-benefit analysis (1.4);
- Identify the general regions of the atmosphere with respect to altitude and the relationship of air pressure to altitude (1.5);
- Interpret air quality data in terms of concentration units (ppm, ppb) and pollution levels, including unreasonableness of "pollution free" levels (1.2–1.3, 1.12, 1.14, 1.15);
- Differentiate among mixtures, elements, and compounds (1.6);
- Understand the differences between atoms and molecules, between symbols for elements and formulas for chemical compounds (1.7);
- Name selected chemical elements and compounds (1.7);
- Write and interpret chemical formulas (1.8);
- Balance chemical equations, including using sphere equation representations (1.9–1.10);
- Discuss the green chemistry initiative (1.11);
- Describe the nature of air quality policies in this country and abroad in terms of their effectiveness in controlling air pollution (1.12–1.13);
- Identify the sources and nature of indoor air pollution (1.14);
- Interpret the nature of air at the molecular level (1.15);
- Use scientific notation and significant figures in performing basic calculations (1.4 and 1.15, respectively).

Questions

The questions in this chapter, as well as those in the remaining chapters, are divided into three categories:

- **Emphasizing Essentials** These questions give you the opportunity to practice the fundamental skills to be developed in the chapter. This set of questions relates most closely to the Your Turns in the chapter. Answers are provided in Appendix 5 for questions whose numbers are in color.
- **Concentrating on Concepts** These questions ask you to focus on the chemical concepts developed in the chapter and their relationships to the topics under discussion. They integrate and apply chemical concepts. This set of questions most closely resembles Consider This activities you have been engaged with throughout the chapter. Answers are provided in Appendix 5 for questions whose numbers are in color.
- **Exploring Extensions** These questions challenge you to go beyond the information presented in the text. They provide an opportunity for extending and integrating the skills, concepts, and communication abilities practiced in the chapter. Some extension questions are closely related to the type of analysis practiced in the *Sceptical Chymist* activities in the chapter. Questions marked with the web icon require using the World Wide Web to obtain further information.

Emphasizing Essentials

1. Calculate the volume of air that a person exhales in an 8-hr day. Assume that each breath has a volume of about 1 L and that the person exhales 15 times a minute.
2. Given that air is 78% nitrogen by volume, how many liters of nitrogen are in 500 L of dry air?
3. A 5.0-L mixture of gases is prepared for photosynthesis studies by combining 0.75 L of oxygen, 4.0 L of nitrogen, and 0.25 L of carbon dioxide. Compare the percentage of carbon dioxide gas with that normally found in the atmosphere.
4. Air contains 9000 ppm (parts per million) argon. Express this value as a percentage.
5. The smoke inhaled from a cigarette contains about 0.04% carbon monoxide, CO. Express this concentration in ppm.
6. The concentration of water vapor in the atmosphere of a tropical rain forest may reach 50,000 ppm. Express this value as a percentage.
7. According to Table 1.2, the percentage of carbon dioxide in inhaled air is *lower* than it is in exhaled air, but the percentage of oxygen in inhaled air is *higher* than in exhaled air. How can you account for these relationships?
8. The permissible limit for ozone for a 1-hr average is 0.12 ppm. If Little Rock, Arkansas registers a reading of 0.15 ppm for 1 hour, by what percent is Little Rock over the limit for atmospheric ozone?
9. Express each of these numbers in scientific notation.
   a. 1500 m, the distance of a foot race
   b. 0.0000000000958 m, the distance between O and H atoms in water

c. 0.0000075 m, the diameter of a red blood cell

d. 150,000 mg of CO, the approximate amount breathed daily

10. Write each of these values in nonscientific notation.

a. 8.5×10^4 g, the mass of air in an average room

b. 1.0×10^7 gallons, the volume of crude oil spilled by the Exxon Valdez

c. 5.0×10^{-3}%, the concentration of CO in the air of a city street

d. 1×10^{-5} g, the recommended daily allowance of vitamin D

11. Express each of these numbers in scientific notation.

a. 72000000 cigarettes; the number of cigarettes smoked per hour in the United States

b. 15000°C; the temperature near the spark plug in an automobile engine

c. 0.000000003 g; the number of grams of the insecticide DDT that dissolves in 1 g of water

d. 0.00022 g; the number of grams of NO_2 that can be detected by smell in 1 m^3 of air

12. Use Figure 1.6 to verify this statement: "Below about 20 km . . . the pressure decreases by about 50% for every 5 km increase in altitude." Does this relationship hold throughout the troposphere?

13. Consider this periodic table.

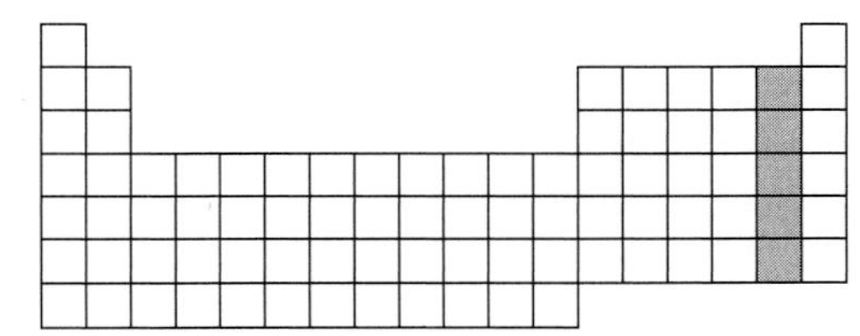

a. What is the group number indicated by the shading on this periodic table?

b. What elements make up this group?

c. What is a general characteristic of the elements in this group?

14. Consider this periodic table.

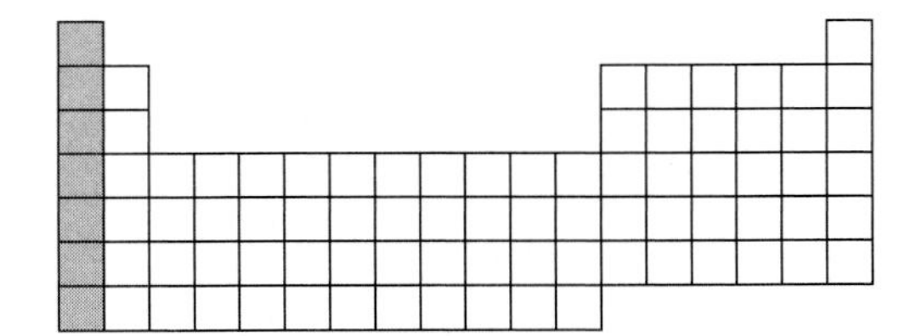

a. What is the group number indicated by the shading on this periodic table?

b. What elements make up this group?

c. What is a general characteristic of the elements in this group?

15. Classify each of these substances as an element, compound, or mixture.

a. a sample of "laughing gas" (dinitrogen monoxide)

b. steam coming from a pan of boiling water

c. a bar of deodorant soap

d. a sample of copper

e. a cup of mayonnaise

f. the helium filling a balloon

16. Name the compounds that contain these pairs of elements.

a. potassium and oxygen

b. aluminum and chlorine

c. sodium and iodine

d. magnesium and bromine

17. Give the correct formula for each of these substances.

a. "laughing gas," chemically named dinitrogen monoxide

b. ozone, used in water purification

c. sodium fluoride, an anti-cavity ingredient in toothpaste

d. carbon tetrachloride, formerly used as a dry-cleaning agent

18. What information does each formula convey about its compound, each a trace component of the atmosphere?

a. CH_2O, formaldehyde

b. H_2O_2, hydrogen peroxide

c. CH_3Br, methyl bromide

19. Write balanced chemical equations to represent these reactions.

a. nitrogen (N_2) reacting with oxygen gas (O_2) to form nitric oxide (NO)

b. ozone (O_3) decomposing into oxygen gas (O_2) and atomic oxygen (O)

c. sulfur (S) reacting with oxygen gas (O_2) to form sulfur trioxide (SO_3)

20. Write balanced sphere equations to represent each of the reactions in question 19.

21. Balance these equations, all of which involve the reaction of ethylene gas, C_2H_4, with oxygen gas, O_2.

a. $C_2H_4(g) + O_2(g) \longrightarrow C(s) + H_2O(g)$

b. $C_2H_4(g) + O_2(g) \longrightarrow CO(g) + H_2O(g)$

c. $C_2H_4(g) + O_2(g) \longrightarrow CO_2(g) + H_2O(g)$

22. Consider the three equations you balanced in question 21. Compare the coefficient for oxygen gas in these equations. How is it related to the products formed in each case?

23. Demonstrate that each of these equations is balanced by counting atoms of each element on either side of the arrow.

a. $2\ C_3H_8(g) + 7\ O_2(g) \longrightarrow 6\ CO(g) + 8\ H_2O(l)$

b. $2\ C_8H_{18}(g) + 25\ O_2(g) \longrightarrow 16\ CO_2(g) + 18\ H_2O(l)$

24. Platinum, palladium, and rhodium are used in automobile catalytic converters.

  a. What is the symbol for each of these metals?

  b. Where are these metals located on the periodic table?

  c. What can you infer about the properties of these metals, given that they are useful in this application?

25. If a room is 6 m long, 5 m wide, and 3 m high, how many milligrams of formaldehyde must be present if the concentration is reported as 40 ppm?

Concentrating on Concepts

26. In Section 1.1, air was referred to as ". . . that invisible stuff. . ." Is this always true? What factors influence if air appears "invisible" or if you can "see" it?

27. In 1.1 Consider This, you calculated the volume of air exhaled in a day. How does this volume compare with the volume of air in your chemistry classroom? Show your calculations. *Hint:* Think ahead about the most convenient unit to use for measuring or estimating the dimensions of your classroom.

28. In 1.3 Consider This, you considered how life on Earth would change if the concentration of oxygen were doubled. Now consider the opposite case; discuss how life on Earth would change if the concentration of O_2 were only 10%. Give some specific examples of how burning, rusting, and most metabolic processes in humans and plants would be affected.

29. Explain why the concentrations of some components in the atmosphere are expressed in percent (parts per hundred) and others are given in ppm (parts per million).

30. Consider this table of data from the U.S. Environmental Protection Agency, Office of Air Quality Planning and Standards. The data indicate the number of days metropolitan statistical areas failed to meet acceptable air-quality standards (Pollutant Standards Index rating over 100).

Air Quality of Selected U.S. Metropolitan Areas, 1992 – 1999

Metropolitan statistical area	1992	1993	1994	1995	1996	1997	1998	1999
Philadelphia, PA	24	51	26	30	22	32	37	32
Phoenix, AZ	13	16	10	22	17	12	17	12

  a. Prepare a visual representation of these data. Use any type of representation that you feel will best convey the information to the general public.

  b. Use your representation to discuss the trends in these two cities from 1992–99.

  c. Can you use your representation to *predict* the air quality in Philadelphia and Phoenix in 2000? Discuss your reasoning.

31. A certain city has an ozone reading of 0.13 ppm for one hour, and the permissible limit is 0.12 for that time. You have the choice of reporting that the city has exceeded the ozone limit by 0.01 ppm or saying that it has exceeded the limit by 8%. What are the advantages of each method?

32. a. Arrange these measurements in order of increasing size: 1 m, 3.0×10^2 m, 5.0×10^{-3} m.

  b. Draw an analogy between these three length measurements and time. Let 1 year be equal to 1 m. How long will the other two measurements be in terms of time expressed in years?

33. Air quality reports are often published in local newspapers, but rarely reported during televised weather reports, unless there is a dangerously high level of air pollution. Why do you think this is the case?

34. If risk is related to public perception, what is your feeling about the relative risks associated with each of these items? Rank them in order of your perception of the *most* risky to the *least* risky. Be prepared to explain your choices to class members.

smoking, using roller blades, eating beef, driving to work, getting a suntan, taking aspirin, drinking tap water, breathing polluted air

35. The cabins of commercial airliners flying at 30,000 feet are pressurized. Buildings in Denver, the "mile-high city," do not need to be pressurized. What is the best explanation for these observations? Figure 1.6 might be helpful in answering this question.

36. In these diagrams, the larger circles represent one kind of atom and the smaller circles represent a different kind. Characterize each of the samples as an element, compound, or mixture and give reasons for your answers.

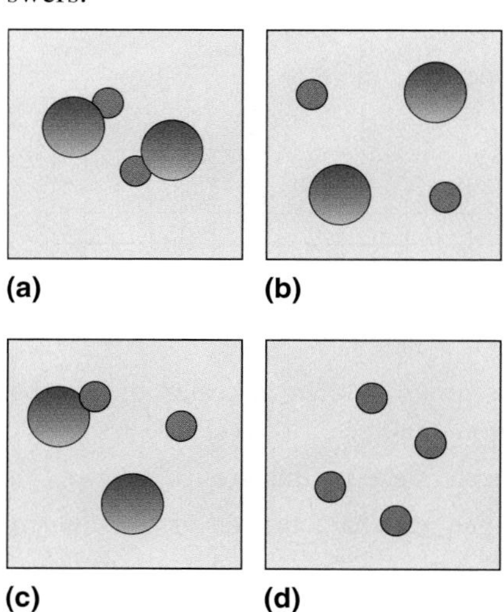

37. Consider this representation of the reaction between nitrogen and hydrogen to form ammonia.

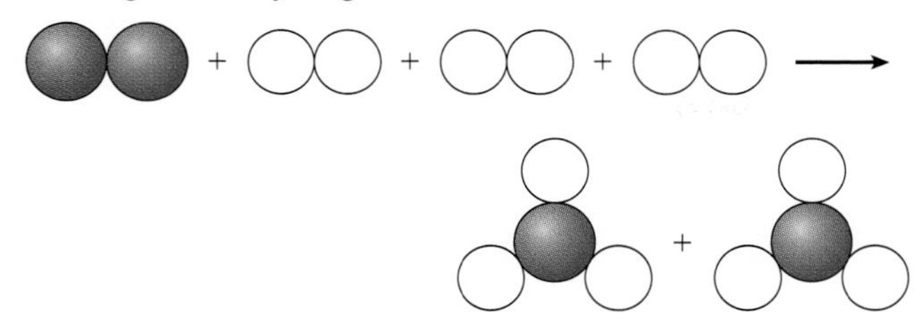

Use it to comment on these questions.

a. Are the mass of reactants and products the same?

b. Are the number of molecules of reactants and of products the same?

c. Are the total number of atoms in the reactants and the total number of atoms in the products the same?

38. Consider the information in Figure 1.13.

a. Control measures have not uniformly decreased all air pollutants during the period 1980 to 1999. Rank the pollutants from the one that shows the *greatest* percent reduction to the one having the *smallest* percentage reduction during this period.

b. What factors help explain the differences in the percent reduction?

39. What correlation might exist between the annual variation in lead concentrations in human blood and the lead compounds in gasoline in selected U.S. cities? Explain your reasoning.

40. Young adults in Beijing, China have gone to bars after work, not for glasses of beer or wine, but for fresh air. These "oxygen bars" provide one half-hour of deep breathing for the equivalent of $6.

a. What does this tell you about air pollution in Beijing?

b. Consider the information in Table 1.10. If you wanted to set up "oxygen bars" in other cities of the world, which would be your first likely markets?

41. Air quality in Santiago, Chile is such a major problem that driving private cars has been severely restricted. Special decals indicate the days particular cars can be driven. Some citizens purchased a second car and obtained a decal for that car. However, the increase in the total number of cars will likely make the pollution problem even worse. Write a letter to a friend in Santiago suggesting a possible solution to this problem and defending your suggestion.

42. The concentration of formaldehyde in *outdoor* air is typically about 0.01 ppm in urban areas, unless conditions are right for smog formation. The level of formaldehyde *indoors* can average 0.1 ppm, which is the level at which most people can smell its pungent odor. What factors can lead to formaldehyde accumulation indoors?

Exploring Extensions

43. The percentage of oxygen gas in the atmosphere (21%) is usually expressed as the volume of oxygen gas relative to the total volume of the atmosphere being considered. The percentage can also be reported as the mass of oxygen gas relative to the total mass of the atmosphere being considered; in this case, it is 23%. Offer a possible explanation for why these two values are not the same.

44. The EPA oversees the Presidential Green Chemistry Challenge Awards. Use the EPA web site to find when the program started and to find the list of the most recent winners of the Presidential Green Chemistry Challenge Award. Pick one winner and summarize in your own words the green chemistry advance that merited the award.

45. Recreational scuba divers usually use compressed air that has the same composition as normal air. A mixture being used is called Nitrox®. What is its composition and why is it being used?

46. Here are some data from the U.S. EPA, Office of Air Quality Planning and Standards. Data indicate the number of days metropolitan statistical areas failed to meet acceptable air-quality standards (Pollutant Standards Index rating over 100).

Air Quality of Selected U.S. Metropolitan Areas, 1992 – 1999

Metropolitan statistical area	1992	1993	1994	1995	1996	1997	1998	1999
Boston, MA	9	6	10	8	2	8	7	5
Denver, CO	11	3	1	2	0	0	5	1
Houston, TX	32	28	38	66	26	47	38	50

a. Do these values show the same types of trends shown in question 30?

b. What factors influence these values? Offer some reasonable explanations, based on your research or knowledge about these cities.

47. An article in *USA Today* on January 12, 1999, is titled "Taking technology from here to the infinitesimal." By the year 2020, the article predicts, "the age of atomic engineering . . ., a type of nanotechnology, will dawn." What does this term imply? What kinds of applications will be possible that are not now part of our technology?

48. Consider this graph showing the dependence of hydrocarbon and ozone concentrations with time for a major metropolitan area.

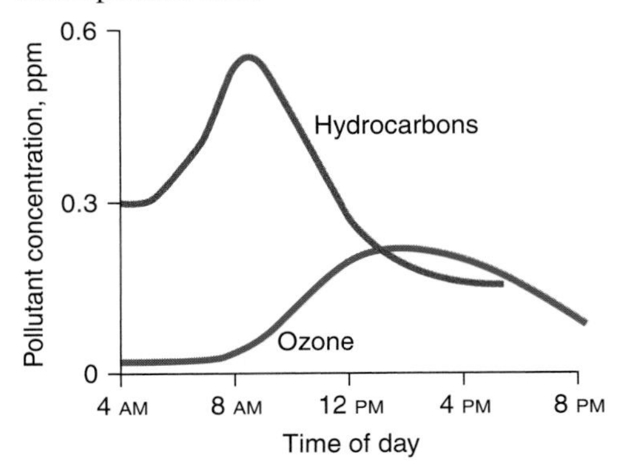

a. Interpret the two curves, explaining what they imply about air pollution in an urban area.

b. Where do you think the curve for NO would fit on this graph?

c. What type of health effects would be felt in this major metropolitan area?

49. A concept web or concept map is a convenient way to represent knowledge and connection among ideas. Concept webs are constructed by joining a word or expression to another one by means of linking words. For example, the atmosphere has three layers, the mesosphere, stratosphere, and troposphere.

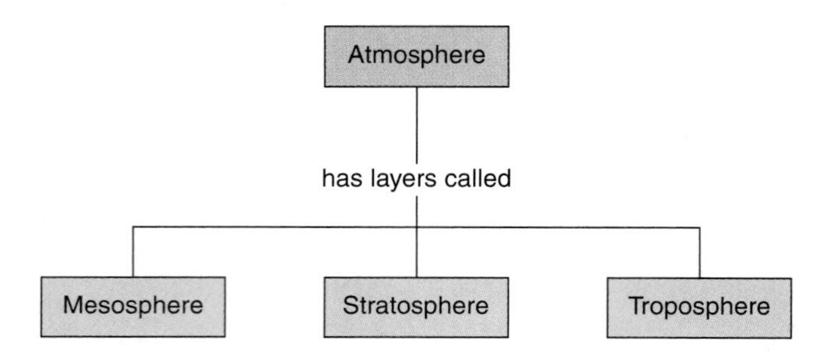

What advantages or disadvantages does this representation have compared with Figure 1.4? Explain your reasoning.

50. Consider this graph showing the effects of carbon monoxide inhalation on humans.

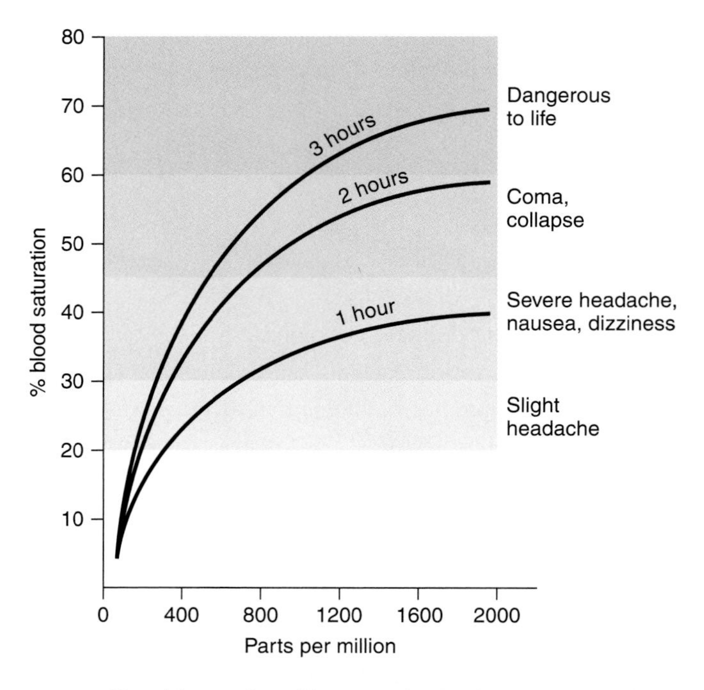

a. Use this graph to illustrate that both the avenue of exposure and the duration of exposure have an effect on CO toxicity in humans. Explain your reasoning.

b. Use the information in this graph to write a paragraph intended to inform purchasers of a home carbon monoxide detection kit about the potential health hazards of carbon monoxide gas.

2

Protecting the Ozone Layer

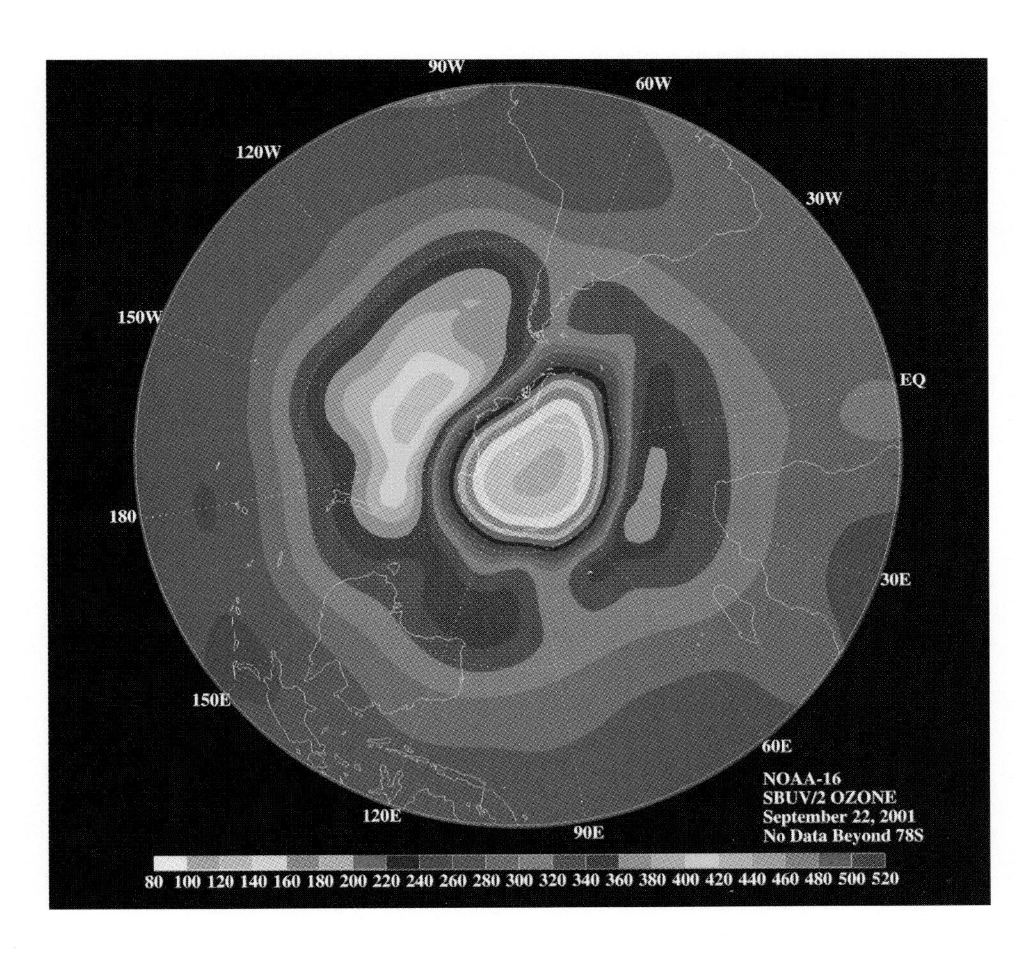

A view of a 2001 Antarctic stratospheric ozone hole (inside the purple color) taken from a National Oceanic and Atmospheric Administration (NOAA) satellite. The maximum size of the hole observed was the largest one yet, extending over about 29 million square kilometers, an area larger than North America's 24 million square kilometers.

"Orbiting above the Earth, an astronaut can look down on our home and see the thin blue ribbon that rims our planet. That transparent blanket—our atmosphere—makes life possible. It provides the air we breathe and regulates our global temperature. And it contains a special ingredient called ozone that filters deadly solar radiation. Life as we know it is possible because of the protection afforded by the ozone layer. Gradually it has become clear to scientists and to governments alike that human activities are threatening our ozone shield. Behind this environmental problem lies a tale of twin challenges: the scientific quest to understand our ozone shield and the debate among governments about how best to protect it . . ."

Daniel Albritton *et al.*

So begins *Our Ozone Shield,* a 1992 Report to the Nation On Our Changing Planet, a publication of the University Corporation for Atmospheric Research in conjunction with the National Oceanic and Atmospheric Administration. The challenges raised in the opening statement, written a decade ago, nevertheless are still with us. Striking images, such as the one that opens this chapter, have been taken by satellites. In June 2003, the National Aeronautics and Space Administration (NASA) will launch a new mission called Earth Observing System (EOS) Aura to gather additional data about changes in the Earth's stratospheric ozone layer. Essential to life on Earth, ozone in the stratosphere remains a subject of close scrutiny by scientists, politicians, and diplomats alike. Although its concentration can vary by season, stratospheric ozone plays a vital role in protecting the Earth's surface and those who live there from damaging solar radiation. Where you live also plays a role in the amount of stratospheric ozone overhead and how well it provides its protective effects.

- NASA's EOS Aura mission will also collect data about tropospheric air quality (Chapter 1) and global warming (Chapter 3).

2.1 Consider This: Ozone Levels Above Your House

What are the current ozone levels in the stratosphere? NASA can provide you with values. In fact, if a satellite is sending back data as you read this, you may be able to get today's ozone level in the stratosphere right above where you live.

a. Use the NASA link at the *Chemistry in Context* web site to access satellite data. Click on your location on the world map or enter your specific latitude and longitude to find the total column ozone amount at your location for today. Request an earlier date if today's data are not available. As you will learn later in the chapter, 320 Dobson units is the average ozone level over the northern U.S. How does your value compare with the average?
b. Again using the NASA link, obtain ozone values for some other parts of the world. How do they compare with the value for your location?

There is now ample experimental evidence that the concentration of ozone in the stratosphere has diminished beyond that expected seasonally from natural causes. A class of compounds called **chlorofluorocarbons (CFCs),** used in air conditioners, refrigerators, propellants, and other applications, has been implicated in the ozone reduction. The story of the stratospheric ozone layer and the role of these chemicals in its depletion are superb examples of chemistry in context. As you will discover in the pages that follow, the "chemistry" includes topics such as atomic and molecular structure, chemical reactions, bond breaking and bond making, the interaction of radiation and matter, the effect of radiant energy on living things, and the laboratory search for CFC substitutes. The "context" is created by the global environmental impact of stratospheric ozone depletion, the widespread use of chlorofluorocarbons, and the economic and political issues associated with achieving international agreement on the phase out of compounds that attack stratospheric ozone.

The 1995 Nobel Prize in chemistry was awarded to Dr. F. Sherwood Rowland, Dr. Mario Molina, and Dr. Paul Crutzen for their ground-breaking research on stratospheric zone depletion. A *New York Times* editorial called the award "A Nobel Prize With Political Punch" in reference to the fact that Rowland, Molina, and Cruzen had to win over international scientists and government officials who were skeptical of the award-winning research that affirmed the stratospheric ozone problem as real. We will consider the types of experimental evidence they used to help detect and understand processes leading to the depletion of the ozone layer in the stratosphere. Despite their successes, many unanswered questions remain: Just how serious is the stratospheric ozone depletion that has already occurred? How dangerous is the associated increase in exposure to ultraviolet radiation? What can be done to halt or correct the problem? Will the proposed measures work, and how much will they cost?

CHAPTER OVERVIEW

In this chapter, we will address the questions raised in the introduction by considering both scientific and societal issues. As we investigate the properties of ozone, we soon find that some knowledge of its molecular structure and the nature of light are needed to understand how ozone acts in the stratosphere to filter the Sun's harmful radiation. A consideration of some fundamental properties of atomic structure is also necessary. These ideas are then used to predict the molecular structures of a number of substances, including ozone. A discussion of sunlight in particular and radiation in general follows. You will find that light is strangely schizophrenic—it can behave both like waves and like little particles of energy. The particulate properties are especially useful in describing how ozone and oxygen molecules absorb ultraviolet radiation and how radiation damages biological materials. The formation and fate of ozone and its distribution in the atmosphere are considered. We then turn to an analysis of the various mechanisms for the depletion of stratospheric ozone, some involving naturally occurring chemicals and others involving synthetic compounds. Among the latter are the chlorofluorocarbons, whose properties and uses are discussed at length. The following section describes the stratospheric interaction of chlorofluorocarbons and ozone. After a discussion of the Antarctic ozone hole, the chapter concludes with a consideration of the social and technical problems associated with reducing chlorofluorocarbon emissions and developing substitutes.

2.1 Ozone: What Is It?

The central substance in this chapter is ozone, an atmospheric gas. If you have ever been near a sparking electric motor or a copying machine, or been in a severe lightning storm, you probably smelled ozone. Its odor is unmistakable, but hard to describe. It is possible to smell concentrations as low as 10 parts per billion (ppb)—10 molecules out of one billion. Appropriately enough, the name "ozone" comes from a Greek word meaning "to smell."

Ozone is oxygen that has undergone rearrangement from the normal diatomic molecule, O_2, to a triatomic form, O_3. A simple chemical equation summarizes the reaction:

$$\text{Energy} + 3\,O_2 \longrightarrow 2\,O_3 \qquad [2.1]$$

We have inserted a reminder that energy must be absorbed for this reaction to occur, which accounts for the fact that ozone forms when oxygen is subjected to electrical discharge, whether from an electric spark or lightning.

Ozone is called an **allotrope** or allotropic form of oxygen. Allotropes are two or more forms of the same element that differ in their molecular or crystal structure, and therefore in their properties. The familiar allotropes of carbon—diamond and graphite—have different crystal structures, as do fullerenes (buckyballs), the recently discovered, but less common, carbon allotrope. Diatomic oxygen, O_2, and triatomic ozone, O_3, obviously differ in molecular structure. This variance is responsible for differences in the physical and chemical properties of the two allotropes. For example, ordinary oxygen (O_2) is odorless. It condenses and changes from a colorless gas to a light blue liquid at −183°C and a pressure of one atmosphere. Ozone is more easily liquefied, changing its physical state from a gas to a dark blue liquid at −112°C. Because ozone is chemically more reactive than oxygen, O_3 is used in the purification of water and the bleaching of paper pulp and fabrics. At one time it was even advocated as a deodorant for air in crowded interiors and continues to be used by hotels to remove residual smoke from rooms.

In the troposphere, the region of the atmosphere in which we live, ozone forms in photochemical smog and other kinds of air pollution. But what is detrimental in one region of the atmosphere may be essential in another. The stratosphere, at an altitude of 20 to 30 km, is where ozone performs most of its filtering function on ultraviolet light from the Sun. That process involves the interaction of matter and radiant energy, and to understand it requires knowledge about both of these fundamental topics. We turn first to a submicroscopic view of matter.

2.2 Atomic Structure and Elementary Periodicity

The chemical and physical properties of oxygen and ozone and the interaction of these allotropes with sunlight are intimately related to the structure of the O_2 and O_3 molecules. Before we can speak about molecular structure, we must consider the atoms from which molecules are formed. You will recall from Chapter 1 or from your previous study that each element consists of its own distinctive, characteristic atoms. During the 20th century, chemists and other scientists made great progress in discovering details about the structure of atoms and the particles that make them up. The physicists have been almost too successful; they have found more than 200 subatomic particles. Fortunately, most chemistry can be explained with only three.

We now know that every atom has at its center a minuscule **nucleus.** This nucleus is composed of particles called **protons** and **neutrons.** Protons are positively charged and neutrons are electrically neutral, but both have almost exactly the same mass. Indeed, the protons and neutrons in the nucleus account for almost all of an atom's mass. Well beyond the nucleus are the **electrons** that define the outer boundary of the atom. An electron has a much smaller mass than a proton or neutron, approximately 1/2000th the mass. Moreover, an electron has a negative electrical charge that is equal in magnitude to that of a proton, but opposite in sign. The charge and mass properties of these particles are summarized in Table 2.1.

In any electrically neutral atom, the number of electrons equals the number of protons. This number of protons is called the **atomic number.** The atomic number is important because it determines the elemental identity of the atom. Each element has its own characteristic atomic number. For example, the simplest atom is hydrogen, and each hydrogen atom contains one proton, and thus has an atomic number of 1. Helium (He) has an atomic number of 2, hence each atom of this element contains two protons. With each successive element, the atomic number increases, right up through element 112, whose atoms contain 112 protons.

The atomic number also indicates the number of electrons in a *neutral* atom. To be electrically neutral, an atom must have the same number of protons (positive charge) as electrons (negative charge). Correspondingly, a neutral hydrogen atom contains one proton and one electron; a neutral helium atom has two protons and two electrons.

Table 2.1

Properties of Subatomic Particles

Particle	Relative Charge	Relative Mass	Actual Mass, kg
Proton	+1	1	1.67×10^{-27}
Neutron	0	1	1.67×10^{-27}
Electron	−1	1/1838 = 0.0005444	9.11×10^{-31}

2.2 Your Turn

Using the periodic table as a guide, specify the number of protons and electrons in a neutral atom of each of these elements.

a. carbon (C) **b.** chlorine (Cl) **c.** calcium (Ca) **d.** chromium (Cr)

Answers

a. 6 protons, 6 electrons **b.** 17 protons, 17 electrons

We wish that we could include a drawing of a typical atom. However, atoms defy such representation, and any depictions in textbooks are at best oversimplifications. Electrons are sometimes pictured as moving in orbits about the nucleus, but the reality is a good deal more complicated and abstract. For one thing, the relative size of the nucleus and the atom create serious problems for the illustrator. If the nucleus of a hydrogen atom were the size of a period on this page, the atom's single electron would most likely be found at a distance of about 10 feet from that period. An atom is thus mostly empty space. Moreover, electrons do not follow specific circular orbits. In spite of what one reads, an atom is really not very much like a miniature solar system. Rather, the distribution of electrons in an atom is best represented by probability and statistics. The nucleus is surrounded by a sort of fuzzy cloud in which electrons are more or less likely to occur.

If this sounds rather vague to you, you are not alone. Common sense and our experience of ordinary things are not particularly helpful in our efforts to visualize the interior of an atom. Instead, we are forced to resort to mathematics and metaphors. The mathematics required (a field called quantum mechanics) can be formidable. Chemistry majors do not normally encounter this field until rather late in their undergraduate study. We cannot fully share with you the strange beauties of the peculiar quantum world of the atom, although we can provide some useful generalizations.

In the periodic table, the elements are arranged in order of increasing atomic number. The periodic table also organizes elements so that those with similar chemical and physical properties fall in the same columns (groups). This arrangement shows that the properties of the elements vary in a regular way with increasing atomic number and have a pattern of repeating periodically. Thus, lithium (Li, atomic number 3), sodium (Na, 11), potassium (K, 19), rubidium (Rb, 37), and cesium (Cs, 55) must share something besides their behavior as highly reactive metals. What fundamental feature accounts for these similar properties?

Today we know that the periodicity of properties *is chiefly the consequence of the number and distribution of electrons* in the atoms of the elements. Because the atomic number represents the number of protons in each atom (and electrons in a neutral atom) of each particular element, properties vary with atomic number. And when properties repeat themselves, it signals a repeat in electronic arrangement.

It can be demonstrated by experiment and calculation that the electrons are arranged in levels (sometimes called shells) about the nucleus. The electrons in the innermost level are the most strongly attracted by the positively charged nucleus. The greater the distance between an electron and the nucleus, the weaker the attraction is between them. We say that the more distant electron is in a higher energy level, which means that the electron itself possesses more energy.

An important feature of these energy levels is the fact that they have maximum electron capacities and are particularly stable when they are fully occupied. The innermost level, corresponding to the lowest energy, can hold only two electrons. The second level has a maximum capacity of eight, and the higher levels are also particularly stable when they contain eight electrons.

Because it is difficult to picture an atom accurately, Table 2.2 is used to represent the arrangement of electrons in the atoms of the first 18 elements. The total number of electrons in each atom is printed in blue; the number of outer electrons is printed in red. The number of electrons in fully filled inner or lower energy levels is given in parentheses. The number of *outer electrons* is particularly important because these electrons account for many of the chemical and physical properties of the corresponding elements.

Lithium

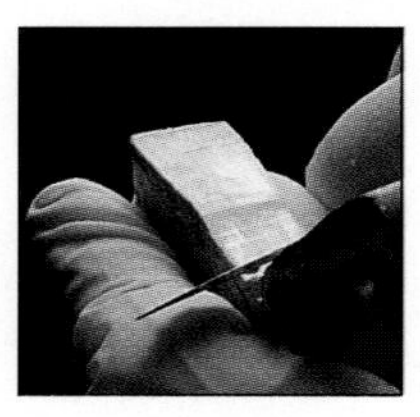

Sodium

Potassium

Rubidium

Figure 2.1
Selected Group 1A Elements

Table 2.2

Electronic Arrangements in Atoms of the First 18 Elements

Group 1A	2A	3A	4A	5A	6A	7A	Noble Gases 8A
1 H 1							2 He 2
3 Li (2) + 1	4 Be 2	5 B 3	6 C 4	7 N 5	8 O 6	9 F 7	10 Ne 8
11 Na (2)+(8)+1	12 Mg 2	13 Al 3	14 Si 4	15 P 5	16 S 6	17 Cl 7	18 Ar 8

- Number *above* the atomic symbol is the number of electrons in a neutral atom. This is the same as the **atomic number,** which gives the number of protons.
- Number *below* the atomic symbol is the number of **outer** electrons.
- Number in *parentheses* indicates electrons in **fully filled energy levels** and applies to entire row.
- Number in each Group designation corresponds to the number of **outer** electrons.

Note, for example, that in spite of the fact that they have different total numbers of electrons, lithium and sodium atoms both have one *outer* electron per atom. This fact explains much of the chemistry that these two alkali metals have in common. It places them in Group 1A of the periodic table (the 1 indicates one outer electron). Moreover, we would be correct in assuming that potassium, rubidium, and the other elements in column 1A of the periodic table also have a single outer electron in each of their atoms. They are all soft metals that react readily with oxygen, water, and a wide range of other chemicals. In fact, chemical reactivity and the bonding that holds atoms together to form molecules and crystals are largely consequences of the number of outer electrons in any element. Figure 2.1 shows photographs of Group 1A elements.

The periodic table is a useful guide to electron arrangement in the various elements. In the elementary families or groups marked "A", the number that heads the column indicates the number of outer electrons in each atom. You have already seen that Group 1A elements are characterized by one outer electron. Similarly, the atoms of the Group 2A elements (the "alkaline earths") all have two outer electrons. The same pattern holds true for all of the A Groups in the periodic table. Group 3A elements have three outer electrons, Group 4A elements have four, and so on across the table. Seven outer electrons characterize the atoms of the "halogens" that make up Group 7A—fluorine (F), chlorine (Cl), bromine (Br), iodine (I), and astatine (At). The next two exercises provide some practice with elements in the A groups.

- The Group number does not *necessarily* indicate the number of outer electrons for elements in B families, where the situation is a bit more complicated.

2.3 Your Turn

Give the number of outer electrons in neutral atoms of each of these elements.

a. oxygen (O) **b.** germanium (Ge) **c.** krypton (Kr) **d.** calcium (Ca)

Answers

a. 6 (Group 6A) **b.** 4 (Group 4A)

2.4 Your Turn

What feature of atomic structure is shared by oxygen (O), sulfur (S), selenium (Se), and tellurium (Te)?

In addition to electrons and protons, atoms also contain neutrons. The one exception is ordinary hydrogen, which consists of one electron and one proton. But even in pure hydrogen, one atom out of 6700 also has a neutron in its nucleus. Recall our earlier statement that most of the mass of any atom is associated with its nucleus. Because both the proton and the neutron have relative masses of almost exactly 1, the relative mass of an atom of this "heavy hydrogen" very nearly equals 2. This form of hydrogen is also called deuterium. It is an example of a naturally occurring **isotope** of hydrogen. Isotopes are two or more forms of the same element (same number of protons) whose atoms differ in number of neutrons, and hence in mass.

Isotopes are identified by their **mass numbers**—the sum of the number of protons and the number of neutrons in an atom. Thus, the mass number of ordinary hydrogen is 1, reflecting the fact that the nucleus contains only one proton. This isotope may be identified as hydrogen-1, or H-1; it can also be represented by the nuclear symbol $^{1}_{1}H$ or even simplified to ^{1}H. The subscript can be omitted in such a symbol because hydrogen *always* has an atomic number of 1, which means it always has 1 proton. Clearly there is redundancy in actually giving the atomic number as well as the elemental symbol, even though it may be convenient for the reader. The mass number, indicated by a superscript, must be given because it can vary for the same element. For example, the nucleus of an atom of deuterium contains one proton plus one neutron and is therefore assigned a mass number of 2. In identifying isotopes, the mass number follows the name or symbol of the element. Thus, deuterium is designated as hydrogen-2, H-2; it can also be represented by $^{2}_{1}H$ or simplified to ^{2}H, where the mass number of 2 is indicated by a superscript. There is also a third isotope of hydrogen, called tritium, whose atoms consist of two neutrons in addition to the one proton and one electron characteristic of all neutral hydrogen atoms. Tritium, a radioactive isotope that is rare in nature, thus has a mass number of 3 (1 proton and 2 neutrons). It can be represented as hydrogen-3, H-3, $^{3}_{1}H$, or ^{3}H. Table 2.3 summarizes this information about the isotopes of hydrogen.

2.5 Your Turn

Specify the number of protons, electrons, and neutrons in a neutral atom.

a. carbon-14 ($^{14}_{6}C$) **b.** uranium-235 ($^{235}_{92}U$) **c.** copper-64 ($^{64}_{29}Cu$) **d.** gold-198 ($^{198}_{79}Au$)

Answers

a. 6 protons, 6 electrons, 8 neutrons **b.** 92 protons, 92 electrons, 143 neutrons

Table 2.3

Isotopes of Hydrogen

Name	Nuclear Symbol	Number of Protons	Number of Neutrons	Sum of Protons and Neutrons
hydrogen or hydrogen-1	$^{1}_{1}H$	1	0	1
deuterium or hydrogen-2	$^{2}_{1}H$	1	1	2
tritium or hydrogen-3	$^{3}_{1}H$	1	2	3

All elements have isotopes, but the number of stable and unstable ones varies considerably. Each element's atomic mass, the number you see on every periodic table, takes the relative abundance of isotopes, as well as their masses, into account. Although the concept of atomic mass is important, we do not require it at this time. Following our general rule of introducing information only as needed, we will defer a discussion of atomic masses to Chapter 3.

- Mass number gives the total number of protons and neutrons in a specific isotope. Atomic mass refers to a weighted average of all isotopes of that element.

2.3 Molecules and Models

After this excursion into the atom, we come to our primary motivation for studying atoms, which is molecular structure. The stability of filled electron shells can be invoked to explain why atoms bond to each other to form molecules. The simplest case is H_2, a diatomic molecule we encountered in Chapter 1. A hydrogen atom has only one electron, but if two hydrogen atoms come together, the two electrons become common property. Each atom effectively has a share in both electrons. The resulting H_2 molecule has a lower energy than two individual H atoms, and consequently the molecule with its bonded atoms is more stable than the separate atoms. The two electrons that are shared constitute what is called a **covalent bond.** Appropriately, the name "covalent" implies "shared strength."

If we represent each atom by its symbol and each electron by a dot, the two individual hydrogen atoms might look something like this:

H · and · H

Bringing the two atoms together yields a molecule that can be represented this way.

H : H

This is called a dot or **Lewis structure,** after Gilbert Newton Lewis (1875–1946), an American chemist who pioneered its use. Lewis structures can be predicted for any molecule by following a few simple steps. We first illustrate the procedure with hydrogen fluoride, HF, a very reactive compound used to etch glass.

1. Starting with the chemical formula of the compound, note the number of outer electrons contributed by each of the atoms (Remember that the periodic table is a useful guide).

1 H·   1 H atom × 1 outer electron per atom = 1 outer electron

1 :F̤̈·   1 F atom × 7 outer electrons per atom = 7 outer electrons

2. Add the outer electrons contributed by the individual atoms to obtain the total number of outer electrons available.

$$1 + 7 = 8 \text{ outer electrons}$$

3. Arrange the outer electrons in pairs. Then distribute them in such a way as to maximize stability by giving each atom a share in enough electrons to fully fill its outer shell—two electrons in the case of hydrogen, eight electrons for most other atoms.

H:F̤̈:

We surrounded the F atom with eight dots, organized into four pairs. The pair of dots between the H and the F represents the electron pair that forms the bond uniting the hydrogen and fluorine atoms. The other three pairs of dots are the three pairs of electrons that are not shared with other atoms and hence not involved in bonding. As such, they are called "nonbonding" electrons or "lone pairs."

When only one pair of shared electrons is involved in a covalent bond, as it is in HF, the linkage is called a **single bond.** A horizontal line often replaces the electron pair forming a single covalent bond. This line connects the symbols for the two atoms.

H—F̤̈:

Sometimes even the nonbonding electrons are removed from a Lewis structure, simplifying the structure still more.

$$\mathrm{H{-}F}$$

Remember that the single line represents one pair of shared electrons. These two electrons plus the six electrons in the three nonbonding pairs mean that the fluorine atom is associated with a total of eight outer electrons, whether or not all the electrons are specifically shown. Remember there are no additional electrons other than the single pair shown around the hydrogen atom, thanks to its small size.

The fact that electrons in many molecules are arranged so that every atom (except hydrogen) shares in eight electrons is called the **octet rule.** This generalization is a useful guide for predicting Lewis structures and the formulas of compounds. Consider the Cl_2 molecule, the diatomic form of elemental chlorine. From the periodic table, we can see that chlorine, like fluorine, is in Group 7A, which means that its atoms each have seven outer electrons. Using the scheme given for HF earlier, we first count and add up the outer electrons for Cl_2.

$2\,:\!\ddot{\underset{\cdot\cdot}{\mathrm{Cl}}}\cdot$ 2 Cl atoms $\times$ 7 outer electrons per atom = 14 outer electrons

For Cl_2 to exist, there must be a bond between the two atoms, which we show by a single line designating a shared electron pair, a single covalent bond. The remaining twelve electrons constitute six nonbonding pairs, distributed in such a way as to give each chlorine atom eight electrons (two bonding and six nonbonding). This meets the octet rule. Accordingly, this is the Lewis structure for Cl_2.

$$:\ddot{\underset{\cdot\cdot}{\mathrm{Cl}}}{-}\ddot{\underset{\cdot\cdot}{\mathrm{Cl}}}:$$

2.6 Your Turn

Use the procedure just outlined to draw the Lewis structures for each of these substances; both species obey the octet rule.

a. HBr **b.** I_2

Answer

a.

1 H· 1 H atom $\times$ 1 outer electron per atom = 1 outer electron

$1\ \cdot\ddot{\underset{\cdot\cdot}{\mathrm{Br}}}:$ 1 Br atom $\times$ 7 outer electrons per atom = 7 outer electrons

The total number of electrons is eight. These are the Lewis structures.

$$\mathrm{H}:\ddot{\underset{\cdot\cdot}{\mathrm{Br}}}: \quad \text{or} \quad \mathrm{H}{-}\ddot{\underset{\cdot\cdot}{\mathrm{Br}}}:$$

So far we have dealt only with molecules having just two atoms. But there are many compounds whose molecules contain more than two atoms. The octet rule is a generalization that applies to many of these compounds as well. Here is another generalization just as useful as the octet rule in helping to predict Lewis structures. In most molecules where there is only one atom of one element bonded to two or more atoms of another element (or elements), *the single atom goes in the center of the Lewis structure.* There are exceptions to these generalizations, but this is a good place to begin to apply the generalizations. We start with a water molecule, H_2O, as an example.

Following the same procedures used for two-atom molecules, we first count and add up the outer electrons.

2 H· 2 H atoms $\times$ 1 outer electron per atom = 2 outer electrons

$1\cdot\ddot{\underset{\cdot\cdot}{\mathrm{O}}}\cdot$ 1 O atom $\times$ 6 outer electrons per atom = 6 outer electrons

Total = 8 outer electrons

We place the O representing the oxygen atom in the center and distribute the eight electrons (dots) around the O, in conformity with the octet rule. Each of the hydrogen atoms

- Each hydrogen atom forms only one bond (two shared electrons). Oxygen, because it can form two bonds, is the central atom.

is bonded to the oxygen atom with a pair of electrons. The remaining four electrons are also placed on the oxygen, but as two nonbonding pairs. This is the result.

$$\mathrm{H{:}\ddot{\underset{..}{O}}{:}H}$$

A quick count confirms that the O is surrounded by eight dots, representing the eight electrons predicted by the octet rule. Alternatively, we could symbolize the water molecule with lines for the single bonds, with or without oxygen's nonbonding electrons.

$$\mathrm{H{-}\ddot{\underset{..}{O}}{-}H} \quad \text{or} \quad \mathrm{H{-}O{-}H}$$

These Lewis representations provide more information than does the chemical formula, H_2O. The formula shows the types and ratio of atoms present, and so does the Lewis structure. In addition, the Lewis structure indicates how the atoms are connected to each other. On the other hand, Lewis structures do not directly reveal the shape of a molecule. From the structures for water given so far, it might appear that the atoms of the water molecule all fall in a straight line. In fact, the molecule is bent. It looks something like this.

(bent H–O–H with nonbonding pairs on O) or (bent H–O–H)

We will return to this discussion of shape in Chapter 3 and see how the Lewis structure can lead to the prediction of this bent structure. We will examine the experimental evidence for the shape of the water molecule in Chapter 5.

Another example of a molecule with more than two bonded atoms is methane, CH_4. Using the rules and generalizations given earlier, and recognizing that carbon is in Group 4A, we can write the Lewis structure of methane.

• The combustion of methane was discussed in Chapter 1.

4 H·	4 H atoms × 1 outer electron per atom = 4 outer electrons
1 ·Ċ· (C with four dots)	1 C atom × 4 outer electrons per atom = 4 outer electrons
	Total = 8 outer electrons

The C representing a carbon atom goes in the center and is surrounded by the eight electrons, giving carbon an octet of electrons. Each of the four hydrogen atoms uses two of the electrons to form a shared pair with carbon, for a total of four single covalent bonds. This gives us the Lewis structure of methane.

• The geometry of the methane molecule will be described in Chapter 3.

$$\begin{array}{c}\mathrm{H}\\ \mathrm{H{:}\ddot{\underset{..}{C}}{:}H}\\ \mathrm{H}\end{array} \quad \text{or} \quad \begin{array}{c}\mathrm{H}\\ |\\ \mathrm{H{-}C{-}H}\\ |\\ \mathrm{H}\end{array}$$

Checking the methane structure we see that the carbon atom has a share in eight electrons, complying with the octet rule. Remember that H can only accommodate a pair of electrons.

2.7 Your Turn

Use the procedure just outlined to draw the Lewis structures for each of these compounds. Both species obey the octet rule.

a. hydrogen sulfide (H_2S) **b.** dichlorodifluoromethane (CCl_2F_2)

Answer

a.

2 H·	2 H atoms × 1 outer electron per atom = 2 outer electrons
1 ·S̈· (S with six dots)	1 S atom × 6 outer electrons per atom = 6 outer electrons

The total number of electrons is eight. The Lewis structures are $\mathrm{H{:}\ddot{\underset{..}{S}}{:}H}$ or $\mathrm{H{-}\ddot{\underset{..}{S}}{-}H}$

In some structures, single covalent bonds do not allow the atoms to follow the octet rule. Consider, for example, the very important gas oxygen, O_2. Here we have 12 outer electrons to distribute, six from each of the Group 6A oxygen atoms. There are not enough electrons to give each of the atoms a share in eight electrons if only one pair is held in common. However, the octet rule can be satisfied if the two atoms share four electrons (two pairs). A covalent bond consisting of two pairs of shared electrons is called a **double bond.** This bond is represented by four dots or by two lines, with or without the nonbonding electrons.

$$\ddot{\underset{\cdot\cdot}{O}}::\ddot{\underset{\cdot\cdot}{O}} \quad \text{or} \quad \ddot{\underset{\cdot\cdot}{O}}=\ddot{\underset{\cdot\cdot}{O}} \quad \text{or} \quad O=O$$

Double bonds are shorter, stronger, and harder to break than single bonds involving the same atoms. The length and strength of the bond in the O_2 molecule correspond to a double bond. However, oxygen has a peculiar property that is not fully consistent with the Lewis structure just drawn. When liquid oxygen is poured between the poles of a strong magnet, it sticks there like iron filings. Such magnetic behavior implies that the electrons are not as neatly paired as the octet rule would suggest. But this little discrepancy is hardly a reason to discard a useful generalization. After all, simple scientific models seldom if ever explain all phenomena, but they can be helpful approximations. There are other common examples in which the straightforward application of the octet rule leads to discrepancies in interpreting experimental evidence. In the best of all scientific worlds, coming across data that do not conform to existing models may lead to the development of even better ones.

For completeness, we need to note that a **triple bond** is a covalent linkage made up of three pairs of shared electrons. The nitrogen molecule, N_2, contains a triple bond. Each Group 5A nitrogen atom contributes five outer electrons for a total of 10. These 10 electrons can be distributed in accordance with the octet rule if six of them (three pairs) are shared between the two atoms, leaving four of them to form two nonbonding pairs, one on each nitrogen atom.

$$:N:::N: \quad \text{or} \quad :N\equiv N: \quad \text{or} \quad N\equiv N$$

The ozone molecule introduces another structural feature. We again start with the octet rule. Each of the three oxygen atoms contributes six outer electrons for a total of 18. These 18 electrons can be arranged in two ways; each way gives a share in eight outer electrons to each atom.

$$\underset{\mathbf{a}}{\ddot{\underset{\cdot\cdot}{O}}::\ddot{O}:\ddot{\underset{\cdot\cdot}{O}}:} \qquad \underset{\mathbf{b}}{:\ddot{\underset{\cdot\cdot}{O}}:\ddot{O}::\ddot{\underset{\cdot\cdot}{O}}}$$

There are experimental techniques for determining molecular structure, and it turns out that neither of the above versions is exactly correct. Structures a and b predict that the molecule should contain one single bond and one double bond. In **a** the double bond is to the left of the central atom; in **b** it is to the right. But experiments reveal that the two bonds in the O_3 molecule are identical in length and strength, being somewhere between a single bond and a double bond. Structures a and b are called **resonance** forms. They represent hypothetical extremes of electron arrangements that do not exist as represented. The actual structure of the ozone molecule is something like a hybrid of the two resonance forms. A double-headed arrow linking the different forms is used to represent the resonance phenomenon.

$$:\ddot{\underset{\cdot\cdot}{O}}-\ddot{O}=\ddot{\underset{\cdot\cdot}{O}} \longleftrightarrow \ddot{\underset{\cdot\cdot}{O}}=\ddot{O}-\ddot{\underset{\cdot\cdot}{O}}:$$

This representation and the word "resonance" seem to imply that the electrons are jumping back and forth between the two arrangements, but in fact this does not happen. Resonance is just another useful concept invented by chemists to describe the complex microworld of molecules.

A closer experimental inspection of that microworld reveals that the O_3 molecule is bent, not linear as the simple Lewis structures just drawn seem to indicate. Lewis structures tell us only what is connected to what, not anything about the geometry. Thus a more accurate picture of the resonance forms is this:

$$\ddot{O}(=O)(-\ddot{O}:) \longleftrightarrow \ddot{O}(-O:)(=\ddot{O})$$

An explanation of why the O_3 molecule is bent will have to wait until Chapter 3. At this point we are more concerned about how bonding in O_2 and O_3 influences their interaction with sunlight.

2.8 Your Turn

Use the octet rule to draw the Lewis structures for each of these compounds.

a. carbon monoxide (CO) **b.** sulfur dioxide (SO_2)

Answer

a. $\cdot\dot{\underset{\cdot}{C}}\cdot$ 1 C atom × 4 outer electrons per atom = 4 outer electrons

$\cdot\ddot{\underset{\cdot\cdot}{O}}\cdot$ 1 O atom × 6 outer electrons per atom = 6 outer electrons

The total number of electrons is 10. The Lewis structures are $:C:::O:$ or $:C{\equiv}O:$ Note that the number of electrons, 10, is the same as the number in the N_2 molecule. It also has a triple bond.

2.4 Waves of Light

To better understand how stratospheric ozone screens out much of the Sun's harmful radiation requires knowing the molecular structure of ozone, as well as understanding some fundamental properties of light. The interaction of sunlight with matter is important, such as in photosynthesis or in the damage high-energy solar radiation can cause in living organisms. Therefore, we turn now to develop an understanding of light.

Every second, five million tons of the Sun's matter are converted into energy, which is radiated into space. The fact that our eyes are capable of detecting different colors is one indication that the radiation that reaches us is not all identical. Prisms and raindrops break sunlight into a spectrum of colors. Each of these colors can be identified by the numerical value of its wavelength. The word correctly suggests that light behaves rather like a wave in the ocean. The **wavelength** is the distance between successive peaks. It is expressed in units of length and symbolized by the Greek letter lambda (λ). Waves are also characterized by a certain **frequency,** the number of waves passing a fixed point in one second. Figure 2.2 shows two waves of different wavelength and frequency for comparison.

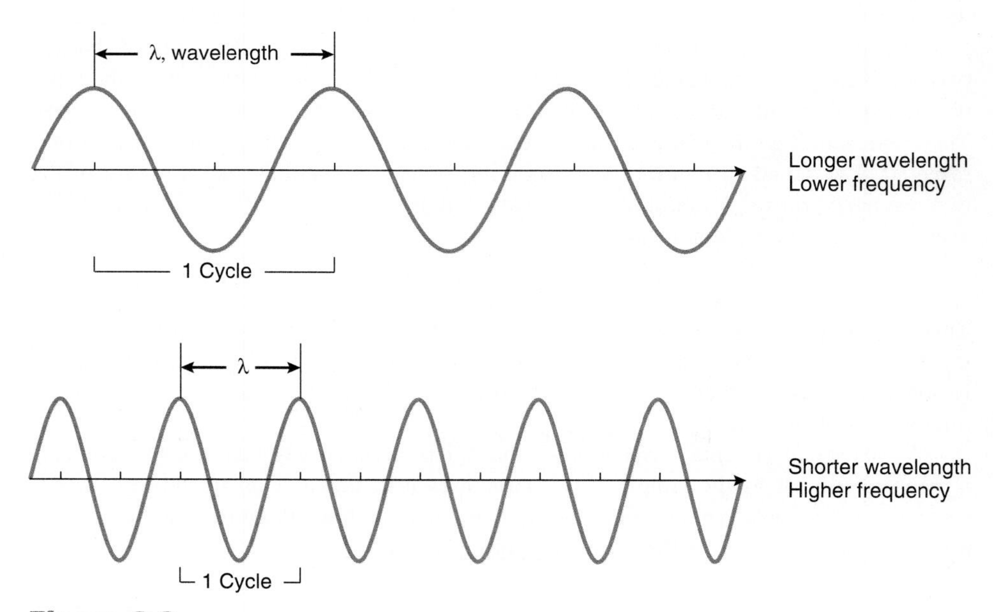

Figure 2.2

Two different waves.

It is both interesting and humbling to realize that out of the vast array of radiant energies, our eyes are sensitive to light in only a very tiny portion of the electromagnetic spectrum—wavelengths between about 700×10^{-9} m (corresponding to red) and 400×10^{-9} m (corresponding to violet). These lengths are very short, so we typically express them in nanometers. One nanometer (nm) is defined as one one-billionth of a meter (m). This is the relationship in symbols.

$$1 \text{ nm} = \frac{1}{1{,}000{,}000{,}000} \text{ m} = \frac{1}{10^9} \text{ m} = 1 \times 10^{-9} \text{ m}$$

We can use this equivalence to convert meters to nanometers; for example, to convert 700×10^{-9} meters to nanometers.

$$\text{wavelength} = \lambda = 700 \times 10^{-9} \cancel{\text{m}} \times \frac{1 \text{ nm}}{1 \times 10^{-9} \cancel{\text{m}}} = 700 \text{ nm}$$

The meters units cancel and we are left with nanometers.

2.9 Your Turn

Green light in the visible part of the spectrum has a wavelength of 500 nm. Express this wavelength in meters.

Answer

500×10^{-9} m or, expressed in scientific notation, 5.00×10^{-7} m.

Another way of quantifying color is to express the waves associated with each color in terms of frequency. If you were watching waves on the surface of a lake or the ocean, you could measure the distance between successive crests (the wavelength). But you could also determine how often the crests passed your point of observation by counting the number in a particular time interval. That would give you the frequency of the waves. The same idea applies to radiation, as shown in Figure 2.5. Frequency and wavelength are related; the shorter the wavelength, the higher the frequency, that is, the greater the number of waves that pass the observer in one second. For any wave, as the frequency increases, the wavelength decreases; their values change in opposite directions. In mathematical terms, wavelength and frequency are inversely proportional.

- As wavelength ↓, frequency ↑.

Instead of reporting frequency as "waves per second," the units are shortened to "per second" and written as 1/s or s^{-1}. This unit is also called a hertz (Hz), perhaps familiar to you from radio station frequencies. A companion unit is the megahertz (mHz), one million hertz, 1×10^6 s^{-1}. Frequency is represented by the Greek letter nu(ν).

The relationship between frequency and wavelength just described in words can be summarized in a simple equation where ν is the frequency and c represents the constant speed at which light and other forms of radiation travel.

$$\text{Frequency} = \nu = \frac{c}{\lambda} \qquad [2.2]$$

In metric units, the speed of light is 3.00×10^8 meters/second or 3.00×10^8 m·s^{-1}. You may be more familiar with the speed of light as 186,000 miles/second. The form of equation 2.2 indicates that wavelength and frequency are *inversely* related: As the value for λ decreases, the value for ν increases, and vice versa. Red light, which has a wavelength of 700 nm or 700×10^{-9} m, has a frequency of 4.29×10^{14} s^{-1}. This value can be calculated using equation 2.2.

$$\text{Frequency} = \nu = \frac{c}{\lambda} = \frac{3.00 \times 10^8 \text{ m·s}^{-1}}{700 \times 10^{-9} \text{ m}} = 4.29 \times 10^{14} \text{ s}^{-1}$$

Violet light has a shorter wavelength (400 nm) and hence a higher frequency (7.50×10^{14} s^{-1}) than red light.

2.10 Your Turn

Green light in the visible part of the spectrum has a wavelength of 500 nm.

a. Calculate the frequency of green light, reporting your answer in s^{-1}.
b. Compare the wavelengths and frequencies for green and red light.

Answer
a. $6.00 \times 10^{14}\ s^{-1}$

2.11 Consider This: Analyzing a Rainbow

Which color in the rainbow (Figure 2.3) has the

a. shortest wavelength? **b.** lowest frequency?

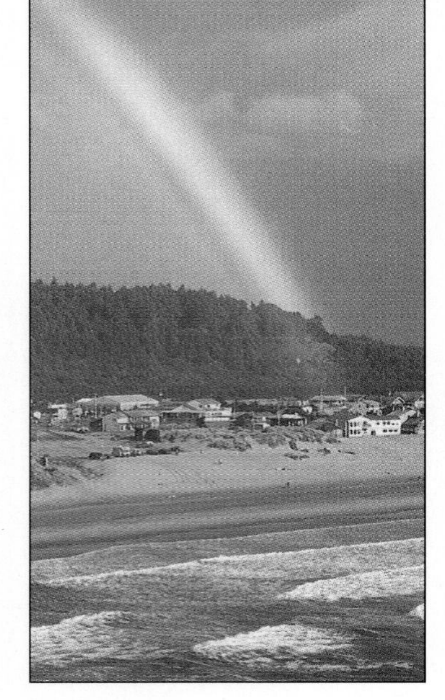

Figure 2.3
A rainbow of color. Water droplets act as prisms to separate visible light into its component colors.

The light that we can see directly is only a narrow band in the electromagnetic spectrum. The **electromagnetic spectrum** is the entire range of radiant energy, and it is a very wide range indeed. All waves in the electromagnetic spectrum travel at the same forward speed, $3.00 \times 10^8\ m \cdot s^{-1}$, although they differ widely in their wavelengths and frequencies. Scientists have devised a variety of detectors that are sensitive to the radiation in various parts of this broad band. As a consequence, we can speak with confidence about the regions of the spectrum that are invisible to our eyes. Figure 2.4 shows the continuum of types of waves of the electromagnetic spectrum and their relative frequencies and wavelengths.

At wavelengths longer than those of red visible light, one first encounters **infrared (IR),** waves that we cannot see, but can certainly feel. They are also known as heat rays. The microwaves used in radar and to cook food quickly have wavelengths on the order of centimeters. At still longer wavelengths (1 m to 1000 m) are the regions of the spectrum used to transmit your favorite AM and FM radio and television programs.

In this chapter we are most concerned with the **ultraviolet (UV)** region, which lies at wavelengths shorter than those of violet. At still shorter wavelengths are the **X rays** used in medical diagnosis and the determination of crystal structure and **gamma rays** that are given off in certain radioactive processes. Some gamma rays can have wavelengths as short as 10^{-16} m. In the next chapter, we will be most concerned with the **infrared (IR)** region of the electromagnetic spectrum. Figure 2.5 will help you keep these important regions of the electromagnetic spectrum in perspective.

2.12 Your Turn

Consider these four types of radiant energy from the electromagnetic spectrum.

infrared microwave ultraviolet visible

a. Arrange them in order of *increasing* wavelength.
b. Arrange them in order of *increasing* frequency.
c. Are the two arrangements the same or different? Explain your reasoning.

Answer
a. ultraviolet, visible, infrared, microwave

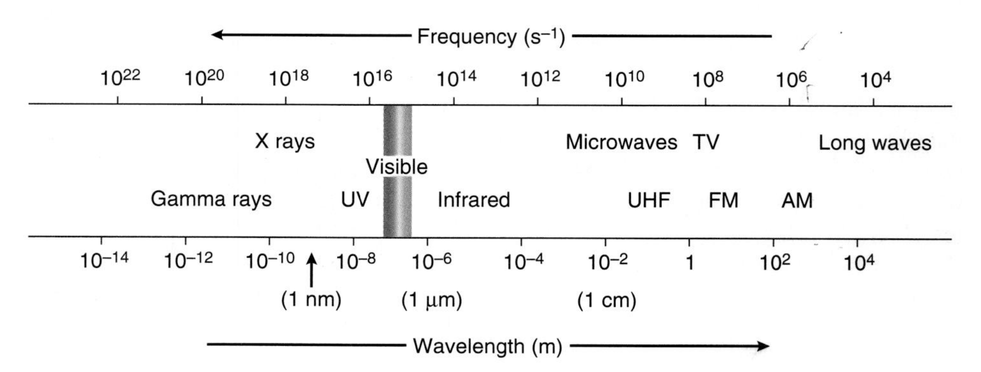

Figure 2.4
The electromagnetic spectrum.

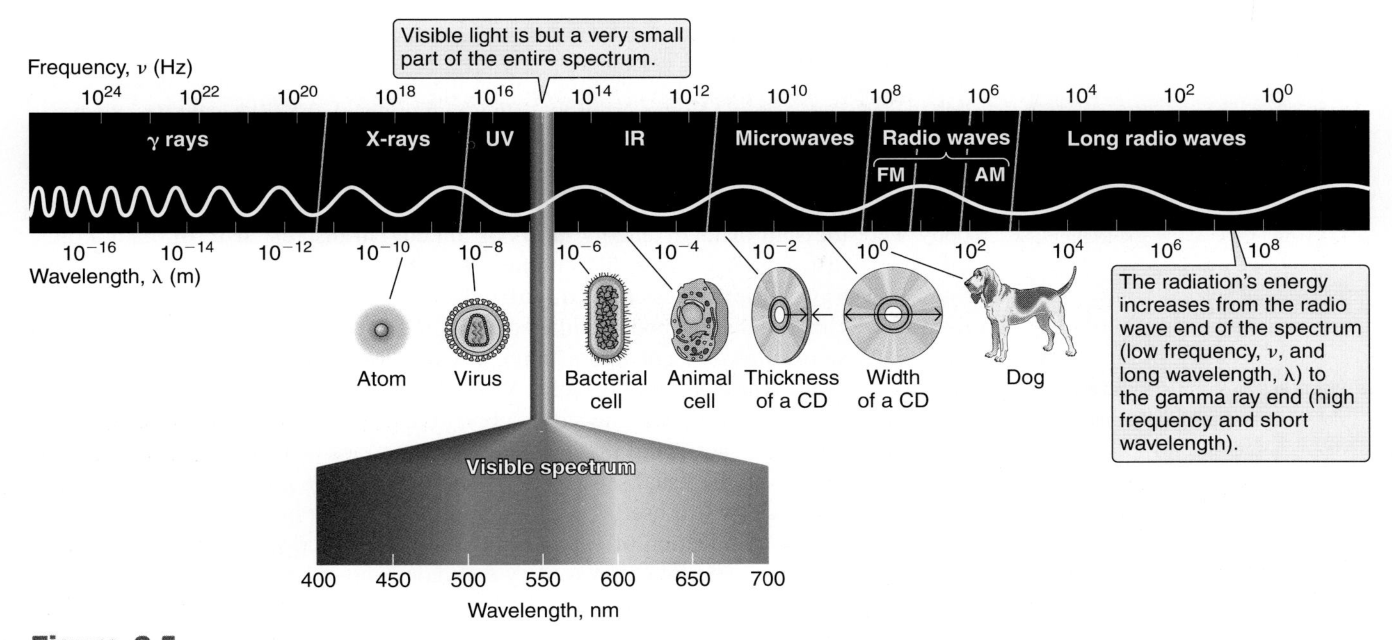

Figure 2.5

Pictorial representations of wavelengths in the electromagnetic spectrum.

Figures Alive!

Our local star, the Sun, emits many types of radiant energy, including infrared, visible, and ultraviolet radiation. However, our Sun does not emit all types with equal intensity. This is evident from Figure 2.6, a plot of the relative intensity of solar radiation as a function of wavelength. The curve represents the spectrum as measured *above* the atmosphere, before there has been opportunity for interaction of radiation with the molecules of the air. The peak indicating the greatest intensity is in the visible region. However, infrared radiation is spread over a much wider wavelength range, with the result that 53% of the total energy emitted by the Sun is radiated to Earth as infrared radiation. This is the major source of heat for the planet. Approximately 39% of the energy comes to us as visible light and only about 8% as ultraviolet. (The areas under the curve give an indication of these percentages.) But in spite of its small percentage, the Sun's UV radiation is potentially the most damaging to living things. To understand why, we must look at electromagnetic radiation in a different light, this time in terms of its energy.

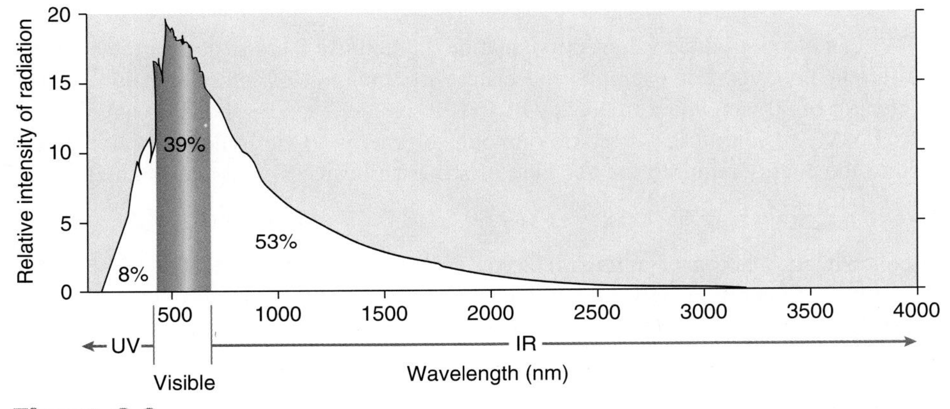

Figure 2.6

Energy distribution of solar radiation above the Earth's atmosphere.

2.5 "Particles" of Energy

The idea that radiation can be described in terms of wave-like character is well established and very useful. However, around the beginning of the 20th century, scientists found a number of phenomena that seemed to contradict this model. In 1909, a German physicist named Max Planck (1858–1947) argued that the shape of the energy distribution curve pictured in Figure 2.6 could only be explained if the energy of the radiating body were the sum of many energy levels of minute but discrete size. In other words, the energy distribution is not really continuous, but consists of many individual steps. Such an energy distribution is called **quantized.** An often-used analogy is that the quantized energy of a radiating body is like steps on a staircase, which are also quantized (no partial steps allowed), not like a ramp, which allows any sized stride. Five years later, in the work that won him his Nobel Prize, Albert Einstein (1879–1955) suggested that radiation itself should be viewed as constituted of individual bundles of energy called **photons.** One can regard these photons as "particles of light," but they are definitely not particles in the usual sense. For example, they have no mass.

• Planck and Einstein were both amateur violinists who played duets together.

The wave model is still useful, even with the new development of the quantum theory to explain the atomization of energy. Both are valid descriptions of radiation. This dual nature of radiant energy seems to defy common sense. How can light be described in two different ways at the same time, both waves and particles? There is no obvious answer to that very reasonable question—that's just the way nature is. The two views are linked in a simple relationship that is one of the most important equations in modern science. It is also an equation that is very relevant to the role of ozone in the atmosphere.

$$\text{Energy, } E = h\nu = \frac{hc}{\lambda} \qquad [2.3]$$

Here E represents the energy of a single photon. It is *directly* proportional to ν, the frequency of radiation and *inversely* proportional to the wavelength, λ. Consequently, as the wavelength of radiation gets shorter, its energy increases; as the energy decreases, the wavelength increases. On the other hand, as the frequency of the radiation increases, so does its energy. These qualitative relationships are summarized in the margin and you may wish to refer to these relationships as you do the next Your Turn. The symbol h in equation 2.3 represents **Planck's constant,** which has a value of 6.63×10^{-34} joule·second (J·s). Note the important fact that a **joule** is a unit of energy.

• As wavelength ↓, frequency ↑, energy ↑.

2.13 Your Turn

Arrange these colors of the visible spectrum in order of *increasing* energy per photon.

a. green **b.** red **c.** yellow **d.** violet

Answer

red < yellow < green < violet

We can also introduce values into equation 2.3 and use them to compare the energy of different photons. For example, the energy of a photon of ultraviolet light with a wavelength of 300 nm and a frequency of $1.00 \times 10^{15}\ s^{-1}$ can be shown to have energy of 6.63×10^{-19} joule (J), a very tiny amount of energy. (One joule is approximately equal to the energy required for one beat of a human heart.) This is the calculation.

$$E = h\nu = (6.63 \times 10^{-34}\ \text{J·s})(1.00 \times 10^{15}\ \text{s}^{-1}) = 6.63 \times 10^{-19}\ \text{J}$$

By contrast, the photon of a 100 mHz FM radio signal with a wavelength of 300×10^{7} nm has an energy of only 6.63×10^{-26} J. Although these energies are very small, there is a significant difference in the energies of the photon of UV radiation and the photon of the radio signal.

UV radiation: 6.63×10^{-19} J per photon

Radio signal: 6.63×10^{-26} J per photon

The energy of a photon of UV radiation is 10^7 or 10 million times larger than the energy of a photon emitted by your favorite radio station. Remember that as wavelength decreases (from radio waves to ultraviolet radiation), the energy per photon of radiation increases.

One consequence of this great difference in energy is the fact that you cannot get a tan from listening to the radio—unless you happen to be listening to it outside in the sunlight. Whether or not your radio is turned on, you are continuously bombarded by radio waves. Your body cannot detect them, but your radio can. As we have just seen, the energy associated with each of the radio photons is very low—about 7×10^{-26} J. This energy is not sufficient to produce a local increase in the concentration of the skin pigment, melanin, to cause tanning. That process involves a quantum jump, an electronic transition that requires approximately 7×10^{-19} J, far more than radio wave photons can supply.

Your body cannot store the 10 million low-energy photons of radio frequency that would be necessary to equal the energy required for the tanning reaction. It is an either/or situation: either a photon has enough energy to cause a specific chemical change or it does not. Photons of ultraviolet radiation of 300 nm or shorter do have sufficient energy to bring about the changes that result in tanning, burning, or in some cases, skin cancer.

All of this may seem relatively unimportant. But it was essentially this line of reasoning, on a different system, that won Einstein his Nobel Prize in 1905. Moreover, this same logic extends to any interaction of electromagnetic radiation and matter.

2.14 Your Turn

Return once more to the green light of 2.9 and 2.10 Your Turns. Calculate the energy of a photon of this radiation, expressing your answer in joules.

Answer
3.98×10^{-19} J

2.6 Radiation and Matter

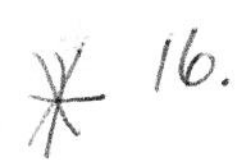

The Sun bombards the Earth with countless photons—indivisible packages of energy. The atmosphere, the surface of the planet, and the Earth's living things all absorb photons. Radiation in the infrared region of the spectrum warms the Earth and its oceans, causing molecules to move, rotate, and vibrate. The cells of our retinas are tuned to the wavelengths of visible light. Photons associated with different wavelengths are absorbed and the energy is used to "excite" electrons in biological molecules. The electrons jump to higher energy levels, triggering a series of complex chemical reactions that ultimately lead to sight. Compared to animals, green plants capture photons in an even narrower region of the visible spectrum (corresponding to red light) and use the energy to convert carbon dioxide and water into food, fuel, and oxygen in the process of photosynthesis.

Remember that as the wavelength of light *decreases*, the energy carried by each photon *increases.* Consequently, the interaction of shorter-wavelength radiation and matter becomes more energetic. Photons in the UV region of the spectrum are sufficiently energetic to eject electrons from atoms and molecules, converting them into positively charged species, rather than leaving them as neutral species. The even shorter UV wavelength photons break bonds causing molecules to come apart. In living things, such changes disrupt cells, and create the potential for genetic defects and cancer. This is shown schematically in Figure 2.7.

It is part of the fascinating symmetry of nature that this interaction of radiation with matter explains both the damage ultraviolet radiation can cause and the atmospheric mechanism that protects us from it. We turn next to understanding the ultraviolet shield provided by oxygen and ozone in our stratosphere.

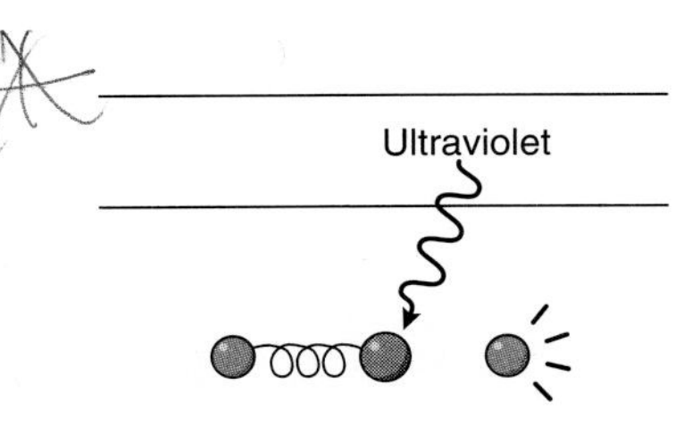

Figure 2.7
Ultraviolet radiation can break chemical bonds.

2.7 The Oxygen/Ozone Screen

The presence of oxygen and ozone in the Earth's stratosphere guarantees that electromagnetic radiation reaching the surface of the planet is different from that emitted by the Sun in some important respects. Much of the ultraviolet light is blocked out by these two allotropes of the same element, oxygen. This is shown schematically in Figure 2.8.

As we noted in Chapter 1, 21% of the atmosphere consists of diatomic oxygen. The forms of life that inhabit our planet are absolutely dependent on the chemical properties of this gas and its interaction with ultraviolet radiation. The strong covalent bond holding the two oxygen atoms together in the O_2 molecule can be broken by the absorption of a photon of the proper radiant energy. The photon excites a bonding electron to a higher energy level, causing the atoms to come apart. But the bond will be broken and the molecule will dissociate only if the photon has energy corresponding to a wavelength of 242 nm or less ($\lambda \leq 242$ nm). This energy is in the ultraviolet region of the spectrum.

$$O_2 + \text{photon} \longrightarrow 2\ O \qquad [2.4]$$

$$\lambda \leq 242 \text{ nm}$$

Because of this reaction, stratospheric oxygen shields the surface of the Earth from high-energy radiation. As green plants flourished on the young planet, they released oxygen to the atmosphere. The increasing oxygen concentration led to more effective interception of ultraviolet radiation. Consequently, forms of life evolved that were less resistant to UV radiation than they would otherwise have been.

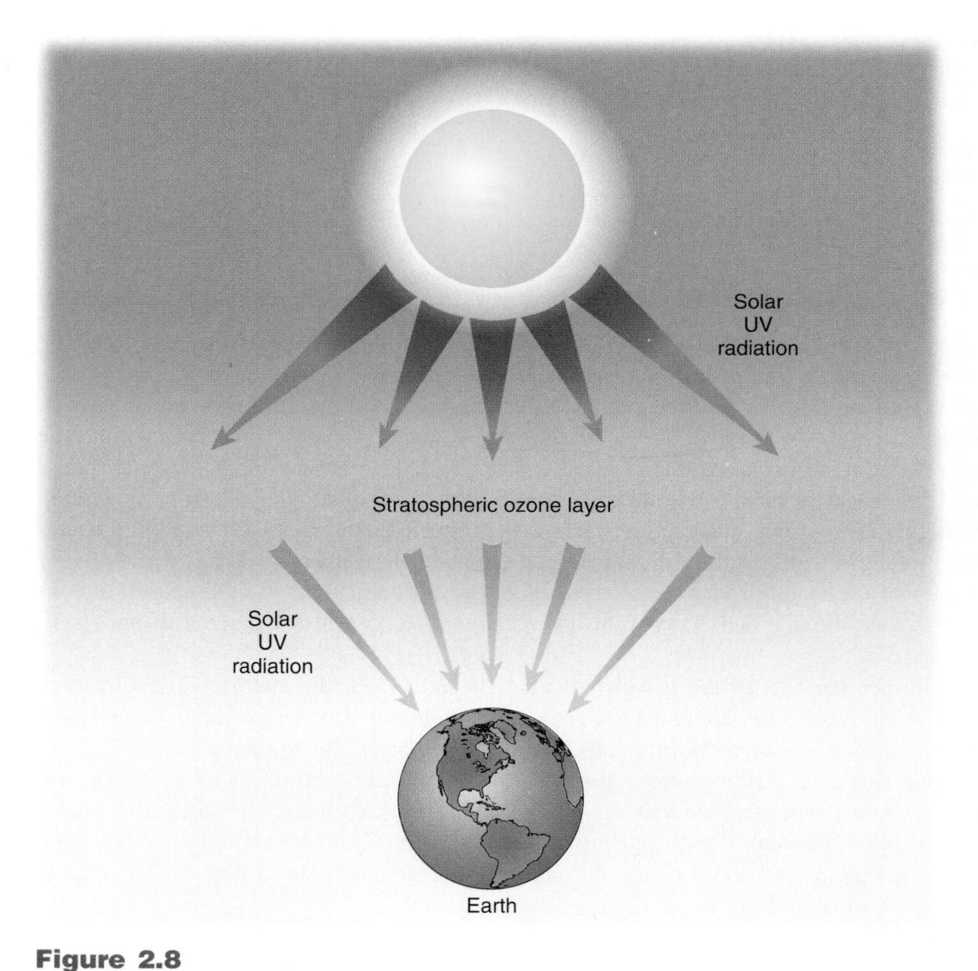

Figure 2.8

Solar ultraviolet radiation comes to Earth. Solar UV radiation is greatly diminished by passing through oxygen and particularly through ozone in the stratosphere.

Diatomic oxygen screens out radiation with wavelengths shorter than 242 nm. If O_2 were the only UV absorber in the atmosphere, the surface of the Earth and the creatures that live on it would still be subjected to damaging radiation in the 242–320 nm range. It is here that ozone plays its protective role. The fact that ozone is more reactive than diatomic oxygen suggests that the O_3 molecule is more easily broken apart than O_2. Recall that the atoms in the O_2 molecule are connected with a strong double bond. Each of the bonds in O_3 is somewhere between a single and double bond in length and in strength. This makes the bonds in O_3 energetically weaker than the double bonds in O_2. Therefore, photons of a lower energy (longer wavelength) should be sufficient to separate the atoms in O_3. This is in fact the case: radiation of wavelength 320 nm or less induces the following reaction.

$$O_3 + \text{photon} \longrightarrow O_2 + O \qquad [2.5]$$

$$\lambda \leq 320 \text{ nm}$$

Because of this reaction and that represented by equation 2.4, only a relatively small fraction of the Sun's UV radiation reaches the surface of the Earth. However, what does arrive can do significant damage.

2.15 Your Turn

Calculate the difference (in joules) between the energy of a 242-nm photon and that of a 320-nm photon.

Answer

2.00×10^{-19} J

2.8 Biological Effects of Ultraviolet Radiation

The consequences of ultraviolet radiation on plants and animals depend primarily on two factors: the intensity of UV radiation and the sensitivity of organisms to that radiation. The vertical scale of Figure 2.9 indicates the quantity of ultraviolet solar energy (expressed in joules) that falls on a surface one square meter in area in one second. The graph shows how this energy varies with wavelength. For any wavelength, the total amount of energy is the product of the number of photons striking the surface and the energy per photon. The rather flat upper curve reveals that the energy input above the

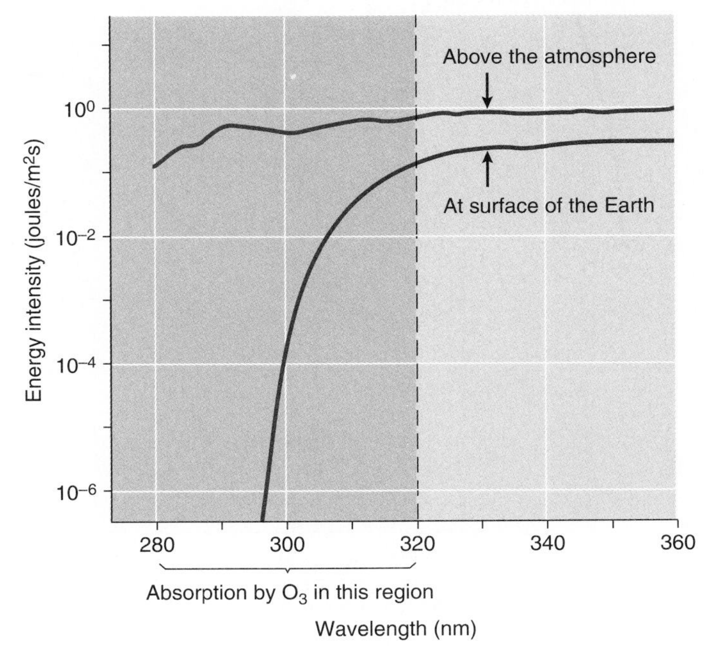

Figure 2.9
Variation of solar energy with wavelength of UV radiation.

atmosphere does not depend significantly on wavelength. If the Earth's atmosphere did not exist, the surface of the planet and the creatures on it would be subjected to these extremely high levels of radiant energy. However, the lower curve indicates that the energy reaching the surface of the Earth varies markedly with wavelength. It starts dropping at 330 nm and falls off sharply as the wavelength decreases, due to the absorption of UV radiation by stratospheric ozone.

In fact, the decrease in UV radiation is a good deal more dramatic than the figure at first suggests. The vertical scale in Figure 2.9 is logarithmic, a method of presenting data that permits the inclusion of a wide range of values. A logarithmic scale is one in which every mark on the axis represents a value (in this case, an energy value) that is one-tenth of that corresponding to the mark immediately above it on the vertical axis. Thus, at 320 nm, where ozone starts absorbing, the energy input to the Earth's surface is 1×10^{-1} or 0.1 joules per square meter per second ($J \cdot m^{-2} \cdot s^{-1}$). At 300 nm, the energy has dropped to 1×10^{-4} or 0.0001 $J \cdot m^{-2} \cdot s^{-1}$.

Highly energetic photons can excite electrons and break bonds in biological molecules, rearranging them and altering their properties. Solar radiation at wavelengths below 300 nm is almost completely screened out by O_2 and O_3 in the stratosphere. This is most fortunate, because radiation in this region of the spectrum is particularly damaging to living things. This relationship is evident from Figure 2.10, where biological sensitivity is plotted versus wavelength. As defined here, biological sensitivity is based on experiments in which the damage to deoxyribonucleic acid (DNA), the chemical basis of heredity, is measured at various wavelengths. In the figure, the biological sensitivity is expressed in relative units, once more on a logarithmic scale. Biological sensitivity at 320 nm is about 10^{-5} or 0.00001 units. But at 280 nm, the sensitivity is 10^0 or 1 unit. This means that radiation at 280 nm is 10^5 or 100,000 times more damaging than radiation at 320 nm. As we have seen, this is because the energy per photon and the potential for biological damage increase as the wavelength decreases.

- A discussion of DNA appears in Chapter 12.
- As wavelength ↓, frequency ↑, energy ↑, DNA damage ↑.

2.16 Your Turn

Arrange these types of radiation in order of *decreasing* energy.

a. gamma rays **b.** microwaves **c.** ultraviolet **d.** visible

Answer

gamma rays, ultraviolet, visible, microwaves

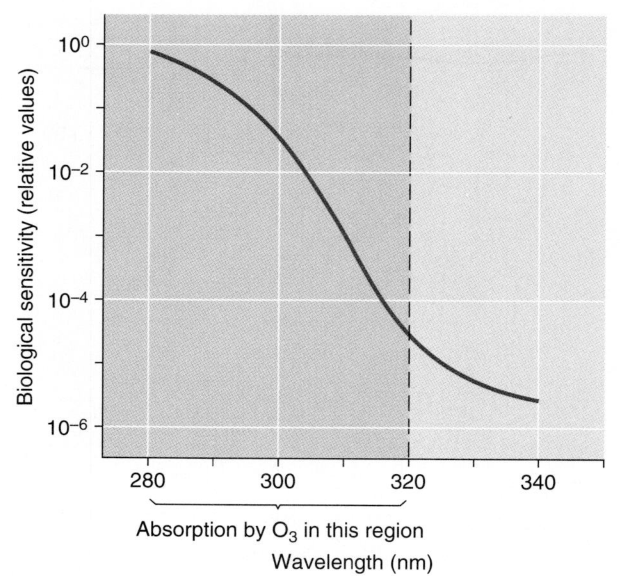

Figure 2.10

Variation of biological sensitivity of DNA with wavelength of UV radiation.

If more high-energy photons reach the surface of the Earth from the Sun, the potential for significant biological damage increases. All evidence shows that the average stratospheric ozone concentration has dropped significantly in the last 20 to 30 years. This phenomenon has been documented in many regions of the world, using data gathered from high-flying aircraft, ground-based systems, and satellites. Later in this chapter we will explore the special reasons why the observed percent ozone depletion is the greatest in the Antarctic. However, stratospheric ozone depletion means that the ability of the atmosphere to screen out UV radiation with wavelengths below 320 nm has decreased. Although this has happened to varying extents in different regions, living things are now exposed to greater intensities of potentially damaging radiation. Scientists have made calculations predicting that a given percent decrease in stratospheric ozone will increase the flux of biologically damaging UV radiation by twice that percentage. For example, a 6% decrease in stratospheric ozone could mean a 12% rise in skin cancer, especially the more easily treated form, non-melanoma skin cancers such as basal cell and squamous cell cancers. This condition is considerably more common among Caucasians than among those with more heavily pigmented skin. People of African and Indian origin are better equipped to withstand the high levels of UV radiation in the intense sunlight that strikes the Earth near the equator.

There is good evidence linking the incidence of non-melanoma skin cancers, the intensity of UV radiation, and latitude. For example, the disease generally becomes more prevalent as one moves farther south in the Northern Hemisphere (Figure 2.11).

2.17 Consider This: Geography of Skin Cancer

Many generations of immigrants have come to the United States. There are fair-skinned Northern Europeans, for example, who have settled in the area around San Antonio, TX; there are many other equally fair-skinned immigrants who have settled in the area around Seattle. Based on Figure 2.11, compare their relative risks of developing skin cancer. Identify several other factors that may affect the risk for any individual in these two populations.

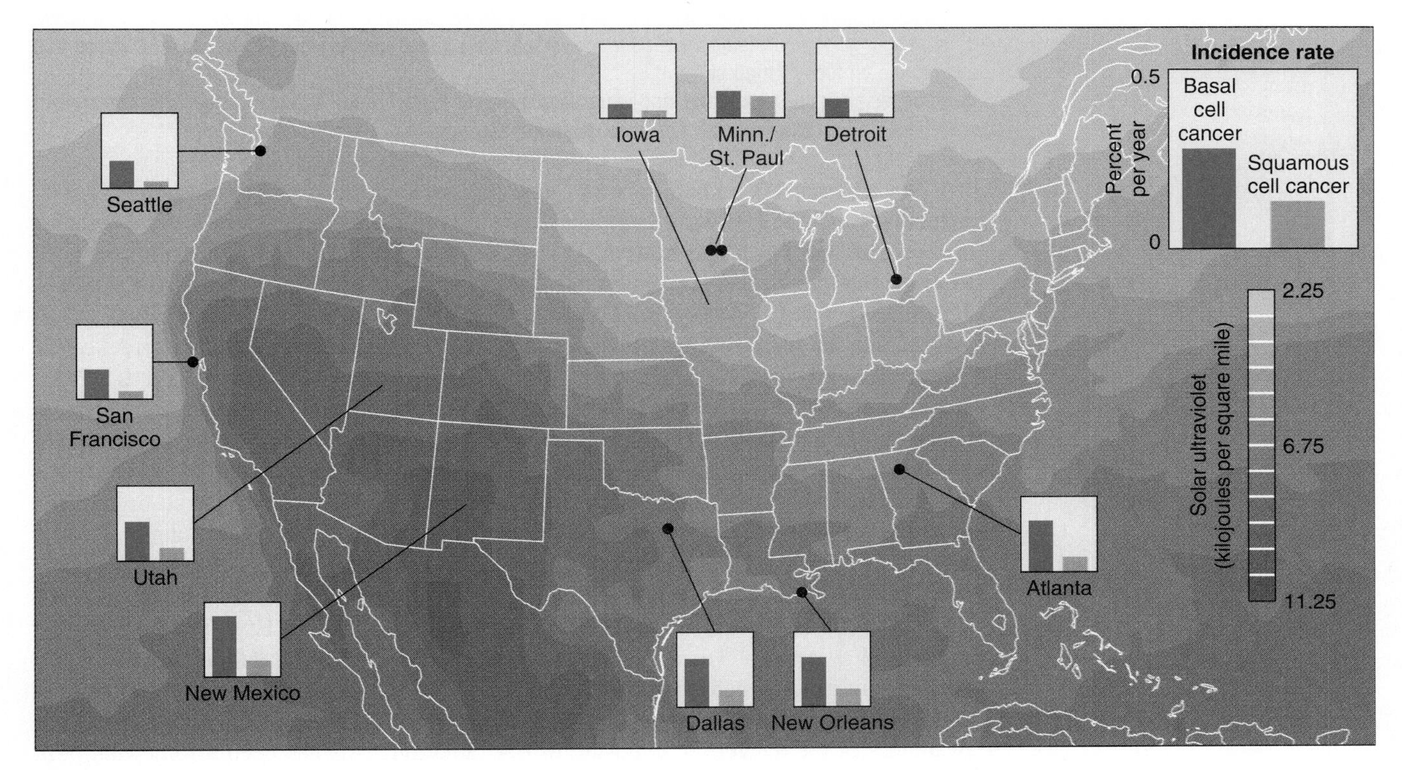

Figure 2.11
Skin cancer risks in selected U.S. locations.

Table 2.4

UV Index

UV Index	Exposure Level	Minutes for some damage to occur in light-skinned people who "never tan"	Minutes for some damage to occur in dark-skinned people who "never tan"
0–2	Minimal	30	>120
3–4	Low	15–20	75–90
5–6	Moderate	10–12	50–60
7–9	High	7–8.5	33–40
10–15	Very high	4–6	20–30

Because of the damage that can be caused by exposure to UV radiation, the National Weather Service issues an ultraviolet index forecast that appears nationally in newscasts, in newspapers, and on the Web. UV Index values range from 0 to 15 and are based on how long it takes for skin damage to occur (Table 2.4). The Index has played an important role in alerting the public to the dangers of increasing exposure to UV radiation.

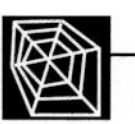

2.18 Consider This: UV Index Forecasts

The UV Index indicates the amount of UV radiation reaching the Earth's surface at solar noon (1 P.M. daylight time).

a. The UV Index depends on the latitude, the day of the year, time of day, amount of ozone above the city, elevation, and the predicted cloud cover. How is the UV index affected by each of these?

b. The UV Index Forecast is available on the Web, compliments of a satellite launched by the National Oceanographic and Atmospheric Administration (NOAA). Search either for "Current UV Index Forecast, NOAA" or go to the *Chemistry in Context* web site for a direct link. Account for the range of values that you see on today's map of the U.S.

c. Surfaces such as snow, sand, and water intensify your exposure to UV radiation, because they reflect it back at you. What outdoor activities might increase your risk from exposure?

Not all solar UV radiation is the same in wavelength and energy (Figure 2.5). The particular range of UV wavelengths and energies influence how much UV solar radiation reaches the Earth and how much damage it can cause. The UV radiation coming from the Sun can be categorized by its wavelength as UV-A (320–400 nm), UV-B (280–320 nm), or UV-C (<280 nm) (Table 2.5). UV-C radiation, with the shortest wavelengths, has the highest energy. Thankfully UV-C radiation is completely screened by stratospheric ozone and oxygen. The ozone layer protects against most of the UV-B radiation, although some UV-B still reaches the Earth's surface. UV-B is particularly effective at damaging DNA and has been linked to weakening the human immune system. It is also implicated in causing eye damage, skin cancers, including non-melanoma basal cell cancer, as well as being responsible for damage to crops and marine organisms. Recent research has linked non-melanoma skin cancers specifically to mutations of the *p*53 gene in skin cells. UV-B photons are sufficiently energetic to directly alter the DNA structure in those cells, causing the mutations. Unlike UV-B or UV-C, the longer wavelength UV-A is not screened by stratospheric ozone or oxygen. Even though UV-A is relatively low energy, it is suspected in the premature wrinkling and aging of skin, as well as skin cancer, including melanomas that can be fatal.

Skin cancer rates continue to rise in all countries (Figure 2.12), despite increased awareness of the dangers of exposure to UV radiation. This has been a somewhat puzzling phenomenon because improvements in early detection and effective treatment have lowered death rates for many cancers, including many skin cancers. Changes in the

Table 2.5

Categories and Characteristics of UV Radiation

Radiation	Wavelength Range	Energy	Comments
UV-A	320–400 nm	Least energetic of these three UV categories	Least damaging, reaches Earth's surface in greatest amount
UV-B	280–320 nm	More energetic than UV-A, less energetic than UV-C	More damaging than UV-A, less damaging than UV-C, most absorbed by ozone in the stratosphere
UV-C	200–280 nm	Most energetic of these three categories	Most damaging of these three, but not a problem because totally absorbed by oxygen and ozone in stratosphere

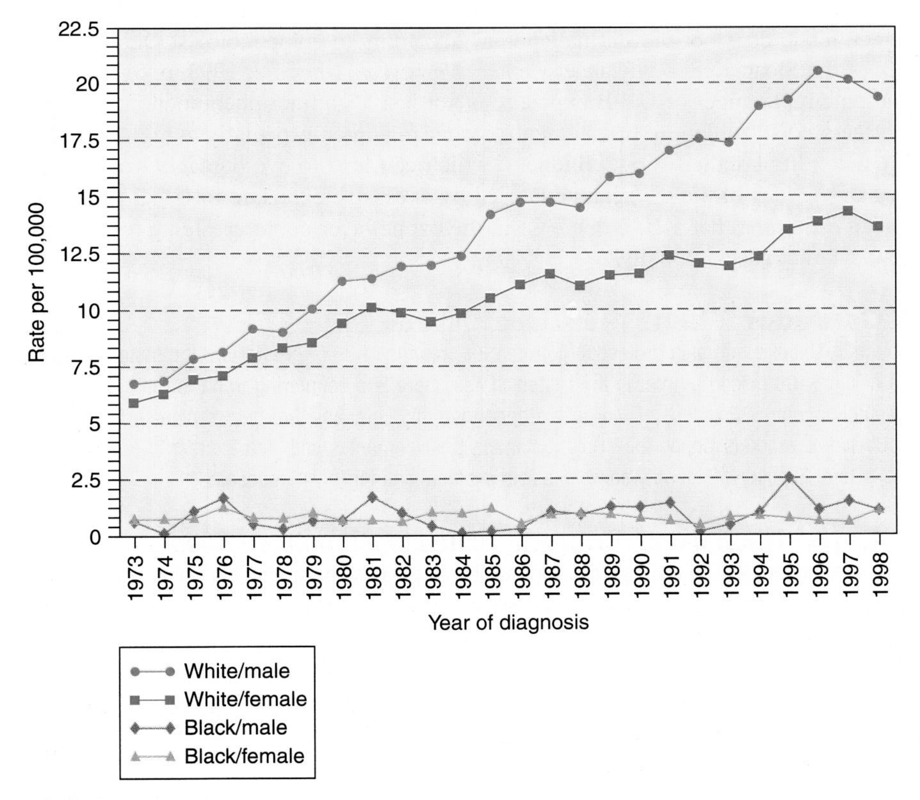

Figure 2.12

Increase in incidence of melanoma skin cancer in the United States, 1973–1998.

(Source of data: Surveillance, Epidemiology, and End Results (SEER) Program of the National Cancer Institute.)

Figure 2.13
It is estimated that up to 60% of Australians will be treated for skin cancer in their lives. Wearing protective sunscreen, such as Blue Lizard suncream, is one way to reduce the risk of skin cancer.

From the makers of Blue Lizard suncream: "Blue Lizard suncream is packaged in a unique bottle that changes colour to blue in UV light. This gives you an extra reminder that the dangers of UV light are still present. This is a deliberate message to reinforce the seriousness of looking after your health."

natural protection afforded by the stratospheric ozone layer are only partially responsible for higher rates of skin cancer. In a study conducted for the National Institutes of Health, researchers were surprised to find that Vermont was the U.S. state with the highest melanoma rate in 1998. Clearly, geographic location cannot be the only factor influencing development of skin cancer. Can this be linked to the availability of outside sports such as skiing in winter and hiking in the summer? Or can this be linked to the fact that Vermont also has a large number of tanning beds per capita? There is mounting evidence that tanning beds, which emit UV-A and UV-B radiation, may increase the risk of skin cancer, even melanomas.

Tanning, either naturally or in a tanning bed, is clearly a risk/benefit activity. Many light-skinned sun worshippers consider a golden tan a cosmetic benefit worth taking a risk for. The risk undoubtedly exists, as evidenced by the fact that about one million new cases of skin cancer occur each year in the United States, almost as many as the total number of cases of all other cancers. Skin cancers can develop many years after repeated, excessive exposure has stopped. Cancers even have been linked to a single episode of extreme sunburn. Fair-haired, fair-skinned individuals are at the highest risk to develop skin cancer from UV-B radiation; fair-skinned whites in Australia have the highest skin cancer rates in the world. Dr. Paul Jelfs, head of the Australian Institute of Health and Welfare, reported that during the 1990s, skin cancer rates in Australia increased by 4.3% per year for men and 1.8% per year for women. The Australian government has acted in several ways to turn this trend around, including banning tanned models from all advertising media. It is also important to note that people of all skin types face greater risk with increased exposure to the sun, not just in Australia.

Some protection against harmful UV rays is offered by sunscreens, products that contain compounds that absorb UV-B to some extent; some sunscreens also contain compounds that absorb UV-A (Figure 2.13). The American Academy of Dermatology recommends a sunscreen with a skin protection factor (SPF) of 15 to 30. But wearing a sunscreen does not mean that you are without risk from the Sun's UV rays. Because sunscreens let you be exposed for a longer time without burning, they may ultimately cause greater UV damage. "Sunscreen alone does not prevent melanoma" according to epidemiologist Dr. Marianne Berwick of the Sloan-Kettering Cancer Center (*Newsweek,* March 2, 1998, p. 61).

The potential danger of UV-B exposure is not just a summer phenomenon. Because snow reflects UV-B radiation, harm can also occur in winter, especially at high altitudes, creating a condition called snow blindness, which can lead to eye damage. Cataracts, a clouding of the lens of the eye, can be caused by excessive exposure to UV-B radiation. It has been estimated that a 10% decrease in the ozone layer could create up to two million new cataract cases globally.

2.19 Consider This: Sunscreen and You

Even though sunblocks with skin protection factors (SPF) ratings of 25 or greater are the fastest-growing segment of the $400 million market in the United States, there is a countering trend to market products low in protection with SPF ratings of 4 or less. Coppertone, the company that helped pioneer SPF ratings and the UV Index, markets Coppertone Gold®. Banana Boat competes with Tan Express® and Hawaiian Tropic sells Total Exposure®. Identify some possible explanations for these actions.

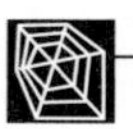

2.20 Consider This: Bronze by Choice—Tanning Salons

The indoor tanning industry maintains a constant public relations campaign that highlights positive news about indoor tanning, promoting it as part of a healthy lifestyle. Countering these claims are the studies published in scientific journals that support the view of dermatologists that there is no such thing as a "safe tan." Investigate at least two web sites that present each point of view and list the specific claims made by each side of this issue. Based on your findings, what are the criteria you would use to decide whether or not to go to an indoor tanning salon?

Those who endure the long nights and short days of northern winters are compensated in general by a level of non-melanoma skin cancer that is only about half that of those who enjoy year-round sunshine. The geographical effect on radiation intensity and skin cancer is, at least to date, much greater than that caused by ozone depletion.

That seemingly reassuring statement should not be interpreted as suggesting that the problem of ozone depletion can be ignored. Human beings are not the only creatures on the globe affected by UV radiation. Our existence is inextricably linked to the entire ecosystem. Increases in UV radiation will bring harm to young marine life, such as floating fish eggs, fish larvae, juvenile fish, and shrimp larvae. Plant growth is suppressed by UV radiation, and experiments have measured the negative impact that increased UV-B radiation has on phytoplankton. These photosynthetic microorganisms live in the oceans where they occupy a fundamental niche in the food chain. Phytoplankton ultimately supply the food for all the animal life in the oceans, and any significant decrease in their number could have a major effect globally. In a report released in September 1999, an international panel of scientists confirmed that exposure to elevated levels of UV-B radiation affected phytoplankton movement (up and down in water) and their motility (moving through water). Without such movement and motility, phytoplankton cannot achieve proper position in the water and are unable to carry out photosynthesis as effectively. Moreover, these tiny plant-like organisms play an important role in the carbon dioxide balance of the planet by absorbing approximately 80% of the atmospheric CO_2 created by human activities. Thus, it is possible that ozone depletion may influence another atmospheric problem—the greenhouse effect, the topic of Chapter 3. There is also experimental evidence of DNA damage in the eggs of Antarctic ice fish.

Decreasing stratospheric ozone and consequences of this reduction are cause for concern and action. But action requires knowledge of the chemistry that occurs 15 miles above the surface of the Earth; we turn next to consider that.

2.9 Stratospheric Ozone: Its Formation and Fate

Every day, 300,000,000 (3×10^8) tons of stratospheric ozone form and an equal mass decomposes. As with any chemical or physical change, totally new matter is neither created nor destroyed but merely changes its chemical or physical form. In this particular case, the overall concentration of ozone remains constant. The process is an example of a **steady state,** a condition in which a dynamic system is in balance so that there is no net change in concentration of the major species involved.

A steady state arises when a number of chemical reactions, typically competing reactions, balance each other. In the case of stratospheric ozone, the steady state is the net result of four reactions that constitute the **Chapman cycle,** named after Sydney Chapman, a physicist who first proposed it in 1929. Consider Figure 2.14 as you work through these equations for the reactions in the four steps of the cycle.

Step 1. Monatomic oxygen formation (decomposition of O_2):

$$O_2 + \text{UV photon} \longrightarrow 2\ O$$

Step 2. Ozone formation (O_2 and O consumption):

$$O_2 + O \longrightarrow O_3$$

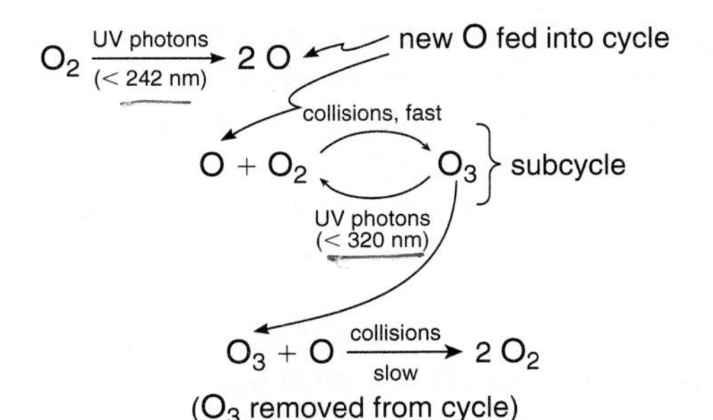

Figure 2.14
The Chapman cycle.

Step 3. Ozone decomposition (O_2 and O formation; opposite of step 2):

$$O_3 + \text{UV photon} \longrightarrow O_2 + O$$

Step 4. Diatomic oxygen formation (O_3 and O conversion):

$$O_3 + O \longrightarrow 2\,O_2 \qquad \text{(slow)}$$

This natural process shows both ozone formation and ozone decomposition. Ozone forms from O_2 via steps 1 and 2, and decomposes back to O_2, the material from which it originates, in steps 3 and 4, thus completing the cycle (Figure 2.14).

You have already encountered the reaction identified as step 1 (equation 2.4). It is the process by which oxygen molecules absorb photons of UV radiation and dissociate into individual oxygen atoms. These reactive atoms tend to combine readily with other atoms and molecules. One such reaction is step 2, which occurs when a monatomic oxygen atom (O) strikes an oxygen molecule (O_2) to generate an ozone molecule, O_3. As we can see from step 3, once an O_3 molecule is generated, it can absorb a photon of UV radiation, causing the molecule to dissociate and regenerate O_2 and O. It is, of course, by means of the reaction in step 3 that ozone screens out UV radiation. Most of the oxygen molecules and atoms formed in step 3 recombine to form ozone molecules via step 2. Occasionally, however, an O_3 molecule collides with an O atom to form two O_2 molecules (step 4). This slow reaction removes the "odd oxygen" species O and O_3 from the cycle. The "lifetime" of a given ozone molecule depends strongly on altitude, ranging from days to years. In the center of the ozone layer, an O_3 molecule can persist for several months before it dissociates into O_2 and O.

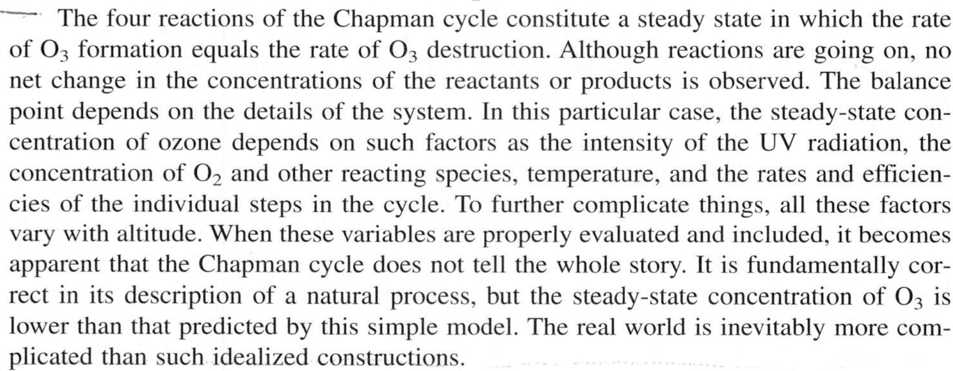

The four reactions of the Chapman cycle constitute a steady state in which the rate of O_3 formation equals the rate of O_3 destruction. Although reactions are going on, no net change in the concentrations of the reactants or products is observed. The balance point depends on the details of the system. In this particular case, the steady-state concentration of ozone depends on such factors as the intensity of the UV radiation, the concentration of O_2 and other reacting species, temperature, and the rates and efficiencies of the individual steps in the cycle. To further complicate things, all these factors vary with altitude. When these variables are properly evaluated and included, it becomes apparent that the Chapman cycle does not tell the whole story. It is fundamentally correct in its description of a natural process, but the steady-state concentration of O_3 is lower than that predicted by this simple model. The real world is inevitably more complicated than such idealized constructions.

2.10 Distribution of Ozone in the Atmosphere

One key to understanding ozone depletion is reliable information about atmospheric ozone concentrations. The total amount of ozone in a vertical column of air of known volume can be determined with relative ease. It is done from the surface of the Earth by measuring the amount of UV radiation reaching a detector; the lower the intensity of the radiation, the greater the amount of ozone. Such data have been collected since the late 1930s. Since the 1970s, measurements of total ozone have also been made from the top of the atmosphere. Satellite-mounted detectors record the intensity of the ultraviolet radiation scattered by the upper atmosphere. The results are then related to the amount of O_3 present. Measuring the ozone concentration at various altitudes is a good deal more difficult. Detectors are carried aloft on airplanes, rockets, balloons, and satellites. The results of some of these measurements are summarized in Figure 2.15.

This plot includes a good deal of important information and is worth careful study. First, note that concentration on the horizontal axis has units of O_3 molecules per cubic meter. Now examine the values along this axis. Here we once more use a logarithmic scale to include a wide range of data. (If a typical linear scale were used to present these data, it would be inconveniently long.) There is a 10-fold increase in concentration between two successive markers. Thus, 10^{15} is 10 times larger than 10^{14}; 10^{16} is 10 times larger than 10^{15}, and 100 times greater than 10^{14}.

• A review of exponents is found in Appendix 2.

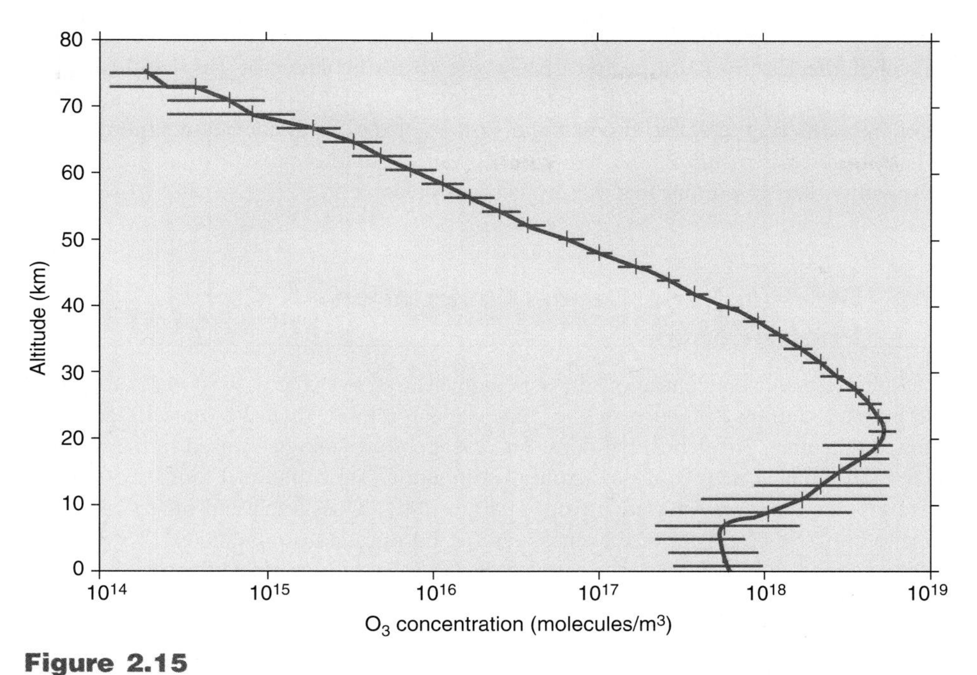

Figure 2.15

Ozone concentrations at various altitudes.

(Source of data: United States Standard Atmosphere, 1976, U.S. EPA.)

To put these concentrations into perspective, recall the Chapter 1 calculation of the concentration of CO molecules in a breath. There we concluded that the average 1-L breath would contain 2×10^{17} CO molecules. This corresponds to 2×10^{20} CO molecules per cubic meter (1 m^3 = 1000 L). Note that according to Figure 2.15, the highest concentration of ozone in the stratosphere is between 10^{18} and 10^{19} O_3 molecules per cubic meter. This means that the maximum stratospheric ozone concentration is about 1% of the maximum tropospheric carbon monoxide concentration.

The horizontal lines that cross the curve in Figure 2.15 are called uncertainty bars or error bars. Each bar is a composite of the individual concentration values obtained in a number of separate measurements. Not all measurements give identical results, and the length of the bar indicates the variability of the data. In some cases, especially at low and at high altitudes, the variation is considerable. The cause of this variability may be limitations in instrument sensitivity, experimental error, or genuine variation in concentration. It is impossible to determine the source of variability from the graph. In this case, most of the scatter is due to seasonal and geographical variation.

The message is that when evaluating experimental results, whether published in a daily newspaper, on the Web, or in a scientific journal, it is important to know the limitations of the data. Just how good are the measurements and how many were performed? Are the results statistically significant? How great is the uncertainty? How much confidence can you have in the conclusions? A political poll reporting that 51% ± 5% of those surveyed favored one candidate and 49% ± 5% favored the other candidate would be of little value. The potential errors (±5%) are greater than the difference between the percentages for the two candidates. Healthy and informed skepticism is a useful attribute in analyzing any experimental evidence or statistical summary.

In Figure 2.15 the curve is drawn through the midpoints of the uncertainty bars—points that represent the averages of the measurements. Even allowing for uncertainties, it is clear that the highest concentration of ozone occurs between 10 and 30 km, with a maximum around 20 km. Roughly 91% of the Earth's ozone is found in the stratosphere, between the altitudes of 10 and 50 km.

Because this 40-km band is so broad, the concept of an ozone layer can be a little misleading. At the altitudes of the maximum ozone concentration, the atmosphere is very thin, so the total amount of ozone is surprisingly small. If all the O_3 in the atmosphere could be isolated and brought to the average pressure and temperature at the surface of the Earth (1.0 atm and 15°C), the resulting layer of gas would be just 0.30 cm thick, or about 1/8 of an inch. On a global scale, this is a minute amount of matter. Yet, this fragile shield protects the surface of the Earth and its inhabitants from the harmful effects of ultraviolet radiation. Because ozone is present in a small and finite quantity, it is important that we protect and preserve it.

2.11 Stratospheric Ozone Destruction—A Global Phenomenon

Stratospheric ozone concentrations have been measured over the past 80 years at ground experimental stations spread over the planet and for more than 20 years by satellite-mounted detectors. Airplanes, balloons, and rockets have also been used in the quest to understand what is happening to ozone in our upper atmosphere. Figure 2.16 shows measurements made in Antarctica from 1980 to 2001. This graph and others to follow report total ozone levels above the surface of the Earth in Dobson units (DU). A Dobson unit corresponds to about one ozone molecule for every billion molecules of air. For our purposes, the precise details of obtaining these measurements are less important than the fact that, at this time, a value of 320 DU represents the average ozone level over the northern hemisphere. A value of 250 DU is typical at the equator.

- The Dobson unit is named after G. M. B. Dobson, a scientist at Oxford University. In 1920, he invented the first instrument to quantitatively measure total atmospheric ozone.

Figure 2.16 shows the dramatic decline in ozone levels observed near the South Pole. The data displayed in Figure 2.16, collected in the fall of every year from 1980 through 2001, illustrate the decline in the minimum amount of ozone detected. Indeed, these changes were so pronounced that, when the British monitoring team at Halley Bay

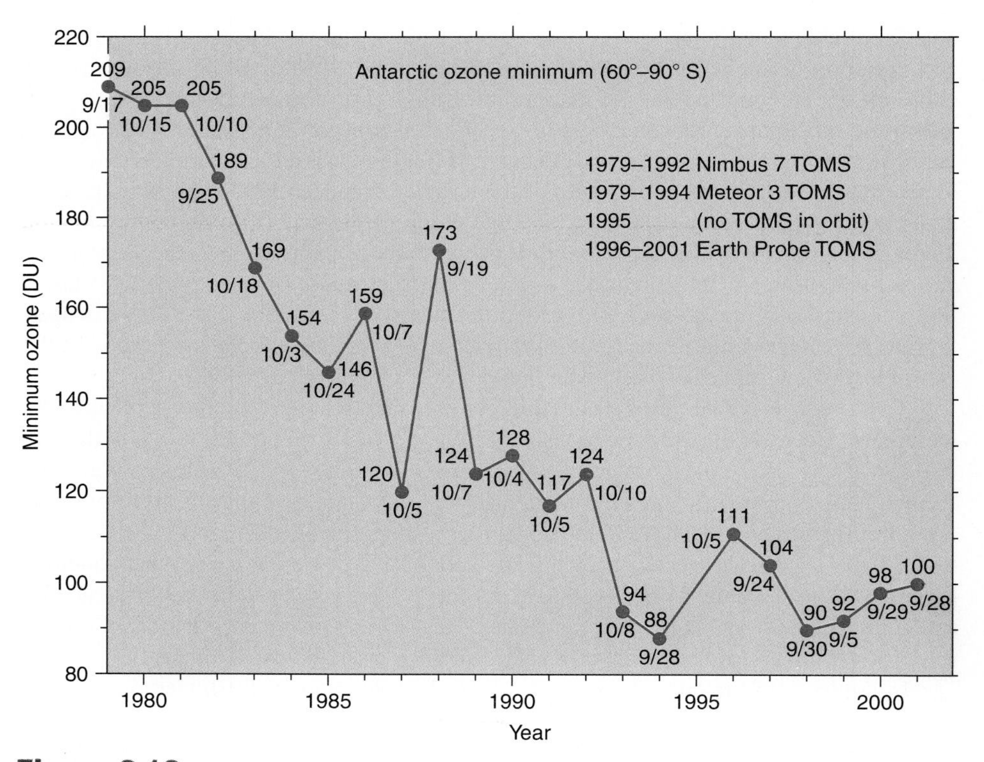

Figure 2.16

Minima in fall stratospheric ozone in Antarctica, 1979–2001. The number above each dot is minimum reading in DU. The number below each dot is the date the minimum occurred that year. TOMS the total ozone-measuring spectrometer, an analytical instrument.

(Source: Climate Prediction Center, NOAA.)

in Antarctica first observed it in 1985, they thought their instruments were malfunctioning. The area covered by ozone levels less than 220 DU, defined as the "ozone hole," was larger in early September 2000, than in any previous year. Total ozone destruction in the hole occurred from an altitude of 15 km to 20 km, consistent with measurements taken in recent years. The minimum total ozone value of 98 DU was not the lowest ever recorded. In early October 1993, the total ozone there dropped to 86 DU, the lowest recorded anywhere in the world in 36 years of measurement, less than 30% of the 1956 value. Keep in mind that there has always been a seasonal variation in ozone concentration over the South Pole, with a minimum in late September or early October, the Antarctic spring. What is unprecedented is the dramatic decrease in this minimum that has been observed over the 40 years.

• The reasons for the Antarctic ozone hole are discussed in Section 2.14.

More importantly, the stratospheric ozone concentrations are lower than those predicted using the simple Chapman cycle mechanism. That by itself is neither cause for alarm nor proof that the lower stratospheric ozone concentrations are the consequence of human intervention. In fact, many of the factors influencing the stratospheric ozone layer are natural in origin. We know that the processes establishing the steady-state concentration of stratospheric ozone are more complicated than originally believed.

For one thing, the natural concentration of stratospheric ozone is not uniform over all parts of the globe. On average, the total O_3 concentration increases the closer one gets to either pole (with the exception of the seasonal "hole" over the Antarctic). The formation of ozone via steps 1 and 2 of the Chapman cycle (p. 71) is triggered when an O_2 molecule absorbs a photon of ultraviolet light. Therefore, ozone production increases with the intensity of the radiation striking the stratosphere, an intensity that is not constant. Intensity varies with the seasons, reaching its maximum (in the Northern Hemisphere) in March and its minimum in October (just the reverse of the Southern Hemisphere). Consequently, stratospheric ozone concentrations also follow this seasonal pattern. In addition, the amount of radiation emitted by the Sun changes over an 11- to 12-year cycle related to sunspot activity. This variation also influences O_3 concentrations, but only by 1–2%. The winds blowing through the stratosphere cause other variations in ozone concentrations, some on a seasonal basis and others over a 28-month cycle. To further complicate matters, seemingly random fluctuations often occur. Finally, it is well established that certain gases from both natural and human-made sources are also responsible for the destruction of stratospheric ozone.

Dramatic TOMS images, such as the one that opens this chapter, are color coded to show stratospheric ozone concentrations in Dobson Units. The violet and purple regions are those where the greatest destruction of O_3 occurred, that is, lowest ozone concentrations. Since the early to mid-1990s, the size of the depleted ozone region annually equals nearly the total area of the North American continent, in some cases exceeding it. The next Consider This gives you the opportunity to examine a series of these images, each taken annually in October.

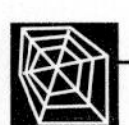

2.21 Consider This: Purple Octobers

NASA satellites provide stratospheric ozone data over time that can be tabulated in a number of ways, including global images, Antarctica ozone minima, and size of the ozone hole. All three are provided at the *Chemistry in Context* web site.

a. Using the web site, look first at the global images centered on Antarctica. Describe what is happening with the passage of time.

b. Now look at the graphs that show the minimum ozone levels and the size of the region affected. What information does each plot give you?

c. Use the information from all three views to write a description of what is meant by the term "ozone hole." In your statement, include references to the region of the globe, area affected, amount of ozone, and time.

The problem of ozone depletion is not limited to Antarctica. As another example, consider Figure 2.17. Scientists have been gathering data in Arosa, Switzerland, for more than 70 years, including satellite measurements started in the early 1980s. This figure shows that ozone levels in this Northern Hemisphere location, although lower than

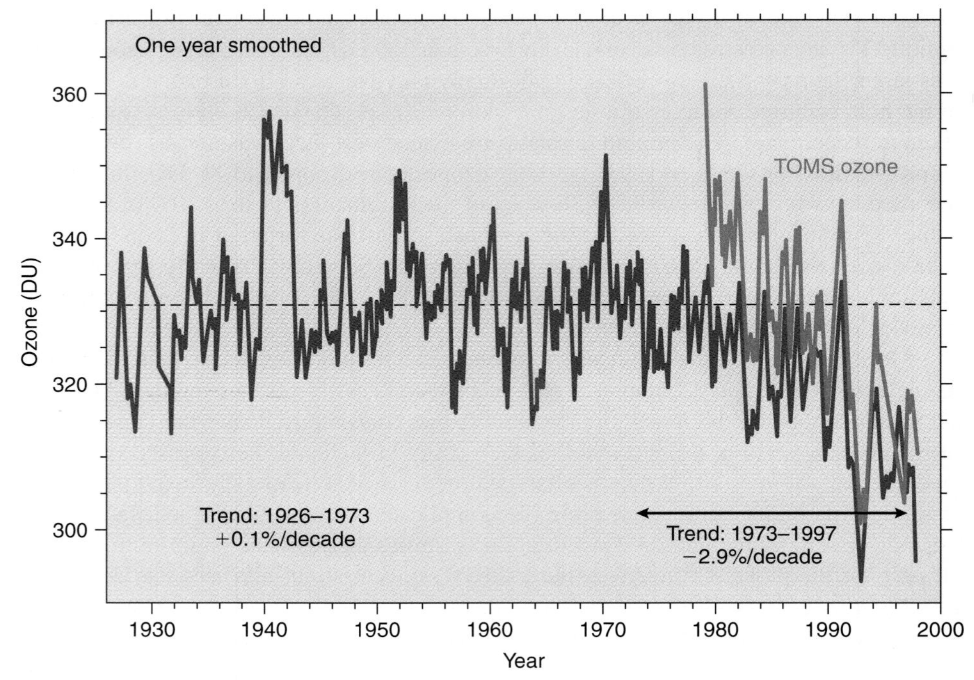

Figure 2.17

Ozone concentrations in Arosa, Switzerland, 1926–1998. The TOMS instrument was carried into the stratosphere by both Earth Probe (EP) satellites and the Japanese Advanced Earth Observations Satellite (ADEOS). DU is Dobson units, the common unit for expressing total ozone in the atmosphere.

(Source: www.epa.gov/docs/ozone/science/arosa.html. U.S. EPA.)

those predicted by the Chapman cycle mechanism, are not, however, not nearly the large percentage decrease documented in Antarctica.

The major naturally occurring cause of ozone destruction, wherever it takes place around the globe, is a series of reactions involving water vapor and its breakdown products. The great majority of the H_2O molecules that evaporate from the oceans and lakes fall back to the surface of the Earth as rain or snow. But a few reach the stratosphere, where the H_2O concentration is about 5 ppm. There, photons of ultraviolet radiation trigger the dissociation of water molecules into hydrogen atoms (H·) and hydroxyl (·OH) free radicals. A **free radical** is an unstable chemical species with an unpaired electron. The unpaired electron is often indicated with a dot, as it is here.

- Free radicals are also discussed in Chapter 9 in conjunction with plastics formation.

$$H_2O + \text{UV photon} \longrightarrow H\cdot + \cdot OH$$

Because of its unpaired electron, a free radical reacts readily. Thus, the H· and ·OH radicals participate in many reactions, including some that ultimately convert O_3 to O_2. It turns out that this is the most efficient mechanism for destroying ozone at altitudes greater than 50 km.

2.22 Your Turn

Can the reaction of H· and ·OH radicals account completely for depletion of ozone in the stratosphere? Why or why not?

Water molecules and their breakdown products are not the only agents responsible for natural ozone destruction. Another is NO (nitrogen monoxide, also called nitric oxide). Most of the NO in the stratosphere is of natural origin. It is formed when nitrous oxide, N_2O, reacts with oxygen atoms. The N_2O is produced in the soil and oceans by microorganisms and gradually drifts up to the stratosphere. There is really little that can or should be done to control this process. It is part of a cycle involving compounds of nitrogen and living things.

- NO has an uneven number of outer electrons, 11 (5 from N and 6 from O), so it has an unpaired electron. As a free radical, NO reacts with additional oxygen to form NO_2 and N_2O_4, other oxides of nitrogen.

However, not all the nitric oxide in the atmosphere is of natural origin; human activities can alter steady-state concentrations. That is why, in the 1970s, chemists became concerned about the increase in NO that would result from developing and deploying a fleet of supersonic transport (SST) airplanes. These planes were designed to fly at altitudes of 15–20 km, the region of the ozone layer. The scientists calculated that much additional NO would be generated by the direct combination of nitrogen and oxygen.

$$\text{energy} + N_2 + O_2 \longrightarrow 2\ NO$$

This reaction requires large amounts of energy, which can be supplied by lightning or the high-temperature jet engines of the SSTs. To evaluate this risk/benefit situation, many experiments and calculations were carried out, leading to predictions about the net effect of a fleet of SSTs. The conclusion was that the risks outweighed the benefits, and the decision was made, partly on scientific grounds, not to build an American fleet. The Anglo-French Concorde is the only commercial plane that operates at this altitude.

Subsequent research has indicated that atmospheric reactions involving NO and other oxides of nitrogen are more numerous and more complicated than originally thought. The National Aeronautics and Space Administration (NASA) is currently sponsoring a project to investigate effects of another generation of high-altitude supersonic aircraft. As you already know from your reading of Chapter 1, a more serious pollution problem involving the oxides of nitrogen occurs at ground level.

- Nitrogen oxides are also prominently featured in Chapter 6 dealing with acid rain.

Even when the effects of water, nitrogen oxides, and other naturally occurring compounds are included in stratospheric models, the measured ozone concentration is still lower than predicted. Measurements worldwide indicate that the ozone concentration has been decreasing over the past 20 years. There is a good deal of fluctuation in the data, but the trend is clear. Stratospheric ozone concentration at mid-latitudes (60° south to 60° north) has decreased by more than 8% in some cases (Figure 2.18). These changes cannot be correlated with changes in the intensity of solar radiation, so we must look elsewhere for a more complete explanation.

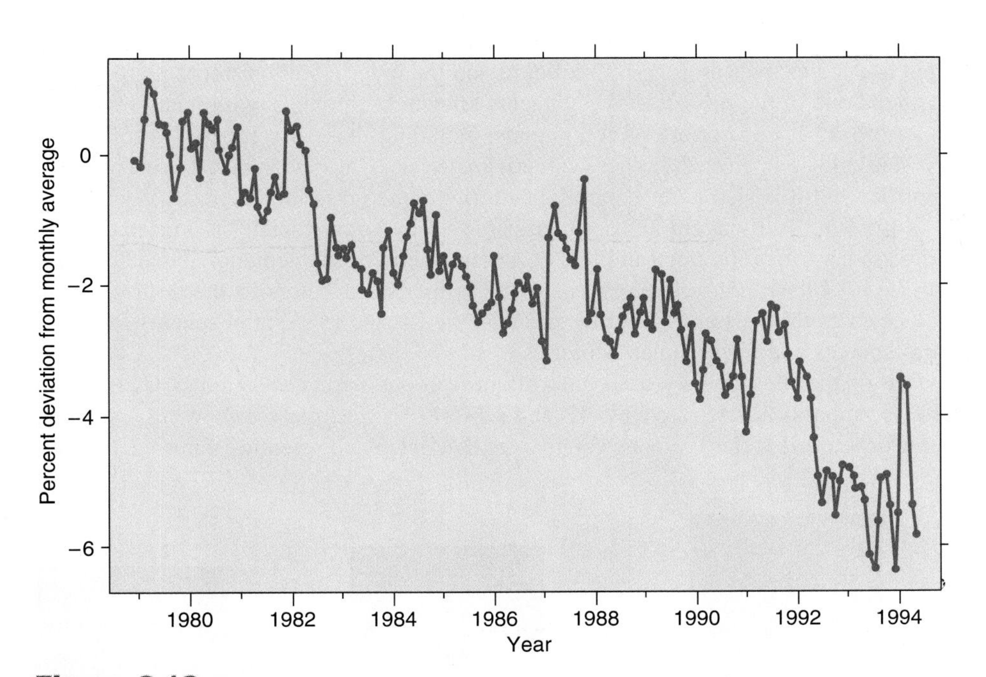

Figure 2.18

Change in stratospheric ozone concentration (60° south to 60° north) expressed as percent deviation from the monthly average, 1979–1995.

Source of data: Scientific Assessment of Ozone Depletion, 1994. Executive Summary, World Meterological Organization Global Ozone Research and Monitoring Project, Report No. 37, Geneva, 1995.

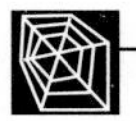

2.23 Consider This: Up and Down the Latitudes

In an earlier exercise, you used the Web to get stratospheric ozone data at a location of your choice (presumably above your head). Now go to NASA's archive of satellite data on stratospheric ozone levels to find out how the values have varied between 1979 and 1992 over the lower Northern Hemisphere latitudes. You may wish to coordinate your efforts with other students, so that together you cover a range of years.

a. Obtain values of stratospheric ozone levels at latitudes from +45 degrees north to +0 (the equator) for the year of your choice. Enter −90° west (the middle of the United States) as the longitude and use the satellite Nimbus-7 and a June 15 date. Obtain readings 5° apart. Make a table of the stratospheric ozone values and compute the average.

b. Compare with others in your class data over these 13 years. Note that you may not always be able to use the average as a meaningful comparison, because satellite data may be missing at some latitudes.

We have seen much evidence for the abnormally large decrease in stratospheric ozone. Although natural processes can cause such changes, they are insufficient to cause or explain the magnitude of the depletion. It is time to turn our attention to understanding chlorofluorocarbons, compounds that are an important factor in stratospheric ozone depletion.

2.12 Chlorofluorocarbons: Properties and Uses

A major cause of stratospheric ozone depletion was uncovered through the masterful scientific sleuthing of 1995 Nobel laureates F. Sherwood Rowland, Mario Molina, and Paul Crutzen, as well as other chemists and physicists. Vast quantities of atmospheric data have been collected and analyzed, hundreds of chemical reactions have been studied, and complicated computer programs have been written to identify the chemical culprit. As with most scientific results, some uncertainties remain, but there is now compelling evidence implicating an unlikely group of compounds—the chlorofluorocarbons (CFCs).

As the name implies, **chlorofluorocarbons** are compounds composed of the elements chlorine, fluorine, and carbon. Fluorine (F) and the more familiar chlorine (Cl) are members of the same chemical family, the halogens. The other halogens are bromine (Br) and iodine (I) (Figure 2.19). These elements appear in a column labeled Group 7A in the periodic table. At ordinary temperature and pressures, fluorine and chlorine exist as gases made up of diatomic molecules, F_2 and Cl_2. Bromine and iodine also form diatomic molecules, but the former is a liquid and the latter a solid at room temperature. Fluorine is one of the most reactive elements known. It combines with many other elements to form a wide variety of compounds, including those in "Teflon" (a trademark of the DuPont Company) and other synthetic materials. Chlorine is best known as a water purifier, but it is also a very important starting material in the chemical industry.

Chlorofluorocarbons do not occur in nature; they are artificially produced. This is an important verification point in the CFC/stratospheric ozone depletion debate because *there are no known natural sources of CFCs.* Other contributors to the destruction of ozone, such as the ·OH and ·NO free radicals, are formed in the atmosphere from both natural sources and from human activities.

Two of the most widely used chlorofluorocarbons have the formulas CCl_2F_2 and CCl_3F, commonly known as CFC-12 and CFC-11, respectively, following a naming scheme developed in the 1930s by chemists at DuPont. Their scientific names and Lewis

Chlorine

Bromine

Iodine

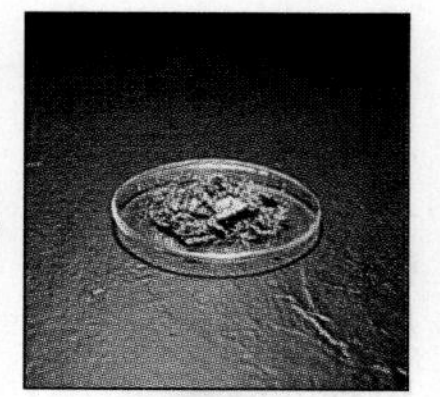

Figure 2.19
Selected Group 7A elements.

Table 2.6

Two Important Chlorofluorocarbons

CFC-11	CFC-12
Freon 11	Freon 12
CCl_3F	CCl_2F_2
trichlorofluoromethane	dichlorodifluoromethane

```
        ..                        ..
       :F:                       :F:
  ..    |    ..             ..    |    ..
 :Cl — C — Cl:             :Cl — C — F:
  ..    |    ..             ..    |    ..
       :Cl:                      :Cl:
        ..                        ..
```

structures are given in Table 2.6. Note that the scientific names for these two compounds are based on methane, CH_4. The prefixes *di-* and *tri-* specify the number of halogen atoms that substitute for hydrogen atoms. (Freon is a trademark of the DuPont Company.)

The introduction of CFC-12 (Freon 12) as a refrigerant in the 1930s was rightly hailed as a great triumph of chemistry and an important advance in consumer safety and environmental protection. This synthetic substance replaced ammonia or sulfur dioxide, two naturally occurring toxic and corrosive refrigerants that made leaks in refrigeration systems extremely hazardous. In many respects, CFC-12 was (and is) an ideal substitute. It has a boiling point in the right range, it is not poisonous, it does not burn, and the CCl_2F_2 molecule is so stable that it does not react with much of anything.

The many desirable properties of CFCs soon led to other uses: as propellants in aerosol spray cans, as the gases blown into polymer mixtures to make expanded plastic foams, as solvents for oil and grease, and as sterilizers for surgical instruments. Similar compounds, in which bromine replaces some of the chlorine or fluorine, have proved to be very effective fire extinguishers. These halons are used to protect property that would be especially vulnerable to water and other conventional fire-fighting chemicals. Thus, they have found applications in electronic and computer installations, chemical storerooms, aircraft, and rare book rooms.

Before we consider the key issue of how CFCs interact with stratospheric ozone, it is important to remember the impact their synthesis has on our lives. Because CFCs are nontoxic, nonflammable, cheap, and widely available, they revolutionized air conditioning, making it readily accessible in the United States for homes, office buildings, shops, schools, even automobiles. Oppressive summer heat and humidity became manageable with the use of low-cost CFCs as coolants. Throughout the American South, beginning in the 1960s and '70s, air conditioning using CFCs helped to spur on the booming growth of cities such as Atlanta, Dallas–Fort Worth, and Houston, to be followed by others such as San Antonio, Austin, Charlotte, Phoenix, Memphis, Orlando, and Tampa. Some of these are now among our nation's most populated metropolitan areas. In effect, a major sociological shift occurred because of CFC-based technology that transformed the economy and business potential of an entire region of the country.

By 1985, the combined annual international production of CFC-11 and CFC-12 was approximately 850,000 tons. Venting of refrigerators and air conditioners, evaporation of solvents, and escape during polymer foaming inevitably released some of this material into the atmosphere. In 1985, the ground-level atmospheric concentration of CFCs was about six molecules out of every 10 billion (0.6 ppb)—a value that has been increasing by about 4% per year. Of course, the fact that stratospheric ozone levels have been decreasing while CFC levels have been increasing does not prove that the two are causally related. However, other evidence suggests that there is a connection. Ironically, it is the very property that makes CFCs so ideal for so many applications—chemical inertness—that ends up posing a threat to the environment.

2.13 Interaction of CFCs with Ozone

Chlorofluorocarbons represent a classic case where an apparent virtue becomes a liability. Many of the uses of these compounds capitalize on their low reactivity. The carbon-chlorine and carbon-fluorine bonds in the CFCs are so strong that the molecules can remain intact for long periods. For example, it has been estimated that an average CCl_2F_2 molecule will persist in the atmosphere for 120 years before it is destroyed. In a much shorter time, typically about five years, many CFC molecules penetrate to the stratosphere with their structures intact.

In 1973, Rowland and Molina, motivated largely by intellectual curiosity, set out to study the fate of these stratospheric CFC molecules. They knew that as altitude increases and the concentrations of oxygen and ozone decrease, the intensity of ultraviolet radiation increases. Therefore, they reasoned that in the stratosphere high-energy photons, such as UV-C, corresponding to wavelengths of 220 nm or less, can break carbon-chlorine bonds. This reaction releases chlorine atoms.

$$CCl_2F_2 + \text{photon} \longrightarrow \cdot CClF_2 + \cdot Cl \qquad [2.6]$$

$$\lambda \leq 220 \text{ nm}$$

Similar reactions occur with other CFCs.

- Recall that free radicals have an unpaired electron, making them very reactive.

Atomic chlorine, ·Cl, is very reactive. A Cl atom has seven outer electrons, six of them paired and one unpaired. In equation 2.6, we emphasize the unpaired electron by writing the atom as ·Cl. The free radical chlorine atom exhibits a strong tendency to achieve a stable octet by combining and sharing electrons with another atom. Rowland and Molina and subsequent researchers hypothesized that this reactivity would result in the following chain of reactions.

First, the chlorine atom pulls an oxygen atom away from the O_3 molecule, forming chlorine monoxide, ·ClO, and leaving an O_2 molecule. Equation 2.7 shows this for two ·Cl atoms reacting with two ozone molecules to form two O_2 molecules and two chlorine monoxide molecules, key reaction intermediates in ozone depletion.

$$2 \cdot Cl + 2\, O_3 \longrightarrow 2 \cdot ClO + 2\, O_2 \qquad [2.7]$$

The ·ClO molecule is another free radical; it has 13 outer electrons (7 + 6). Recent experimental evidence indicates that 75–80% of stratospheric ozone depletion involves ·ClO joining to form ClOOCl, as shown in equation 2.8.

$$\cdot ClO + \cdot ClO \longrightarrow ClOOCl \qquad [2.8]$$

The ClOOCl decomposes in the two-step sequence shown in equations 2.9a and 2.9b.

- Blue is used to indicate reactants, red for products. Species occurring as reactants and products can be cancelled.

$$ClOOCl + \text{UV photon} \longrightarrow \cdot ClOO + \cdot Cl \qquad [2.9a]$$

$$\cdot ClOO + \text{UV photon} \longrightarrow \cdot Cl + O_2 \qquad [2.9b]$$

Note that ·Cl, ·ClO, and ClOOCl appear on both sides of these equations. We can treat this series of chemical equations as if they were mathematical equations. We subtract the ·Cl, the ·ClO, and the ClOOCl species from both sides of the chemical equations, or if you prefer, we cancel the ·Cl, the ·ClO, and the ClOOCl species. What remains is the net equation showing the conversion of ozone into oxygen gas.

$$\text{Net equation: } 2\, O_3 \longrightarrow 3\, O_2 \qquad [2.10]$$

Thus, the interaction of ozone with atomic chlorine provides a pathway for the destruction of ozone.

2.24 Consider This: A New Proposal—The Deep Space Connection

Two Canadian scientists proposed that high-energy radiation, sometimes called "cosmic rays", from deep space may be responsible for releasing chlorine free radicals from CFCs in the stratosphere. A publication in the August 13, 2001, issue of *Physical Review Letters* suggests that these cosmic rays can penetrate ice clouds, knocking electrons loose. These energetic electrons interact with CFCs to liberate the active chlorine atoms. How does this new proposal differ from our current understanding of the mechanism for production of chlorine free radicals?

The fact that ·Cl appears as a reactant (equation 2.7) and a product (equations 2.9a and 2.9b), is important. This indicates that ·Cl is both consumed and regenerated in the cycle, so there is no net change in its concentration. Such behavior is characteristic of a **catalyst,** a chemical substance that participates in a chemical reaction and influences its speed without undergoing permanent change. Atomic chlorine acts catalytically by being regenerated and recycled to remove more ozone molecules. On average, a single ·Cl atom can catalyze the destruction of as many as 1×10^5 O_3 molecules before it is carried back to the lower atmosphere by winds.

Interestingly, the currently accepted mechanism for ozone destruction by CFCs in the stratosphere was not the one first proposed by Rowland and Molina. Their initial hypothesis was that ·Cl reacted with O_3 to form ·ClO and O_2. The second step proposed was that ·ClO reacted with oxygen atoms to form O_2 and regenerate ·Cl atoms.

$$\cdot Cl + O_3 \longrightarrow \cdot ClO + O_2$$

$$\cdot ClO + O \longrightarrow \cdot Cl + O_2$$

Although this mechanism did not prove to be the correct one in the stratosphere, it did provide a reasonable explanation for why a limited number of chlorine atoms could be recycled and therefore be responsible for the destruction of a large number of ozone molecules. As is often true in science, hypotheses need to be recast in light of experimental evidence.

Atomic chlorine can also become incorporated into stable compounds that do not react to destroy ozone. Hydrogen chloride, HCl, and chlorine nitrate, $ClONO_2$, are two of these "safe" compounds that are quite readily formed at altitudes below 30 km. Thus, chlorine atoms are fairly effectively removed from the region of highest ozone concentration (about 20 km). Maximum ozone destruction by chlorine atoms appears to occur at about 40 km, where the normal ozone concentration is quite low.

Rowland, a professor at the University of California at Irvine, and Molina, then a post-doctoral fellow in Rowland's laboratory, published their first paper (two pages) on chlorofluorocarbons and ozone depletion in 1974 in the scientific journal *Nature.* At about the same time, other scientists were obtaining the first experimental evidence of stratospheric ozone depletion and CFCs in the stratosphere. Since then, the correctness of the Rowland-Molina hypothesis has been well established. Perhaps the most compelling evidence for the involvement of chlorine and chlorine monoxide in the destruction of stratospheric ozone is presented in Figure 2.20. These two graphs contain two

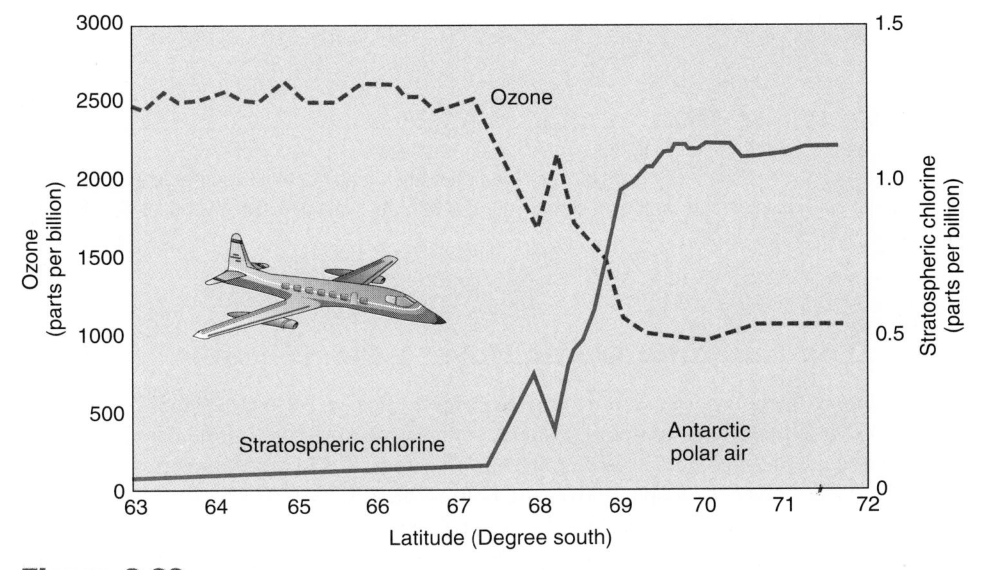

Figure 2.20

Antarctic O_3 and ·ClO concentrations. Measurements of ozone and reactive chlorine from a flight into the Antarctic ozone hole, 1987.

(Source of data: United Nations Environment Program. Taken from www.unep.ch/ozone/oz-story/sld008.shtml.)

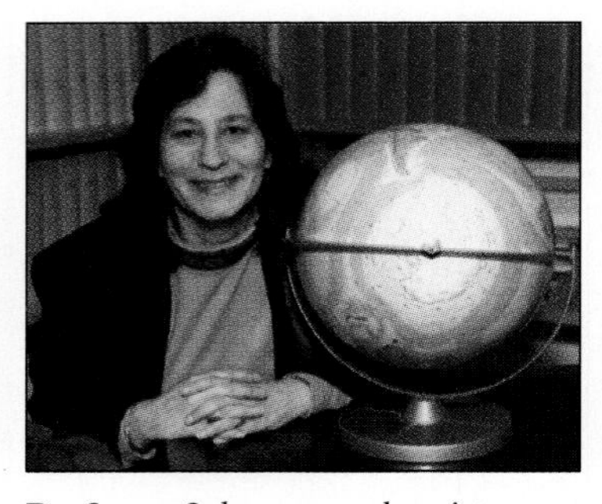

Dr. Susan Solomon, a chemist, headed the team that first gathered stratospheric ·ClO and ozone data over Antarctica. The data solidified the causal connection between CFCs and the ozone hole. She was just 30 years old at the time.

plots of Antarctic data: one of O_3 concentration and the other of ·ClO concentration. Both are plotted versus the latitude at which samples were measured. *As stratospheric O_3 concentration decreases, the ·ClO concentration increases;* the two curves mirror each other almost perfectly. The major effect is a decrease in ozone and an increase in chlorine monoxide as the South Pole is approached. Because ·ClO, ·Cl, and O_3 are linked by equation 2.7, the conclusion is compelling. Figure 2.20 is sometimes described as the "smoking gun," the clinching evidence. James G. Anderson of Harvard University, who conducted these measurements, has recently refined his instrumentation so that he can detect pollutants in concentrations as low as one part in 10 trillion, the equivalent of the area of a postage stamp in an area 10 times the size of Texas.

Although most of the attention has focused on chlorofluorocarbons, it is important to remember that not all of the chlorine implicated in stratospheric ozone destruction comes from CFCs. The atmosphere currently contains about 4 ppb of chlorinated carbon compounds capable of reaching the ozone layer. The comparable concentration in 1950 was 0.8 ppb, at least implying that the source of some of the chlorine is from human activities. On the other hand, some critics, including radio talk-show hosts and some politicians, have argued that essentially all the chlorine in the stratosphere comes from natural sources such as seawater and volcanoes. However, the majority of atmospheric scientists agree that most chlorine from natural sources is in water-soluble forms. Therefore, any natural chlorine-containing substances are washed out of the atmosphere by rainfall, long before they would reach the stratosphere. Of particular significance are the data gathered by NASA and by international researchers that high concentrations of HCl and HF (hydrogen fluoride) always occur together. Although some of the HCl might conceivably arise from a variety of natural sources, the only reasonable origin of significant stratospheric HF is CFCs. The conclusion is that most of the HCl comes from the same synthetic source.

2.25 Consider This: Talk Radio Opinion

"And if prehistoric man merely got a sunburn, how is it that we are going to destroy the ozone layer with our air conditioners and underarm deodorants and cause everybody to get cancer? Obviously we're not . . . and we can't . . . and it's a hoax. Evidence is mounting all the time that ozone depletion, if occurring at all, is not doing so at an alarming rate."* Consider the first thing you would ask this talk-show host about these statements. Remember that you need to formulate a short and focused question to get any airtime!

*Limbaugh, R. 1993. *See, I Told You So.* New York: Pocket Books.

2.26 Consider This: Graffiti with a Message

a. What was the source of humor in this cartoon when it originated in the mid-1970s?
b. Is this cartoon still relevant to the problem of ozone depletion today? Explain your reasoning.
c. If you are so inclined, create your own cartoon dealing with the issue of ozone layer depletion. Be sure that the chemistry is correct!

2.14 The Antarctic Ozone Hole: A Closer Look

A particularly intriguing question is why the greatest losses of stratospheric ozone have occurred over Antarctica and not elsewhere, such as over the North Pole. Evidence suggests that a special mechanism is operative in that region. This mechanism is related to the fact that the lower stratosphere over the South Pole is the coldest spot on Earth. From June to September, during the Antarctic winter, circulator winds blowing around the South Pole prevent warmer air from entering the region. Temperatures get as low as −90°C. Under these conditions, the small amount of water vapor present freezes into thin stratospheric clouds, called polar stratospheric clouds, of ice crystals. The clouds have also been found to contain sulfate particles and droplets or crystals of nitric acid. Atmospheric scientists believe that chemical reactions occurring on the surface of these cloud particles convert otherwise safe, that is, non-ozone depleting, molecules like $ClONO_2$ and HCl, to more reactive species such as HOCl and Cl_2. When the Sun comes out in late September or early October to end the long Antarctic night, the solar radiation breaks down the HOCl and Cl_2, releasing ·Cl atoms. The destruction of ozone, which is catalyzed by these atoms, accounts for the missing ozone. Notice the conditions needed for the hole to form: extreme cold and no wind for an extended period to permit ice crystals to provide a surface for the reactions; darkness followed by rapidly increasing levels of sunlight.

The seasonal variations of the ozone hole (1955–1995) over Antarctica are shown in Figure 2.21. Note the rapid ozone decline during spring at the South Pole (September–early November) compared to the summer (January–March). As the sunlight warms the stratosphere, the ice clouds evaporate, halting the chemistry that occurs on the ice crystals. Moreover, air from lower latitudes flows into the polar region, replenishing the depleted ozone levels. Thus, by the end of November, the hole is largely refilled. However, annual repetitions of this process could result in a general lowering of ozone concentrations in the Antarctic region and eventually across the entire atmosphere unless successful corrective actions are taken.

Although the deepest decrease in the ozone layer over Antarctica occurs during the spring, creating the ozone hole, recent discoveries by British Antarctic Survey researchers, based on improved detection instruments, indicate that the ozone depletion may begin earlier, as early as midwinter at the edges of the Antarctic, including over populated southern areas of South America.

There is already evidence that the ozone reduction over the Southern Hemisphere is greater than would be predicted solely on the basis of the mid-latitude chlorine cycle. Australian scientists believe that wheat, sorghum, and pea production have already been lowered as a result of increased ultraviolet radiation. We have already noted that Australian health officials observed significant increases in skin cancers despite a very

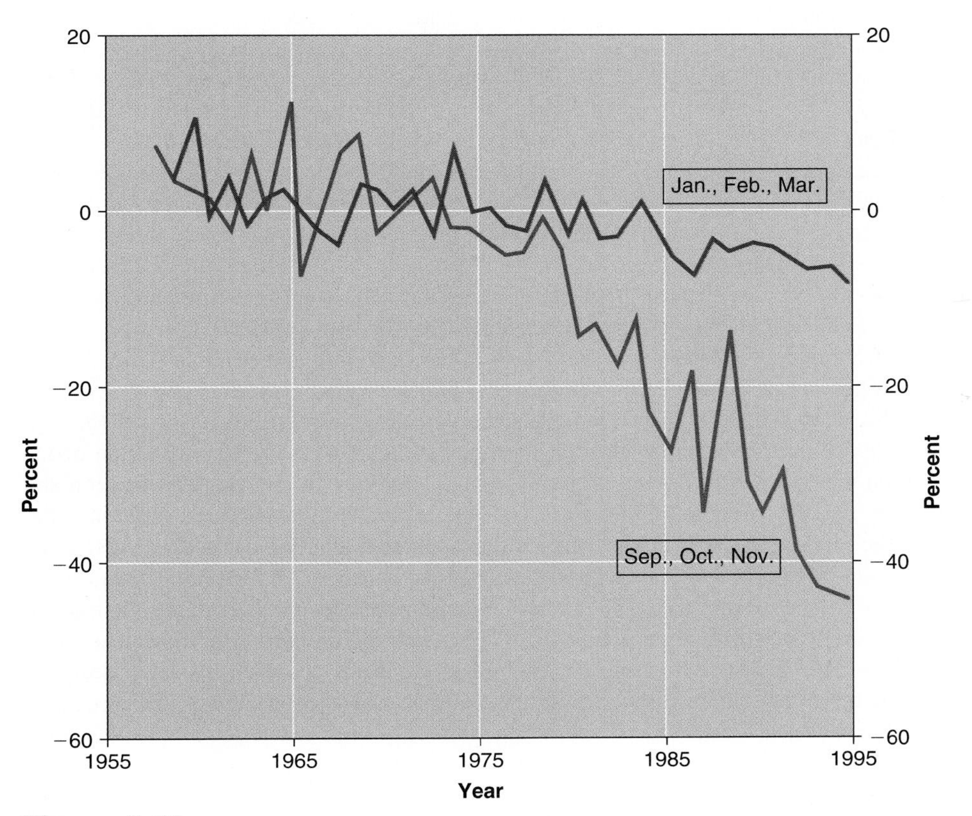

Figure 2.21
Deviations from average ozone concentrations over Antarctica.

- Decreased stratospheric ozone over the South Pole leads to increased UV-B levels reaching the Earth, causing increased skin cancer rates in Australia.

active public health campaign to alert the population to the danger of exposure to ultraviolet radiation. Ultraviolet alerts have even been issued in Australia. Similar effects are also being felt in southern Chile in the area around Punta Arenas, and on the island of Tierra del Fuego at the southernmost tip of South America. Lidia Amarales, Chile's health minister, has warned the 120,000 residents of Punta Arenas not to be out in the sun between 11 A.M. and 3 P.M. during the fall, when ozone depletion reaches its peak. As reported by BBC News in October 2000, Morales said, "If people have to leave their homes, they should wear high-factor suncreams, UV protective sunglasses, wide-brimmed hats and clothing with long sleeves." Despite these warnings, not all of the residents of this area can afford to take these precautions. "I have to go to buy bread and scarcely have money for that, so forget the sunglasses and suncream" said Adriana Cerpa. Those who live in such a stratospheric ozone-depleted area near the South Pole may show the first consequences of ozone destruction in that region.

The dramatic seasonal loss of stratospheric ozone near the South Pole raises the logical question of whether such a phenomenon also occurs at the North Pole. The effect, if it occurs, should be observed in the North during March–April rather than September–October because of the reversal of the seasons in the two hemispheres. Because a far larger fraction of the Earth's people live north of 50° north latitude than live south of 50° south latitude, the possibility of a northern ozone hole has important implications.

Measurements have indicated that some of the same ozone-depleting conditions operating at the South Pole also apply over the North Pole. Potentially destructive chemical species such as $\cdot ClO$ have been detected. In fact, the highest stratospheric $\cdot ClO$ concentration ever observed, 1.5 ppb, was measured in January 1992 by the Second Airborne Arctic Stratospheric Expedition. Information gathered by the Finnish Meteorological Institute indicates that Arctic ozone reached a record low during the winter of 1994–95. More evidence of Arctic ozone depletion is continually gathered from satellite measurements conducted by the United States National Oceanic and Atmospheric Administration. Local losses when unusually cold winters occur can be up to 50% at some altitudes. The location of the depletion also can shift to be the strongest over Britain, such as happened in 1994–1995. Measurements taken aboard NASA's ER-2 plane showed that 60% of the ozone above the Arctic disappeared between January and mid-March 2000. Dr. Owen B. Toon, one of the NASA project scientists making flights over the artic region, attributes this startling loss to the persistence of polar stratospheric clouds during that month. In a May 2000 NASA news release, he stated: "We found that the clouds lasted longer during the 1999–2000 winter than during past winters, allowing greater ozone depletion over the Arctic." This loss had the effect of causing a slight depletion in ozone in the Northern Hemisphere as well, in an area stretching through northern Europe, Canada, and the United States. In the spring of 2000, however, the deviation from expected levels was greatest over the Canadian and Russian Artic.

The general extent of the depletion in the Northern Hemisphere is not as severe as in the Southern Hemisphere. Even the observed downward trend observed through 1994–1995 has reversed in more recent years. Scientists have not classified the ozone depletion over the North Pole as a "hole," but are carefully monitoring the location and intensity of UV-B radiation being received. One reason for the observed difference between the total ozone changes in the two hemispheres is that the atmosphere above the North Pole usually is not as cold as that over its Southern Hemisphere counterpart. Although polar stratospheric clouds have been observed in the Arctic, the air trapped over the Arctic generally begins to diffuse out of the region before the Sun gets bright enough to trigger as much ozone destruction as has been observed in Antarctica. The fact that the stratosphere above the North Pole reached record low temperatures in 1994–95 may have been a factor in the uncommonly high ozone depletion. Whatever the combination of causes, scientists are giving the situation their close attention. A map of stratospheric ozone over the Northern Hemisphere from March 2001 is shown in Figure 2.22. Note that there are regions in the lower northern latitudes that show readings 5–10% higher than the expected 320 DU. During the winter and spring of

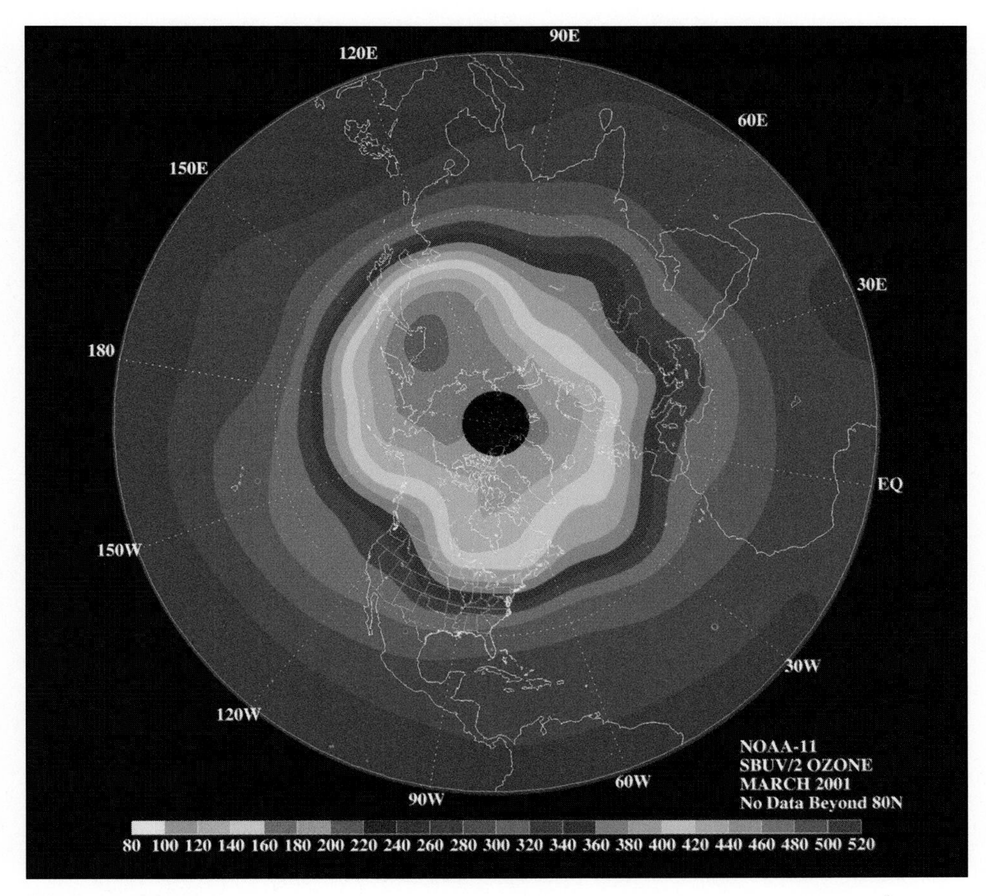

Figure 2.22
Total ozone in the Northern Hemisphere, March 2001.
(Source: NOAA satellite image; March 2001.)

2000–2001, total ozone values over the Artic region were generally higher than the expected average. There were some colder than average days in January and February with low total ozone, but even these readings were still well above comparable values in the 1980s.

2.27 Consider This: Comparison of Northern Hemisphere Ozone Maps

Figure 2.22 shows total ozone detected in the Northern Hemisphere in March 2001. Another way to present this information is to map the total ozone percent difference between March 2001 and the March average from 1979–86. Visit the *Chemistry in Context* web site for direct access to the NOAA Climate Prediction Center's site and obtain this map. Compare these two maps and offer some possible advantages for using each type of map.

2.15 Response to a Global Crisis

About 82% of stratospheric chlorine comes from synthetic compounds, such as CFCs, which are implicated in enlarging the ozone hole. The response to the threat of ozone destruction has been surprisingly rapid and reasonably effective. The first steps toward reversing ozone depletion were taken by individual countries. For example, the use of chlorofluorocarbons in spray cans was banned in North America in 1978, and their use as foaming agents for plastics was discontinued in 1990. The problem of CFC production and subsequent release, however, is a global one, and it requires international cooperation.

In 1985, in response to the first experimental evidence of an ozone hole, a number of world governments participated in the Vienna Convention on the Protection of the Ozone Layer. Through action taken at the convention, these nations committed themselves to protecting the ozone layer and to conduct scientific research to better understand atmospheric processes. A major breakthrough came with the signing, in 1987, of the Montreal Protocol on Substances That Deplete the Ozone Layer. The participating nations agreed to reduce CFC production to one-half the 1986 levels by 1998. An important provision of the agreement was to plan for future meetings to revise goals as scientific knowledge evolved. Therefore, with growing knowledge of the cause of the ozone hole and the potential for global ozone depletion, atmospheric scientists, environmentalists, chemical manufacturers, and government officials soon agreed that the original Montreal Protocol was not sufficiently stringent. In 1990, representatives of approximately 100 nations met in London and decided to ban the production of CFCs by the year 2000. Even that phaseout time was further accelerated in the amendments enacted in Copenhagen in 1992 and again in Montreal in 1997. The Beijing Amendment in 1999 added bromochloromethane to the schedule for phaseout, and revised controls on consumption of HCFCs. The production of CFCs and other fully halogenated CFCs are to be eliminated by 2010 by all parties to the Montreal Protocol, no matter the basic domestic economic needs. The production of other ozone-depleting compounds was identified for elimination between 2002 and 2010. These include including fire-fighting halons (carbon-fluorine-bromine compounds), the solvent carbon tetrachloride (CCl_4), and methyl bromide (CH_3Br), a widely used agricultural fumigant. Developing nations have until 2015 to discontinue using methyl bromide.

- HCFCs, developed as alternatives to CFCs, are discussed in Section 2.16.

The United States and 140 other countries agreed to a complete halt in CFC manufacture after December 31, 1995. Figure 2.23 indicates that the decline in global CFC production has been dramatic. By 1996, production of CFCs had dropped to 1960 levels. Production and consumption of CFCs fell by 86% overall between 1986 and 1996, and by 95% in industrialized countries by the end of 1998, the last year for which full data are available. Halon production and consumption fell by 70% between 1986 and 1996 and by 84% by 1998. Production of the originally controlled halons fell by 99.8% by the end of 1998. The atmospheric concentration of chlorine is known to have peaked in the mid-1990s and is now declining. It is estimated that without the international

- CFCs used as propellants in medical inhalers, such as those used by asthmatics, are exempt from the ban on CFCs for other uses.

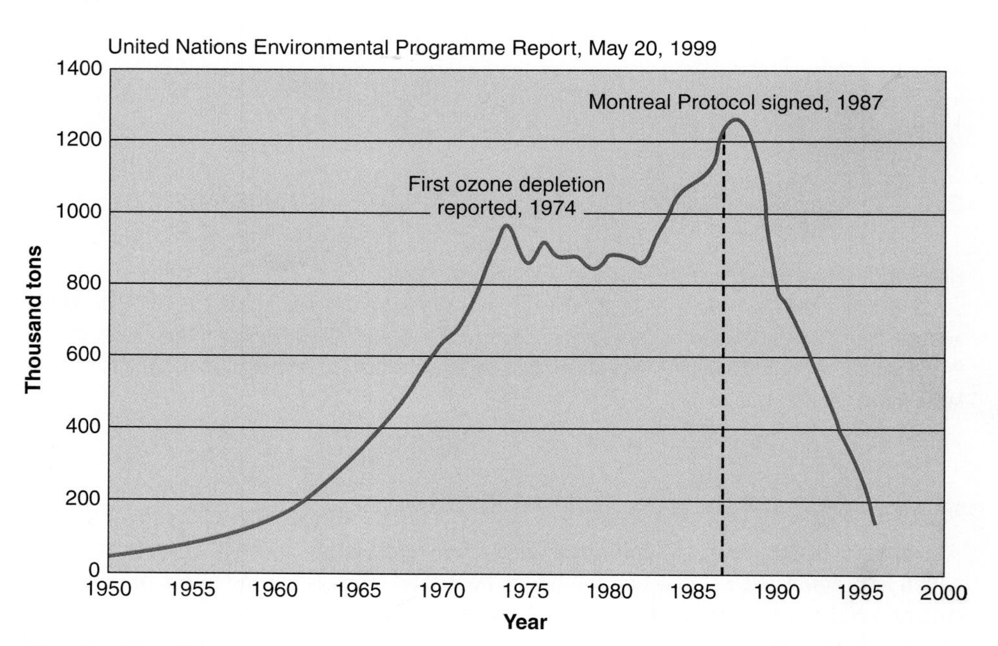

Figure 2.23

Global production of CFCs, 1950–1996.

(Source: United Nations Environment Program.)

action brought to bear by the Montreal Protocol, stratospheric abundances of chlorine would have tripled by the middle of the 21st century (2050).

Although these protocols set dates for the halt of CFC production, the sale of existing stockpiles of CFCs will remain legal until phaseout dates in the future. One reason for this is that in the United States alone, 140 million car air conditioners and the majority of home air conditioners are designed to use these compounds. Nevertheless, the U.S. government is promoting conversion to less harmful substitute refrigerants by imposing a tax of $5.35 per pound on chlorofluorocarbons. As a result, the price of CFCs has risen from $1 per pound in 1989 to over $15 at this time. This situation has tempted bootleggers to smuggle CFCs into the United States, largely from the Russian Federation, China, Eastern Europe, and Mexico. China now holds the dubious honor of being the leading supplier of black-market CFCs. According to law enforcement officers, CFCs are second only to illicit drugs as the most lucrative illegal import. In 1997 alone, "freon busts" by the U.S. Justice Department led to the confiscation of nearly 12 million pounds of illegal CFCs, bringing in fines of $38 million.

One of the largest CFC smuggling operations was uncovered in 1999, when federal agents from the EPA enforcement unit intercepted a shipment of six large industrial refrigeration units from Venezuela to the United States. These units only require 3–4 lbs of CFC-12 each for operation, but they had been altered to each hold more than 2500 lbs of CFC-12. It is still legal in Venezuela to use previously produced CFCs and evidently, the price was right. The street value in the U.S. of just this shipment was approximately $600,000, making the economic incentive clear. This shipment turned out to be just one part of a very large Freon smuggling ring, but the federal crackdown on the black market in CFCs was successful in this case.

Despite this trend to develop a global black market in CFCs, a new study reported in the May 2000 issue of *Atmospheric Environment*, concludes that illegal trade in CFCs is only a small threat to ozone layer recovery at this time. What this profit-driven smuggling threatens is the efforts to recover, recycle, and destroy these banned materials. Furthermore, Duncan Brack of the Royal Institute of International Affairs in London points out that the CFC problem must be solved if environmental crime is to be taken seriously in the global arena. Challenges such as these must be successfully met if progress brought about by the Montreal Protocol is to be sustained.

2.16 Substitutes for the Future

Where do we go from here? Atmospheric scientists feel that it is very likely that the atmosphere will eventually rid itself of the ozone-destroying agents, but it will take a long time, given the 100-year lifetime of many CFCs. Analysis of trends in atmospheric chlorine levels indicates that stratospheric chlorine peaked at 4.1 ppb in the late 1990s and then diminished slowly. This is taken as evidence that the Montreal Protocol and its amendments have slowed the release of CFCs and related ozone-depleting materials. But we are not completely in the clear. Scientists estimate that even under the most stringent international controls on the use of ozone-depleting chemicals, the stratospheric chlorine concentration would not drop to 2 ppb until 2050 or 2075. This value is significant because the Antarctic ozone hole first appeared when chlorine levels reached 2 ppb.

To be sure, some have proposed attempting to scrub the atmosphere of chlorine and some chlorine-containing compounds by intentionally introducing other chemicals. For example, a researcher at the University of California, Los Angeles, suggested injecting electrons into the stratosphere. He postulated that these electrons would react with chlorine atoms, converting them into negatively charged chloride ions (Cl^-). But a recent report argued that this approach would not be effective. Another proposal advocated adding propane (C_3H_8) to the stratosphere, reasoning that the hydrocarbon would react with chlorine atoms and remove them from the ozone-depletion cycle. The author of the scheme estimated that 50,000 tons of C_3H_8 would have to be added each year, at an annual cost of $100 million. This strategy has also been discredited.

Figure 2.24

A solution to ozone depletion?

Nor can we adopt the solution proposed by the industrialist in Sidney Harris's cartoon (Figure 2.24). The authors of this book calculated that it would take about 17 million planeloads of ozone to replenish 10% of the stratospheric ozone. Sherwood Rowland has estimated that "the energy that would be needed to move the ozone up [to the stratosphere] is about $2\frac{1}{2}$ times all of our current global power use." Even if we could temporarily replace the lost ozone, the steady-state cycle would soon re-establish itself.

Although we cannot undo what already has been done, we can stop doing it. Key to the process will be the success of chemists in finding replacements for CFCs. No one advocates the return to ammonia and sulfur dioxide in home refrigeration units. In designing replacement molecules, chemists are concentrating on compounds similar to the CFCs. The assumption is that the substitute molecules will include one or two carbon atoms, at least one hydrogen atom, and several chlorine and/or fluorine atoms. The rules of molecular structure limit the options. For example, each carbon atom forms single bonds to four other atoms in the molecules under consideration.

The chemical and physical properties of all compounds depend on elementary composition and molecular structure. In synthesizing substitutes for CFCs, chemists must weigh three undesirable properties—toxicity, flammability, and extreme stability—and attempt to achieve the most suitable compromise. Compounds containing only carbon and fluorine (fluorocarbons) are neither toxic nor flammable, and they are not decomposed by ultraviolet radiation, even in the stratosphere. Consequently, they would not catalyze the destruction of ozone. This would be ideal, were it not for the fact that the undecomposed fluorocarbons would eventually build up in the atmosphere and contribute to the global warming effect by absorbing infrared radiation.

Introducing hydrogen atoms in place of one or more halogen atoms reduces molecular stability and promotes destruction of the compounds at low altitudes, long before they enter the ozone-rich regions of the atmosphere. However, too many hydrogen atoms increase flammability and too many chlorine atoms seem to increase toxicity. For these reasons, chloroform, $CHCl_3$, would not be a good substitute. Moreover, if a hydrogen atom replaces a halogen atom, the total mass of the molecule is decreased. This results

Table 2.7

Two Important Hydrochlorofluorocarbons

HCFC-22	HCFC-141b
$CHClF_2$	$C_2H_3Cl_2F$
Chlorodifluoromethane	Dichlorofluoroethane

```
      :F:                  H  :Cl:
       |                   |   |
   H—C—F:              H—C—C—Cl:
       |                   |   |
     :Cl:                  H  :F:
```

in a decrease in boiling point making the compounds less ideal for use as refrigerants. A boiling point in the -10 to $-30°C$ range is an important property for a refrigerant. Therefore, the relationship between composition, molecular structure, boiling point, and proposed use must be considered along with toxicity, flammability, and stability.

Fortunately, chemists already know a good deal about how these variables are related, and they have used this knowledge to synthesize some promising replacements for CFCs. Table 2.7 shows the formulas, names, and structures for these substitute materials. The current substitutes are the hydrochlorofluorocarbons (HCFCs). These compounds of hydrogen, chlorine, fluorine, and carbon decompose in the troposphere more readily than CFCs, and hence do not accumulate to the same extent in the stratosphere. HCFC-22 ($CHClF_2$) is the most widely used HCFC, being used in air conditioners and in the production of foamed fast-food containers. Its ozone-depleting potential is about 5% that of CFC-12 and its estimated atmospheric lifetime is only 20 years, compared to 111 years for CFC-12. HCFC-141b ($C_2H_3Cl_2F$) is also used to form foam insulation and more than 250 million pounds of this HCFC were produced worldwide in 1996 for this purpose. But, because HCFCs have some adverse effects on the ozone layer, they are regarded as only an interim solution to the problem. The 1992 Copenhagen amendments to the Montreal Protocol call for a halt in the manufacture of these compounds by 2030.

In the long run, hydrofluorocarbons (HFCs), compounds of hydrogen, fluorine, and carbon, may be more suitable. HFC-134a (CF_3CH_2F), with a boiling point of $-26°C$, could prove to be the substitute of choice for CFC-12. HFC-134a has no chlorine atoms to interact with ozone, and its two hydrogen atoms facilitate its decomposition in the lower atmosphere without making it flammable under normal conditions. Other potential replacement refrigerants include hydrocarbons such as propane (C_3H_8) and isobutane (C_4H_{10}). Hydrocarbons do not destroy ozone, but they are flammable and can contribute to global warming. This interconnection between ozone depletion and global warming mechanisms warrant additional study as more is learned about global atmospheric modeling.

2.28 Your Turn

Draw the structural formula of HFC-134a, given that its formula is CF_3CH_2F.

Hint: You may wish to refer to the structural formula for the two-carbon HCFC-141b given in Table 2.6.

2.29 Consider This: Air Conditioning Your Car

a. How does the air conditioning system fluid of a car differ in a pre-1994 motor vehicle and a post-1994 vehicle?

b. If you need to have your car's air conditioning system's refrigerant recharged, what will it cost you? Will the cost be the same for a pre-1994 car as for a post-1994 car? Why or why not?

c. If you are considering buying a new or used car, what information about the air cooling system fluid will you need to know?

Halons are compounds composed of carbon, fluorine, and bromine atoms used in fire fighting and other applications. Long hailed as clean, nonconductive, safe, and extremely effective, replacement of halons phased out by the Montreal Protocol proved to be a challenge. Halons are also implicated in the greenhouse effect and therefore subject to control by the Kyoto Protocol as well (see Chapter 3). Using a green chemistry approach, Pyrocool Technologies synthesized a halon substitute. The product, Pyrocool FEF, is a foam tested to be environmentally benign yet more effective than halons in fighting fires, even large-scale fires such as those on oil tankers and jet airplanes (Figure 2.25). Pyrocool Technologies won a 1998 Presidential Green Chemistry Challenge Award for this development. Many other companies, nationally and internationally, are working on the challenge of replacing halon compounds.

The phaseout of CFCs is not without major economic considerations. At its peak, the annual worldwide market for CFCs reached $2 billion, but that was only the tip of a very large financial iceberg. In the United States alone, chlorofluorocarbons were used in or used to produce goods valued at about $28 billion per year. Even today, over $100 billion worth of equipment, probably including your refrigerator and automobile air conditioner, rely on CFCs. Although the conversion to CFC replacements has had some additional costs associated with it, the overall effect on the U.S. economy has been minimal. Companies that produce refrigerators, air conditioners, insulating plastics, and other goods have adapted to using the new compounds. Some substitutes for CFC refrigerants are less energy efficient, hence increasing energy consumption by several percent. But the conversions provide a market opportunity for innovative syntheses using green chemistry to produce environmentally benign substances.

On a domestic level, the political dimension of CFC regulation raises many issues. What agency establishes the limits? Where is the legislation enacted? Is this a national, state, or local affair? Who will enforce the regulations? What limitations and what time constraints are reasonable, responsible, possible? How much testing is necessary before replacement compounds can be introduced? How can the country be confident that those making the political, legal, and economic decisions are getting the best scientific advice and interpreting it correctly? There are no easy answers to these questions, maybe not even any right or wrong answers, but 2.30 Consider This gives you an opportunity to struggle with some of them.

Figure 2.25
Pyrocool FEF.

2.30 Consider This: Environmental Legislation and States' Rights

In 1995, the Arizona legislature, in defiance of the federal government, passed a law permitting the manufacture and use of CFCs within the state. In defining the law, state representative Robert N. Blendu made the following statement: "Before we ask people to spend millions and millions of dollars in Arizona to replace the Freon in their equipment, we need proof [CFCs] are harmful. We heard testimony on both sides of the issue, and it's only a matter of opinion that CFCs are bad." The Arizona law is a symbolic protest, because federal law banning the manufacture of CFCs supersedes it. On the other hand, the U.S. EPA has ruled that its federal regulations governing CFCs do not preempt the rights of states and cities to enact legislation with stricter, more stringent controls. At first sight, this appears to be an unequal application of states' rights. Write an essay in which you present arguments either for or against a system that prohibits state laws that are more lenient than federal environmental regulations, but allows states to set stricter limits.

Developing countries face another set of economic problems and priorities. Chlorofluorocarbons have played an important role in improving the quality of life in the industrialized nations. Few would be willing to give up the convenience and health benefits of refrigeration or the comfort of air conditioning. It is understandable that millions of people over the globe aspire to the lifestyle of the industrialized nations. As an example, over the past decade, the annual production of refrigerators in China has increased from 500,000 to eight million. But, if the developing nations are banned from using the relatively cheap CFC-based technology, they may not be able to afford alternatives. "Our development strategies cannot be sacrificed for the destruction of the environment caused by the West," asserts Ashish Kothari, a member of an Indian environmental group.

In recognition of these legitimate expectations, the Montreal Protocol and its adjustments have established a more lenient timetable for developing countries to phase out ozone-depleting substances. These nations were not expected to begin cutting back on the use of CFCs and halons until 1999, and a complete halt on production is not required until 2010.

As a result of these international agreements, CFC consumption by industrialized countries dropped between 1986 and 2000. But, during the same interval, use of these compounds by the developing world increased. This suggests that the use of CFCs by developing nations will continue to grow during the next decade. Without further restrictions, total CFC emissions by developing countries might easily equal one million tons by 2010. Although the result may be industrial and economic progress for one segment of the world's population, it hardly represents progress in the protection of the environment.

Recently, environmentalists have urged that the phase out schedule for developing countries should be accelerated, but such action is rife with political complications and may well require the infusion of funds from the industrialized nations. There is a precedent. Both India and China refused to sign the original Montreal Protocol because they felt that it discriminated against developing countries. To gain the participation of these highly populated nations, the industrially developed nations created a special fund in 1990. Ten donor countries (Austria, Denmark, Finland, Germany, Italy, Japan, Norway, Sweden, United Kingdom, and U.S.) committed $19 million dollars to the Russian Federation alone to help them close production facilities for CFCs and halons by the year 2000. The U.S. contributed nearly one-fourth of the total to this fund. The World Bank oversees the funds. In 2000, it approved 20 full-project grants, totaling $266 million, and 17 medium-sized grants just over $13 million. The World Bank feels that these investments are having a significant impact in helping countries phase out the use of ozone-depleting substances.

- In 1991, China signed the Montreal Protocol and has begun to take steps to phase out the production and use of CFCs. Starting in 1998, it banned the industrial use of CFCs as aerosol propellants.

Developing nations apply to this special fund for grants to underwrite specific projects that lead to discontinuation of CFC use. But even this amount of money is insufficient to cover the costs of conversion and phaseout in many nations. To make matters worse, a number of industrialized nations are behind in their payments to the fund, and the United States has not always been supportive of foreign aid. Without financial assistance, the developing nations may not be able or willing to meet a more stringent timetable for discontinuing their use of substances that deplete the ozone layer. Clearly, an understanding of chemistry is necessary to protect the ozone layer, but it is not sufficient. And thus, the second part of the twin challenge—the debate among governments about how best to protect the stratospheric ozone layer—continues in the global political arena.

2.31 Consider This: Equity and National Rights

Since the 1950s, the U.S. had the luxury of using CFCs with apparently little concern for the long-range consequences, until recently. Do we now have the right to deny this privilege to countries that are presently starting to develop? Write an essay in which you address these issues.

CONCLUSION

Chemistry is intimately entwined with the story of ozone depletion. Chemists created the chlorofluorocarbons whose near-perfect properties only relatively recently revealed their dark side as predators of stratospheric ozone. Chemists worked internationally to discover the mechanism by which CFCs destroy this ozone and warned of the dangers of increasing ultraviolet radiation. And chemists will continue to synthesize the substitutes necessary to ultimately replace CFCs. But the issues involve more than just chemistry. The "Action on Ozone" report for 2000 from the Ozone Secretariat of the United Nations Environmental Program, sums up the global experience with ozone depletion story in this manner: "Perhaps the most important feature of the ozone regime is the way in which it has brought together an array of different participants in pursuit of a common end. Scientists have provided the information, with steadily increasing degrees of precision, on the causes and effects of ozone depletion. Industry, responding to the stimulus provided by the control measures, has developed alternatives far more rapidly and more cheaply than initially thought possible, and has participated fully in the debates over further phaseout. NGOs (nongovernmental organizations) and the media are the essential channels of communication, and education, with the peoples of the world in whose name the measures have been taken. . . . Governments have worked well together in patiently negotiating agreements acceptable to a range of countries with widely varying circumstances, aims, and resources—and showed courage and foresight in putting the precautionary principle into effect before the scientific evidence was entirely clear." These actions will be lessons to keep in mind as we turn to our next topic, the Chemistry of Global Warming.

Chapter Summary

Having studied this chapter, you should be able to:

- Describe the chemical nature of ozone, the ozone layer, and factors affecting its existence (2.1, 2.7, 2.9, 2.11)
- Apply the basics of atomic structure, that is, protons, neutrons, and electrons to particular elements (2.2)
- Translate an element's atomic number to its position in the periodic table (2.2)
- Write electron distributions, by levels, for elements in A groups (2.2)
- Differentiate atomic number from mass number; apply the latter to isotopes (2.2)
- Write Lewis dot structures using the octet rule; interpret such structures in terms of the nature of their bonds (2.3)
- Describe the electromagnetic spectrum in terms of frequency, wavelength, and energy, and use appropriate calculations to determine these quantities (2.4–2.5)
- Interpret graphs related to wavelength and energy, radiation and biological damage, and ozone depletion (2.4, 2.8, 2.11)
- Understand how the ozone layer protects against harmful ultraviolet radiation (2.7–2.8)
- Discuss the interaction of radiation with matter and changes caused by such interactions, including biological sensitivity and the use of the UV index (2.6–2.8)
- Differentiate among the energies and biological effects of UV-A, UV-B, and UV-C radiation (2.8)

- Appreciate the complexities of collecting accurate data for stratospheric ozone depletion and interpreting them correctly and unambiguously (2.9–2.13)
- Understand the Chapman cycle (2.9) and the role of nature (2.11) and the role of CFCs (2.12–2.16) in stratospheric ozone depletion, that is, ozone hole formation (2.10, 2.14)
- Evaluate articles on green chemistry alternatives to stratospheric ozone-depleting compounds and recognize that market forces determine the success of these innovations (2.16)
- Summarize the political dimensions of CFC regulation and the Montreal Protocol and its amendments (2.15–2.16)

Questions

Emphasizing Essentials

1. The text states that ozone can be detected in concentrations as low as 10 ppb. Will you be able to detect the odor of ozone in any of these air samples?

a. 0.118 ppm ozone, a concentration reached in the troposphere

b. 25 ppm ozone, a concentration reached in the stratosphere

2. Which of these pairs are allotropes?

a. diamond and graphite

b. water, H_2O, and hydrogen peroxide, H_2O_2

c. white phosphorus, P_4, and red phosphorus, P_8

3. Using the periodic table as a guide, specify the number of protons and electrons in a neutral atom of each of these elements.

a. oxygen (O) **b.** nitrogen (N)

c. magnesium (Mg) **d.** sulfur (S)

4. Consider this periodic table.

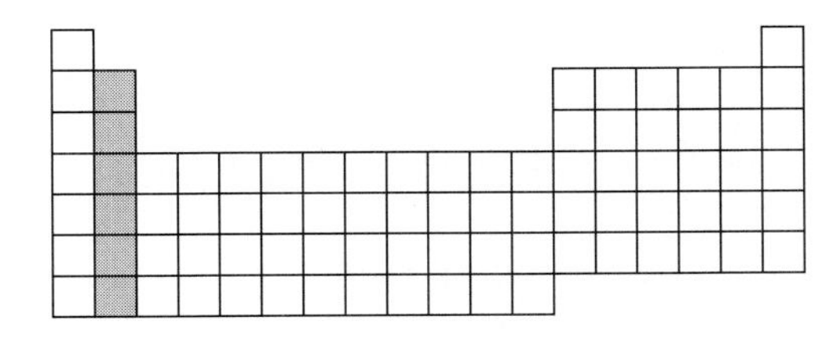

a. What is the group number indicated by the shading?

b. What elements make up this group?

c. What is the number of electrons for each element in this group?

d. What is the number of outer electrons for each element of this group?

5. Using the periodic table as a guide, give the name and symbol of the element that has the given number of protons in the nucleus of its atoms.

a. 2

b. 19

c. 29

6. Give the number of protons, neutrons, and electrons in each of these isotopes.

a. oxygen-18 (O-18, $^{18}_{8}O$)

b. sulfur-35 (S-35, $^{35}_{16}S$)

c. uranium-238 (U-238, $^{238}_{92}U$)

7. Give the symbol showing the atomic number and the mass number for the element that has

a. 9 protons and 10 neutrons (an isotope used in nuclear medicine)

b. 26 protons and 30 neutrons (the most stable isotope of this element)

c. 86 protons and 136 neutrons (the radioactive gas found in homes)

8. Give the number of protons, neutrons, and electrons in each of these isotopes.

a. bromine-82 (Br-82, $^{82}_{35}Br$)

b. neon-19 (Ne-19, $^{19}_{10}Ne$)

c. radium-226 (Ra-226, $^{226}_{88}Ra$)

9. Write the electron dot structures for each of these elements.

a. calcium

b. nitrogen

c. chlorine

d. helium

10. Assuming that the octet rule applies, write Lewis structures for each of these compounds. Start by counting the number of available outer electrons. (a) Write the complete electron dot structure; (b) then write the structure representing shared pairs with a dash, showing nonbonding electrons as dots.

a. CCl_4 (carbon tetrachloride, a substance formerly used as a cleaning agent)

b. H_2O_2 (hydrogen peroxide, a mild disinfectant; the atoms are bonded in this order: H to O to O to H)

c. H_2S (hydrogen sulfide, a gas with the unpleasant odor of rotten eggs)

11. Assuming that the octet rule applies, write Lewis structures for each of these compounds. Start by counting the number of available outer electrons. (a) Write the complete electron dot structure; (b) then write the structure representing shared pairs with a dash, showing nonbonding electrons as dots.

a. N_2 (nitrogen gas, the major component of the atmosphere)

b. HCN (hydrogen cyanide, a molecule found in space and used in some "death" chambers)

c. N_2O (nitrous oxide, "laughing gas"; the atoms are bonded in this order: N to N to O)

d. CS_2 (carbon disulfide, used to kill rodents; the atoms are bonded in this order: S to C to S)

12. Consider these two waves representing different parts of the electromagnetic spectrum.

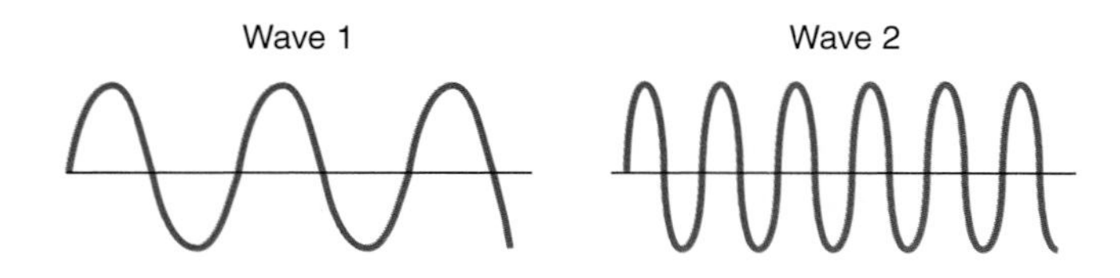

a. How do these two waves compare in wavelength?

b. How do these two waves compare in frequency?

c. How do these two waves compare in forward speed?

13. Use Figure 2.4 to specify the region of the electromagnetic spectrum where radiation of each wavelength is found. *Hint:* Change each wavelength to meters before making the comparison.

a. 2.0 cm

b. 400 nm

c. 50 μm

d. 150 mm

14. Calculate the frequency that corresponds to each of the wavelengths in question 13.

15. Calculate the energy of a photon for each wavelength in question 13. Which wavelength possesses the most energetic photons?

16. Arrange these types of radiation in order of *increasing* energy per photon.
gamma rays, infrared radiation,
radio waves, visible light

17. The microwaves in home microwave ovens have a frequency of 2.45×10^9 s^{-1}. What is the wavelength of these waves in meters?

18. Ultraviolet radiation coming from the Sun is categorized by wavelength into three bands. These are UV-A, UV-B, and UV-C. Arrange these three bands in order of their increasing:

a. wavelengths

b. energies

c. potential for biological damage

19. Consider the Chapman cycle in Figure 2.14. Rewrite each step of the cycle using a sphere equation to illustrate the changes that take place.

20. Cl, NO_2, ClO, and HO are all free radicals that catalyze atmospheric ozone depletion.

a. Count the number of outer electrons available and then draw a Lewis structure for each of these species.

b. What characteristic is shared by these free radicals that makes them so reactive?

21. In Chapter 1 the role of nitric oxide, NO, in forming photochemical smog was discussed. What role, if any, does NO play in stratospheric ozone depletion? Are NO sources the same in the troposphere and in the stratosphere?

22. Which graph shows how measured increases in UV-B radiation correlate with percent reduction in the concentration of ozone in the stratosphere over the South Pole? All data were taken between February 1991 and December 1992 and reported in *Science* in 1993.

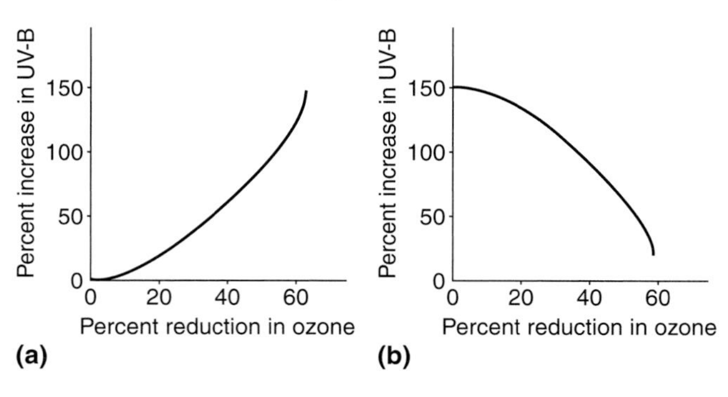

23. a. Most CFCs that have been used are based on methane, CH_4, or ethane, C_2H_6. Use structural formulas to represent these two compounds.

b. Substituting chlorine and/or fluorine for hydrogen atoms, how many different CFCs can be formed from methane?

c. Which of the compounds you wrote in part b have been most successful?

d. Why weren't all of these compounds equally successful?

24. a. How were the original measurements of increases in chlorine monoxide and the stratospheric ozone depletion over the Antarctic gathered?

b. How are these measurements made today?

Concentrating on Concepts

25. The allotropes oxygen and ozone differ in molecular structure. What differences does this produce in their properties, uses, and significance?

26. Why is it possible to detect the pungent odor of ozone after a lightning storm or around electrical transformers?

27. How do *allotropes* of oxygen and *isotopes* of oxygen differ? Explain your reasoning.

28. Consider the Lewis structure for SO_2. How is it similar to or different from the Lewis structure for ozone?

29. It is possible to write three resonance structures for ozone, not just the two shown in the text. Verify that all three structures satisfy the octet rule, and offer an explanation as to why the triangular structure is not used.

$$:\ddot{O}:\ddot{O}::\ddot{O}: \leftrightarrow :\ddot{O}::\ddot{O}:\ddot{O}: \longleftrightarrow \text{triangular } O_3$$

30. The average length of an oxygen-to-oxygen single bond is 132 pm. The average length of an oxygen-to-oxygen double bond is 121 pm. What do you predict the oxygen-to-oxygen bond lengths will be in ozone? Will they all be the same? Explain your predictions.

31. The equation $E = \frac{hc}{\lambda}$ indicates the relationships among energy, frequency, and wavelength for electromagnetic radiation. Which variables are directly related and which are inversely related?

32. Which of these forms of electromagnetic radiation from the Sun has the lowest energy and therefore the least potential for damage to biological systems? infrared radiation, ultraviolet radiation, visible radiation, radio waves

33. Why is it that you cannot get a suntan from standing in front of your radio in your living room or dorm room?

34. The morning newspaper reports a UV index of 6.5. What should that mean to you as you plan your daily activities?

35. UV-C has the shortest wavelengths of all UV radiation and therefore the highest energies. All the reports of the damage caused by UV radiation focus on UV-A and UV-B radiation. Why is the focus of attention not on the damaging effects that UV-C radiation can have on our skin?

36. If all the 3×10^8 tons of stratospheric ozone that are formed every day are also destroyed every day, how is it possible for stratospheric ozone to offer any protection from ultraviolet radiation?

37. Explain how the small changes in ClO concentrations in Figure 2.20 (measured in parts per billion) can cause the much larger changes in O_3 concentrations (measured in parts per million).

38. Development of the stratospheric ozone hole has been most dramatic over Antarctica. What set of conditions exist over Antarctica that help to explain why this area is well-suited to studying changes in stratospheric ozone concentration?

39. The free radical $CF_3O\cdot$ is produced during the decomposition of HFC-134a.

a. Propose a Lewis structure for this free radical.

b. Offer a possible reason why this free radical does not cause ozone depletion.

40. Consider this graph from the World Meteorological Organization.

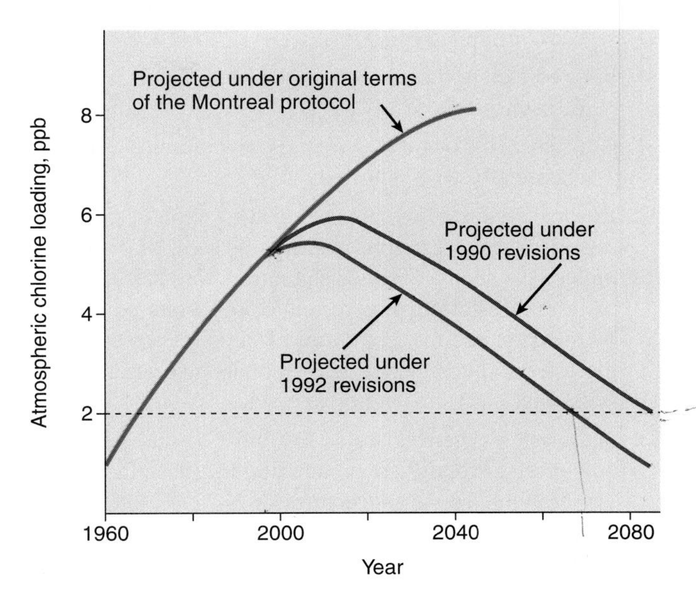

Note: The original Montreal Protocol was signed in 1987 and amended in 1990 in London, in 1992 in Copenhagen, and again in 1997 in Montreal.

a. Scientists from the World Meteorological Organization think that if atmospheric chlorine loading in the atmosphere were reduced to 2 ppb, shown by the dashed line on the graph, the ozone hole in the Antarctic would disappear. According to the predictions based on the 1992 revisions to the Montreal Protocol, what is the approximate total time span (in years) that the Antarctic ozone hole will be in existence?

b. Imagine that a local newspaper published this graph with the headline "Montreal Protocol Right on Target—Ozone Depletion Problem Solved by the Year 2002." Does this headline correctly convey the information in the graph? Explain your reasoning.

Exploring Extensions

41. Consider this periodic table.

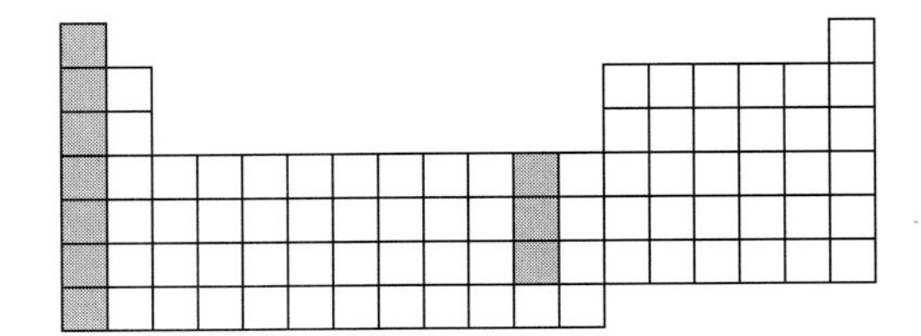

Two groups are highlighted; the first is Group 1A and the second is Group 1B. The text states that although A groups have very regular patterns, the "situation gets a bit more complicated with the B families."

Researching in other resources, find out which of these predictions becomes more complicated.

a. predicted number of electrons for each element in each group

b. predicted number of outer electrons for each element of each group

c. predicted formula when each element combines chemically with chlorine

42. Resonance structures can be used to explain the bonding in charged groups of atoms as well as in neutral molecules, such as ozone. The nitrate ion, NO_3^-, has one additional electron plus the outer electrons contributed by nitrogen and oxygen atoms. That extra electron gives the ion its charge. Draw the resonance structures, verifying that each obeys the octet rule.

43. Although oxygen exists as O_2 and O_3, nitrogen exists only as N_2. Propose an explanation for these facts. *Hint:* Try drawing a Lewis structure for N_3.

44. Consider Figure 2.15 showing ozone concentrations at various altitudes.

a. What does this graph tell you about the concentration of ozone as you travel upward from the surface of the Earth? Write a brief description of the trends shown in the graph, describing the location of the ozone layer.

b. The y-axis in this graph starts at zero. Why doesn't the x-axis start at zero?

45. It has been suggested that the term "ozone screen" would be a better descriptor than *ozone layer* to describe ozone in the stratosphere. What are the advantages and disadvantages to each term?

46. The effect a chemical substance has on the ozone layer is measured by a value called its *ozone-depleting potential,* ODP. This is a numerical scale that estimates the lifetime potential stratospheric ozone that could be destroyed by a given mass of the substance. All values are relative to CFC-11, which has an ODP defined as equal to 1.0. Use those facts to consider these questions.

a. What factors do you think will influence the ODP value for a chemical? Why?

b. Most CFCs have ODP values ranging from 0.6 to 1.0. What range do you expect for HCFCs? Explain your reasoning.

c. What ODP values do you expect for HFCs? Explain your reasoning.

47. Recent experimental evidence indicates that OCl· initially reacts to form Cl_2O_2.

a. Predict a reasonable Lewis structure for this molecule. Assume the order of atom linkage is Cl-O-O-Cl.

b. What impact does this evidence have on understanding the mechanism for the catalytic destruction of ozone by ClO?

48. The chemical formulas for individual CFCs, such as CFC-11 (CCl_3F), can be figured out from their code numbers. A quick way to interpret the code number for CFCs is to add 90 to the number. In this case, 90 + 11 = 101. The first number in this sum is the number of carbon atoms, the second is the number of hydrogen atoms, and the third is the number of fluorine atoms. CCl_3F has one carbon, no hydrogens, and one fluorine atom. All remaining bonds are assumed to be chlorine until carbon has the required four single covalent bonds to satisfy the octet rule.

a. What is the chemical formula for CFC-12?

b. What is the code number of CCl_4?

c. Will this "90" method work for HCFCs? Use HCFC-22, which is $CHClF_2$, to explain your answer.

49. As halons are being phased out under the conditions of the Montreal Protocol, are halon substitutes ready to take their place? Halon substitutes such as Pyrocool FEF are effective and environmentally benign. Are they also economically successful? Use the resources of the Web to assess the scientific and economic success of halon substitutes.

50. One outcome of regulations regarding ozone-depleting substances has been the development of a black market for CFCs.

a. Why is this a significant problem for the United States?

b. Has anyone ever been penalized by the United States for purchasing or possessing illegal CFCs?

3

The Chemistry of Global Warming

Construction of the Trans-Amazonian Highway in Brazil has opened millions of square kilometers of previously inaccessible land to development and the expansion of mining, logging, and cattle industries. The resulting deforestation of nearly 14% of the Brazilian Amazon has created environmental impact in two ways: (1) Felled trees cannot carry out photosynthesis, therefore not remove atmospheric CO_2. (2) Burning the cut trees adds CO_2 to the atmosphere. This chapter considers the role of CO_2 in global warming. (Source: Canadian Forestry Advisors Network, CFAN. http://www.rcfa-cfan.org/english/issues.12-5.html.)

"In May 2000, at the InterAcademy Panel (IAP) meeting in Tokyo, 63 scientific academies of science from all parts of the world issued a statement in which they noted that 'global trends in climate change . . . are growing concerns' and pledged themselves to work for sustainability—meeting current human needs while preserving the environment and natural resources needed by future generations. It is now evident that human activities are already contributing adversely to global climate change. Business as usual is no longer a viable option."

Editorial: "The Science of Climate Change"
Science, 18 May 2001, p. 1261

"We know that greenhouse gases are accumulating in Earth's atmosphere, causing surface temperatures to rise," said committee chair Ralph Cicerone, chancellor, University of California at Irvine. "We don't know precisely how much of this rise to date is from human activities, but based on physical principles and highly sophisticated computer models, we expect the warming to continue because of greenhouse gas emissions."

Press release: "Leading Climate Scientists Advise White House on Global Warming." Committee Report: *Climate Change Science: An Analysis of Some Key Questions.*
National Academy of Sciences, 6 June 2001

- The 1997 Kyoto Protocol and its subsequent amendments established goals for worldwide reduction of emissions of greenhouse gases into the atmosphere.

"When President George W. Bush yanked support for the Kyoto Protocol . . . and called for more research to reduce the "uncertainties" about global warming, many policy-makers and scientists worldwide let out a collective groan. Bush's stance, they said, was just another excuse for inaction. Many climate scientists believe that, despite the admittedly large uncertainties, their current knowledge merits action."

Richard A. Kerr, News Focus: Climate Change
"Major Challenges for Bush's Climate Initiative."
Science, 13 July 2001, p. 199

These quotes focus attention on the hot topic of global warming and raise a number of questions. What is global warming, and has global warming actually been demonstrated? How could an acceleration of global warming cause climate change, and how could this affect our lives? Are answers available to questions about the causes of global warming? Why is international cooperation in considering questions of global warming essential? How can we address some of the complex political and economic questions related to climate change? What is the current status of the Kyoto Protocol? What is it about carbon dioxide that makes it the major player in the debate about global warming? After all, CO_2 is an essential component of the atmosphere—a gas that all animals exhale and green plants absorb. Finding answers to these and other questions will lead us on a journey into the land of chemistry and its interaction with public policy.

3.1 Consider This: Science Behind the Policy

Policy decisions on global warming are based on scientific research and careful assessment of environmental and economic effects. The United Nations and the World Health Organization sponsor the Intergovernmental Panel on Climate Change (IPCC). In the United States, the Global Change Research Program (USGCRP) carries out climate change research. What is the membership of each agency? What are the major responsibilities and activities of each agency? How do these two groups carry out their work? How do they interact? Use a search engine or the direct links provided at the *Chemistry in Context* web site to learn about these two groups and their work.

CHAPTER OVERVIEW

The first two sections of this chapter describe the greenhouse effect and its relationship to the evolution of the Earth and its atmosphere. Central to the issue of global warming is the Earth's energy balance and the molecular mechanism by which carbon dioxide and other compounds absorb the infrared radiation emitted by the planet. Some knowledge of molecular structure and shape is necessary to understand this mechanism. Therefore, the next sections develop a general method for predicting molecular geometry and relating that geometry to infrared-induced vibrations. The two sections that follow make it clear that most of the CO_2 in the atmosphere is of natural origin, but increased human contributions are the chief cause of the current concern about the enhanced greenhouse effect. These concerns have a significant quantitative component; we need numbers to help assess the seriousness of the situation. That need justifies several sections in which we introduce and illustrate some fundamental chemical concepts, including atomic and molecular mass, Avogadro's number, and the mole concept. Examples and exercises demonstrate how the important ideas of mass and moles are

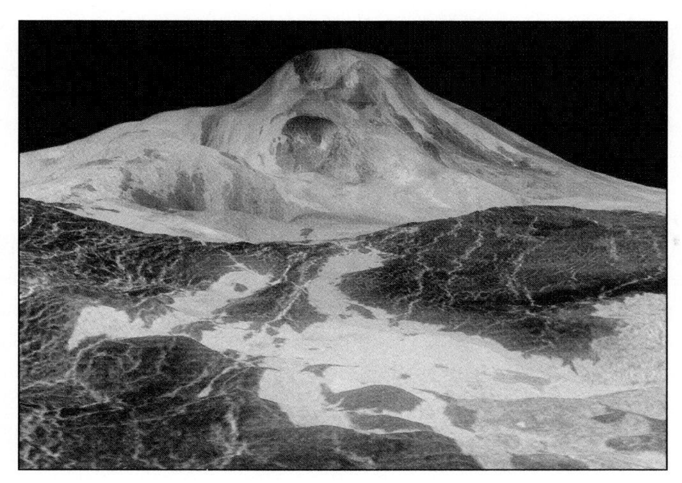

Figure 3.1
Computer-generated three-dimensional image of Maat Mons, a volcano on Venus, created from radar data collected by the space probe Magellan.

related. Thus armed, we return to another look at several greenhouse gases. A discussion of predictions based on computer modeling of the climate leads to an assessment of the current situation. The chapter ends with some suggested answers to the all-important questions about global warming: "What can we do? What should we do?" Recognizing that global warming has international implications, should we and can we, as nations, collectively address global warming through initiatives such as the Kyoto Conference protocols? Finally, how does the issue of climate change compare with the issue of stratospheric ozone depletion considered in the previous chapter?

3.1 In the Greenhouse

The brightest and most beautiful body in the night sky, after our own moon, is considered by many to be Venus (Figure 3.1). It is ironic that the planet named for the goddess of love is a most unlovely place by earthly standards. Spacecraft launched by the United States and the former Soviet Union have revealed a desolate, eroded surface with an average temperature of about 450°C (840°F). The atmosphere surrounding Venus has a pressure 90 times greater than that of the Earth, and it is 96% carbon dioxide, with clouds of sulfuric acid. It makes the worst smog-bound day anywhere on Earth seem like a breath of country air. The beautiful blue-green ball we inhabit has an average annual temperature of 15°C (59°F). The point of this little astronomical digression is that both Venus and Earth are warmer than one would expect based solely on their distances from the Sun and the amount of solar radiation they receive. If distance were the *only* determining factor, the temperature of Venus would average approximately 100°C, the boiling point of water. The Earth, on the other hand, would have an average temperature of −18°C (0°F) and the oceans would be frozen year around.

3.2 Consider This: Science Fiction Story

A number of successful writers of science fiction began their careers as majors in a natural science. Their best work reveals a sound understanding of scientific phenomena and principles. Often a good science fiction story assumes a slightly different scientific reality than the one we know. For example, *Dune*, by Frank Herbert, takes place on a desert planet.

Here is an opportunity to exercise your imagination in a different climate. Make the assumption that the planet has an average temperature of −18°C (0°F). What would human life be like? Write a brief description of a day on a frozen planet. (Residents of northern climates should have a great advantage in this exercise.)

The composition of the atmosphere is central to understanding why our planet is about 33°C warmer than we would expect, considering the amount of solar energy reaching its surface and its distance from the Sun. The moderating effect is primarily due to two of the minor constituents of the atmosphere: water vapor and carbon dioxide. There is a sort of wonderfully harmonious symmetry in the fact that the two compounds that keep our planet warm enough to sustain life are also among the essential ingredients of all living things.

The idea that atmospheric gases might somehow be involved in trapping some of the Sun's heat was first proposed around 1800 by the French mathematician and physicist, Jean-Baptiste Joseph Fourier (1768–1830). Fourier compared the function of the atmosphere to that of the glass in a "hothouse" (his term) or greenhouse. Although he did not understand the mechanism or know the identity of the gases responsible for the effect, his metaphor has persisted. Some 60 years later, John Tyndall (1820–1893) in England experimentally demonstrated that carbon dioxide and water vapor absorb heat radiation. In addition, he calculated the warming effect that would result from the presence of these two compounds in the atmosphere. Given the perspective of time and additional research to augment Fourier's and Tyndall's work, we now know three things. Carbon dioxide absorbs heat, the concentration of CO_2 in the atmosphere has increased over the past 150 years, and the Earth's average temperature has not remained constant. The major question to consider is: Are these three items related?

3.2 The Testimony of Time

In the 4.5 billion years that our planet has existed, its atmosphere and climate have varied widely. Evidence from the composition of volcanic gases suggests the concentration of carbon dioxide in the early atmosphere of the Earth was perhaps 1000 times greater than it is today. Much of the CO_2 dissolved in the oceans became incorporated in rocks such as limestone, which is calcium carbonate, $CaCO_3$. High concentrations of carbon dioxide all those years ago also made possible the most significant event in the history of our planet—the development of life on Earth. Even though the Sun's energy output was 25–30% less than it is today, the ability of CO_2 to trap heat kept the Earth sufficiently warm to permit the development of life. As early as three billion years ago, the oceans were filled with primitive plants such as cyanobacter. Like their more sophisticated descendants, these simple plants were capable of **photosynthesis.** They were able to use chlorophyll to capture sunlight and use this energy to combine carbon dioxide gas and water, forming molecules such as glucose that are more complex.

$$6\,CO_2 + 6\,H_2O \xrightarrow{\text{chlorophyll}} \underset{\text{glucose}}{C_6H_{12}O_6} + 6\,O_2$$

Photosynthesis dramatically reduced atmospheric CO_2 concentration and increased the amount of O_2 present. The microbiologist Lynn Margulis has called this "the greatest pollution crisis the Earth has ever endured." We, and all past and future generations, are the unknowing beneficiaries of this long-ago pollution crisis. The increase in oxygen concentration helped make possible the evolution of animals. But even 100 million years ago, in the age of dinosaurs and well before humans walked the Earth, the average temperature is estimated to have been 10–15°C warmer than it is today, and the CO_2 concentration is assumed to have been considerably higher.

How do we know such numbers? Reasonably reliable evidence is available about temperature fluctuations during the past 200,000 years—only yesterday in geological terms. Deeply drilled cores from the ocean floor give us a slice through time. The number and nature of the microorganisms present at any particular level indicate the temperature at which they lived. Supplementing this, the alignment of the magnetic field in particles in the sediment provides an independent measure of time.

Other relevant information comes from the analysis of ice cores. The Soviet drilling project at the Vostok Station in Antarctica has yielded over a mile of ice cores taken

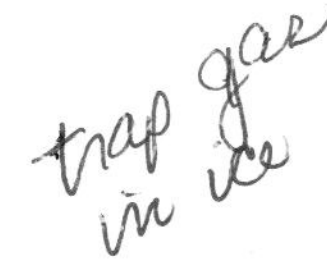

Figure 3.2

Scientists use data from ice cores to reconstruct greenhouse gas concentration and temperature information going back as far as 160,000 years ago.

from the snows of 160 millennia. The ratio of deuterium, ^2H, to hydrogen, ^1H, in the ice can be measured and used to estimate the temperature at the time the snow fell. Water molecules containing atoms of ordinary hydrogen (mass number 1) are lighter than molecules of "heavy water," which contain deuterium (mass number 2). The lighter H_2O molecules evaporate more readily than the heavier ones. This means that there is relatively more ordinary hydrogen and less deuterium in the water vapor of the atmosphere than in the oceans. The rain or snow that condenses from atmospheric water vapor also reflects this enrichment of ^2H. The ratio of ^2H to ^1H in precipitation further varies with average temperature. Higher temperatures tend to increase the deuterium/hydrogen ratio in the rain or snow. This is the key to estimating ancient temperatures by the analysis of the isotopic composition of ice cores. In addition, the bubbles of air trapped in the ice can be analyzed for carbon dioxide and other gases. Figure 3.2 shows drilling for these icy record keepers of the past.

- Isotopes of hydrogen were discussed in Section 2.2.

Both carbon dioxide concentration and temperature data are incorporated in Figure 3.3. The upper curve (corresponding to the scale on the left) is a plot of parts per million of carbon dioxide in the atmosphere versus time over a span of 160,000 years. The lower plot and the right-hand scale indicate how the average global temperature has varied over the same period. For example, the figure shows that 20,000 years ago, during the last ice age, the average temperature of the Earth was about 9°C below the 1950–1980 average. At the other extreme, a maximum temperature (just over 16°C) occurred approximately 130,000 years ago.

What is particularly striking about Figure 3.3 is that temperature and carbon dioxide concentration follow the same patterns. When the CO_2 concentration was high, the temperature was high. Other measurements show that periods of high temperature have also been characterized by high atmospheric concentrations of methane (CH_4). Such correlations do not necessarily *prove* that elevated atmospheric concentrations of CO_2 and CH_4 caused the temperature increases. Presumably, the converse could have taken place. But these compounds are known to trap heat, and there is no doubt that they can and do contribute to global warming.

To be sure, other mechanisms are also involved in the periodic fluctuations of global temperature. Temperature maxima seem to come at roughly 100,000-year intervals, with interspersed major and minor ice ages. Over the past million years, the Earth has experienced 10 major periods of glacier activity and 40 minor ones. Some of this temperature variation is probably caused by minor changes in the Earth's orbit, which affect the distance of the Earth to the Sun and the angle with which sunlight strikes the planet. However, this hypothesis cannot fully explain the observed temperature fluctuations. It

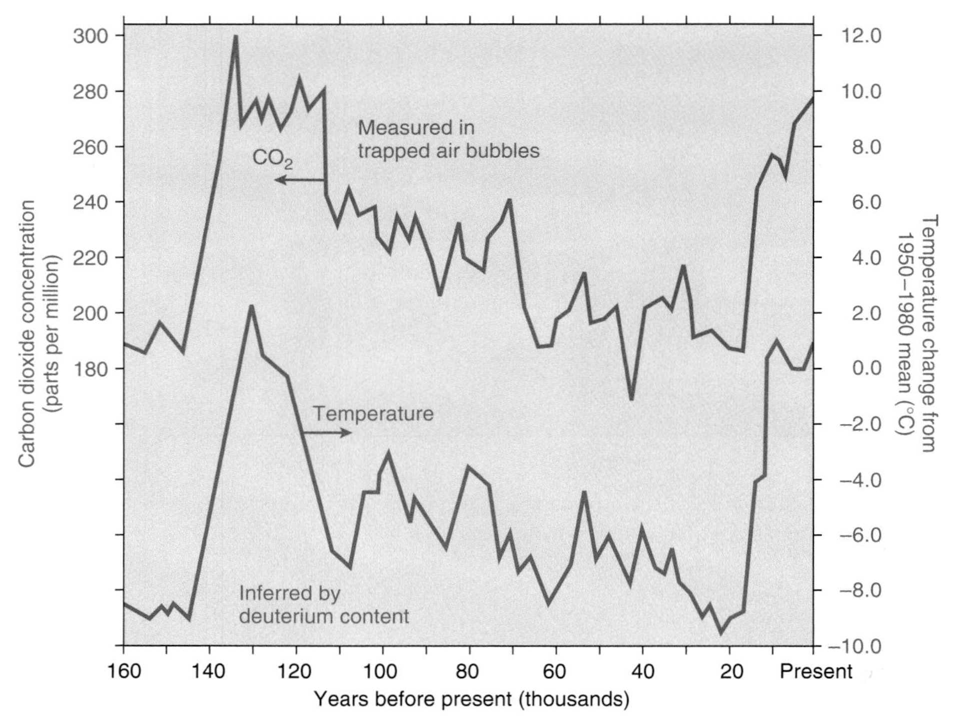

Figure 3.3
Atmospheric CO_2 concentration and average global temperature over 160,000 years (ice core data).

is likely that the orbital effects are coupled with terrestrial events such as changes in reflectivity, cloud cover, airborne dust, and carbon dioxide and methane concentration. These factors can diminish or enhance the orbital-induced climatic changes. The feedback mechanism is complicated and not well understood. One thing is clear: The Earth is a far different place in the 2000s than it was at the time of our last temperature maximum 130,000 years ago. Our ancestors had discovered fire by then, but they had not learned to exploit it as we have.

3.3 The Earth's Energy Balance

The major source of the Earth's energy is the Sun. About half the radiant energy that strikes our atmosphere is either reflected or absorbed by the molecules that make up this envelope of air. You know from your study of Chapter 2 that oxygen and ozone intercept much of the ultraviolet radiation. The rays that do reach the surface of the planet are largely in the visible and infrared (heat) regions of the spectrum. This radiation is absorbed by the Earth, and as a result, the continents and oceans are warmed. The current average temperature of the planet, about 15°C (59°F), is much higher than the −270°C of outer space. Consequently, the Earth acts like a global radiator, radiating heat to its frigid surroundings.

Figure 3.4 is a schematic representation of the Earth's energy balance. The widths of the arrows are roughly proportional to energy flow. Thus, the rate at which energy escapes the surface of the Earth is over twice the rate at which the planet directly absorbs energy from the Sun. This means that the Earth acts a little like an extravagant person, spending money faster than he or she earns it. If this process were the only one occurring, the person would soon be deeply in debt and the Earth would be a very cold place. Fortunately, the planet has something few people do—a sort of built-in forced savings account. Almost 84% of the heat it radiates is absorbed by gases in the atmosphere and

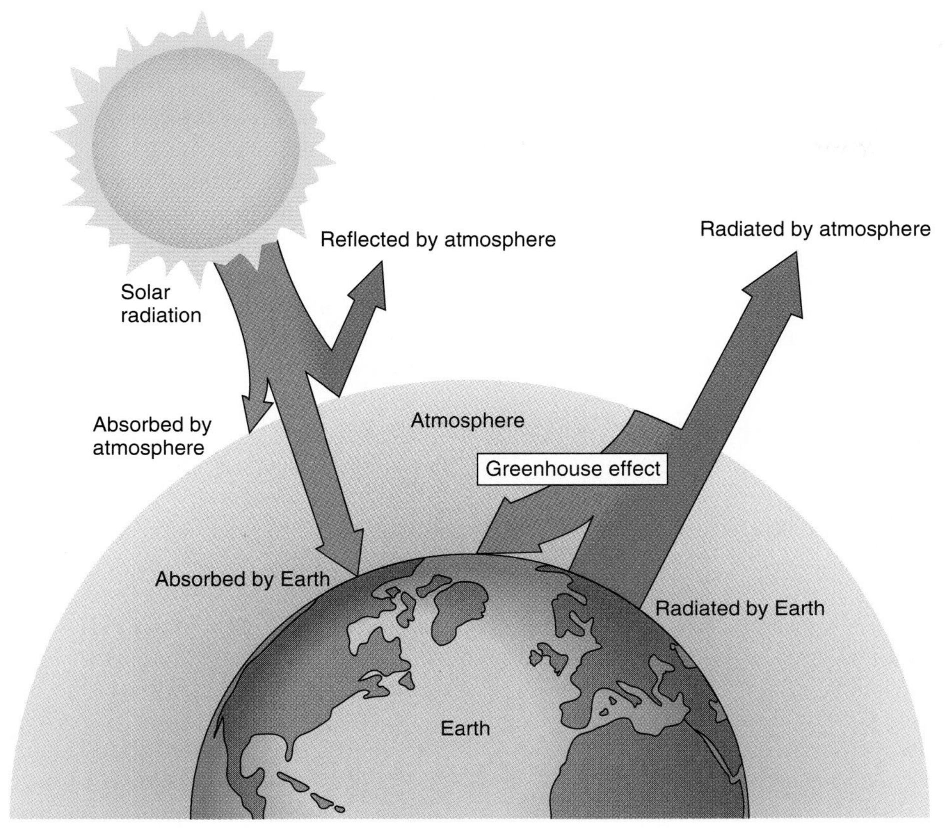

Figure 3.4
The Earth's energy balance. The width of each arrow is roughly proportional to energy flow.

 **Figures Alive!**

then reradiated back to the Earth's surface. Because of this exchange, the books are balanced and the total energy input from the Sun balances the energy output from the Earth. A steady state is established, with more or less constant average terrestrial temperatures.

- Steady-state disturbance was also important in the Chapter 2 discussion of ozone depletion.

The "more or less" is the reason for the current concern over global warming. What if more than 84% of the energy radiated is returned to the Earth? The term **global greenhouse effect** refers to the return of 84% of the energy radiated from the surface of the Earth. The term **enhanced greenhouse effect** is often used to refer to an energy return of **greater** than 84%. You likely experienced an enhanced greenhouse effect in your automobile if you left your car windows closed on a sunny day. The windows allow the UV and visible light in sunlight to pass through into the car, where some of the radiation is absorbed. Part of this absorbed energy is radiated in the car as infrared radiation, which, because of its longer wavelength, cannot pass back through the windows to the outside world. The infrared radiation (heat) is trapped inside the car, causing it to warm up. There is no circulation of the hot air in your car until you open the car doors and windows or turn on the air conditioner.

- In sunny climates, temperatures in a closed car can quickly exceed 120°F. For this reason, small children or pets should not be left in cars under these conditions.

3.3 Consider This: Clear or Cloudy?

Why does the temperature drop more on a clear night than on a cloudy night? Explain your reasoning.

The "windows" of the atmospheric greenhouse are made of molecules that are transparent to visible light but absorb in the infrared region of the spectrum. Carbon dioxide, water, methane, and the molecules of several other atmospheric compounds act in this way and are called *greenhouse gases*. They permit the radiation coming from the Sun to pass through but trap much of the heat emitted by the Earth.

Obviously, the greenhouse effect is essential in keeping our planet habitable for the species that have evolved here. However, if some CO_2 in the atmosphere is a good thing, more is not necessarily better. An increase in the concentration of this infrared absorber will very likely mean that more than 84% of the radiated energy will be returned to the Earth's surface, with an attendant increase in average temperature. Back in 1898, the Swedish chemist Svante Arrhenius (1859–1927), estimated the extent of this effect. He calculated that doubling the concentration of CO_2 would result in an increase of 5–6°C in the average temperature of the planet's surface. Writing in the *London, Edinburgh, and Dublin Philosophical Magazine* to announce his findings, Arrhenius colorfully described the phenomenon: "We are evaporating our coal mines into the air." At the end of the 19th century, the Industrial Revolution was already well under way in Europe and America, and it was "picking up steam" as well as generating it (and CO_2 also).

3.4 Consider This: Evaporating Coal Mines

Although the Arrhenius statement about "evaporating our coal mines into the air" certainly was effective in grabbing attention in 1898, what process do you think he really was referring to in discussing the amounts of CO_2 being added to the air? What is your reasoning?

Hint: Consider Figure 3.15 later in this chapter.

Recent trends in atmospheric carbon dioxide and average global temperatures are important for assessing the current and future status of the greenhouse effect. There is compelling evidence that CO_2 concentrations have increased significantly in the past century. The best data are those acquired at Mauna Loa in Hawaii. Figure 3.5 presents values from Antarctic ice cores taken from 1860 to the 1950s, and then adds the Mauna Loa data to show the continuation of the trend. The zigzag line from 1960 is a consequence of seasonal variation, but the general increase in average annual values from 315 ppm to about 370 ppm is clear. Later in this chapter we will learn why scientists believe that much of the added carbon dioxide has come from the burning of fossil fuels.

- Projections of changes in CO_2 based on computer modeling will be discussed in Section 3.11.

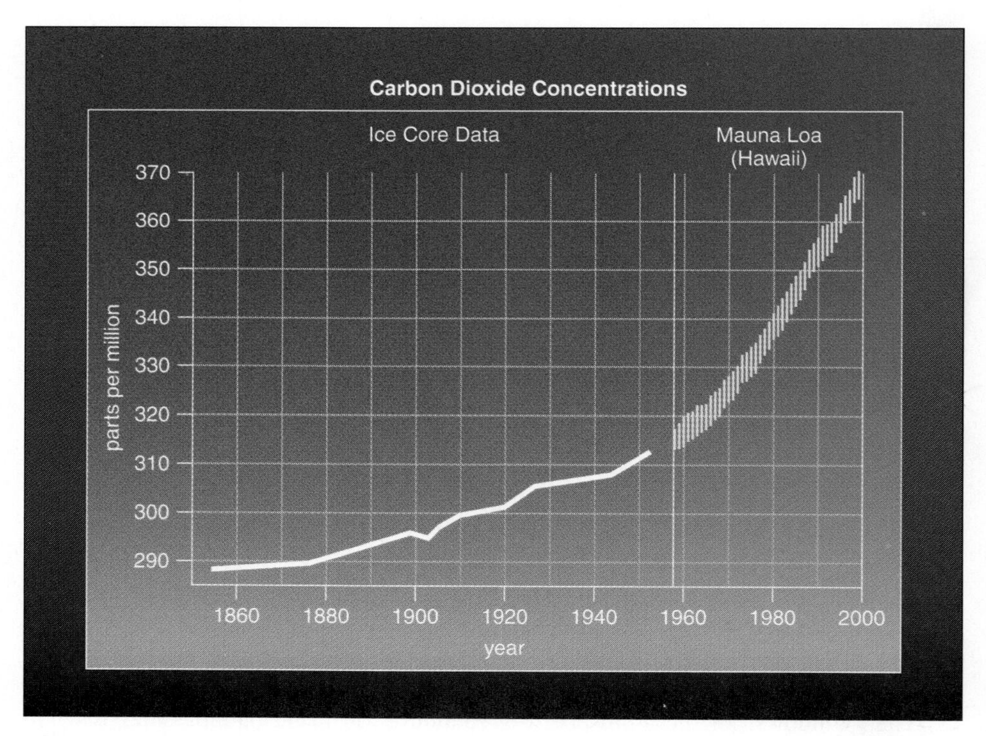

Figure 3.5

The atmospheric level of carbon dioxide has risen from 280 ppm to 370 ppm since 1860.

(From "The Greenhouse Effect and Historical Emissions," figure 4. Taken from http://clinton2.nara.gov/Initiatives/Climate/greenhouse.html.)

3.5 Consider This: The Cycles of Mauna Loa

The text states the zigzag pattern observable in Figure 3.5 is due to "seasonal variation."

a. What is the trend within each year?
b. Propose some reasons to explain seasonal changes in CO_2 concentrations on Mauna Loa.

3.6 Sceptical Chymist: Checking the Facts on CO_2 Increases

a. A recent government report states that the atmospheric level of CO_2 has increased 30% since 1860. Use the data in Figure 3.5 to either prove or disprove this statement.
b. A global warming skeptic states that the percent increase in the atmospheric level of CO_2 since 1957 has been only about half as great as the percent increase from 1860 to the present. Comment on the accuracy of that statement and how it could affect policy on global warming.

Other measurements indicate that during the past nearly 120 years, the average temperature of the planet has increased by somewhere between 0.4 and 0.8°C. Figure 3.6 shows the changes in the air temperature at the Earth's surface from 1880 to 2000. Some scientists have correctly pointed out that a century or two is an instant in the 4.5 billion-year history of our planet. They caution restraint in reading too much into what may be short-term temperature fluctuations. In fact, while some areas such as in Alaska and northern Eurasia have warmed by up to 6°C, cooling has occurred in the North Atlantic and in the central North Pacific. Short-term changes in atmospheric circulation patterns are thought to cause some observed temperature anomalies. Nevertheless, most credible researchers agree that there seems to be a trend. From Figure 3.6 we see that the average temperature of the Earth is about 0.6°C higher than it was in 1880. Whether this temperature increase is a consequence of the increased CO_2 concentration cannot be concluded with absolute certainty. Nevertheless, experimental evidence implicates carbon dioxide from human-related sources as a cause of recent global warming.

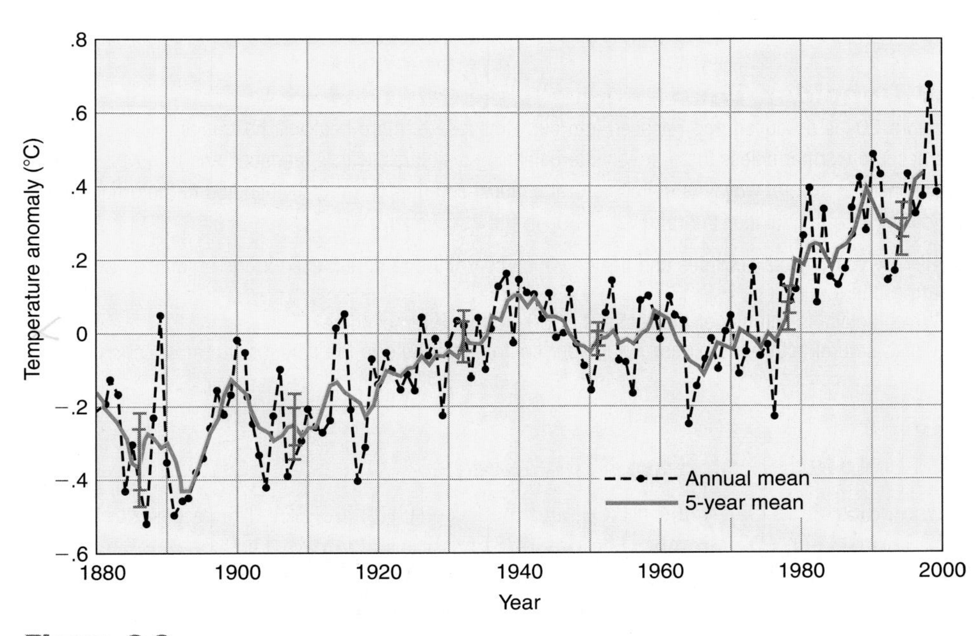

Figure 3.6
Global surface temperature change. Note the variability in temperatures from year to year, as well as the longer-term trends. Many scientists conclude that only since about 1970 is there a demonstrated upward climb in global temperatures.

(Taken from http://nside.org/NASA/SOTC/Intro.html.)

3.7 Consider This: Conclusion Justified?

Do you think the comment made in the cartoon is justified? Why or why not?

Pepper . . . and Salt

"This winter has lowered my concerns about global warming . . . "

When temperature measurements are extrapolated into the future, predictions made by Arrhenius must be revised downward. Current estimates are that doubling the CO_2 concentration will result in a temperature increase of between 1.0 and 3.5°C. If and when that doubling will occur depends, to a considerable extent, on the human beings who inhabit this planet. We are a long way from the out-of-control hothouse of Venus, but we face difficult decisions. These decisions are not going to be easy, but they will be better informed with an understanding of the mechanism by which greenhouse gases interact with radiation to create the greenhouse effect. For that we must again assume a submicroscopic view of matter.

3.8 Consider This: Adding to the CO_2 Level

As more CO_2 is produced and released into our atmosphere, more heat will be trapped, and our planet will become warmer unless there are counterbalancing measures. CO_2 is produced and removed from our atmosphere in several ways. As members of the global community, our activities add to the production of CO_2. They also may hinder the removal of CO_2 in the atmosphere.

a. Review your typical activities and decide which contribute to an increase in the amount of CO_2 in the atmosphere.
b. Which activities can hinder the removal of CO_2 from the atmosphere?
c. Which natural activities that are *not* in your control contribute to the amount of atmospheric CO_2?

3.4 Molecules: How They Shape Up

Carbon dioxide (CO_2), water (H_2O), and methane (CH_4) are greenhouse gases; nitrogen (N_2) and oxygen (O_2) are not. The obvious question is "why?" The not-so-obvious answer has to do with molecular structure and shape. When you encountered Lewis structures in Chapter 2, geometry was not the main consideration. The octet rule provides a generally reliable method for predicting bonding in molecules. Moreover, in the case of diatomic molecules, it also predicts molecular geometry. In molecules such as O_2 and N_2, the shape is unambiguous. In a molecule with two atoms, the atoms can only be in a straight line.

$$:N:::N: \quad \text{or} \quad :N{\equiv}N: \quad \text{or} \quad N{\equiv}N \quad \text{and} \quad :\ddot{O}::\ddot{O}: \quad \text{or} \quad :\ddot{O}{=}\ddot{O}: \quad \text{or} \quad O{=}O$$

With molecules of three or more atoms, differences in molecular geometry become possible. Fortunately, knowing where the outer electrons are located provides insight into molecular shape. Therefore, the first step in predicting molecular shape is to write the Lewis structure for the molecule. If the octet rule is obeyed throughout the molecule, each atom (except hydrogen) will be associated with four pairs of electrons. Some molecules include nonbonding lone-pair electrons, but all molecules contain some bonding electrons or they would not be molecules! These bonding electrons can be grouped in one or more pairs to form one or more single bonds. In other molecules, the bonding electrons are involved in double bonds consisting of two pairs of electrons, or in triple bonds made up of three pairs of electrons. A basic rule of electricity is that unlike charges attract and like charges repel. Negatively charged electrons are attracted to a positively charged nucleus in every case. However, the electrons all have the same charge and therefore are found in space as far from each other as possible while still maintaining their attraction to the oppositely charged nucleus. This leads to the important point that groups of negatively charged electrons will repel each other. *The most stable arrangement is one in which the mutually repelling electron groups are as far away from each other as possible.* This electronic arrangement determines the atomic arrangement and the shape of the molecule.

We illustrate this stepwise procedure for predicting molecular structure with methane, CH_4, one of the greenhouse gases.

1. **Determine the number of outer electrons associated with each atom in the molecule.** The carbon atom (atomic number 6, Group 4A) has four outer electrons; each of the four hydrogen atoms contributes one electron. There is a total of $4 + (4 \times 1)$ or 8 outer electrons.

2. **Arrange the outer electrons and the atoms in pairs in such a way as to satisfy the octet rule. This may require single, double, and/or triple bonds.** The eight outer electrons in a CH_4 molecule are arranged around the central carbon atom in four bonding pairs, each pair connecting the carbon atom to a hydrogen atom. This is the Lewis structure.

```
  H               H
  ..              |
H:C:H    or   H — C — H
  ..              |
  H               H
```

   This structure seems to imply that the CH_4 molecule is flat or planar. This is only because we are restricted to the two dimensions of a sheet of paper. The architecture of molecules is three dimensional, and we must look into that third dimension.

3. **Assume that the most stable molecular shape is the one in which the bonding or nonbonding electron groups attached to any atom are as far from each other as possible, within the constraints of bonding.**

The four electron pairs around the carbon atom in CH_4 repel each other, and in their most stable arrangement they are as far from each other as they can be and still form C—H bonds. Furthermore, because a hydrogen atom is attached to each pair of electrons, the four hydrogen atoms also are as far from each other as possible. This means that the shape of a CH_4 molecule is similar to the base of a folding music stand. The four C—H bonds correspond to the three evenly spaced legs and the vertical shaft of the stand. The angle between each pair of bonds is 109.5°. This shape is said to be tetrahedral, because the hydrogen atoms correspond to the corners of a **tetrahedron,** a four-cornered figure with four equal triangular sides. This shape has been experimentally confirmed. Indeed, the tetrahedral structure is one of the most common atomic arrangements in nature, particularly in carbon-containing molecules.

The legs and shaft of a folding music stand approximate the bond angles in a tetrahedral molecule like methane.

The representation in Figure 3.7a is a generally accepted method used by chemists to convey the three-dimensional structure of methane. The wedge-shaped line

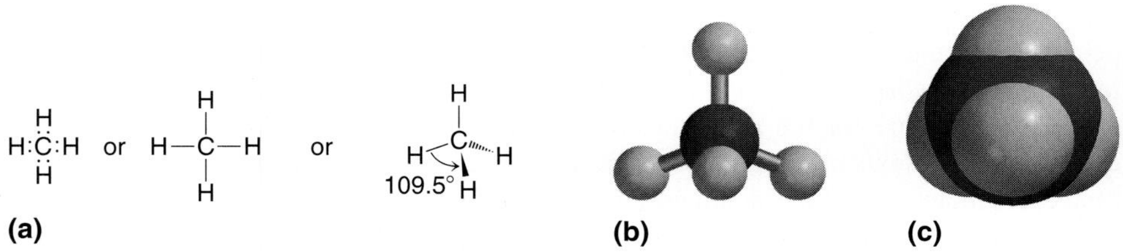

(a) (b) (c)

Figure 3.7

(a) Lewis structure and structural formula; (b) Ball-and-stick model of methane; (c) Space-filling model illustration of the compound.

represents a bond that is coming out of the paper at an angle that is generally toward the reader, the dashed wedge represents a bond pointing away from the reader, and the solid lines are assumed to be in the plane of the paper. This is an improvement over the two-dimensional structure, but the best way to visualize molecules is with models, as in Figure 3.7b,c. Your instructor will certainly show you such models, and may give you an opportunity to use them yourself. Another way to understand the shape of molecules is to use a molecular modeling program, as you will do in 3.11 Consider This.

- CFCs were discussed in Section 2.12.

We can apply the same set of steps with CCl_3F, a CFC that can act as a greenhouse gas (Figure 3.8). Using step 1, we determine from the molecular formula and the periodic table that there are a total of 32 outer electrons: four for carbon (Group 4A), seven for fluorine (Group 7A), and seven for *each* chlorine (Group 7A). Applying step 2 reveals that carbon is the central atom and the other four atoms are bonded to it by four single covalent bonds (8 electrons; four shared electron pairs). This satisfies the octet rule for carbon. The remaining 24 electrons serve as nonbonding (unshared; lone) electron pairs on the other bonded atoms, thus achieving an octet around each one.

Like the four electron pairs around carbon in CH_4, the four shared electron pairs in CCl_3F also repel each other so as to be as far apart from each other as possible. The resulting molecular shape is also tetrahedral, with the fluorine and chlorine atoms at the corners of a tetrahedron.

In some molecules, the central atom is surrounded by an electron octet, but not all of the electrons are bonding electron pairs. Some are nonbonding (lone) pairs, as in ammonia, NH_3, a refrigerant gas that was largely replaced by CFCs (Figure 3.9).

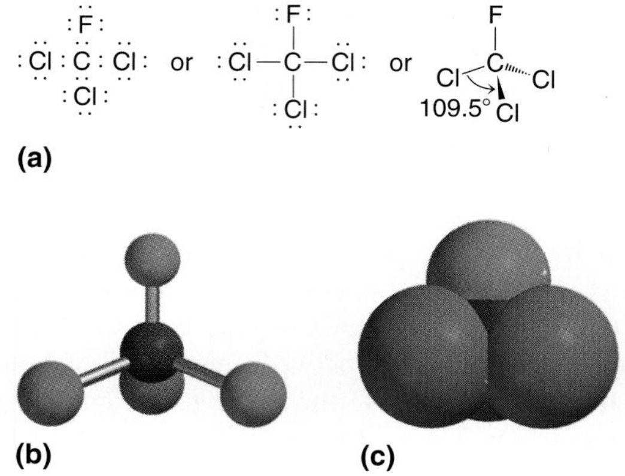

(a)

(b) (c)

Figure 3.8

(a) Lewis structure and structural formula; (b) Ball-and-stick model of CCl_3F; (c) Space-filling model illustration of the compound.

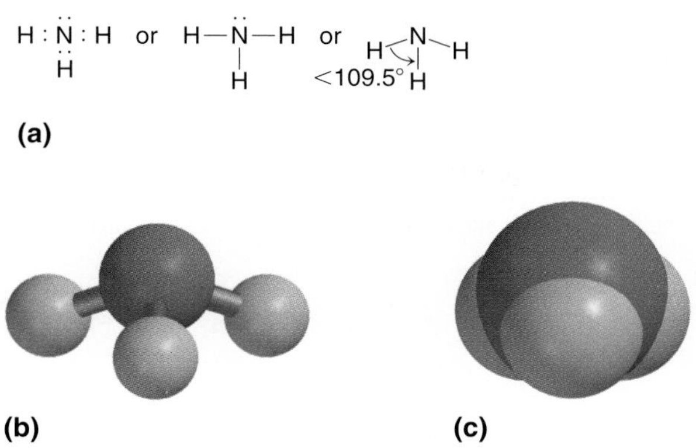

H : N̈ : H or H—N̈—H or H—N—H
H H $<109.5^\circ$ H

(a)

(b) **(c)**

Figure 3.9

(a) Lewis structure and structural formula; (b) Ball-and-stick model of ammonia; (c) Space-filling model illustration of the compound.

The nonbonding pair occupies greater space than a bonding pair. Consequently, it repels the bonding pairs somewhat more strongly than the bonding pairs repel each other. This stronger repulsion forces the bonding pairs closer to each other creating an H—N—H angle slightly less than the predicted 109.5° of a regular tetrahedron. The experimental value is close to this, 107.5°, indicating that our model is reasonably reliable.

In describing the shape of a molecule, we do it in terms of the atoms, not the electrons. The hydrogen atoms form a triangle with the nitrogen atom above them at the top of the pyramid. Thus, ammonia is said to be a triangular pyramid; it has a **triangular pyramidal structure** (Figure 3.9).

Water, another naturally occurring greenhouse gas, illustrates yet another type of molecular shape. There are eight outer electrons: one from each hydrogen atom plus six from the oxygen (Group 6A). Its Lewis structure (Figure 3.10) discloses how the eight electrons on the central oxygen atom are distributed—two pairs involved in bonding and two lone pairs.

If these four pairs of electrons are arranged so that they are as far apart as possible, the distribution will be similar to that in methane, and we might predict water to have a tetrahedral shape, with a 109.5° H—O—H bond angle. Unlike the four bonding pairs in methane, water has two bonding pairs and two nonbonding pairs. The repulsion

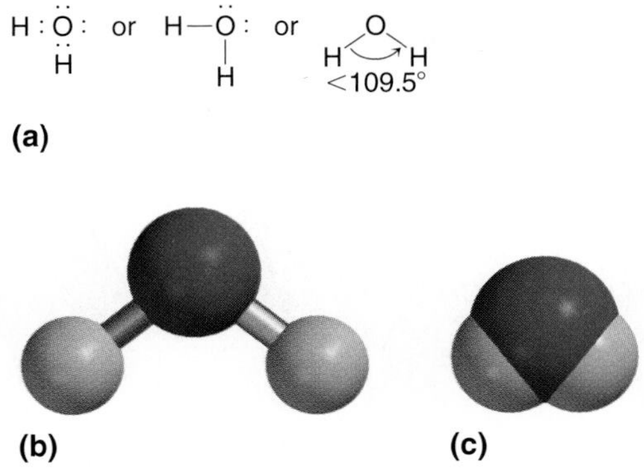

H : Ö : or H—Ö : or O
H H H H
$<109.5^\circ$

(a)

(b) **(c)**

Figure 3.10

(a) Lewis structure and structural formula; (b) Ball-and-stick model of water; (c) Space-filling model illustration of the compound.

between the two nonbonding pairs and their repulsion of the bonding pairs cause the bond angle to be less than 109.5°. Experiments indicate a value of approximately 104.5°. Thus, a water molecule is said to have a *bent* shape (Figure 3.10).

3.9 Your Turn

Using the strategies just described, predict and sketch the shape of each of these molecules.

a. CCl_4 (carbon tetrachloride)
b. CCl_2F_2 (Freon-12; dichlorodifluoromethane)
c. H_2S (hydrogen sulfide)

Answer

a. The first step in predicting a structure is to write the correct Lewis structure. Each of the four chlorine atoms has seven outer electrons, and the carbon atom has four. The chlorine atoms bond to the central carbon atom, forming four single bonds. Each bond is a shared pair of electrons. Thus, each atom is surrounded by 8 electrons.

The bonding electron pairs and the attached chlorine atoms arrange themselves so that their separation is maximized. There are no nonbonding pairs on the central carbon atom. It follows that the shape of a carbon tetrachloride molecule is tetrahedral—the same as a methane molecule:

:Cl:
:Cl:C:Cl: or :Cl—C—Cl: or Cl, C, Cl, Cl, Cl (109.5°)
:Cl:

Carbon dioxide is an example of an important greenhouse gas. A count of outer electrons gives a total of 16. Four electrons are contributed by the carbon atom and six from each of the two oxygen atoms. If only single bonds are involved, there are not enough electrons to provide eight electrons for each atom. That would require 20 electrons. However, the octet rule will be obeyed if the central carbon atom shares 4 electrons (a double bond) with each of the oxygen atoms. This means that two double bonds are formed.

In the CO_2 molecule, two groups of four electrons each are associated with the central atom. These groups of electrons repel each other, and the most stable configuration provides the furthest separation of the negative charges. This occurs when the angle between them is 180° and the molecule is **linear.** The model predicts that all three atoms in a CO_2 molecule will be in a straight line. This is, in fact, the case. Here is an instance where the simple Lewis structure reveals the correct molecular geometry, as shown in Figure 3.11.

We applied the idea of electron pair repulsion to molecules in which there are four groups of electrons (CH_4, CCl_3F, NH_3, and H_2O) and two groups of electrons (CO_2). Electron pair repulsion also applies reasonably well to molecules that include three, five, or six groups of electrons. In most molecules, the electrons and atoms are arranged to keep the separation of the electrons at a maximum. This logic accounts for the bent

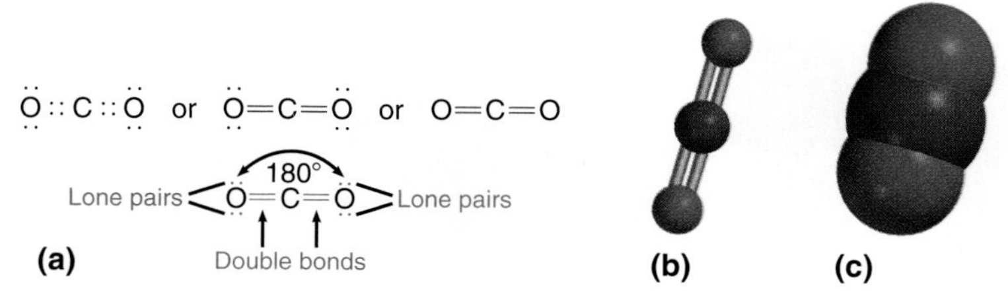

Figure 3.11
(a) Lewis structure and structural formula; (b) Ball-and-stick model of carbon dioxide; (c) Space-filling model illustration of the compound.

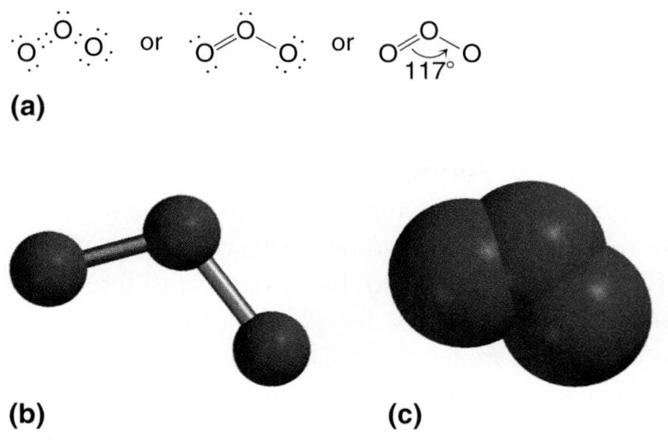

Figure 3.12
(a) Lewis structure and structural formula; (b) Ball-and-stick model of ozone; (c) Space-filling model illustration of the compounds.

• These structures for ozone show only one resonance form. See Section 2.3 for a discussion of resonance.

shape we associated with the ozone molecule in Chapter 2. Remember that according to the octet rule, the O_3 molecule (18 total outer electrons) contains a single bond and a double bond. Remember also that the central oxygen atom carries a nonbonding lone pair of electrons. Thus, there are three groups of electrons on this central atom: the pair that makes up the single bond, the two pairs that constitute the double bond, and the lone pair. These three groups of negatively charged particles repel each other, and the minimum energy of the molecule corresponds to the furthest separation of these electron groups. This will occur when the electron groups are all in the same plane and at an angle of about 120° from each other. We predict, therefore, that the O_3 molecule should be bent, and the angle made by the three atoms should be approximately 120°. The fact that experiment shows the angle to be 117° is confirmation of the general utility of this method (Figure 3.12).

The 3.10 Your Turn that follows gives you an opportunity to apply the method to two more molecules of atmospheric importance.

3.10 Your Turn

Using the strategies just described, predict and sketch the shapes of these molecules:

a. SO_2 (sulfur dioxide) **b.** SO_3 (sulfur trioxide)

Hint: Note the periodic table family resemblance between SO_2 and O_3.

3.11 Consider This: Chiming in on Molecules

Three-dimensional representations of molecules can be viewed on the Web with the aid of CHIME, a free plug-in that you can download and install. Following the links on the *Chemistry in Context* web site, use CHIME to view the molecules considered in this section. Has your mental picture of these molecules changed after seeing the 3-D representations? Explain.

3.5 Vibrating Molecules and the Greenhouse Effect

Now that we know the molecular shapes of some important greenhouse gases, we can turn to the important phenomenon of how these molecules interact with infrared radiation. When a molecule absorbs a photon, it responds to the added energy. You already learned in Chapter 2 that if the photon corresponds to the UV region of the spectrum, it has sufficient energy to disrupt the arrangement of electrons within the molecule. This can cause covalent bonds to break, as in the dissociation of O_2 and O_3 by UV-B and UV-C radiation.

Radiation in the infrared region of the spectrum is not sufficiently energetic to cause such molecular disruption. However, a photon of IR radiation can enhance the vibrations in a molecule. The covalent bonds holding atoms together can be thought of as springs, and the atoms can move back and forth. Depending on the molecular structure, only certain vibrations are permitted, and each of these vibrations has a characteristic set of permissible energy levels. The energy of the photon must correspond exactly to the vibration energy of the molecule for the photon to be absorbed. This means that different molecules absorb IR radiation at different wavelengths and thus vibrate at different energies.

We illustrate these ideas with the carbon dioxide molecule, representing the atoms as balls and the bonds as springs. A molecule of CO_2 can vibrate in the four ways pictured in Figure 3.13. The arrows indicate the direction of motion. Vibrations **a** and **b** are called stretching vibrations. In vibration **a,** the central carbon atom is stationary and the oxygen atoms move back and forth (stretch) in opposite directions. Alternatively, the oxygen atoms can move in the same direction and the carbon atom in the opposite direction (vibration **b**). Vibrations **c** and **d** look very much alike. In both cases, the molecule bends from its normal linear shape. The bending counts as two vibrations because it can occur in either of two planes.

In any molecule, the amount of energy required to cause vibration depends on the nature of the motion, the "stiffness" and strength of the bonds, and the masses of the atoms that move. If you have ever examined a spring or played with a Slinky toy, you have probably observed that more energy is required to stretch a spring than to bend it. Similarly, more energy is required to stretch a CO_2 molecule than to bend it. This means that the more energetic photons—corresponding to shorter wavelengths—are needed to excite stretching vibrations **a** or **b** than those for bending vibrations **c** or **d.** The two bending motions (**c** and **d**) are both stimulated when the molecule absorbs IR radiation with a wavelength of 15.00 micrometers (μm).

- A micrometer is equal to one-millionth of a meter: $1\ \mu m = 1 \times 10^{-6}$ m.

Vibration **b** requires more energy; it will occur only if radiation of wavelength of 4.26 μm is absorbed. Vibrations **b, c,** and **d** account for the greenhouse properties of carbon dioxide. It turns out that vibration **a** cannot be triggered by the direct absorption of IR radiation. For such absorption to occur, the overall electrical charge distribution in the molecule must change during the vibration. In a CO_2 molecule, the average concentration of electrons is greater on the oxygen atoms than on the carbon atom. This means that the oxygen atoms carry a partial negative charge relative to the carbon atom. As the bonds stretch, this charge distribution alters. Because of the linear shape and symmetry of the molecule and vibration **a,** the changes in charge distribution cancel each other and no infrared absorption occurs.

The infrared energies absorbed or transmitted by molecules can be measured with an instrument called an infrared spectrometer. Heat radiation from a glowing filament is passed through a sample of the compound to be studied, in this case gaseous carbon dioxide. A detector measures the amount of radiation, at various frequencies, transmitted by the sample. High transmission means low absorbance, and vice versa. This information is recorded on a chart, where radiation intensity is plotted versus wavelength. The result is called the *infrared spectrum* of the compound. Figure 3.14 is the infrared spectrum of CO_2, obtained in just this way. There are two steep valleys where the intensity of the

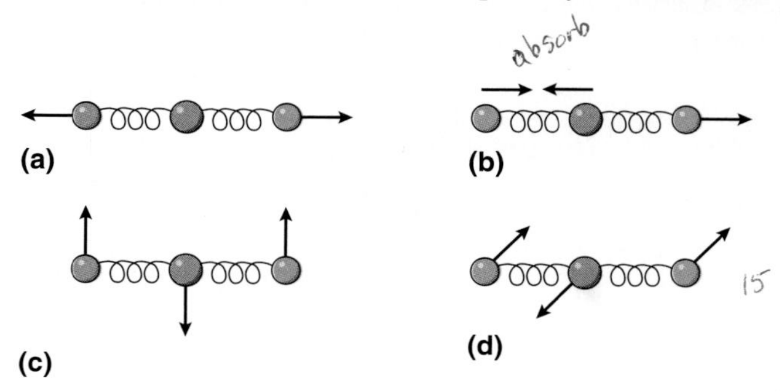

Figure 3.13

Molecular vibrations in CO_2. Each spring represents a C═O bond.

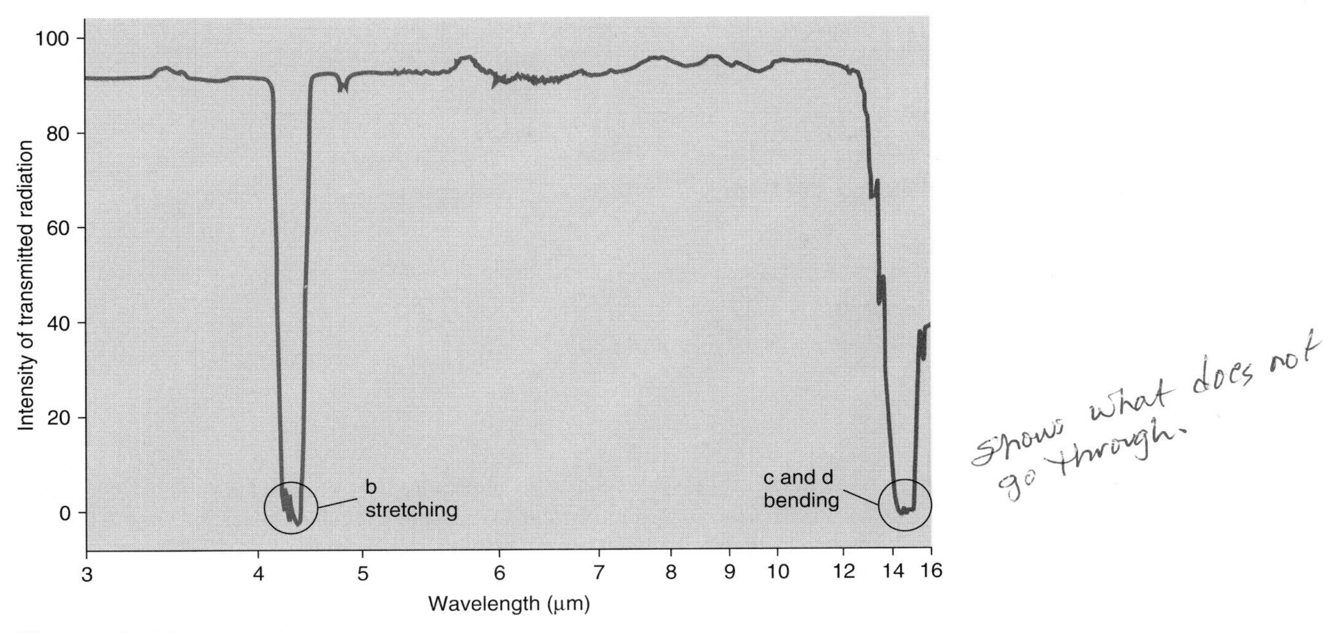

Figure 3.14

Infrared spectrum of carbon dioxide. Maximum absorbencies of IR energy occur at 4.26 μm and 15.00 μm. Figure 3.13 depicts stretching and bending modes.

transmitted radiation drops almost to zero. This means that most of the radiation is absorbed by the CO_2 molecules. Note that these absorbencies occur at 4.26 and 15.00 μm, as predicted. The same phenomenon occurs in the atmosphere. Carbon dioxide molecules absorb infrared energy at these wavelengths. They vibrate for a while and then re-emit the energy and return to their normal unexcited or "ground" state. This is how carbon dioxide captures and returns the infrared radiation coming from the surface of the Earth. This is what makes carbon dioxide a greenhouse gas.

Any molecule that can vibrate in response to the absorption of infrared radiation is potentially a greenhouse gas. There are many such substances. Carbon dioxide and water are the most important in maintaining the temperature of the Earth. However, methane (CH_4), nitrous oxide (N_2O), ozone (O_3), and chlorofluorocarbons (such as CCl_3F) are among the other substances that help retain planetary heat. Figure 3.15, the infrared

- Spectroscopy is the field of study that examines matter by passing electromagnetic energy through a sample. Wavelengths absorbed and/or transmitted are changed to electrical signals in the detector, producing a pattern called a spectrum or a spectrograph.

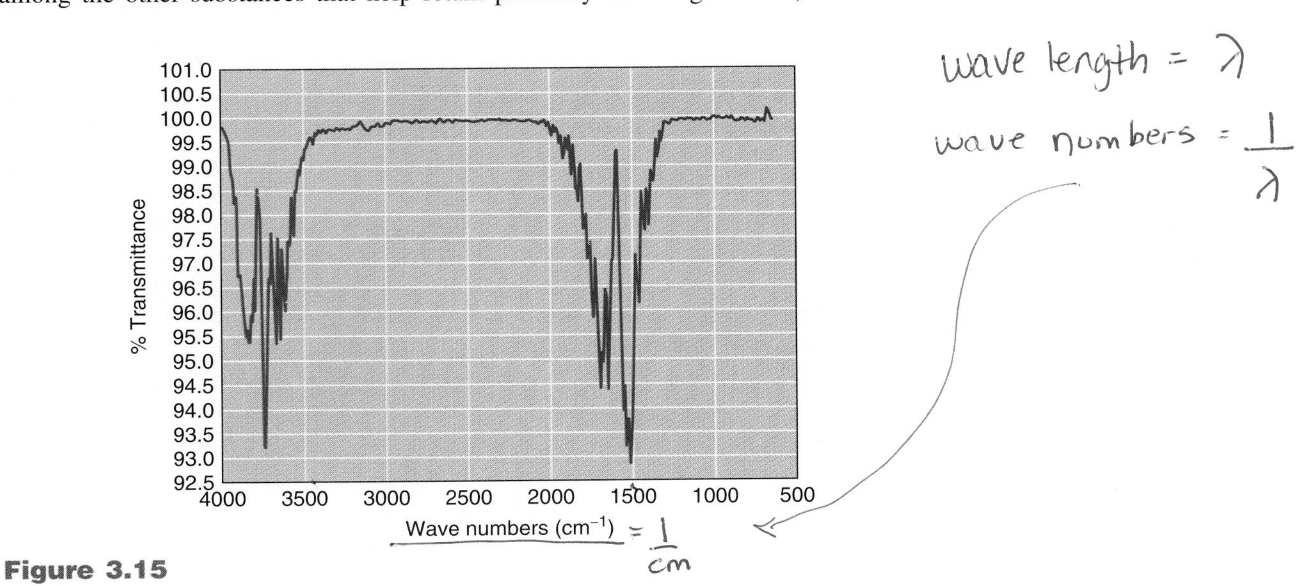

Figure 3.15

Infrared spectrum of water vapor. Maximum absorbancies of IR energy occur at 2.66 μm (3776 cm^{-1}) and 6.27 μm (1595 cm^{-1}).

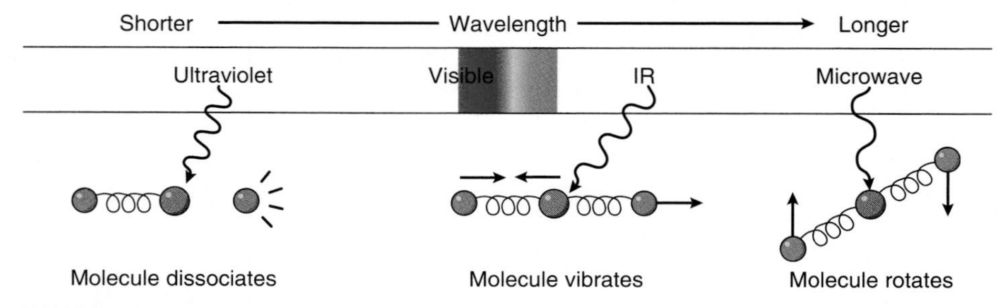

Figure 3.16

Molecular response to types of radiation.

spectrum of water vapor, shows that H_2O molecules absorb infrared radiation equivalent to 2.66 and 6.27 μm. Diatomic N_2 and O_2 are not greenhouse gases. Although molecules consisting of two identical atoms do vibrate, the overall electrical charge distribution does not change during these vibrations. Hence, these molecules do not absorb infrared radiation.

3.12 Consider This: Bending and Stretching Water Molecules

Consider the two "peaks" in the IR spectrum of water shown in Figure 3.15. Which do you predict represents bending vibrations and which represents stretching? Explain the basis of your predictions.

Hint: Compare the IR spectrum of CO_2 with that for H_2O.

You have encountered two responses of molecules to radiation. Highly energetic photons with high frequencies and short wavelengths (such as UV radiation) can break up molecules. The less energetic photons of infrared light cause many molecules to vibrate. Both processes are depicted in Figure 3.16, but the figure also includes another response of molecules to radiant energy that is probably a good deal more familiar to you. This response happens in a microwave oven. The radiation generated in such a device is of relatively long wavelength, about a centimeter. This means that the energy per photon is quite low. This energy is insufficient to cause a molecule to vibrate or dissociate, but it is enough to set the molecule spinning. Microwave ovens are tuned to generate radiation that causes water molecules to rotate. As the H_2O molecules absorb the photons and spin more rapidly, the resulting friction warms up the leftovers. The same region of the spectrum is used for radar. Beams of microwave radiation are sent out from a generator. When the beams strike an object such as an airplane, the microwaves bounce back and are detected by a sensor.

The practical consequences of the interaction of radiation and matter are immense, but there is another application of great significance in our understanding of nature. Spectroscopy provides a means of studying atomic and molecular structure. Electronic, vibrational, and rotational energy are all quantized: Only certain energy levels are permitted. No matter what region of the spectrum is employed, spectroscopy reveals differences between energy levels. Using the appropriate mathematical model, scientists can translate these energy differences into information about bond lengths, bond strengths, and bond angles. The assurance with which chemists describe the invisible is a consequence of looking through a spectroscopic window into atoms and molecules.

3.6 The Carbon Cycle: Contributions from Nature and Humans

In his book *The Periodic Table*, the late chemist, author, and World War II concentration camp survivor Primo Levi, wrote eloquently about carbon dioxide.

"This gas which constitutes the raw material of life, the permanent store upon which all that grows draws, and the ultimate destiny of all flesh, is not one of the principal components of air but rather a ridiculous remnant, an 'impurity' thirty times less abundant than argon, which nobody even notices. . . . [F]rom this ever renewed impurity of the air we come, we animals and we plants, and we the human species, with our four billion discordant opinions, our millenniums of history, our wars and shames, nobility and pride."

In the essay from which this quotation is taken, Levi traces a brief portion of the life history of a carbon atom from a piece of limestone (calcium carbonate, $CaCO_3$) where it lies "congealed in an eternal present," to a CO_2 molecule, to a molecule of glucose in a leaf, and ultimately to the brain of the author. And yet that is not the final destination. "The death of atoms, unlike our own," writes Levi, "is never irrevocable." That carbon atom, already billions of years old, will continue to persist into the unimagined future.

This marvelous continuity of matter, a consequence of its conservation, is beautifully illustrated by the carbon cycle. Even without Primo Levi's poetic gifts, the story is fascinating, one that is important to understand to see the danger of the cycle being changed by human's activities. It is certain that without the proper functioning of the carbon cycle, every aspect of life on Earth could undergo dramatic change.

3.13 Consider This: Understanding the Natural Carbon Cycle

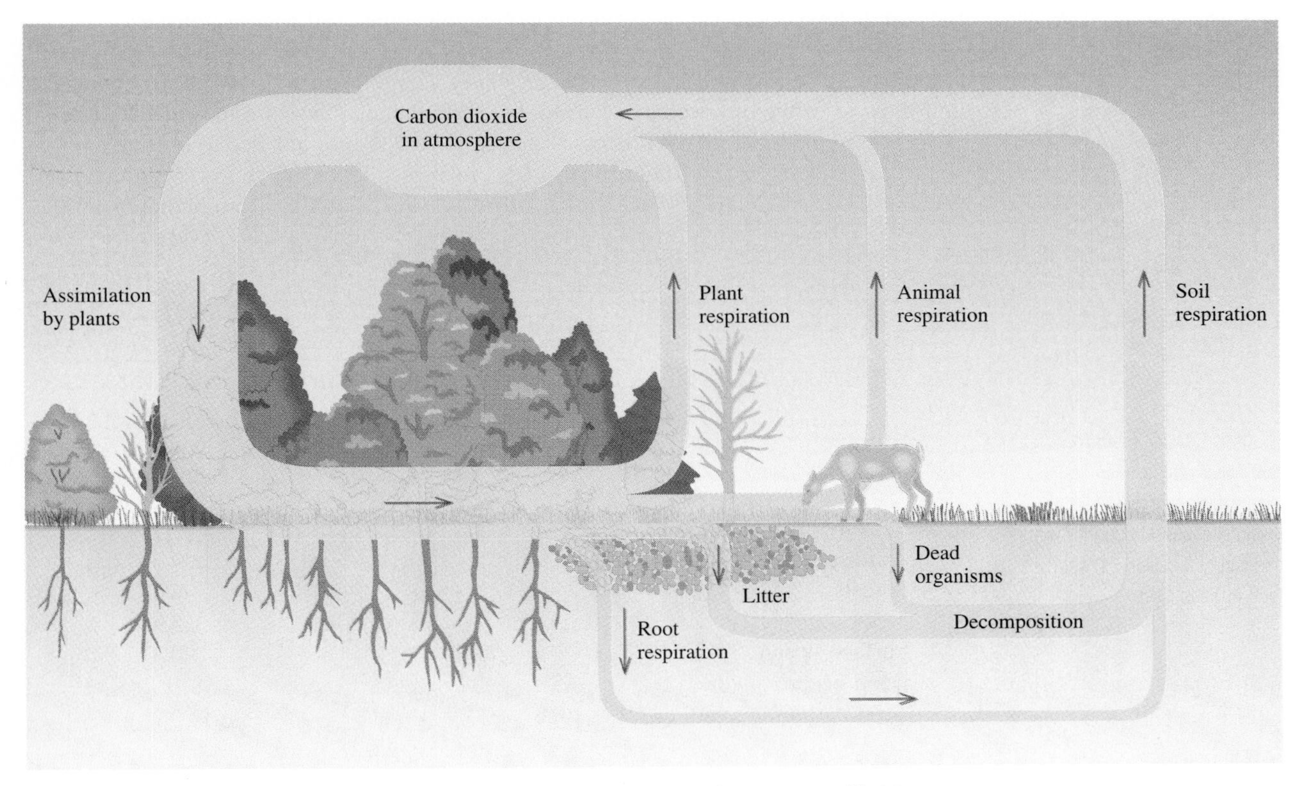

a. This figure illustrates several steps in the natural carbon cycle. Which processes add CO_2 to the atmosphere and which remove it?

b. What steps in the natural carbon cycle are *not* shown in this figure?

The carbon cycle is a dynamic, steady state system that returns as much CO_2 to the atmosphere as is extracted. Plants die and decay, releasing CO_2. Other plants enter the food chain where their complex molecules are broken down into CO_2, H_2O, and other simple substances. Animals exhale CO_2, carbonate rocks decompose, and carbon dioxide escapes through the vents of volcanoes. And the cycle goes on and on. Michael B.

McElroy of Harvard University estimated, "The average carbon atom has made the cycle from sediments through the more mobile compartments of the Earth back to sediments, some 20 times over the course of Earth's history." Carbon dioxide in the air today may have come from the campfires of Attila the Hun, more than a thousand years ago.

Where is carbon on the Earth? The world's carbon supply is widely distributed, as shown in Table 3.1. The relative amount of carbon found in each location is part of the overall story of the carbon cycle. Changes to that cycle affect the distribution of carbon noted in Table 3.1.

3.14 Your Turn

a. Which is the largest reservoir for carbon on Earth?
b. What percentage of the total fossil fuel carbon reservoir is from coal?
c. What percentage of the Earth's total carbon reservoir is in fossil fuels?

As members of the animal kingdom, we *Homo sapiens* participate in the carbon cycle along with our fellow creatures. But we do more than our share; we do more than simply inhale and exhale, ingest and excrete, live and die. We have developed processes that permit us to significantly perturb the system. Because of our involvement, the carbon cycle is out of balance. The Industrial Revolution, which began in Europe in the late 18th century, was fueled largely by coal. Coal was used to power steam engines in mines, factories, locomotives, ships, and later, electrical generators. The subsequent discovery and exploitation of vast deposits of petroleum made possible the development of automobiles and other types of transportation. To a very considerable extent, the Industrial Revolution was a revolution in energy sources and energy transfer.

The result, however, is a large-scale transfer of carbon stored in fossil fuels to carbon dioxide released into the atmosphere. Every year, human activity releases more carbon dioxide to the atmosphere from burning fossil fuels than is removed from the atmosphere by natural cycles. Processes that remove CO_2 are not necessarily as rapid as needed, so the amounts of atmospheric CO_2 increase. Table 3.2 shows the sources and sinks of CO_2 from anthropogenic origin, that is, sources arising from human activities. Table 3.2 presents an optimistic picture that atmospheric, oceanic, and terrestrial sinks

- A Gigaton (Gt) is a billion metric tons, or about 2200 billion pounds. For comparison, a fully loaded 747 jet airplane weighs about 800,000 lbs. It would take nearly 3 million 747s to have a total mass of 1 Gt.

Table 3.1

The Earth's Carbon Reservoirs

	Size (Gt carbon/year)
Reservoir	
Atmosphere	750
Forests	610
Soils	1,580
Surface ocean	1,020
Deep ocean	38,100
Total carbon, excluding fossil fuels	42,060
Fossil fuels	
Coal	4,000
Oil	500
Natural gas	500
Total fossil fuel	5,000
Overall total, all sources	47,060

Source: *Climate Change*, Intergovernmental Panel on Climate Change (IPCC), as quoted in *Consequences, The Nature & Implications of Environmental Change*, Vol. 4, No. 1.

Know *

Table 3.2

Human Perturbations to the Global Carbon Budget

	Flux (Gt carbon/year)
CO_2 sources	
Fossil fuel combustion and cement production	5.5 ± 0.5
Tropical deforestation	1.6 ± 1.0
Total anthropogenic emissions	7.1 ± 1.1
CO_2 sinks	
Storage in the atmosphere	3.3 ± 0.2
Uptake by the ocean	2.0 ± 0.8
Northern Hemisphere forest regrowth	0.5 ± 0.5
Other terrestrial sinks (CO_2 fertilization, nitrogen fertilization, climatic effects)	1.3 ± 1.5
Total sinks for CO_2	7.1 ± 1.1

Source: *Climate Change*, Intergovernmental Panel on Climate Change (IPCC), as quoted in *Consequences, The Nature & Implications of Environmental Change*, Vol. 4, No. 1.

- A carbon "sink" is a natural storage place in the environment that removes carbon from another part of its cycle. "Flux" refers to the amount of carbon moving through the environment in a year.

will match the sources. Note, however, that nearly half of CO_2 from human-based sources remains in the atmosphere, leading to the observed rise in CO_2 concentration in recent years.

3.15 Consider This: Uncertainty–A Part of the Problem

Data in Table 3.2 are reported with a plus and minus value that indicates the level of uncertainty in the information. Uptake of CO_2 by the oceans is one of the most difficult sink values to determine accurately.

a. Offer some possible reasons why the value for uptake of CO_2 by oceans has a high uncertainty.
b. What could be the effect on global warming if the value for uptake by the oceans is too low? Too high?
c. Which is the most uncertain anthropogenic source of CO_2? Offer some possible reasons.

As the generation of energy and the consumption of fossil fuels increased, so did the quantity of combustion products such as carbon dioxide released to the atmosphere. Since 1860, the CO_2 concentration has increased from 290 ppm to 370 ppm and the current rate of increase is about 1.5 ppm per year. At the present level, fossil fuels containing more than 5 Gt of carbon are burned annually, but not necessarily for the same purpose. Most fossil fuel–based CO_2 comes from power utilities (35%) and transportation (31%); much less, for example, comes from home and commercial heating (Figure 3.17).

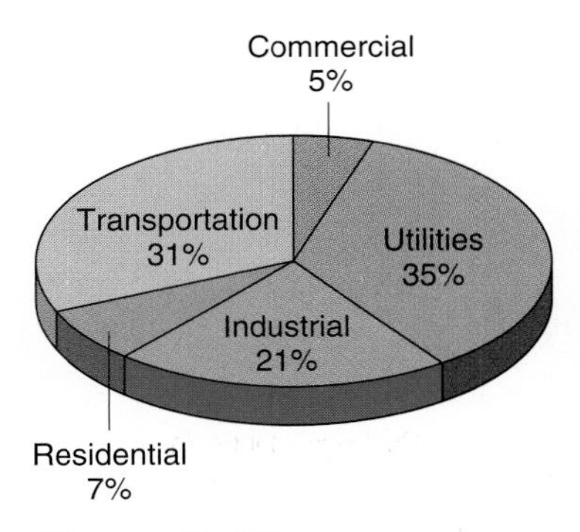

Figure 3.17
CO_2 emission sources in the U.S.
(Source of data: DOS/EPA.)

3.16 Consider This: The CO_2 Emissions–Implications for Policy

Figure 3.17 gives the sources of CO_2 emissions from fossil fuel consumption. These values also have implications for personal action and for setting control policies.

a. As an individual, which sources of CO_2 can you control? Explain your reasoning.
b. Do you think that national priorities for controlling CO_2 emissions are set based on the rank order of percentages in this figure? Why or why not? Explain your reasoning.

Table 3.2 also reveals that deforestation by burning releases 0.6–2.6 Gt of carbon (in the form of CO_2) to the atmosphere each year. It is estimated that 2.00×10^6 km^2 of rain forest has been cut down or burned globally from 1980 to 1995. This is an area equal to the combined size of Mexico or Indonesia. Brazil continues as the country with the greatest loss of forest, with an annual change of 2.6×10^4 km^2 a year. As a result, trees, which are very efficient absorbers of carbon dioxide, are removed from the cycle.

If the wood is burned, vast quantities of CO_2 are generated; if it is left to decay, that process also releases carbon dioxide, but more slowly. Even if the lumber is harvested for construction purposes and the land is replanted in cultivated crops, the loss in CO_2 absorbing capacity may approach 80%. The systematic deforestation is not a new phenomenon nor is it limited to tropical forests. The focus in the 20th century shifted from the heavily deforested regions of Europe and North America to the tropical rainforests of Central and South America, Africa, and Asia. There are actually more trees in the U.S. now than there were in colonial times and in the later 1800s, although the same cannot be said for European countries.

The total quantity of carbon dioxide released by the human activities of deforestation and burning fossil fuels is 6.0–8.2 Gt per year. About half of this is recycled into the oceans and the biosphere. The remainder stays in the atmosphere as CO_2, adding between 3.1 and 3.5 Gt of carbon per year to the existing base of 750 Gt noted in Table 3.1. We are concerned primarily with this *increase* in atmospheric carbon dioxide, because the *excess* carbon dioxide is implicated in global warming. Therefore, it would be useful to know the mass (Gt) of CO_2 added to the atmosphere each year. In other words, what mass of CO_2 contains 3.3 Gt, the midpoint between 3.1 Gt and 3.5 Gt, of carbon? Other books might just give the answer, 12 Gt. But here the urge to work through a thoughtful explanation of the 12 Gt is too great. Rather than make the assertion and leave it at that, we will demonstrate where that number comes from. This will require another scenic tour into the land of chemistry, one you will find useful elsewhere in this text.

- Remember that the natural "greenhouse effect" makes life on Earth possible. Problems occur when the amount of greenhouse gases *increases* faster than the sinks can accommodate this increase. This is sometimes called the enhanced greenhouse effect.

3.7 Weighing the Unweighable

To solve the problem just posed, we need to know the mass fraction or mass percent of carbon in carbon dioxide. This information can be obtained experimentally by burning a weighed sample of carbon in oxygen and capturing and weighing the carbon dioxide formed. Alternatively, we could decompose a known mass of CO_2 and weigh the carbon and oxygen formed. A third approach, the one we use, is to calculate an answer based on the formula of the compound. In doing this calculation, we illustrate some fundamental chemical concepts, including the conservation of matter. Regardless of the source of carbon dioxide, its formula is stubbornly the same, CO_2. Thus the mass percent of carbon in CO_2 is also unwavering. As you work through Sections 3.7–3.9, keep in mind that we are seeking a value for that percentage.

The procedure requires the use of the atomic masses of the elements involved. But this raises an important question: How much does an individual atom weigh? Recall from Chapter 2 that most of the mass of an atom is attributable to the neutrons and protons in the nucleus. Thus, elements differ in atomic mass because their atoms differ in composition. Rather than use absolute masses of individual atoms, chemists have found it convenient to employ relative atomic masses—in other words, to relate all atomic masses to some convenient standard. The internationally accepted atomic mass standard is carbon-12, the isotope that makes up 98.9% of all carbon atoms. Carbon-12 has a mass number of 12 because each atom has a nucleus consisting of 6 protons and 6 neutrons plus 6 electrons outside the nucleus. The mass of one of these atoms is arbitrarily assigned a value of exactly 12 atomic mass units (amu). We can thus define the **atomic mass** of an element as the average mass of an atom of that element as compared to an atomic mass of exactly 12 amu for carbon-12. Because atoms are so very small, an atomic mass unit is an extremely small unit: 1 amu $= 1.66 \times 10^{-24}$ g.

You will note that the periodic table in your text shows that the atomic mass of carbon is 12.011, not 12.000. This is not an error; it reflects the fact that carbon exists naturally as three isotopes. Although C-12 predominates, 1.1% of carbon is C-13, with six protons and *seven* neutrons per atom and an isotopic mass very close to 13. In addition, natural carbon contains a trace of C-14, whose nuclei consist of six protons and *eight* neutrons. The tabulated atomic mass value of 12.011 is a weighted average that takes into consideration the masses of the three isotopes of carbon and their natural abundances. This isotopic distribution and this average atomic mass characterize carbon obtained from any chemical source—a graphite ("lead") pencil, a tank of gasoline, a loaf of bread, or a lump of limestone.

Carbon-14 is a radioactive isotope that plays a key role in determining the origin of the increasing atmospheric carbon dioxide. In all living things, one out of 10^{12} carbon atoms is a C-14 atom. A plant or animal constantly exchanges CO_2 with the environment, and this maintains the C-14 concentration in the organism at a constant level. However, when the organism dies, the carbon-14 is no longer replenished as it undergoes radioactive decay to form nitrogen-14. This means that after the death of the organism, the concentration of C-14 decreases with time. Coal and oil are the fossilized remains of plant life that died millions of years ago. Hence, the level of C-14 is extremely low in fossil fuels, and in the carbon dioxide released when fossil fuels burn. Experiments show that there has been a recent decrease in the concentration of C-14 in atmospheric CO_2. This strongly suggests that the origin of the added carbon dioxide is indeed the burning of fossil fuels, a decidedly human activity.

Most of the atomic masses listed in the periodic table are experimentally obtained values that correspond to averages reflecting individual isotopic masses and the natural distribution of those isotopes. Even so, the atomic masses of many elements closely approximate whole numbers. There are two reasons for this. In the first place, the nucleus of any atom consists of a whole number of particles. Each of these neutrons and protons has a relative mass of very nearly one atomic mass unit. Second, for many elements one isotope dominates and is by far the most plentiful. Therefore, the experimentally determined atomic mass will be close to the sum of protons and neutrons in the most plentiful isotope. For example, this is the case with nitrogen (N). The most abundant isotope of nitrogen is N-14, with seven protons and seven neutrons in each atomic nucleus. The mass number of this isotope is thus 14, a number that corresponds closely to the average atomic mass of 14.0067 for the naturally occurring isotopic mixture. Other elements with atomic masses close to whole numbers are oxygen (O, atomic mass 15.9994), argon (Ar, 39.948), uranium (U, 238.03), and as we have seen, carbon. When the tabulated atomic mass differs significantly from a whole number, as it does for chlorine (Cl, 35.453) or copper (Cu, 63.546), it indicates that the natural distribution involves sizeable concentrations of two or more isotopes. For some of the artificially produced heavy elements at the end of the periodic table, the atomic mass reported is the mass number of the most plentiful isotope.

- Number of protons + number of neutrons = mass number; see Section 2.2 to review this relationship.

3.17 Your Turn

Gold (Au) has an atomic mass of 196.967 and an atomic number of 79.

a. What is the number of protons, neutrons, and electrons in a neutral atom of the most common isotope, Au-197?

b. Gold-198 is a naturally radioactive isotope that is medically useful in diagnosing liver function. How does the number of protons, neutrons, and electrons in a neutral atom of Au-198 compare with that of Au-197?

3.18 Your Turn

Lithium (Li) has an atomic mass of 6.941. The isotope Li-7 makes up 92.5% of naturally occurring lithium. Lithium has only one other naturally occurring isotope. Predict the mass number of this isotope. Explain your answer.

With that bit of digression about isotopes and their relation to atomic masses as background information, we return to the matter at hand—the masses of atoms and particularly the atoms in CO_2. Not surprisingly, it is impossible to weigh a single atom because of its extremely small mass. A typical laboratory balance can detect a minimum mass of 0.1 mg; that corresponds to 5×10^{18} carbon atoms, or 5,000,000,000,000,000,000 carbon atoms. An atomic mass unit is far too small to measure in a conventional chemistry laboratory. The gram is the chemist's mass unit of choice. Therefore, scientists use exactly 12 g of carbon-12 as the reference for the atomic masses of all the elements. **Atomic mass** (or atomic weight) *can thus be alternately defined as the mass (in grams) of the same number of atoms that are found in exactly 12 g of carbon-12.* The number of atoms in exactly 12 g of C-12 is called *Avogadro's number*, after an Italian scientist with the impressive name of Lorenzo Romano Amadeo Carlo Avogadro di Quaregna e di Ceretto (1776–1856). (His friends called him Amadeo.) Note that Avogadro's name consists of 52 letters. His number, if written out, requires 24 numerals: 602,000,000,000,000,000,000,000. It is more compactly written in scientific notation as 6.02×10^{23}.

Avogadro's number may be shorter than his name, but it is so large that about the only way to hope to comprehend it is through analogies. For example, one Avogadro's number of regular-sized marshmallows, 6.02×10^{23} of them, would cover the surface of the United States to a depth of 650 miles. Or, if you are more impressed by money than marshmallows, assume 6.02×10^{23} pennies were distributed evenly among the more than 6 billion inhabitants of the Earth. Every man, woman, and child could spend one million dollars every hour, day and night, and half of the pennies would still be left unspent at death. And remember, this is the number of atoms in 12 g of carbon—a tablespoonful of soot.

Because one Avogadro's number of carbon-12 atoms has a mass of exactly 12 g, the mass of an equal number of oxygen atoms should correspond to the atomic mass of oxygen (Figure 3.18). All we need to do is count out 6.02×10^{23} atoms and weigh them. We could use some help with this assignment, so suppose we enlist all the human beings currently alive and set them each counting—one oxygen atom per second per person, 24 hours a day, 365 days a year. Even at that rate, things do not look good. It would take almost 4 million years for all of us to complete the job. Fortunately, we do not need to count the atoms. There are fairly accurate ways of estimating the number. And remember, we can be off by 5×10^{18} atoms and never notice the difference in mass on a laboratory balance! Without going into the experimental details, the result is that one Avogadro's number of oxygen atoms weighs 15.9994 g. This means that the atomic mass of oxygen is 15.9994. This value is consistent with the structure of the oxygen atom. By far the most common isotope of oxygen is O-16, with atoms consisting of 8 protons and 8 neutrons.

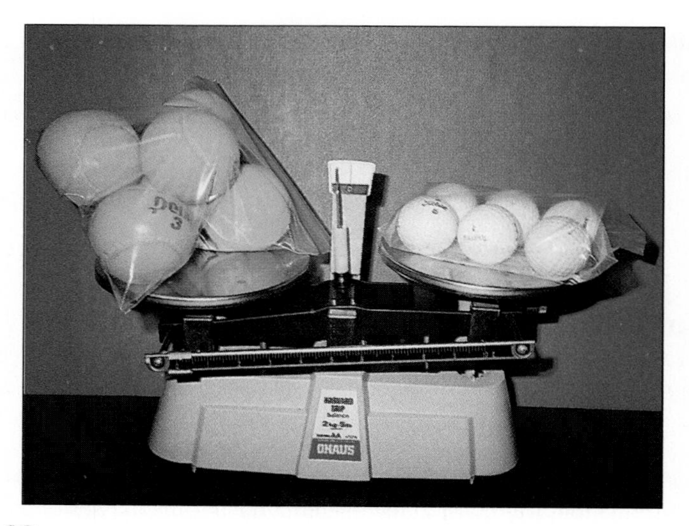

Figure 3.18

Relative masses. Like atoms of different elements, the masses of a tennis ball and a golf ball differ. Six tennis balls have a greater mass than six golf balls. The number of balls is the same in both cases—six in each bag.

A knowledge of Avogadro's number and the atomic mass of any element permits us to calculate the average mass of an atom of that element. Thus, the mass of 6.02×10^{23} oxygen atoms is 15.9994 g. To find the mass of one oxygen atom, we simply divide.

$$\frac{15.9994 \text{ g oxygen}}{6.02 \times 10^{23} \text{ oxygen atoms}} = \frac{2.66 \times 10^{-23} \text{ g oxygen}}{\text{oxygen atom}}$$

With few exceptions, chemists never work with individual atoms and molecules. We manipulate trillions at a time. Therefore, practitioners of this art need to measure matter with a sort of chemist's dozen—a very large one, indeed. To learn about it, read on . . . but only after stopping to practice your new skill.

3.19 Your Turn

a. Calculate the average mass (in grams) of an individual atom of nitrogen.
b. Calculate the mass (in grams) of five trillion nitrogen atoms.
c. Calculate the mass (in grams) of 6×10^{15} nitrogen atoms.

Answer

a. $$\frac{14.0067 \text{ g nitrogen}}{6.02 \times 10^{23} \text{ nitrogen atoms}} = \frac{2.33 \times 10^{-23} \text{ g nitrogen}}{\text{nitrogen atom}}$$

3.8 Of Molecules and Moles

An Avogadro's number, 6.02×10^{23} of anything, is called a **mole,** a term derived from the Latin word to "heap" or "pile up." It is a chemist's way of counting. Usually, the mole is used to count atoms, molecules, electrons, or other small particles. Thus, one mole of carbon atoms consists of 6.02×10^{23} C atoms, one mole of oxygen gas (O_2) is made up of 6.02×10^{23} oxygen molecules, and 1 mole of carbon dioxide molecules corresponds to 6.02×10^{23} carbon dioxide molecules.

Moles are fundamental to chemistry because chemistry involves the interaction of individual atoms and molecules. As you already know from Chapter 1, chemical formulas and equations are written in terms of atoms and molecules. For example, reconsider the equation for the reaction of carbon and oxygen.

$$C + O_2 \longrightarrow CO_2$$

In Chapter 1, we interpreted this expression as stating that one atom of carbon combines with one molecule of diatomic oxygen to yield one molecule of CO_2. The equation reflects the ratio in which the particles interact. Thus, it would be equally correct to say that 10 carbon atoms react with 10 oxygen molecules (20 oxygen atoms) to form 10 carbon dioxide molecules. Or, putting the reaction on a grander scale for that matter, we could say 6.02×10^{23} C atoms combine with 6.02×10^{23} O_2 molecules (12.0×10^{23} oxygen atoms) to yield 6.02×10^{23} CO_2 molecules. The last statement is equivalent to saying: "one *mole* of carbon plus one *mole* of diatomic oxygen yields one *mole* of carbon dioxide." The point is that *the numbers of atoms and molecules taking part in a reaction are proportional to the numbers of moles of the same substances.* The atomic ratio reflected in a chemical formula or the molecular ratio in a molecular equation is identical to the molar ratio. Accordingly, just as there are two oxygen *atoms* for each carbon *atom* in a CO_2 *molecule*, there are also two *moles* of oxygen atoms for each *mole* of carbon atoms in a *mole* of carbon dioxide. The ratio of two oxygen atoms to one carbon atom remains the same regardless of the number of carbon dioxide molecules, as summarized in Table 3.3.

In the laboratory and the factory, the quantity of matter required for a reaction is usually measured by mass or weight. The mole is a way to simplify matters (and matter) by relating number of particles and mass. Central to this approach is **molar mass,** defined as the mass of one Avogadro's number of whatever particles are specified. In chemistry, molar masses are almost always expressed in grams. Thus, the mass of a mole of carbon atoms, rounded to the nearest tenth of a gram, is 12.0 g. Similarly, a mole of

Table 3.3

Counting Atoms and Molecules
$C + O_2 \longrightarrow CO_2$

# Carbon Atoms	# Oxygen Atoms	# Carbon Dioxide Molecules
1	2	1
2	4	2
5	10	5
10	20	10
100	200	100
1000	2000	1000
6.02×10^{23} (1 mole)	$2\ (6.02 \times 10^{23})$ (2 moles)	6.02×10^{23} (1 mole)

oxygen atoms has a mass of 16.0 g. But we can also speak of a mole of O_2 molecules. Because there are two oxygen atoms in each oxygen molecule, there are two moles of oxygen atoms in each mole of molecular oxygen, O_2. Consequently, the molar mass of O_2 is 32.0 g—twice the molar mass of O. Some books refer to this as the molecular mass or molecular weight of O_2, emphasizing its similarity to atomic mass or atomic weight.

The same logic for the molar mass of O_2 applies to compounds of two or more elements, which brings us, at last, to the composition of carbon dioxide. The formula, CO_2, and the molecular structure reveal that each molecule contains one carbon atom and two oxygen atoms. Scaling up by 6.02×10^{23}, we can say that each mole of CO_2 consists of 1 mole of C and 2 moles of O atoms (Table 3.3). But remember that we are interested in the mass composition of carbon dioxide—the number of grams of carbon per gram of CO_2. This requires the molar mass of carbon dioxide, which we obtain by adding the molar mass of carbon to twice the molar mass of oxygen:

$$\text{Molar mass } CO_2 = (1 \times \text{molar mass C}) + (2 \times \text{molar mass O})$$

Substituting numerical values for the molar masses of the elements gives the desired result.

$$\left(1\ \cancel{\text{mole C}} \times \frac{12.0\text{ g C}}{1\ \cancel{\text{mole C}}} = 12.0\text{ g C}\right)$$

$$+\left(2\ \cancel{\text{mole O}} \times \frac{16.0\text{ g O}}{1\ \cancel{\text{mole O}}} = 32.0\text{ g O}\right)$$

$$\text{Molar mass } CO_2 = 44.0\text{ g/mole } CO_2$$

This procedure is routinely used in chemical calculations, where molar mass is an important property. Some examples are included in the next activity. In every case, you count the number of moles of the constituent elements in one mole of the compound, multiply the number of moles of each element by the corresponding elementary molar mass (the atomic mass in grams), and add the result.

3.20 Your Turn

Calculate the molar mass of each of these substances important in atmospheric chemistry.

a. O_3 (ozone)
b. NO (nitrogen monoxide or nitric oxide)
c. Freon-11; trichlorofluoromethane

Answer

a. $\frac{16.0 \text{ g O}}{1 \text{ mol O}} \times \frac{3 \text{ mol O}}{1 \text{ mol } O_3} = \frac{48.0 \text{ g O}}{1 \text{ mol } O_3}$

- When used with a number, the term "mole" is commonly abbreviated "mol".

3.9 Manipulating Moles and Mass with Math

You may recall that several pages ago we set out to calculate the mass of carbon dioxide that could be produced from burning 3.3 Gt of carbon. We at last have all the pieces necessary to solve the problem. To do so, we use the quantitative compositional information implicit in the formula of a compound, in this case CO_2. Because 44.0 g CO_2 contain 12.0 g C, we can readily find the ratio (by mass) of carbon to carbon dioxide. Out of every 44.0 g of CO_2, 12.0 g are C and the remaining 32.0 g oxygen. The carbon-to-carbon dioxide ratio is $\frac{12.0 \text{ g C}}{44.0 \text{ g } CO_2}$. This mass ratio holds for all samples of carbon dioxide, and we can use it to calculate the mass of carbon in any known mass of carbon dioxide. For example, we could compute the number of grams of C in 100.0 g CO_2 by setting up a proportion in this manner.

- The question to be answered is: What mass of CO_2 contains 3.3 Gt of carbon?

$$\frac{? \text{ g C}}{100.0 \text{ g } CO_2} = \frac{12.0 \text{ g C}}{44.0 \text{ g } CO_2}$$

Next, rearrange the equation using standard algebraic procedures (This process is referred to as "cross multiplication").

$$? \text{ g C} \times 44.0 \text{ g } CO_2 = 12.0 \text{ g C} \times 100.0 \text{ g } CO_2$$

Solving for the number of grams C yields the desired result.

$$? \text{ g C} = \frac{12.0 \text{ g C} \times 100.0 \text{ g } CO_2}{44.0 \text{ g } CO_2} = 27.3 \text{ g C in } 100.0 \text{ g } CO_2$$

Note that carrying along the labels, "g CO_2" and "g C," helps you do the calculation correctly. In the center part of the expression, "g CO_2" appears in the top (numerator) and the bottom (denominator). Hence, they can be canceled, and you are left with the desired label, "g C." This is a useful strategy in solving many problems. The fact that there are 27.3 grams of carbon in 100.0 grams of carbon dioxide is equivalent to saying that the mass percent of C in CO_2 is 27.3%. Alternatively, CO_2 is 27.3% C by mass.

To find the mass of carbon dioxide that contains 3.3 gigatons (Gt) of carbon, we use the same mass ratio and a similar approach. We could convert 3.3 Gt to grams, but it is not necessary. As long as we use the same units for the mass of C and the mass of CO_2, the same numerical ratio holds and is still $\frac{12.0 \text{ Gt C}}{44.0 \text{ Gt } CO_2}$. But there is one important difference, this time we are solving for the mass of CO_2, not the mass of C.

$$\frac{3.3 \text{ Gt C}}{? \text{ Gt } CO_2} = \frac{12.0 \text{ Gt C}}{44.0 \text{ Gt } CO_2}; \ ? \text{ Gt } CO_2 \times 12.0 \text{ Gt C} = 3.3 \text{ Gt C} \times 44.0 \text{ Gt } CO_2$$

$$? \text{ Gt } CO_2 = \frac{3.3 \text{ Gt C} \times 44.0 \text{ Gt } CO_2}{12.0 \text{ Gt C}} = 12 \text{ Gt } CO_2 \text{ from } 3.3 \text{ Gt C}$$

Once again the labels cancel and the answer comes out in the desired form, Gt CO_2.

Our innocent question, "What is the mass of carbon dioxide added to the atmosphere each year from burning fossil fuels?" has finally been answered: 12 gigatons. Of course, our not-so-hidden agenda was to demonstrate the problem-solving power of chemistry and to introduce five of its most important ideas: atomic mass, molecular mass, Avogadro's number, mole, and molar mass. The next few activities provide opportunities to practice your skill with these concepts and manipulations.

3.21 Your Turn

a. Calculate the mass ratio of sulfur (S) in sulfur dioxide (SO_2).
b. Find the mass percent of S in SO_2.
c. Calculate the mass ratio and the mass percent of N in N_2O.

Answers

a. The mass ratio is found by comparing the number of grams of sulfur to the molar mass of SO_2.

$$\frac{32.1 \text{ g S}}{64.1 \text{ g } SO_2} = \frac{0.501 \text{ g S}}{1.00 \text{ g } SO_2} \text{ is the mass ratio.}$$

b. To find the mass percent, multiply the mass ratio by 100.

$$\frac{0.501 \text{ g S}}{1.00 \text{ g } SO_2} \times 100 = 50.1\% \text{ S}$$

3.22 Your Turn

a. It is estimated that volcanoes globally release about 19 million metric tons of SO_2 per year. Calculate the mass of sulfur in this amount of SO_2.
b. If 142 million metric tons of SO_2 are released per year by fossil fuel combustion, calculate the mass of sulfur released.

Answer

a. The mass ratio of S to SO_2 is known from 3.21 Your Turn, so it is the easiest way to approach this problem.

$$\frac{32.1 \times 10^6 \text{ metric tons S}}{64.1 \times 10^6 \text{ metric tons } SO_2} \times 19 \times 10^6 \text{ metric tons } SO_2 = 9.5 \times 10^6 \text{ metric tons S}$$

Note: There is no need to change the unit of SO_2 from million metric tons to grams or any other unit. The mass ratio expressed in grams to grams is in the same proportion as if expressed in any other comparable unit, as long as the unit is the same.

If you know how to apply these ideas, you have gained the ability to evaluate critically media reports about releases of carbon or carbon dioxide (and other substances as well) and judge their accuracy. One can either take such statements on faith or check their accuracy by applying mathematics to the relevant chemical concepts. Obviously, there is insufficient time to check every assertion, but we hope that readers develop questioning and critical attitudes toward all statements about chemistry and society, even those found in this book.

3.23 Sceptical Chymist: Checking Carbon from Cars

A clean-burning automobile engine will emit about five pounds of carbon in the form of carbon dioxide for every gallon of gasoline it consumes. The average American car is driven about 12,000 miles per year. Using this information, check the statement that the average American car releases its own weight in carbon into the atmosphere each year. Note the assumptions you make in solving this problem. Compare your list, and your answer, with those of your classmates.

3.10 Methane and Other Greenhouse Gases

Concerns about global warming are based primarily, but not solely on increases in atmospheric CO_2. There are several other gases of concern, all of which have been increasing because of human activities. For example, methane, CH_4, is in much lower concentration in the atmosphere than CO_2. Methane is also approximately 30 times more effective than CO_2 in its ability to trap infrared energy. Fortunately, CH_4 is quite readily converted to less harmful chemical species and has a relatively short average atmospheric lifetime of 12 years. This means that most of the CH_4 added to the air in a given year will be gone from the atmosphere 12 years later. Compare that situation with CO_2, a gas with a lifetime ranging from decades to hundreds—and some say thousands—of years. Atmospheric concentration of CH_4 at this time is relatively low, but its current

Table 3.4

Changes in Greenhouse Gases Since Preindustrial Times

	CO_2	CH_4	N_2O
Preindustrial conc.	280 ppm	0.70 ppm	0.28 ppm
2000 conc.	370 ppm	1.8 ppm	0.31 ppm
Rate of conc. change	1.5 ppm/yr	0.010 ppm/yr	0.0008 ppm/yr
Atmospheric lifetime (yr)	50–200	12	120

level of 1.8 ppm is estimated to be more than twice that before the Industrial Revolution. Table 3.4 gives a comparison of changes in methane concentration with those of carbon dioxide and nitrous oxide, N_2O.

- The compound N_2O has the common name nitrous oxide. Its proper (systematic) name is dinitrogen monoxide.

Methane comes from a wide variety of sources. Most of them are natural, but they have been magnified by human activities. For example, because CH_4 is a major component of natural gas, some has always leaked into the atmosphere from rock fissures. But the exploitation of these deposits and the refining of petroleum have led to increased emissions. Similarly, CH_4 has always been released by decaying vegetable matter. Its early name, "marsh gas," reflects this origin. Any human activity that contributes to similar conditions leads to increased CH_4 release. Thus, the decaying organic matter in landfills and from the residue of cleared forests generates CH_4. Methane formed in the main New York City landfill is used for residential heating, but at most landfills it simply escapes into the atmosphere.

Another major source of CH_4 is agriculture, particularly cultivated rice paddies. Rice is grown with its roots under water where anerobic ("oxygen-poor") bacteria produce methane, which is released into the atmosphere. Additional agricultural methane comes from an increasing number of cattle and sheep. The digestive systems of these ruminants contain bacteria that break down cellulose. In the process, methane is formed and released through belching and flatulence—about 500 L per cow per day. The ruminants of the Earth release a staggering 73 million metric tons of methane each year. Even termites, which carry on similar chemistry in their guts, generate methane. And there is more than half a ton of termites for every man, woman, and child on the planet.

There is a possibility that global warming may exacerbate the release of methane from ocean mud, bogs, peat lands, and even the permafrost of northern latitudes. In these areas, a substantial amount of methane appears to be trapped in "cages" made of water molecules. The methane trapped in this way is referred to as methane hydrate. As the temperature increases, the escape of CH_4 becomes more likely. The Commonwealth Scientific and Industrial Research Organization (CSIRO) has been conducting a series of ocean core drillings to gather evidence about methane hydrate and its role in global warming. Its findings link periods of historic global warming with the release of methane (Figure 3.19).

The complex details of the generation and fate of atmospheric methane make it difficult to speak with certainty about its future effect on the average temperature of the planet. Current predictions are that methane's effect will be less pronounced than temperature changes caused by CO_2, with methane adding only perhaps a few tenths of a degree to the average temperature of the Earth. This is in sharp contrast with the major effect predicted for CO_2, expected to cause a temperature rise of 1.0–3.5°C by the end of this century.

Nitrous oxide, also known as "laughing gas", is used as an inhaled anesthetic for dental and medical purposes. In the atmosphere, it is less useful. There, a typical N_2O molecule persists for about 120 years, absorbing and emitting infrared radiation. Over the past decade, atmospheric concentrations of the compound have shown a slow but steady rise. Major anthropogenic sources of N_2O are synthetic fertilizers and the burning of biomass. The majority of N_2O molecules in the atmosphere come from the

Figure 3.19
Icy methane (frozen methane hydrate). CSIRO researchers found evidence to support a theory that an abrupt warming of the Earth 55 million years ago was caused by the sudden release of previously frozen methane from the ocean floor. The Ocean Drilling Program obtained this sample of still frozen methane hydrate from the continental shelf off the coast of Florida.

bacterial removal of nitrogen from soils. Agricultural practices, again linked to population pressures, can exacerbate this nitrogen removal from soils.

In addition to its role in the greenhouse effect, nitrous oxide contributes to stratospheric ozone depletion, as discussed in Chapter 2. Near the surface of the Earth, however, the reactions of nitrogen oxides and hydrocarbons (like methane) lead to the production of ozone. Ozone itself can also act like a greenhouse gas, but its efficiency depends very much on altitude. It appears to have its maximum warming effect in the upper troposphere (around 10 km). Depletion of ozone has a cooling effect in the stratosphere and it may also promote slight cooling at the surface of the Earth. Chlorofluorocarbons (CFCs), already implicated in the destruction of stratospheric ozone, also absorb infrared radiation.

One important consideration in all of this is that not all greenhouse gases are equally effective in absorbing infrared radiation. This effectiveness is quantified by the **greenhouse factor,** a number that represents the relative contribution of a molecule of the indicated substance to global warming. Carbon dioxide is assigned the reference value of 1; all other greenhouse gases are indexed with respect to it. Values of the greenhouse factor for seven common atmospheric trace gases and their average concentrations in the troposphere are given in Table 3.5. Note from the table that one molecule of the

Table 3.5

Greenhouse Factors for Some Common Atmospheric Trace Constituents

Substance	Greenhouse Factor	Tropospheric Abundance (%)
CO_2	1 (assigned value)	3.7×10^{-2}
CH_4	30	1.8×10^{-4}
N_2O	160	3.1×10^{-5}
O_3	2000	4.0×10^{-6}
CCl_3F	21,000	2.6×10^{-8}
CCl_2F_2	25,000	5.2×10^{-8}

chlorofluorocarbon CCl_2F_2 has the same global warming effect as 25,000 CO_2 molecules. However, the tropospheric abundances of most highly effective absorbers of IR radiation are very low.

Figure 3.20 shows the relative amounts of carbon equivalents contributed by the different sources of all greenhouse gas emissions in the U.S.

3.24 Your Turn

a. Is the tropospheric abundance of methane given in Table 3.5 the same as that in Table 3.4? Show your reasoning.

b. The tropospheric abundance of ozone is 4.0×10^{-6}%. What is the concentration expressed in ppm? Again, show your work.

c. Comment on why the values in Table 3.5 are expressed in percent tropospheric abundance rather than ppm.

3.11 Climatic Modeling

The previous paragraphs have merely hinted at the complexity of atmospheric chemistry. To accurately model global climate, one must also include a number of often incompletely understood astronomical, meteorological, geological, and biological factors. Among these are variations in the intensity of the Sun's radiation as a consequence of sunspot activity, winds and air circulation patterns, cloud cover, volcanic activity, dust and soot, aerosols, shifting sea ice and glaciers, the oceans, and the extent and nature of living things, especially human beings. Urban areas create their own "heat islands" further increasing the difficulty of climate modeling on a local rather than global scale. The situation is greatly complicated by the fact that many of these variables are interrelated and cannot be studied independently as in a controlled experiment. Dr. Michael

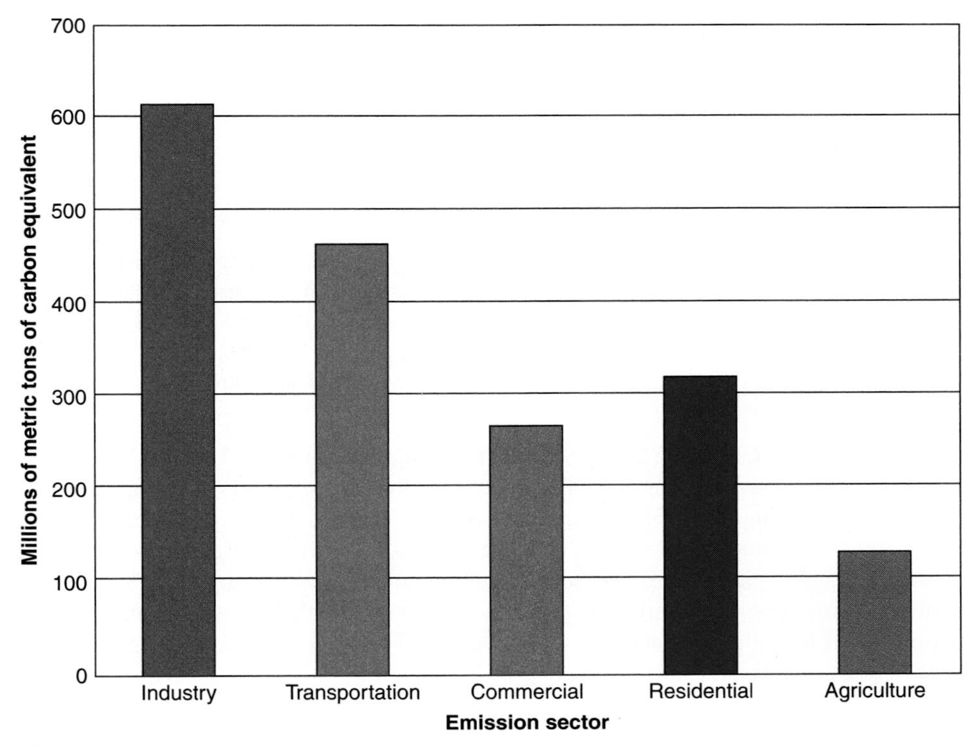

Figure 3.20

Greenhouse gas emission sources in the U.S., 1998.

(Source: U.S. EPA.)

Schlesinger, who directs climate research at the University of Illinois, remarked: "If you were going to pick a planet to model, this is the *last* planet you would choose." Despite all these difficulties, policy decisions must be made on the best possible models. The usefulness and limitations of all models needs to be clearly understood and the level of uncertainty in them quantified. These are tall tasks, to be sure, but essential ones for making informed assessments of climate change.

For example, we know from the solubility properties of most gases that increasing the temperature of the oceans will decrease the solubility of CO_2, thus releasing more of it into the atmosphere. An increase in the temperature of the oceans may promote the growth of tiny photosynthetic plants called phytoplankton, and hence increase CO_2 absorption. But the result could be just the opposite. Water in a warmer ocean will not circulate as well as it does now, which may inhibit plankton growth and CO_2 fixing. Decreased snow and ice cover, which would attend global warming, would lower the amount of sunlight reflected from the Earth's surface. The resultant increase in absorbed radiation would promote a further increase in temperature.

A warmer Earth would presumably mean that the tree line would move north, bringing with it added CO_2 absorbing capacity. Countering this, related reductions in rainfall might turn areas that are currently covered by vegetation into deserts, thus reducing carbon dioxide absorption. Global warming would also cause more water to evaporate, increasing the average relative humidity and thus adding to the greenhouse effect. More clouds would form, but their influence cannot be generalized. It seems that high clouds contribute little to the greenhouse effect and reflect sufficient sunlight so that they have a net cooling effect on the surface of the Earth. Low clouds have a net warming effect.

In spite of such formidable problems and sometimes countervailing effects, scientists are developing computer programs to model the Earth's climate. As supercomputers have become more powerful, models have become more sophisticated. The oceans are represented as a multilayer circulating system and the model atmosphere is assumed to contain 10 or more interacting layers. Typically, the surface of the planet is divided into about 10,000 cells, not enough to provide detailed predictions, but sufficient to include general patterns of weather development. One test of these simulations of global climate is how well they predict the 0.6 ± 0.2°C temperature increase observed over the last century when CO_2 concentrations increased by 25–30%. Most models estimate a temperature increase of about twice that actually measured. This suggests that certain relevant factors may have been omitted or that some variables may have been incorrectly weighted.

3.25 Consider This: Climate Questions

If you visit climate-modeling sites on the Web, you may be deluged with technical terms and numerical analyses. A good place to begin your understanding of climate modeling is to visit the National Climatic Data Center (NCDC), billed as "the world's largest active archive of weather data." A direct link to the NCDC is provided at the *Chemistry in Context* web site. What types of data do the NCDC provide? Propose two or three questions that you might like to investigate using data provided by NCDC.

One group of researchers, led by Benjamin Santer of Lawrence Livermore National Laboratory, has found that predictions agree more closely with observations if the model includes the cooling effect of atmospheric aerosols. These aerosols consist primarily of tiny particles of ammonium sulfate, $(NH_4)_2SO_4$, that form from sulfur dioxide released by natural or artificial sources. These particles promote global cooling by reflecting and scattering sunlight. In addition, they serve as nuclei for the condensation of water droplets and hence cloud formation. Thus, aerosols counter the effects of greenhouse gases. The temporary drop in average global temperature that followed the eruption of Mount Pinatubo in 1991 may well have been the consequence of the large volume of sulfur dioxide released by the volcano. The Third Assessment Report issued in early 2001 by the Intergovernmental Panel on Climate Change (IPCC) reaffirms the conclusion that the evidence strongly supports the position that human activity is the cause of the increase in average global temperature observed over the last century. Figure 3.21 shows how the current data match the most complete computer models.

Much of the heat radiated by the greenhouse gases may be going into the oceans, which act as a thermal buffer. Some climatic models may underestimate the amount of heat absorbed by the oceans. Although the oceans are very important in moderating the temperature of the planet, there are limits to their capacity to do so. It is also instructive that some climatologists have found that if greenhouse gases are *omitted* from their computer models, predictions *underestimate* observed temperature increases.

Given the complexity of the global system, considerable uncertainty is associated with climate models. There is no wonder, then, that experts sometimes disagree. First of all is the matter of projected levels of greenhouse gases. The *rate* of their emission currently increases by about 1.5% per year. This is largely a consequence of growing global population, agricultural production, and industrialization. The population of the planet has tripled in this century and it is expected to double or triple again before reaching a plateau sometime in the next century. Industrial production is 50 times what it was 100 years ago. In the next 50 years, it will probably grow to 5 or 10 times what it is today. Most of this growth has been powered by the combustion of fossil fuels. Every year,

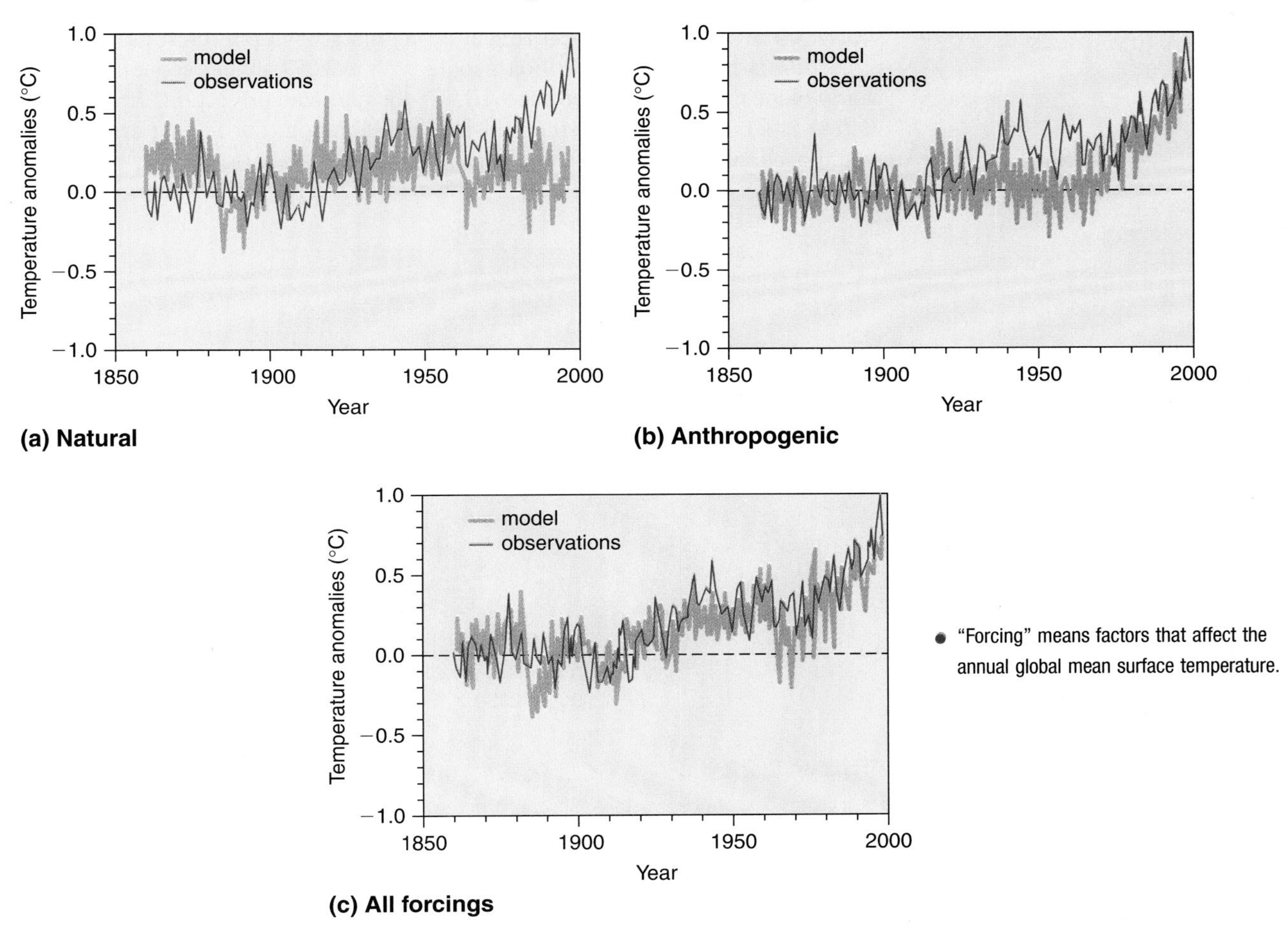

• "Forcing" means factors that affect the annual global mean surface temperature.

Figure 3.21

Simulated annual global mean surface temperatures. Simulating the Earth's temperature variations and comparing the results to measured changes can provide insight into the underlying causes of the major changes.

(Source: Intergovernmental Panel on Climate Change (IPCC) Report, 2001)

2–3% more energy is generated than in the previous one, and most of it comes from burning coal. If these rates of fuel consumption continue, the atmospheric concentration of CO_2 will be twice its 1860 level sometime between the years 2030 and 2050.

All models predict that this doubling will increase the average global temperature, but the magnitude of that increase is variously estimated between 1.4 and 4.8°C. Many predictions, including the most recent one by the 2001 Intergovernmental Panel on Climate Change (IPCC), fall in the range of 1.4 and 5.8°C (2.5 and 10.4°F) by the end of the century. Such an increase in average global temperature might seem trivial, but it is not. A temperature drop of 5–9°F (2.8–5.0°C) is the difference between the current average global temperature and that of a much chillier epoch, the last ice age, about 20,000 years ago.

Projected increases of CO_2 developed by the IPCC are shown in Figure 3.22. This graph depicts atmospheric carbon dioxide concentrations since 1990 and several projections of CO_2 concentrations. The projections are based on differing scenarios, each assuming that no dedicated efforts are made to diminish greenhouse gas emissions. Because increased numbers of people translate into increased energy use and greater greenhouse gas emissions through burning fossil fuels, the scenarios incorporate population growth rates. A significant uncertainty in any such projections is trying to predict whether the rate of population growth will stabilize during the next hundred years; during the 20th century worldwide population increased from two billion people to approximately six billion today. The low-end scenario assumes a world population in the year 2100 of 6.4 billion and an annual economic growth rate of 1.2%. The mid-range projection is based on 11.3 billion people with a 2.3% annual economic growth rate, nearly twice that of the low-end scenario. In the high-end projection, the year 2100 population is 11.3 billion as in the mid-range calculation, but annual economic growth would occur at 3.0%.

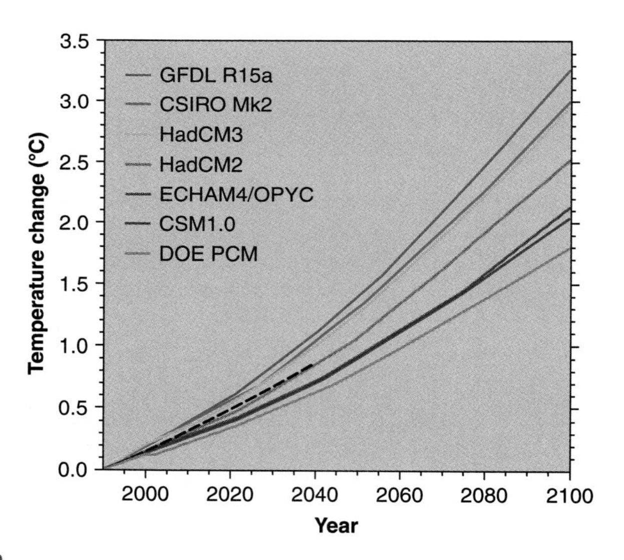

Figure 3.22

Projecting CO_2 concentrations. Several different models have been created by various teams in several countries to make climate change projections. The heavy dashed line and blue plume show the median and 5–95% range of anthropogenic warming 1991–2041 under a high-end scenario. (The high-end scenario predicts a population of 11.3 billion in 2100 and an annual economic growth rate of 3.0%.) Other lines are the result of projections produced by other countries or agencies. All models seek to reconcile complex climate simulations with recent observed climate change information.

GFDL is a geophysical fluid dynamic laboratory model from NOAA; CSIRO is an Australian model; Had is from the Hadley Centre, Bracknell, UK; ECHAM is a German model from the Max Planck Institute; CSM is a Canadian model; and DOE is a U.S. Department of Energy model. *Science*, Vol. 293, 20 July 2001, p.431.

Table 3.6

Judgmental Estimates of Confidence

Term Used	Probability That a Result is True
Virtually certain	> 99%
Very likely	90–99%
Likely	66–90%
Medium likelihood	33–66%
Unlikely	10–33%
Very unlikely	1–10%

Source: *Summary for Policymakers, A Report of Working Group I of the Intergovernmental Panel on Climate Change*, Shanghai: IPCC, January 1, 2001.

Policymakers must evaluate all climate change data over time to determine what trends are due to natural variability and what are the result of human activity. To help policy makers and the public at large better understand the inherent uncertainty and reliability of the data, the many hundreds of scientists from all over the world who participated in preparing the 2001 IPCC report used words that quantified the probability that their estimates of change are valid. Table 3.6 gives the scientists' definitions.

Using this language, the conclusions reported by the IPCC in 2001 are shown in Table 3.7. One clear effect of observed climatic changes is shown in Figure 3.23.

Table 3.7

IPCC Conclusions

Very Likely	Likely	Very Unlikely
The 1990s were the warmest decade and 1998 the warmest year since 1861.	Temperatures in the Northern Hemisphere during the 20th century are *likely* to have been the highest of any century during the past 1000 years.	The observed warming over the past 100 years is due to climate variability alone, providing new and even stronger evidence that changes must be made to stem the influence of human activities.
Higher maximum temperatures are observed over nearly all land areas.	Arctic sea-ice thickness declined about 40% during late summer to early autumn in recent decades.	
Snow cover decreased about 10% since the 1960s (satellite data); in the 20th century there was a reduction of about two weeks in lake and river ice cover in the mid- and high-latitudes of the Northern Hemisphere. (independent ground-based observations).	An increase in rainfall, similar to that in the Northern Hemisphere, has been observed in tropical land areas falling between 10°N and 10°S.	
Increased precipitation has been observed in most of the Northern Hemisphere continents.	Increased summer droughts are *likely* in a few areas.	

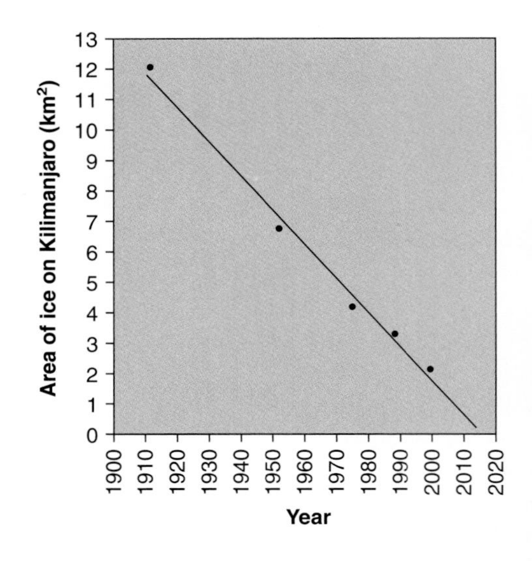

Figure 3.23

The snows of Kilimanjaro. In 1912, Kilimanjaro in Tanzania had significant snow cover on its peak. By 2001, about 82% of the massive ice field atop Kilimanjaro has been lost. If the measured rate of retreat of the snow cover continues unchanged, the snow cover will have disappeared by 2015.

(Source: Science Vol 293 6 July 2001, p 47.)

Changes in sea ice, snow, and glaciers can all contribute to changes in sea level. Global mean sea level is projected to rise by 9–88 cm (3.5–34.6 in.) between 1990 and 2100. Predictions of rising sea levels, which have been scaled back from earlier IPCC predictions of 13–92 cm (5.1–36.2 in.), are caused by thermal expansion of warmer water as well as by melting of frozen precipitation. Sea level rises of 9–88 cm would endanger New York, New Orleans, Miami, Venice, Bangkok, Taipei, and other coastal cities. Millions of people might have to relocate. Dr. Richard Williams of the U.S. Geological Survey states: "If the ice sheets on Greenland melted, that alone could raise sea level by 20 feet." It is far from certain that major increases in sea level will occur. Even if they do, they will take place over many years, providing considerable time for preparation and protection.

Some climatologists also feel that an increase in the average temperature of the oceans could cause more weather extremes including storms, floods, and droughts. In the Northern Hemisphere, summers are predicted to be drier and winters wetter. The regions of greatest agricultural productivity could change. Drought and high temperatures could reduce crop yields in the American Midwest, but the growing range might extend farther into Canada. It is also possible that some of what is now desert could get sufficient rain to become arable. Many of these predictions have not yet been substantiated and some have been called into question. One region's loss may well become another locale's gain, but it is too early to tell with any certainty.

In other respects, we may all be losers in a warmer world. Recently, physicians and epidemiologists have attempted to assess the costs of global warming in terms of public health. An increase in average temperatures is expected to increase the geographical range of mosquitoes, tsetse flies, and other insects. The result could be a significant increase in diseases such as malaria, yellow fever, and sleeping sickness in new areas, including Asia, Europe, and the United States. Indeed, it has been suggested that the deadly 1991 outbreak of cholera in South America is attributable to a warmer Pacific Ocean. The bacteria that cause cholera thrive in plankton. The growth of plankton and bacteria are both stimulated by higher temperatures.

Even more uncertainty is associated with the regional weather patterns predicted by the various models. Most climate models lack enough detail to predict changes in small-scale phenomena. Thunderstorms, tornadoes, hail, and lightning are not included in most climate models. One of the more controversial forecasters is James Hansen of NASA. Hansen has estimated that doubling the concentration of greenhouse gases would mean that New York City could expect 48 days a year with temperatures above 90°F instead of the current 15. In Dallas, the number of days per year with temperatures above 100°F would increase from 19 to 87. It is important to note that many scientists have questioned Hansen's estimates, and they have not been endorsed by IPCC.

3.26 Consider This: Winners and Losers

If significant global warming occurs, some countries will probably be winners and some losers. Identify three nations that would most likely benefit from a warmer Earth and three that would face serious problems. State the reasons for your selections, including the gains and losses you anticipate.

3.12 Has the Greenhouse Effect Already Started?

The answer to this question is most definitely "yes." Recall that without the greenhouse effect the average temperature of the Earth would be about −18°C. Under such circumstances, we would not be here, or perhaps more correctly, we would be very different creatures. But, as generally asked, the question implies, Has the *enhanced* greenhouse effect started on a global basis? The question becomes whether or not the average temperature of the planet increased as a consequence of human activity. Is global warming, the amplification of the greenhouse effect by human activities, occurring? Here the answer is somewhat less certain. Therefore, this is an excellent opportunity for the Sceptical Chymist to exhibit his or her skills of inquiry.

To better address the issues, it would be well for us to review the status of our knowledge. We do so by making some statements and then attempting to assess their accuracy. In this, we are guided by the 2001 report from the IPCC, the Intergovernmental Panel on Climate Change.

1. *Carbon dioxide contributes to an elevated global temperature.* Definitely true and supported by much experimental evidence, including the average temperatures of the Earth and Venus. The mechanism for global warming, the absorption of infrared radiation by vibrating molecules, is well understood and widely accepted.
2. *The concentration of carbon dioxide in the atmosphere has been increasing over the past century.* Definitely true. Analytical data strongly support this statement.
3. *The increase in atmospheric carbon dioxide over the past century is a consequence of human activity.* Very likely. Carbon isotope ratio measurements strongly suggest that at least part of the increase is attributable to human activities such as increased burning of fossil fuels and cutting of forests.
4. *Average global temperature has increased during the past century.* Virtually certain. The data are consistent with this interpretation, indicating a rise of about 0.6°C (±0.2°C) over the 20th century. Computer models as well as experimental data from ice core samples, tree ring measurements, and coral growth rates point to the 20th century as the warmest since 1400. Measurement methods have changed during this period, which might call the conclusion into some question. Also, a century may be too short a time to reveal genuine temperature trends. Annual temperature trends are shown in Figure 3.24.

(a) Annual temperature trends, 1901 to 2000

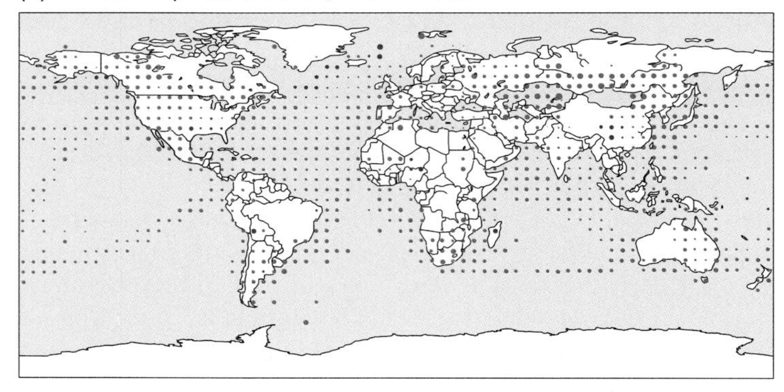

(b) Annual temperature trends, 1976 to 2000

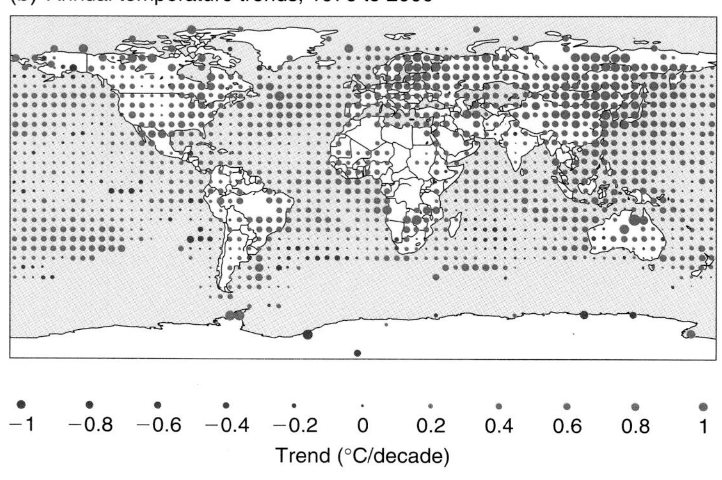

Figure 3.24

Annual temperature trends 1901–2000 and 1976–2000. Annual temperatures that are increasing are represented in red and decreasing ones in blue. 2001 Technical Summary of the Intergovernmental Panel on Climate Change, Working Group 1 Report, National Academy of Sciences, Washington, D.C.

(Source: 2001 Technical Summary of the IPCC Working Group 1 Report.)

5. *Carbon dioxide and other gases generated by human activity are responsible for this temperature increase.* May be true. This causal connection is not unambiguously established. It is possible that there are other, natural causes for the measured temperature increase. Thus far, climate scientists have created increasingly sophisticated computer models linking the temperature increase and the increase in the concentration of greenhouse gases. The evidence implicating CO_2 from human sources is growing, but remains circumstantial. One of the most influential groups studying climate change is the IPCC, a collection of 2500 world-leading international scientists and technical experts whose mission is to assess climate change research in a balanced way through peer review. In its 2001 report, the IPCC does not equivocate on the matter of global warming; it takes the position that humans have caused untoward climatic changes: "The balance of evidence suggests a discernible human influence on global climate." It goes on to say "With the growth in atmospheric concentrations of greenhouse gases, influence with the climate system will grow in magnitude, and the likelihood of adverse impacts from climate change that could be judged dangerous will become greater."
6. *The average global temperature will continue to increase as anthropogenic emissions of greenhouse gases increase.* Uncertain. This statement assumes that the (probable) increase in average global temperature observed over the last century has been caused by the (very likely) increase in human-generated CO_2 and other gases. As we have

seen, this cause-and-effect relationship has not been unambiguously established. Extrapolations into the future are even more uncertain because of the complexities of the global system.

All of us have a strong temptation to address the issues surrounding global warming and climate change by resorting to anecdote and personal experience. But such arguments can be misleading. It does not necessarily follow that the widespread North American drought of 1988, floods of 1993, or the summer heat waves of the 1990s were evidence of a global warming trend. Fluctuations in temperature and precipitation occurring over short periods of time are common and correspond to variations in *weather* patterns. They may not signal large-scale and long-range changes that shape the *climate*. On the other hand, the fact that 10 of the 12 years between 1986 and 1998 have been the warmest on record may be significant and a predictor of things to come; 1998 has been the warmest to date since records have been kept. You may want to refer to the cartoon in 3.7 Consider This to see if your response has changed to the statement made there.

3.27 Consider This: Global Warming in the News

Given the rate at which new information about global warming is being generated, some parts of this book were almost certainly out of date before it was printed. Consult the Web to find two documents on global warming that were published in the last calendar year. For each, give the title, author or source, URL, and the date last updated. Summarize the new information that you found. If this information is different from your textbook, cite the differences.

3.28 The Sceptical Chymist: Global Warming Skeptics

Some people conclude that based on the evidence, human activities have amplified the greenhouse effect; others do not. Find out what such skeptics have to say about the topic. You can locate organizations that take a lukewarm view of global warming by searching for "global warming" and "skeptics." Many search engines allow "wildcards" such as *. Thus, by typing skeptic*, you will bring up sites that include related words such as skeptics, skeptical, or skepticism.

Locate a global warming skeptics web site and write a short description about the information found there. Include the URL, the title of the site, its sponsor, and the date the web site was last updated. Summarize three points made on the web site in opposition to the concept that there are anthropogenic contributions to global warming.

3.13 What Can We Do? What Should We Do?

Given these uncertainties, what is the current status about the possibility of global warming? The debate over climate change has subtly shifted in the last five years. The focus is no longer on whether or not there is an observable increase in global temperatures. The majority of climate scientists agree that global warming is occurring. Instead, the focus now is on understanding the causes of such increases and means of preventing projected changes. There are two related questions. What *can* we do and what *should* we do about the possibility of significant climate change caused by global warming? One thing is clear: Given the recent results from improved climate modeling, we will start seeing climatic changes within a decade or so. But can we prudently wait that long, or is prompt action essential? Whether or not to act, and how to act are not just scientific issues. What determines our response is a complicated mix of science, perception of risk, societal values, politics, and economics.

There are three lines of thought for dealing with what we can and should do, although such a classification system is somewhat arbitrary and unfair when applied to a complicated issue. Simply put, there are: (1) those who would act now; (2) those who would study more; and (3) those who would not act at all, believing that climate change is inevitable.

The most obvious action strategy for dealing with global warming would be to reduce our reliance on fossil fuels. Such action would be very difficult, not only because this energy source is so important to our modern economy but because of its

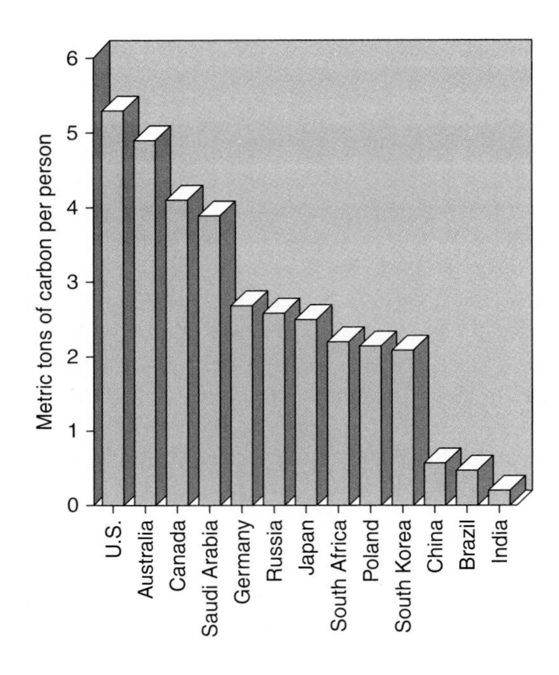

Figure 3.25

Per capita CO_2 emissions for 1998 (measured as metric tons of carbon per year) for selected nations.

(Source of data: Carbon Dioxide Information Analysis Center).

international dimensions. Although the developing countries may well become the major producers of carbon dioxide and other greenhouse gases in the not-too-distant future, the developed countries have a prodigious lead. The annual carbon output (in the form of CO_2) of the United States is about 5.3 tons for each of its inhabitants, compared to a worldwide average only one-fifth as great (Figure 3.25). On a per capita basis the United States leads the world for developed countries, creating a mandate for the country to help lead the way in reducing emissions, but to do so without sacrificing the economy. Comparable per capita values for China and India are about 0.6 and 0.2 tons, respectively. Even so, the Peoples' Republic of China ranks second behind the United States in *total* carbon dioxide emissions from fossil fuels. If China were to succeed in raising its per capita gross national product to only 15% of the U.S. figure, the increase in CO_2 production would approximately equal the current American annual emissions from coal. The IPCC has estimated that by 2010, the developing countries plus the former Soviet Union nations will produce more than half the world's CO_2. It is unrealistic for the developed countries to expect the nations of the Third World to abandon their hopes for economic growth and become good, nonpolluting global citizens.

- From 1996 to 2000, CO_2 emissions from fossil fuel combustion in China decreased by 9%. By contrast, from 1995 to 1999, CO_2 emissions from fossil fuel combustion in western Europe increased by 4.5%, in the U.S. by 6.3%, and in Japan by 3.0%.

3.29 Consider This: The Top Emitters

It is no secret which countries are emitting the highest amounts of CO_2. You can access a list of the top fossil fuel CO_2 emitters on the Web. This list is provided by the Carbon Dioxide Information Analysis Center and gives carbon dioxide emissions both per country and per capita (person). Three sets of data are available: regional, national, or global basis.

a. This text just cited the United States and Peoples' Republic of China as the leaders in *total* CO_2 emissions from fossil fuels. What countries rank third, fourth, and fifth in the most recent total CO_2 data? Whose emissions are relatively low?

b. Now look at the emissions per capita. Which countries have the highest emissions per person? Do the top five differ per capita by very much? Why or why not?

To just keep using fossil fuels at the same rate until we run out of them is not a reasonable approach to decreasing our reliance on them. It is already too late for this to be a viable option. There are significant time lags between production of CO_2, its accumulation in the atmosphere, and the eventual return of the CO_2 into an environmental sink. Thus, waiting until we run out of fossil fuels to take action will not solve the problem. In the interim, ever-increasing amounts of CO_2 will be added to the atmosphere. Morris Adelman, now professor emeritus at MIT, has been making this point for over 30 years. From a realistic perspective, continuing improvements in technology and the drive of economic development will mean that we will continue to depend heavily on fossil fuels into the near future. Currently, more than 85% of the world's energy needs are supplied by fossil fuels, making it difficult to develop sufficient alternate technological capacity soon enough to mitigate climate change. There also is the issue of cost. The costs to reduce atmospheric carbon dioxide would be huge if only alternate technologies such as solar and wind power are considered. While these are essential parts of the energy mix, they cannot be the only answers.

There are those, including President George W. Bush, who advocate delaying action regarding global warming because they feel that further study is needed. During the 2000 presidential campaign, candidates George W. Bush and Al Gore expressed the same opinion, agreeing that carbon dioxide is a pollutant and that legislation should be passed to help reduce CO_2 emissions from power plants. When elected President, George W. Bush's proposed energy policy no longer included mandatory reductions. The Bush administration argues that any shift from coal to natural gas would mean higher prices for electricity, a change that could diminish our economy. Others argue that uncertainties in our predictive powers and climate models are so great that money and effort would be wasted now to undertake preventive or ameliorative action. Without more knowledge, we run the risk of making more mistakes.

Understandably, some greeted President Bush's reversed position with acclaim and others with dismay. Frederick L. Webber, president and CEO of the American Chemistry Council, a chemical industry trade group, characterized President Bush's move as "an important element of a sound, long range national energy strategy that will help achieve a reliable and affordable energy supply." A different view was expressed by the UN Environment Program Executive Director, Klaus Topfer: "Without U.S. leadership, effective global action on climate change may not be possible." Representative Sherwood L. Boehlert (R-NY), chair of the House Science Committee, stated that he is "profoundly disappointed. Capping CO_2 was in Bush's campaign energy plan, and I would like to know what happened in the interim." The "not yet" attitude is also found unacceptable by Dr. Michael Oppenheimer, chief scientist of the Environmental Defense Fund, who stated: "Scientific uncertainties, which are substantial, of course, [regarding global warming] are not a reason to put off action. In fact, we only have one Earth to experiment on."

In marked contrast to the "business as usual" and "wait and study" schools of thought are the "do anything but do something fast" activists. The danger with such an approach is that some of the proposed cures may be worse than the disease. A number of emergency responses have been suggested, and most of them have been discredited. For example, some have proposed that we follow the example of Mount Pinatubo and release sulfur-containing compounds that would generate cooling sulfate aerosols in the atmosphere. But critics have pointed out that the equivalent of 300 Pinatubo eruptions would be required each year to counteract the projected temperature increases. The not-so-desirable associated effects would include global acid rain, serious ozone loss, and severe air pollution. Another approach suggested fertilizing southern oceans with iron to increase the growth of phytoplankton. According to the plan, these primitive green plants would thrive, absorbing vast quantities of carbon dioxide from the atmosphere. This strategy was even tested and found wanting. Iron did increase phytoplankton growth, but the growth of zooplankton, the animals that eat the phytoplankton, kept pace; there was no significant change in net CO_2 absorption.

Others suggest some very different approaches involving various ways of putting the carbon dioxide back into the Earth, known as *carbon sequestration*. Carbon sequestration often refers to planting trees as environmental sinks that absorb CO_2. It also includes capturing carbon dioxide emitted from stationary sources—an electric power

- Sequestration means keeping something apart. If carbon dioxide is properly sequestered, it cannot reach the troposphere and contribute to global warming.

plant, for example. The CO_2 would be liquefied and pumped deep into the ocean. Proponents of such carbon sequestration technology point out that much of the carbon dioxide now emitted eventually finds its way into the oceans anyway. Carbon sequestration is currently being implemented off the coast of Norway where the Sleipner natural gas rig pumps carbon dioxide 1000 m below the ocean's surface, a project driven by concern for global warming and financial incentive. Norway's Statoil oil company is projected to save millions of dollars by using carbon sequestration. The company built an $80 million dollar at-sea facility to separate CO_2 from natural gas for two reasons: first because they could not sell their natural gas to European customers without first removing carbon dioxide from it, and second because sequestering the CO_2 avoids a stiff "carbon tax" imposed by Norway. This tax would have cost Statoil about $50 for every ton of CO_2 emitted, thus saving about $50 million a year in taxes. Critics cite experimental evidence that increased oceanic carbon dioxide could damage coral reefs.

Oil companies propose separating the CO_2 from the natural gas that accompanies drilling for oil and pumping the CO_2 directly back underground either into nearby depleted natural gas fields, or use it to flush residual crude oil from shrinking reservoirs. This method could be especially useful in the case of a huge Indonesian natural gas deposit. The deposit is principally CO_2 that, if released, would equal about 0.5% of all carbon dioxide emitted globally from fossil fuel burning. Also, in over 65 oil wells in the U.S. companies are successfully injecting the carbon dioxide underground to make the drilling operation more efficient. These approaches and those previously suggested must be evaluated thoroughly because, although some beneficial and benign forms of climate engineering may be possible, we would be well advised to proceed with caution.

Still others look at the magnitude of the potential problems associated with global warming and conclude that nothing can be humanly done to halt or reverse the process. Their outlook is that the generation of energy, and with it carbon dioxide, is an essential feature of modern industrial life, and we must therefore learn to live with its consequences and begin adapting to our warm new world.

You are encouraged to debate and discuss these and other options. Clearly, your opinions and the evidence you marshal to support them are important. However, the authors of this text hope that there will be at least some advocates who develop a compromise strategy out of the best characteristics of the extreme positions just described. There are, after all, likely to be elements of truth in each position.

3.30 Consider This: Three Reactions to Global Warming

The text identifies three possible reactions to the problem of global warming: act now, continue to study it, and take no action at all. The most effective response will take place in the shortest amount of time if human and financial resources are directed toward a single unified plan. Review the three positions and their consequences. Discuss the situation with others and prepare to state and defend your position to the rest of the class.

3.31 Consider This: The Massachusetts Model

In April 2001, Governor Jane Swift issued a rule requiring the major coal- and oil-fired power plants in Massachusetts to meet stringent new air emissions standards, including those for emission of carbon dioxide. Power plants must limit carbon dioxide emissions by an amount equal to 10% of what generators have been discharging in the state, but they have freedom in how they do it.

a. What are some of the means proposed to meet this new rule?
b. Could the Massachusetts model be useful in other parts of the United States?

3.32 Consider This: Warming Mars?

Mars is a very chilly place with an average annual temperature of 55 °C below zero. It has been proposed that greenhouse gases that cause problems on Earth might provide the means to make Mars a more comfortable place for humans to colonize in the future. In the process, we could solve a perplexing climate problem on Earth and learn more about Mars at the same time. Find out about this research by using a search engine or use the direct links provided at the *Chemistry in Context* web site. What problems are associated with this rather far-out approach?

3.14 The Kyoto Protocol on Climate Change

It has been more than a century since Ahrrenius first proposed that carbon dioxide emissions could accumulate in the atmosphere and lead to global warming. The world may have been slow in responding, but the last decade has seen considerable progress towards global understanding of the need to start formulating plans for action. The stage was set in 1990 when the IPCC certified the scientific basis for the greenhouse effect. Nearly 10,000 participants from 161 countries at the Kyoto Conference in 1997 established goals to stabilize and reduce atmospheric greenhouse gas concentrations at environmentally more responsible levels. To meet these goals, binding emission targets based on five-year averages were set for 38 developed nations to reduce their emissions of six greenhouse gases from 1990 levels. Accomplishing these goals between the years 2008 to 2012 would potentially decrease emissions from industrialized nations overall by about 5%. Under the Kyoto Conference protocol, the United States was expected to reduce emissions to 7% below its 1990 levels, the European Union nations 8%, and Canada and Japan 6%. No new binding emission targets were established for developing countries, a contentious issue then and now. Developed nations are permitted to trade emission credits to meet their targets. That is, countries that have emissions lower than their targets can sell the residual amounts to countries exceeding their targets. Developed nations also can receive further credits, as early as 2000, for investments and projects to help developing countries reduce their emission of greenhouse gases through better technologies. The gases regulated include carbon dioxide, methane, nitrous oxide, hydrofluorocarbons (HFCs), perfluorocarbons (PFCs), and sulfur hexafluoride.

- An example of a perfluorocarbon (PFC) is C_2F_6. It has a lifetime of 10,000 years in the troposphere, but its concentration is very low, 4.0×10^{-10}%.

- Sulfur hexafluoride (SF_6), used for electrical insulation in transformers, has a tropospheric lifetime of 3200 years. It is nearly 24,000 times more potent as a greenhouse gas than CO_2, but its atmospheric concentration is very low, 3.5×10^{-10}%.

3.33 Consider This: Kyoto Conference Humor

What is the humor in this cartoon? Would everyone find it amusing? Explain your reaction to this cartoon, including whether or not you feel it is trying to communicate a certain point of view.

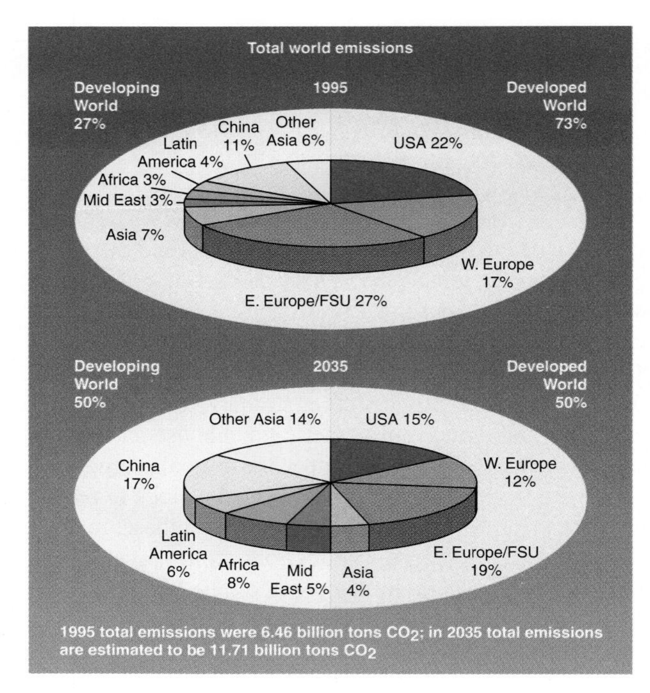

Figure 3.26

Total world CO_2 emissions, 1995, 2035. 1995 total emissions were 6.46 billion tons CO_2. In 2035, total emissions are estimated to be 11.71 billion tons CO_2. Notice how the relative contributions of the developed and developing world change.

(Source: Taken from http://clinton2.nara.gov/initiatives/Climate/Figure 6.gif)

Although there is no deadline for implementation of the Kyoto Protocol, the agreement will not take effect until 90 days after at least 55 countries that account for at least 55% of total 1990 CO_2 emissions have ratified it. This has not happened, although as of July 20, 2001, 84 countries have signed and 37 ratified the treaty. In 2002, the treaty was ratified by the European Union as a bloc and by Japan, but not by the U.S.. Former President Clinton made it clear while president that he would not submit the treaty to the U.S. Senate for ratification until key developing countries agreed to limit their greenhouse gas emissions. This step is important because the rates of greenhouse gas emissions of developing nations are increasing more than those of industrialized countries and expected to grow even faster (Figure 3.26). Now President George W. Bush stated he does not support the agreements reached in Kyoto, calling it "fatally flawed" because it does not require developing nations to meet target restrictions on CO_2 emissions. Prospects of U.S. ratification are nil. What does this mean for the future of global climate control?

3.34 Consider This: Developing Countries; Developed Countries

Consider how your position on controlling emissions of carbon dioxide would change if you were a student in a developing country rather than in the United States. To gather some information to help your consideration, pick a country in the developing world. Then use a search engine or the direct access provided on the *Chemistry in Context* web site to find out if that country has signed and ratified the Kyoto Protocol. Also find the total tons of carbon dioxide emitted and the per capita emission for that country. Compare these data to those of the United States and comment on the differences and their possible effect on policy.

International summit talks on climate control held in 2000 at The Hague broke down over issues of the proper "carbon credit" for sequestration of carbon in forests, and penalties proposed for failing to meet goals. In July 2001, delegates from 179 nations met in Bonn, Germany to once again address changes in rules governing the Kyoto Protocol. After three days of marathon negotiating, new agreements were reached. The agreement reached by 178 countries, the United States not among them, was largely the product of give and take among Japan, Australia, Canada, and the European Union. Japan's role was particularly crucial given its position having the largest economy after the United States. Olivier Deleuze, the energy and sustainability secretary of Belgium, shared the view of many European leaders in saying that many provisions were weakened or could be otherwise criticized. "But," he said, "I prefer an imperfect agreement that is living than a perfect agreement that doesn't exist."

The agreement reached in Bonn in 2001 requires the 38 developed countries to reduce their combined annual greenhouse gas emissions by 2012 to 5.2% below levels measured in 1990. The United States was not party to the agreement. If these targets are not met, stiff penalties will be imposed. Japan and other countries were given permission to use existing managed forests to count for carbon credits as carbon sinks, although this remains a point of contention. The 1997 specific targets remain unchanged for developed countries. Under this new agreement developing countries still do not need to restrict their emissions, a major point of contention for the United States representatives. A system for buying and selling credits earned by reducing carbon dioxide emissions was formalized. The European Union has pledged $410 million a year to help developing countries adapt to climate change and to improve their technology infrastructure.

Many diverse interests, both economic and environmental, were part of the compromise reached in July 2001 in response to a general consensus that the risk of total collapse of the Kyoto protocol was just too great unless an agreement was reached at the meeting. Because the U.S. chose not to participate, some think that the consensus achieved at the meeting is a bit hollow. Given that the U.S. accounts for 25% of greenhouse gas emissions worldwide, many international policy makers feel that the U.S. needs to be a major player on the world stage to reduce CO_2 emissions. In 2000, U.S. emissions of CO_2 increased by 3.1%, the second highest increase in the decade. In Japan, CO_2 emissions have risen by 17% since 1990.

3.35 Consider This: Greenhouse Gas Guidelines

Suppose you have been appointed a U.S. delegate to the next meeting of the climate control talks based on the Kyoto Protocol and its amendments. What position would you advocate that the U.S. take in future talks? Support your economic and environmental views with facts and figures.

What else can we do? Humans can intervene by rectifying their prior assaults on the planet. Deforested areas can be replanted, converting a CO_2 source back to a CO_2 sink. We can conserve energy by making changes in our lifestyle, with no significant impact on the quality of our lives. And, while promoting the growth of other species, we must be aware of the fact that many of our environmental problems are related to the expanding human population. Unfortunately, it is unlikely that these preventive measures, though necessary, will be sufficient to avert a temperature increase. We must also be prepared to meet and mitigate future climate changes. This includes protecting arable soil, improving water management, prudently using agricultural technology and agricultural chemicals, maintaining global food reserves, and establishing an effective mechanism for disaster relief.

No matter what happens, continued careful study is essential to improve our options. It is a response that carries little risk and the potential for great benefit. Extensive environmental monitoring is necessary to provide more reliable climatic and meteorological data. Appropriate technology must be developed and transferred to those who need it. And the public must be educated and informed. Complete evidence and absolute certainty will never be ours in such matters, but intervention based on the best available scientific and technical information is called for.

3.15 Global Warming and Ozone Depletion

Global warming and ozone depletion are important environmental issues that involve the atmosphere, and both are much in the news. There are enough apparent similarities

Table 3.8

Global Warming and Ozone Depletion: Some Characteristics

	Global Warming	Ozone Depletion
Region of atmosphere involved	Mostly troposphere	Stratosphere
Major substances involved	CO_2, CH_4, N_2O	O_3, O_2, CFCs
Interaction with radiation	IR radiation absorbed by molecules, which vibrate and remit energy to Earth	UV radiation absorbed by molecules, which are dissociated into smaller fragments
Nature of problem	Increasing concentrations of greenhouse gases are apparently increasing average global temperature	Decreasing concentration of O_3 is apparently increasing exposure to UV radiation
Source of problem	Release of CO_2 from burning fossil fuels, deforestation; CH_4 from agriculture	Release of CFCs from refrigeration, foaming agents, solvents; CFCs release Cl· which destroys O_3
Possible consequences	Altered climate and agricultural productivity, increased sea level	Increased incidence of skin cancer, damage to phytoplankton
Possible responses	Decrease use of fossil fuels and discontinue deforestation	Eliminate use of CFCs and find suitable replacements
International control	Kyoto Protocol	Montreal Protocol

between the two phenomena that the casual reader of newspaper accounts may mix them up. Sometimes, the authors of the articles themselves get confused. One aim of this text is to avoid such mix-ups, and for that reason it is probably a good idea to conclude this chapter by summarizing some of the important differences between the greenhouse effect and ozone destruction. We do so in Table 3.8. Such a tabulation is an invitation to oversimplification, but it can be a useful reminder of some of the important aspects of these two environmental problems. The 3.36 Consider This provides you an opportunity to express your informed opinion about their relative significance.

3.36 Consider This: Air Quality, Ozone Depletion, or Global Warming

Now that you have studied air quality (Chapter 1), stratospheric ozone depletion (Chapter 2), and global warming (Chapter 3), which one do you believe poses the most serious problem? Discuss your reasons with others and draft a one-page report on this question.

CONCLUSION

Is our world warming to unintended climatic effects? To assess and reverse such effects, much will depend on the quality of information gathered and how it is used to make sound economic and environmental decisions. This chapter and the two that preceded it

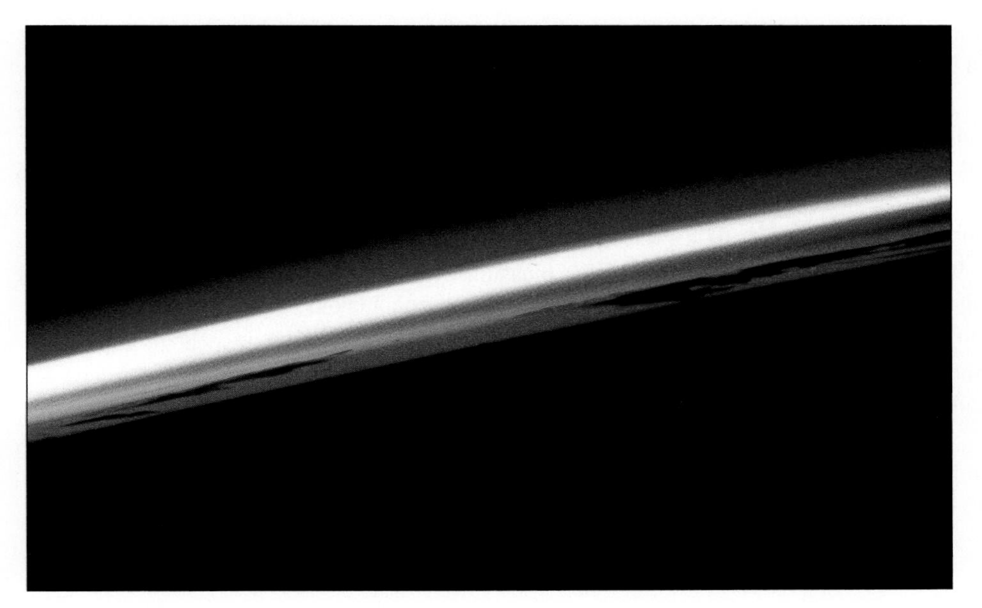

Figure 3.27

An astronaut's view of the Earth's atmosphere from an altitude of 128 miles. The black regions at the top and bottom of the picture are outer space and the Earth, respectively. The outermost layer of the atmosphere appears blue. The reddish streak immediately below the white band, at an altitude of 20–27 km, is caused by ash and sulfuric acid particles from the eruption of Mount Pinatubo in 1991. The region closest to the surface of the Earth is dark red because the sunlight is scattered by dust, smoke, and water vapor.

disclosed that our atmosphere is both fragile and yet more robust than it appeared to the German astronaut, Ulf Merbold. When he saw an astronaut's view of the Earth's atmosphere, such as is seen in Figure 3.27, he was awestruck and terrified at the same time.

> *"For the first time in my life I saw the horizon as a curved line. It was accentuated by a thin seam of dark blue light—our atmosphere. Obviously this was not the ocean of air I had been told it was so many times in my life. I was terrified by its fragile appearance."*
>
> Ulf Merbold

Nevertheless, during the 20th century, the air on which our very existence depends was subjected to repeated assaults. The fact that most of these environmental insults were unintentional and, in some cases, the unexpected consequences of social progress does not alter the problems we face. We have only recently recognized the potential harm that air pollution, stratospheric ozone depletion, and global warming can bring to our personal, regional, national, and global communities. To reverse the damage already done and to prevent more, all these communities must respond with intelligence, compassion, commitment, and wisdom. It is instructive that even in the absence of threats such as global warming, much of what has been advocated in the preceding section would be sound, prudent, and responsible stewardship of our planet.

Chapter Summary

Having completed this chapter you should be able to:

- Realize the difference between the Earth's natural greenhouse effect and the enhanced greenhouse effect (3.1)
- Explain the methods used to gather evidence for global warming (3.2)
- Understand the mechanism by which global warming occurs (enhanced greenhouse effect), the chief role played by carbon dioxide in it, and the nature of the carbon cycle (3.3–3.6)

- Relate Lewis structures and molecular geometry to absorption of particular radiation, for example, infrared (3.4–3.5)
- Know which molecular species are greenhouse gases because of their molecular structures; write such Lewis structures (3.4)
- Summarize the contributions human activities make to the carbon cycle and through it, to global warming (3.6)
- Apply atomic weight and Avogadro's number data to molar mass and mole calculations (3.7–3.9)
- Recognize the contributions and limitations made by computer models of climatic change and the current limitations to such models (3.11)
- Consider the national and global implications of a 1.4–5.8°C rise in the Earth's average temperature (3.11–3.13)
- Describe ways in which carbon dioxide emissions can be reduced (3.13)
- Assess the factors involved in global warming (3.1, 3.3, 3.4, 3.10–3.15)
- Explain world and U.S. policy concerning the Kyoto Protocol (3.14)
- Read and hear news articles on global warming with some measure of confidence in your ability to interpret the accuracy and conclusions of such reports
- Take an informed position with respect to issues surrounding global warming

Questions

Emphasizing Essentials

1. Concentrations of CO_2 in the early atmosphere of the Earth were much higher than today. What happened to this CO_2?

2. a. Write and balance an equation for the photosynthetic conversion of CO_2 and H_2O to form glucose, $C_6H_{12}O_6$, and O_2. Show all states for reactants and products, and include any necessary conditions for the reaction to take place.

b. Demonstrate that the equation is balanced by counting atoms of each element on either side of the arrow.

c. Is the number of molecules on either side of the equation the same? Why or why not?

3. Using the analogy of a greenhouse to understand the energy radiated by the Earth, what are the "windows" of the greenhouse made of?

4. Consider Figure 3.3.

a. How does the concentration of CO_2 in the atmosphere at present compare with the concentration of CO_2 20,000 years ago? How does the present concentration of CO_2 compare with the concentration 120,000 years ago?

b. Compared with the 1950–1980 mean temperature of the atmosphere, how does the temperature at present compare? How does the temperature 20,000 years ago compare? How does each of these values compare with the average temperature 120,000 years ago?

c. Do your answers to parts a and b indicate causation, correlation, or no relation? Explain.

5. Understanding the Earth's energy balance is essential to understanding the issue of global warming. For example, the solar energy striking the surface of the Earth averages 169 watts per square meter, but the energy leaving the surface of the Earth averages 390 watts per square meter. Why isn't the Earth cooling rapidly?

6. Explain each of the observations.

a. The inside of a car left in the sun may become hot enough to endanger the lives of pets or small children.

b. Clear winter nights tend to be colder than cloudy ones.

c. There is a much wider daily temperature variation in the desert than in a moist environment.

7. Using the Lewis structures for H_2 and H_2O as examples, why is it that the Lewis structure for H_2 allows an unambiguous prediction of molecular geometry whereas this is not the case for H_2O?

8. Use a molecular model kit to make a methane molecule, CH_4. (If a kit is not available, this model can be made using Styrofoam balls or gumdrops to represent the atoms and toothpicks to represent the bonds.) Prove that the hydrogen atoms are farther from each other in a tetrahedron than they would be if the molecular structure of methane were square planar.

9. Use the stepwise procedure given in the text to predict the shape of each of these molecules.

a. CH_2Cl_2

b. CO

c. HCN

d. PH_3

10. a. Write the Lewis structure for H_3COH, which is methanol or wood alcohol.

b. Based on this structure, predict the hydrogen-to-carbon-to-hydrogen bond angle. Explain the reason for your prediction.

c. Based on this structure, predict the hydrogen-to-oxygen-to-carbon bond angle. Explain the reason for your prediction.

11. a. Write the Lewis structure for H_2CCH_2, ethene, a simple hydrocarbon.

b. Based on this structure, predict the hydrogen-to-carbon-to-hydrogen bond angle. Explain the reason for your prediction.

c. Sketch the molecule showing the predicted bond angles.

12. The text states that a UV photon can break chemical bonds but that an IR photon can cause only vibration in the bonds.

a. Calculate the energy associated with each of these processes by assuming a wavelength of 320 nm for the UV photon and a wavelength of 5000 nm for the IR photon.

b. What is the ratio of the energy that breaks bonds to the energy that causes vibration?

13. Here are three different modes of vibration of a water molecule. Imagine the atoms being connected by bonds that act as springs. Each vibration can be seen by moving each atom in the direction indicated and then back again.

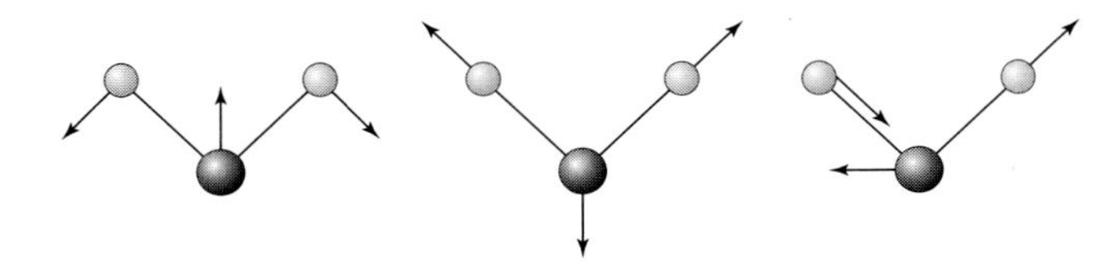

Can any of these modes of vibration contribute to the greenhouse effect? Explain your reasoning.

14. If a carbon dioxide molecule interacts with certain photons in the IR region, the molecule vibrates. For CO_2, the major wavelengths of absorption occur at 4.26 μm and 15.00 μm.

a. What is the energy for each of these IR photons?

b. What happens to the energy in the vibrating CO_2 species?

15. What effect would each of these changes have on global warming?

a. volcanic eruptions

b. CFCs in the troposphere

c. CFCs in the stratosphere

16. One of the biochemical processes that releases carbon dioxide to the atmosphere is the fermentation of sugar to produce alcohol. For example, when glucose, $C_6H_{12}O_6$, undergoes fermentation, ethanol (C_2H_5OH) and CO_2 are produced. Yeast is used to catalyze this conversion. Write a balanced chemical equation for this reaction.

17. Consider Figure 3.17.

a. What is the major source of CO_2 emission from fossil fuel combustion?

b. What are some of the alternatives for each of the major contributors to CO_2 emissions?

18. Silver (Ag) has an atomic mass of 107.870 and an atomic number of 47.

a. What is the number of protons, neutrons, and electrons in a neutral atom of the most common isotope, Ag-107?

b. How do the numbers of protons, neutrons, and electrons in a neutral atom of Ag-109 compare with those of Ag-107?

19. There are just two naturally occurring isotopes of silver. If silver-107 accounts for 52% of natural silver, what is the mass number of the other isotope of silver?

20. a. Calculate the average mass (in grams) of an individual atom of silver.

b. Calculate the mass (in grams) of 10 trillion silver atoms.

c. Calculate the mass (in grams) of 5.00×10^{45} silver atoms.

21. Calculate the molar mass of each of these substances important in atmospheric chemistry.

a. H_2O

b. CCl_2F_2 (Freon-12)

c. CO

22. a. Calculate the mass ratio and mass percent of oxygen in H_2O.

b. Calculate the mass ratio and the mass percent of fluorine in CCl_2F_2.

c. Calculate the mass ratio and the mass percent of carbon in CO.

23. The total mass of carbon in living systems is estimated to be 7.5×10^{17} g. Given that the total mass of carbon on the Earth is estimated to be 7.5×10^{22} g, what is the concentration of carbon atoms on the Earth? Report your answer in percent and in ppm.

24. Consider the information presented in this graph.

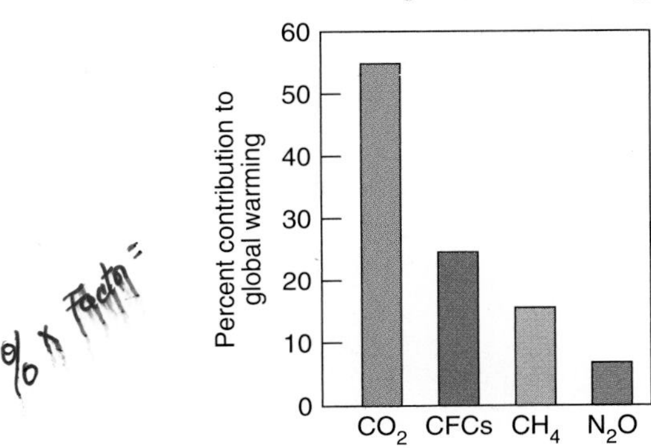

a. Which substance makes the largest contribution to global warming?

b. Use these percentages together with the greenhouse factors given in Table 3.5 to calculate the net effect of each of these gases on global warming. Which gas has the largest net effect?

25. Use the data in Table 3.5 to rank these gases in order of decreasing greenhouse factor: CH_4, CO_2, CCl_3F, N_2O, H_2O, CCl_2F_2, and O_3.

26. Consider the information presented in Table 3.4. Calculate these changes.

a. What is the percent increase in CO_2 when comparing 2000 concentrations with preindustrial concentrations?

b. Considering CO_2, CH_4, and N_2O, which has shown the greatest percentage increase when comparing 2000 concentrations with preindustrial concentrations?

Concentrating on Concepts

27. The text makes a distinction between the *correlation* of two events and the *causation* of one by the other. Identify each of these pairs as an example of correlation, causation, or no relationship. Explain your reasoning.

a. metric tons of coal burned	metric tons of CO_2 emitted
b. national per capita income	per capita emission of CO_2
c. number of cigarettes smoked per day	incidence of lung cancer
d. number of bonds between two oxygen atoms	length of oxygen-to-oxygen bond
e. building a greenhouse	raising beautiful tropical plants
f. buying a pair of roller blades	breaking your leg

28. Why do people sometimes bring living plants, rather than cut flowers, to help a friend along the road to recovery from an illness?

29. Given that direct measurements of the Earth's atmospheric temperature over the last several thousands of years are not available, how can scientists know the fluctuations in the temperature in the past?

30. Consider Figure 3.5 showing atmospheric concentrations of CO_2 at Mauna Loa, and Figure 3.6, showing temperature changes on the Earth's surface. Why is the pattern with the trend so regular in Figure 3.5, but not as regular in Figure 3.6?

31. Consider this graph showing sources of CO_2 emissions in the United States.

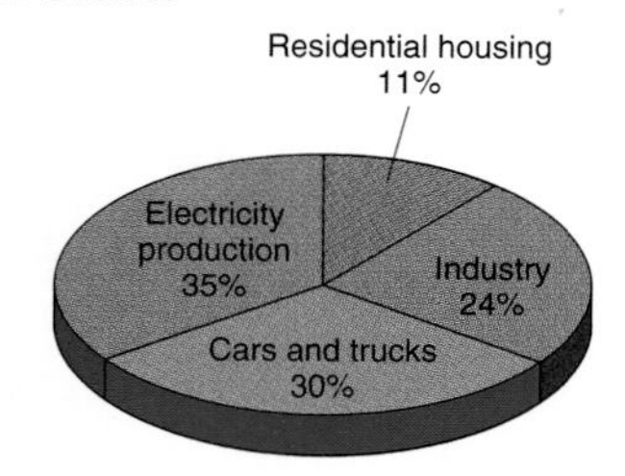

a. If you want to cut down your personal contribution to CO_2 emissions, what changes can you make that will be most effective? Explain your reasoning.

b. Will the entire quantity of CO_2 emitted end up in the atmosphere? Why or why not?

32. Why is it said that Lewis dot structures show linkages (what is hooked to what) but they do not show shape? Explain what is meant by this statement, using the molecule H_2S as an example.

33. The molecule BF_3 is triangular planar with 120° fluorine-to-boron-to-fluorine bond angles, whereas the NF_3 molecule is triangular pyramidal with fluorine-to-nitrogen-to-fluorine bond angles of about 103°. Account for the different geometries and bond angles. *Hint:* Boron is too small to obey the octet rule and is stable with only six electrons around it in the Lewis structure.

34. Carbon dioxide gas and water vapor both absorb IR radiation. Do they also absorb visible radiation? Offer some evidence based on your everyday experiences to help explain your answer.

35. How do you think that the energy required to cause IR-absorbing vibrations in CO_2 would change if the carbon and oxygen atoms were bonded with single bonds rather than with double bonds?

36. Explain why food placed in a plastic container is quickly warmed in a microwave oven, while the container warms much more slowly.

37. Consider Table 3.1. Compare forests, oceans, and fossil fuels in the order of the increasing amounts of carbon they contain. Give reasons for your rankings.

38. Why is the atmospheric lifetime of a potential greenhouse gas important?

39. CO_2 gas has a greater density than N_2 gas. Why don't these gases settle out into layers in the atmosphere?

40. One of the first radar devices developed during World War II used microwave radiation of a specific wavelength that triggers the rotation of water molecules. Why was this design not successful?

41. William McKibben suggests that 73 million metric tons of CH_4 are produced by the Earth's ruminants, such as cattle and sheep, each year. How many metric tons of carbon are present in this mass of CH_4?

42. a. The January 16, 1998 issue of *Science* magazine reported that 1997 was the warmest year of this century. Does this prove that theories about global warming are correct? Explain your reasoning.

b. Evidence has been reported that certain butterflies have recently changed their migration patterns and are moving farther north. Does this prove that theories about global warming are correct? Explain your reasoning.

Exploring Extensions

43. Figure 3.6 shows the temperature increase on the Earth's surface from 1880 to 2000. Imagine you are in charge of extending this graph to include the present year. What kind of information would you need and how might you gather such data? *Hint:* Consider the source of the data in Figure 3.6.

44. If a water molecule interacts with certain photons in the IR region, the molecule will vibrate (Figure 3.16). This equation can be used to represent that interaction, in which the species with the * represents the vibrating molecule.

$$H_2O + h\nu \longrightarrow H_2O^*$$

a. The maximum IR absorbance for water vapor occurs at a wavelength of 6.6×10^{-6} m. Calculate the energy associated with an IR photon represented by this equation.

b. Which absorption peak represents higher energy?

c. Using the * notation, write an equation representing a vibrating H_2O molecule passing its energy to an H_2O molecule that is not vibrating.

45. Data taken over time show an increase in CO_2 in the atmosphere. The large increase in the combustion of hydrocarbons since the Industrial Revolution is often cited as a reason for the increasing levels of CO_2. However, an increase in water-vapor has *not* been observed during the same time period. Remembering the general equation for the combustion of a hydrocarbon, does the difference in these two trends *disprove* any connection between human activities and global warming?

46. a. Assuming that only single bonds are present, how many outer electrons are needed to form a three-atom molecule formed from atoms of **X, Y,** and **Z**? Assume that **X, Y,** and **Z** follow the octet rule.

b. What is the shape of the molecule formed from **X, Y,** and **Z** in part a? What is the **X**-to-**Y**-to-**Z** bond angle? Explain your reasoning.

c. How will the answers to parts **a** and **b** change if you remove the restriction on using only single bonds? Explain.

47. Consider this information about three greenhouse gases. These data are from the United Nations Intergovernmental Panel on Climate Change, representing the change from preindustrial time to 1994. Then compare these data with those given in Table 3.4, which bring us up to date with the present.

Gas	Preindustrial Concentration	1994 Concentration	Atmospheric Lifetime
CO_2	280 ppm	358 ppm	50–200 yr
N_2O	275 ppb	311 ppb	120 yr
CH_4	700 ppb	1721 ppb	12 yr

Use these data to write a commentary for your local newspaper explaining which gases are experiencing the greatest percent increases. Explain some of the reasons for the observed increases, and explain how knowing the atmospheric lifetime of a greenhouse gas is an important piece of information for setting control strategies for greenhouse gases. How has this information changed from 1994 to 2000?

48. The per capita CO_2 emissions for several countries are shown in Figure 3.25. This table gives the per capita income for some of those countries.

U.S.	\$22,470
Australia	\$18,054
Canada	\$19,400
Japan	\$19,100
China	\$360
India	\$380

a. What is the relationship between the per capita CO_2 emissions and the per capita income? Offer some reasonable explanation for the relationship.

b. Considering the relationship between per capita income and per capita CO_2 emissions, what are the policy implications for implementing the Kyoto agreement?

49. The CNN business news web site reported this headline on November 9, 1998. "Some U.S. businesses warming to global emissions treaty." What type of businesses do you predict are supporting the Kyoto agreement? Which are most likely not to be supportive? What changes in business support for the Kyoto agreement have taken place in the last year? Research this aspect of U.S. support for the Kyoto agreement, and write a report summarizing your findings.

50. The atmospheric problems described in Chapters 2 and 3 stimulate different responses. The evidence for ozone-depletion resulted in the Montreal Protocol, which included a schedule for decreasing the production of ozone-depleting chemicals. The evidence for global warming led to the Kyoto agreement, which calls for targeted reductions in greenhouse gases. Suggest some reasons why the question of ozone depletion was dealt with by the world community before taking up the question of global warming.

4

Energy, Chemistry, and Society

Motor vehicles in the U.S. use prodigious amounts of energy daily from burning gasoline.

"America in the year 2001 faces the most serious energy shortage since the oil embargoes of the 1970s. The effects are already being felt nationwide . . . A fundamental imbalance between supply and demand defines our nation's energy crises."

From "Reliable, Affordable, Environmentally Sound Energy for America's Future"

"This [the report] has a lot of opportunities for mischief from the energy producers and no real solid commitments for the green components."

David G. Hawkins, senior lawyer for the National Resources Defense Council. *New York Times,* May 17, 2001, p. A1

"This report relies heavily on the heavy polluters of the past and looks to ease important environmental protections."

Senator Harry Reid (D-NV), ranking Democrat, Senate Environment and Public Works Committee, *New York Times,* May 18, 2001, p. A4

• "As a country, we have demanded more and more energy. But we have not brought on line supplies needed to meet that demand . . . We can explore for energy, we can produce energy and use it, and we can do so with a decent regard for the natural environment." Vice-President Richard Cheney, "Reliable, Affordable, Environmentally Sound Energy for America's Future."

These stark statements are from and about the May 2001 report "Reliable, Affordable, Environmentally Sound Energy for America's Future." The report was produced by the National Energy Policy Development Group, a task force established by President George W. Bush and headed by Vice-President Richard Cheney to develop a national energy policy. Written at a time when Californians were experiencing rolling blackouts and sharp increases in electricity costs, the report goes on to note that our energy needs will likely far exceed expected levels of domestic energy production if such production increases at only the same rate as it had during the 1990s. If allowed to continue, this imbalance between production and demand would, according to the report, "inevitably undermine our economy, our standard of living, and our national security." Among its 105 recommendations, the report encourages greater use of fossil-based energy resources and exploration for them, as well as the reexamination of federal policies to allow greater use of nuclear energy.

Critics see the report as a plan for developing new energy supplies rather than encouraging energy-saving strategies and allowing the search for fossil-fuel supplies to take priority over environmental concerns. Responses critical to the report came from a variety of sources.

At its June 2001 meeting, the Board of Directors of the American Chemical Society, the world's largest scientific society, a nonprofit scientific and educational organization chartered by Congress, issued a statement regarding national energy policy. The statement begins:

> "Our nation's energy policy needs to be revisited in light of energy shortages and continued concerns about the adverse effects of energy consumption and our nation's growing reliance on imported energy. Any change to our national energy policy will affect the mix of energy sources we use and the infrastructure we have for generating, selling, and distributing energy. Policy choices on every energy extraction and use will impact the environment and local communities. Congress should take advantage of the current concerns to develop a comprehensive and coherent policy response to the many challenges we face . . . The policy challenges we face can be met if the nation sets long-term goals to develop and deploy a sustainable energy supply, to use energy efficiently, and to conserve where feasible."

Clearly, energy sources, production, and policy have the attention of politicians, environmentalists, industrialists, scientists, and consumers. The continuing availability of adequate energy at reasonable cost is vital to us and others around the world. Energy is a common thread that runs through the first three chapters of this book. The Earth's finite energy reserves create the probability of energy shortages, a compelling reason to attend to issues related to energy. Our lives are intimately connected to this familiar yet elusive concept. The time has obviously come to take a closer look at energy.

CHAPTER OVERVIEW

To consider energy, we need to understand the meanings of some fundamental concepts such as energy, work, and heat. We begin with these essential ideas and energy units. The first law of thermodynamics is a generalization that describes some of the constraints governing the generation and use of energy. This law links energy, work, and heat. With this foundation we turn to a consideration of energy sources and uses in the past, the present, and the future. Insight into the sources of energy requires a look at what happens at the molecular level when a chemical reaction occurs. We describe these energetic transformations in qualitative terms and then use bond energies to calculate the energy changes associated with typical combustion reactions. Most of the energy currently used in home and industry comes from fossil fuels, so the topics of coal and petroleum are explored in some depth and detail. But the supplies of these fuels are limited, so the text then turns to discuss substitutes and additives from various sources, including renewable biological sources such as corn.

Mechanisms for transforming energy come under scrutiny next. Such mechanisms are subject to the second law of thermodynamics, another natural constraint. We introduce the second law to explain the inescapable inefficiencies of energy transformation. Along the way we develop the concept of entropy, a measure of disorder and an indication of the directionality of natural change. The chapter concludes with some observations on the importance of conserving energy and fuels.

4.1 Energy: Hard Work and Hot Stuff

Energy is one of those words that everyone uses, but whose precise meaning is not well understood. Unfortunately, the dictionary definition, the capacity to do work, doesn't help much because work is another common but poorly defined concept. To a scientist, work is done when movement occurs against a restraining force, and it is equal to the force multiplied by the distance over which the motion occurs. Thus, when you lift a book against the force of gravity, you are doing work. When you read a book without moving it, you are not, strictly speaking, working. On the other hand, you are again doing something that will not happen by itself, and that does require energy.

The source of much of the work done on our planet is another familiar form of energy—heat. The formal definition sounds a little strange: **heat** is that which flows from a hotter to a colder body. But a child once burned knows the meaning of heat. Our understanding of temperature is also based on experience, and we know that temperature and heat are not the same thing. But a standard definition of temperature sounds awkward and circular: **Temperature** is a property that determines the direction of heat flow. When two bodies are in contact, heat always flows from the object at the higher temperature to that at the lower temperature. Consider that a burning match and a bonfire are at the same high temperature. But, as experience tells us, the bonfire produces far more heat than the match. Heat is a consequence of motion at the molecular level. When matter, for example, liquid water in a pan, absorbs heat, its molecules move more rapidly. Temperature is a statistical measure of the average speed of that motion. Hence, temperature rises as the amount of heat energy in a body increases.

Consider two containers of water, each at room temperature (25°C). One contains 100 mL of water; the other holds 200 mL of water. Because the two water samples are at the same temperature, the average speed of their water molecules is the same, but their heat content is not the same. Starting at the same lower temperature, it takes twice as much heat energy to raise the temperature of 200 mL of water to 25°C than it does to reach that temperature for 100 mL. Therefore, the 200 mL of water has twice the heat energy than the smaller volume of water.

Before we turn to matters of energy demand, we need a unit in which to express it. Historically, there have been many, but recently there has been an international agreement to make the common unit the **joule.** To put this unit into context, one joule (1 J) is approximately equal to the energy required to raise a 1 kg (approximately 2 lb) book 10 cm (4 in.) against the force of gravity. On a more personal basis, each beat of the human heart requires about 1 J of energy. As the name implies, one kilojoule (1 kJ) is equal to 1000 J.

Much of the published data relating to energy are reported in calories, not in joules. The **calorie** was introduced with the metric system in the late 18th century as a measure of heat. Originally, the calorie was defined as the amount of heat necessary to raise the temperature of exactly one gram of water by one degree Celsius. It has been redefined as exactly 4.184 J. Calories are perhaps most familiar when used to express the

- The **calorie** was more specifically defined as the amount of heat required to raise the temperature of one gram of pure liquid water from 14.5°C to 15.5°C.

energy released when foods are metabolized. The values tabulated on package labels and in diet books are, in fact, kilocalories (1 kcal = 1000 cal = 1 **Calorie**); when Calorie is written with a capital C, it generally means kilocalorie (Figure 4.1). Thus, the energetic equivalent of a donut is 425 Cal (425 kcal, 425,000 calories). For most purposes we will use joules and kilojoules in this chapter, but when it seems more appropriate or more easily understandable, we express energy in calories or kilocalories. We will not worry about British Thermal Units (BTU), ergs, or foot-pounds, but you are cautioned that the world also expresses energy in these units.

Figure 4.1
Food labels have the energy listed in Calories or kilojoules.

4.1 Your Turn

a. Convert to kilojoules the 425 kcal (425 food Calories) released when a donut is metabolized.
b. Calculate the number of 2-lb books you could lift to a shelf 6 ft off the floor with the amount of energy from metabolizing one donut.
c. A 12-oz can of a soft drink has an energy equivalent of 92 kcal. Convert the energy released when metabolizing the soft drink, expressing the answer in kJ.
d. Assume that you use this energy to lift concrete blocks that weigh 22 lb (10 kg) each. How many blocks could you lift to a height of 4 ft with this quantity of energy?

Answers

a. Find the relationship necessary for this calculation. You will find 1 kcal is equivalent to 4.184 kJ.

$$425 \text{ kcal} \times \frac{4.184 \text{ kJ}}{1 \text{ kcal}} = 1.78 \times 10^3 \text{ kJ}$$

b. Earlier the text states that 1 J is approximately equal to the energy required to raise a 2-lb book a distance of 4 in. against Earth's gravity. We can use this information to calculate the number of 2-lb books that could be lifted 6 ft. First note that 6 ft is 72 in. Next, calculate the energy (joules) required to lift a 2-lb book:

$$72 \text{ in.} \times \frac{1 \text{ J}}{4 \text{ in.}} = 18 \text{ J}$$

Then, express this value in kilojoules.

$$18 \text{ J} \times \frac{1 \text{ kJ}}{10^3 \text{ J}} = 0.018 \text{ kJ}$$

This value allows the final calculation to be made.

$$1.78 \times 10^3 \text{ kJ} \times \frac{1 \text{ book}}{0.018 \text{ kJ}} = 9.9 \times 10^4 \text{ books}$$

In round numbers, about 100,000 books! Lots of exercise is required to work off one donut.

4.2 The Sceptical Chymist: Checking Assumptions

A simplifying (and erroneous) assumption was made in doing the calculations in parts **b** and **d** of the previous Your Turn. What was the assumption, and is it reasonable? Is the answer based on this assumption too large or too small? Give a reason for your answer.

4.2 Energy Conservation and Consumption

Strictly speaking, energy is not consumed. The **first law of thermodynamics,** also called the **law of conservation of energy,** states that energy is neither created nor destroyed. Energy is often transformed as it is transferred, but the energy of the universe is constant. However, energy sources such as coal, oil, and natural gas are consumed. In the United States, we burn a prodigious quantity of these fossil fuels to generate a huge amount of energy. Figure 4.2 compares our annual per capita energy use with that of other parts of the world. The energy comes from many different sources, but in the bar graph it is expressed as if all of it were generated from oil. Thus, in 1999 the energy share of the average North American was roughly 9.3 tons of oil, a quantity that yields about 116 million kcal. In India, the amount of energy derived from commercial fuel

- Energy content of fuels
  1 lb wood (oak) = 7.6×10^3 kJ
  1 lb gasoline = 2.2×10^4 kJ

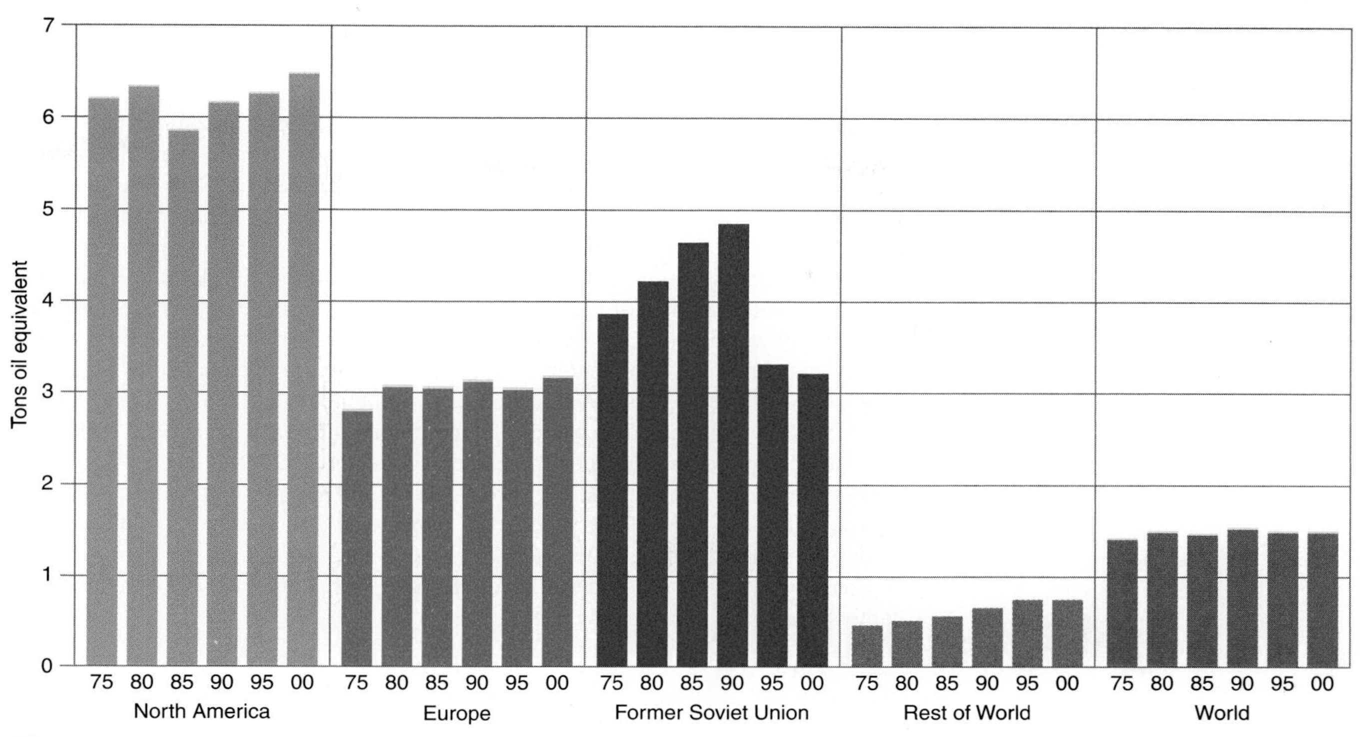

Figure 4.2

Annual per capita energy consumption, 2000. "Energy consumption per capita at the global level has shown almost no growth over the last 20 years. Per capita consumption in North America remains at more than twice the levels of Europe and the Former Soviet Union and at almost 10 times the level in the rest of the world."

(Source of data: Data from BP Statistical Review of World Energy 2001. *June 2000, British Petroleum Company.)*

sources corresponded to about 0.30 tons of oil per person per year, or less than 3% of the values for the United States and Canada. Perhaps an equal quantity of India's energy was derived from traditional fuels such as wood, grass, or animal dung, but the international energy imbalance remains staggering. It is no coincidence that the nations with the highest consumption in Figure 4.2 are industrialized and wealthy, and that those with the lowest are struggling with poverty. Energy appears to drive industrial and economic progress, and gross national product correlates well with energy production and use. So do life expectancy, infant mortality, and literacy.

The great burst of energy consumption is of relatively recent origin. Two million years ago, before our primitive ancestors learned to use fire, the sole source of energy available to an individual was that of his or her own body. Earliest hominids probably consumed the equivalent of 2000 kcal per day and expended most of it finding food. This roughly corresponds to the energy output of a 100-watt light bulb burning for 24 hours. The discovery of fire and the domestication of beasts of burden increased the energy available to an individual by about six times. Hence, we estimate that about 2000 years ago, a farmer with an ox or donkey had roughly 12,000 kcal at his or her disposal each day. The Industrial Revolution brought another five- or sixfold increase in the energy supply, most of it from coal via steam engines. Yet another energy jump occurred during the 20th century. By the end of the 20th century, the total energy used in the United States (from all sources and for all purposes) corresponded to about 650,000 kcal per person per day. This translates to an annual equivalence of 65 barrels of oil or 16 tons of coal for each American. Or, to put it in human terms, the energy available to each resident of the United States would require the physical labor of 130 workers. Yet, there are still people on the planet whose energy use and lifestyle closely approximate those of 2000 years ago.

The history of increasing energy consumption is closely related to changing energy sources and the development of devices for extracting and transforming that energy. Figure 4.3 displays the average American energy consumption from a variety of sources over a 150-year period. The data start in 1850 and are projected to 2000. The graph indicates that wood was originally the major energy source in the United States, and it continued to be until around 1890, when it was surpassed by coal. Coal provided more than 50% of the nation's energy from then until about 1940. By 1950, oil and gas were the source of more than half of the energy used in this country. Falling water has long been used to power mills and, more recently, to generate electricity, but it provides only a small percentage of our total energy output. Nuclear fission, once hailed as an almost limitless source of energy, has not achieved its full potential for a variety of reasons.

• Nuclear fission is discussed in Chapter 7.

Geothermal, wind, and solar sources are combined into a sliver marked "Other" in Figure 4.4a. This pie chart indicates the percentage of the U.S. 1999 energy consumption derived from various sources. Figure 4.4b is a similar representation of world energy consumption for 1999. These global data indicate an 8% reliance on traditional energy sources and relatively less dependence on oil than is characteristic of the United States. Some of the currently underutilized alternate energy sources will be considered in Chapters 7 and 8.

4.3 Consider This: U.S. Sources of Energy Over Time

Many changes in our sources of energy are represented in Figure 4.3. Concentrating on the period from 1950 to 2000 shown in the figure, which sources of energy have shown steady growth and which have not? Propose some possible reasons for the observed trends.

4.4 Consider This: Now and in the Future

Dr. Ronald Breslow, a Columbia University chemistry professor and a recent president of the American Chemical Society, speaking at a February 2001 symposium on "Sustainability Through Science," re-

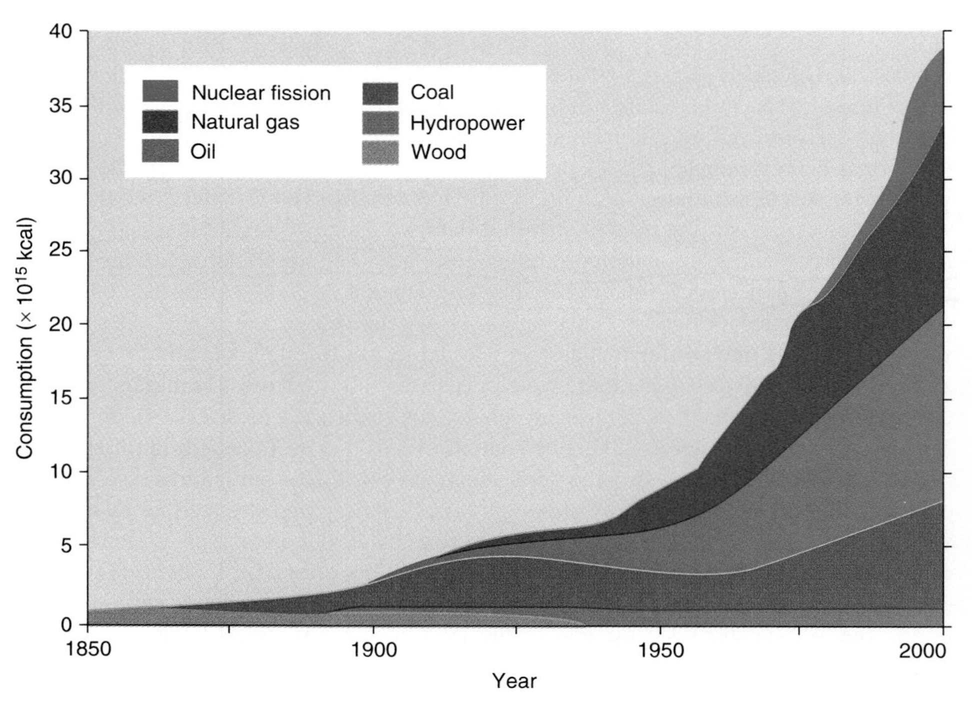

Figure 4.3

Annual United States energy consumption from various sources, 1850–2000.

(Source of data: Department of Energy/EIA.)

marked: "Succeeding generations are going to curse us for burning their future raw materials, and they are right. Not only are we using up valuable resources—petroleum and coal—but we are adding pollution and carbon dioxide which may be contributing to global warming." Comment on Dr. Breslow's remarks.

4.3 Energy: Where From and How Much?

At a time when the nation is seeking new sources of energy, it is reasonable to ask what makes some substances such as coal, gas, oil, or wood usable as fuels, while many others are not. To find an answer, we must consider the properties of fuels and how energy is released from them. The most common energy-generating chemical reaction is burning, or combustion. **Combustion** is the combination of the fuel with oxygen to form product compounds. In such a chemical transformation, the **potential energy** (on a molecular scale, the energy related to the positions of atoms and molecules) of the reactants is greater than that of the products. Because energy is conserved, the difference in energy is given off, primarily as heat.

We illustrate the process with the combustion of methane, CH_4, the principal component of natural gas, a major home heating fuel. The products are carbon dioxide and water. In Chapter 1 you encountered the reaction represented by this equation.

$$CH_4(g) + 2\ O_2(g) \longrightarrow CO_2(g) + 2\ H_2O(g) + \text{energy} \quad [4.1]$$

The reaction is said to be **exothermic**—a term applied to any chemical or physical change accompanied by the release of heat. The quantity of heat energy released in a combustion reaction such as this can be experimentally determined with a device called a calorimeter (Figure 4.5). Not surprisingly, the amount of heat generated depends on the amount of fuel burned. Therefore, a known mass of fuel and an excess of oxygen

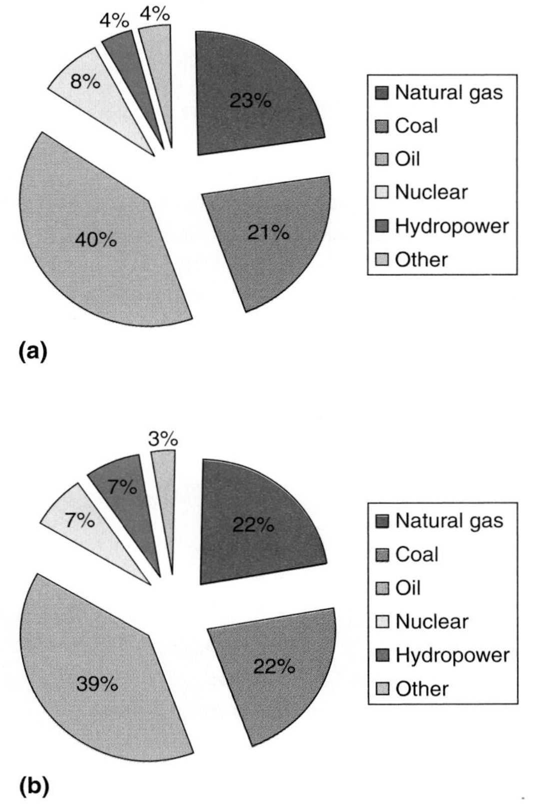

Figure 4.4
Energy consumption from various sources, 1999: (a) United States, (b) World.
(Source of data: Department of Energy/EIA.)

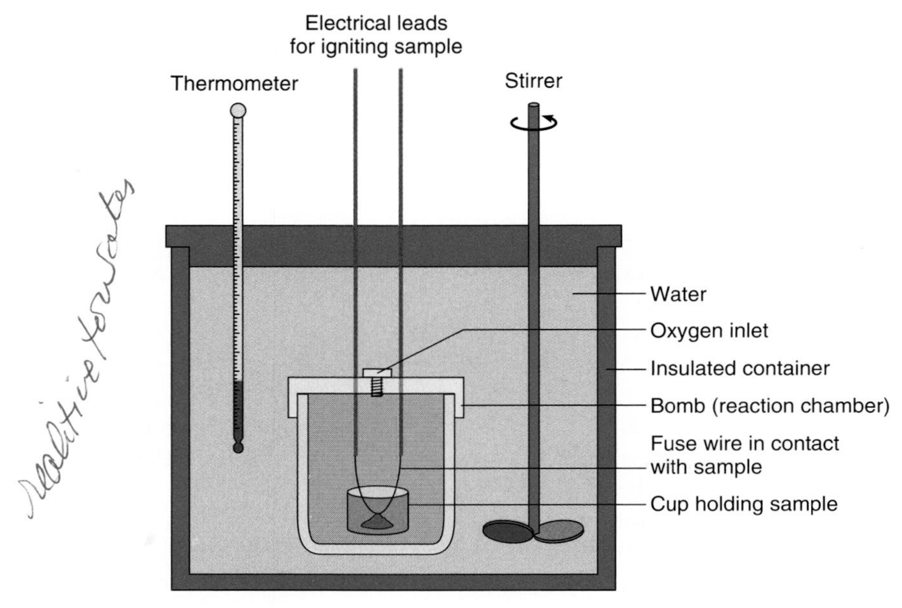

Figure 4.5
Schematic drawing of a bomb calorimeter.

are introduced into a heavy-walled stainless steel "bomb." The bomb is then sealed and submerged in a bucket of water. The reaction is initiated with an electrical current that burns through a fuse wire. The heat evolved by the exothermic reaction flows from the bomb to the water and the rest of the apparatus. As a consequence, the temperature of the entire calorimeter system increases. The quantity of heat given off by the reaction can be calculated from this temperature rise and the known heat absorbing properties of the calorimeter and the water it contains. The greater the temperature increase, the greater the quantity of energy evolved.

- Heats of combustion, by convention, are tabulated as positive values even though all combustion reactions *release* heat.

Experimental measurements of this sort are the source of most tabulated values of heat of combustion. As the name suggests, the **heat of combustion** is the quantity of heat energy given off when a specified amount of a substance burns in oxygen. Heats of combustion are typically reported as positive values in kilojoules per mole (kJ/mole), kilojoules per gram (kJ/g), kilocalories per mole (kcal/mole), or per gram (kcal/g). The energy equivalents of various foods are also usually determined by calorimetry. In the case of methane, experiment shows its heat of combustion to be 802.3 kJ. This means that 802.3 kJ of heat are given off when one mole of $CH_4(g)$ reacts with two moles of $O_2(g)$ to form one mole of $CO_2(g)$ and two moles of $H_2O(g)$ (see equation 4.1). We can also calculate the number of kilojoules released when one gram of methane is burned. The molar mass of CH_4, calculated from the atomic masses of carbon and hydrogen, is 16.0 g/mole. The heat of combustion per gram of methane gas (kJ/g) is obtained as follows:

$$\frac{802.3 \text{ kJ}}{1 \cancel{\text{mole CH}_4}} \times \frac{1 \cancel{\text{mole CH}_4}}{16.0 \text{ g CH}_4} = 50.1 \text{ kJ/g CH}_4$$

- $\text{Energy}_{\text{products}} - \text{Energy}_{\text{reactants}} < 0$ for an exothermic reaction.

The fact that heat is evolved signals that there is a decrease in the energy of the chemical system during the reaction. In other words, the reactants (methane and oxygen) are at higher energy than the products (carbon dioxide and water). The burning of methane is thus somewhat like a waterfall or a falling object. In all these changes, potential energy decreases and is manifested in some other form of energy (heat, sound, etc.). This decrease is signified by the negative sign that is traditionally attached to the energy change for *all* exothermic reactions. For the combustion of methane, the energy change is listed as −802.3 kJ/mole. Figure 4.6 is a schematic representation of this process. The downward arrow indicates that the energy associated with 1 mole of $CO_2(g)$ and 2 moles of $H_2O(g)$ is less than the energy associated with 1 mole of $CH_4(g)$

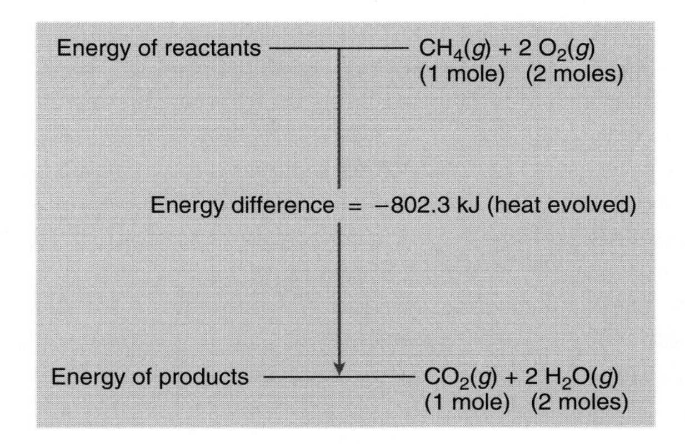

Figure 4.6
Energy difference in an exothermic reaction.

and 2 moles of $O_2(g)$. The energy difference between the products and the reactants is thus a negative quantity, as is the case for all exothermic reactions. In the combustion of methane, the energy difference is −802.3 kJ.

4.5 Your Turn

According to information in this section, the heat of combustion of methane is 802.3 kJ/mol. Methane is usually sold by the standard cubic foot, SCF. One SCF contains 1.25 moles of methane. Calculate the energy (in kJ) that is released by burning 1.00 SCF of methane.

Answer
1000 kJ released

We still need to adequately explain the origin of the energy released in an exothermic reaction. To do that, we investigate the structure of the molecules involved. We already encountered all the reactants and products in the combustion of methane. Consequently, we can write Lewis structures for all of the molecular species.

$$CH_4 + 2\ \ddot{\underset{\cdot\cdot}{O}}{=}\ddot{\underset{\cdot\cdot}{O}} \longrightarrow \ddot{\underset{\cdot\cdot}{O}}{=}C{=}\ddot{\underset{\cdot\cdot}{O}} + 2\ H{-}\ddot{\underset{\cdot\cdot}{O}}{-}H \qquad [4.2]$$

The reaction represented by this and any other chemical equation is a rearrangement of atoms. It involves the breaking and making of chemical bonds. Energy is required to break bonds, just as energy is required to break wood or tear paper. Bond breaking is thus an **endothermic** process, a term applied to any chemical or physical change that absorbs energy. On the other hand, the formation of chemical bonds is an exothermic process in which energy is released. The overall energy change associated with a chemical reaction depends on the net effect of the bond breaking and bond making. If the energy required to break the bonds in the reactants (endothermic) is greater than the energy released (exothermic) when the products form, the reaction is *endothermic*; energy is absorbed. If, on the other hand, the situation is reversed, the exothermic bond-making energy of the products is greater than the endothermic bond breaking in the reactants, then the net energy change is *exothermic*; energy will be released by the reaction.

Endothermic Reaction	Exothermic Reaction
$Energy_{products} > Energy_{reactants}$	$Energy_{products} < Energy_{reactants}$
Energy change is positive	Energy change is negative
Energy is absorbed	Energy is released

The potential energy associated with any specific chemical species, for example, a CH_4 molecule, is in part a consequence of the interaction of the atoms via chemical bonds. When methane or any other fuel burns in oxygen, the energy released in bond formation exceeds the energy absorbed in bond breaking. The net result is the evolution of energy, mostly in the form of heat. Another way to look at such exothermic reactions is as a conversion of reactants involving weaker bonds (for example, CH_4 and O_2) to products involving stronger bonds (CO_2 and H_2O). In general, the products are more stable and less reactive than the starting substances.

Although the chemical reactions used to generate energy are all exothermic, many naturally occurring endothermic reactions absorb energy as they occur, such as photosynthesis. You already encountered two that are very important in atmospheric chemistry. One is the decomposition of O_3 to yield O_2 and O, and the other is the combination of N_2 and O_2 to yield two molecules of NO. Both reactions require energy, which can be in the form of electrical discharge, high-energy photons, or high temperatures. It is possible to experimentally determine the energy changes associated with many reactions—exothermic or endothermic. But sometimes it is easier to calculate values. We illustrate the process in the next section.

4.4 Calculating Energy Changes in Chemical Reactions

Hydrogen provides a lot of "bang for the buck" because of the large amount of energy per gram produced when it burns. Thus, there is much interest in hydrogen as a fuel. But, there is a catch. Hydrogen would become a plentiful fuel if an economical way could be developed to decompose water into its elements, hydrogen and oxygen (Figure 4.7). Methods currently are available to carry out the decomposition, but energy requirements and costs are too high to make hydrogen competitive with fossil fuels.

We can calculate the total energy change associated with the combustion of hydrogen to form water, as represented by equation 4.3.

$$2\ H_2(g) + O_2(g) \longrightarrow 2\ H_2O(g) + \text{energy} \qquad [4.3]$$

To simplify matters for such calculations, we assume that all the bonds in the reactant molecules are broken and then the individual atoms are reassembled into the product molecules. In fact, the reaction does not occur that way. But we are interested in only the overall or net change, not the details. Therefore, we proceed with our convenient fiction and see how well our calculated result agrees with the experimental value.

The numbers we need in the computation are given in Table 4.1, a listing of the bond energies associated with a large variety of covalent bonds. **Bond energy** is the amount of energy that must be absorbed to break a specific chemical bond. Thus, be-

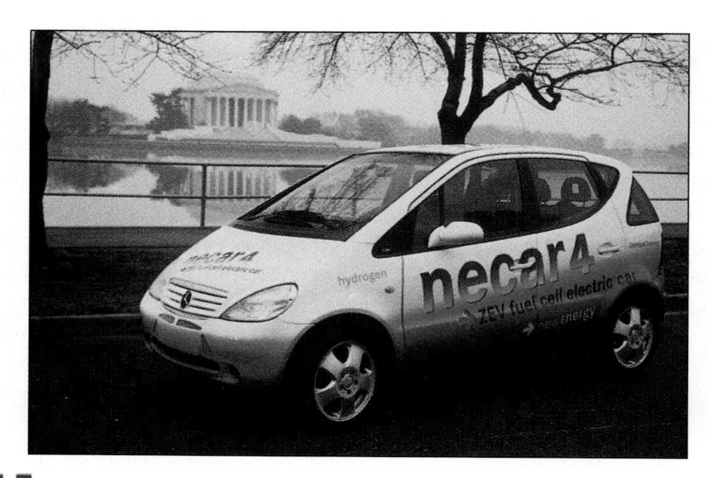

Figure 4.7

The Daimler-Benz NeCar (New Electric Car), a hydrogen-fueled vehicle. The car moves by the energy released when hydrogen (a "clean fuel") and oxygen react to form water.

Table 4.1

Bond Energies (in kJ/mole)

	H	C	N	O	S	F	Cl	Br	I
Single bonds									
H	436								
C	416	356							
N	391	285	160						
O	467	336	201	146					
S	347	272	—	—	226				
F	566	485	272	190	326	158			
Cl	431	327	193	205	255	255	242		
Br	366	285	—	234	213	—	217	193	
I	299	213	—	201	—	—	209	180	151

Multiple bonds

C=C	598	C=N	616	C=O	803 in CO_2
C≡C	813	C≡N	866	C≡O	1073
N=N	418	O=O	498		
N≡N	946				

cause energy must be absorbed, breaking bonds is an endothermic process, and all the bond energies in Table 4.1 are positive. Obviously, the amount of energy required depends on the number of bonds broken—more bonds take more energy. Typically, bond energies are expressed in kilojoules per mole of bonds. Note that element symbols appear across the top of Table 4.1 and down the left side. The number at the intersection of any row and column is the energy (in kilojoules) needed to *break a mole of bonds* linking the atoms of the two elements thus identified. For example, the bond energy of an H—H bond, as in the H_2 molecule, is 436 kJ/mole. Similarly, the energy required to break one mole of O=O is 498 kJ, as noted from the bottom part of the table. Bond energies for other double bonds, as well as triple bonds, are also given in the table.

Because we are doing energetic bookkeeping, we need to keep track of the energy change involved in each step and whether the energy is taken up or given off. To do this, we assume that energy that is absorbed carries a positive sign, like a deposit to your checkbook. On the other hand, energy given off is like money spent; it bears a negative sign. Bond energies are positive because they represent energy absorbed when bonds are broken. But the formation of bonds releases energy, and hence the associated energy change is negative. For example, the bond energy for the O=O double bond is 498 kJ/mole. This means that when one mole of O=O bonds is broken, the energy change is +498 kJ; correspondingly, when one mole of O=O bonds is formed, the energy change is −498 kJ.

Now we are finally ready to apply these concepts and conventions to the burning of hydrogen gas, H_2. First we need to determine how many moles of bonds are broken and how many moles of bonds are formed. We can do so using Lewis structures relating them to the balanced equation 4.4 for this combustion reaction:

$$2\ \mathrm{H{-}H} + \ddot{\underset{\cdot\cdot}{\mathrm{O}}}{=}\ddot{\underset{\cdot\cdot}{\mathrm{O}}} \longrightarrow 2\ \mathrm{H{-}\ddot{\underset{\cdot\cdot}{O}}{-}H} \quad [4.4]$$

Remember that chemical equations are written in terms of moles. In this case, equation 4.4 indicates "2 moles of $H_2(g)$ plus 1 mole of $O_2(g)$ yields 2 moles of gaseous water (water vapor)." But, to use bond energies, we need to count the number of moles of *bonds* involved. Because each H_2 molecule contains one H—H bond, 1 mole of H_2 must contain 1 mole of H—H bonds. Similarly, equation 4.4 indicates that 1 mole of O_2 contains 1 mole of O═O bonds. Each mole of water contains 2 moles of H—O bonds; thus, 2 moles of water contain 4 moles of H—O bonds. Therefore, we now have the total number of moles of bonds to be broken (2 moles of H—H and 1 mole of O═O) and those to be formed (4 moles of H—O). These number of bonds are then *multiplied* by the representative bond energy, using the appropriate sign convention (+ for bonds broken; − for bonds made).

Number of moles	Moles of bonds per mole of molecules	Energy	Energy Change (energy absorbed for bond breaking)
Bonds Broken in Reactants			
2 mole H—H	1 mole H—H per mole H_2	+436 kJ	2 × (+436 kJ) = +872 kJ
1 mole O═O	1 mole O═O per mole O_2	+498 kJ per mole O═O	1 × (+498 kJ) = +498 kJ

Total energy, bonds broken (energy *absorbed*) = (+872 kJ) + (+498 kJ)
= +1370 kJ

Number of moles	Moles of bonds per mole of molecules	Energy	Energy Change (energy released by bond making)
Bonds Made in Products			
2 moles H_2O	2 moles H—O per mole of H_2O	−467 kJ per mole H—O	4 × (−467 kJ) = −1868 kJ

Total energy, bonds made (energy *released*) = −1868 kJ
Consequently, the *overall* energy change in breaking bonds and forming new ones is (+1370 kJ) + (−1868 kJ) = −498 kJ.

A schematic representation of this calculation is presented in Figure 4.8. The energy of the reactants, 2 H_2 and O_2, is set at zero, chosen as an arbitrary but convenient value. The green arrows pointing upward signify energy absorbed to break bonds and convert the reactant molecules into individual atoms: 4 H and 2 O. The black arrow pointing downward represents energy released as these atoms are reconnected with new bonds to form the product molecules: 2 H_2O. The heavy red arrow corresponds to the net energy change of −498 kJ signifying that the overall combustion reaction is strongly exothermic. The *release* of heat corresponds to a *decrease* in the energy of the chemical system, which explains why the energy change is *negative.* The energy released when hydrogen burns is impressive (Figure 4.9).

We can also use bond energies from Table 4.1 to calculate the energy change for the combustion of methane.

$$CH_4(g) + 2\ O_2(g) \longrightarrow CO_2(g) + 2\ H_2O(g) + \text{Energy}$$

$$\text{H—C(H)(H)—H} + 2\ \ddot{O}{=}\ddot{O} \longrightarrow \ddot{O}{=}C{=}\ddot{O} + 2\ H{-}\ddot{O}{-}H$$

One mole of methane contains four moles of C—H bonds, each with a bond energy value of 416 kJ. Breaking two moles of O═O bonds requires 996 kJ (2 × 498 kJ). Bonds formed in the products are two moles of C═O bonds in one mole of CO_2 (2 × −803 kJ), and four moles of H—O bonds in two moles of water (4 × −467 kJ). Note again that bond formation is exothermic and the associated bond energies have minus signs. These energy changes can be summarized as follows.

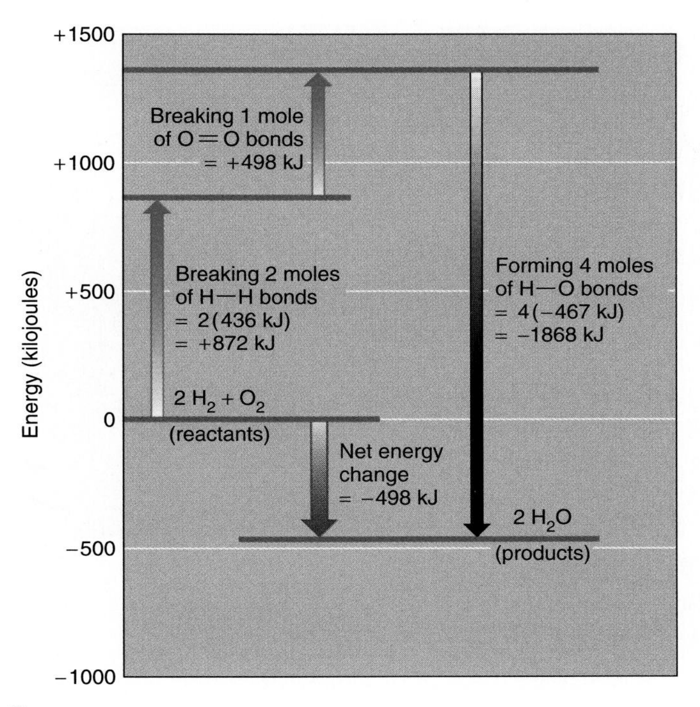

Figure 4.8

The energetics of the combustion of hydrogen to form water.

 **Figures Alive!**

Number of moles	Moles of bonds per mole of molecules	Energy	Energy Change (energy absorbed for bond breaking)
Bonds Broken in Reactants			
1 mole CH_4	4 moles of C—H bonds per 1 mole of CH_4	+416 kJ per mole C—H bonds	4 × (+416 kJ) = +1664 kJ
2 moles O_2	1 mole O=O bonds per 1 mole O_2	+498 kJ per mole O=O bonds	2 × (+498 kJ) = +996 kJ

Number of moles	Moles of bonds per mole of molecules	Energy	Energy Change (energy released with bond making)
Bonds Formed in Products			
1 mole of CO_2	2 moles of C=O bonds per 1 mole CO_2	−803 kJ per mole C=O bonds	2 × (−803 kJ) = −1606 kJ
2 moles of H_2O	2 moles of H—O bonds per 1 mole H_2O (4 moles of H—O bonds per 2 moles of H_2O)	−467 kJ per mole H—O bonds	4 × (−467 kJ) = −1868 kJ

Total energy change in breaking bonds = (+1664 kJ) + (+996 kJ) = +2660 kJ
Total energy change in making bonds = (−1606 kJ) + (−1868 kJ) = −3474 kJ
Net energy change = (+2660 kJ) + (−3474 kJ) = −814 kJ

Figure 4.9
The Hindenburg, a German airship inflated with hydrogen, was destroyed in a spectacular fire as it prepared to dock at Lakehurst, NJ in 1937.

Heats of combustion, by convention, are listed as positive values. Thus, the heat of combustion of methane calculated using bond energies is +814 kJ.

- Generally, experimental values differ slightly from those calculated using bond energies. The experimental heat of combustion of methane is 802.3 kJ; the calculated value, 814 kJ, differs by 1.5%.

The energy changes we just calculated from bond energies, −506 kJ for the burning of one mole of hydrogen and −814 kJ for one mole of methane combustion, compare favorably with the experimentally determined values. This agreement justifies our rather unrealistic assumption that all the bonds in the reactant molecules are first broken, then all the bonds in the product molecules are formed. This is not at all what actually happens. But the energy change that accompanies a chemical reaction depends on the energy *difference* between the products and the reactants, not on the particular process, mechanism, or individual steps that connect the two. This is an extremely powerful idea for understanding chemical energetics and doing related calculations. Even so, not all calculations come out as well as this one did. For one thing, the bond energies of Table 4.1 apply only to gases, so calculations using these values agree with experiment only if all the reactants and products are in the gaseous state. Moreover, tabulated bond energies are really average values. The strength of a bond depends on the overall structure of the molecule in which it is found; in other words, on what else the atoms are bonded to. Thus, the strength of an O—H bond is slightly different in HOH (H_2O), HOOH (H_2O_2), and CH_3OH. Nevertheless, the procedure illustrated here is a useful way of estimating energy changes in a wide range of reactions. The approach also helps illustrate the relationship between bond strength and chemical energy.

This analysis also helps clarify why the H_2O or CO_2 formed in combustion reactions cannot be used as fuels. There are no substances into which these compounds can be converted that have stronger bonds and are lower in energy; we cannot run a car on its exhaust.

4.6 Your Turn

Use the bond energies in Table 4.1 to calculate the heat of combustion of propane, C_3H_8 (LP or "bottled gas"). Report your answer both in kJ/mole of C_3H_8 and kJ/g C_3H_8. This is the equation for the reaction, written with structural formulas.

```
    H   H   H
    |   |   |
H — C — C — C — H + 5 O=O ⟶ 3 O=C=O + 4   O
    |   |   |                            /   \
    H   H   H                           H     H
```

Hint: Note that there are 8 moles of C—H bonds, 2 moles of C—C bonds, 5 moles of O=O bonds, 6 moles of C=O bonds and 8 moles of O—H bonds involved in the reaction.

Answer

Energy change = −2024 kJ/mole C_3H_8 or −46.0 kJ/g C_3H_8

Heat of combustion = 2024 kJ/mole C_3H_8 or 46.0 kJ/g C_3H_8.

4.7 Your Turn

Use the bond energies in Table 4.1 to calculate the heat of combustion of ethanol, C_2H_5OH, one of the components of "gasohol" fuel. This is the molecular structure of ethanol.

```
    H   H
    |   |   ..
H — C — C — O — H
    |   |   ..
    H   H
```

Answer

Energy change = −1281 kJ/mole C_2H_5OH or −27.8 kJ/g C_2H_5OH

4.8 Your Turn

Use the bond energies in Table 4.1 plus information from Chapter 2 to explain why the ultraviolet radiation absorbed by O_2 has a shorter wavelength than the UV radiation absorbed by O_3.

4.5 Getting Started: Activation Energy

Just because two substances can react in an exothermic process does not mean that they will do so, even if they are in intimate contact. For example, if you turn on the gas jet at a Bunsen burner, methane and oxygen will be present in a potentially combustible mixture. But they will not react unless a spark, a flame, or some other source of energy is supplied (Figure 4.10). This turns out to be a fortunate feature of matter. Wood, paper, and many other common materials are energetically unstable and capable of exothermic conversion to water, carbon dioxide, and other simple molecules, but they do not suddenly burst into flame.

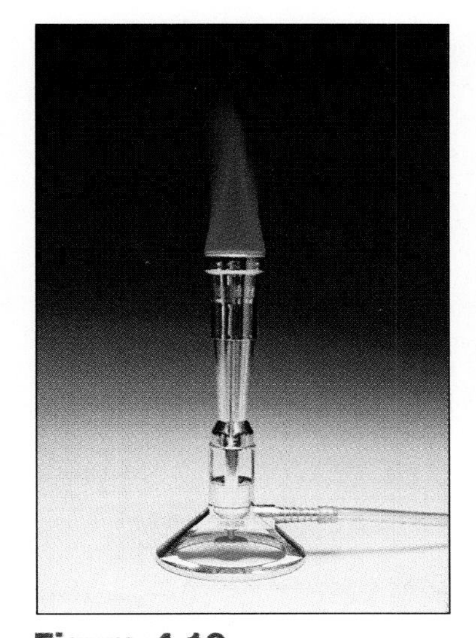

Figure 4.10
Methane from a laboratory burner ignites to produce a very hot flame.

The energy necessary to initiate a reaction is called its **activation energy.** Figure 4.11 is a schematic of the energy changes that might occur in a typical exothermic reaction. It looks a little like the cross section of a hill. The activation energy corresponds to a peak over which a boulder must be pushed before it will roll downhill. Although

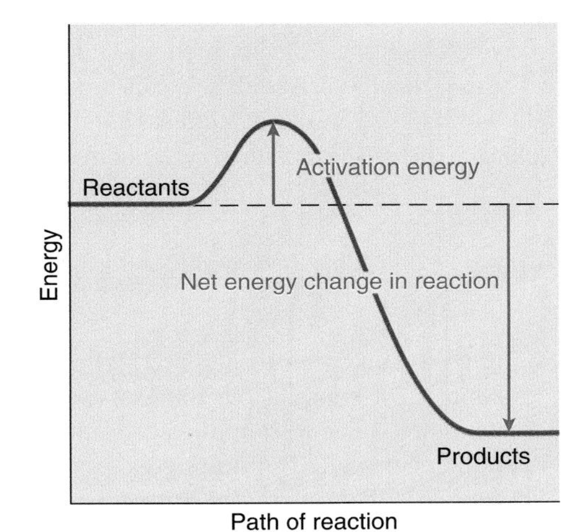

Figure 4.11
Energy-reaction pathway diagram.

energy must be expended to get the reaction (or the boulder) started, a good deal more energy is given off as the process proceeds to a lower potential energy state. Generally, reactions that occur rapidly have low activation energies; slower reactions have higher activation energies.

Activation energy is also involved in another aspect of chemical reactions that determines whether a given substance can be used as a fuel. Useful fuels react at rates that are neither too fast nor too slow. Slow reactions are of little use in producing energy because the energy is released over too long a time. For example, it would not be very practical to try to warm your hands over a piece of rotting wood, even though the overall reaction is similar to burning and forms CO_2 and H_2O. On the other hand, fast reactions can release energy too rapidly to be put to convenient use. In fact, such reactions often lead to explosions because the gases produced expand rapidly.

One way to speed up the rate of a reaction is to divide the fuel into small particles. This principle is used in fluidized-bed power plants in which pulverized coal is burned in a blast of air. The fine coal dust is quickly heated to the kindling point and the large surface area means that oxygen reacts rapidly and completely with the fuel. The combustion actually occurs at a lower temperature than that required to ignite larger pieces of coal. As a result, the generation of nitrogen oxides is minimized. If finely divided limestone (calcium carbonate) is mixed with the powdered coal, sulfur dioxide is also removed from the effluent gas. Thus, the amount of pollution is reduced while the efficiency of coal combustion is enhanced.

- Explosions occur in grain elevators (storage tanks) when an inadvertent spark or flame causes finely powdered grain to burn explosively.

Increasing temperature also increases the rates at which reactions occur. The added heat energy helps the reactants over the activation energy barrier. Catalysts, including those used in automobile catalytic converters (Section 1.12) and in petroleum refining (Section 4.8), increase reaction rates by providing alternate reaction pathways with lower activation energies.

4.9 Consider This: The Striking Case of a Single Match

It is not unusual for the electrical power to go out in a residential area during a summer thunderstorm. Imagine that the power went out in your house just as you were about to cook dinner. You scrambled around in the camping gear looking for an alternative to your electric stove. You found a portable gas stove, one match, a small grill, and a bag of charcoal briquettes. Because you have only one match and no other sources of fire, you must choose between using the portable gas stove or the charcoal grill. Which will you choose and why? Be sure to justify your decision based on what you have learned about combustion.

4.6 There's No Fuel Like an Old Fuel

Coal, oil, and natural gas possess many of the properties needed in a fuel. Therefore, most of the energy that drives the engines of our economy comes from these remnants of the past. In a very real sense, these fossil fuels are sunshine in the solid, liquid, and gaseous state. The sunlight was captured millions of years ago by green plants that flourished on the prehistoric planet. The reaction is the same one that is carried out by plants today.

$$2800 \text{ kJ} + 6\ CO_2(g) + 6\ H_2O(l) \xrightarrow{\text{chlorophyll}} \underset{\text{glucose}}{C_6H_{12}O_6(s)} + 6\ O_2(g) \qquad [4.5]$$

This conversion of carbon dioxide and water to glucose and oxygen is endothermic. It requires the absorption of 2800 kJ of sunlight per mole of $C_6H_{12}O_6$ or 15.5 kJ/g glucose formed. The reaction could not occur without the absorption of energy and the participation of a green pigment molecule called chlorophyll. The chlorophyll interacts with photons of visible sunlight and uses their energy to drive the photosynthetic process.

- $\dfrac{2800 \text{ kJ}}{1 \text{ mole glucose}} \times \dfrac{1 \text{ mole glucose}}{180 \text{ g glucose}} = 15.5 \text{ kJ/g glucose formed}$

You are already aware of the essential role of photosynthesis in the initial generation of the oxygen in the Earth's atmosphere, in maintaining the planetary carbon dioxide balance, and in providing food and fuel for creatures like us. In our bodies, we run the preceding reaction backward, like living internal combustion engines.

$$C_6H_{12}O_6(s) + 6\ O_2(g) \longrightarrow 6\ CO_2(g) + 6\ H_2O(l) + 2800\ \text{kJ} \qquad [4.6]$$

We extract the 2800 kJ released per mole of glucose "burned" and use that energy to power our muscles and nerves, though we do not do it with perfect efficiency (see 4.2 Sceptical Chymist). The same overall reaction occurs when we burn wood, which is primarily cellulose, a polymer composed of repeating glucose units. In equation 4.6, an *exo*thermic reaction, energy is shown as a product (2800 kJ) because energy is *given off* by the reaction. In an *endo*thermic reaction, energy is noted as a reactant because the reaction *absorbs* energy. Many chemical equations are written without including energy as a reactant or product. Including the energy in an equation is a way of emphasizing the energy change associated with the reaction.

- Polymers are discussed in Chapter 9.

When plants die and decay, they are also largely transformed into CO_2 and H_2O. However, under certain conditions, the glucose and other organic compounds that make up the plant only partially decompose and the residue still contains substantial amounts of carbon and hydrogen. Such conditions arose at various times in the prehistoric past of our planet, when vast quantities of plant life were buried beneath layers of sediment in swamps or on the ocean bottom. There, these remnants of vegetable matter were protected from atmospheric oxygen, and the decomposition process was halted. However, other chemical transformations occured in Earth's high-temperature and high-pressure reactor. Over millions of years, the plants that captured the rays of a young Sun were transmuted into the fossils we call coal and petroleum.

- "You will die but the carbon will not; its career does not end with you . . . it will return to the soil, and there a plant may take it up again in time, sending it once more on a cycle of plant and animal life." Jacob Bronowski, *Biography of an Atom—And the Universe*

4.7 Coal: Black Gold

The great exploitation of fossil fuels began with the Industrial Revolution, about two centuries ago. The newly built steam engines consumed large quantities of fuel, but in England, where the revolution began, wood was no longer readily available. Most of the forests had already been cut down. Coal turned out to be an even better energy source than wood because it yields more heat per gram. Burning 1 gram of coal releases approximately 30 kJ, compared to 10–14 kJ per gram of wood. This difference in heat of combustion is a consequence of differences in chemical composition. When wood or coal burn, a major energy source is the conversion of carbon to carbon dioxide. Coal is a better fuel than wood because it contains a higher percentage of carbon and a lower percentage of oxygen and water.

Coal is a complex mixture of compounds that naturally occurs in varying grades. Although coal is not a single compound, it can be approximated by the chemical formula $C_{135}H_{96}O_9NS$. This formula corresponds to a carbon content of 85% by mass. The carbon, hydrogen, oxygen, and nitrogen atoms come from the original plant material. In addition, samples of coal typically contain small amounts of silicon, sodium, calcium, aluminum, nickel, copper, zinc, arsenic, lead, and mercury. Soft lignite or brown coal is the lowest grade. The vegetable matter that makes it up has undergone the least amount of change, and its chemical composition is similar to that of wood or peat. Consequently, the heat of combustion of lignite is only slightly greater than that of wood (Table 4.2). The higher grades of coal, bituminous and anthracite, have been exposed to higher pressures in the Earth. In the process, they lost more oxygen and moisture and have become a good deal harder—more mineral than vegetable. The percentage of carbon has increased, and with it, the heat of combustion. Anthracite has a particularly high carbon content and low concentrations of sulfur, both of which make it the most desirable grade of coal. Unfortunately, the deposits of anthracite are relatively small and the United States supply is almost exhausted. We now rely most heavily on bituminous and subbituminous coal.

Generally speaking, the less oxygen a compound contains, the more energy per gram it will release on combustion because it is higher up on the potential energy scale.

Table 4.2

Fuel Value of Various U.S. Coals

Type of Coal	State of Origin	Heat Content (kJ/g)
Anthracite	PA	30.5
Bituminous	MD	30.7
Subbituminous	WA	24.0
Lignite (brown coal)	ND	16.2
Peat	MS	13.
Wood	—	10.4–14.1

This explains why burning one mole of carbon to form carbon dioxide yields about 40% more energy than obtained from burning one mole of carbon monoxide. To be sure, coal is a mixture, not a compound, but the same principles apply. Anthracite and bituminous coals consist primarily of carbon. Their heat of combustion is, gram for gram, about twice that of lignite, which contains a much lower percentage of carbon.

The global supply of coal is large and it remains a widely used fuel, but it is not without some serious drawbacks. Coal is difficult to obtain, and underground mining is dangerous and expensive. In *Invention and Technology,* Summer 1992, Mary Blye Howe reported that since 1900 more than 100,000 workers have been killed in American coal mines by accidents, cave-ins, fires, explosions, and poisonous gases. Many thousands more have been injured or incapacitated by respiratory diseases. If the coal deposits lie sufficiently close to the surface, surface or strip mining can be used. In this method, the overlying soil and rock are stripped away to reveal the coal seam, which is then removed by heavy machinery.

Much care is necessary in mining to prevent serious environmental deterioration. Current regulations require the replacement of earth and topsoil and the planting of trees and vegetation at mine sites. But, in the past, these regulations were not in place to prevent the great holes in the Earth and heaps of eroding soil that still dot regions of abandoned strip mines. Once the coal is out of the ground, its transportation is complicated by the fact that it is a solid. Unlike gas and oil, coal cannot be pumped unless it is finely divided and suspended in a water slurry.

Perhaps the most widely discussed drawback of coal is that it is a dirty fuel. It is, of course, physically dirty, but its dirty combustion products may be more serious. The unburned soot from countless coal fires in the 19th and early 20th centuries blackened buildings and lungs in many cities. Less visible but equally damaging are the oxides of sulfur and nitrogen formed when certain coals burn. If these compounds are not trapped, they can contribute to the acid precipitation that forms the subject for Chapter 6. In addition, coal suffers from the same drawback of all fossil fuels: The greenhouse gas carbon dioxide is an inescapable product of its combustion.

In spite of these less-than-desirable properties, the world's energy dependence on coal will likely increase rather than decrease. The recoverable world supply of coal is estimated as 20 to 40 times greater than world petroleum reserves. As the latter become depleted, reliance on coal will increase, unless alternative energy sources are developed. It is, however, possible that coal will not be burned in its familiar form, but rather converted to cleaner and more convenient liquid and gaseous fuels. That is a subject for a subsequent section, after we consider the properties of petroleum.

4.10 Your Turn

a. Assuming the composition of coal can be approximated by the formula $C_{135}H_{96}O_9NS$, calculate the mass of carbon (in tons) contained in 1.5 million tons of coal. This is approximately the quantity of coal that might be burned by a power plant in one year.

b. What mass of CO_2 will be produced by the complete combustion of 1.5 million tons of this coal?
c. Compute the amount of energy (in kJ) released by burning this mass of coal. Assume the process releases 30 kJ per gram of coal. Recall that 1 ton = 2000 lb and that 1 lb = 454 g.

Answers

a. Start by calculating the approximate molar mass of coal from the given formula.

$$135 \text{ mole C} \times \frac{12.0 \text{ g C}}{1 \text{ mole C}} = 1620 \text{ g C}$$

$$96 \text{ mole H} \times \frac{1.0 \text{ g H}}{1 \text{ mole H}} = 96 \text{ g H}$$

$$9 \text{ mole O} \times \frac{16.0 \text{ g O}}{1 \text{ mole O}} = 144 \text{ g O}$$

$$1 \text{ mole N} \times \frac{14.0 \text{ g N}}{1 \text{ mole N}} = 14 \text{ g N}$$

$$1 \text{ mole S} \times \frac{32.1 \text{ g S}}{1 \text{ mole S}} = 32.1 \text{ g S}$$

The sum of these values gives the molar mass of $C_{135}H_{96}O_9NS$ = 1906 g/mole. Note that there are 1620 g C in every 1906 g of coal. The mass-to-mass relationship stays the same as long as the same mass unit is used for both; the ratio is just as useful expressed in tons.

$$\text{Mass C} = 1.5 \times 10^6 \text{ tons } C_{135}H_{96}O_9NS \times \frac{1620 \text{ tons C}}{1906 \text{ tons } C_{135}H_{96}O_9NS}$$

$$= 1.3 \times 10^6 \text{ tons C} = 1.3 \text{ million tons C}$$

b. 4.8 million tons
c. 4.1×10^{13} kJ

4.8 Petroleum: Black Liquid Gold

Children in the average American city or town would be hard-pressed to find lumps of coal. Indeed, the children may never have seen coal, but they have undoubtedly seen gasoline. Around 1950 petroleum surpassed coal as the major energy source in the United States. The reasons are relatively easy to understand. Petroleum, like coal, is partially decomposed organic matter, but it has the distinct advantage of being liquid. It is easily pumped to the surface from its natural, underground reservoirs, transported via pipelines, and fed automatically to its point of use. Moreover, petroleum is a more concentrated energy source than coal, yielding approximately 40–60% more energy per gram. Typical figures are 48 kJ/g for petroleum and 30 kJ/g for coal.

The major fuel component extracted from petroleum (crude oil) is gasoline. Although it has been extracted from petroleum since the mid-1800s, gasoline became valuable and important only with the advent of the automobile and the internal combustion engine early in the 20th century. That fuel-engine partnership has led to our seemingly insatiable appetite for gasoline. In 1998, 125 billion gallons of gasoline were burned in more than 203 million American automobiles, SUVs, and light trucks traveling an astounding 2.6 trillion miles, the equivalent of 1400 round trips to the Sun. However, our capacity to consume gasoline has far outstripped our ability to produce it from crude oil extracted in this country. With 5% of the world's population, we consume 25% of the oil produced worldwide. By the 1970s, the United States was producing less than half of the crude oil it required to power its automobiles and factories, heat its homes, and lubricate its machines (Figure 4.12).

Our fragile dependency on oil from abroad continues today, increasing from 4.3 million barrels per day in 1985 to 10 million barrels per day in 2000. Our dependency on foreign oil rose from 35% in 1973 to 52% in 2000. But, over the past decade, we have shifted our sources of imported oil from the politically volatile Middle East;

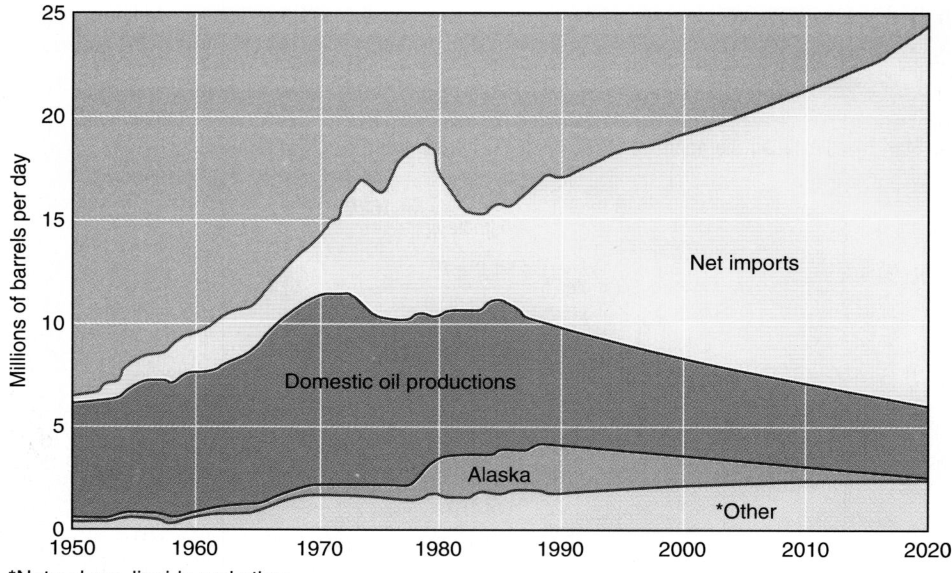

Figure 4.12

U.S. oil production and imports. At present, more than 50% of the total oil used in the United States is imported, and projections show oil imports will continue to increase.

sources in the Western Hemisphere now constitute nearly 55% of our imported oil (Figure 4.13). Saudi Arabia provides the bulk of oil imported from the Middle East.

4.11 Consider This: Shipping Oil

Figure 4.13 gives regional sources of U.S. oil imports in 2000. Use the Web to find the 10 countries that supply the most oil to us.

a. List the countries in order of the amount of oil provided (first to tenth).
b. Which non-Western Hemisphere countries are in this list?
c. Which countries, if any, surprised you as sources of our imported oil? Why?

As a nation, our voracious appetite for oil was met by consuming an average of nearly 20 million barrels of oil *daily* in 2000. Two-thirds of this was consumed for transportation. Unlike coal, however, crude oil is not ready for immediate use when it is extracted from the ground. Crude oil must first be refined—a process that has given gainful employment to many chemists and chemical engineers (and quite a few others). It has also provided an amazing array of products. Petroleum is a complex mixture of thousands of different compounds. The great majority are **hydrocarbons,** molecules consisting of only hydrogen and carbon atoms. The hydrocarbon molecules can contain from one to as many as 60 carbon atoms per molecule. Concentrations of sulfur and

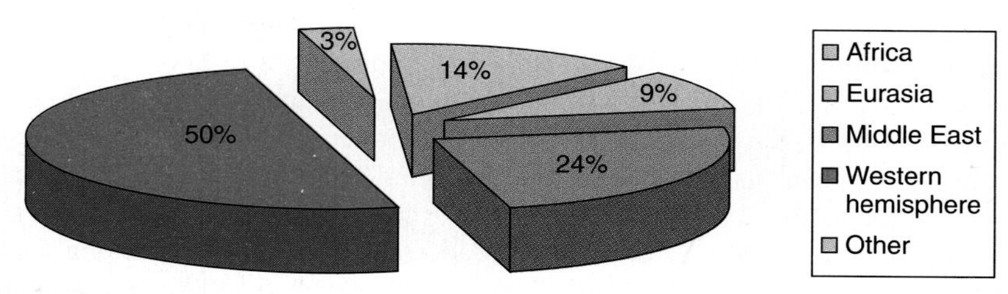

Figure 4.13

Regional sources of oil imported by the U.S., 2000.

(Source of data: National Energy Policy Development Group Report, figure 8.3.)

other contaminating elements are generally quite low, minimizing polluting combustion products.

The oil refinery has become a symbol of the petroleum industry (Figure 4.14). In the refining process, the crude oil is separated into individual compounds or, more often, into **fractions** that consist of compounds with similar properties. This fractionation is accomplished by distillation. **Distillation** is a purification or separation process in which a solution is heated to its boiling point and the vapors are condensed and collected. The petroleum is pumped into an industrial-sized retort or still, and the mixture is heated. As the temperature increases, the components with the lowest boiling points are the first to vaporize. The gaseous molecules escape from the liquid and move up a tall distillation column or tower. There the cooled vapors recondense into the liquid state, only this time in a much purer condition. By varying the temperature of the still and the fractionating column, a petroleum engineer can regulate the boiling point range of the fractions distilled and condensed. Higher temperatures mean higher boiling compounds. All fractions are not produced in equal proportions; market demand dictates which are maximized and which are decreased.

Figure 4.15 illustrates a distillation tower and lists some of the fractions obtained. They include gases such as methane, liquids such as gasoline and kerosene, waxy solids, and a tarry asphalt residue. Note that the boiling point goes up with increasing number of carbon atoms in the molecule and hence with increasing molecular mass and number of electrons. Heavier, larger molecules with an abundance of electrons are attracted to each other with stronger intermolecular forces than are lighter, smaller molecules having fewer electrons. Higher temperatures are required to vaporize the compounds of higher molecular mass to overcome their stronger intermolecular forces.

Figure 4.14
An oil refinery, symbol of the petroleum industry.

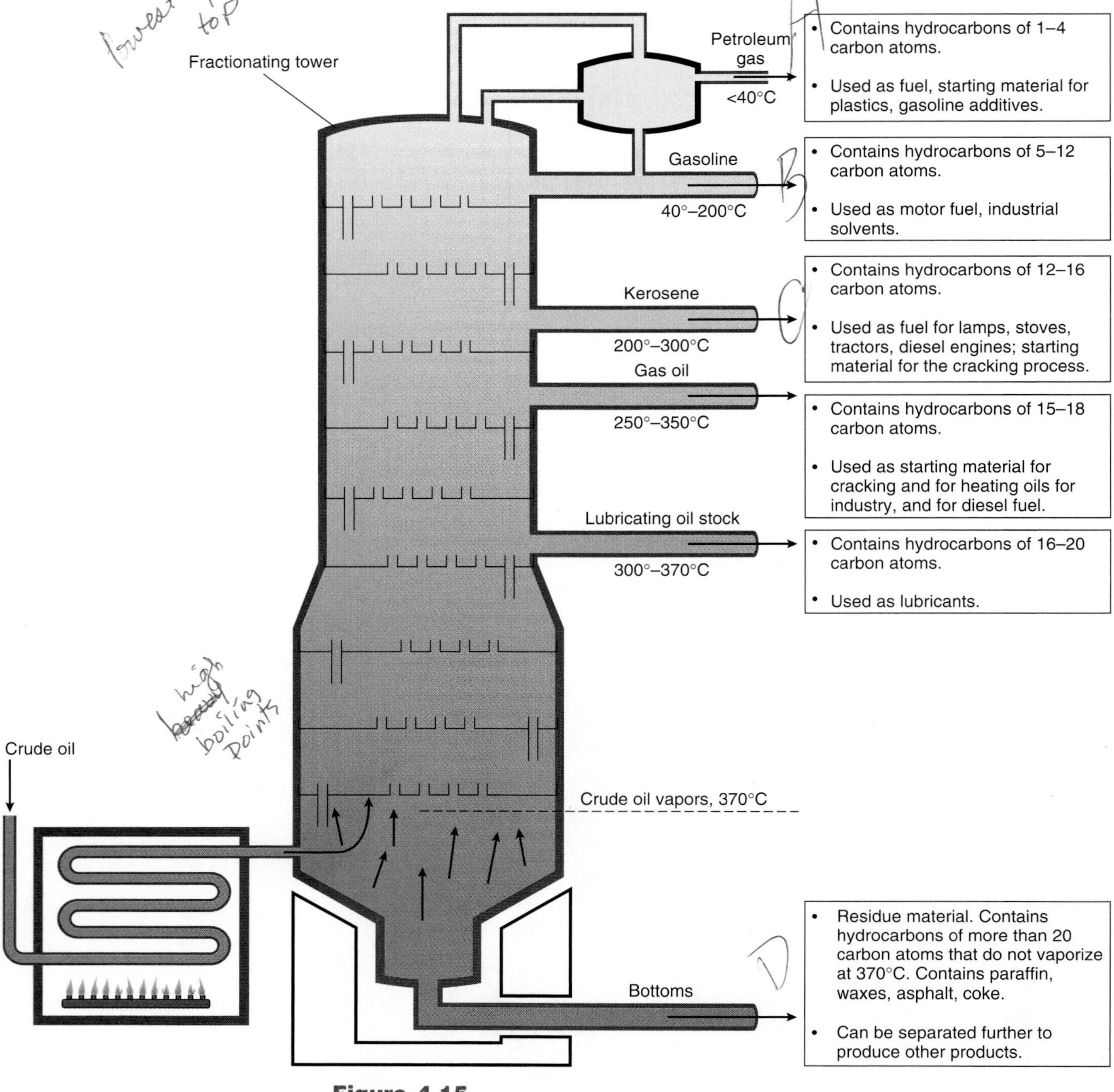

Fractionating tower

Petroleum gas <40°C
- Contains hydrocarbons of 1–4 carbon atoms.
- Used as fuel, starting material for plastics, gasoline additives.

Gasoline 40°–200°C
- Contains hydrocarbons of 5–12 carbon atoms.
- Used as motor fuel, industrial solvents.

Kerosene 200°–300°C
- Contains hydrocarbons of 12–16 carbon atoms.
- Used as fuel for lamps, stoves, tractors, diesel engines; starting material for the cracking process.

Gas oil 250°–350°C
- Contains hydrocarbons of 15–18 carbon atoms.
- Used as starting material for cracking and for heating oils for industry, and for diesel fuel.

Lubricating oil stock 300°–370°C
- Contains hydrocarbons of 16–20 carbon atoms.
- Used as lubricants.

Crude oil

Crude oil vapors, 370°C

Bottoms
- Residue material. Contains hydrocarbons of more than 20 carbon atoms that do not vaporize at 370°C. Contains paraffin, waxes, asphalt, coke.
- Can be separated further to produce other products.

Figure 4.15

Diagram of a fractional distillation tower and various fractions.

Because of differences in properties, the various fractions distilled from crude oil have different uses. Indeed, the great diversity of products obtained has made petroleum a particularly valuable source of matter and energy. The lowest boiling components are gases at room temperature, and are used as "bottled gas" and other fuels. Sometimes flames at the tops of refinery towers signal that the gas is being burned off in what seems to be an unnecessary waste of a valuable resource. The gasoline fraction, containing hydrocarbons with 5 to 12 carbon atoms per molecule, is particularly important to our automotive civilization. Efforts at designing and mass producing self-propelled vehicles were largely unsuccessful until petroleum provided a convenient and relatively safe liquid fuel. The kerosene fraction is somewhat higher boiling, and it finds use as a fuel in

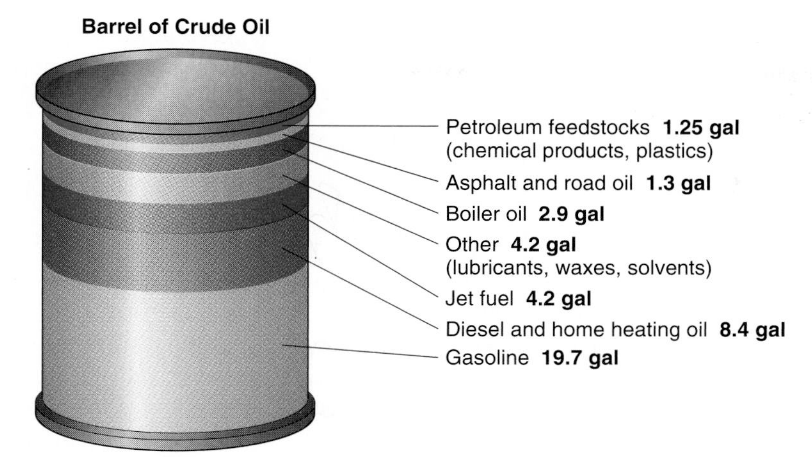

Figure 4.16
End uses for products from the refining of 1 barrel (42 gallons) of crude oil.

diesel engines and jet planes. Still higher boiling fractions are used to fire furnaces and as lubricating oils.

The refining of a barrel of crude oil (42 gallons) provides an impressive array of products, the vast majority of which (20 gal) is gasoline (Figure 4.16). A staggering 35 gal of the 42 gal in a barrel of crude oil is simply burned for heating and transportation. The remaining 7 gal are for nonfuel uses, including only 5 quarts (1.25 gal) set aside to serve as nonrenewable starting materials (reactants, commercially called *feedstocks*) to make the myriad of plastics, pharmaceuticals, fabrics, and other carbon-based industrial products so common in our society.

It is appropriate that a discussion of petroleum also should include natural gas. This fuel, which is mostly methane, currently provides heat for two-thirds of the single-family homes and apartment buildings in the United States. Recently, there has also been increased interest in using natural gas as an energy source for generating electricity and for powering cars and trucks. A distinct advantage of natural gas is that it burns much more completely and cleanly than other fossil fuels. Because of its purity it releases essentially no sulfur dioxide. Natural gas emits only very low levels of unburned volatile hydrocarbons, carbon monoxide, and nitrogen oxides; and it leaves no residue of ash or heavy metals. Moreover, on a per joule of energy basis, natural gas produces 30% less carbon dioxide than oil and 43% less carbon dioxide than coal.

4.12 Consider This: The Changing Mix of Products from a Barrel of Crude Oil

Chemical research has contributed to increasing the amount of gasoline derived from a barrel of crude oil. For example, in 1904 a barrel of crude oil produced 4.3 gal gasoline, 20 gal kerosene, 5.5 gal fuel oil, 4.9 gal lubricants, and 7.1 gal miscellaneous products. By 1954, the mix was 18.4 gal gasoline, 2.0 gal kerosene, 16.6 gal fuel oil, 0.9 gal lubricants, and 4.1 gal miscellaneous products. Compare these values with those shown in Figure 4.16 and offer some reasons why the mix of products has changed over time.

4.9 Manipulating Molecules

Research has shown that many of the compounds distilled from crude oil are not ideally suited for the desired applications. Nor does the normal distribution of molecular masses of the distilled fractions correspond to the prevailing use pattern. For example, the demand for gasoline is considerably greater than that for higher boiling fractions. Gasoline that comes directly from the fractionating tower represents less than 50% of the original crude oil. Heavier and lighter crude oil fractions can be manipulated to form still more gasoline. Therefore, chemistry is used to rearrange large molecules by breaking them into smaller ones suitable to be used in gasoline, a process called **cracking.** For example, a hydocarbon with 16 carbons can be cracked into two almost equal fragments,

$$C_{16}H_{34} \longrightarrow C_8H_{18} + C_8H_{16} \quad [4.7]$$

or into different sized ones.

$$C_{16}H_{34} \longrightarrow C_5H_{12} + C_{11}H_{22} \quad [4.8]$$

Note that the numbers of carbon and hydrogen atoms are unchanged from reactants to products; the larger reactant molecules simply have been rearranged to form smaller, more economically important molecules. When cracking is achieved by heating the starting materials, the process is called thermal cracking. However, valuable energy can be saved if catalysts are used to speed up the molecular breakdown at lower temperatures in an operation called catalytic or cat cracking. The catalysts employed in this process are chemically similar to the ion exchangers used in water softening (see Chapter 5). Important cracking catalysts were developed by research chemists at Mobil Oil and at Union Carbide.

If there is an excess of small molecules and a need for intermediate-sized ones, such as those in gasoline, the former can be catalytically combined to form the latter.

$$4\ C_2H_4 \longrightarrow C_8H_{16} \quad [4.9]$$

The refining process can also rearrange the atoms within a molecule. It turns out that not all the molecules with a single chemical formula are necessarily identical. For example, octane, an important component of gasoline, has the formula C_8H_{18}. Careful analysis discloses that there are 18 different compounds with this formula. Although the chemical and physical properties of these forms are similar, they are not identical. For example, the substance called *octane* has a boiling point of 125°C, whereas that of the compound commonly known as *isooctane* is 99°C. Different compounds with the same formula are called **isomers.** Isomers differ in molecular structure—the way in which the constituent atoms are arranged. This is like the letters a, e, and t: they can be organized to form three different recognizable words—ate, eat, and tea. The structures of octane and isooctane are illustrated below.

```
                                                  H
                                                  |
                                               H—C—H  CH3
                                               H  |  H  |  H
                                               |  |  |  |  |
                                            H—C—C—C—C—C—H
  H  H  H  H  H  H  H  H                       |  |  |  |  |
  |  |  |  |  |  |  |  |                       H  |  H  H  H
H—C—C—C—C—C—C—C—C—H                            H—C—H
  |  |  |  |  |  |  |  |                          |
  H  H  H  H  H  H  H  H                          H
```

octane isooctane

The molecules of both isomers consist of 8 carbon atoms and 18 hydrogen atoms, but these atoms are arranged differently in the two compounds. In octane all the carbon atoms are in an unbranched ("straight") line; in isooctane the carbon chain is branched. Chapter 10 includes more information about isomers and how to interpret these structures.

Both octane and isooctane have essentially the same heat of combustion, but the latter compound ignites much more easily. In a well-tuned car engine, gasoline vapor and air are drawn into a cylinder, compressed by a piston, and ignited by a spark. But compression alone is often enough to ignite the fuel before the spark occurs. This premature firing is called **preignition** (Figure 4.17). **Knocking** occurs after the spark ignites the fuel, but before the flame has burned all of the fuel in the cylinder. Knocking and preignition reduce the efficiency of the engine because the energy of the exploding and expanding gas is not applied to the pistons at the optimum time. Extensive knocking can cause engine damage.

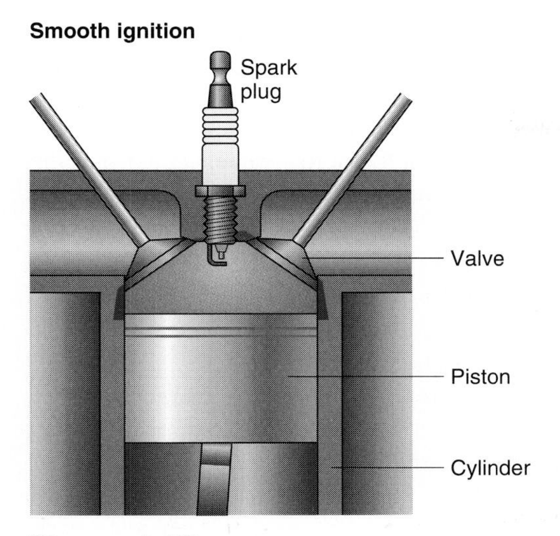

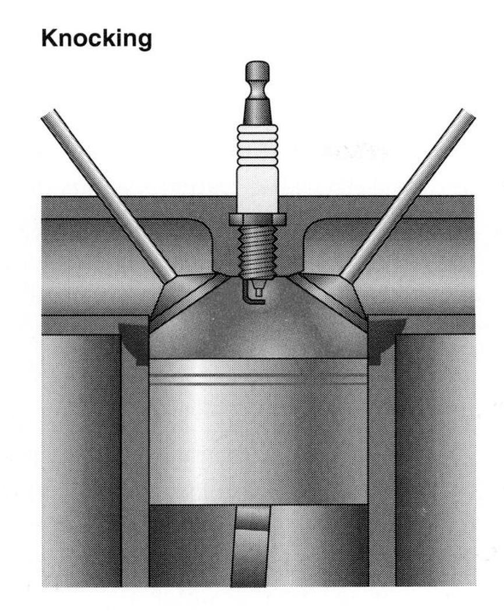

Figure 4.17
Smooth ignition and knocking.

Unlike octane, isooctane is very resistant to preignition. Isooctane is the standard of excellence for rating the tendency of fuels to knock, though some compounds are even better. The performance of isooctane in an automobile engine has been measured and arbitrarily assigned an octane rating of 100. On this scale, octane has an octane rating of −20 (Table 4.3). When you go to the gasoline pump and fill up with 87 octane, the gasoline has the same knocking characteristics as a mixture of 87% isooctane (octane number 100) and 13% heptane (octane number zero). Higher-grade gasolines are also available: 89 octane (regular plus) and 92 octane (premium); these contain a greater percentage of higher octane compounds (Figure 4.18).

Figure 4.18
Gasoline is available in 87, 89, and 92 octane.

It is possible to rearrange or "reform" octane to isooctane, thus greatly improving its performance. This is accomplished by passing octane over a catalyst consisting of rare and expensive elements such as platinum (Pt), palladium (Pd), rhodium (Rh), or iridium (Ir). Reforming isomers to improve octane rating has become particularly important because of the nationwide efforts to ban lead from gasoline. Lead does not occur naturally in petroleum. But in the 1920s, the compound tetraethyl lead, $Pb(C_2H_5)_4$, was first added to gasoline to reduce knocking and increase octane rating in more powerful engines (Figure 4.19). The strategy proved successful, but it was not without environmental consequences. In a very short time, automobile internal combustion engines became a major source of lead, emitted into the environment as volatile lead oxide through cars' tailpipes. Lead is a heavy metal poison, with cumulative neurological effects that are particularly damaging to young children. Therefore, major efforts have

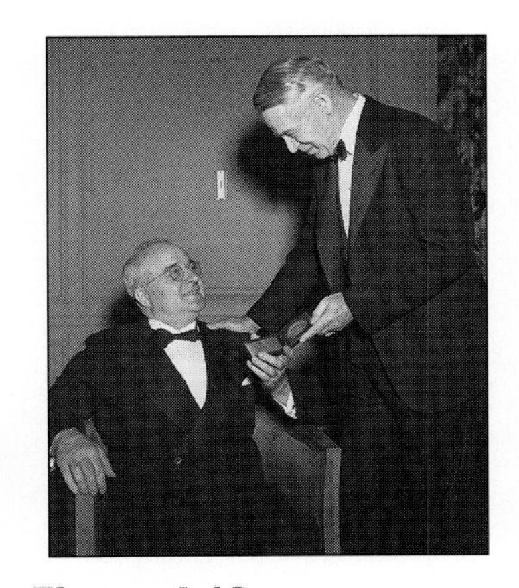

Figure 4.19
Dr. Thomas Midgley (seated) receiving the Willard Gibbs Medal from the American Chemical Society President Dr. Harry N. Holmes. The medal was awarded to Midgley for his research that developed, among other things, tetraethyl lead and CFCs.

Table 4.3

Octane Ratings of Several Substances

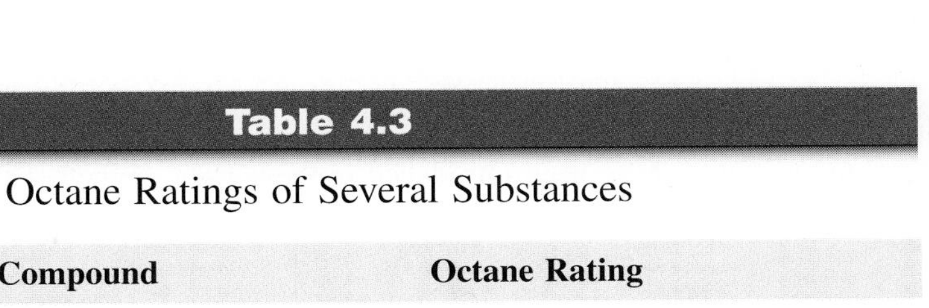

Compound	Octane Rating
Octane	−20
Heptane	zero
Isooctane	100
Methanol	107
Ethanol	108
MTBE	116

been launched in the United States to discontinue the practice of adding lead compounds to motor fuel. Since 1976, all new cars and trucks sold in the United States have been designed to run on unleaded gasoline. The results have been dramatic. In 1970, approximately 200 thousand metric tons of lead were released into the atmosphere. Over the next 30 years, lead emissions dropped to less than 5% of that value. On balance, this decrease has been environmentally beneficial, but there have been associated costs, as the next activity suggests. About 50 countries worldwide still allow leaded gasoline containing 0.8 g of lead per liter, about 75% of the amount previously permitted in the United States.

4.13 Consider This: Leaded versus Unleaded

Modern car engines designed to burn unleaded gasoline are somewhat less efficient than older engines designed to burn leaded fuel. It can therefore be argued that the switch to unleaded gasoline contributed to greater fuel consumption and air pollution by unburned exhaust residues. Moreover, some of the hydrocarbons introduced into gasoline when lead was phased out have been identified as possibly causing cancer. Some countries are still facing the switch from leaded to unleaded gasoline. To help with their decision making, draw up a list of risks and benefits associated with leaded and unleaded gasoline, and indicate the additional information you would need to appropriately weigh these risks and benefits.

4.10 Newer Fuels: Oxygenated and Reformulated Gasolines

Eliminating the use of tetraethyl lead as an octane enhancer required finding substitutes for it so as to produce gasoline with sufficiently high octane ratings to meet the requirements of modern automobile engines. Several substitutes are used, including ethanol and MTBE (methyl-*tertiary*-butyl ether), each with an octane rating greater than 100.

```
     H         CH3              H   H
     |    ..   |                |   |    ..
H ─ C ─ O ─ C ─ CH3       H ─ C ─ C ─ O ─ H
     |    ..   |                |   |    ..
     H         CH3              H   H

        MTBE                    ethanol
```

These additives also are used in **oxygenated gasolines,** which are blends of gasolines with oxygen-containing compounds such as MTBE, ethanol, or methanol (CH_3OH). Because they already contain oxygen, oxygenated gasolines burn more cleanly by producing less carbon monoxide than their nonoxygenated counterparts, thereby reducing CO emissions. Cities with excessive wintertime carbon monoxide emissions are required by the Clean Air Act Amendments of 1990 to use oxygenated gasolines that contain 2.7% oxygen by weight. Oxygenated gasoline is also required in the 40 U.S. cities with the highest air pollution. Questions about the efficacy of MTBE and oxygenated gasolines in cold weather have been raised in Alaska and Wisconsin. Some data indicate an actual decrease in air quality and an increase in respiratory health problems since oxygenated gasolines have been introduced in these states. Whether there is a causal relationship among these factors is still being studied.

- Oxygenated gasoline is required to be sold in Los Angeles, New York, Baltimore, Chicago, Hartford, Houston, Milwaukee, Philadelphia, Sacramento, and San Diego, cities with very bad smog.

4.14 Your Turn

The molecular formula of MTBE is $C_5H_{12}O$; that of ethanol is C_2H_6O. Calculate the percent (by mass) of oxygen in

a. MTBE
b. ethanol

Hint: See Section 3.8 for a review of such calculations.

4.15 Consider This: Ethanol and Politics

The Clean Air Act Amendments of 1990 contains an amendment that requires gasoline to contain 2% of oxygenates such as ethanol. The amendment was championed by Senators Tom Daschle of South Dakota and Bob Dole of Kansas. Why did Senators Daschle and Dole have so much interest in the amendment?

During the 1990s, reformulated gasolines were developed to meet the mandate of the Clean Air Act of 1990 and its amendments. **Reformulated gasolines (RFGs)** are oxygenated gasolines that also contain a lower percentage of certain more volatile hydrocarbons such as benzene found in nonoxygenated conventional gasoline. RFGs cannot have greater than 1% benzene (C_6H_6) and must be at least 2% oxygenates. Because of their composition, reformulated gasolines evaporate less easily than conventional gasolines, and produce less carbon monoxide emissions. The more volatile hydrocarbons (benzene, etc.) in conventional gasoline also are involved in tropospheric ozone formation, especially in high-traffic metropolitan areas. To diminish ozone levels in these areas, RFGs have been required since 1995 in the nine cities in the United States with the highest ozone levels, along with voluntary compliance in nearly 90 other metropolitan areas with ozone levels above the standard. One of the goals of the Clean Air Act Amendments of 1990 is to improve air quality by reducing the amount of carbon monoxide pollution formed during combustion of automobile fuels. Because they already contain oxygen, reformulated gasolines burn more cleanly by producing less carbon monoxide than nonoxygenated fuels.

- The molecular structure of benzene is discussed in Section 10.3.

The use of RFGs and oxygenated gasolines exemplifies a risk-benefit situation. The potential benefits are considerable. For example, using RFGs in the San Francisco Bay area could prevent 40 tons of hydrocarbon emissions and 160 tons of CO daily, the equivalent of removing over half a million vehicles from the road. If proportional reductions are realized in other cities across the country, the impact on air quality improvement truly would be significant. Yet, RFGs and oxygenated gasolines may not be risk free; they have not been used long enough for possible long-term adverse effects, if any, to arise. Currently, risk levels are presumed to be low, with benefits outweighing risks. But the usefulness of RFGs and oxygenates to abate CO during the winter in cold climates has been called into question in several areas. Commenting on MTBE use, Dr. Sandra Mohr, a National Research Council committee member and occupational physician at the New Jersey Environmental and Occupational Health Sciences Institute, reported in the *Trenton Times*: "From a health effects point of view, we can't say there are any long-term health consequences [to MTBE use] but we recommend a cost-benefit analysis of the wintertime oxygenated fuel program."

Although MTBE is now a major component of oxygenated and reformulated gasolines, its continued use for this purpose is now in doubt. In California, MTBE has leaked from underground gasoline storage tanks at gas stations into ground water. In March 1999, California's governor decided to phase out the use of MTBE in gasoline in that state by the end of 2002. Ten other states, many in the Northeast, also plan to phase out or restrict the use of MTBE.

MTBE notice on a California gasoline pump.

4.16 Consider This: What's Up with MTBE?

Use the Web to find this information.

a. The status of the MTBE phase out in California
b. Which other states are phasing out or plan to phase out MTBE in gasoline
c. How the phase out plans for these states resemble or differ from that in California

4.17 Consider This: Your Contribution to Air Quality

According to the EPA, driving a car is "a typical citizen's most 'polluting' daily activity."

a. Do you agree? Why or why not?
b. What pollutants do cars emit?

Information on automobile emissions provided by the EPA (together with the information in your text) can help you fully answer this question. Check the *Chemistry in Context* web site for a direct link.

c. Reformulated gasolines (RFGs) play a role in reducing emissions. Where in the country are RFGs required? Check the current list published on the Web by the EPA.

d. Explain which emissions reformulated gasolines are supposed to lower.

4.11 Seeking Substitutes

Because the world's coal supply far exceeds the available oil reserves, there is interest in converting coal into gaseous and liquid fuels that are identical with or similar to petroleum products. As a matter of fact, some of the appropriate technology is quite old. Before large supplies of natural gas were discovered and exploited, cities were lighted with water gas. This is a mixture of carbon monoxide and hydrogen, formed by blowing steam over hot coke (the impure carbon that remains after volatile components have been distilled from coal):

$$\underset{\text{coke}}{C(s)} + H_2O(g) \longrightarrow \underbrace{CO(g) + H_2(g)}_{\text{water gas}} \qquad [4.10]$$

This same reaction is the starting point for the Fischer-Tropsch process for producing synthetic gasoline. The carbon monoxide and hydrogen are passed over an iron or cobalt catalyst, which promotes the formation of hydrocarbons. These can range from the small molecules of gases like methane, CH_4, to the medium-sized molecules (containing five to eight carbon atoms) typically found in gasoline. This process, developed during the 1930s in Germany, is economically feasible only where coal is plentiful and cheap, and oil is scarce and expensive. This is the case in South Africa, where 40% of gasoline is obtained from coal. In the future, such technology may also become competitive in other parts of the world.

Concerns about the dwindling supply of petroleum have also led to the use of renewable energy sources. This generally means **biomass**—materials produced by biological processes. One such source, wood, was much touted during the 1970s energy crisis. But the energy demands of our modern society cannot possibly be met by burning wood. Burning trees would also destroy effective absorbers of carbon dioxide while adding that greenhouse gas and other pollutants to the atmosphere. In some parts of the country, the use of wood-burning stoves has been severely curtailed because the smoke and soot particles produced have a negative effect on air quality.

Ethanol (C_2H_5OH or CH_3CH_2OH) is another alternative fuel produced from renewable biomass. It is formed by a method known since ancient times, the fermentation of starch and sugars in grains such as corn. Enzymes released by yeast cells catalyze the reaction typified by this equation.

$$\underset{\text{glucose}}{C_6H_{12}O_6} \xrightarrow{\text{enzymes}} 2\ C_2H_5OH + 2\ CO_2 \qquad [4.11]$$

Ethanol can also be prepared commercially in large quantities by the reaction of water (steam) with ethylene, C_2H_4 (CH_2CH_2).

$$CH_2CH_2(g) + H_2O(g) \longrightarrow CH_3CH_2OH(l)$$

When the second method is used to produce ethanol for oxygenated fuels, any residual water must scrupulously be removed so that it will not create problems in an automobile engine. Making ethanol by fermenting grains creates a fuel that, unlike gasoline, is renewable because crops can continue to be planted. Thus, the source of the fuel can be replaced.

The burning of ethanol, as in the following equation, releases 1367 kJ per mole of C_2H_5OH.

$$C_2H_5OH(l) + 3\ O_2(g) \longrightarrow 2\ CO_2(g) + 3\ H_2O(l) + 1367\ \text{kJ} \qquad [4.12]$$

The energy output of 1376 kJ per mole of ethanol burned corresponds to 29.7 kJ/g. This value is less than the 47.8 kJ/g produced by burning C_8H_{18}, because the ethanol is already partially oxidized. Nevertheless, ethanol is being mixed with gasoline to form "gasohol." At the usual concentration of 10% ethanol, gasohol can be used without modifying standard automobile engines. Higher concentrations of alcohol require changes in design. Of the 13 million vehicles in Brazil, more than 4 million use pure ethanol, made from fermenting sugar cane juice. Most of the rest of Brazilian cars operate on a mixture of ethanol and gasoline.

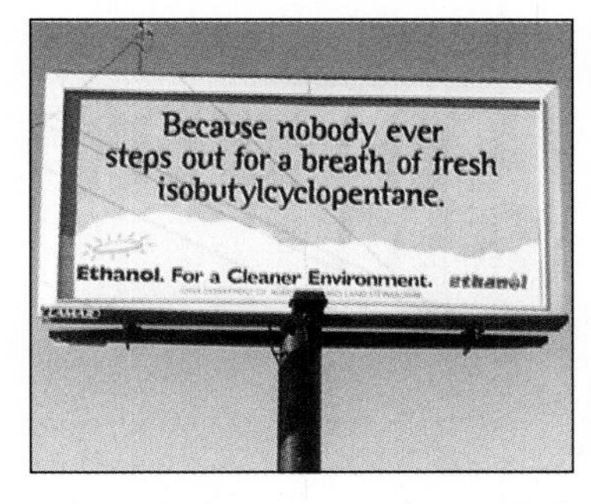

Billboard advertisement for gasohol (Iowa Corn Grower's Association).

- Racing engines have been modified to run on pure methanol (methyl alcohol), CH_3OH.

However, ethanol as a fuel is not without its drawbacks and critics. There are those who point to the fact that a gram of ethanol does not produce as much energy as a gram of gasoline. In addition, those opposed to ethanol as a fuel question whether valuable farmland, normally used to grow crops such as corn that feed people and animals, should be used to produce grain for ethanol. Others claim that ethanol burns incompletely to form acetaldehyde, CH_3CHO, a component of urban smog, and a compound that can cause allergic reactions in some people.

In these statements about ethanol, we see some of the basic issues raised in this chapter. An energy-intensive society such as the United States seeks ways to achieve what to some may be incompatible goals—improving air quality while continuing to consume increasing amounts of fossil fuels to provide relatively low-cost energy for transportation and other uses. But the days of plentiful (and consequently cheap) fuel and energy are obviously limited. The planet has a finite supply of coal, oil, and gas. Colin Campbell and Jean Laherrère, in their 1998 *Scientific American* essay "The End of Cheap Oil," suggest that world production of conventional oil will peak within the next decade to be followed by a permanent decline. The fact that previous forecasts of major energy crises have not developed thus far is by no means reason for complacency.

Whether ethanol will make a significant contribution to energy production depends on other factors, especially agriculture and politics. The great variety of alcoholic beverages indicates that C_2H_5OH can be prepared by fermentation from almost any plant product—corn, wheat, barley, rice, sugar beets, sugar cane, grapes, apples, potatoes, dandelions, and so on. But these sources also serve as food for humans or other animals. Therefore, the use of agricultural products for the production of fuel must depend on supply and demand, surpluses and shortages. Currently, the United States produces a significant surplus of corn and other grains that could be converted to ethanol. But, in a recent paper, Bernard Gilland, a civil engineer, estimated that meeting only 10% of the world's current primary energy demand with alcohol would require that one-quarter of the world's cropland be removed from food and feed production. Clearly, there are limits to the amount of energy we can obtain from the fermentation of crops.

The use of ethanol as a petroleum substitute or gasoline oxygenating agent has become a political hot potato in the United States. The issues involve not only ethanol, but other oxygen-containing compounds such as MTBE. Since 1999, California has sought a waiver of the Clean Air Act requirement for oxygenates as a way of eliminating MTBE from gasoline, arguing that oil refiners have better and cheaper ways to blend cleaner-burning gasoline without using oxygenates. Under the Clinton administration the EPA decided to allow the waiver, convinced that doing so would reduce emissions; however, President Clinton did not complete action on the waiver. In June 2001, the EPA, under the George W. Bush administration, denied the waiver, saying: "After an extensive analysis, the Agency concluded that there is significant uncertainty over the change in emissions that would result from a waiver. California has not clearly demonstrated what the impact on smog would be from a waiver of the oxygen mandate." The EPA ruled that, if California banned the use of MTBE, the state would have to replace it specifically with ethanol. If ethanol replaces MTBE in California, estimates are that the state would use about 580 million gallons of ethanol for fuel per year, roughly one-third of the ethanol now produced in the U.S. In anticipation of a shift to a greater use of ethanol as a fuel, the nation's ethanol producers are building their capacity to produce 3.5 billion gallons by 2003.

4.18 Consider This: To MTBE or Not to MTBE? That Is the Question

Senator Tom Harkin (D-Iowa), chair of the Senate Agriculture Committee, has been a foe of California's request for a waiver of the Clean Air Act requirement for oxygenates. He predicts a need of 500 million additional gallons of ethanol per year to meet the replacement.

a. What might be the basis for the Senator's opposition?
b. Does his prediction of 500 million additional gallons of ethanol per year seem reasonable? Why?
c. What would be the likely position of the National Corn Growers Association on this replacement? Check this by going to its web site. Does it agree with your original answer?

4.19 Sceptical Chymist: Keeping Track of Ethanol

There are 2962 grams of ethanol in a gallon of ethanol. California has approximately 17 million passenger vehicles. Assume that 400 million of the 500 million gallons predicted by Senator Harkin are to be used by these vehicles. Also, assume that the ethanol is used exclusively in gasohol. Ten gallons of gasohol contains 1 gal of ethanol and 9 gal of gasoline. Show calculations that will support or refute Senator Harkin's estimate.

The battle lines regarding the use of ethanol as a fuel are clearly drawn, largely on the basis of self-interest. Supporting the greater use of ethanol are the EPA, Archer Daniels Midland (ADM, a huge agribusiness), and over 20 farm groups, including the National Corn Growers Association (Figure 4.20). In opposition are the American Petroleum Institute, the petroleum refiners and gasoline companies, and the Sierra Club. U.S. senators stake out positions on the issue depending on whether they are from agricultural states or ones tied closely to oil. There is a good deal at stake—among other things, 100 million to 200 million bushels of corn per year. The pro-petroleum faction responds that, even with extensive farm support programs, ethanol is significantly more expensive per gallon and per kilojoule than conventional gasoline. Amid all the lobbying, the claims and counterclaims, the charges and the rebuttals, it is frustratingly difficult to find the facts, even the scientific ones. For example, some researchers have argued that when all aspects of production and processing are included, burning ethanol actually contributes more CO_2 to the environment than burning petroleum-based fuel. A

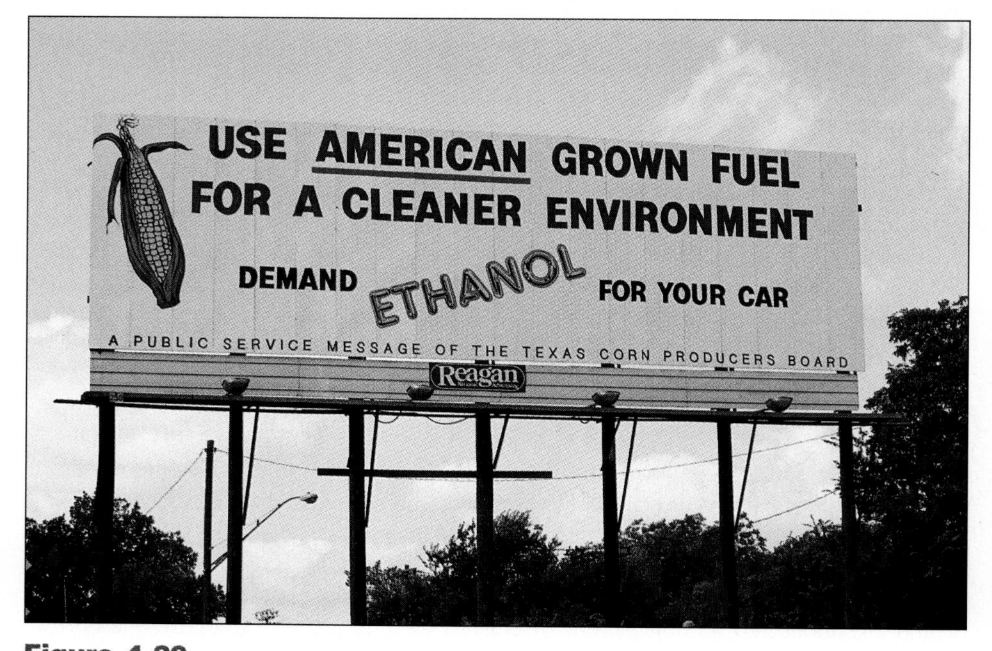

Figure 4.20
An advertisement for gasohol made using ethanol.

related issue is the amount of energy required to produce a gallon of ethanol. The Sun is not the only source involved. Energy is required to plant, cultivate, and harvest the corn; to produce and apply the fertilizers; to distill the alcohol from the fermented mash; and to manufacture tractors. Hard, accurate data are difficult to get, but some sources claim that more energy goes into producing a gallon of ethanol than can be obtained by burning it.

4.20 Consider This: Gasohol

Unlike fuels obtained from petroleum, a nonrenewable source, ethanol can be produced from renewable resources. One method of producing ethanol is through the fermentation of crops such as sugar cane, potatoes, and corn.

a. Draw up a list of reasons that argue for using crops for food and also a list of reasons for converting crops to ethanol for use in fuels.
b. What factors determine how you decide which is the optimal use of these crops?

Yet another potential energy source is a commodity that is cheap, always present in abundant supply, and always being renewed—garbage. No one is likely to design a car that will run on orange peels and coffee grounds, but approximately 140 power plants in the United States do just that. One of these, pictured in Figure 4.21, is the Hennepin County Resource Recovery Facility in Minneapolis, Minnesota. Hennepin County produces about one million tons of solid waste each year. About one-third of that is burned in the waste-to-energy facility. The heat evolved is used to generate electricity via a process described in the next section. At full capacity, the Minneapolis plant produces 37,000 kJ per second, enough energy to meet the needs of 40,000 homes. One truckload of garbage (about 27,000 lb) generates the same quantity of energy as 21 barrels of oil.

This resource recovery approach, as it is sometimes called, simultaneously addresses two major problems—the growing need for energy and the growing mountain of waste. The great majority of the trash is converted to carbon dioxide and water and no supplementary fuel is needed. The unburned residue is disposed of in landfills, but it represents only about 10% of the volume of the original refuse. Although some citizens have expressed concern about gaseous emissions from garbage incinerators, the incinerator's stack effluent is carefully monitored and must be maintained within established

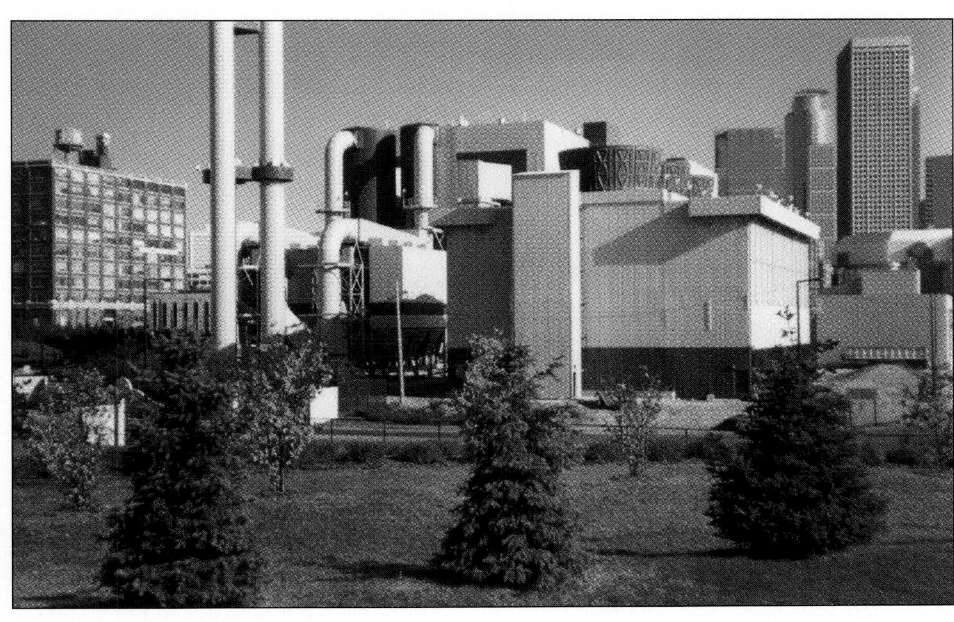

Figure 4.21
Hennepin County Resource Recovery Facility, a garbage-burning power plant.

limits. Both Japan and Germany are making considerably greater use of waste-to-energy technology than the United States.

Perhaps the ultimate example of using waste as an energy source is provided by methane generators. Rural China and India have over one million reactors in which animal and vegetable wastes are fermented to form biogas. This gas, which is about 60% CH_4, can be used for cooking, heating, lighting, refrigeration, and generating electricity. The technology lends itself very well to small-scale applications. The daily manure from one or two cows can generate enough methane to meet most of the cooking and lighting needs of a farm family. Two-thirds of China's rural families use biogas as their primary fuel.

4.21 Consider This: Hennepin County Waste Burning Plant

The Hennepin County Resource Recovery Facility in Minneapolis was the subject of a great deal of controversy for the county residents. The idea of generating usable energy from trash sounds wonderful, until the facility is built in your neighborhood. This is an issue faced by homeowners and residents in the area surrounding the plant. To address residents' concerns, an open meeting between the residents and representatives of the plant was held. Managers from the plant, engineers, and representatives of the state pollution control agency made presentations. Prepare a list of questions that you, as a resident in this area, would like to see addressed at such a meeting.

4.12 Transforming Energy

Essentially all the fuels we have been considering in this chapter—coal, oil, alcohol, and garbage—give up their energy through combustion. They are burned to generate heat. For the most part, however, heat is not the form in which the energy is ultimately used. Heat is nice to have around on a cold winter day, but it is a cumbersome form of energy. It is dangerous if uncontrolled, difficult to transport, and hard to harness for other purposes. The industrialization of the world's economy began only with the invention of devices to convert heat to work. Chief among these was the steam engine, developed in the latter half of the 18th century. The heat from burning wood or coal was used to vaporize water, which in turn was used to drive pistons and turbines. The resulting mechanical energy was used to power pumps, mills, looms, boats, and trains. The smoke-belching mechanical monsters of the English midlands soon replaced humans and horses as the primary source of motive power in the Western world.

A second energy revolution occurred early in the 1900s with the commercialization of electrical power. Today, one-third of the energy produced in the United States is electrical. Most of it is generated by the descendants of those early steam engines. Figure 4.22 illustrates a modern power plant. Heat from the burning fuel is used to boil water, usually under high pressure. The elevated pressure serves two purposes: It raises the boiling point of the water and it compresses the water vapor. The hot, high-pressure vapor is directed at the fins of a turbine. As the gas expands and cools, it gives up some of its energy to the turbine, causing it to spin like a pinwheel in the wind. The shaft of the turbine is connected to a large coil of wire that rotates within a magnetic field. The turning of this dynamo generates an electric current—a stream of electrons that represents energy in a new and particularly convenient form. Meanwhile, the water vapor leaves the turbine and continues in its closed cycle. It passes through a heat exchanger where a stream of cooling water carries away the remainder of the heat energy originally acquired from the fuel. The water condenses into its liquid state and reenters the boiler, ready to resume the energy transfer cycle.

This process of energy transformation can be summarized in the three steps diagrammed in Figure 4.23. Potential energy in the chemical bonds of fossil fuels is first converted to *heat energy*. The heat released in combustion is absorbed by the water vaporizing it to steam. This heat is then transformed into *mechanical energy* in the spinning turbine that turns the generator that changes the mechanical energy into *electrical energy*. In compliance with the first law of thermodynamics, energy is conserved

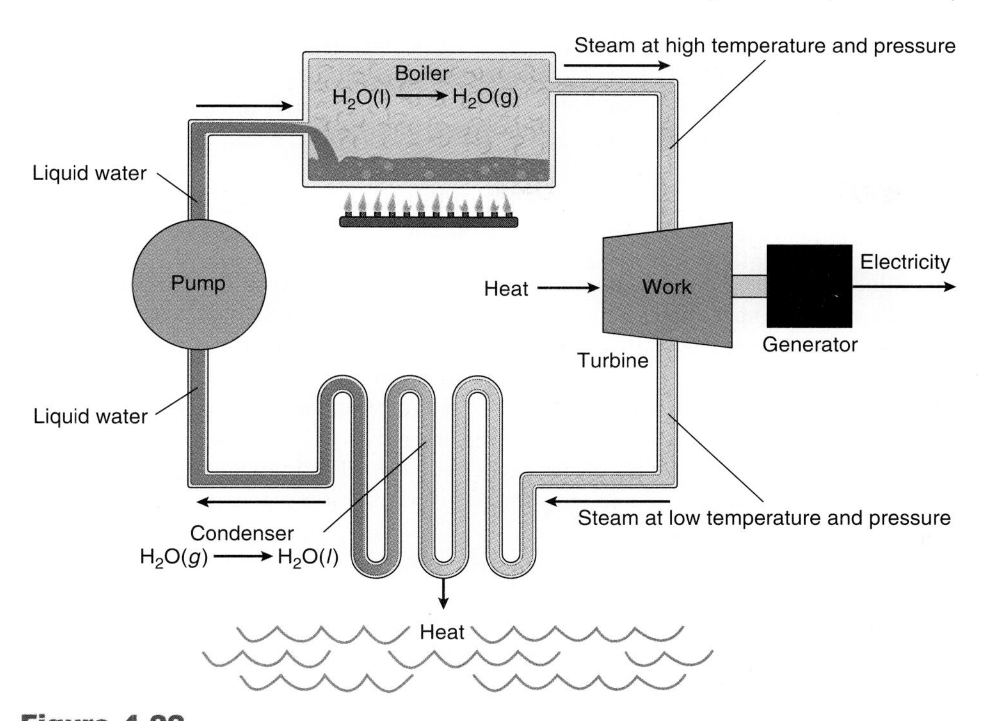

Figure 4.22
Diagram of a power plant for the conversion of heat to work to electricity.

throughout these transformations. To be sure, no new energy is created, but none is lost, either. We may not be able to win, but we can at least break even . . . or can we?

4.13 Energy and Efficiency

The last question is not as facetious as it might sound. In fact, we cannot break even. No power plant, no matter how well designed, can completely convert heat into work. Inefficiency is inevitable, in spite of the best engineers and the most sincere environmentalists. There are, of course, energy losses due to friction and heat leakage that can be corrected, but these are not the major problems. The chief difficulty is nature; more specifically, the nature of heat and work.

The *maximum theoretical efficiency* with which a power plant can convert heat to work depends on the difference between the highest temperature to which the water vapor is heated and the lowest temperature to which the condensed water is cooled. The mathematical expression for the efficiency is given by this equation.

$$\text{Efficiency} = \frac{\text{highest temp} - \text{lowest temp}}{\text{highest temp}}$$

It is important to note that the temperatures in this equation are expressed in the **absolute** or **Kelvin scale.** Zero on the Kelvin scale is absolute zero or −273°C, the lowest temperature possible. In fact, it cannot quite be attained. To convert a temperature from the Celsius scale to the Kelvin scale, you simply add 273.

$$\text{Temperature (in Kelvins)} = \text{Temperature (in degrees Celsius)} + 273$$

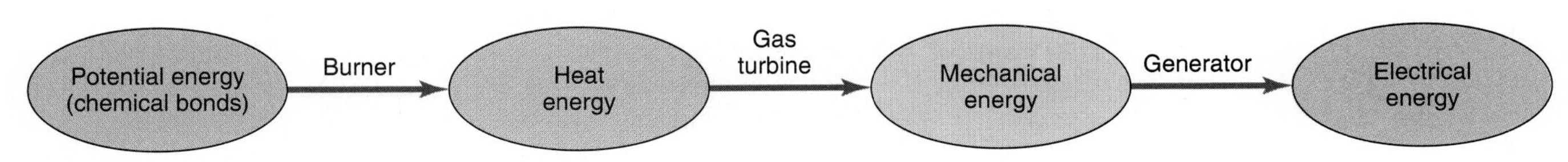

Figure 4.23
Energy transformation in an electric power plant. each step you lose some energy.

A well-designed modern fossil-fuel-burning power plant operates between a high temperature of about 550°C and a low temperature of 30°C. This means that, on the Kelvin scale, highest temp = 550° + 273 = 823 K, and lowest temp = 30° + 273 = 303 K. It follows that the efficiency is 63%.

$$\text{Efficiency} = \frac{823\text{ K} - 303\text{ K}}{823\text{ K}} = \frac{520\text{ K}}{823\text{ K}} = 0.63 \times 100 = 63\%$$

This value of 0.63 is the maximum theoretical efficiency of a power plant operating between these two temperatures. It means that at best only 63% of the heat that is obtained from the burning fuel is actually converted to work. The remainder is discharged to the cooling water, which consequently warms up. In addition, other inefficiencies are associated with friction, loss over long-distance power transmission lines, and so forth. Table 4.4 lists the efficiencies of a number of steps in energy production. The overall efficiency is the product of the efficiencies of the individual steps: The individual efficiencies are multiplied. The net result is that today's most advanced power plants operate at an overall efficiency of only about 42%.

Consider, for example, the case of electrical home heating, sometimes touted as being clean and efficient. We will assume that the electricity is produced by a methane-burning power plant with a maximum theoretical efficiency of 60%. The efficiencies of the boiler, turbine, electrical generator, and power transmission lines are given in Table 4.4; converting the electrical energy back into heat in the home is 98% efficient.

Because efficiencies are multiplicative, to find the overall efficiency of the electricity generation-to-home heating sequence, we multiply the efficiencies of the individual steps, expressed as their decimal equivalents.

$$\begin{aligned}\text{Overall efficiency} &= \text{efficiency of (power plant)} \times \text{(boiler)} \times \text{(turbine)} \times \\ &\quad \text{(electrical generator)} \times \text{(power transmission)} \times \text{(home heat)} \\ &= 0.60 \times 0.90 \times 0.75 \times 0.95 \times 0.90 \times 0.98 = 0.34\end{aligned}$$

The overall efficiency of 0.34 indicates that only 34% of the total heat energy derived from the burning of methane at the power plant is available to heat the house. If the electrically heated house requires 3.5×10^7 kJ (a typical value for a northern city in January), how much methane (grams) has to be burned at the power plant to furnish the heat needed? The combustion of 1 gram of methane releases 50.1 kJ. Remember that only 34% of the energy from the burned methane is available to heat the house. So, because of inefficiencies, far more methane has to be burned than the amount to release 3.5×10^7 kJ. The total quantity of heat that must be used can be calculated.

$$\text{Heat needed} = \text{Heat used} \times \text{efficiency}$$

$$3.5 \times 10^7\text{ kJ} = \text{Heat used} \times 0.34$$

$$\text{Heat used} = \frac{3.5 \times 10^7}{0.34}\text{ kJ} = 1.0 \times 10^8\text{ kJ}$$

Table 4.4

Some Typical Efficiencies in Power Production

Maximum theoretical efficiency	55–65%
Efficiency of boiler	90%
Mechanical efficiency of turbine	75%
Efficiency of electrical generator	95%
Efficiency of power transmission	90%

Because each gram of burning methane yields 50.1 kJ, the mass of methane that must be burned to furnish 1.0×10^8 kJ is 2.0×10^6 g of methane.

$$1.0 \times 10^8 \text{ kJ} \times \frac{1 \text{ g CH}_4}{50.1 \text{ kJ}} = 2.0 \times 10^6 \text{ g CH}_4$$

You can compare the efficiencies of heating a home with electricity or a natural gas furnace by completing the following Sceptical Chymist activity.

4.22 Sceptical Chymist: Clean Electric Heat

Is electric heat clean and efficient? The electricity must first be generated, usually by a fossil-fuel power plant.

a. The house could also have been heated directly with a gas furnace burning methane at 85% efficiency. Calculate the number of grams of methane required in January using this method of heating.

Answer

Because the only inefficiency is that of the furnace, we can do a calculation similar to that just done but using 0.85 as the efficiency. The energy required is 4.1×10^7 kJ, which corresponds to 8.2×10^5 g of methane. This is only 41% of the methane needed to provide the heat electrically.

b. On the basis of your answer to part a and the discussion preceding this exercise, comment on the claim that electric heat is "clean" and efficient.

4.23 Your Turn

A coal-burning power plant generates electrical power at a rate of 500 megawatts; that is, 500×10^6 or 5.00×10^8 joules per second. The plant has an overall efficiency of 0.375 for the conversion of heat to electricity.

a. Calculate the total quantity of electrical energy (in joules) generated in one year of operation and the total quantity of heat energy used for that purpose.

Answer

1.58×10^{16} J generated; 4.20×10^{16} J used

b. Assuming the power plant burns coal that releases 30 kJ/g, calculate the mass of coal (in grams and metric tons) that will be burned in 1 year of operation (1 metric ton = 1×10^3 kg = 1×10^6 g).

Answer

1.40×10^{12} g; 1.40×10^6 metric tons

4.14 Improbable Changes and Unnatural Acts

Although we asserted that heat cannot be completely converted into work, we offered little evidence and no explanation for this fact of nature. To illustrate the problem, push this book off the desk, and wait for it to come back up by itself. Be prepared to wait quite a while! The idea of a book picking itself off the ground and rising against the force of gravity is so bizarre, it is unbelievable. In fact, it is just unlikely—extremely unlikely. To understand why, let us examine the process in a little more detail.

You probably recall that a book resting on a table has potential energy by virtue of its position above the floor. When the book is dropped, the potential energy is converted to kinetic energy, the energy of motion. When the book strikes the floor, the kinetic energy is released with a bang. Some of it goes into the shock wave of moving air molecules that transmit the sound. Most of the energy goes to increase the motion of the atoms and molecules in the book and in the floor beneath it. This microscopic motion is the origin of what we call *heat*. A careful measurement would show that the book and the floor and the air immediately around them are all very slightly warmer than they were before the impact. Energy has been conserved, but it has also been dissipated. Heat or **thermal energy** is characterized by the random motion of molecules. They move chaotically in all directions.

Now consider what would be necessary for the book to rise by itself, and thus to work against the force of gravity. All the molecules in the book would have to move upward at the same time. At that instant, all the molecules in the floor under the book would also have to move in an upward direction, giving the book a little shove. Needless to say, such agreement among the 10^{25} or so molecules involved is very unlikely. Yet, such a change is necessary to convert heat into work—to transform random thermal motion into uniform motion. The first law of thermodynamics may confidently assert that all forms of energy are equal, but the fact remains that some forms of energy are more equal than others! The chaotic, random motion that is heat is definitely low-grade energy.

4.15 Order versus Entropy

Both the inability of a power plant to convert heat into work with 100% efficiency or for dropped books to spontaneously rise are both manifestations of the same law of nature, the **second law of thermodynamics.** There are many versions of the second law, but all describe the directionality of the universe. One version states that it is impossible to completely convert heat into work without making some other changes in the universe. Another observes that heat will not of itself flow from a colder to a hotter body. The falling book provides another version of the second law. That, too, is a transformation of ordered kinetic energy into random heat energy. Like all naturally occurring changes, it involves an increase in the disorder or randomness of the universe. This randomness in position or energy level is called **entropy.** The most general statement of the second law of thermodynamics is **the entropy of the universe is increasing.** This means that organized energy, the most useful kind for doing work, is always being transformed into chaotic motion or heat energy.

A helpful way to look at the increase in universal entropy that characterizes all changes is in terms of probability. Disordered states are more probable than ordered ones, and natural change always proceeds from the less probable to the more probable. Let's suppose you define perfect order as a beautifully organized sock drawer, all the socks matched, folded, and placed in rows. This would represent a condition of zero entropy (no randomness). If you are like most people, this is probably a rather unlikely arrangement. It certainly did not occur by itself; it took work to organize the socks. Without the continuing work of organization, it is quite possible that, over the course of a week or a month or a semester, the entropy and disorder of that sock drawer will increase. The point is that there are lots of ways in which the socks can be mixed up, and therefore disorder is more probable than order. Conversely, it is not very likely that you will open your drawer some morning and find that the previously jumbled socks are in perfect order and the entropy in that particular part of the universe has suddenly and spontaneously decreased without any external intervention. That sort of change from disorder to order is essentially what is involved in the conversion of heat to work. Professor Henry Bent has estimated that the probability of the complete conversion of one calorie of heat to work is about the same as the likelihood of a bunch of monkeys typing Shakespeare's complete works 15 quadrillion times (15×10^{15}) in succession without a mistake.

Perhaps by now some Sceptical Chymist in the class has objected that there are many earthly instances in which order increases. Sock drawers do get organized; power plants convert heat to work; dropped objects get picked up; water can be decomposed into hydrogen and oxygen; refrigerators transfer heat from a colder to a hotter body; and students learn chemistry. All of these are "unnatural" events—**nonspontaneous** in the vocabulary of thermodynamics. They will not occur by themselves; they require that work be done by someone or something. An input of energy is necessary to reduce the entropy and increase the order. And in every case, the work that is done generates more entropy somewhere in the universe than it reduces in one small part of the universe. Even when entropy appears to decrease in a spontaneous change, for example, the freezing of water at temperatures below 0°C, there are balancing increases in entropy. In this particular case, the heat given off by the freezing water adds to the disorder of its surroundings. In short, when the entire universe is considered, entropy always increases. THERE IS NO FREE LUNCH!

One word of caution: you need to be a little careful about the scientific meaning of "spontaneous" and "nonspontaneous." Spontaneous is often taken to mean a process that occurs all by itself, with no apparent initiation, as in the tabloid headline "SLEEPING MAN BURSTS INTO FLAME." In scientific usage, a spontaneous change is one that *could* occur; in other words, it is thermodynamically possible. But it might not take place all by itself because a large activation energy barrier must be overcome to start the reaction. Let us return to our sleeping man. Human beings are thermodynamically unstable with respect to combustion products such as carbon dioxide and water. So the burning of a human is a spontaneous change in the scientific sense of that term. But, fortunately for us, the activation energy barrier for that process is so high that we do not have to worry about bursting into flames without any provocation. In the case of the reported sleeping man, it would be a good idea to look for an outside agent, perhaps a disgruntled friend, who might have helped him over that barrier with a can of gasoline and a match.

4.24 Consider This: Humpty Dumpty

All around us there are examples of the natural tendency for things to get messed up. Some have even been enshrined in literature: "All the king's horses and all the king's men, couldn't put Humpty together again." Cite some examples of your own.

4.25 Consider This: Entropy Decrease–Entropy Increase

During midterm time, many students become very serious about their studying, and, for hours on end, will concentrate on the plays of Shakespeare or the causes of World War II. This decrease in intellectual entropy is often associated with an increase in the entropy of the student's room. Identify another process in which entropy appears to decrease but is actually coupled with an increase in entropy elsewhere in the universe.

4.16 The Case for Conservation

A fundamental feature of the universe is that energy and matter are conserved. However, the process of combustion converts both energy and matter to less useful forms. For example, the energy stored in hydrocarbon molecules is eventually dissipated as heat when those molecules are converted to carbon dioxide and water—essential compounds, to be sure, but unusable as fuels. As residents of the universe, we have no choice; we must obey its inexorable laws. Nevertheless, there are many options within those constraints. One of the most important is to make a human contribution to the conservation of energy and matter.

The planet's store of fossil fuels is finite, of course, although our appetites for them seem infinite (Figure 4.24). Worldwide demand for oil is increasing at more than 2% per year. Over the past 10 years, energy use is up in Latin America (32%), Africa (27%), and Asia (30%). In the mid-1990s, China became an oil importer rather than an exporter as it had been. Projections are for China to increase oil imports from 1 million barrels per day currently to 5 to 8 million barrels per day in the next two decades, predominantly from Middle East sources. Oil forecasts by the U.S. Energy Information Administration project a 60% increase in global demand for oil by 2020 to a whopping 40×10^9 barrels per year. To meet this growth, the market share of oil from Middle Eastern countries will likely rise beyond 30%, approaching the levels that produced the oil crises of 1974 and again in 1979. The conventional oil reserves (2.183×10^{12} barrels) are calculated to last 43 years at the current rate of consumption. In addition, enough petroleum is locked up in heavy crude oil, bitumen, and oil shale to provide for another 170 years, though this reserve will be more difficult and more expensive to

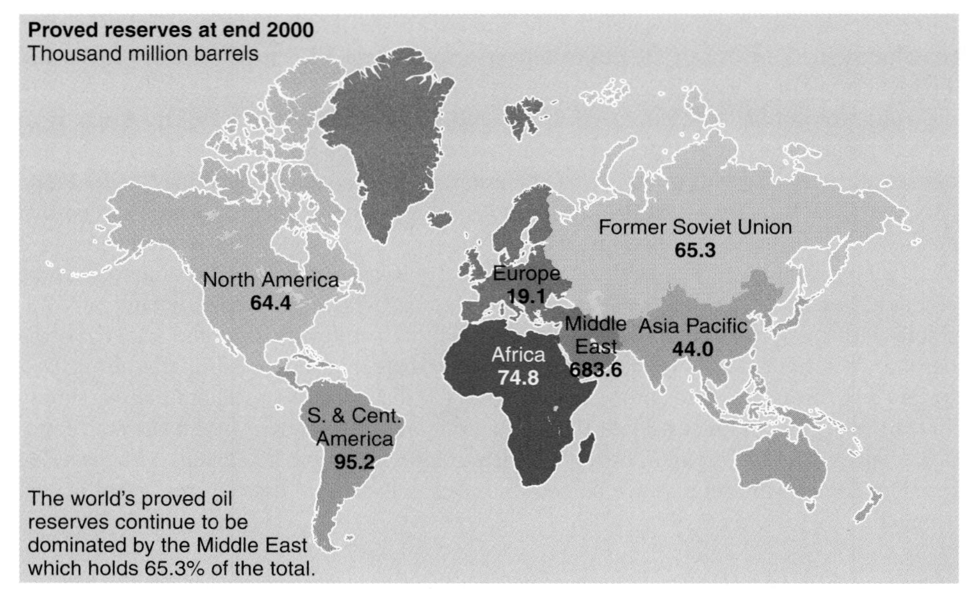

Figure 4.24

World proven reserves of oil (billions of barrels).

(Source: BP Statistical Review of World Energy 2001, *June 2000. Taken from www.bp.com/centres/energy/world_stat>rev/oil/reserves.asp.)*

extract. Global coal reserves appear to be considerably greater, but they too are limited. And in any event, we can be certain that the world's energy consumption will not remain at its current level.

Oil is the prime energy source for all regions of the world except for the former Soviet Union, where the prime fuel is natural gas (Figure 4.25). The Organization for Economic Co-operation and Development, an organization of western European nations, has estimated that between 1990 and 2010 world energy consumption will increase by 50%, oil consumption will increase by 40%, coal consumption will increase by 45%, and natural gas consumption will increase by 66%. One consequence of this increased use of energy will be a 50% rise in global CO_2 emissions. Not surprisingly, the greatest increases are expected to occur in the developing countries that, by 2010, will account for over half of the energy consumed. Growing populations, migration to the cities, and industrialization will drive up the energy demand of the Third World to unprecedented highs. The pattern has already been established. In China, energy utilization in 1993 was 22 times what it was in 1952. From 1990 to 1999, electricity consumption in China nearly doubled (97%). The construction of the immense and controversial Three Gorges dam and hydroelectric power station in China is one attempt by that country to meet its exploding electrical energy demands (Figure 4.26).

- To continue to meet such increases, it is estimated that China will need to open one medium-sized power plant every week for a year.

The demands of conventional power plants for coal, oil, and gas are voracious. But fossil fuels are so important as feedstocks for chemical synthesis that it is a great waste to burn them. Late in the 19th century, Dmitri Mendeleev, the great Russian chemist who proposed the periodic table of the elements, visited the oil fields of Pennsylvania and Azerbaijan. He is said to have remarked that burning petroleum as a fuel "would be akin to firing up a kitchen stove with bank notes." Mendeleev recognized that oil could be a valuable starting material for a wide variety of chemicals and the products made from them. But he would, no doubt, be amazed at the fibers, plastics, rubber, dyes, medicines, and pharmaceuticals currently produced from petroleum. Yet, we continue to ignore Mendeleev's warning and burn nearly 85% of the oil pumped from the ground. This is clearly a risk-benefit situation.

- The use of nuclear power, as discussed in Chapter 7, is another strategy to reduce the dependency on fossil fuels for energy.

In short, the arguments for conserving energy and the fuels that supply it are compelling. Fortunately, some promising strategies are available, and considerable savings have already been realized. Although energy production and fuel consumption have increased since the 1974 oil crisis, there has also been a significant increase in the

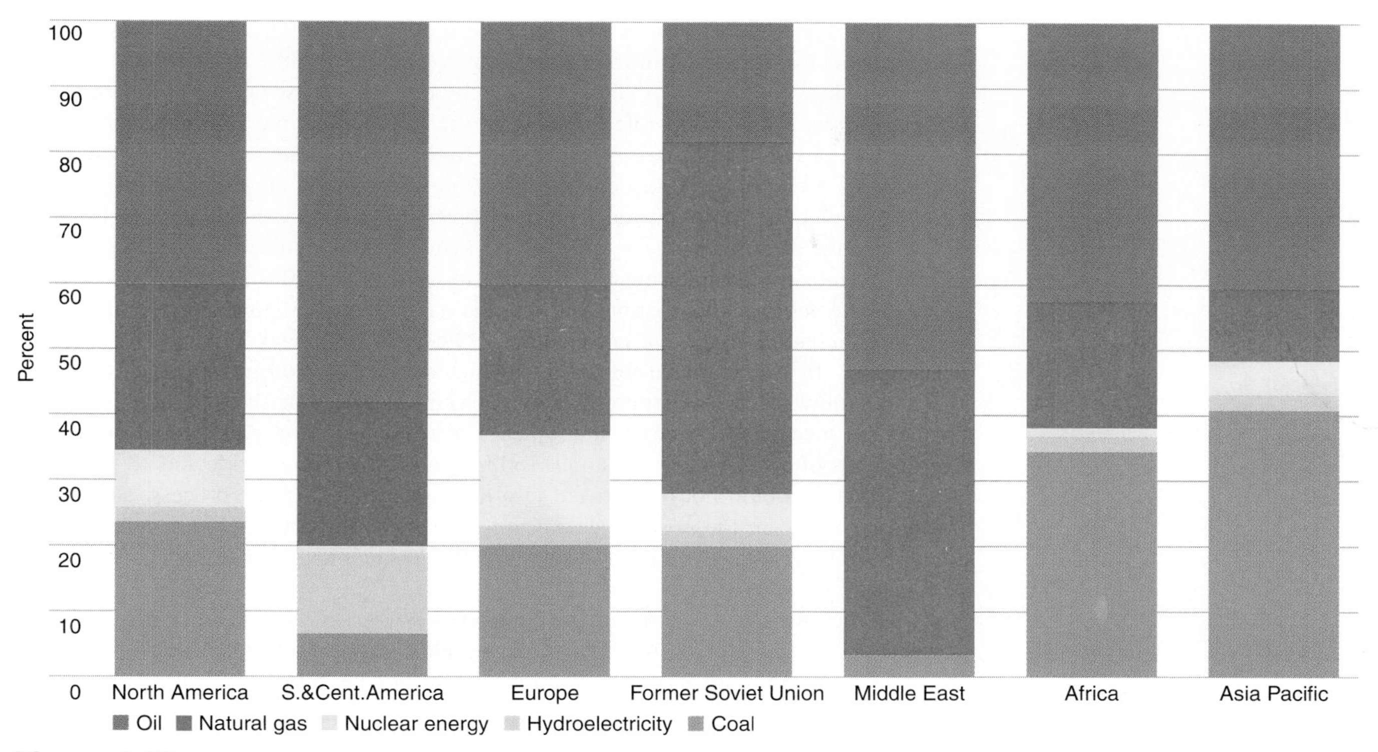

Figure 4.25

Regional energy consumption pattern, 2000. (green) Oil, (red) Natural gas, (yellow) Nuclear energy, (blue) Hydroelectricity, (purple) Coal. Oil remains the largest single source of energy in all regions of the world, except for the former Soviet Union where natural gas is the prime fuel.

(Source: BP Statistical Review of World Energy 2001, *June 2000. Taken from www.bp.com/centres/energy/world_stat>rev/oil/reserves.asp.)*

Figure 4.26

The Three Gorges Dam will supply China with significant hydroelectric power.

efficiency with which the fuels are used. The production of electricity by power plants is the major use of energy in the United States, making up 38% of the total. The conversion of heat to work is, of course, limited by the second law of thermodynamics, but power plants currently operate well below the thermodynamic maximum efficiency. Better design will bring power plants closer to that upper limit, perhaps to overall efficiencies of 50 or 60%. A particularly appealing approach is an integrated system that uses "waste" heat from a power plant to warm buildings.

Once the electricity is generated, great savings can be realized in its use. Estimates of the technically feasible savings in electricity range from 10 to 75%. The wide range in these predictions is worrisome, but specific data are encouraging. For example, in an article in *Scientific American* in September 1990, Arnold P. Fickett, Clark W. Gellings, and Amory B. Lovins made the following statement: "If a consumer replaces a single 75-watt bulb with an 18-watt compact fluorescent lamp that lasts 10,000 hours, the consumer can save the electricity that a typical United States power plant would make from 770 pounds of coal. As a result, about 1600 pounds of carbon dioxide and 18 pounds of sulfur dioxide would not be released into the atmosphere." In the process, about $100 would be saved in the cost of generating electricity. Improvements in the design of electric motors and refrigeration units also hold considerable potential for increased efficiency.

It is noteworthy that some utility companies now promote consumer education and provide financial inducements for conserving electricity. Sophisticated economic planning, new financing arrangements, and pricing policy are all part of efforts to save energy. One particularly important concept is "payback time," the period necessary before a private consumer, an industry, or a power company recaptures in savings the initial cost of a more efficient refrigerator, manufacturing process, or power plant.

4.26 The Sceptical Chymist: Light Bulbs Revisited

The quotation from Fickett, Gellings, and Lovins provides a marvelous opportunity for the Sceptical Chymist to apply her or his knowledge of chemistry. For example, let us check their assertion that replacement of a 75-watt (W) bulb with an 18-W fluorescent lamp will save the electricity made from 770 lb of coal. That is, assume that 770 pounds of coal, containing 65% carbon and 2% sulfur, will not be burned.

First, we note that the difference in the rate of energy consumption of the two bulbs is 75 W − 18 W = 57 W or 57 J/s. The total projected energy savings over the life of the bulb (10,000 hr) is obtained in this way:

$$\text{Energy savings} = 10{,}000 \text{ hr} \times \frac{60 \text{ min}}{1 \text{ hr}} \times \frac{60 \text{ s}}{1 \text{ min}} \times \frac{57 \text{ J}}{1 \text{ s}} = 2.05 \times 10^9 \text{ J}$$

The energy comes from coal, and coal typically yields 30 kJ/g or 30×10^3 J/g. To determine the mass of coal that must be burned to obtain 2.05×10^9 J, the following operation is performed.

$$\text{Mass coal} = 2.05 \times 10^9 \text{ J} \times \frac{1 \text{ g coal}}{30 \times 10^3 \text{ J}} \times \frac{1 \text{ lb coal}}{454 \text{ g coal}} = 150 \text{ lb coal}$$

This is a significant discrepancy from the quoted value of 770 lb. Possibly Fickett, Gellings, and Lovins made an error, or perhaps they made an assumption that we neglected. Explore the latter possibility, and suggest what the assumption might have been.

4.27 Your Turn

Now it is your turn to exercise your skepticism and your computational skills by checking the other two claims. Calculate the mass of CO_2 and SO_2 that would *not* be released into the atmosphere if a 75-W bulb were replaced by an 18-W compact fluorescent lamp. Assume that 770 lb of coal, containing 65% carbon and 2% sulfur, would not be burned.

4.28 The Sceptical Chymist: Shedding Light on Assumptions

Fickett, Gellings, and Lovins also claim that the bulb replacement we have been discussing would save about $100 in the cost of generating electricity. What value are they assuming for the cost of electricity?

Electricity is generally priced per kilowatt-hour (kWh), so we need to know the number of kWh saved over the 10,000 hr lifetime of the bulb. First we multiply the 57 watts saved by 10,000 hr:

$$57 \text{ watts} \times 10{,}000 \text{ hr} = 570{,}000 \text{ watt hr (Wh)}$$

Then we convert the answer to kilowatt-hours, recognizing that 1 kWh = 1000 Wh:

$$570{,}000 \text{ Wh} = \frac{1 \text{ kWh}}{1000 \text{ Wh}} = 570 \text{ kWh}$$

If 570 kWh of electricity cost $100, as the writers imply, what is the cost per kilowatt-hour? Once you have calculated the answer, find the cost of electricity to consumers in your city. Compare the results.

Recent advances in information technology and data processing also made possible sizeable energy savings. "Smart" office buildings or homes feature a complicated system of sensors, computers, and controls that maintain temperature, airflow, and illumination at optimum levels for comfort and energy conservation. Similarly, the computerized optimization of energy flow and the automation of manufacturing processes have brought about major transformations in industry. Over the past 20 years, industrial production in the United States has increased substantially, but the associated energy consumption has actually gone down. A case in point is the low-pressure, gas-phase process developed by Union Carbide chemical engineers for making polyethylene, which is the world's most common plastic. This new process uses only one-quarter the energy required by previous high-pressure methods. Although the capital investment associated with such conversions is often substantial, consumers and manufacturers may ultimately enjoy financial savings and increased profits. Mention should also be made of the energy conservation that results from recycling materials, especially aluminum. Because of the high energy cost required to extract aluminum metal from its ore, recycling the metal yields an energy saving of about 70%. To put things in perspective, you could watch television for three hours on the energy saved by recycling just one aluminum can.

One final area where energy conservation has a direct impact on lifestyle is transportation. About 20% of the total energy used and one-half of the world's oil production goes to power motor vehicles. But even here, we are making some progress in conservation. From the mid-1970s to the early 1990s, gasoline consumption in the United States dropped by one-half. Much of this saving was attributable to lighter-weight vehicles, thanks to the use of new materials, and to new engine designs. Yet this environmentally friendly trend is under assault as the 21st century begins. Since 1988 there has been an overall decline in vehicle fuel economy. It currently stands at 24.0 miles per gallon (mpg), the lowest since 1980 and 1.9 mpg less than the 25.9 mpg highest value obtained in 1988. A major factor for the decline in fuel economy has been the significant increase the proportion of vehicles classified as **light-duty trucks**—sport utility vehicles (SUVs), vans, and pickup trucks—which now command a substantial part (46%) of the traditional car market.

Two decades ago, the average light-duty truck and car weighed about the same and had similar engines. That is no longer the case; light-duty trucks are now more than 15% heavier and have nearly 80% more horsepower than those of 20 years ago. Although car manufacturers must meet a federally mandated 27.5 mpg average fuel consumption for cars, the requirement is only 20.7 mpg for light-duty trucks. SUVs are classified as light trucks, not as cars. Therefore, SUVs need not meet the higher mpg requirements for cars. The lower mpg standards for trucks and SUVs have allowed their weight and horsepower to increase. A full-size SUV or a pickup truck emits much more CO_2 than even a large automobile (SUV, 80% more; pickup truck, 67% more). Passenger cars have an average fuel economy of 28.1 mpg; that of light-duty (pickup) trucks is only 20.1 mpg (Figure 4.27). SUVs average 20.0 mpg; vans and minivans, 22.5 mpg.

Senator Diane Feinstein (D-CA) introduced a bill that would require SUVs and pickup trucks to meet the same fuel economy standards as automobiles by 2007. Feinstein asserts: "Simply put, this is the single most effective action we can take to

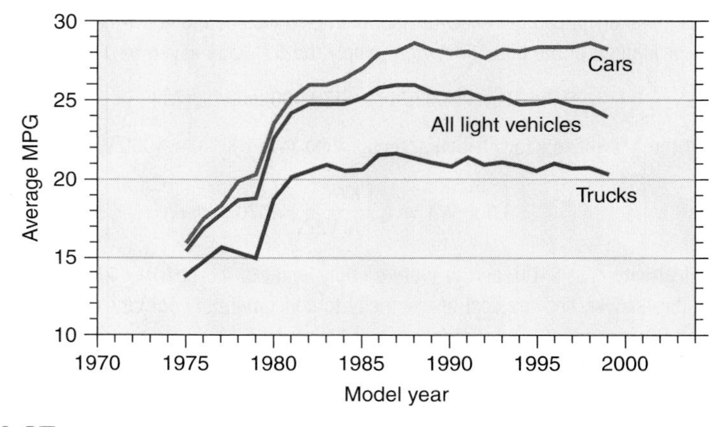

Figure 4.27
Fuel economy by model year and vehicle type.
(Source: U.S. EPA. Taken from www.epa.gov/otaq/cert/mpg/retrends.)

limit our dependency on foreign oil, to save consumers at the pump, and to reduce global warming." Ford Motor Company has pledged to increase the fuel economy of its SUVs by 25% by the 2005 model year, which would result in its vehicles able to get 23 mpg, up from the current 18 mpg.

On the plus side, research to develop cars with higher gas mileage continues. The 2000 Honda Insight is the most fuel efficient vehicle sold in the U.S. since 1975, with a rating of 61 mpg in the city and 70 mpg on the highway. The two-seater Insight is a hybrid vehicle that uses a combination of gasoline and battery power, like the Toyota Prius, which gets 52 mpg (city) and 45 mpg (highway) (Figure 4.28). Such hybrid vehicles are discussed in Chapter 8. Methane and propane are being used to power a growing number of cars and trucks, and Chapter 8 explores such alternative energy sources as hydrogen, electric batteries, fuel cells, and photovoltaic cells.

Figure 4.28
The Toyota Prius is a gasoline/battery hybrid car.

Such potential improvements in fuel economy are impressive, but the fact remains that the automobile is an energy-intensive means of transportation. A mass transit system is far more economical, provided it is heavily used. In Japan, 47% of travel is by public transportation, compared to only 6% in the United States. Of course, Japan is a compact country with a high population density. The great expanse of North America is not ideally suited to mass transit, although some regions in the United States, such as the population-dense Northeast and some other metropolitan areas, are. And one must also reckon with the long love affair between Americans and their automobiles.

4.29 Consider This: Getting People to Travel Together: Mass Transit

Advocates propose mass transit as a way to reduce fossil fuel consumption in the U.S. With a few notable exceptions, mass transit has not been widely adopted.

a. List the reasons why mass transit has not been implemented more widely.
b. What would it take for mass transit to become used to a greater degree?
c. What other creative measures could be taken to reduce the use of personal vehicles?

4.30 Consider This: Gasoline Rationing

Imagine that you are transported to the future at a time when you must pay $4.00 per gallon for gasoline. Your budget allows $250 per month for gasoline. Assume that your vehicle averages 20 miles per gallon. Describe the number of people in your family, how many of them go to school or work each day, and how far each must travel. Use this information to prepare a detailed gasoline budget of how you and your family would use the $250 worth of gasoline in a month.

4.31 Consider This: The Price of Gasoline

Oil is a valuable resource, even beyond its use for home heating and gasoline. If we continue to use our petroleum supply to extract gasoline from it, we may lose our starting materials to make other petroleum-based products, such as many pharmaceuticals and plastics. Up to now, voluntary conservation of gasoline has not been effective. The government could force more conservation by rationing gasoline or by heavily taxing it. If our government increased the price of gasoline to $4.00 per gallon (a price typical of that in western Europe and Japan), sales of gasoline likely would drop. This price would have serious consequences for the American work force because the price of a gallon of gasoline would be raised compared to the current minimum hourly wage.

Suppose a bill has been introduced in Congress to raise the price of gasoline to $4.00 per gallon. Draft a letter to a friend at another college, either supporting or protesting the bill. Include the reasons for your position and your opinion on what should be done with this new revenue if the bill is passed.

CONCLUSION

To a considerable extent, choice ultimately influences what technology can do to conserve energy. As individuals and as a society, we must decide what sacrifices we are willing to make in speed, comfort, and convenience for the sake of our dwindling fuel supplies and the good of the planet. The costs might include higher taxes, more expensive gasoline and electricity, fewer and slower cars, warmer buildings in summer and cooler ones in winter, perhaps even drastically redesigned homes and cities. During the 1970s a series of energy crises occurred because of a dramatic rise in the cost of imported crude oil, principally from the Middle East. Although we have broadened our sources of imported crude oil to others than solely the Middle East, our reliance on imported fossil fuels remains high, regardless of their sources. This ongoing dependence keeps alive the specter of whether supply and demand factors for crude oil could precipitate another energy crisis, perhaps on a global basis. One thing seems to be clear. The best time to examine our options, our priorities, and our will is before we face another full-blown energy crisis. Quite obviously energy, chemistry, and society are closely intertwined. This chapter is an attempt to untangle them.

Chapter Summary

Having studied this chapter, you should be able to:

- Distinguish between energy and heat, and be able to convert among energy units (joules, kilocalories, Calories)
- Describe the factors related to the United States' dependency on fossil fuels for energy (4.2)
- Apply the terms exothermic, endothermic, and activation energy to chemical systems (4.3–4.5)
- Interpret chemical equations and basic thermodynamic relations to calculate heats of reaction, particularly heats of combustion (4.4)
- Use bond energies to describe the energy content of materials (4.4)
- Evaluate the risks and benefits associated with petroleum, coal, and natural gas as fossil fuel energy sources (4.7–4.8)
- Relate energy use to atmospheric pollution and global warming (4.7, 4.8)
- Understand the physical and chemical principles associated with petroleum refining (4.8–4.9)
- Describe "octane rating" and how refining, leaded gasoline, ethanol, and MTBE relate to it (4.9)
- Discuss approaches to alternative (supplemental) automobile fuels (4.9–4.11)
- Describe why reformulated and oxygenated gasolines are used (4.10)
- Relate the energy potentially available from a process with the efficiency of that process (4.13)
- Use entropy as a concept to explain the second law of thermodynamics (4.15)
- Take an informed stand on what energy conservation measures are likely to produce the greatest energy savings (4.16)
- With confidence, examine news articles on energy crises and energy conservation measures to interpret the accuracy of such reports (4.16)

Questions

Emphasizing Essentials

1. **a.** What is the origin of fossil fuels?

   b. Name some examples of fossil fuels.

   c. Are fossil fuels a renewable resource?

2. Consider Figure 4.12 showing U.S. oil production and oil imports.

   a. Calculate the percentage of total oil production and imports supplied by domestic oil production in 1970 and in 1990. Also calculate the predicted value for 2010.

   b. How are these values changing with time?

   c. How have the sources of oil changed from 1970 to 1990 to what is predicted for 2010?

3. The reaction of water vapor with ethylene is one way to produce ethanol for use as a gasoline additive:

$$CH_2CH_2 + H_2O \longrightarrow CH_3CH_2OH$$

   a. Rewrite this equation using Lewis structures.

   b. Was it necessary to break *all* the chemical bonds in the reactants to form the product ethanol? Give a reason for your answer.

4. Consider the water in each of these two containers.

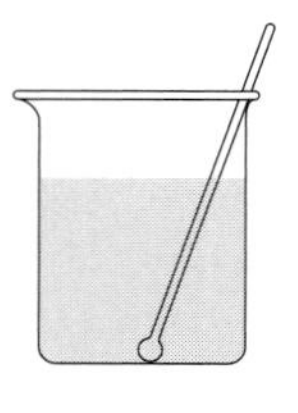

Container 1
80 g H_2O
70°C

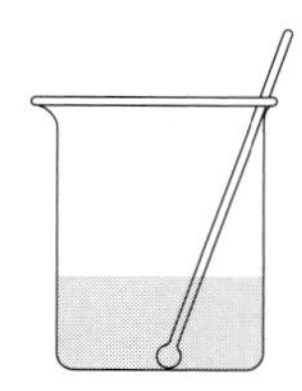

Container 2
40 g H_2O
70°C

The temperature of the water is the same in each of the containers. Is the heat content of the water the same in each of these containers? How do you know?

5. The Calorie, used to express food heat values, is the same as a kilocalorie of heat energy. If you eat a chocolate bar from the United States with 600 Calories of food energy, how does the energy compare with eating a Swiss chocolate bar that has 3000 kJ of food energy? (*Note:* 1 kcal = 4.184 kJ)

6. A single serving bag of Granny Goose Hawaiian Style Potato Chips® has 70 Calories. Assuming that all of the energy from eating these chips goes towards keeping your heart beating, how long can the energy from these chips sustain a heartbeat of 80 beats per minute? *Note:* 1 kcal = 4.184 kJ and each human heart beat requires 1 J of energy.

7. The text states that an energy consumption of 650,000 kcal per person per day is equivalent to an annual personal consumption of 65 barrels of oil or 16 tons of coal. Use this information to calculate the amount of energy available in each of these quantities.

   a. one barrel of oil

   b. one gallon of oil (42 gallons per barrel)

   c. one ton of coal

   d. one pound of coal (2000 pounds per ton)

8. Use the information in question 7 to find the ratio of the quantity of energy available in one pound of coal to that in one pound of oil. *Hint:* One pound of oil has a volume of 0.56 quarts.

9. Use Figure 4.4 to compare the sources of U.S. energy consumption with the sources of world energy consumption. Arrange each in order of decreasing percentage and comment on the relative rankings.

10. Equation 4.1 represents the complete combustion of methane.

    a. Write a similar chemical equation for the complete combustion of ethane, C_2H_6.

    b. Represent this equation with Lewis structures.

    c. Represent this reaction with a sphere equation.

11. The heat of combustion for ethane, C_2H_6, is 52.0 kJ/g. How much heat will be released if one mole of ethane undergoes combustion?

12. **a.** Write the chemical equation for the complete combustion of heptane, C_7H_{16}.

    b. Given that the heat of combustion for heptane is 4817 kJ/mole, how much heat will be released if 250 kg of heptane undergo complete combustion?

13. Figure 4.6 shows energy differences for the combustion of methane, which is an exothermic chemical reaction. The combination of nitrogen gas and oxygen gas to form nitric oxide, NO, is an example of an endothermic reaction:

$$180 \text{ kJ} + N_2(g) + O_2(g) \longrightarrow 2\ NO(g)$$

Sketch an energy difference diagram for this reaction.

14. From your personal experience, predict whether each of these processes is endothermic or exothermic. Give a reason for each prediction.

    a. a charcoal briquette burns

    b. water evaporates from your skin

    c. ice melts

    d. wood burns

15. Use the bond energies in Table 4.1 to estimate the energy change associated with this reaction.

$$2\,C{\equiv}O + O{=}O \longrightarrow 2\,O{=}C{=}O$$

16. Draw a diagram like Figure 4.8 for the reaction in question 15.

17. Use the bond energies in Table 4.1 to explain why

    a. chlorofluorocarbons, CFCs, are so stable

b. it takes less energy to release Cl atoms than F atoms from CFCs

18. Use the bond energies in Table 4.1 to calculate the energy changes associated with each of these reactions. Remember to consider the Lewis structures for each of the reactants and products. Indicate which reactions are endothermic and which are exothermic.

 a. $2\ C_5H_{12}(g) + 11\ O_2(g) \longrightarrow 10\ CO(g) + 12\ H_2O(l)$

 b. $H_2(g) + Cl_2(g) \longrightarrow 2\ HCl(g)$

 c. $N_2(g) + 3\ H_2(g) \longrightarrow 2\ NH_3(g)$

19. Use the bond energies in Table 4.1 to calculate the energy changes associated with each of these reactions. Indicate which reactions are endothermic and which are exothermic.

 a. $H_2(g) + O_2(g) \longrightarrow H_2O_2(g)$

 b. $2\ H_2(g) + O_2(g) \longrightarrow 2\ H_2O(g)$

 c. $2\ H_2(g) + CO(g) \longrightarrow CH_3OH(g)$

20. Pentane, C_5H_{12}, has a boiling point of 36.1°C. Octane, C_8H_{18}, has a boiling point of 125.6°C. Will pentane and octane be gases at room temperature (25°C)?

21. Consider this reaction representing the process of cracking.

$$C_{16}H_{34} \longrightarrow C_5H_{12} + C_{11}H_{22}$$

 a. What bonds are broken and what bonds are formed in this reaction? Use Lewis structures to help answer this question.

 b. Use that information and Table 4.1 to calculate the energy change during this cracking reaction.

22. How many isomers are there for butane, C_4H_{10}? Write the Lewis structure for each isomer. *Hint:* Be careful not to repeat isomers. Remember that Lewis structures show how atoms are linked, but not their spatial arrangement.

23. A premium gasoline available at most stations has an octane rating of 92.

 a. What does the octane rating tell you about the knocking characteristics of this gasoline?

 b. What does this tell you about whether the fuel contains oxygenates?

24. Assume that three power plants have been proposed, operating at the temperatures given here.

Plant I	$T_{high} = 1200°C$	$T_{low} = 0°C$
Plant II	$T_{high} = 600°C$	$T_{low} = 20°C$
Plant III	$T_{high} = 450°C$	$T_{low} = -150°C$

 a. Calculate the maximum efficiency of each plant.

 b. Identify the factors that affect the efficiency.

 c. Discuss the practical limits that govern such efficiencies. Which plant is most likely to be built?

25. Which is the better analogy for a state of high entropy—an unopened deck of playing cards or a plate of cooked spaghetti? Explain your reasoning.

Concentrating on Concepts

26. How can you explain the difference between temperature and heat to a friend? Use some practical, everyday examples to help your friend understand. Assume your friend has not taken a chemistry course.

27. One risk of depending on oil imports is the shortage of gasoline in the event of unfavorable international events. Does a gasoline shortage affect only individual motorists? What are some of the ways that a gasoline shortage could affect your life?

28. How is the statement that "energy is neither created nor destroyed in a chemical reaction" related to the law of conservation of energy and the first law of thermodynamics?

29. A friend tells you that hydrocarbons containing larger molecules are better fuels than those containing smaller molecules.

 a. Use these data, together with appropriate calculations, to discuss the merits of this statement.

Hydrocarbon	Heat of Combustion
Octane, C_8H_{18}	5450 kJ/mol
Butane, C_4H_{10}	2859 kJ/mol

 b. Considering your answer to part a, do you expect the heat of combustion per gram of candle wax, $C_{25}H_{52}$, to be more or less than the heat of combustion per gram of octane? Do you expect the molar heat of combustion of candle wax to be more or less than the molar heat of combustion of octane? Justify your predictions.

30. Halons are synthetic chemicals similar to CFCs, but they also include bromine. Although halons are excellent materials for fire fighting, they are more effective at ozone depletion than CFCs. The structural formula for halon-1211 is

```
        ..
       :F:
        |
  ..    |    ..
 :Cl — C — Br :
  ..    |    ..
        |
       :F:
        ..
```

 a. Which bond in this compound is broken most easily? How is that related to the ability of this compound to interact with ozone?

 b. C_2HClF_4 is a compound being considered as a replacement for halons as a fire extinguisher. Draw the Lewis structure for this compound and identify the bond broken most easily. How is that related to the ability of this compound to interact with ozone?

31. During the distillation of petroleum, kerosene and hydrocarbons with 12–18 carbons used for diesel fuel will be located at position C marked on this diagram.

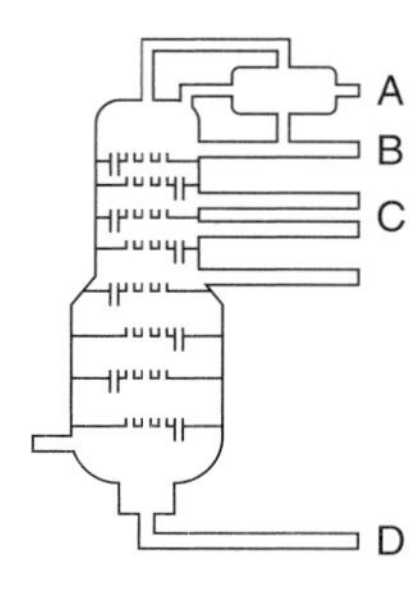

a. Separating hydrocarbons by distillation depends on the hydrocarbons having differences in a specific physical property. Which property is that?

b. How will the number of carbon atoms in the hydrocarbon molecules separated at A, B, and D compare with those separated at position C? Explain the basis for your prediction.

c. How will the uses of the hydrocarbons separated at A, B, and D differ from those separated at position C? Explain your reasoning.

32. Imagine you are at the molecular level, looking at what happens when liquid ethene, C_2H_4, boils. Consider a collection of four ethene molecules, representing each molecule with this sphere formula.

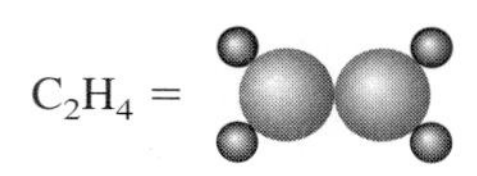

a. Draw a representation of ethene in the liquid state and then in the gaseous state. How will the collection of molecules change?

b. Estimate the temperature at which this transition from liquid to gas is taking place. What is the basis for your estimation?

33. It seems inefficient that petroleum first undergoes separation by distillation and then some fractions undergo further change through cracking.

a. Explain why cracking is necessary.

b. Hydrocarbons undergo physical changes during distillation in a fractionating tower. Does cracking also involve physical changes? Explain your reasoning.

34. Catalysts are used to speed up cracking reactions in oil refining and allow them to be carried out at lower temperatures.

a. Draw a sketch similar to Figure 4.11, in which you illustrate the energy changes for such a reaction in the absence and in the presence of the catalyst. Explain how your sketch illustrates the effect of the catalyst.

b. What examples of catalysts were given in the first two chapters of this text?

35. Consider these three structural formulas representing octane, C_8H_{18}; all hydrogen atoms and all C—H bonds have been omitted for simplicity.

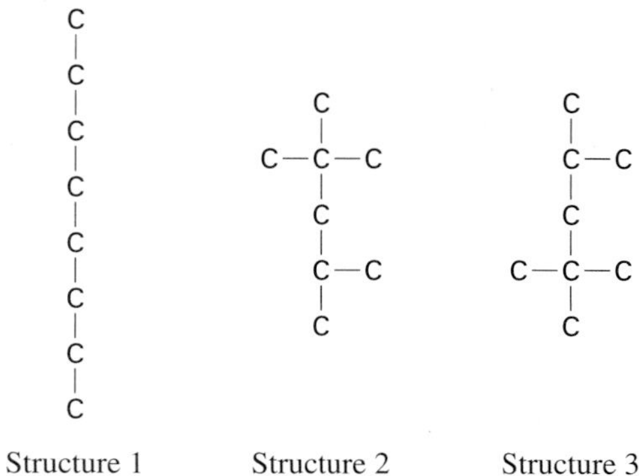

a. Fill in the missing hydrogen atoms and demonstrate that these structures all represent C_8H_{18}.

b. Are any representations duplicated here? How can you tell?

c. Obtain a model kit and make one of the structures. What are the carbon-to-carbon-to-carbon bond angles in the structure?

d. If you make a different one of the structures, will the carbon-to-carbon-to-carbon bond angles change? Why or why not?

e. Write the structural formula of at least one more isomer of octane.

36. How is the growth in oxygenated gasolines related to

a. restrictions on the use of lead in gasoline?

b. federal and state air quality regulations?

37. In each of these pairs, select the substance that has the greater entropy. Explain the reasons behind your choice.

a. $H_2O(g)$ at 100°C and $H_2O(l)$ at 100°C

b. a solid piece of iron and an equal mass of iron powder

c. peanuts or an equal mass of peanut butter

38. State whether the entropy increases or decreases in each of these reactions or processes.

a. Liquid water is converted to ice.

b. Solid sodium chloride is dissolved in water.

c. Hydrocarbons with 16 carbons are cracked into smaller hydrocarbons.

39. One mole of diamond has an entropy value of 2.4 J/K at 25°C; 1 mole of methanol, $CH_3OH(l)$, has an entropy value of 127 J/K at 25°C. What generalization can be drawn from these two values?

40. a. Do oxygenated fuels have a higher energy content than nonoxygenated fuels? Explain your reasoning.

b. Do oxygenated fuels have a higher entropy than comparable nonoxygenated fuels? Explain your reasoning.

Exploring the Extensions

41. The text states that reformulated gasolines, RFGs, burn more cleanly by producing less carbon monoxide than nonoxygenated fuels. What is the evidence that supports this statement?

42. Consider this diagram.

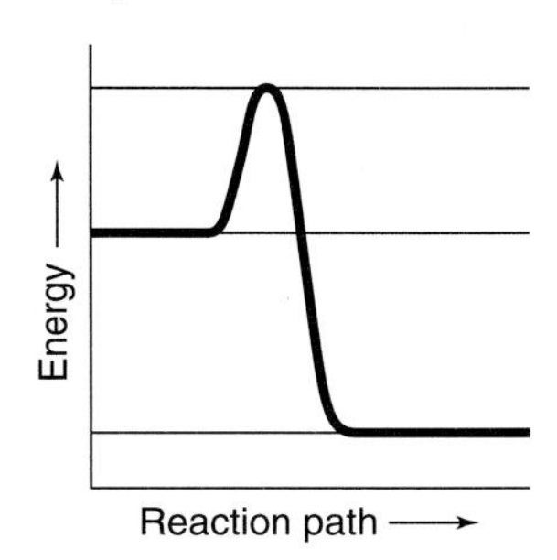

a. Does this representation show an exothermic or endothermic reaction? Explain your reasoning by commenting on the shape of the curve.

b. Sketch this type of energy diagram for the type of reaction, exothermic or endothermic, *not* shown in the diagram.

43. Bond energies such as those in Table 4.1 are sometimes found by "working backwards" from heats of reaction. A reaction is carried out and the heat absorbed or evolved is measured. From this value and known bond energies, other bond energies can be calculated. For example, the energy change associated with this reaction is +81 kJ.

$$NBr_3(g) + 3\ H_2O(g) \longrightarrow 3\ HOBr(g) + NH_3(g)$$

Use this information and the values found in Table 4.1 to calculate the energy of the N—Br bond. Assume all atoms other than hydrogen obey the octet rule.

44. Explain why it is possible to use a distillation tower to separate a mixture of hydrocarbons into different fractions, but it is not possible to separate seawater, also a complex mixture, into all of its different fractions.

45. The text states that both octane and isooctane have essentially the same heat of combustion. How is that possible if they have different structures? Explain your thinking.

46. Why do you think that countries are willing to go to war over energy issues, but not over other environmental issues? Write a brief op-ed piece for your school newspaper discussing this issue.

47. What relative advantages and disadvantages are associated with using coal and with using oil as energy sources? Which do you see as the better fuel for the 21st century? Give reasons for your choice.

48. What are the advantages and disadvantages of replacing gasoline with renewable fuels such as ethanol? Indicate your personal position on the issue and state your reasoning.

49. In December 1998 the EPA announced it would create a panel of experts from the public health and scientific communities, the automotive fuels industry, water utilities, and state and local governments to review the use of MTBE and other oxygenates added to gasoline. Find out more about why representatives of these groups have been included, and what the panel was asked to accomplish. Has the panel issued a final report at this time?

50. C. P. Snow, a noted scientist and author, wrote an influential book called *The Two Cultures,* in which he stated the following: "The question, 'Do you know the second law of thermodynamics?' is the cultural equivalent of 'Have you read a work of Shakespeare's?'" How do you react to this comparison? Discuss these questions in light of your own educational experiences.

5

The Water We Drink

Many types of bottled water are available to consumers.

"Americans have long felt safe drinking their water out of the tap, and with good reason. For the most part, public water supplies and even well water throughout the country are highly unlikely to make anyone acutely ill. The occasional failures of public systems to operate properly are few and attract national attention. . . . But in recent years, the concern has been mounting about less obvious potential hazards in drinking water, especially residues of pesticides and industrial wastes, lead from old pipes, and compounds formed by chlorine and organic matter that are believed to be cancer-causing. The concern about water safety has prompted millions of Americans to reject the water that comes straight from the tap, resulting in two new growth industries: bottled water and filtration systems."

Jane E. Brody
New York Times, July 18, 2000

A "Right-to-Know" Report about a community's water supply.

So begins an article entitled "On Tap or Bottled, Pursuing Purer Water" that appeared in the personal health column of the *New York Times.* Similar questions about water quality may be raised about your water supply, even about the water supply for your college or university. You may even have received a Right-To-Know Report, also called a Consumer Confidence Report. The Safe Drinking Water Act as amended in 1996 mandates that such reports be delivered once a year to consumers of water from community water systems, public or private. As a person studying chemistry, you are in a good position to understand the meaning of the measurements being reported and the standards of quality that must be met. You might be asked to help friends or family understand the measurements and conclusions reached in a report they might receive about their water quality.

Such reports raise a number of questions that we address in this chapter. What is a regulated chemical and what are its maximum levels allowed in drinking water? How do such chemicals get into drinking water? Should we be concerned about any of them? Who establishes the rules? What are the major provisions of the Safe Drinking Water Act of 1996? What options are there for alternative sources of drinking water? In particular, why is bottled water cited as the most likely alternative? Are all water filtration systems equally effective?

Water is, indeed, a special compound. In spite of its commonness, this colorless liquid called water is amazing stuff. The noted anthropologist and essayist Loren Eiseley speaks poetically of the wonders of it: "If there is magic on this planet, it is contained in water . . . Its substance reaches everywhere; it touches the past and prepares the future; it moves under the poles and wanders thinly in the heights of the air." Arguably, water is the most important chemical compound on the face of the Earth. In fact, it covers about 70% of that face, giving the planet the lovely blue color in the famous "blue marble" photos taken from outer space by astronauts. Water is essential to all living species; without it humans would die within a week. Our bodies are approximately 60% water, blood is at least 50% water, and a human brain is an astonishing 77% water! Water is so important to life as we know it that speculation about life elsewhere in the universe hinges first and foremost on the availability of water. Water refreshes us, dominates weather systems, provides for many types of recreation on and in it, and even gives us aesthetic and relaxing pleasures.

• See the opening of Chapter 1 for a "Blue Marble" photo.

CHAPTER OVERVIEW

Although we generally take water for granted, it is a remarkable chemical compound with unique properties that account for its essential life-supporting role. In this chapter, we will consider water from the perspective of those who drink it. First a question of aesthetics: What makes a glass of water pleasing to the eye and to the palate? There is more to water, however, than can be seen or tasted. Unseen impurities in water, depending on their identities and amounts, can impart a crisp, fresh taste or produce an unpleasant illness. And so, we next look at water as a solvent and at some of the things that may be dissolved in drinking water. How much of a substance dissolves in water makes concentration an important part of the story. The concentrations of substances dissolved in water can be expressed in several ways, including descriptions of the extremely low concentrations. To better comprehend aqueous solutions and why some substances dissolve in water while others do not, we will relate the properties of water molecules through such concepts as electronegativity, polarity, and hydrogen bonding.

Water dissolves various kinds of compounds, including those containing positive and negative ions, as well as some molecular compounds not composed of ions. We will consider the solubility of both types of substances and how it affects their presence in drinking water. Because the quality of drinking water is regulated by federal and state legislation, we will examine how drinking water is tested against standards, and how water is treated to make it potable (safe for drinking). Three case studies will give us the chance to look more closely at water quality. We examine the effect of calcium on water hardness, lead in water, and a particular category of compounds called trihalomethanes that are formed during water purification by chlorination. The questions posed early in the chapter about choices between tap water and bottled water are revisited. Finally, because most of the world's people do not have good access or choices for safe drinking water, we look briefly at ways to purify water, including home "filtration" systems, deionization, distillation, and reverse osmosis. And so, with water glasses or bottles raised, we extend the invitation to "take a drink."

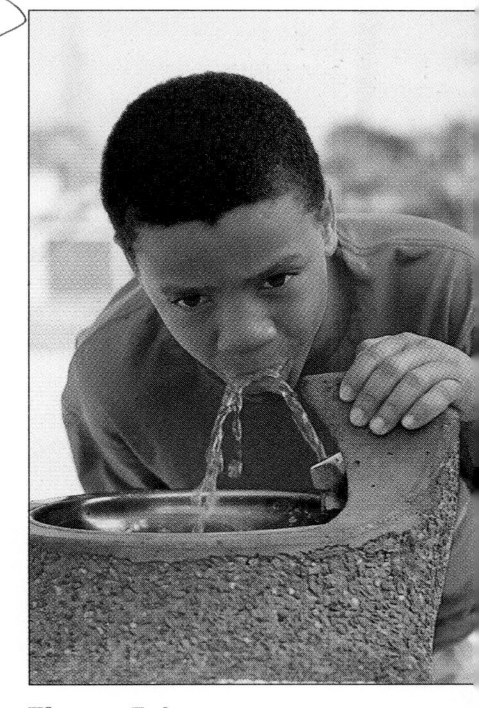

Figure 5.1
We usually take safe drinking water for granted.

5.1 Take a Drink of Water

Chapter 1 began with an invitation to "Take a breath of air." Without thinking, we do it automatically about 12–16 times a minute. Without air, we die in less than 15 minutes. Furthermore, we generally don't have any choice about *what* air we breathe; we must rely on that which surrounds us. On the other hand, we generally have choices with water: about *how frequently* we drink, *how much* we drink, and about the *source* of the water. Tap water, water from a lake or stream, bottled water, or water in bottled beverages are among the choices. If potable water is not available, we are in danger; without it, we would die in a few days.

Nothing could be more familiar than this clear, colorless, and (usually) tasteless liquid. And yet, we generally take water and water quality for granted in most regions of this country (Figure 5.1). Unless there is a water emergency brought on by drought or contamination of our municipal water supply, we seldom think about where the water comes from, what it contains, how pure it is, or how long the supply will last. We turn on a faucet for a drink or a shower and simply expect a sufficient quantity of water to come flowing out of the tap. The 5.1 Consider This activity asks you to think a bit more about the water you drink.

5.1 Consider This: Take a Drink of Water

Obtain a glass of tap water and answer these questions.

a. Carefully describe the water—its taste, odor, and appearance. What do you like or dislike about the water?
b. Make a list of five qualities that you want to have in your drinking water.
c. What concerns, if any, do you have about the safety of the tap water or possible long-term health effects of unsafe water?

Bottled water is a very popular beverage.

5.2 Bottled Water

Most Americans obtain their drinking water from a water faucet or a drinking fountain. This marvelous liquid is remarkably inexpensive, costing only about 1/10 of a penny per quart. But not everyone drinks tap water. An increasing number of Americans are drinking bottled water instead. Indeed, the possible need for turning to bottled water was mentioned in the quotation opening this chapter. Because of its growing popularity, bottled water will be a subtheme of this chapter.

Bottled water is big business and is now the fastest growing segment of the beverage industry in the world. Current estimates, based on information provided by the International Bottled Water Association and independently evaluated in a study conducted by the World Wildlife Fund, are that annual revenues worldwide from the sale of bottled water are over 35 billion U.S. dollars per year. Sales in the U.S. alone have now topped $6 billion annually. Growth rates in selling bottled water exceed 10% yearly, far outstripping rates seen for soft drinks, fruit beverages, or beer. In Europe, bottled water is now the biggest selling "soft drink." It is a common sight on campuses worldwide to see students carrying bottled water. Consumers 18 to 24 years old are the major users of bottled water, particularly pint-sized plastic bottles of water.

You have only to walk down a grocery aisle in the U.S., Canada, Europe, or Asia to take a tour of bottled water brands. Advertisements for bottled water tout its purity, often using words or images that conjure up nature, purity, and pristine beauty. For example, the label for Dansani bottled water, the brand owned by Coca-Cola, makes these statements.

> *"Purified Water. Enhanced with Minerals for a Pure, Fresh Taste."*

The label for LeBleu UltraPure Drinking Water conjures up images of pure drinking water while explaining the natural water cycle.

> *"When the air was clean, nature did a perfect job of making fresh, pure water. The sun evaporated water from lakes, streams, and oceans, then winds blew it to the mountains where it condensed into rain and snow. Now we can duplicate this purity nature once so freely gave us."*

The web sites for bottled water also try to convey images of purity and natural processes. This is Evian's statement.

> *"Every drop of Evian Natural Spring Water begins as rain and snow falling high in the pristine and majestic French Alps."*

DENNIS THE MENACE

"HEY, JOEY. FORGET THE LEMONADE... THIS IS WHERE THE BIG MONEY IS."

Figure 5.2

Although highly popular, bottled water is also very expensive, relative to tap water. Typically, bottled water costs from $1.00 to $2.00 or more per quart, which is approximately 1000 times more expensive than the same volume of tap water. Bottled water is far more expensive, drop for drop, than milk, soft drinks, or even West Texas crude oil (Figure 5.2). Are there important reasons why consumers are willing to ante up so much more for bottled water? One of the goals of this chapter is to help you learn enough about drinking water quality to make intelligent choices. Are there compelling reasons to drink bottled water?

5.2 Consider This: Bottled Water and You

Why do people buy bottled water, despite its high cost relative to tap water?

a. List what you perceive to be the advantages and the disadvantages of drinking bottled water.
b. Prioritize your lists in decreasing order of importance (most to least) for your personal decision about whether to drink bottled water.

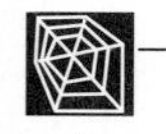

5.3 Consider This: Finding Out About Bottled Water

If you search for "bottled water" on the Web, you will get over a million "hits." Select two sites to explore, one provided by a supplier and the other provided as a source of consumer information. The former may flood you with statistics about the benefits of bottled water; the latter may raise questions, such as "Is bottled water safer?" or "Is it worth the cost?" For each site, list the title, author, URL, and two things that you learned about water from the site.

- Growth of the bottled water industry has had a direct impact on CO_2 emissions (Chapter 3), energy costs for production and transportation (Chapter 4), and plastics production (Chapter 9).

5.3 Where Does Drinking Water Come From?

What journey does water take to get from its natural source to your tap or bottle? Water is widely distributed on planet Earth (Figure 5.3). On the surface, it is found in oceans,

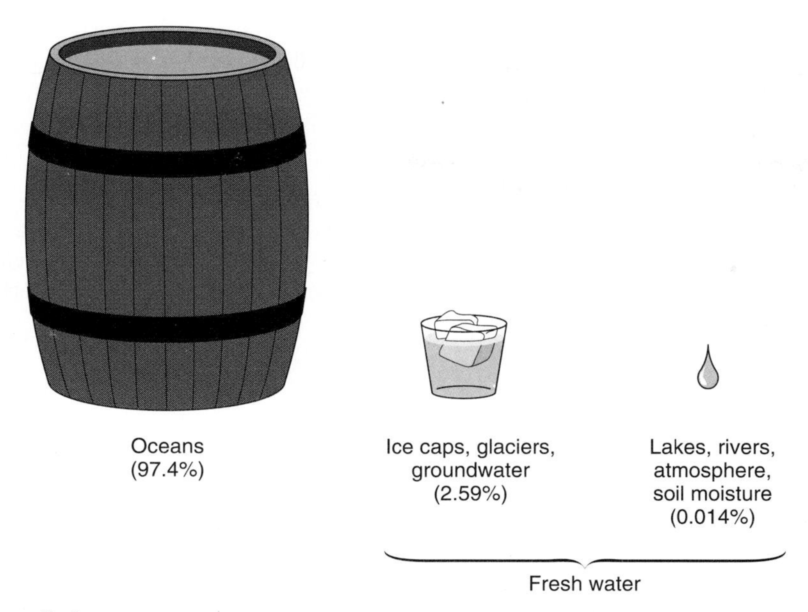

Figure 5.4
Distribution of water on Earth.

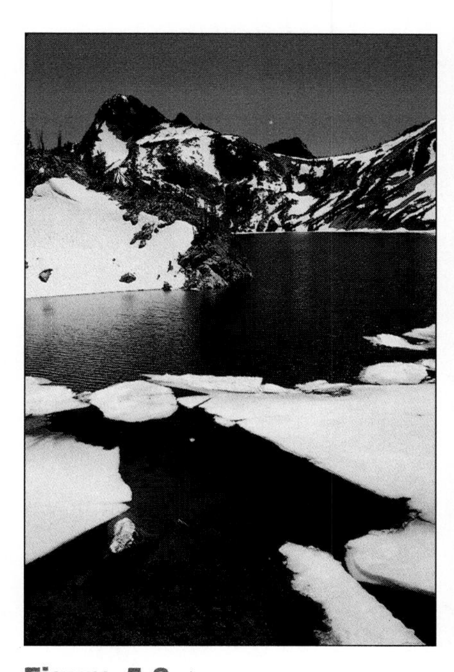

Figure 5.3
Lakes and reservoirs provide much of our drinking water.

lakes, rivers, snow, and glaciers (Figure 5.4). In the atmosphere it exists as water vapor and as tiny droplets in clouds that replenish surface water by means of rain and snow.

Water is also found underground in **aquifers,** great pools of water trapped in sand and gravel 50–500 ft below the surface. Some aquifers are enormous, such as the Ogallala Aquifer in the center of the United States that underlies parts of eight states from South Dakota to Texas. Aquifers underlying many other regions of the country also can provide an important source of water. Keeping these underground resources free from contamination is an important consideration in conserving sources of drinking water. If an aquifer becomes contaminated, it may take decades to become clean again.

Water that can be made suitable for drinking normally comes from either surface water or groundwater. **Surface water**—lakes, rivers, and reservoirs—frequently contains substances that must be removed before it can be used as drinking water. By contrast, **ground water** is that pumped from wells that have been drilled into underground

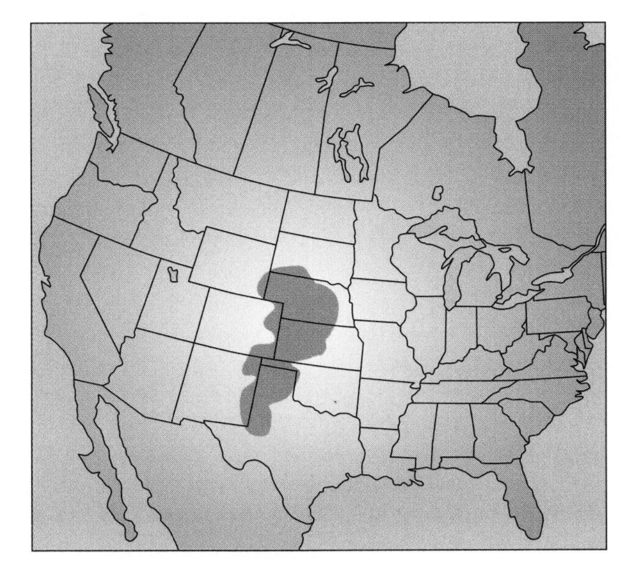

The Ogallala Aquifer is shown in dark blue on this map.

aquifers, and is usually free of harmful contaminants. Large-scale water supply systems for cities tend to rely on surface water resources; smaller cities, towns, and private wells tend to rely on ground water, the source of drinking water for a little over half the U.S. population.

5.4 Consider This: Your Drinking Water Source

Do you know where your drinking water comes from? You can find out by using the Safe Drinking Water Information System (SDWIS) of the EPA web site either directly or through the direct link on the *Chemistry in Context* web site. Search for the source of your home's drinking water by entering the geographic area in which you live. Then answer these questions.

a. What is the name of the water system that services your home?
b. What is the primary water source for that system?
c. What information is given about the quality of that system's drinking water?

5.5 Consider This: Water Quality in Your State

Each state has different concerns about its surface and ground water.

a. Draw up a list of the issues in your state. For example, you might mention agricultural runoff, leaking storage tanks, or pollution from other human activities.
b. Read the state fact sheet provided by the Office of Water at the EPA through the direct link at the *Chemistry in Context* web site. What does the sheet say about the surface and ground water quality in your state?
c. The EPA notes that some states have lakes, rivers, and streams that support no aquatic life. What percent of the surface water in your state falls in this category?

The Earth was not always as wet as it is now. Scientists believe that much of the water now on this planet originally was spewed as vapor from thousands of volcanoes that pocked the Earth. The vapor condensed as rain and the process repeated over the ages. Water molecules cycled from sea to sky and back again. About three billion years ago, primitive plants, and then animals, extracted water from and contributed water to the cycle, a cycle that continues today. During an average year, enough precipitation falls on the continents to cover all the land area to a depth of more than 2.5 ft. As we are well aware, this precipitation does not fall all at once, nor is it distributed uniformly. The wettest place on record is Mount Waialeale on Kauai, Hawaii, which receives an annual average rainfall of 460 inches. In contrast is the 0.03 inches of rain per year average in Arica, Chile. Data have been gathered there for the past 61 years, and for 14 of those years, it did not rain at all! Closer to home, an average of 1.5×10^{13} liters of water falls daily on the continental United States—enough to fill about 400 million swimming pools.

This sounds like a great deal of water, but on a global scale, the amount of *fresh* water is relatively quite small. The great majority of the Earth's water, 97.4% of the total, is in the oceans, water that is undrinkable without expensive purification (Figure 5.4). The remaining 2.6% is all the fresh water we have, but almost all of the world's fresh water is permanently frozen in glaciers and polar ice caps. *Only about 0.01% of the Earth's total water is conveniently located in lakes, rivers, and streams as fresh water.* Consequently, the world's drinking water supplies are quite limited, varying widely depending on locale. In the United States, 80% of the fresh water is used to irrigate crops and cool electrical power plants.

5.4 Water as a Solvent

A major reason why we must consume water is because it is an excellent solvent for many of the chemicals that make up our bodies, as well as for a wide variety of other substances. A **solvent** dissolves other substances. Substances that dissolve in a solvent are referred to as **solutes.** The resulting mixture is a **solution,** a homogeneous mixture of uniform composition. Solutions in which water is the solvent are called **aqueous solutions.** Later in this chapter we will examine why certain kinds of substances dissolve

Table 5.1

Importance of Water as a Solvent in Living Organisms and in the Environment

- Blood plasma is an aqueous solution containing a variety of life-supporting substances.
- Inhaled oxygen dissolves in blood plasma in the lungs where O_2 combines with hemoglobin.
- Blood carries dissolved CO_2 to the lungs to be exhaled from the body.
- Water transports nutrients into all the cells and organs of our body.
- Water can transport *toxic* substances into, within, and out of living organisms.
- Water helps to maintain a chemical balance in the body by flushing wastes away.
- Water-soluble toxic substances, such as some pesticides, lead ions, and mercury ions, can be spread through the environment.
- Water may reduce the concentrations of pollutants to safe levels by dilution and/or by carrying them away.
- Rainwater carries substances, including acid rain, from the atmosphere down to Earth.

• Ions are discussed in Section 5.11

in water and others do not. For now, we simply note that a remarkable variety of substances can dissolve in water and that this has important consequences for living organisms as well as for the environment. Table 5.1 summarizes some examples of water acting as a solvent.

5.6 Your Turn

Based on your experience, which of these substances dissolves in water? In describing the solubility, use terms such as very soluble, partially soluble, or insoluble.

a. salt
b. sugar
c. sidewalk chalk
d. rubbing alcohol
e. cooking oil
f. aspirin

5.5 How Pure Is Drinking Water?

The word "pure" does not mean the same thing to everyone. Regardless of its source, drinking water is rarely, if ever, just pure H_2O. You can be assured that almost certainly it contains other substances. For example, a label on Evian bottled water includes the information in Table 5.2.

Of these seven items, all but one of them, silica, will be discussed in this chapter. We will also find that the items listed on the label are themselves not chemical compounds, but ions, parts of compounds. The number given with each item indicates how much of that substance (in milligrams) is present in one liter of Evian water. This raises

Table 5.2

Mineral Composition of Evian, mg/L

Calcium	78	Bicarbonates	357
Magnesium	24	Sulfates	10
Silica	14	Chlorides	4
		Nitrates (N)	1

Table 5.3

Mineral Composition of Tap Water, mg/L

Calcium	66	Sulfate	42
Magnesium	24	Chloride	48
Sodium	18	Nitrate	6
		Fluoride	1

a reasonable question: Should we be concerned about any of these substances and its amount? Calcium ions, for example, have a definite health benefit in producing stronger bones. Milk, not Evian water, is the preferred source for calcium ions; you would have to drink 4 liters of Evian water to get the same amount of calcium ions as that in one 8-oz glass of milk. In contrast, nitrate ions, depending on their concentration, can be a health problem, especially for infants. The other substances listed for Evian bottled water are not likely to be a health problem. Elsewhere on the label it is noted that sodium (sodium ions), a health issue for some people, is present at less than 5 mg per bottle.

- 1.00 liter = 1.06 quart.

Such composition information can also be obtained for tap water supplied by municipal water companies. A typical analysis of tap water in a Midwest home revealed the information in Table 5.3.

5.7 Your Turn

One 500-mL bottle of Evian water provides 4% of the recommended daily requirement of calcium.

a. Use the label information to find the number of milligrams of calcium recommended per day.
b. How many 500-mL bottles of Evian water would you have to drink to obtain your total daily supply of calcium?

Perhaps you have never considered drinking a glass of water as a risk/benefit act, yet it is. We usually consider water that has been chemically analyzed and treated has important benefits with very low risk. Fortunately, that is most often the case. But, however useful the information on either the bottled water label or in the tap water analysis may be, it is incomplete. The information indicates nothing about whether other substances, if any, are present in the water. It also does not indicate how much of each of the other substances is present or whether such substances pose health concerns at those concentrations. For example, even though a tiny amount of lead is found in almost all water samples, it is usually in such low amount as not to be a health problem. If the water has been chlorinated to purify it, the water almost certainly has trace amounts of some chlorination by-products. Indeed, we rarely stop to think about what trace amounts of substances may be in the water, because we tend to assume that the water is safe to drink. Overwhelmingly, this is a valid assumption. In part, this is because extensive federal and state regulations and standards govern water quality to protect the public. Most bottled water is regulated as well, often by self-imposed industry standards.

- Chlorination and its by-products will be examined in detail later in this chapter.
- Absence of evidence is not the same as evidence of absence. A substance may be present, but in undetectable amounts. It just means that it can't be detected at current levels of analytical chemical technology. You may recall that this same point was made for air pollutants in Chapter 1.

In assessing the health-giving or risk-taking aspects of drinking water, it is not sufficient to know only *what substances* are present in the water and *how toxic* they are. We also need to know *how much* of each substance is present in a particular amount of the water. In other words, we need to be able to understand what is meant by the concentration of a solute and the usual ways of expressing it. And so, we turn to this topic.

5.6 Solute Concentration in Aqueous Solutions

First introduced in Chapter 1 in relation to air quality, the concept of concentration warrants review and elaboration here in terms of substances dissolved in water. The composition of the major components of air was expressed in terms of the *percentages*, whereas the much lower concentrations of toxic air pollutants (ozone, sulfur dioxide, etc.) were given in *parts per million*. Although concentrations of air components might be a bit hard to visualize, solute concentrations in water solutions are more familiar and more easily imagined. For example, when a cooking recipe asks you to dissolve 1 teaspoon of an ingredient in 1 cup of water, a solution of a specific concentration results: 1 tsp per cup (tsp/cup). Note that you would have the same 1 tsp/cup concentration if you also dissolved 2 tsps of the ingredient in 2 cups of water, 4 tsps in 4 cups, or 1/2 tsp in 1/2 cup. Even though you used larger or smaller quantities of the ingredient, the number of cups of water increased or decreased proportionally. Therefore, the **concentration**—*the ratio of amount of ingredient to amount of water*—would be the same in each case: 1 tsp *per* 1 cup (tsp/cup). Solute concentrations in water solutions follow the same pattern as in the recipe, but are expressed in different units. We use four ways of expressing concentration: percent; parts per million; parts per billion; and molarity. Each unit has particular application in various circumstances.

4 ways

Percent The simplest and most familiar way of expressing concentration is percent, defined as *parts per hundred*, that is, parts of solute per 100 parts of solution. A solution containing 5 g of sodium chloride in 100 g of solution would be a five percent (5%) solution. Hydrogen peroxide solutions, often found as an antiseptic in medicine cabinets, are usually 3% hydrogen peroxide, which indicates that they contain 3 grams of hydrogen peroxide in 100 grams of solution (or 6 grams in 200 grams of solution, etc.).

Ppm and ppb Concentrations of substances in drinking water are normally far lower than 1% (1 part per hundred). Correspondingly, different units are used to express such low concentrations. **Parts per million (ppm)** is the most common way of expressing the concentration of a solute in drinking water. A 1 ppm solution of calcium in water contains 1 gram of calcium in 1 million (1,000,000) grams of water. The same concentration, 1 ppm, could be applied to a solution with 2 grams of calcium in 2 million grams of water, 5 g in 5 million g of water, or 5 mg (0.005 g) in 5000 g of water. Although parts per million is a very useful concentration unit, measuring one million grams of water is not very convenient. Therefore, we look to find an easier but equivalent way to establish ppm. We find it using the **liter,** the volume occupied by 1000 g of water at 4°C, as the volume unit, rather than 1 million g of water. Proportionally, we use milligrams of solute, not grams, to express ppm in an alternative but equivalent way: *1 ppm of any substance in water equals 1 mg of that substance per liter of water.*

- The quart, not the liter, is a more familiar measure of volume for Americans, although not for the rest of the world. A liter is just a bit larger than a quart.
- 1 ppm = 1 mg/L

$$1 \text{ ppm} = \frac{1 \text{ g solute}}{1{,}000{,}000 \text{ g water}} = \frac{1 \text{ mg solute}}{1{,}000 \text{ g water}} = \frac{1 \text{ mg solute}}{1 \text{ L water}}$$

Drinking water often contains various substances naturally present at concentrations in the ppm range, as illustrated on the Evian bottled water label. Certain toxic water pollutants may also be present in the ppm concentration range. For example, the acceptable limit for nitrate, often found in well water in agricultural areas, is 10 ppm; and the limit for fluoride is 4 ppm.

Because some pollutants are of concern at concentrations much lower than even parts per million, they may be reported as **parts per billion (ppb).** One part per billion of mercury in water means 1 gram of mercury in 1 billion grams of water. In more convenient terms, this means 1 *microgram* (1×10^{-6} g, abbreviated as 1 μg) of mercury in 1 liter (10^3 g) of water. For example, the acceptable limit for mercury in drinking water is 2 ppb.

- 1 ppb = 1 μg/L

$$2 \text{ ppb mercury} = \frac{2 \text{ g mercury}}{1 \times 10^9 \text{ g water}} = \frac{2 \times 10^{-6} \text{ g mercury}}{1 \times 10^3 \text{ g water}} = \frac{2\ \mu\text{g mercury}}{1 \text{ L water}}$$

One ppm is a very small concentration. Several analogies to a concentration of 1 ppm were given in Section 1.2, including the statement that 1 ppm corresponds to 1 second in nearly 12 days. A similar analogy can be offered for parts per billion: 1 ppb corresponds to 1 second in 33 years, or approximately 1 inch on the circumference of the Earth.

5.8 Your Turn

a. If 80 μg of lead were detected in 5 L of water, what would be the concentration of lead? Express your answer in both ppb and ppm.

b. If the maximum lead concentration in drinking water allowed by the federal government is 15 ppb, is the sample in part **a** in compliance with federal limits? Explain.

• The concept of moles was introduced in Section 3.8.

Molarity Another useful concentration unit in chemistry is molarity, which is based on the mole, chemists' favorite way of measuring matter. **Molarity,** abbreviated M, is defined as the number of moles of solute present in one liter of solution. Written more compactly, molarity is (moles of solute)/L solution or simply moles/L. The great advantage of molarity is that a 1 molar (1 M) solution of *any* solute contains exactly the same number of chemical units (atoms or molecules) as any other 1 molar solution. The *mass* of solute may vary depending on the molar mass, but the *number* of chemical units will be the same for all 1 M solutions. Methods of chemical analysis of water (Section 5.15) frequently use molarity. It is also particularly useful for chemical reactions involving solutions, a topic explored more fully in the next chapter. For now, we simply want to develop some familiarity with molarity itself.

• The molar mass of NaCl, 58.5 g, is the sum of the mass of 1 mole of sodium, 23.0 g, plus 1 mole of chlorine, 35.5 g. Calculating molar masses was covered in Section 3.8.

As an example, consider a solution of sodium chloride in water. The molar mass of NaCl is 58.5 g; therefore, one mole of NaCl weighs 58.5 grams. If we dissolve 58.5 grams of NaCl in some water and then add enough water to make exactly 1.00 L of solution, we will have a 1.00 molar NaCl solution (Figure 5.5). But, there are many ways to make a 1-molar NaCl solution. For example, we could use one-tenth of a mole of NaCl (5.85 g) in one-tenth of a liter (0.100 L = 100 mL). Another possibility, among many others, would be to use 2 moles NaCl (117.0 g) in 2 liters of solution.

$$1 \text{ M NaCl} = \frac{2 \text{ mole NaCl}}{2 \text{ L solution}} = \frac{1 \text{ mole NaCl}}{1 \text{ L solution}} = \frac{0.5 \text{ mole NaCl}}{0.5 \text{ L solution}} = \frac{0.1 \text{ mole NaCl}}{0.1 \text{ L solution}}$$

5.9 Your Turn

a. Consider a 1.5 M NaCl solution and a 0.15 M NaCl solution. How many moles of solute are present in 500 mL of each solution?

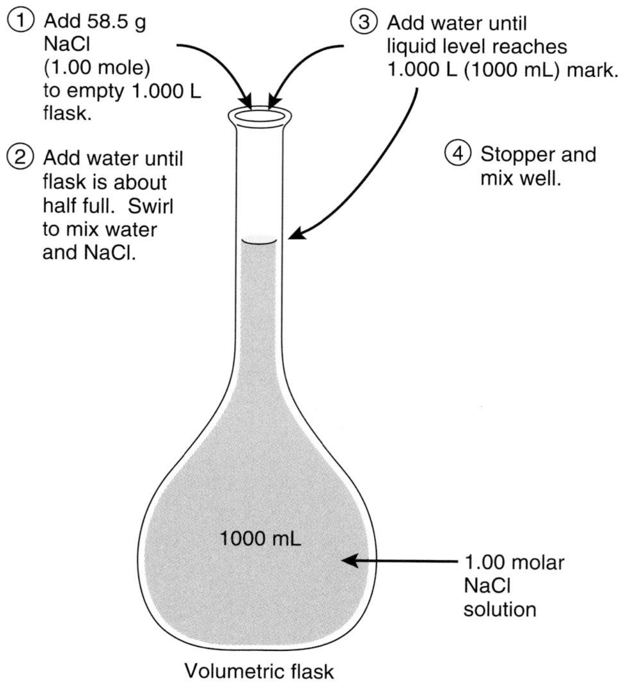

Figure 5.5
Preparing a 1.00 M NaCl solution.

b. A solution is prepared by adding enough water to 0.5 mole of NaCl to form 250 mL of solution. A second solution is prepared by adding enough water to 0.6 mole of NaCl to form 200 mL of solution. Which solution is more concentrated? Explain your reasoning.

Now that we have some idea about the importance of drinking water and some of the substances that may be present in it, we shift to a more detailed examination of this liquid at the molecular level. Our aim is to understand its unique properties, including its behavior as a solvent.

5.7 H_2O: Surprising Stuff—Molecular Structure and Physical Properties

The many and varied uses of water are a consequence of its properties. Not only is water an effective solvent for a wide range of materials, but it also has a number of unusual properties. Water is so ubiquitous that it has become the standard for many of the units of modern science, including the Celsius temperature scale, the kilogram, and the calorie. Yet, for all of that, the physical properties of water are quite peculiar, and we are very fortunate that they are. If water were a more conventional compound, we would be very different creatures.

This most common of liquids is full of surprises. First, it is noteworthy that water is a liquid and not a gas at room temperature (about 25°C) and normal atmospheric pressure. The molar mass of water is 18.0 g/mole. Almost all other compounds with molar masses in that range are gases under these conditions of temperature and pressure. Consider three common atmospheric gases—N_2, O_2, and CO_2—whose molar masses are 28, 32, and 44 g/mole, respectively. All have molar masses greater than that of water, yet they are gases to breathe rather than liquids to drink.

- Generally speaking, as molar mass increases in a series of similar compounds, the boiling point also increases.

Moreover, not only is water a liquid under these conditions, it has an anomalously high boiling point of 100°C. This temperature is one of the reference points for the Celsius temperature scale. The other is the freezing point of water, 0°C. And when water freezes, it exhibits another—bizarre property—it expands. Most liquids contract when they solidify. These and other unusual properties derive from water's chemical composition and molecular structure. Table 5.4 compares several properties of water with other substances and offers some reasons why these properties are significant for our lives.

Table 5.4

Unusual Properties of Water

Property	Comparison with Other Substances	Importance in Physical and Biological Environments
Specific heat (4.184 J/g·°C)	Highest of all common liquids and solids except NH_3	Affects climate; moderates temperatures in organisms and the environment
Heat of fusion (freezing) (331 J/g)	Highest of all molecular solids except NH_3	Releases enormous amounts of heat on freezing; saves crops from freezing by spraying them with liquid water
Heat of vaporization (2250 J/g)	Highest of all molecular substances	Releases very large quantities of thermal energy when condensed; storms can result

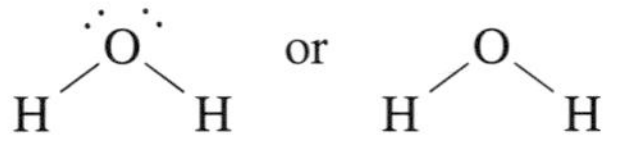

Figure 5.6

Alternative representations of a water molecule.

Group							
1A	2A	3A	4A	5A	6A	7A	8A
H 2.1							He —
Li 1.0	Be 1.5	B 2.0	C 2.5	N 3.0	O 3.5	F 4.0	Ne —
Na 0.9	Mg 1.2	Al 1.5	Si 1.8	P 2.1	S 2.5	Cl 3.0	Ar —

Figure 5.7

Electronegativity values for the first 18 elements.

To understand the chemical and physical properties of water, we need information about its chemical composition and molecular structure. The chemical composition is known to practically everyone. Indeed, the formula for water, H_2O, is very likely the world's most widely known bit of chemical information. Recall from Chapter 2 that water is a covalent molecule (Section 2.3). The oxygen atom is at the center of the bent molecule, attached by covalent bonds to two hydrogen atoms. Each of the two O—H bonds consists of one pair of electrons shared between the hydrogen and oxygen atoms (Figure 5.6).

It turns out that the bonding electrons are not shared equally between oxygen and hydrogen atoms. Experimental evidence indicates that the oxygen atom attracts the shared electron pair more strongly than does the hydrogen atom. To use the appropriate technical term, oxygen is said to have a higher electronegativity than hydrogen. **Electronegativity** is a measure of an atom's attraction for the electrons it shares in a covalent bond. The greater the electronegativity, the more an atom attracts bonding electrons to itself. Figure 5.7 shows a periodic table of electronegativity values for the first 18 elements.

- The periodic table was introduced in Section 1.6 and is found on the inside front cover.

An examination of Figure 5.7 reveals some useful generalizations about electronegativity. The highest electronegativity values are associated with nonmetallic elements such as fluorine and chlorine. These halogens, members of Group 7A, have atoms with seven outer electrons. Recall (Section 2.3) that each of these atoms has a strong tendency to bond with another atom in such a way as to acquire a share in an additional electron, thus completing a stable octet of electrons. A similar argument explains the high electronegativity values of other nonmetals. For example, oxygen, with six outer electrons per atom, also exhibits a powerful attraction for shared electrons. Conversely, the lowest electronegativity values are associated with the metals found in Groups 1A and 2A. Atoms of these metallic elements have much weaker attractions for electrons than do nonmetals. In general, electronegativity values *increase* as you move across a row of the periodic table from *left to right* (from metals to nonmetals) and *decrease* as you move *down* a group of the table.

According to Figure 5.7, the electronegativity of oxygen is 3.5; that of hydrogen is 2.1. Because of these electronegativity differences, the shared electrons are actually pulled by the more electronegative oxygen to itself and away from the less electronegative hydrogen. This unequal sharing gives the oxygen end of the O—H bond a partial negative charge and the hydrogen end a partial positive charge. Because the bond has oppositely charged ends or poles, it is said to be a **polar covalent bond.** Polar bonds arise whenever atoms of differing electronegativities are covalently bonded to each other. The greater the electronegativity differences of the elements involved, the more polar the bond.

- A polar covalent bond is an example of an *intramolecular* force, a force that exists *within* a molecule.

Many of the unique properties of water are a consequence of its molecular shape and the polarity of its bonds. Both are shown in the drawing of an H_2O molecule in Figure 5.8. Arrows are used to indicate the direction in which the electron pairs are displaced. The δ^+ and δ^- symbols indicate partial positive and partial negative charges, respectively. Note that the hydrogen atoms are partially positive and that a partial negative charge appears to be concentrated in the two nonbonding pairs of electrons on the oxygen atom.

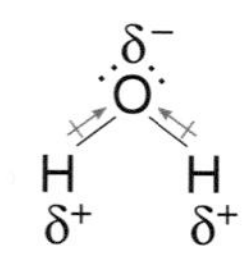

Figure 5.8

H_2O, a polar covalent molecule.

5.10 Your Turn

For each pair, identify the more polar bond. To which of the atoms in the bond will the electron pair be more strongly attracted?

a. H—F and H—Cl
b. N—H and O—H
c. N—O and S—O

Hint: Use the electronegativity values given in Figure 5.7.

5.8 Hydrogen Bonding

Polar bonds can be used to understand the unusually high boiling point of water. Consider what happens when two water molecules approach each other. Because opposite charges attract, one of the partially positively charged hydrogen atoms of one water molecule is attracted to one of the regions of partial negative charge associated with the two nonbonding electron pairs of the other water molecule. This is an *inter*molecular attraction, one that occurs *between* molecules. The fact that each H_2O molecule has two hydrogen atoms and two nonbonding pairs of electrons increases the opportunities for intermolecular attraction. Each of the two nonbonding electron pairs on oxygen can form a loose, long-distance kind of attraction, called a **hydrogen bond,** to a hydrogen atom of a neighboring water molecule. Similarly, each of the two hydrogen atoms in a water molecule can attract a nonbonding electron pair from an adjacent water molecule to form hydrogen bonds. Thus, a single water molecule can simultaneously hydrogen bond to as many as four other water molecules, as pictured in Figure 5.9.

- Hydrogen bonds are *inter*molecular forces, those occurring *between* molecules.

Hydrogen bonds, such as those between water molecules, are only about one-tenth as strong as the covalent bonds that connect atoms together *within* molecules. Nevertheless, as intermolecular forces go, hydrogen bonds are quite strong. Their strength is reflected in the relatively high boiling point of water (100°C) and the large amount of energy required to convert liquid water into vapor (2250 J/g or 4.05×10^4 J/mole). To boil water, the H_2O molecules must be separated from their relatively close contact in the liquid state and moved into the gaseous state, where they are much farther apart. In other words, their intermolecular hydrogen bonds must be broken. If the hydrogen bonds in water were weaker, water would have a much lower boiling temperature and require less energy to boil. If water had no hydrogen bonding at all, it would boil at about −75°C, making life as we know it very uncomfortable, if not impossible. Although water is wet, boiling points may seem dry and impersonal until you reflect on the fact that our bodies are over 60% water. Because of hydrogen bonding, almost all of our body's water, whether in cells, blood, or other body fluids, is in the liquid state, well below the boiling point. Our very existence depends on hydrogen bonding; without it, we would be a gas!

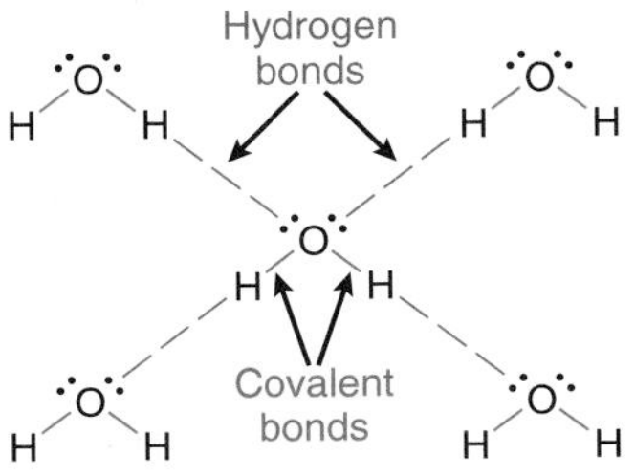

Hydrogen bonds

Covalent bonds

Figure 5.9
Hydrogen bonding in water (distances not to scale).

 **Figures Alive!**

5.11 Consider This: Bonds Within and Between Water Molecules

Use Figure 5.9 to help explain what bonds are broken when water boils. Draw a diagram to help illustrate your understanding.

Hint: Diagrams and sphere equations were used extensively in Chapter 1 to help illustrate chemical reactions. Start with molecules of water in the liquid state, then show what happens to those molecules when water boils.

It should be noted that hydrogen bonds are not restricted to water. There is evidence for similar intermolecular attraction in many molecules that contain hydrogen atoms covalently bonded to oxygen, nitrogen, or fluorine atoms. Thus, ammonia, NH_3, and hydrogen fluoride, HF, also have unusually high boiling points, but lower than water. The net effect of hydrogen bonding in these compounds is weaker than in water. Hydrogen bonding is also important in stabilizing the shape of large biological molecules, such as proteins and nucleic acids. In proteins, which are major components of skin, hair, and muscle, hydrogen bonding occurs between hydrogen atoms and oxygen or nitrogen atoms. The coiled, double-helical structure of DNA (deoxyribonucleic acid) is stabilized by thousands of hydrogen bonds formed between particular segments of the linked DNA strands. So in this respect, too, hydrogen bonding plays an essential role in the life process.

- The molecular structures of proteins and nucleic acids are discussed in Chapters 11 and 12, respectively.

Hydrogen bonding also explains why ice cubes and icebergs float in water. Ice is a regular array of water molecules in which every H_2O molecule is hydrogen bonded to four others. The pattern is shown in Figure 5.10. Note that the pattern includes a good deal of empty space in the form of hexagonal channels. When ice melts, this regular array begins to break down and individual H_2O molecules can enter the open channels. As a result, the molecules in the liquid state are, on the average, more closely packed than in the solid state. Thus, a volume of one cubic centimeter (1 cm^3) of liquid H_2O contains more molecules than 1 cm^3 of ice. Consequently, liquid water has a greater mass per cubic centimeter than ice. This is simply another way of saying that the *density* of liquid water is greater than that of ice.

- One cubic centimeter (1 milliliter) is about the same volume as one-fifth of a teaspoon.

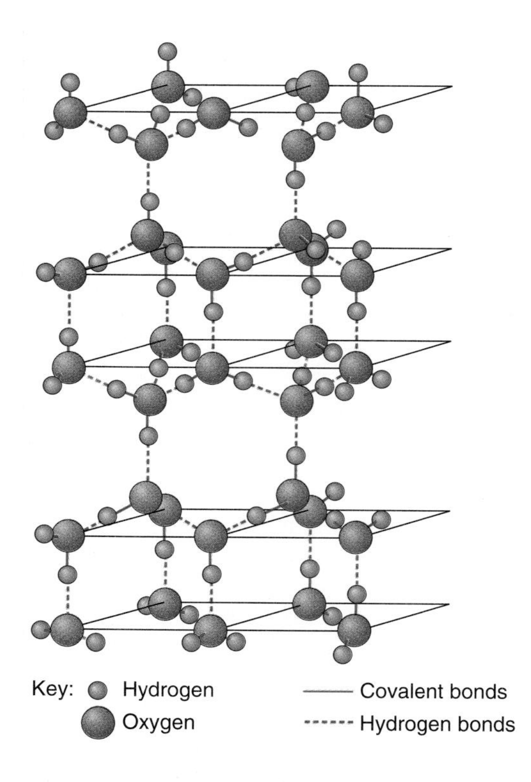

Figure 5.10

The hydrogen-bonded lattice structure of the common form of ice. Note the open channels between "layers" of water molecules, which cause ice to be less dense than liquid water.

The **density** of a substance is defined as its mass per unit of volume. For scientific purposes, mass is given in grams and the "unit of volume" is one cubic centimeter, which is identical to one milliliter (mL). One cubic centimeter of liquid water weighs 1.00 g. In other words, its density is 1.00 g/cm^3 or 1.00 g/mL. On the other hand, 1 cm^3 of ice weighs 0.92 g, so its density is 0.92 g/cm^3 or 0.92 g/mL. People often confuse mass with density. For example, you may hear someone say that iron is heavy or that lead is very heavy. Pieces of iron and lead are indeed often quite heavy, but it is more accurate to say that iron has a high density (7.9 g/mL) and that lead has an even higher density (11.3 g/mL). On the other hand, popcorn has a low density and we are likely to say that a bag of popcorn feels very light.

For the great majority of substances, the solid state is more dense than the liquid. The fact that water shows the reverse behavior means that lakes freeze from the top down, not the bottom up. This topsy-turvy behavior is convenient for aquatic plants, fish, and ice skaters. On the other hand, it is not so convenient for people whose water pipes and car radiators burst when the water inside them expands as it freezes.

5.12 Consider This: Oil and Water

Relative densities have practical consequences for water when it mixes (but does not dissolve) with other substances in the environment. Crude oil has a density of approximately 0.8 g/cm^3. What implications does this have for cleaning oil spills in the ocean?

Finally, we want to examine another of water's unusual properties, namely its uncommonly high capacity to absorb and release heat. On a global scale, this property helps determine worldwide climates. By absorbing vast quantities of heat, the oceans and the droplets of water in clouds help mediate global warming. The specific details of these processes are among the uncertainties that complicate efforts to model global warming accurately. We do know that heat is absorbed when water evaporates from seas, rivers, and lakes. Heat is also released when water condenses as rain or snow. These changes between solid, liquid, and gaseous forms of water create the great thermal engine that helps drive weather patterns in the short term and regulates climates over longer periods of time. But, when not changing its physical state, liquid water absorbs more energy than the ground if equal masses are used. This happens because water has a higher capacity to store heat than do rocks and dirt. As a consequence, when the weather turns colder, the ground has less stored heat to lose than the water and therefore, cools more quickly. The water retains more heat and is able to provide more warmth for a longer time to the areas bordering it. Such properties should be familiar to anyone who has ever jumped into a warm lake or pool on a cool day.

- Global warming was discussed in Chapter 3.

A quantitative measure of a substance's capacity to absorb heat is called its **specific heat.** Specific heat is defined as the quantity of heat energy that must be absorbed to increase the temperature of one gram of the substance by one degree Celsius. The specific heat of liquid water is 1.00 cal/g·°C, which means that one calorie of energy will raise the temperature of one gram of liquid water by 1°C. In fact, the calorie was originally defined in this manner. Conversely, when the temperature of one gram of liquid water falls 1°C, one calorie of heat is given off. Because of the relationship between calories and joules, the specific heat of water can also be expressed as 4.18 J/g·°C, which is equivalent to 1.00 calorie (see Chapter 4).

This 4.18 J/g·°C (1.00 cal) may not sound like a very large value, but liquid water has one of the highest specific heats of any known liquid. Because of this, it is an exceptional coolant used to carry away excess heat in chemical industry, power plants, and the human body. Most other compounds have significantly lower specific heats. For example, liquid benzene, C_6H_6, an organic solvent, has a specific heat of only 0.406 cal/g·°C, less than half that of water. Most solid substances have lower specific heats than that of water. Metals are particularly notable for their low specific heats, most being below 0.1 cal/g·°C, less than 10% that of water. For this reason metals can be heated easily to high temperatures and, conversely, cool quickly when the heat source is removed.

The reason for this difference is again associated with structure. The unusually high heat capacity of water is a consequence of strong hydrogen bonding and the resultant degree of order that exists in the liquid. Temperature is a measure of molecular motion—the higher the temperature, the greater the average motion. When molecules are strongly attracted to each other, a good deal of energy is required to overcome these intermolecular forces and enable the molecules to move more freely. Such is the case with water. On the other hand, intermolecular forces are much weaker in non-hydrogen-bonded liquids, such as benzene, and they are much easier to overcome. Consequently, the specific heats for such compounds are lower.

5.13 Consider This: Showering Heat on Yourself

The high specific heat of water has important consequences for energy consumption or conservation in residences, where large amounts of energy are required to heat water for bathing and washing clothes and dishes. Suppose that the water enters the water heater in your residence at 20°C (68°F), and the heater is set to heat the water to 50°C (122°F).

a. How many calories of heat energy are needed to heat the 100 L of water that are used in a typical 5-minute shower?

b. To conserve energy, the heater is reset to 40°C (104°F). How many calories of heat energy would be saved during that 5-minute shower?

Answer

a. $100 \text{ L} \times \frac{10^3 \text{ g}}{1\text{L}} \times \frac{1 \text{ cal}}{\text{g}\cdot°\text{C}} \times (50°\text{C} - 20°\text{C}) = 3.0 \times 10^6 \text{ cal}$

5.9 Water As a Solvent: A Closer Look

One of the most important properties of water was discussed in Section 5.4, namely that water is an excellent solvent for a wide variety of substances. A great deal of chemistry occurs in water solutions. Because aqueous solutions are so important, we need some understanding of *how* substances dissolve in water.

"Sugar water" and "salt water" are examples of two main classes of aqueous solutions. A significant difference between the two can be demonstrated with a conductivity meter, such as pictured in Figure 5.11. Two wires attach a battery, or another source of electricity, to a light bulb. As long as the two separate wires do not touch, the bulb will not light; there is no completed electrical circuit. If the separated ends are placed into pure "distilled" water or a solution of sugar in distilled water, the bulb will not be illuminated. However, if the separate wires are placed into an aqueous solution of salt, the bulb illuminates. Perhaps the light has also gone on in the mind of the experimenter! Pure water or a solution of sucrose in water do not conduct electricity and therefore do not complete the electrical circuit; the light does not glow. Sugar and other nonconducting solutes are called **non-electrolytes.** On the other hand, a water solution of salt

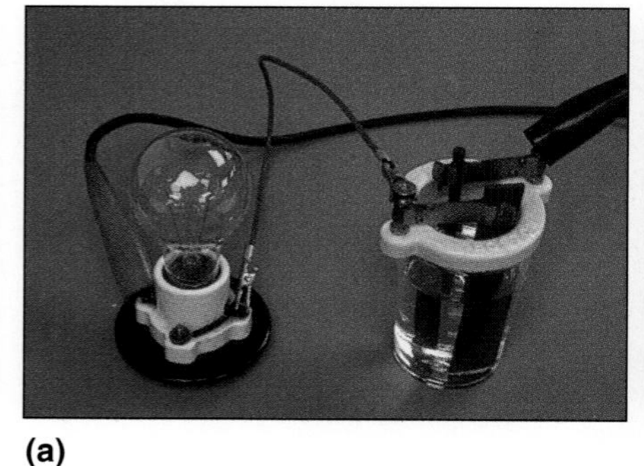

(a)

(b)

(c)

Figure 5.11

Conductivity in water and water solutions.
(a) Distilled water (nonconducting)
(b) Sucrose dissolved in water (nonconducting)
(c) NaCl dissolved in water (conducting)

(sodium chloride) is an electrical conductor and the light bulb lights. Sodium chloride and other conducting solutes are classified as **electrolytes.** What accounts for this difference in properties between electrolytes and nonelectrolytes? That is the next topic we consider.

5.10 Ions and Ionic Compounds

The flow of electric current involves the transport of electrical charge. Therefore, the fact that aqueous sodium chloride solutions conduct electricity suggests they contain electrically charged species. These species are called **ions,** from the Greek for "wanderer." When solid sodium chloride dissolves in water, its positively charged ions (Na^+) called **cations,** and negatively charged ions (Cl^-) called **anions,** separate. As these ions wander about in solution, they transport electrical charge.

- Cations are positively charged; anions are negatively charged.

It may be a little surprising to learn that Na^+ and Cl^- ions exist in the solid crystals of salt such as those in a saltshaker, as well as in a water solution of salt. Solid sodium chloride is a three-dimensional cubic arrangement of sodium and chloride ions occupying alternating positions. The attractions between cations and anions in the crystal are called **ionic bonds** and hold the crystal together. In an **ionic compound,** such as NaCl, there are no true covalently bonded molecules, only positively charged cations (Na^+) and negatively charged anions (Cl^-). Each Na^+ ion is surrounded by six oppositely charged Cl^- ions. Likewise, each Cl^- ion is surrounded by six positively charged Na^+ ions. A single, tiny crystal of sodium chloride consists of many billions of Na^+ and Cl^- ions held together in the arrangement shown in Figure 5.12.

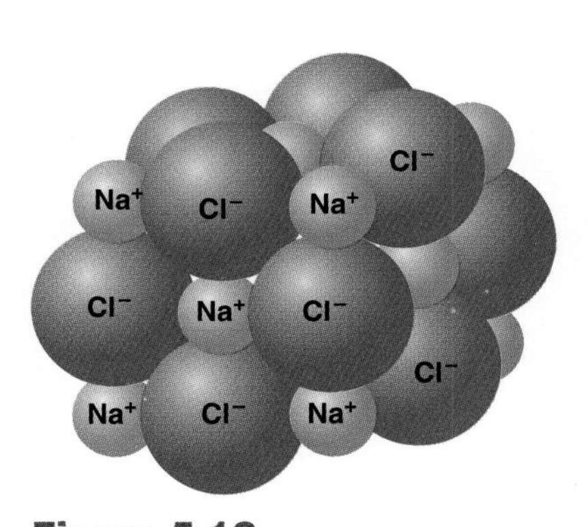

Figure 5.12
The arrangement of Na^+ and Cl^- ions in a crystal of sodium chloride.

Thus far we have described the structure and some of the properties of ionic compounds, but we have not explained *why* certain atoms lose or gain electrons to form ions. Not surprisingly, the answer involves the distribution of electrons within atoms. Recall that a sodium atom, with an atomic number of 11, has 11 electrons and 11 protons. Sodium is in Group 1A and each sodium atom has only one electron in its *outer* energy level. This electron is rather loosely attracted to the nucleus and can be easily removed from the atom by absorbing a small amount of energy. When this happens, the Na atom is transformed into a Na^+ ion by losing an electron (e^-), a process represented by the equation 5.1.

$$Na \longrightarrow Na^+ + e^- \qquad [5.1]$$

A Na^+ ion has a +1 charge because it contains the 11 protons originally present in the Na atom, but only 10 electrons. These 10 electrons are in a configuration that is essentially the same as the 10 electrons in an atom of the inert element neon (Ne). Table 5.5 compares the sodium atom and the sodium ion.

A Na^+ ion, like a Ne atom, has two *inner* electrons and eight *outer* electrons. This is a particularly stable arrangement. We may generalize by saying that *metallic elements tend to form cations by losing electrons.*

By contrast, a chlorine atom has a tendency to *gain* an electron. The electrically neutral Cl atom includes 17 electrons and 17 protons. It has seven *outer* electrons. Because of the stability associated with eight outer electrons, it is energetically

Table 5.5

Comparison of a Sodium Atom with a Sodium Ion

Sodium Atom	Sodium Ion
Na	Na^+
11 protons	11 protons
11 electrons	10 electrons
Net charge: zero	***Net*** charge: +1

favorable for a Cl atom to acquire an extra electron, such as one from a sodium atom, to become a Cl^- ion. In this ion there are 18 electrons and 17 protons; thus the net charge is -1. This information is shown in Table 5.6.

$$Cl + e^- \longrightarrow Cl^- \quad [5.2]$$

Because elemental chlorine consists of diatomic Cl_2 molecules, we can also write this gain in electrons in the following fashion.

$$Cl_2 + 2\ e^- \longrightarrow 2\ Cl^- \quad [5.3]$$

In general, *nonmetallic elements—those on the right-hand side of the periodic chart* (except the noble gases, Group 8A)—*gain electrons to form anions.*

When sodium metal and Cl_2 gas react, electrons are transferred from sodium atoms to chlorine atoms with the release of a considerable amount of energy. The result is the aggregate of Na^+ ions and Cl^- ions known as sodium chloride. In the formation of an ionic compound, such as sodium chloride, the electrons are actually transferred from one atom to another, not simply shared, as they would be in a covalent compound.

Is there *evidence* for electrically charged ions in pure sodium chloride? Experimental tests show that crystals of sodium chloride do not conduct electricity, but when these crystals are melted, the resulting liquid conducts electricity. This provides evidence that Na^+ and Cl^- ions from the solid NaCl also exist in the liquid state, without the presence of water. Crystals of NaCl and other ionic compounds are hard and brittle. When hit sharply, they shatter rather than being flattened, as would be true for a substance consisting of molecules with weak forces between them. This suggests the existence of strong forces that extend throughout the ionic crystal. Literally speaking, there is no such thing as a specific, localized "ionic bond" analogous to covalent bonds in molecules. Rather, there is a generalized ionic bonding that holds together a large assembly of ions.

Having established that sodium chloride is an ionic compound, we next ask what other elements form ions and ionic compounds. Electron transfer to form cations and anions, respectively, is likely to occur between metallic elements (those elements in the left and middle blocks of the periodic table) and nonmetallic elements (those elements on the right side of the periodic table, except the noble gases, Group 8A). Sodium, lithium, magnesium, and other metallic elements have a strong tendency to give up electrons and form positive ions. On the other hand, chlorine, fluorine, oxygen, and other nonmetals have a strong attraction for electrons and readily gain electrons to form negative ions. Therefore, *ionic compounds are formed when elements from opposite sides of the periodic table react.* Potassium chloride (KCl) and sodium iodide (NaI) are two of many such compounds. Because ordinary table salt (NaCl) is such an important example of an ionic compound, chemists frequently refer to other ionic compounds simply as "salts," which are crystalline solids.

The idea of ion formation, where atoms gain or lose electrons, can help explain and even predict the formulas of many compounds. Back in Chapter 1 you were told that the formula of the compound formed from calcium and chlorine is $CaCl_2$, but no explanation was given. The reason should now be apparent. An atom of calcium, a member of Group 2A, readily loses its two outer electrons to form a Ca^{2+} ion.

Table 5.6

Comparison of a Chlorine Atom with a Chloride Ion

Chlorine Atom	Chloride Ion
Cl	Cl^-
17 protons	17 protons
17 electrons	18 electrons
Net charge: zero	***Net*** charge: -1

Calcium atom	Calcium ion
·Ca·	Ca^{2+}

Chlorine, as we have already seen, forms Cl^- ions. For the electrical charges to be balanced, two Cl^- ions are required for each Ca^{2+} ion. Hence, the formula of calcium chloride is $CaCl_2$. In an ionic compound, the *sum* of the positive charges equals the *sum* of the negative charges. Your Turns 5.14 and 5.15 give you opportunities to apply similar logic to other elements and compounds.

5.14 Your Turn

Predict the charge on the ion that will form from each of these atoms. Draw the Lewis structure for each atom and for its ion, clearly labeling the charge on the ion.

a. Br **b.** Mg **c.** O **d.** Al

Hint: Use the periodic table to determine the number of outer electrons. That in turn will help you determine how many electrons must be lost or gained to achieve stability with an octet of electrons.

Answer

a. Bromine is in Group 7A of the periodic table and gains one electron to form a stable ion with a charge of -1, just as was the case for chlorine. These are the Lewis structures.

:B̤̈r· and $[:\ddot{\underset{..}{Br}}:]^-$

5.15 Your Turn

Predict the formulas of the ionic compounds that would be formed by the combination of each pair of elements.

a. Ca and Br **b.** K and F **c.** Li and O **d.** Sr and Br

Hint: The sum of the positive charges must match the sum of the negative charges for the compound to be neutral.

Answer

a. Ca forms Ca^{2+} ions and Br forms Br^- ions. To be electrically neutral, the compound must have two Br^- ions for each Ca^{2+} ion. The formula for calcium bromide is $CaBr_2$.

Some ionic compounds include **polyatomic ions,** ions that are themselves made up of more than one atom or element. A case in point is sodium sulfate, Na_2SO_4. This compound consists of Na^+ and SO_4^{2-} ions. In the sulfate ion, the four oxygen atoms are covalently bonded to a central sulfur atom in a symmetric tetrahedral arrangement. Counting the electrons in the Lewis structure (Figure 5.13) reveals that there are 32 electrons, two more than the 30 valence electrons provided by one neutral sulfur atom (six) and four neutral oxygen atoms (24). The "extra" two electrons give the sulfate ion a charge of -2.

Table 5.7 is an alphabetical list of some common polyatomic ions. Most are anions, but polyatomic cations are also possible, as in the case of the ammonium ion, NH_4^+. Note that some elements (carbon, sulfur, and nitrogen) form more than one polyatomic anion with oxygen.

The rules for predicting formulas of compounds containing polyatomic ions and naming them are similar to those that apply to simple compounds of two elements. Consider, for example, aluminum sulfate, a compound used in water purification. It is formed from Al^{3+} and SO_4^{2-} ions. Like all ionic compounds, this compound must be electrically neutral; thus the sum of the positive charges must equal the sum of the negative charges. This requires that two Al^{3+} ions combine with three SO_4^{2-} ions: $2(+3)$ cancels $3(-2)$; the formula of aluminum sulfate must be $Al_2(SO_4)_3$. Note that the subscript 3 applies to the entire SO_4^{2-} ion that is enclosed in parentheses. The formula of the compound thus represents two Al ions along with three sulfate ions containing a total of three S atoms and 12 O atoms. As is always the case with ionic compounds, the positive ion is named first.

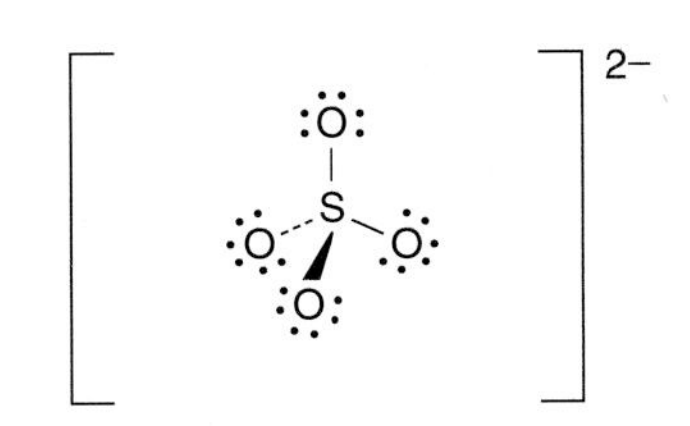

Figure 5.13
Lewis structure of the sulfate ion, SO_4^{2-}.

Table 5.7

Some Common Polyatomic Ions

Name	Formula	Name	Formula
acetate	$C_2H_3O_2^-$	nitrite	NO_2^-
bicarbonate*	HCO_3^-	phosphate	PO_4^{3-}
carbonate	CO_3^{2-}	sulfate	SO_4^{2-}
hydroxide	OH^-	sulfite	SO_3^{2-}
hypochlorite	OCl^-	ammonium	NH_4^+
nitrate	NO_3^-		

*Also called the hydrogen carbonate ion.

5.16 Your Turn

Give the correct name for each of these compounds:

a. KNO_3 **b.** $(NH_4)_2SO_4$ **c.** $NaHCO_3$ **d.** $CaCO_3$ **e.** $Mg_3(PO_4)_2$

Answers

a. Potassium nitrate. This name is just the combination of the names of the two ions, potassium and nitrate.

b. Ammonium sulfate. *Note:* When naming ionic compounds, it is not necessary to use the prefixes such as *di-* or *tri-* as we did when naming covalently bonded compounds such as carbon dioxide or dinitrogen pentoxide.

5.17 Your Turn

Write the formula for each of these compounds.

a. calcium hypochlorite (used in bleaches)
b. lithium carbonate (treatment of bipolar disorders)
c. potassium nitrate (matches and fireworks)
a. barium sulfate (medical X rays)

Answer

a. $Ca(OCl)_2$ As in 5.15 Your Turn, the *sums* of the negative OCl^- ions and the positive charges of Ca^{2+} ions must be equal. Therefore, two hypochlorite ions are needed to equal the charge on a calcium ion, balancing the charges at 2− and 2+.

5.11 Water Solutions of Ionic Compounds

Now we are in a position to understand one of the most important properties of ionic compounds, namely why many are quite soluble in water. Recall from Section 5.7 that water molecules are polar: They have both a partial positive side (hydrogen) and a partial negative side (oxygen). When a solid sample of an ionic compound is placed in water, the polar H_2O molecules are attracted to the individual ions. The partial negatively charged oxygen atom of a water molecule is attracted to the positively charged cations in the crystal. At the same time, hydrogen atoms in H_2O, with their partial positive charges, are attracted to the negatively charged anions of the solute. Thus, the ions are surrounded by water molecules, which diminishes the anion-cation attraction in the solid. The substantial attraction between the ions and H_2O molecules results in the surrounding water molecules literally plucking the ions out of the solid and into solution. In dissolving, the ionic compound separates into its component cations and anions. Equation 5.4 and Figure 5.14 represent this process for sodium chloride and water:

$$NaCl(s) + H_2O(l) \longrightarrow Na^+(aq) + Cl^-(aq) \quad [5.4]$$

The (aq) in the equation indicates that the ions are present in an *aqueous* solution.

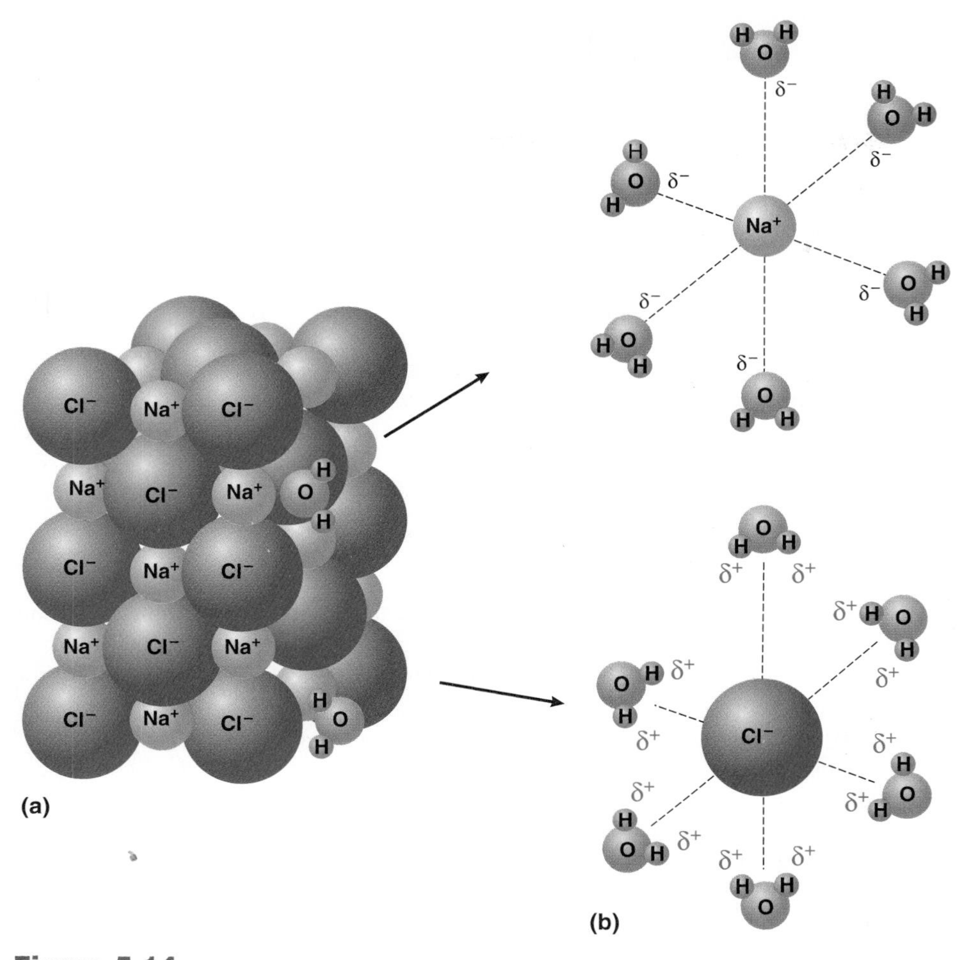

Figure 5.14
Dissolving sodium chloride in water.

When compounds containing polyatomic ions dissolve in water, the polyatomic ions remain intact. For example, when sodium sulfate dissolves in water, the sodium ions and sulfate ions separate, but the SO_4^{2-} ions remain intact.

$$Na_2SO_4(s) + H_2O(l) \longrightarrow 2\ Na^+(aq) + SO_4^{2-}(aq) \qquad [5.5]$$

What has just been described for sodium chloride and sodium sulfate dissolving in water is true for many other ionic compounds. Indeed, this behavior is so common that the chemistry of ionic compounds is largely that of their behavior in aqueous solutions. Conversely, almost all naturally occurring water samples contain various amounts of ions. Even our body fluids contain significant concentrations of ions.

5.18 Consider This: Electricity and Water Don't Mix

Small electrical appliances, such as a hair dryer or curling iron, carry a warning label prominently advising the consumer of the hazard associated with using the appliance near water. Why is this a problem if water does not conduct electricity? What is the best course of action if a plugged-in hair dryer does accidentally fall into a sink full of water?

We conclude the discussion of ionic compounds and their solutions with a brief excursion into the relative solubility of various ionic compounds. In principle, the dissolving process in water, as just described, ought to be true for any ionic compound. Indeed, many ionic compounds are highly soluble in water. But some are at best only slightly soluble, and some have extremely low solubility in water. The reasons for this

Table 5.8

Generalizations about the Solubility of Ionic Compounds in Water

Note: "Insoluble" means that the compounds have extremely low solubility in water (less than 0.01 M). All ionic compounds have at least a very small solubility in water.
All **sodium, potassium,** and **ammonium (NH_4^+)** compounds are soluble.
All **nitrates** are soluble.
Most **chlorides** are soluble (except silver, mercury, and lead chlorides).
Most **sulfates** are soluble (except strontium, barium, and lead sulfate).
Most **carbonates** are insoluble (except those with Group 1A or NH_4^+ cations).
Most **hydroxides** and **oxides** are insoluble (except those with Group 1A or NH_4^+ cations).
Most **sulfides** are insoluble (except those with Group 1A or NH_4^+ cations).

range of behavior would take us too far afield. Basically, they involve the sizes and charges of the ions, how strongly the ions attract each other, and how strongly the ions are attracted to water molecules. Nevertheless, a few generalizations are quite useful for predicting the solubility of common ionic compounds (Table 5.8).

It is possible to use Table 5.8 and a periodic table of the elements to predict the solubility (or insolubility) of many compounds. For example, calcium nitrate, $Ca(NO_3)_2$, is predicted to be soluble in water because all nitrates are soluble. On the other hand, calcium carbonate, $CaCO_3$, is likely insoluble because most carbonates are insoluble, and calcium is not one of the exceptions for carbonates. By similar reasoning, copper hydroxide, $Cu(OH)_2$, is insoluble, but copper sulfate, $CuSO_4$, is soluble.

5.19 Your Turn

Use the solubility generalizations in Table 5.8 to predict the solubility of these compounds.

a. ammonium nitrate, NH_4NO_3 (used in fertilizers)
b. sodium sulfate, Na_2SO_4 (used as an additive in detergents)
c. mercury sulfide, HgS (the mineral cinnabar)
d. aluminum hydroxide, $Al(OH)_3$ (used in some antacid tablets)

Answer

a. Soluble. All nitrates and all ammonium compounds are soluble.

The landmasses on the earth are made up largely of minerals consisting of ionic compounds that have extremely low solubility in water. If that were not the case, most would have dissolved long ago. Table 5.9 summarizes some environmental consequences of the differing solubility of minerals and other substances in water.

5.12 Covalent Compounds and Their Solutions

From the previous discussion, you might get the impression that only ionic compounds dissolve in water. But, other kinds of compounds dissolve as well. Common experience tells us that sucrose (ordinary table sugar) dissolves readily in water. But sucrose contains no ions; it is a **covalent** or **molecular compound.** Like water, carbon dioxide, chlorofluorocarbons, and many of the other compounds you have been reading about, sucrose molecules consist of covalently bonded atoms. The formula for sucrose is $C_{12}H_{22}O_{11}$ and it exists as individual molecules consisting of 45 atoms, as shown in Figure 5.15.

Table 5.9

Environmental Consequences of Solubility

- Sodium chloride dissolves from the land and washes into the sea. Thus, oceans are salty and ocean water cannot be used for drinking water without expensive purification.
- Agricultural fertilizers contain nitrates. Because all nitrates are soluble in water, runoff water from fertilized fields carries nitrates into surface and ground water. At high enough concentration, they may create a health hazard because of the toxicity of nitrate ion.
- Most metals (iron, copper, zinc, chromium, etc.) exist on Earth as insoluble sulfide (S^{2-}) or oxide (O^{2-}) minerals; iron ore (iron oxide, Fe_2O_3) is a common example. If these minerals were soluble in water, they would have been largely dissolved out of the ground and washed to sea long ago.
- Waste piles left from mining operations remain a visual as well as chemical blight on the landscape. They often contain small amounts of toxic ions, such as lead and mercury, which have low solubility. However, these ions may be leached very slowly from the waste into rivers and lakes where they may contaminate water supplies.

When sugar dissolves in water, its molecules become uniformly dispersed among the H_2O molecules. As in all true solutions, the mixing is at the most fundamental level of the solute and solvent—the molecular or ionic level. The $C_{12}H_{22}O_{11}$ molecules remain intact and do not separate into ions. Evidence for this is the fact that aqueous sucrose solutions do not conduct electricity, as was shown in Figure 5.11. However, the sugar molecules do interact with the water molecules. In fact, solubility is always promoted when there is a net attraction between the solvent molecules and the solute molecules or ions. This suggests a general solubility rule: *Like dissolves like.* Compounds with similar chemical composition and molecular structure tend to form solutions with each other. The intermolecular attractive forces between similar molecules are high, promoting solubility. Dissimilar compounds do not dissolve in each other.

- "Like dissolves like" is a useful generalization.

Consider, for example, three familiar covalent (molecular) compounds, all of which are highly soluble in water: sucrose; ethylene glycol (the main ingredient in antifreeze); and ethanol (ethyl alcohol, the "grain alcohol" found in alcoholic beverages). Like all alcohols, they contain an —OH group (Figures 5.16 and 5.17.)

We start with the simplest, ethanol, C_2H_5OH. Its oxygen atom is covalently bonded to a hydrogen atom and to a carbon atom. The —OH group of a C_2H_5OH molecule can form hydrogen bonds with H_2O molecules (Figure 5.17). This hydrogen bonding causes

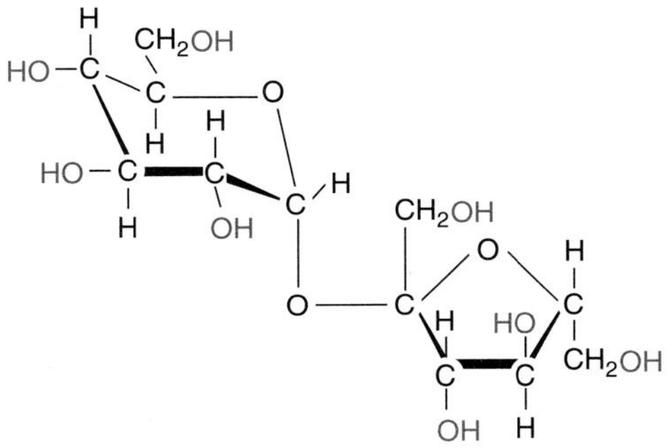

Figure 5.15
Molecular structure of sucrose.

Ethanol

```
    H   H
    |   |   ..
H — C — C — O — H
    |   |   ..
    H   H
```

Ethylene glycol

```
            H   H
    ..      |   |   ..
H — O — C — C — O — H
    ..      |   |   ..
            H   H
```

Figure 5.16

Structures of ethanol and ethylene glycol.

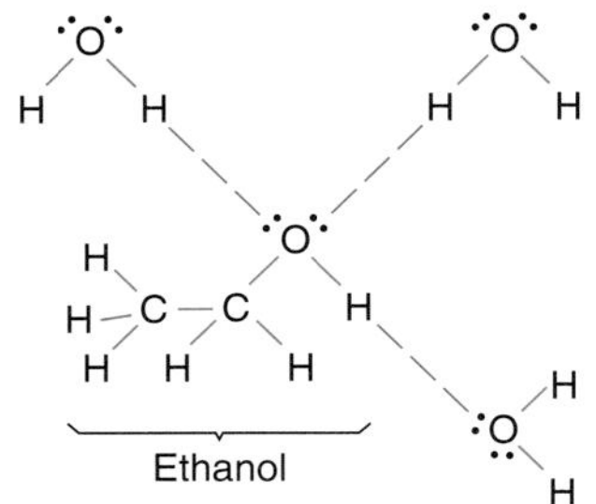
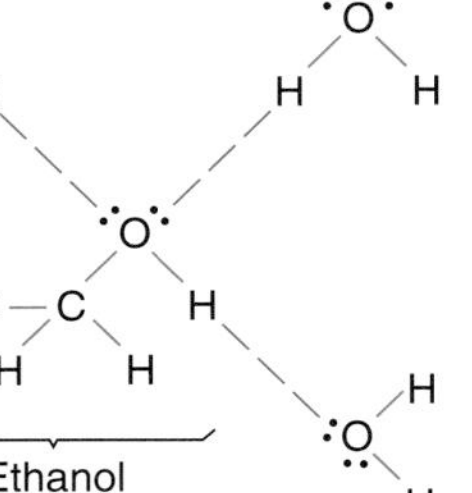

Figure 5.17

Hydrogen bonding of ethanol with water.

— covalent bond

— — — hydrogen bond

water and ethanol to have a great affinity for each other, a conclusion consistent with the fact that they form solutions in all proportions. Ethylene glycol is also an alcohol with two —OH groups available for hydrogen bonding with H_2O. Therefore, ethylene glycol is highly water soluble, a necessary property for an antifreeze ingredient.

5.20 Your Turn

Sketch a diagram to show hydrogen bonding between ethylene glycol and water.

Finally, we consider sucrose, the compound that introduced this section. Examination of its structure (Figure 5.15) discloses that the sucrose molecule contains eight —OH groups and three additional oxygen atoms that can also participate in hydrogen bonding. This accounts for the high solubility of sugar in water.

5.21 Consider This: Three-Dimensional Representations of Molecules

Three-dimensional representations of molecules can be viewed on the Web using CHIME, a free plug-in that you can download and install. Three-dimensional representations of ethanol, ethylene glycol, and sucrose are there. Use these molecular representations to identify the places in each compound where hydrogen bonding occurs. Has your mental picture of these molecules changed after seeing these 3-D representations? Explain.

On the other hand, molecular compounds that differ in composition and molecular structure do not attract each other strongly. It has often been observed that "oil and water don't mix." They don't mix because they are structurally very different; unlike compounds don't like each other, which is just the reverse of "like dissolves like." Water is a highly polar compound, whereas oil consists of nonpolar hydrocarbon compounds. When placed in contact, they remain apart in separate layers (Figure 5.18). Even if shaken vigorously, the oil and water return to their own layers. But oily, nonpolar compounds generally dissolve readily in hydrocarbons or chlorinated hydrocarbons. For this reason, the latter compounds have often been used in dry cleaning solvents.

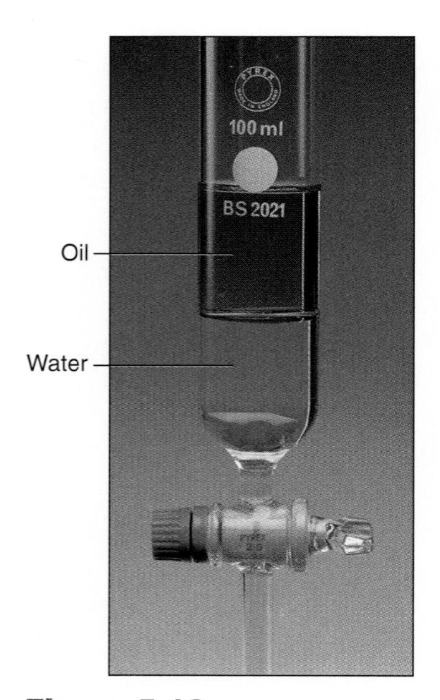

Figure 5.18

Oil and water do not dissolve in each other.

The tendency of nonpolar compounds to mix with other nonpolar substances affects how fish and animals store certain highly toxic substances such as PCBs (polychlorinatedbiphenyls) or the pesticide DDT. PCB and DDT molecules are nonpolar, and so when fish absorb them from water, the molecules are stored in body fat (which is also nonpolar) rather than in the blood (which is a highly polar water solution).

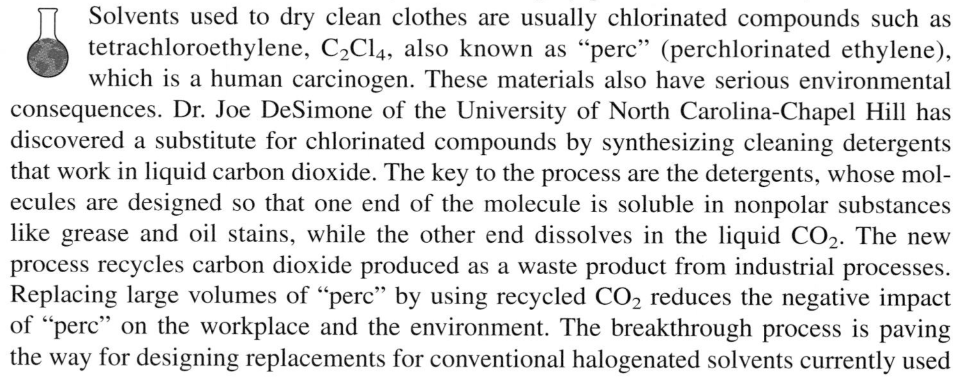

Solvents used to dry clean clothes are usually chlorinated compounds such as tetrachloroethylene, C_2Cl_4, also known as "perc" (perchlorinated ethylene), which is a human carcinogen. These materials also have serious environmental consequences. Dr. Joe DeSimone of the University of North Carolina-Chapel Hill has discovered a substitute for chlorinated compounds by synthesizing cleaning detergents that work in liquid carbon dioxide. The key to the process are the detergents, whose molecules are designed so that one end of the molecule is soluble in nonpolar substances like grease and oil stains, while the other end dissolves in the liquid CO_2. The new process recycles carbon dioxide produced as a waste product from industrial processes. Replacing large volumes of "perc" by using recycled CO_2 reduces the negative impact of "perc" on the workplace and the environment. The breakthrough process is paving the way for designing replacements for conventional halogenated solvents currently used in manufacturing and industries making coatings. For his work, Professor DeSimone received the 1997 Presidential Green Chemistry Challenge Award.

5.13 Protecting Our Drinking Water: Federal Legislation

We can now apply to drinking water what we know about the structure and properties of pure water and aqueous ionic solutions. What dissolves in drinking water determines its quality and the potential for adverse health effects. Keeping public water supplies safe has long been recognized as an important public health issue. In 1974, the U.S. Congress passed the Safe Drinking Water Act (SDWA) in response to public concern about findings of harmful substances in drinking water supplies. The aim of the SDWA, as amended in 1996, is to provide public health protection to all Americans who get their water from community water supplies (over 200 million people). Contaminants that may be health risks are regulated by EPA as required by the SDWA. The EPA sets legal limits for such contaminants that reflect knowledge about health effects and risk calculations (Table 5.10). These limits also take into account the practical realities of the concentration of contaminants likely to be present in drinking water sources, and the ability of water supply utilities to remove the offending contaminants by using available technology.

For each contaminant, the EPA has established a health goal, or **Maximum Contaminant Level Goal (MCLG).** This is the level, expressed in ppm or ppb, at which a person weighing 70 kg (154 lb) could drink two liters (about two quarts) of water containing the contaminant every day for 70 years without suffering any ill effects. Each MCLG includes built-in safety factors for uncertainties in the standardizing data and for individual differences in sensitivity to the contaminant. An MCLG is not a legal limit with which water systems must comply; it is based solely on considerations of human health. For known cancer-causing substances (carcinogens), the EPA has set the health goal at zero, under the assumption that *any* exposure to the substance could present a cancer risk.

The mere presence of a contaminant does not necessarily mean a serious health problem. For a problem to exist, the concentration of the contaminant must exceed the legal limit, expressed in ppm or ppb, referred to as a **Maximum Contaminant Level (MCL).** The EPA sets legal limits for each contaminant as close to the MCLG as possible, keeping in mind the practical realities of technical and financial barriers that may make it difficult to achieve the goals. Except for contaminants regulated as carcinogens, for which the MCLG is zero, most legal limits and health goals are the same. Even when they are less strict than the MCLGs, the MCLs provide substantial public health protection.

Table 5.10

MCLGs and MCLs (in ppm) for Selected Pollutants in Drinking Water

Pollutant	MCLG	MCL
Cadmium (Cd^{2+})	0.005	0.005
Chromium (Cr^{3+}, CrO_4^{2-})	0.1	0.1
Lead (Pb^{2+})	0	0.015
Mercury (Hg^{2+})	0.002	0.002
Nitrate (NO_3^-)	10	10
Benzene (C_6H_6)	0	0.005
Trihalomethanes ($CHCl_3$, etc.)	0	0.080

5.22 Consider This: Understanding MCLGs and MCLs

Most people are unfamiliar with these terms from the Safe Drinking Water Act. Assume you are making a presentation in another class to explain what these acronyms mean and how the information helps to safeguard our drinking water. Prepare a short outline of what you will say. Be prepared to answer questions from the audience, particularly dealing with why MCLs were not set to zero for all carcinogens.

Because of improved detection and quantitative analytical methods, the number of regulated contaminants in drinking water increases each time Congress updates the legislation. Lower limits for MCL values have been established as more accurate risk information becomes available. Currently, there are more than 80 regulated contaminants, which fall into several major categories: metals (such as cadmium, chromium, copper, mercury, and lead), a few nonmetallic elements (e.g., fluoride and arsenic), pesticides, industrial solvents, compounds associated with plastics manufacturing, and radioactive materials. Depending on the particular contaminant, MCLs vary from around 10 ppm to less than 1 ppb. Some contaminants interfere with liver or kidney functions. Others can affect the nervous system if ingested over a long period at levels consistently above the legal limit (MCL). Pregnant women and infants are at particular risk for some contaminants because of their effects on a developing fetus or the digestive system of an infant. In Section 5.15 we look at case studies for two contaminants, lead and trihalomethanes (THMs).

In addition to contaminants that can pose chronic health problems, other substances in drinking water present acute health risks. For example, nitrate (NO_3^-) and nitrite (NO_2^-) ions limit the blood's ability to carry oxygen. Even when consumed in tiny doses, these ions cause immediate health effects for infants. Therefore, the EPA limit for nitrate and nitrite ions in drinking water specifically protects infants. Another acute health risk is biological, not chemical—from bacteria, viruses, and other microorganisms including *Cryptosporidium* and *Giardia.* News media warnings announcing a "boil water emergency" are typically the result of a "total coliform" violation. Coliforms are a broad class of bacteria, most of which are harmless, that live in the digestive tracts of humans and other animals. The presence of high coliform concentration in water usually indicates that the water treatment or distribution system is not working properly. Diarrhea, cramps, nausea, and vomiting, the symptoms of coliform-related illness, are not serious for a healthy adult, but can be life threatening for the very young, the elderly, or those with weakened immune systems.

5.23 Consider This: A Drink of Water—What Is In It?

Table 5.10 is merely a starting point for the wealth of information available about six possible pollutants in drinking water. At the EPA Office of Ground Water and Drinking Water, you can access a consumer fact sheet on each of these pollutants, as well as dozens more. A consumer version and a technical version are available, and the latter is recommended. Look up a pollutant listed in Table 5.10 to find out how the pollutant gets into the water supply and how you would know if it were in your drinking water. Is your state listed as one of the top states that releases the contaminant?

Note: Arsenic, cadmium, lead, chromium, mercury, and nitrate/nitrite ions are found under the section on Inorganic Chemicals. Benzene is listed under Volatile Organic Chemicals. No trihalomethane such as $CHCl_3$ currently is listed, but you can find a variety of other chlorinated compounds found in water, such as CCl_4 and CH_2Cl_2.

5.24 Consider This: MTBE in Ground Water

The gasoline additive MTBE was discussed in Chapter 4 as one of the mandated oxygenates in certain metropolitan areas to reduce air pollution. Concern about the accumulation of MTBE in surface and groundwater, with the potential to compromise the quality of drinking water supplies, has led to some changes in policy in some areas of the U.S. Consult the *Chemistry in Context* web site or other resources suggested by your instructor to find the current status of the use of MTBE.

In addition to the Safe Drinking Water Act, there is also federal legislation to control pollution of surface waters, including lakes, rivers, and coastal areas. The Clean Water Act (CWA), passed by Congress in 1972 and amended several times, provided the foundation for dramatic progress in reducing surface water pollution over the past three decades. The CWA establishes limits on the amounts of pollutants that industries can discharge into surface waters, resulting in actions that have removed over one billion pounds of toxic pollution from U.S. waters every year. Improvements in surface water quality have at least two major beneficial effects: They reduce the amount of clean-up needed for public drinking water supplies, and they result in a healthier natural environment for aquatic organisms. In turn, a healthier aquatic ecosystem has many indirect benefits for humans. In keeping with the new trend toward "green chemistry," industries are finding ways to convert these waste materials into useful products, as well as to design processes "up front" so that they neither use nor produce substances that degrade water quality.

5.25 Consider This: *Cryptosporidium*

As of January 1, 2002, EPA's surface water treatment rules require systems using surface water or groundwater under the direct influence of surface water to remove or deactivate 99% of *Cryptosporidium*.

a. What is *Cryptosporidium*?
b. What are the sources of this contaminant in drinking water?
c. What are the potential health effects from drinking water contaminated with *cryptosporidium?*
d. Why was this rule issued in 2000, but not required until 2002?

5.14 How Is Drinking Water Treated to Make It Safe?

A large supply of water is no guarantee that it is fit to drink. Coleridge's shipwrecked Ancient Mariner knew this all too well, surrounded as he was by "Water, water every where, nor any drop to drink." So how is water treated to make it **potable,** that is, fit for human consumption?

The Safe Drinking Water Act requires that standards of purity and safety for public water supplies be established and enforced by the U.S. EPA. The first step in a typical municipal treatment plant for treating water to be used for drinking is to pass the water through a screen that excludes larger objects both natural (fish and sticks) and artificial (tires and beverage cans). The usual next step is to add two chemicals, aluminum sulfate, $Al_2(SO_4)_3$, and calcium hydroxide, $Ca(OH)_2$. These compounds react to form a sticky gel of aluminum hydroxide, $Al(OH)_3$, which collects suspended clay and dirt particles on its surface.

- Calcium hydroxide is known as slaked lime.

$$Al_2(SO_4)_3(aq) + 3\ Ca(OH)_2(aq) \longrightarrow 2\ Al(OH)_3(s) + 3\ CaSO_4(aq) \qquad [5.6]$$

The $Al(OH)_3$ gel settles slowly carrying with it the suspended particles down into a settling tank. Any remaining particles are removed as the water is filtered through gravel and then sand.

The next step, disinfection to kill disease-causing organisms, is the most crucial one for making drinking water safe. In the United States, this is most commonly done by **chlorination.** Chlorine is usually added in one of three forms: chlorine gas, Cl_2; sodium hypochlorite, NaOCl; or calcium hypochlorite, $Ca(OCl)_2$. The antibacterial agent generated in solution by all three substances is hypochlorous acid, HOCl. The degree of chlorination is adjusted so that a very low concentration of HOCl, between 0.075 and 0.600 ppm, remains in solution to protect the water against further bacterial contamination as it passes through the pipes to the user.

- NaOCl is used in Clorox™ and other brands of laundry bleach. $Ca(OCl)_2$ is commonly used to disinfect swimming pools.

Before chlorination was used, thousands died in epidemics spread via polluted water. In a classic study, John Snow was able to trace a mid-1800s cholera epidemic in London to water contaminated with the excretions of victims of the disease. A more contemporary example occurred in Peru in 1991. This cholera epidemic was traced to bacteria in shellfish growing in estuaries polluted with untreated fecal matter. The bacteria found their way into the water supply, where they continued to multiply because of the absence of chlorination.

Chlorination, however, is not without some drawbacks. The taste and odor of residual chlorine may be objectionable to some and is a reason commonly cited as why people drink bottled water or use home water filters to remove residual chlorine at the tap. A possibly more serious drawback is the reaction of residual chlorine with other substances in the water to form by-products at potentially toxic levels. The most widely publicized of these are the trihalomethanes (THMs) such as chloroform, $CHCl_3$, which are the subject of a case study (Section 5.15.3).

Many European and a few U.S. cities use gaseous ozone (O_3) to disinfect their water supplies. Chapter 1 discussed tropospheric ozone as a serious air pollutant. Chapter 2 described the beneficial effects of the stratospheric ozone layer. In water treatment, the toxic property of ozone is used for a beneficial purpose. The degree of antibacterial action necessary can be achieved with a smaller concentration of ozone than chlorine, and ozone is more effective than chlorine against water-borne viruses. But ozonation is more expensive than chlorination and becomes economical only for large water treatment plants. An additional major drawback of ozone is that it decomposes quickly and hence does not protect the water from contamination after the water leaves the treatment plant. Consequently, a low dose of chlorine is added to ozonated water as it leaves the treatment plant.

Another disinfection method gaining in popularity is the use of ultraviolet (UV) radiation. In Chapter 2 it was pointed out that UV radiation is dangerous for living species, including bacteria. Ultraviolet disinfection is very fast; there are no residual by-products, and it is economical for small installations (including rural homes with potentially unsafe well water). Like ozonation, UV disinfection does not protect the water from contamination after it leaves the treatment site unless a low dose of chlorine is added.

5.26 Consider This: Risk/Benefit—To Chlorinate or Not to Chlorinate?

Chlorination of water supplies has risks as well as benefits associated with it. Prepare to debate the issues involved. In preparing for the debate, list the major advantages and disadvantages of chlorination. Take a position either in favor or opposed to chlorination. If opposed, be ready to discuss alternative methods of disinfection. If in favor, be prepared to offer reasons why alternatives are not as effective as chlorination.

Depending on local conditions, one or more additional purification steps may be carried out at the water treatment facility after disinfection. Sometimes the water is sprayed into the air to remove volatile chemicals that create objectionable odors and taste. If the water is sufficiently acidic to cause problems such as corrosion of pipes or leaching of heavy metals from pipes, calcium oxide (lime) is added to partially neutralize the acid. If there is little natural fluoride in the water supply, many municipalities add about 1 ppm of fluoride (as NaF) to protect against tooth decay. In water, sodium fluoride, NaF, dissociates into $Na^+(aq)$ and $F^-(aq)$ ions. In teeth, fluoride ions are incorporated into a calcium compound called fluorapatite, which is more resistant to dental decay than apatite, the usual tooth material.

5.27 Consider This: Purifying Water Away From Home

How can you purify water when you are hiking? Use the resources of the Web to explore some of the possibilities. What are the relative costs and effectiveness of these alternatives? Are any of the methods similar to those used to purify municipal water supplies? Why or why not?

5.15 Case Studies of Water Quality

It should be clear by now that water is an excellent solvent for many different substances. Some solutes in water, such as oxygen, can be beneficial. Some solutes can change the properties of water. Some are highly toxic and are cause for concern. In this section, we consider three case studies, each of which deals with a different aqueous impurity. Our aim is to understand the nature of the impurity, how it gets into water, how its concentration in water can be measured, how serious a health threat it poses, and what can or should be done to reduce the impurity.

5.15.1 Case Study: Hard Water

Our first case study is of hard water, in which the properties of water are altered by the solutes present. The culprits in this case are some common aqueous ions—principally calcium, Ca^{2+}, and magnesium, Mg^{2+}—frequently present at high concentrations in hard water. Although they pose no health threats for most folks, these ions can be a considerable nuisance and a reason for increased costs unless removed from solution. Water containing Ca^{2+} can form a hard, insoluble deposit of $CaCO_3$ in water heaters, cooking ware, pipes, and industrial equipment. If you live in a hard water region, you have probably noticed a white deposit inside a teakettle or other utensil used to heat water. In hot water pipes, the buildup can cause serious interference with water flow (Figure 5.19). In water heaters, the buildup interferes with heat transfer, resulting in wasted energy and greater cost for water heating.

Water hardness is a property commonly reported for drinking water supplies. **"Hard" water** contains calcium, magnesium, and occasionally iron ions, in the form of their chloride, bicarbonate, and sulfate compounds. In contrast, **"soft" water** contains few of these metal ions. Because calcium ions, Ca^{2+}, are generally the largest contributors to hard water, hardness is usually expressed in parts per million of calcium carbonate by mass. This method of reporting does not mean that the water sample actually contains $CaCO_3$ at the indicated concentration. Rather, it specifies the mass of solid $CaCO_3$ that could be formed from the Ca^{2+} in solution, provided sufficient CO_3^{2-} ions were also present:

$$Ca^{2+}(aq) + CO_3^{2-}(aq) \longrightarrow CaCO_3(s) \qquad [5.7]$$

- Table 5.8 indicates that calcium carbonate is not very soluble in water.

Thus, a hardness of 10 ppm indicates that 10 mg of $CaCO_3$ could be formed from the Ca^{2+} ions present in 1 L of water.

The source of hard water is limestone rock, which is composed of calcium carbonate or a mixture of calcium carbonate and magnesium carbonate. Limestone was formed from ancient inland seas in which calcium carbonate slowly deposited over millions of years. When surface or ground water flows over or through limestone rock, a small amount of calcium and magnesium carbonate dissolves in the water. Hard water results when sufficient magnesium and calcium ions dissolve. Water hardness in the United States varies from nearly 0 ppm in mountainous regions with mostly granite rock to over 400 ppm in parts of the Midwest, where limestone is prevalent.

5.28 Your Turn

Write the formulas of:

a. calcium bicarbonate
b. magnesium sulfate
c. magnesium chloride

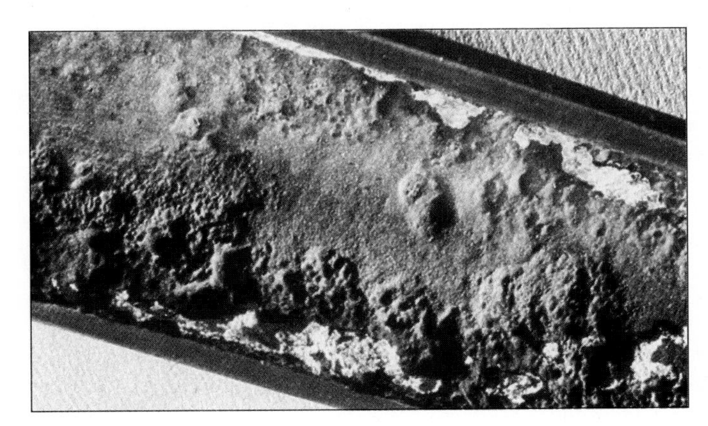

Figure 5.19
A pipe with hard-water scale buildup.

The usual analytical method for measuring water hardness utilizes a **titration** in which a measured volume of hard water is reacted with an ethylenediaminetetraacetate ($EDTA^{2-}$) solution whose molarity is accurately known. In the analysis, calcium ions in the hard water react with a chemical known by the shorthand name "EDTA."

• EDTA is actually in a solution of sodium EDTA, Na_2EDTA.

$$Ca^{2+}(aq) + EDTA^{2-}(aq) \longrightarrow CaEDTA(aq) \quad [5.8]$$

To carry out the analysis, a solution containing a known concentration of $EDTA^{2-}$ is added slowly to a water sample until an indicator in the solution changes color when the number of moles of $EDTA^{2-}$ added exactly equals the number of moles of calcium in the water sample. Note from equation 5.8 that 1 mole of $EDTA^{2-}$ is required for each mole of Ca^{2+} ions in the solution. The volume of water, the volume of $EDTA^{2-}$ solution needed, and the molarity of the EDTA are used to calculate the molarity of calcium ions in the sample. Two methods of performing such a titration are shown in Figure 5.20.

For example, suppose that during a titration 27.3 mL of 0.0100 M $EDTA^{2-}$ were added to a 100.0 mL water sample before the indicator changed color. The number of moles of Ca^{2+} in the water sample can be determined from the moles of $EDTA^{2-}$ used in the titration. Exactly 1 L of 0.0100 M $EDTA^{2-}$ contains 0.0100 mole $EDTA^{2-}$; if 27.3 mL (0.0273 L) of it were used,

$$\text{Moles of } EDTA^{2-} \text{ used} = 0.0273\ \not{L} \times \frac{0.0100 \text{ mole EDTA}}{1\ \not{L}}$$

$$= 2.73 \times 10^{-4} \text{ mole } EDTA^{2-}$$

Equation 5.8 points out that 1 mole of $EDTA^{2-}$ reacts with one mole of Ca^{2+}. Therefore, because 2.73×10^{-4} mole $EDTA^{2-}$ were used, the water sample contained 2.73×10^{-4} mole $CaCO_3$, because 1 mole of Ca^{2+} reacts with 1 mole of CO_3^{2-} (equation 5.7).

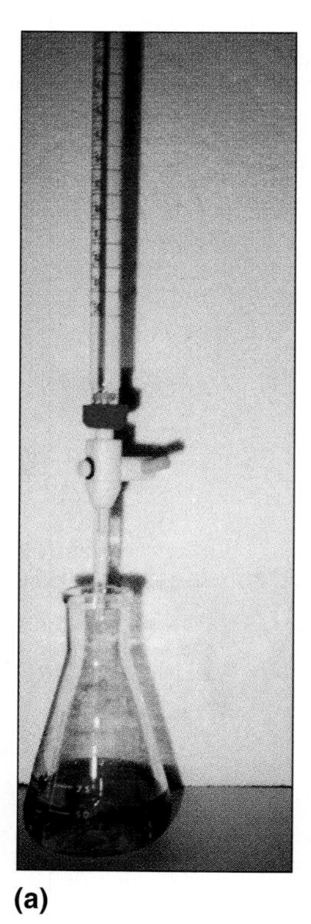

(a)

(b)

Figure 5.20

A titration of hard water with EDTA: (a) a titration using a buret; (b) a smaller-scale titration using a Beral-type pipet and a well plate.

(Courtesy of Conrad Stanitski.)

To express the water hardness of this sample as ppm (mg/L), we must first convert the 2.73×10^{-4} mole of Ca^{2+} to an equivalent number of moles of $CaCO_3$ and then to milligrams of $CaCO_3$:

$$\text{Moles } CaCO_3 = 2.73 \times 10^{-4} \,\cancel{\text{mole } Ca^{2+}} \times \frac{1 \text{ mole } CaCO_3}{1 \,\cancel{\text{mole } Ca^{2+}}}$$

$$= 2.73 \times 10^{-4} \text{ mole } CaCO_3$$

The molar mass of $CaCO_3$ is 100 g/mole (by adding the masses of Ca, C, and three O).

$$\text{mg } CaCO_3 = 2.73 \times 10^{-4} \,\cancel{\text{mole } CaCO_3} \times \frac{100 \,\cancel{\text{g } CaCO_3}}{1 \,\cancel{\text{mole } CaCO_3}} \times \frac{1000 \text{ mg } CaCO_3}{1 \,\cancel{\text{g } CaCO_3}}$$

$$= 27.3 \text{ mg } CaCO_3$$

The final step is expressing the concentration in ppm (mg/L) in the 100.0 mL (0.100 L) sample.

$$\frac{27.3 \text{ mg } CaCO_3}{0.100 \text{ L}} = 273 \text{ mg } CaCO_3/\text{L} = 273 \text{ ppm hardness}$$

Probably the most common manifestation of water hardness is the way in which calcium ions and magnesium ions interfere with the effectiveness of soaps. Magnesium and calcium are members of the same chemical family, Group 2A, of the periodic table. Their ions, Mg^{2+} and Ca^{2+}, share the tendency to react with soap to form an insoluble compound that separates from solution. This insoluble compound is the stuff of bathtub rings and the scum deposited on clothing washed in hard water. Because much of the soap is tied up in the precipitate, more soap is required to form suds and cleanse things in hard water than in soft water.

As shown in Figure 5.21, a soap "molecule" contains two parts: a sodium ion (Na^+) and a long hydrocarbon chain with a negatively charged ionic end (the "soap ion"). In aqueous solution, soap releases Na^+ ions and the negative soap ions. Soaps are used for cleaning because their long, nonpolar hydrocarbon tails dissolve readily in materials that are predominantly nonpolar, such as grease, chocolate, or gravy. The negatively charged ionic ends stick out of the surface of a grease globule because ionic substances are not soluble in nonpolar media. Thus, the grease becomes covered with negative charges. These negatively charged grease particles interact favorably with the partially positive regions of water molecules, are solubilized, and get carried away with the rinse water.

- The action of soap exemplifies the generalization that "like dissolves like" (Section 5.12).

The insoluble precipitate (bathtub ring and scum on laundered clothes) arises when soap ions interact with Ca^{2+} or Mg^{2+} ions. Two of the negatively charged "soap" ions

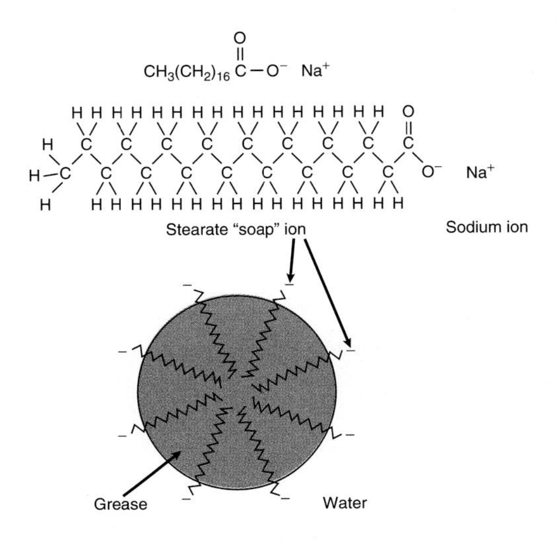

Figure 5.21

Soap and its interaction with grease.

react with each of the Mg^{2+} or Ca^{2+} ions to form a scummy precipitate. One obvious way to avoid the formation of the precipitate is to remove the calcium and magnesium ions, in other words, to "soften" the water. Softening can be accomplished by adding sodium carbonate (washing soda), Na_2CO_3, together with the soap. The carbonate ions (CO_3^{2-}) react with the Ca^{2+} to form insoluble calcium carbonate that is rinsed away (equation 5.7). Other water-softening compounds, such as sodium tetraborate or borax ($Na_2B_4O_7$) and trisodium phosphate (Na_3PO_4), work in a similar fashion. Calgon™ water softener contains hexametaphosphate ions, $P_6O_{18}^{6-}$. These ions are very effective in tying up calcium and magnesium ions as large, soluble ions.

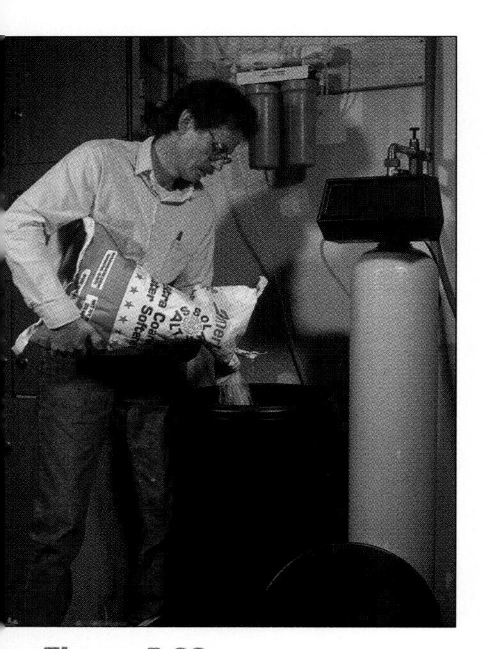

Figure 5.22
Adding salt to recharge an ion exchange water softener.

Another way to soften hard water is to remove Ca^{2+} and Mg^{2+} ions before they get to the washing machine or shower. **Ion exchange** is a process that accomplishes the conversion of hard water to soft water. A water softener typically contains a **zeolite,** a clay-like mineral made up of aluminum, silicon, and oxygen. These atoms are bonded into a rigid, three-dimensional structure bearing many negative charges. All these charges must be balanced by positive charges, usually supplied by Na^+ ions associated with the zeolite. However, when water containing Ca^{2+}, Mg^{2+}, or Fe^{3+} is passed through the zeolite, these ions displace the Na^+ ions because the multiply charged ions are more strongly attracted to the negatively charged zeolite than are the singly charged Na^+ ions. In other words, "hard water" ions (calcium, magnesium, iron) are exchanged for "soft water" sodium ions. If we represent the zeolite as Z, we can write an equation representative of the process.

$$\underset{\substack{\text{zeolite} \\ \text{(sodium form)}}}{Na_2Z(s)} + \underset{\text{(in hard water)}}{Ca^{2+}(aq)} \longrightarrow \underset{\substack{\text{zeolite} \\ \text{(calcium form)}}}{CaZ(s)} + \underset{\text{(in soft water)}}{2\,Na^+(aq)} \qquad [5.9]$$

The Na^+ ions rather than Ca^{2+} ions flow through the tank and into the pipes of the residence. Sodium ions do not interfere with the function of soap or result in the buildup of scale, and so the problem of hardness has been left behind on the zeolite ion exchanger. When the exchanger becomes saturated with Mg^{2+}, Ca^{2+}, and other undesirable ions, it is back-flushed with a concentrated solution of sodium chloride. This is easily accomplished by simply adding salt to the ion exchange unit, as shown in Figure 5.22. The high Na^+ ion concentration of this solution displaces the Mg^{2+} and Ca^{2+} from the zeolite, reversing equation 5.9, and they are flushed down the drain as $MgCl_2$ and $CaCl_2$. The ion exchanger is left in its fully charged sodium form, ready to soften more hard water. Note that zeolite softening adds sodium ions to the water, something that is not beneficial to those on restricted sodium diets.

5.29 Consider This: Water Hardness in Your Area

Find information about the water hardness in your part of the country.

a. How "hard" or "soft" is the water in your area? Were you surprised at the answer?
b. Assuming that water softening is necessary, what are the options for it in your area?
c. Are any health risks associated with using hard water? With using soft water?
d. Has EPA set standards for any of the ions responsible for hard water? Explain?

5.15.2 Case Study: Lead in Drinking Water

Our second case study deals with lead, one of the most serious pollutants potentially found in drinking water. The case study illustrates that standards for acceptable levels of pollutants are of little value unless accurate methods are available to measure the concentrations of pollutants. Newer analytical methods, in turn, have made possible more careful studies of pollutant concentrations in water and their toxicological effects. Chemical analysis of water involves a variety of analytical methods, depending on the impurity in question.

All the heavier metallic elements near the lower right side of the metals in the periodic table are toxic. Several of them, including lead (Pb), mercury (Hg), and cadmium

(Cd) are frequently encountered in drinking water. The metals themselves are not soluble in water. Only their positive ions—Pb^{2+}, Hg^{2+}, and Cd^{2+}—form ionic compounds that are soluble and of concern. Because Pb^{2+} is the most common of these three and poses the most serious health risk due to its widespread occurrence, we examine the lead story in some detail. The source of the problem frequently arises within individual homes. Unless proper precautions are taken, lead from drinking water can have serious long-term health effects, especially tragic for young children.

5.30 Consider This: Lead, Mercury, or Cadmium in Your Drinking Water

Find out whether lead, mercury, or cadmium ions are a significant problem in drinking water where you live or on your campus. You might begin with the map of local drinking water systems provided by EPA's Office of Ground Water and Drinking Water. Your local water utility company or state drinking water program should be able to provide information as well.

a. If these ions are present, what are some likely sources of them?
b. Are the concentrations of these ions above the MCLG or MCL?

In its metallic form, lead is very dense (50% more dense than iron or steel). Because lead is an abundant, soft, and easily worked metal that does not rust, it has been used since ancient times for water pipes and roofs. Romans were likely the first to use lead for water pipes and as a lining for wine casks. Some historians attribute lead poisoning from such extensive use as a major factor contributing to the fall of the Roman Empire.

- The symbol Pb comes from the Latin name for lead, "plumbum," the origin of our word plumbing.

In more modern times, most U.S. homes built before 1900 had lead water pipes, now replaced by copper or plastic ones. Until 1930, lead pipes were commonly used to connect homes to public water mains. There is no accurate way of knowing how many people suffered permanent health damage from living in residences with lead pipes. But, there are a few recorded cases of fatalities caused by lead poisoning in which the victim over many years habitually prepared a morning beverage using the "first draw" of water that had been standing in lead pipes overnight.

An interesting chemical connection exists between water hardness and how much lead dissolves from a lead pipe. Hard water forms a protective coating of calcium carbonate inside lead pipes that prevents water from coming into direct contact with the lead. On the other hand, with naturally soft water no such protective coating forms, and a small amount of lead (as Pb^{2+} ions) can dissolve in the water. Soft water also tends to be more acidic, which causes additional lead to dissolve. The effect is most severe in hot water pipes.

Some Pb^{2+} can get into drinking water even where there are no lead pipes. Solder used to join copper pipes contains 50–75% lead. Some drinking fountains have a holding tank to store chilled water, and the seams in the tank and connections from it to the fountain may be made with lead-based solder. Water for drinking fountains may stand in the tank for many hours, thus providing more contact time for lead from the solder to dissolve into the water.

When ingested, lead causes severe and permanent neurological problems in humans. This is particularly tragic for children, who may suffer mental retardation and hyperactivity as a result of lead exposure, even at relatively low concentrations. Severe exposure in adults causes irritability, sleeplessness, and irrational behavior, including loss of appetite and eventual starvation. Unlike many other toxic substances, lead is a cumulative poison and is not transformed into a nontoxic substance. Once it enters the body, it accumulates in bones and the brain.

Lead toxicity is a particular problem for children because lead ions, Pb^{2+}, are normally incorporated rapidly into bone along with calcium ions, Ca^{2+}. In children, who have less bone mass than adults, the lead remains in the blood longer, where it can damage cells, especially in the brain. Besides lead in drinking water, young children are exposed to large amounts of lead from chewing on lead paint. This is especially the case in older houses where the paint is chipped and flaking. A national program monitoring blood lead levels in children is aimed at identifying children at risk. Health officials are required to investigate cases in which children are known to exceed the currently acceptable blood level of 15 μg/dL (micrograms per deciliter). The EPA estimates that one of six U.S. children under six years of age has a blood lead level above this limit.

• 1 deciliter (dL) = 0.1 L

5.31 Your Turn

Two samples of drinking water were compared for their lead content. One had a concentration of 20 ppb and the other had a concentration of 0.003 mg/L.

a. Explain which one contains the higher concentration of lead.
b. Compare each sample to the current acceptable limit.

Since the 1970s, the federal government has had regulations for acceptable levels of lead in water and foods. These limits have gradually become more restrictive with the development of better analytical methods for measuring extremely low concentrations and as more has been learned about the health effects of lead. Lead is so widespread in the environment that older measurements suffered from unintentional contamination of both the equipment and the reference standards. Until recently, the maximum contaminant level (MCL) for lead ion in drinking water was 15 ppb. In 1992, the U.S. EPA converted this to an "action level," meaning that the EPA will take legal action if 10% of tap water samples exceed 15 ppb. The hazard from lead is so great that the EPA has established a maximum contaminant level goal (MCLG) of 0, even though lead is not a carcinogen.

• MCLs and MCLGs are discussed in Section 5.13.

The good news is that there is very little lead in most public water supplies. It is estimated that lead exceeding allowable limits is present in less than 1% of public water supply systems and they serve less than 3% of the U.S. population. Most lead in drinking water comes from corrosion of plumbing systems, not from the source water itself. When lead is reported, consumers are advised to take simple steps to minimize exposure, such as letting water run before using and using only cold water for cooking.

5.32 Consider This: Regulating Arsenic in Drinking Water

Another toxic metal that can find its way into public water supplies is arsenic. Early in January 2001, the Clinton administration issued a 10-ppb standard for arsenic in drinking water, replacing the standard of 50 ppb set in 1962. The Bush administration soon after recalled the rule before it could take effect, thus reverting to the 50 ppb standard, a controversial decision.

a. What was the reasoning behind each administration's decision?
b. What has been the response to each administration's decision?
c. Determine whether the 50 ppb is still the standard for arsenic.

The limits for lead in drinking water are meaningless unless there are reliable analytical methods for measuring lead accurately in the low-ppb range. The almost universal method for lead analysis utilizes a variation of the spectrophotometric technique of analysis, which relies on absorption of light by a colored species in solution. A substance that forms a colored compound with Pb^{2+} is added to the water sample. The deeper the color, the greater the concentration of the species, say lead (or iron), in the sample. The intensity of the color is then measured with a **spectrophotometer** (Figure 5.23). Light

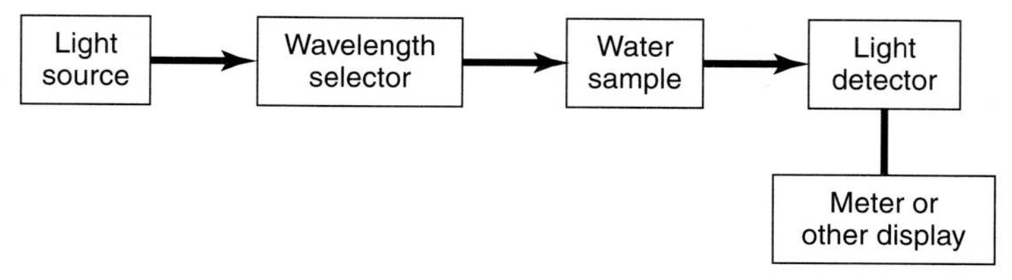

Figure 5.23
Simplified diagram of a spectrophotometer.

of the desired wavelength passes through the sample and strikes a special detector where the light intensity is converted into an electrical voltage. The voltage is displayed on a meter or sent to a computer or other recording device. The amount of light absorbed by the solution, and which therefore does not reach the detector, is proportional to the concentration of the species of interest in the sample. The higher the concentration of the species, the more light that is absorbed by the sample.

- If the species is colorless, a chemical may be added to the sample to produce a colored species in the solution.

To relate the concentration of the species (lead, iron, etc.) analyzed in a water sample to absorbance data taken by the spectrophotometer, an analyst first prepares a **calibration graph.** The graph is made by measuring the absorbances of a set of reference water samples containing known concentrations of the species to be analyzed. An example of a calibration graph for iron analysis is shown in Figure 5.24. Iron concentration is shown on the horizontal axis and absorbance at a wavelength of 505 nm is shown on the vertical axis. If a water sample gives an absorbance measurement of 0.45, an analyst can use that value to read directly from the graph that the concentration of iron is 2.7 ppm (see dashed lines, Figure 5.24). Looking at the horizontal scale on the graph, it is apparent that measurements of iron in water can be made down to low-ppm concentrations.

- The nature of light and wavelength was discussed in Section 2.4. The use of a spectrophotometer to measure infrared radiation was described in Section 3.5.

5.33 Your Turn

Use Figure 5.24 to estimate the concentration of iron in these water samples. If the calibration graph cannot be used, state why not and suggest a possible way around the problem.

a. Absorbance 0.60
b. Absorbance 0.05
c. Absorbance 1.5

Answer
a. The concentration of iron is approximately 2.6 ppm.

Figure 5.24 illustrates a caution about water analysis: The accuracy of the analysis is only as good as the accuracy of the calibration graph. Note that all the reference samples do not lie on an absolutely straight line. Rather, there is some bit of uncertainty in each of the measurements, which contributes to a small uncertainty in the analysis of any water sample.

The specific method used for lead analysis is called **atomic absorption spectrophotometry.** A small water sample is vaporized at very high temperature into a beam of UV light coming from a lead-containing lamp. Radiation unique to lead atoms is emitted from the hot lead atoms in the lamp and absorbed by lead atoms in the vaporized water sample. The sample's absorbance is then compared to a calibration graph to determine the Pb^{2+} concentration in the water. Conventional versions of the equipment can measure lead ion concentrations down to about 50 ppb, but that is not good enough

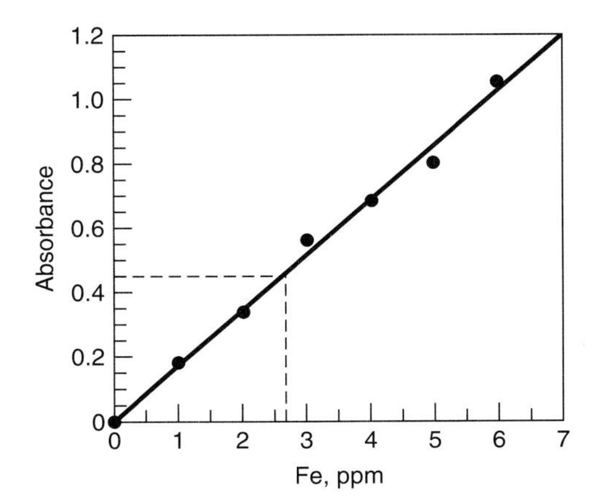

Figure 5.24
Calibration graph for spectrophotometric analysis of iron in water.

to meet the present MCL of 15 ppb. A more sophisticated and expensive atomic absorption spectrophotometer can measure lead at well below 1 ppb.

5.34 Consider This: Shifting Limits for Lead

Before 1962, the recommended limit for lead in drinking water was 100 ppb. In 1962, the limit was lowered to 50 ppb. In 1988, the MCL was lowered to 15 ppb, where it remains today.

a. Suggest some reasons for this trend. Give other examples where legal limits have changed over time.
b. In addition to the current action level of 15 ppb of lead in tap water in residences, the EPA recommends that source water from water utilities should contain no more than 5 ppb lead, and water in school drinking fountains should contain no more than 20 ppb. Suggest possible reasons for these differences.

5.35 Consider This: Policy Trade-offs for Lead in Drinking Water

Establishing more stringent regulations on lead in drinking water may be desirable to protect public health—especially the health of young children and the elderly—but it is costly.

a. Identify some of the costs associated with stricter limits.
b. Which costs do you think are justified?
c. Who should pay these costs? Give the reasons behind your opinions.

5.15.3 Case Study: Trihalomethanes in Drinking Water

The third case study deals with substances known collectively as *trihalomethanes* (THMs), the most important of which is chloroform, $CHCl_3$. Trihalomethanes are derivatives of methane, CH_4. Figure 5.25 gives two representations of the chloroform molecule.

- Halogens are Group 7A elements in the periodic table. The common halogens are fluorine, chlorine, bromine, and iodine.

- Humic acids are complex and variable materials resulting from partial decomposition of plant or animal matter; they form the organic portion of soils.

Other trihalomethanes contain, as the name implies, any three "halogen" atoms from Group 7A of the periodic table. The most significant trihalomethanes, in addition to $CHCl_3$, are $CHCl_2Br$, $CHClBr_2$, and $CHBr_3$.

Chloroform and other THMs are by-products of drinking water chlorination. Recall from Section 5.14 that chlorination leaves some residual HOCl in the water. This HOCl can lead to formation of THMs by its reaction with humic acids, which are breakdown products of plant or animal materials. Humic acids are almost always present in surface waters, and therefore formation of THMs in chlorinated surface water is unavoidable. However, the concentration of THMs in drinking water is normally far less than 1 ppm. But, even in very low concentrations, they contribute an unpleasant taste to chlorinated water. The presence of THMs can sometimes be detected by their hospital-like odor, especially in a hot shower. Recent EPA standards require municipal water treatment facilities to reduce the concentration of humic acids in water prior to its chlorination, which will decrease the likelihood of THMs formation.

The primary health concern is that chloroform, a THM, is suspected of causing liver cancer. There is some epidemiological evidence of slightly greater cancer rates (including bladder and rectal cancer) in people living in communities with chlorinated drinking water compared to those that do not. The current MCL for total THMs established in 1998 by the EPA is 80 ppb (0.080 ppm), down from the previous level of 100 ppb. Most drinking water samples meet that standard. The national average for THMs is 51 ppb for municipal drinking water that comes from surface water. Well water has a lower THM concentration because it usually contains little or no plant material and is not chlorinated. Although the EPA limit (MCL) is 100 ppb, the goal (MCLG) is 0 ppb.

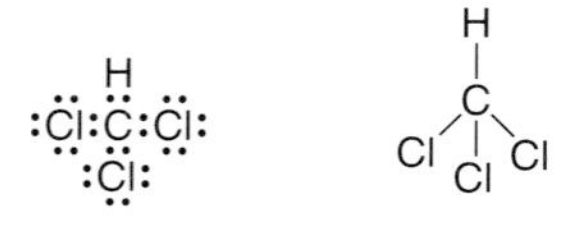

Figure 5.25
Lewis structures of chloroform, $CHCl_3$.

5.36 Consider This: Health Risk From Chloroform

Chloroform has the potential to cause cancer. But, chloroform is also an effective cough suppressant that has been used in many over-the-counter medications. Until recently, many children's cough syrups contained several percent of chloroform. To what extent does this provide convincing evidence that chloroform in drinking water is not a health hazard and therefore that chlorination of drinking water is safe?

A reliable analytical method is needed to measure THMs in the ppb range. The method of choice for THMs is called **gas chromatography (GC),** a powerful technique for measuring trace amounts of various molecular substances in water. In GC, molecular solutes in water are first extracted from a large sample of water into a smaller volume of a nonpolar liquid such as octane. This extraction concentrates the solutes to be analyzed. A very small portion of the extract is injected into a flowing gas stream that passes down a long, heated tube coated with an absorbing material; a detector is at the far end (Figure 5.26a). Components in the sample move down the tube at various rates, thus reaching the detector at different times. The signal from the detector is displayed on a recorder, as shown in Figure 5.26b; each peak corresponds to a different substance. For each plotted peak, the time required for the substance to reach the detector identifies the substance, while peak area measures its concentration. Gas chromatography is the normal analytical method for trace amounts of a wide variety of toxic substances in drinking water. These include pesticides, PCBs, dioxins, industrial solvents, and gasoline or other petroleum products.

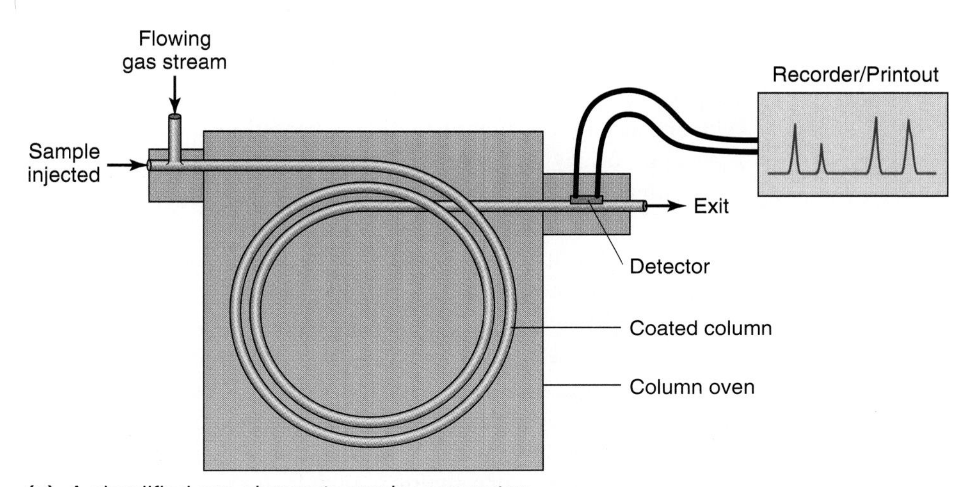

(a) A simplified gas chromatography apparatus.

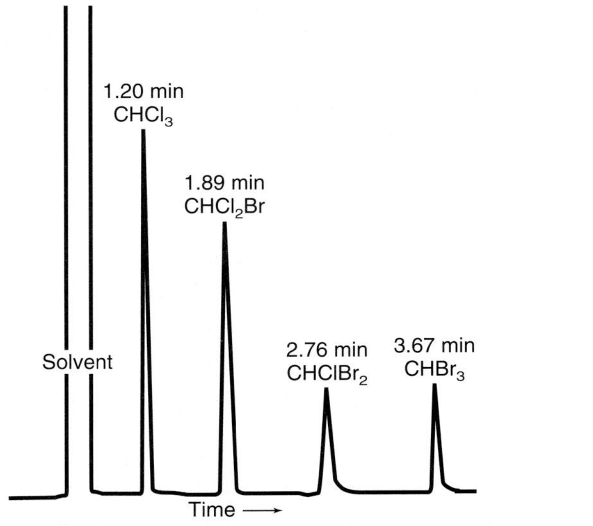

(b) Gas chromatography analysis of a mixture of THMs.

Figure 5.26

Trihalomethanes in chlorinated drinking water create a classic risk/benefit situation. On the one hand, chlorination of drinking water is an efficient, inexpensive, effective method that greatly reduces the health risk from bacteria and other disease-causing organisms in the water. On the other hand, chlorination results in the formation of relatively low concentrations of THMs that may cause cancer.

5.16 Consumer Choices: Tap Water, Bottled Water, and Filtered Water

You should now have a sufficient chemical background to answer some questions about drinking water sources posed early in the chapter and to make informed choices about the water you drink.

Tap Water

1. Is safe drinking water generally available in the United States? The answer, a resounding "yes," is due to high standards mandated by federal regulations for drinking water provided by public water supply utilities. As important, the treatment technology is available to achieve these high standards; without such technology, standards would be merely hollow gestures. Very few people in our country suffer acute illness from drinking contaminated water unless they are using water from a private well that has not been properly tested. It should be noted that the Safe Drinking Water Act Amendments of 1996 enhanced protection, including increased requirements for notifying consumers promptly of any problems with water safety.
2. Is tap water "pure" water? Certainly not; it almost surely contains small amounts of sodium, calcium, magnesium, chloride, sulfate, and bicarbonate ions, as well as trace amounts of other ions. Tap water also contains dissolved air, which is a mixture that includes N_2, O_2, CO_2, and air-borne particles.
3. What *problems* are likely to exist? In some parts of the country, extremely hard water can cause problems, at least economically. Water softening options are readily available, if desired. Some tap water may contain dangerous Pb^{2+} concentrations, although lead is normally a problem only in buildings with lead pipes, and then only if the water is naturally soft. Other heavy metal elements, such as mercury and cadmium, may be present at dangerous concentrations, although this is extremely unlikely. Chlorinated tap water from surface water sources will contain a small amount of residual chlorine. It may also contain small amounts of trihalomethanes, byproducts of chlorination. Depending on its source, the water may contain low concentrations of mercury, nitrate, pesticide residues, PCBs, and industrial solvents. By now you should understand that the presence of such substances in drinking water is not likely a cause for alarm. Rather, the crucial question is "How much?" If pollutant concentrations are below the maximum contaminant levels (MCLs), the EPA regards the water as safe, with an adequate margin of safety.

Bottled Water

This chapter began with a look at bottled water, drunk by people for a variety of reasons: taste, convenience, and/or the belief that it is healthier than tap water. Whatever its source, bottled water is expensive. We can raise the same questions about bottled water as those asked about tap water.

1. Is it safe? The same laws that apply to public water supplies do not regulate bottled water. However, it is highly regulated by other both government- and industry-imposed regulations. Considered a food, bottled water is regulated by the Food and Drug Administration (FDA). Bottled water must meet standards of quality, comply with labeling regulations, and meet good manufacturing practices. A provision of the SDWA amendments of 1996 requires the FDA to develop bottled water standards that are equal to EPA drinking water standards. In years past, critics have questioned the safety of bottled water. However, more than 85% of bottled water currently sold in

the United States is produced by member companies of the International Bottled Water Association (IBWA), whose member companies must meet higher water quality standards than those imposed by the FDA (Figure 5.27). Springs and underground aquifers that do not require disinfection are the principal sources of bottled water. If disinfection is required, it is done with ozone or UV radiation, rather than with chlorine, thus leaving no objectionable taste and no THMs. In addition, most bottled water is subjected to either filtration, reverse osmosis, or distillation (see Section 5.17). The absence of chlorine, THMs, and the probable absence of various trace pollutants found in surface water provide much of the argument for bottled water as a healthier alternative to tap water. In the majority of all bottled water sold in the U.S., the source is municipal tap water that has been subjected to further purification. Interestingly enough, if the municipal water meets processing standards allowing it to be labeled "distilled" or "purified", the water does not need to divulge its municipal tap water source.

2. Is bottled water pure? Because bottled water often comes from springs or wells, we can be sure that it contains ions, dissolved as the water percolates through the surrounding rocks. In fact, bottled water from some well-known spas, such as Bath in England, Baden-Baden in Germany, and White Sulfur Springs in West Virginia, contains relatively large amounts of calcium and other ions, as well as dissolved carbon dioxide. In a few cases, dissolved hydrogen sulfide gas provides a characteristic "sulfur" odor, thought to be a positive virtue by some bottled water connoisseurs.

Filtered Water

Water filter units are a readily available alternative to bottled water. These units generally attach to a kitchen or bathroom faucet and remove most trace pollutants. The units simultaneously employ two purification techniques. The first is "activated carbon," a special form of charcoal with a very high surface area that absorbs most of the

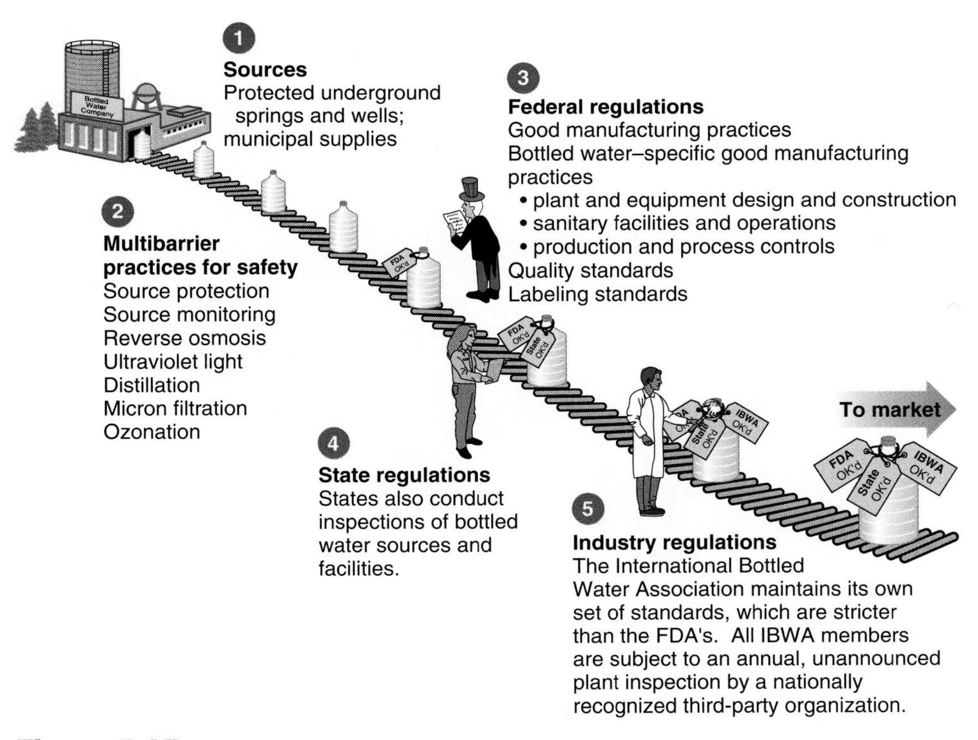

Figure 5.27

Bottled water's path to market. The International Bottled Water Association illustrates the process its members' products follow from the source to the consumer's satisfaction. Federal, state, and industry regulations guarantee safety and quality.

molecular solutes including residual chlorine, THMs, pesticide residues, solvents, and other similar substances. The second component is an ion exchange resin (Section 5.15.1) that removes heavy metal ions (chiefly lead and copper) and hard-water calcium and magnesium ions. Water from such a unit is free of objectionable taste and odor and should be free of most hazardous substances. More correctly, the use of filters reduces the concentrations of such substances to extremely low values, well below the concentrations of concern for human health. Filtered water typically costs only 20% as much as bottled water.

5.37 Consider This: Evaluating Bottled Water

In 5.2 Consider This, you were asked to list and then rank the advantages and the disadvantages associated with drinking bottled water. Having now studied this chapter, check your list. Would the order of importance be the same? Explain how your reasoning may have changed based on the information and understanding gained in the study of this chapter.

5.17 International Needs for Safe Drinking Water

Those who live in the United States are privileged to have drinking water choices available. We can select from tap, bottled, or filtered water, all generally of high quality. Such is not the case for people in most of the rest of the world. The reality is that more than a billion people (one in six), principally in developing nations, lack access to safe drinking water. About 1.8 billion people do not have adequate sanitary facilities. One estimate, made by *Scientific American,* is that it would cost $68 billion dollars over the next 10 years to provide safe water and decent sanitation facilities to everyone. Lack of access to safe water poses a particular risk to infants and young children. Whereas bottled water is a discretionary option for many in the United States, the majority of the world's population does not have that option. Figure 5.28 shows how access to clean water in urban areas varies worldwide.

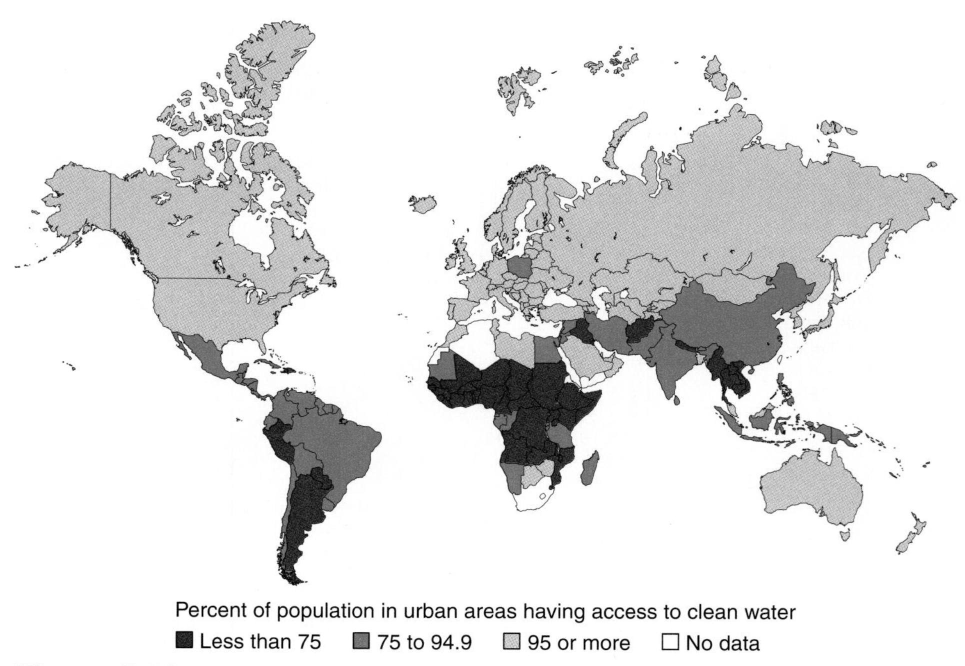

Figure 5.28

Access to safe drinking water varies widely across the world.

Source: The World Resources Institute, data based on surveys of national governments. http://www.sciam.com/1197issue/1197scicit5.html

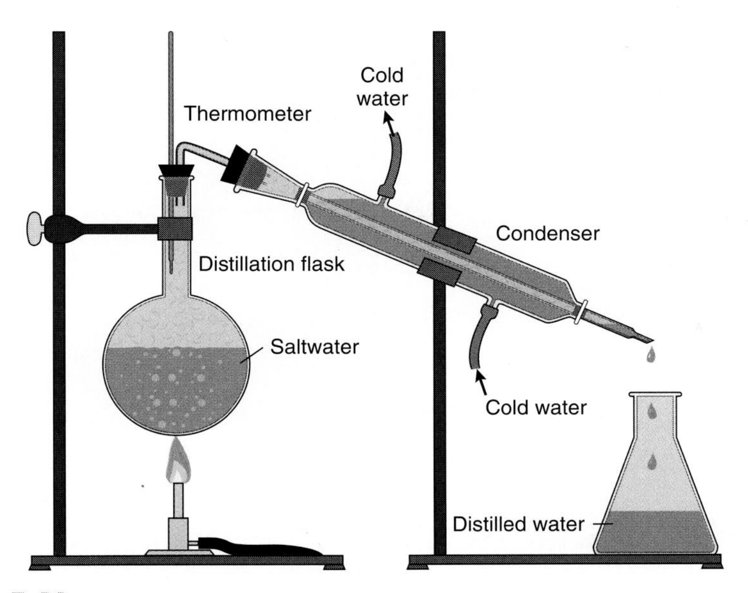

Figure 5.29
Water purification by distillation.

For those living in arid regions, such as the Middle East, fresh water is scarce. Seawater is readily available in many such areas, but its high salt concentration makes it unfit for human consumption. Coleridge's Ancient Mariner is more than just a poetic fantasy; it is a physiological reality. Ocean water contains 3.5% salt compared to only about 0.9% salt in body cells. Consequently, seawater can be drunk only after most of the salt is removed. Fortunately, there are ways to do this, but they require large amounts of energy. Collectively, the methods are known as **desalination,** a broad general term describing any process that removes ions from salty water.

One desalination method is distillation, an old and rather common way of purifying water for laboratory and other uses. Distilled water is used in steam irons, car batteries, and other devices whose operation can be impaired by dissolved ions. **Distillation** is remarkably simple—a liquid is evaporated and then condensed, just as in the natural hydrologic cycle. An apparatus such as that shown in Figure 5.29 is used. Impure water is put into a flask, pot, or other container and heated to its boiling point, 100°C. As the water vaporizes, it leaves behind most of its dissolved impurities. The water vapor passes through a condenser where it cools and reverts back into a liquid, now free of contaminants. Not surprising, the product is usually called "distilled water." If distillation is done very carefully, extremely pure water, with no detectable amounts of contaminants, is produced.

Notice that to distill water, it must be heated to its boiling temperature (100°C) and then heated further to convert the liquid water to steam. Each step requires a large energy input. Energy is required for the distillation of any liquid, but recall from Section 5.8 that water has an unusually high specific heat and an unusually large amount of heat required for evaporation. Both result from the uniquely extensive hydrogen bonding in water. High energy costs for water purification by distillation suggests that it is economically practical only for countries or regions with abundant and cheap energy. Ion exchange, described in Section 5.15.1, is another method for desalination, although it is not very practical for large-scale desalination.

Another desalination technique gaining in popularity is *reverse osmosis.* To understand this method, we need to know that **osmosis** is the natural tendency for a solvent (water in this case) to move through a membrane from a region of higher solvent concentration to a region of lower solvent concentration. This tendency to equalize concentrations is involved in many biological processes. In this particular instance, the net effect is that cells lose water rather than gain it. However, osmosis can be reversed. If

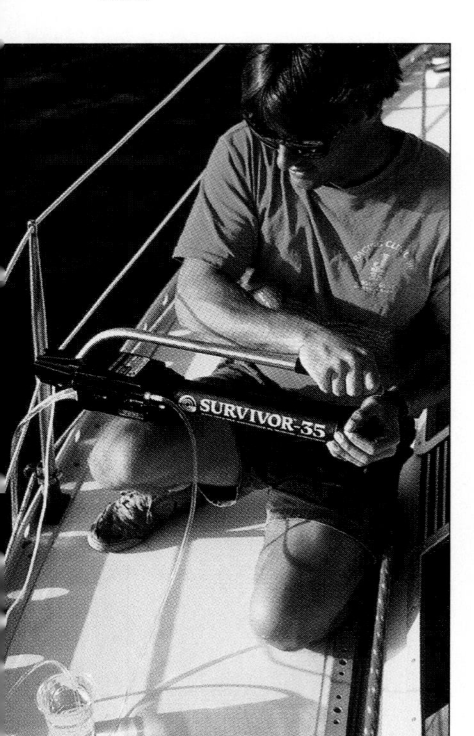

Figure 5.31
A small reverse osmosis apparatus for converting seawater to potable water.

- Solar-powered reverse osmosis desalination units providing 400 L/day were developed at Murdoch University in Perth, Australia. Twenty-five units are now in operation in Australia and Asia. Larger units providing 15,000 L/day are being tested.

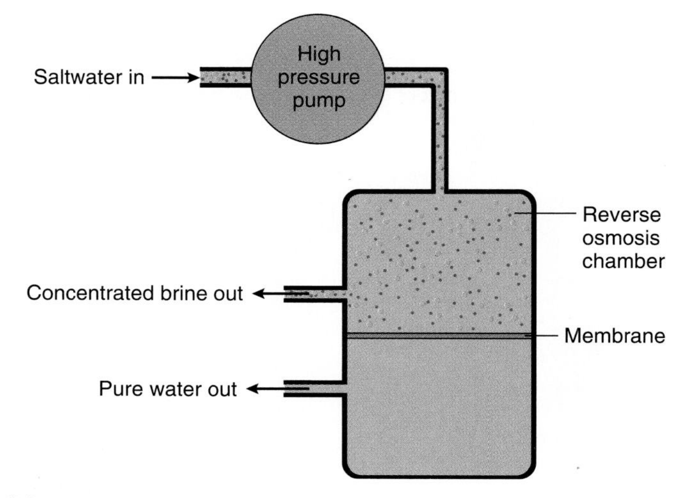

Figure 5.30
Water purification by reverse osmosis.

sufficient pressure is applied to the salt water side, water molecules can be forced through the membrane leaving ions behind. Figure 5.30 is a schematic representation of this process.

The world's largest desalination plant, located at Jubail, Saudi Arabia, provides 50% of that country's drinking water using reverse osmosis to desalinate Persian Gulf water. Although most such installations are in the Middle East, the number of reverse osmosis plants is increasing in the United States. Florida has over 100 reverse osmosis desalination facilities, including the one that furnishes the city of Cape Coral with 15 million gallons of fresh water every day from brackish underground supplies. Small reverse osmosis installations are used in spot-free car washes and individual units are available for boaters. Figure 5.31 shows a small unit, suitable for use on a sailboat. It must be pointed out, however, that reverse osmosis desalination is too expensive for use in most developing nations.

CONCLUSION

This chemical compound H_2O is a very unusual substance, with many unique properties that contribute to its life-supporting role. Like the air we breathe, water is central to life, and we humans require large quantities of it. We take for granted that our drinking water, whether from the tap or bottle, typically is free of harmful contaminants. This chapter has focused almost exclusively on the quality of drinking water—its sources, substances dissolved in it, and potential contaminants and their concentrations. Federal and state regulations help make our drinking water safe. In three case studies, we considered how particular substances in water can be analyzed and treated. In the next chapter, we examine rainwater and the ways in which substances dissolved in rain can adversely affect the environment.

Chapter Summary

Having studied this chapter you should be able to:

- Describe the desirable properties of drinking water (5.1)
- Explain some of the reasons why bottled water is so popular (5.2)
- Recognize the sources and distribution of water (5.3)

- Discuss why water is such an excellent solvent for ionic and some covalent compounds (5.4, 5.11, 5.12)
- Describe the factors involved in providing pure drinking water (5.5, 5.14)
- Use concentration units: percent, ppm, ppb, and molarity (5.6)
- Relate the molecular structure of water to its properties (5.7, 5.8)
- Discuss the relationship between the properties of water and its structure (5.7)
- Describe the specific heat of water and compare it to that of other substances (5.8)
- Understand how electronegativity and bond polarity are related to the structure of water (5.8)
- Describe hydrogen bonding and its importance to the properties of water (5.8)
- Describe how the density of water is related to its molecular structure (5.8)
- Predict ion formation and the formulas for ionic compounds, including those with common polyatomic ions (5.10)
- Discuss the Maximum Contaminant Level Goal (MCLG) and the Maximum Contaminant Level (MCL) established by the EPA to ensure water quality (5.13)
- Discuss how drinking water is made safe to drink (5.14)
- Relate chlorination with water purification (5.14)
- Know the causes and effects of water hardness (5.15.1)
- Cite water softening methods (5.15.1)
- Describe atomic absorption spectrophotometry and gas chromatography as methods for analyzing contaminants in water (5.15.2, 5.15.3)
- Compare and contrast tap water, bottled water, and filtered water in terms of water quality (5.16)
- Understand distillation and reverse osmosis (5.17)

Questions

Emphasizing Essentials

1. a. The text states that currently over three billion gallons of bottled water are sold each year in the United States, more than 10 times that sold 20 years ago. Calculate the volume of bottled water sold 20 years ago.

   b. What percentage increase has there been in the sale of bottled water over the last 20 years?

2. a. What is an aquifer?

   b. Why is it important to prevent unwanted substances from reaching a clean aquifer?

3. If a 500-L drum of water represented the world's total water supply, how many liters would be water actually available for drinking in the United States? *Hint:* See Figure 5.4.

4. Based on your experience, predict the solubility of each of these substances in water. Use terms such as very soluble, partially soluble, or not soluble. Cite some supporting evidence for your prediction.

   a. orange juice concentrate

   b. liquid clothes washing detergent

   c. household cleanser

   d. chicken broth

   e. chicken fat

5. a. A certain bottled water lists a calcium concentration of 55 mg/L. What is its calcium concentration expressed in ppm?

   b. How does this concentration compare with that for Evian listed in Table 5.2?

6. A certain vitamin tablet contains 162 mg of calcium and supplies 16% of the recommended daily amount of calcium required by a person on a typical 2000-Calorie diet. How many 500-mL bottles of Evian bottled water would you have to drink each day to obtain the same mass of calcium? *Hint:* See 5.7 Your Turn and Table 5.2.

7. The acceptable limit for nitrate, often found in well water in agricultural areas, is 10 ppm. If a water sample is found to contain 350 μg/L, does it meet the acceptable limit? Show a calculation to support your answer. NO SAME meaning 10 ppm vs 350ppm

8. One reagent bottle on the shelf in a laboratory is labeled 12 M H_2SO_4 and another is labeled 12 M HCl.

   a. How does the number of moles of H_2SO_4 in 100 mL of 12 M H_2SO_4 solution compare with the number of moles of HCl in 100 mL of 12 M HCl solution?

   b. How does the number of grams of H_2SO_4 in 100 mL of 12 M H_2SO_4 solution compare with the number of grams of HCl in 100 mL of 12 M HCl solution?

9. Both methane, CH_4, and water are compounds in which hydrogen atoms are bonded with a nonmetallic element. Yet, methane is a gas at room temperature and pressure and water is a liquid. Offer a molecular explanation for the difference in properties.

10. Explain why the term "universal solvent" is applied to water.

11. Consult Figure 5.7 to help answer this question.

    a. Calculate the electronegativity difference between each pair of atoms.

    N and C

    S and O

    N and H

    S and F

b. A single covalent bond forms between the atoms in each pair. Identify the atom that attracts the electron pair in the bond more strongly.

c. Arrange the bonds in order of increasing polarity.

12. NaCl is an ionic compound, but chlorine and silicon are joined by covalent bonds in $SiCl_4$.

a. Use Figure 5.7 to determine the electronegativity difference between chlorine and sodium, and between chlorine and silicon.

b. What correlations can be drawn about the difference in electronegativity between bonded atoms and their tendency to form ionic or covalent bonds?

c. How can you explain on the molecular level the conclusion reached in part b?

13. Consider a molecule of ammonia, NH_3.

a. Write the Lewis structure of NH_3.

b. Are there polar bonds in NH_3?

c. Is the NH_3 molecule polar? *Hint:* Consider the geometry of the ammonia molecule.

d. Predict the solubility of NH_3 in water.

14. This diagram represents two water molecules in a liquid state. What kind of bonding force does the arrow indicate? Is this an *inter*molecular or *intra*molecular force?

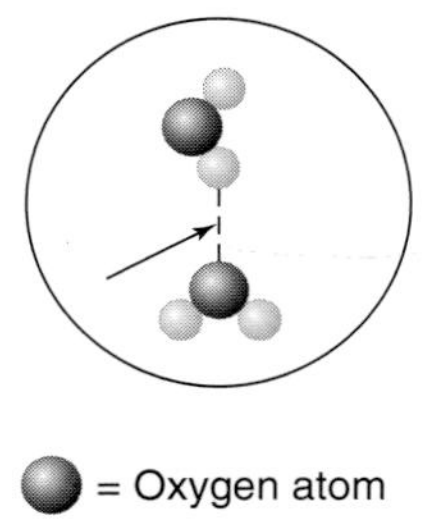

= Oxygen atom

= Hydrogen atom

15. The density of liquid water at 0°C is 0.9987 g/cm^3; the density of ice at this same temperature is 0.917 g/cm^3.

a. Calculate the volume occupied at 0°C by 100. g of liquid water and by 100. g of ice.

b. Calculate the percentage increase in volume when 100. g of water freezes at 0°C.

16. Consider these liquids.

Liquid	Density, g/mL
dishwashing detergent	1.03
maple syrup	1.37
vegetable oil	0.91

a. If you pour equal volumes of these three liquids into a 250-mL graduated cylinder, in what order will you add the liquids to create three separate layers? Explain your reasoning.

b. If an unknown liquid were poured into the cylinder and it formed a layer that was on the bottom of the other three layers, what can you tell about one of the properties of the unknown liquid?

c. What would happen if a volume of water equal to the other liquids were poured into the cylinder in part a and then the contents are mixed vigorously? Explain.

17. Why is there the possibility of a water pipe breaking if the pipe is left full of water during extended frigid weather?

18. Calculate the quantity of heat absorbed (+) or released (−) during each of these changes.

a. 250 g of water (about 1 cup) is heated from 15°C to 100°C

b. 500 g of water is cooled from 95°C to 55°C

c. 5 mL of water at 4°C is warmed to 44°C

19. Solutions can be tested for conductivity using this type of apparatus.

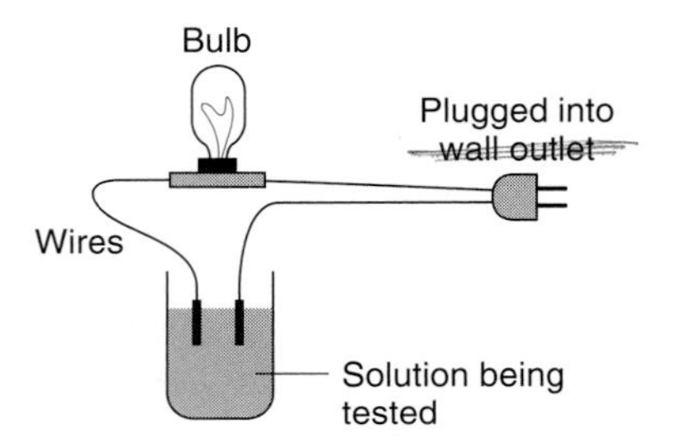

Predict what will happen when each of these dilute solutions is tested for conductivity.

a. $CaCl_2(aq)$

b. $C_2H_5OH(aq)$

c. $H_2SO_4(aq)$

Explain your predictions briefly.

20. Predict the ion most likely to be formed by each of these atoms. Use the Lewis structure of each atom and its corresponding ion to show how the ion obeys the octet rule. *Hint:* Consider Tables 5.5 and 5.6.

a. Cl

b. Ba

c. S

d. Li

e. Ne

21. Predict the formula and give the name of the ionic compound formed by the reaction of each pair of elements.

a. Na and S

b. Al and O

c. Ga and F

d. Rb and I

e. Ba and Se

22. Name each compound.

a. $KC_2H_3O_2$

b. $Ca(OCl)_2$

c. LiOH

d. Na_2SO_4

23. Based on the generalizations in Table 5.8, which compounds are likely to be water soluble?

a. $KC_2H_3O_2$

b. $Ca(NO_3)_2$

c. LiOH

d. Na_2SO_4

24. This represents the structural formula of a soap:

```
      H H H H H H H H H H H H H H H H O
  H    \/ \/ \/ \/ \/ \/ \/ \/ ||
H—C    C  C  C  C  C  C  C  C  C
  H  C  C  C  C  C  C  C  C     O⁻  Na⁺
      /\ /\ /\ /\ /\ /\ /\ /\
      H H H H H H H H H H H H H H H H
```

How is the ability of this soap to remove grease from clothes related to the structure of this soap?

25. Explain why desalination, despite its proven technological effectiveness, is not used more widely to produce potable drinking water.

Concentrating on Concepts

26. We take water for granted. How can you explain to a friend that we should value water as a unique substance, one essential for life?

27. Why is the concentration of calcium often given on the label for bottled water?

28. The label on Evian bottled water lists a magnesium concentration of 24 mg/L. The label of a popular brand of multivitamins lists the magnesium content as 100 mg per tablet. Which do you think is a better source of magnesium? Explain your reasoning.

29. A new sign is posted at the edge of a favorite fishing hole that says "Caution: Fish from this lake may contain over 1.5 ppm Hg." Explain to a fishing buddy what this unit of concentration means, and why the caution sign should be heeded.

30. This periodic table contains four elements identified by numbers.

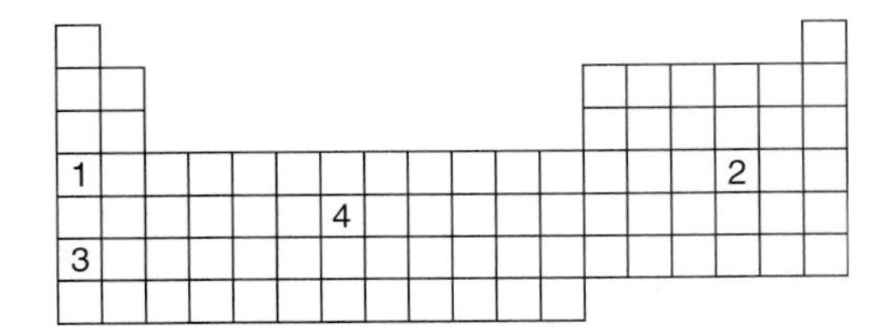

a. Based on trends within the periodic table, which of the four elements do you expect to have the highest electronegativity value? Explain.

b. Based on trends within the periodic table, rank the other three elements in order of decreasing electronegativity values. Explain your ranking.

31. A diatomic molecule **XY** that contains a polar bond *must* be a polar molecule. However, a triatomic molecule $\mathbf{XY_2}$ that contains a polar bond *does not necessarily* form a polar molecule. Use some examples of real molecules to help explain this difference.

32. Imagine you are at the molecular level, looking at what happens when gaseous water condenses.

a. Consider a collection of four water molecules, using

to represent each molecule. Draw a representation of water in the gaseous state and then in the liquid state. How does the collection of molecules change?

b. Discuss what happens to the bonding at the molecular level when water condenses to a liquid.

c. What happens at the molecular level when water changes from a liquid to a solid?

33. Hydrogen bonding has been offered as a reason why ice cubes and icebergs float in water. Consider ethanol, C_2H_5OH.

a. Draw its Lewis structure and use it to predict if pure ethanol will exhibit hydrogen bonding.

b. A cube of solid ethanol sinks rather than floats in liquid ethanol. Explain this behavior in view of your answer in part **a.**

34. The unusually high heat capacity of water is very important to regulating our body temperature and keeping it within a normal range despite time, age, activity, and environmental factors. Consider some of the ways that the body produces heat, and some of the ways that it loses heat. How would these functions differ if water had a much lower heat capacity?

35. How do generalizations about solubility influence water quality? Suppose that you are in charge of regulating an industry in your area that manufactures agricultural pesticides. How will you decide if this plant is obeying necessary environmental controls? What criteria affect the success of this plant?

36. Health goals for contaminants in drinking water are expressed as MCLG, or Maximum Contaminant Level Goals. Legal limits are given as MCL or Maximum Contaminant Levels. How are MCLG and MCL related for a given contaminant?

37. Explain how you would use Figure 5.24 to determine the absorbance of a solution containing 0.002% dissolved iron.

38. Borax, $Na_2B_4O_7$, can be used to soften hard water. Write an equation to represent borax reacting with calcium ions in hard water.

39. An ion exchange resin totally charged with Na^+ ions holds 0.35 g of Na^+ per gram of resin. Use this information to determine the mass of water with a hardness of 40 ppm that could be deionized by 1 g of this resin.

40. Water quality in the chemistry building on a campus was continuously monitored because testing indicated water from drinking fountains in the building had dissolved lead levels above those established by the Safe Drinking Water Act.

a. What is the likely major source of the lead in the drinking water?

b. Does the chemical research carried out in this chemistry building account for the elevated lead levels found in the drinking water? Why or why not?

Exploring Extensions

41. Most people turn on the water tap with little thought about where the water comes from. In 5.4 and 5.5 Consider This, you investigated the source of your drinking water. Now take a more global view. Where does drinking water come from in other areas of the world? Investigate the source of drinking water in a desert country, in a developed European country, and in an Asian country. How do these sources differ?

42. Is there any such thing as "pure" drinking water? Discuss what is implied by this term, and how its meaning might change in different parts of the world.

43. One of the large aquifers in the United States is under the pine barrens of New Jersey.

a. Where are the pine barrens in New Jersey?

b. Why is this aquifer under increased political pressures?

44. In the mid-1990s, researchers in Canada and Australia reported that consumption of drinking water with more than 100 ppb aluminum can lead to neurological damage, such as memory loss and perhaps to a small increase in the incidence of Alzheimer's disease. Has further research substantiated these findings? Find out more about this topic, and write a brief summary of your findings. Be sure to cite the sources of your information.

45. The text states that hydrogen bonds are only about one-tenth as strong as the covalent bonds that connect atoms within molecules. Check out that statement with this information. Hydrogen bonds vary in strength from about 4 to 40 kJ/mole. Given that the hydrogen bonds between water molecules are at the high end of this range, how does the strength of a hydrogen bond between water molecules compare to the strength of an hydrogen-to-oxygen covalent bond within a water molecule? *Hint:* Consult Table 4.1 for covalent bond energies.

46. The text states that mass and density are often confused.

a. What do you think the term "heavy metal" implies when talking about elements on the periodic table?

b. Compare the scientific definitions of this term that you may find in different sources, and discuss whether each definition is related to relative density or to relative mass.

47. We all have the amino acid glycine in our bodies. This is its structural formula.

```
       H  :O:
  ..   |   ||    ..
H—N—C—C—O—H
   |   |         ..
   H   H
```

a. Is glycine a polar or nonpolar molecule? Use electronegativity differences to help answer this question.

b. Can glycine exhibit hydrogen bonding? Explain your answer.

c. Predict the solubility of glycine in water. Explain the reasons for your prediction.

48. How hard is the water in your local area? One way to answer this question is to determine the number of water softening companies in your area. Use the resources of the Web, as well as ads in your local newspapers and yellow pages, to find out if your area is targeted for marketing water softening devices.

49. Some areas have a higher than normal amount of trihalomethanes, THMs, in drinking water. Suppose that you are considering moving to such an area. Write a letter to the local water district asking relevant questions to be answered before deciding to move to that area.

50. PCBs (polychlorinatedbiphenyls) are very useful chemicals that may end up in the wrong place, causing long-term damage to birds and mammals. What are the uses of PCBs that made them desirable, and what are some of the negative effects of these materials?

6

Neutralizing the Threat of Acid Rain

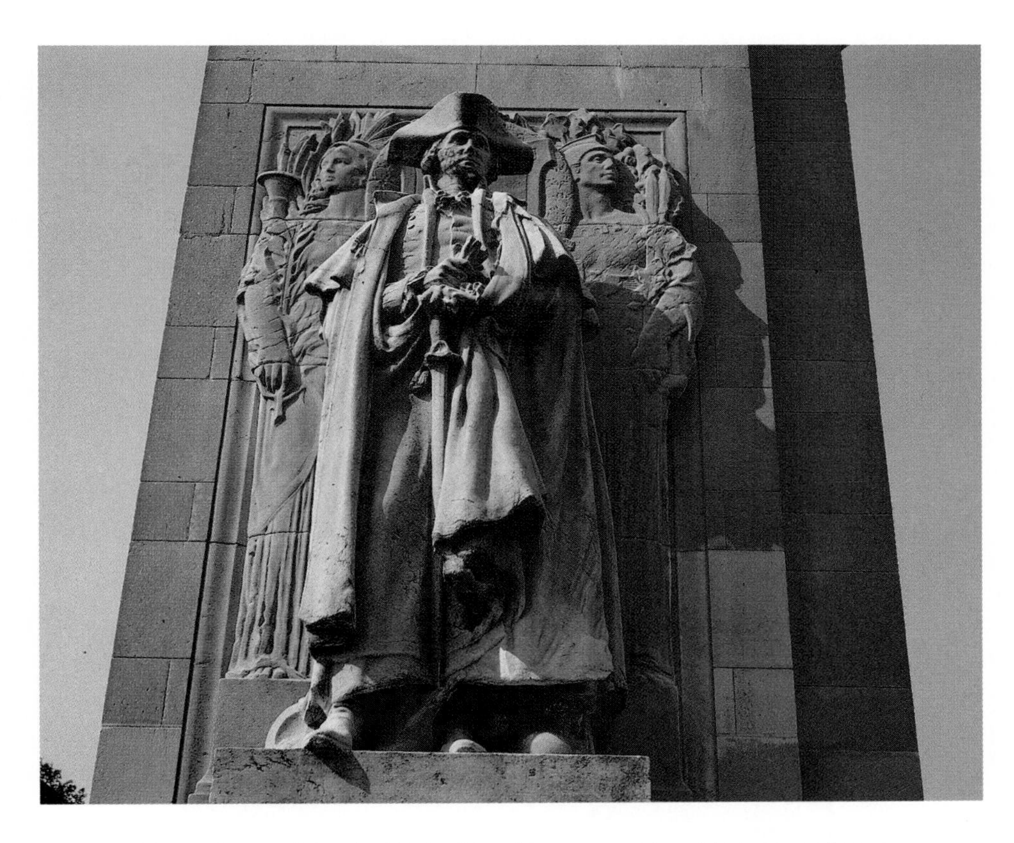

Acid rain damage to statuary.

"The environmental and public health problems caused by acid rain have affected us for several decades. We have, however, started on the path to recovery."

Progress Report on the EPA Acid Rain Program,
November 1999

"That's the thing about acid rain: the technology for reducing it has existed for a long time. But polluters and their congressional representatives have dug in their heels against the remedy. The upshot is that forests and lakes in the Northeast are dying so utility customers upwind of them can save a few cents on their electric bills."

November 22, 2000
Opinion in the *Syracuse News*

"In the 1990s, the businesses, politicians, and public took a collective sigh of relief and said 'that problem is over' and it's not. . . . Acid rain continues to have a significant ecological effect."

Gene E. Likens, *The Boston Globe*, March 26, 2001, page A1

What is the forecast for acid rain in the 21st century? Will those who live downwind of the acidic rains, snows, and mists still battle the power plants in neighboring states that produce the acid-rain forming pollutants? Will people now rest secure in the knowledge that we have lowered acidic emissions? And will fragile ecosystems show a timely recovery from the damage that has been done? No matter what the forecast holds, the years ahead promise us another chapter in the ongoing story of acid rain.

To understand the recent history of acid rain in the United States, think in terms of decades. The 1970s and 1980s stood as a period of passionate concern over our forests and lakes, with outcries over the damage to fragile ecosystems. In 1980, the National Acid Precipitation Assessment Program (NAPAP) was created by Congress and in 1990 released a comprehensive study on the causes and effects of acid rain. That year also marked the beginning of more stringent air quality legislation. The 1990 Clean Air Act Amendments, passed by the joint efforts of Congress and then-President George H. W. Bush, included the creation of the Acid Rain Program. Phase I of the Clean Air Act Amendments commenced in 1995, when 110 electric utility plants in 21 states began phased reductions of emissions that produced acid rain. And the 2000s began expectantly, with Phase II cuts in emissions by these and more players, large and small. Although we clearly have achieved reductions now, we still await the signs of ecological recovery.

- Comments about air being "acidic" were made as early as the beginning of the 1700s.

As you might surmise, our knowledge of acid rain predates these recent legislative acts by over a century. The acidity of rain apparently was first studied in detail in 1852 by a British chemist named Angus Smith. Twenty years later, he wrote a book entitled *Air and Rain*, but the book and Smith's ideas soon fell into obscurity. Then, in the 1950s, the effects of acid rain were rediscovered by scientists working in the northeastern United States, Scandinavia, and the English Lake District.

Reports of damage attributed to acidic precipitation grew dramatically over the next three decades. Dozens of books, scientific papers, and popular articles described the damage already observed and made dire predictions of what was yet to come (Figure 6.1). Similar reports came from every part of the world. Lakes in Norway and Sweden

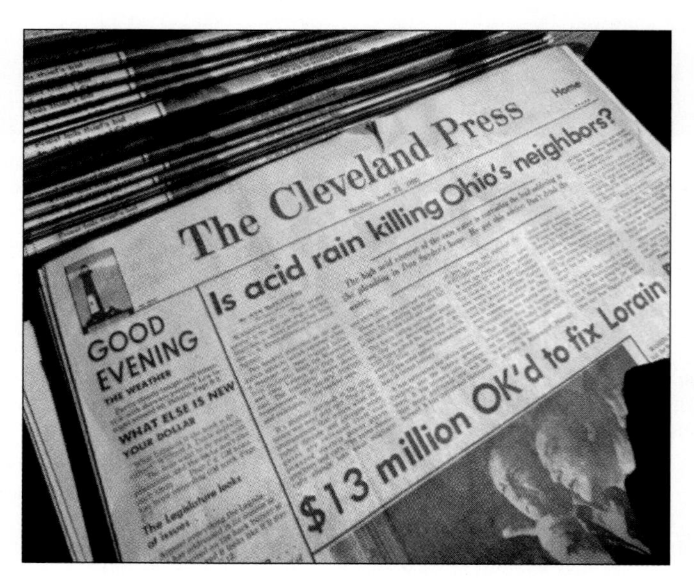

Figure 6.1
Acid rain has been in the news for some time (The Cleveland Press June 23, 1980).

were reported as being effectively "dead," without fish or any other living things. Trees in northern Germany had been stripped of their leaves. The sculptures adorning the exteriors of cathedrals in Europe and prehistoric sites in Mexico and Central America were eroding away. In all these instances, acid rain was blamed as one of the major causes of the damage.

Acid rain differs in scope from the issues discussed in earlier chapters. Ambient air quality, as we saw in Chapter 1, is largely a *localized* problem and most serious in cities and urban areas. The destruction of stratospheric ozone and the greenhouse effect, as described in Chapters 2 and 3, are both *global* atmospheric phenomena. Acid rain, however, tends to be *regional* in character; that is, it falls on areas neighboring its source. Acidic gases that originate in the Midwest, especially in the Ohio River valley, are carried to the northeast by prevailing winds and fall as acidic rain or snow on New York, New England, and eastern Canada. They are also carried to the southeast, producing acidic precipitation in Tennessee and North Carolina. In Europe, acids generated in Germany, Poland, and the United Kingdom are carried northward into Norway and Sweden.

The problem of acid rain is compounded by what may be the environmental paradigm of our times. By trying to use a technological fix for one problem we inadvertently create another. Taller smokestacks were built to eject pollutants high into the atmosphere, where they could be carried away and thus improve local air quality; in effect, an "out of sight out of mind" policy. But what goes up must come down, and it does somewhere, sometimes hundreds of miles away. Thus, potential local pollution problems are converted into regional ones. As acid rain does not respect state or national boundaries, it becomes entangled with intense political controversies and accusations. Once again, we are all caught in the same web.

6.1 Consider This: The Clean Air Act

If you are willing to work your way through over 400 pages of text, read the 1990 version of the Clean Air Act. A far friendlier approach is to browse through "The Plain English Guide to the Clean Air Act" provided by the EPA. Search for this document on the Web or use the link provided at the *Chemistry in Context* web site. The document explains the role of the federal government and that of the states. How do these roles differ? Summarize the Clean Air Act's program to reduce acid air pollutants.

CHAPTER OVERVIEW

Acid rain probably raises a number of questions in your mind: What exactly is an acid, and how acidic is the rain? What makes rain acidic? How does acid rain damage trees, fish, and marble statues? And perhaps most importantly, how do emissions into the atmosphere contribute to acid rain? These are primarily scientific questions, and in the pages that follow we will provide information that can help you answer them.

We begin by considering acids and bases and their properties, then look at acidic and basic solutions to find the ions responsible for these properties. The concept of pH is introduced as a measure of acidity, so that we can discuss the pH of rain. Precipitation with low pH values (high acidity) generally falls in regions where the atmospheric concentrations of sulfur oxides and nitrogen oxides are high. These gases are traced back to their sources: coal-burning power plants and gasoline-burning automobiles, respectively. Then, we turn to the effects of acid precipitation on materials, visibility, human health, lakes and streams, and trees and forests.

Interwoven with these largely scientific issues are the social and political factors that have made the public debate over acid rain as sour as the rain itself. Controversy surrounds the interpretation of the effects of acid rain and the best ways to deal with these effects. We consider several strategies, each with its own price tag. The financial impact of acid rain and the money required to curtail it introduce important economic dimensions. One nationwide pollution solution is that offered by the 1990 federal Clean Air Act Amendments. This legislation and its amendments have already reduced the concentrations of the acidifying oxides. A discussion of this legislation and its current and potential impact brings the chapter to a close.

Figure 6.2
Lemons and other citrus fruits such as limes, oranges, and grapefruits contain citric and ascorbic acids.

6.1 What Is an Acid?

The topic of acid rain brings together water, the subject of Chapter 5, and the atmospheric pollutant gases introduced in Chapter 1. Quite obviously, we need to define acids to understand this linkage. Like many other chemical concepts, acids can be defined either by their observable properties or conceptually, by theories of chemical structure. Chemists usually use both types of definitions, and we shall do so here.

Historically, chemists identified acids by their common properties—sour taste, color changes with indicators, and reactions with carbonate-containing materials. Although tasting is not a safe way to test for chemicals, you undoubtedly know the sour taste of vinegar and of citrus fruit (Figure 6.2); both foods contain acids. You may even be a fan of the incredibly sour candies containing acids that pucker your mouth (Figure 6.3). Other tests for acids rely on their chemical reactivity. A familiar example is the litmus test. Litmus, a vegetable dye, changes from blue to pink in the presence of acids. Indeed, the litmus test is so well known that it is used as a figure of speech in our culture, and a newspaper article might read: "The litmus test for this piece of legislation . . ." Another simple chemical test for acids is to add them to a carbonate-containing material such as marble or eggshell. Doing so causes fizzing to occur due to the release of carbon dioxide bubbles. We will return to this reaction later, after our conceptual definition of an acid.

- Carbonate is a polyatomic ion, CO_3^{2-} (Section 5.10).

From the standpoint of chemical structure, **acids** are substances that release hydrogen ions, H^+, in aqueous solution. You will recall from Chapter 5 that an ion is any atom or group of atoms bonded together that has a net electric charge. Atoms and molecules normally have equal numbers of protons (positively charged) and electrons (negatively charged), so each atom or molecule is electrically neutral. But if electrons are gained or lost, the atom or molecule becomes charged negatively or positively, respectively. For example, a hydrogen atom consists of one electron and one proton and is electrically neutral. If the electron is lost, the hydrogen atom acquires a positive charge and is designated as H^+. Since all that remains is a proton, the hydrogen ion is sometimes referred to as a proton.

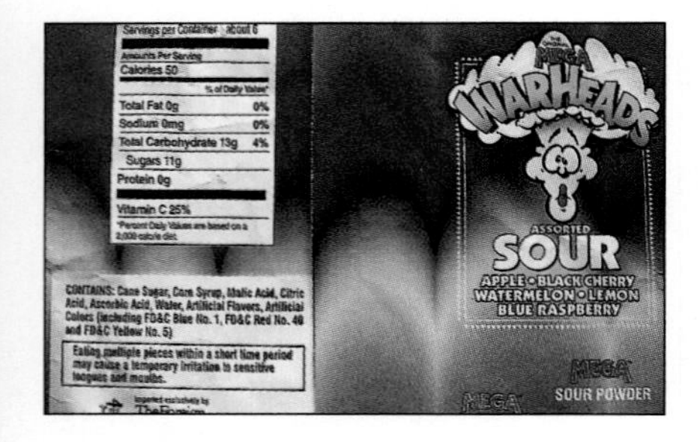

Figure 6.3
Check the label of a sour candy. The candy likely contains malic, citric, or ascorbic acid.

There is a slight complication with the definition of acids as substances that release H^+ ions (protons) in aqueous solutions. H^+ ions are much too reactive to exist by themselves. They always attach to something else, such as water molecules. As an example, consider hydrogen chloride gas, which at room temperature, is made up of covalent HCl molecules. This gas is quite soluble in water and dissolves in water to break apart forming two ions, H^+ and Cl^-.

- The (*aq*) notation, short for *aqueous,* represents a species dissolved in water.

$$HCl(g) \longrightarrow H^+(aq) + Cl^-(aq) \qquad [6.1]$$

Essentially no undissociated HCl molecules remain in solution, and we say that HCl is an acid that ionizes or dissociates completely. When dissolved in water, each HCl donates a proton (H^+) to an H_2O molecule, forming an H_3O^+ ion known as the **hydronium ion,** and leaving a Cl^- (chloride) ion. The overall reaction is

$$HCl(g) + H_2O(l) \longrightarrow H_3O^+(aq) + Cl^-(aq) \qquad [6.2]$$

The resulting solution represented in both equations 6.1 and 6.2 is called hydrochloric acid and has the characteristic properties of an acid because of the presence of H_3O^+

ions. Chemists often simply write H^+ when referring to acids (such as in equation 6.1), but understand this to mean H_3O^+ in aqueous solutions.

- The Lewis dot structure for the hydronium ion is $\left[\begin{array}{c} \text{H} \\ \text{H}:\ddot{\text{O}}:\text{H} \end{array}\right]^+$. It obeys the octet rule.

6.2 Your Turn

Write chemical equations showing the ionization (dissociation) of one hydrogen ion from one molecule of each of these acids.

a. HI (hydroiodic acid)
b. HNO_3 (nitric acid)
c. H_2SO_4 (sulfuric acid)
d. H_3PO_4 (phosphoric acid)

Answer
d. $H_3PO_4(aq) \longrightarrow H^+(aq) + H_2PO_4^-(aq)$

6.3 Consider This: Are All Acids Harmful?

The word "acid" may conjure up all sorts of pictures in your mind, but there are many acids in the foods we eat and the beverages we drink. Check the labels of processed foods and beverages. List some of the acids found in common foods and beverages. What do you think is the function of the acid in each item on your list?

Figure 6.4
Oven cleaner contains NaOH.

6.2 Bases

No discussion of acids would be complete without mentioning their chemical opposites—bases. For our purposes, we define a **base** (or alkali) as any compound that produces hydroxide ions, OH^-, in aqueous solution. Bases have their own characteristic properties attributable to the presence of OH^-. Unlike acids, bases lack any appeal as food items as they generally taste bitter. When dissolved in water, bases have a slippery, soapy feel. Common examples of bases are aqueous solutions of the gas ammonia, NH_3, and of sodium hydroxide, NaOH, sometimes known as lye. The cautions on household drain cleaners that contain lye (Figure 6.4) give dramatic warning that some bases, like some acids, can cause severe damage to eyes, skin, and clothing.

We can visualize how hydroxide ions are produced in solution from sodium hydroxide, an ionic compound. When solid NaOH dissolves in water, the sodium and hydroxide ions separate from the crystalline arrangement of Na^+ and OH^- ions.

$$NaOH(s) \longrightarrow Na^+(aq) + OH^-(aq) \qquad [6.3]$$

However, in an aqueous solution of ammonia, NH_3, the source of the OH^- ions is less obvious until we note this reaction.

$$NH_3(g) + H_2O(l) \longrightarrow \underset{\text{ammonium ion}}{NH_4^+(aq)} + OH^-(aq) \qquad [6.4]$$

Here, a hydrogen ion (H^+) is transferred from a water molecule to an ammonia molecule to form an ammonium ion (NH_4^+) and a hydroxide ion. Reaction 6.4 occurs only to a limited extent; that is, only tiny amounts of the two ions are formed in an aqueous solution of ammonia.

- The ammonium ion, NH_4^+, is formed in a manner analogous to the hydronium ion, H_3O^+.

6.4 Your Turn

Write chemical equations showing the release of ions as each of these solid bases dissolves in water.

a. KOH **b.** LiOH **c.** $Ca(OH)_2$

6.3 Neutralization: Bases Are Antacids

The reaction of an acid and a base is called **neutralization.** We illustrate this familiar process with hydrochloric acid and sodium hydroxide:

$$HCl(aq) + NaOH(aq) \longrightarrow NaCl(aq) + H_2O(l) \qquad [6.5]$$

In this case, the products of neutralization are sodium chloride and water. The corrosive acid and the caustic base offset each other chemically and here produce two neutral compounds; that is, ones that are neither acidic nor basic.

In a neutralization reaction, the hydrogen ions from an acid combine with the hydroxide ions from a base to form water molecules. The chemical reaction can be represented by equation 6.6.

- In cooking we use an acid-base neutralization reaction when we put lemon juice, which contains an acid, on fish. The acid neutralizes the ammonia-like compounds that produce a "fishy smell."

$$\underset{\text{(from an acid)}}{H^+(aq)} + \underset{\text{(from a base)}}{OH^-(aq)} \longrightarrow H_2O(l) \quad [6.6]$$

What happened to the sodium and chloride ions? Recall from equations 6.1 and 6.3 that HCl and NaOH completely dissociate into ions when dissolved in water, and we can rewrite equation 6.5 to show this:

$$H^+(aq) + Cl^-(aq) + Na^+(aq) + OH^-(aq) \longrightarrow Na^+(aq) + Cl^-(aq) + H_2O(l) \quad [6.7]$$

- Recall from Section 5.10 that NaCl is an ionic compound. It dissolves in water to produce sodium ion (Na^+) and chloride ion (Cl^-).

Notice that the Na^+ and Cl^- ions don't take part in the neutralization reaction and remain unchanged on both sides of the equation. Cancelling them in equation 6.7 produces equation 6.6.

6.5 Your Turn

Write chemical equations showing the reaction of each acid and base pair. First write the complete balanced equation, and then rewrite it in ionic form.

a. $HBr(aq)$ and $Ba(OH)_2(aq)$
b. $H_2SO_4(aq)$ and $NaOH(aq)$
c. $H_3PO_4(aq)$ and $Mg(OH)_2(aq)$

Answer

a. $2\ HBr(aq) + Ba(OH)_2(aq) \longrightarrow BaBr_2(aq) + 2\ H_2O(l)$
$2\ H^+(aq) + 2\ Br^-(aq) + Ba^{2+}(aq) + 2\ OH^-(aq) \longrightarrow Ba^{2+}(aq) + 2\ Br^-(aq) + 2\ H_2O(l)$
$2\ H^+(aq) + 2\ OH^-(aq) \longrightarrow 2\ H_2O(l)$
or by cancelling the 2s on both sides:
$H^+(aq) + OH^-(aq) \longrightarrow H_2O(l)$

- **Acidic** solution: $M_{H^+} > M_{OH^-}$
- **Neutral** solution: $M_{H^+} = M_{OH^-}$
- **Basic** solution: $M_{H^+} < M_{OH^-}$

Complete neutralization requires equal concentrations of H^+ and OH^-. This condition exists in pure water or in a neutral solution such as NaCl dissolved in pure water. In contrast, in **acidic solutions,** the concentration of H^+ ions is greater than that of OH^- ions. As you might guess, the concentration of OH^- ions is greater than that of H^+ ions in **basic solutions.**

It may seem strange that acidic solutions contain OH^- ions and likewise that basic solutions contain H^+ ions. A simple, useful, and very important relationship exists between the concentration of hydrogen ions and the concentration of hydroxide ions in any aqueous solution. When these are expressed in units of molarity, M_{H^+} and M_{OH^-}, and multiplied together, the product is a constant, with a value of 1×10^{-14} at 25°C. We represent this as

$$(M_{H^+})(M_{OH^-}) = 1 \times 10^{-14} \quad [6.8]$$

This mathematical expression tells us that the concentrations of H^+ and OH^- are linked to each other. When H^+ increases, OH^- decreases, and vice versa. Both ions are always present in aqueous solutions.

Knowing the value of either M_{H^+} or M_{OH^-}, we can use equation 6.8 to calculate the other. For example, if a rain sample has a H^+ concentration of 1×10^{-5} M, then the OH^- concentration can be calculated.

$$(M_{H^+})(M_{OH^-}) = 1 \times 10^{-14}$$

$$(1 \times 10^{-5}\ M_{H^+})(M_{OH^-}) = 1 \times 10^{-14}$$

$$M_{OH^-} = 1 \times 10^{-14} / 1 \times 10^{-5} = 1 \times 10^{-9}$$

The hydroxide ion concentration (1×10^{-9} M) is smaller than the hydrogen ion concentration (1×10^{-5} M). Therefore, the solution is acidic.

In pure water or a neutral solution, the molarities of hydrogen and hydroxide ions are equal, each with a value of 1×10^{-7} M. Applying equation 6.8, we can see that $(M_{H^+})(M_{OH^-}) = (1 \times 10^{-7})(1 \times 10^{-7}) = 1 \times 10^{-14}$.

- The product $(M_{H^+})(M_{OH^-})$ is somewhat temperature dependent. The value 1×10^{-14} holds true only at 25°C.

6.6 Your Turn

Classify these solutions as acidic, neutral, or basic at 25°C. Then, for parts a and c, calculate M_{OH^-}, and for **b**, calculate M_{H^+}.

a. $M_{H^+} = 1 \times 10^{-4}$
b. $M_{OH^-} = 1 \times 10^{-6}$
c. $M_{H^+} = 1 \times 10^{-10}$

Answer
a. The solution is acidic and has $M_{OH^-} = 1 \times 10^{-14} / 1 \times 10^{-4} = 1 \times 10^{-10}$.

6.7 Your Turn

List all of the ions present in order of decreasing concentration, starting with the most abundant for each of these solutions. Classify each as acidic, basic, or neutral.

a. $KOH(aq)$
b. $HNO_2(aq)$
c. $H_2SO_3(aq)$
d. $Ca(OH)_2(aq)$

Answer
d. $OH^-(aq) > Ca^{2+}(aq) > H^+(aq)$. Note there are two hydroxide ions for every calcium ion. $Ca(OH)_2(aq)$ dissolves to produce a basic solution where the concentration of $OH^-(aq)$ is greater (by far) than the concentration of $H^+(aq)$.

We now need a convenient way of reporting how acidic or basic a solution is. As you will see, the pH scale is just such a tool. It relates the acidity of a solution to its H^+ concentration.

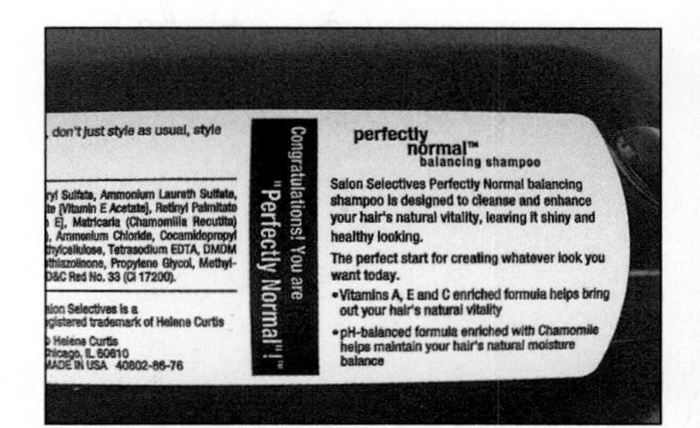

Figure 6.5
Soaps tend to be basic, which can be irritating to skin. Personal care products, such as shampoo, may have their pH adjusted so that they are closer to neutral, not basic.

6.4 Introducing pH

The letters pH show up just about every day. You may see pH in test kits for the soil in your garden, in advertisements for personal care products (Figure 6.5) and of course, in articles about acid rain. To understand such labels and articles, it is necessary to understand pH and its significance. The notation, always a lower case p and an upper case H written together, stands for "power of hydrogen." In simplest terms, **pH** is a number between 0 and 14 (occasionally a bit lower or higher) that indicates the acidity of a solution. The pH of a solution can be readily obtained with the help of a pH meter.

- It is possible for the pH of a sample to be lower than zero and higher than 14, but those cases are rare.

The pH scale has a particular quirk. As the pH value *decreases*, the acidity *increases.* For example, a sample of rainwater with a pH of 5.0 is more acidic than one with a pH of 6.0. Furthermore, as pH changes, the change in acidity is exponential. As the pH *decreases* by one unit, the H^+ concentration *increases* by a factor of 10. Thus, a rain sample with a pH of 5.0 has 10 times *more* H^+ and is 10 times *more* acidic than one with a pH of 6.0.

A pH of 7.0, midpoint on the pH scale of zero to 14, separates acidic from basic solutions. Solutions with a pH of less than 7.0 are acidic; those with a pH greater than 7.0 are described as alkaline or basic; and those with a pH of 7.0 are **neutral.** Stomach acid is highly acidic and can have a pH less than 2. How does the acidity of milk compare with this? Check Figure 6.6 to see pH values for a number of common substances. Note that rain naturally is slightly acidic, with a pH value between 5 and 6. *Acid rain has an even lower pH value*, hence the term **acid rain.**

- We will examine the reasons for the pH of acid rain in Section 6.5.

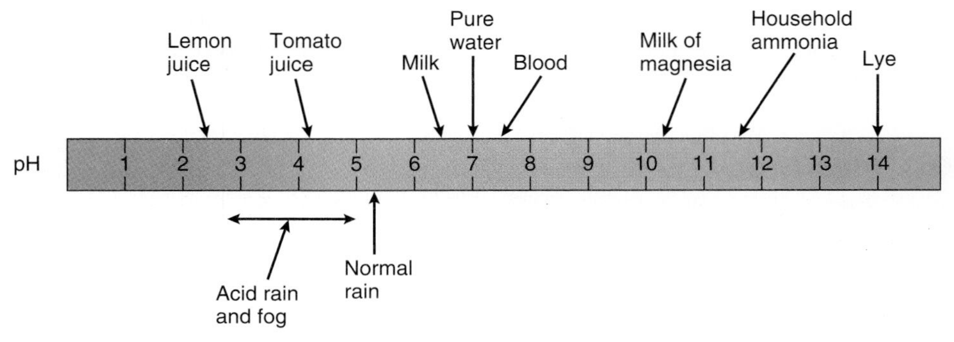

Figure 6.6

The pH of common substances. As pH increases, acidity decreases.

 **Figures Alive!**

6.8 Your Turn

For each pair, state which is more acidic and give the difference in hydrogen ion concentrations.

a. A sample of rain water with pH = 5 and a sample of lake water with pH = 4.
b. A tomato juice sample with pH = 4.5 and milk with pH = 6.5.

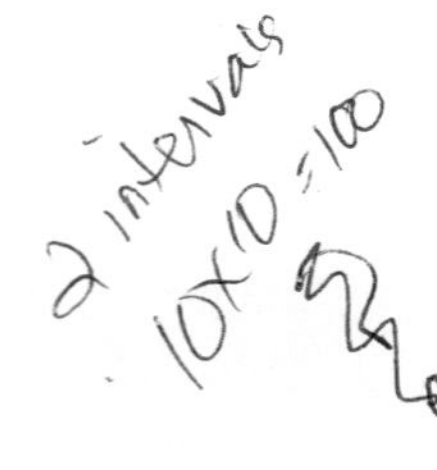

Answer

b. The juice with pH 4.5 is 100 times more acidic and has 100 times more H^+ than the milk with a pH of 6.5.

6.9 Consider This: On The Record

A legislator from a midwestern state is on record as making an impassioned speech in which he argued that the environmental policy of the state should be to bring the pH of rain all the way down to zero. Assume that you are a legislative aide to this legislator and draft a memo to your boss to save him from public embarrassment.

Natural water is typically slightly acidic, with a pH of about 6. On the other hand, pure water has a pH of 7.0 (Figure 6.6). It is neutral—neither acidic nor basic; $M_{H^+} = M_{OH^-}$. So the obvious inference is that well water and normal rain are not pure H_2O. You will soon see what "impurity" makes them acidic.

To many people, the term "acid" has a bad connotation, implying something dangerous or highly corrosive. To be sure, some acids such as "battery acid" (sulfuric acid) have such properties. But these acids have very high H^+ concentrations and correspondingly low pH values. You may be surprised to learn how many acids we eat, drink, and produce through metabolism. The naturally occurring acids in foods contribute distinctive tastes, contain much lower concentrations of H^+, and are much poorer H^+ donors than sulfuric acid. For example, the sharp taste of vinegar (pH ~2.5) comes from acetic acid. The tangy taste of some apples (McIntosh, pH ~3.3) is from malic acid, and the sour taste of lemons (pH ~2.3) is from citric acid. Tomatoes are well known for their acidity, but in fact are usually less acidic (pH ~4.5) than most fruits. Yogurt (pH ~4.0) gets its sour taste from lactic acid, and cola soft drinks (pH ~2.6) contain several acids, including phosphoric acid.

6.10 Consider This: Acidity of Foods

a. List vinegar, tomatoes, lemons, apples, cola, pure water, and yogurt in approximate order of increasing acidity. See Figure 6.6.
b. List any four foods of your own choosing in approximate order of pH. Then look up the pH values on the Web, using the link provided at the *Chemistry in Context* web site or another of your own choosing.
c. Why do you suppose there are so few foods with pH greater than 7?

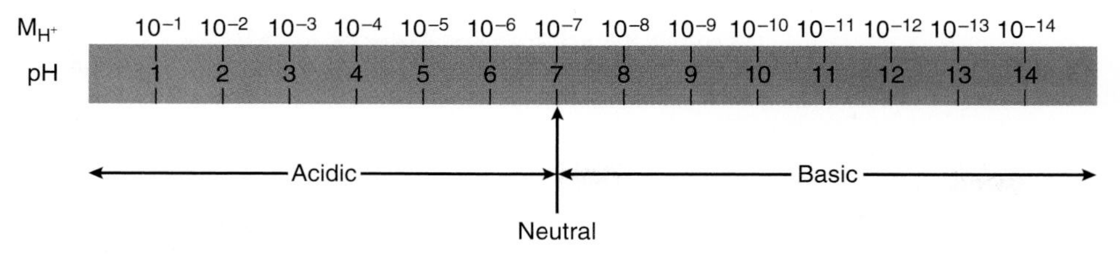

Figure 6.7

The relationship between pH and H^+ ion concentration (molarity). As pH increases, H^+ decreases.

Figures Alive!

Earlier in this section, we pointed out that a change in the pH meant an exponential change in the acidity. We now can demonstrate this using molarity values. For example, a solution in which $M_{H^+} = 0.0001 = 1 \times 10^{-4}$ has a pH of 4. By contrast, a solution of pH = 5 is less acidic with only 1/10 the concentration of hydrogen ion, or $M_{H^+} = 0.00001 = 1 \times 10^{-5}$. This same pattern involving powers of 10 throughout the entire pH scale is shown in Figure 6.7. Notice also that for whole number pH values, the pH is the exponent with its sign changed. For example, if $M_{H^+} = 1 \times 10^{-3}$, then pH = 3. A more detailed description of the relation between pH and M_{H^+} is provided in Appendix 3.

Having established the pH scale as a measure of acidity, we now apply the concept of pH to acid rain and its causes.

6.5 Measuring the pH of Rain

Before we describe the process for determining the pH of rain, it is important to realize that rain is only one of several ways that acids can be delivered to the surface and waters of the Earth. Snow obviously is another, hence the term **acid precipitation.** Even more inclusive is the term **acid deposition,** which includes fog and mist, cloud-like suspensions of microscopic water droplets often more acidic and damaging than acid rain. In addition, acidic deposition includes acidic gases and the acidic solid particles that sometimes settle out on surfaces during dry weather. This "dry deposition" has been shown to be almost as important as the wet deposition of the acids in rain, snow, and fog. It is also more difficult to measure.

The pH of a rain sample or other aqueous solution is usually determined with a pH meter. This device includes a special probe capped with an H^+-sensitive membrane that is immersed in the sample. H^+ ions in a sample create a voltage across the membrane. The meter measures this voltage and converts it to pH, which is indicated on a dial or digital display.

It is fairly easy to measure the pH of rain samples, although certain precautions are necessary to obtain reliable results. The use of scrupulously clean collection containers is crucial, and the containers must be placed high enough to prevent "soil splash" that would contaminate rain samples collected at ground level. The pH meter also must be calibrated carefully with solutions of known pH to ensure that it is operating correctly.

The pH of rain have been collected at selected sites in the United States and Canada since about 1970. A more systematic study has been under way since 1978, with over 225 sites at which weekly samples are collected. The pH is measured immediately and then all samples are sent to a central laboratory in Illinois for further analysis. Figure 6.8 was prepared from such data. It is a map showing the pH of precipitation during 1999. The different colored regions represent areas of the country with average pH values within a given range. Because this map contains a great deal of useful information, we will return to it several times in this chapter.

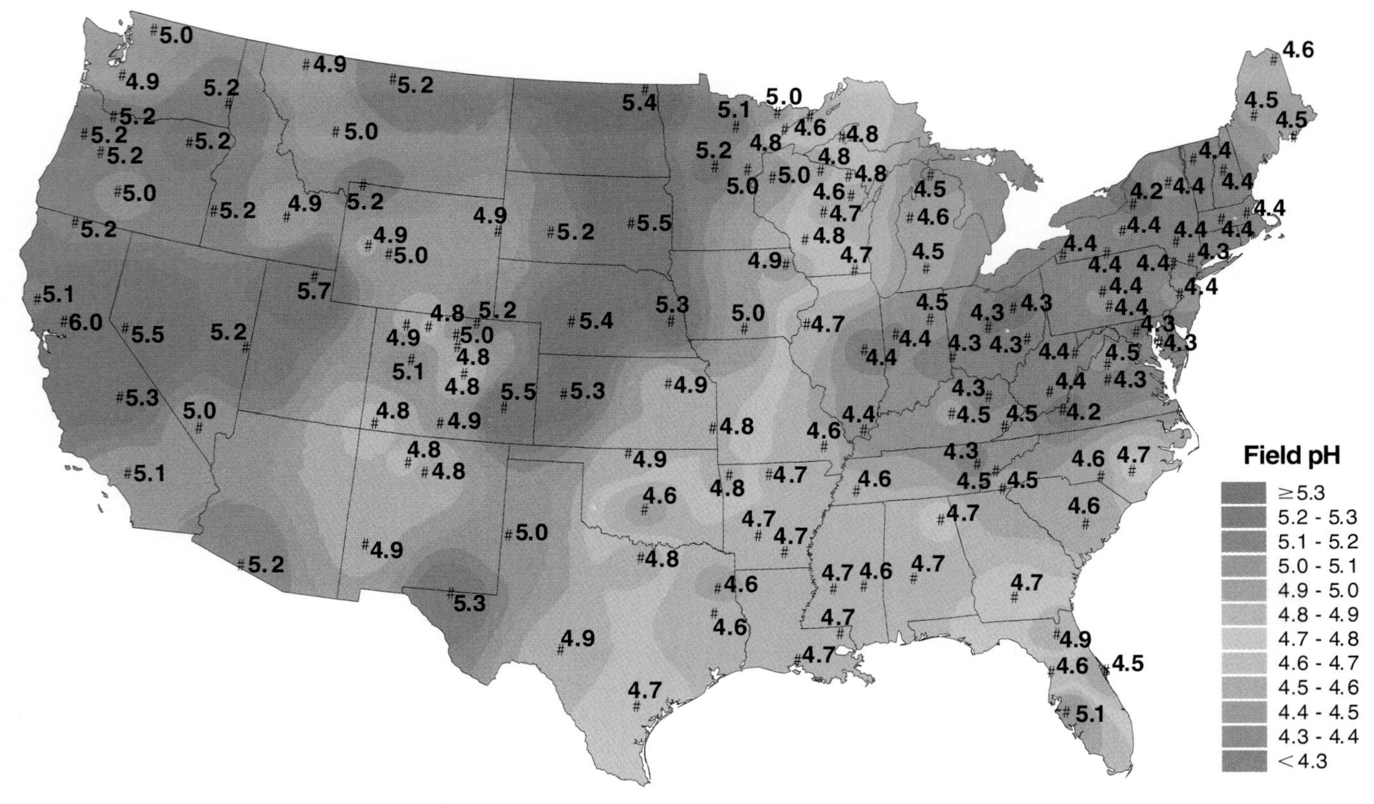

Figure 6.8

Hydrogen ion concentration as pH from measurements made at the Central Analytical Laboratory, 1999.

(Source: National Atmospheric Deposition Program National Trends Network. Taken from http://nadp.sws.uiuc.edu)

From the data in Figure 6.8, it appears that all rain is at least slightly acidic. At first thought, this might seem surprising. If rain is pure water (as we tend to assume), we would expect it to be neutral and have a pH of 7.0. But pure unpolluted rain always contains a small amount of dissolved carbon dioxide, CO_2, from air. Recall that CO_2 is a natural component of the Earth's atmosphere and is removed from the atmosphere when it dissolves in water. Carbon dioxide dissolves to a slight extent in water and reacts with it to produce a slightly acidic solution containing H^+ and HCO_3^- ions:

- At higher pressures, larger amounts of carbon dioxide can be forced to dissolve in water. The result is carbonated water or "soda water" (pH ~4.7).

- $HCO_3^-(aq)$ is the hydrogen carbonate ion (or bicarbonate ion), a polyatomic ion encountered in Chapter 5.

$$CO_2(g) + H_2O(l) \longrightarrow H^+(aq) + HCO_3^-(aq) \qquad [6.9]$$

Reaction 6.9 occurs only to a limited extent; that is, only tiny amounts of H^+ and HCO_3^- are formed. But the hydrogen ions formed are enough to account for most of the acidity in "pure" rainwater. At 25°C, a sample of water exposed to the normal atmospheric concentration of carbon dioxide has a pH of 5.6.

Figure 6.6 indicated that normal rain has a pH of about 5.3. It follows that CO_2 cannot be the sole source of H^+ in rainwater. Small amounts of other natural acids, including formic acid and acetic acid, are almost always present in rain and contribute to its acidity. However, even these additional acids cannot account for the fact that rain frequently may have a pH significantly below 5.3 (Figure 6.8). We are now ready to search for the source of this extra acidity.

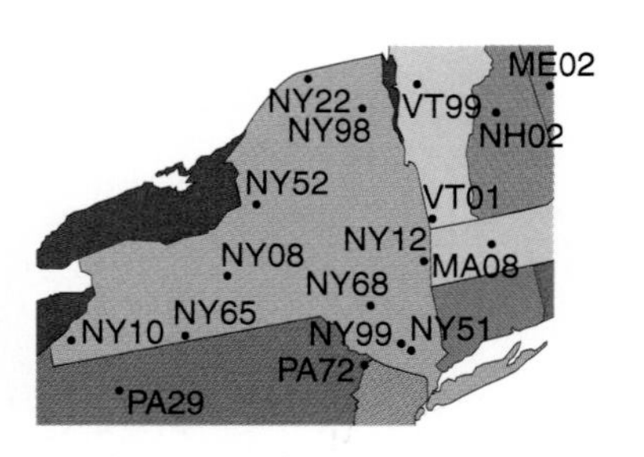

The locations of the precipitation monitoring stations in the Northeast.

(Source: National Atmospheric Deposition Program National Trends Network. Taken from http://nadp.sws.uiuc.edu)

6.11 Consider This: The Rain in Maine . . . or New York or Vermont

All 50 states, as well as Puerto Rico and the Virgin Islands, have one or more precipitation monitoring sites. They are all part of the National Atmospheric Deposition Program/National Trends Network (NADP/NTN). Some have been collecting data since the 70s, and the NADP/NTN now posts the data on the Web. Search or use the link at the *Chemistry in Context* web site to answer these questions.

a. How many monitoring sites are in your state? For a site of your choice, what agency operates it and what agency funds it? Use the "trend data" to create a plot of the pH values of rain over the past few years. What is a typical value? What is the trend?

b. How does the precipitation in your state compare to others? Make a prediction and then look up the data for another state of your choice. Again look at the trend data.

6.6 In Search of the Extra Acidity

According to Figure 6.8 the most acidic rain falls in the eastern third of the United States, with the region of lowest pH being roughly the states along the Ohio River valley. The extra acidity must be originating somewhere in this heavily industrialized part of the country. Analysis of rain for specific compounds confirms that the chief culprits are the oxides of sulfur and nitrogen, which are sulfur dioxide (SO_2), sulfur trioxide (SO_3), nitrogen monoxide (NO), and nitrogen dioxide (NO_2). These compounds are collectively designated SO_x and NO_x and sometimes are called "sox" and "nox."

If this interpretation of the origins of acid precipitation is correct, the geographical regions with the most acidic rain should also be heavy emitters of sulfur and nitrogen oxides. That relationship is generally confirmed by an examination of the maps in Figure 6.9. Emissions of sulfur dioxide are highest in regions where there are many coal-fired electric power plants, steel mills, and other heavy industries that rely on coal. Allegheny County, in western Pennsylvania, is just such an area, and in 1990 it had the dubious distinction of leading the United States in atmospheric SO_2 concentration. Although power plants also generate nitrogen oxides, the highest NO_x emissions are generally found in states with large urban areas, high population density, and heavy automobile traffic. Therefore, it is not surprising that in 1990 (and still today) the highest levels of atmospheric NO_2 were measured over Los Angeles County, the car capital of the country.

The circumstantial evidence linking acid precipitation with the oxides of sulfur and nitrogen appears compelling, but at this stage the Sceptical Chymist should be raising an important question. Given the definition of an acid as a substance that contains and releases H^+ ions in water, how can SO_2, SO_3, NO, and NO_2 qualify? None of these compounds even contains hydrogen! The objection is a sensible one. The explanation is that SO_2 and NO_x react with water to release H^+ ions. Although they are not acids themselves, the oxides of sulfur and nitrogen are acid anhydrides, literally "acids without water." When an *acid anhydride* is added to water, an acid is generated in solution. For example, sulfur dioxide dissolves in water and reacts with the water to form sulfurous acid.

$$\underset{\text{sulfur dioxide}}{SO_2(g)} + H_2O(l) \longrightarrow \underset{\text{sulfurous acid}}{H_2SO_3(aq)} \quad [6.10]$$

Similarly, sulfur trioxide reacts with water to yield sulfuric acid:

$$\underset{\text{sulfur trioxide}}{SO_3(g)} + H_2O(l) \longrightarrow \underset{\text{sulfuric acid}}{H_2SO_4(aq)} \quad [6.11]$$

• Reactions 6.10 and 6.11 are analogous to the reaction of CO_2 with water.

Sulfuric acid then dissociates to yield H^+ and HSO_4^- (hydrogen sulfate) ions.

$$H_2SO_4(aq) \longrightarrow H^+(aq) + \underset{\text{hydrogen sulfate ion}}{HSO_4^-(aq)} \quad [6.12a]$$

The hydrogen sulfate ion further dissociates to yield H^+ and SO_4^{2-} (sulfate) ions.

$$\underset{\text{hydrogen sulfate ion}}{HSO_4^-(aq)} \longrightarrow H^+(aq) + \underset{\text{sulfate ion}}{SO_4^{2-}(aq)} \quad [6.12b]$$

• Not all of the HSO_4^- ions are converted to SO_4^{2-} ions.

The overall result is that sulfuric acid dissociates to yield two H^+ ions and a sulfate ion.

$$H_2SO_4(aq) \longrightarrow 2\ H^+(aq) + SO_4^{2-}(aq) \quad [6.12c]$$

In a similar but a bit more complicated way, NO_2 can yield nitric acid, HNO_3, which dissociates into H^+ and NO_3^- ions.

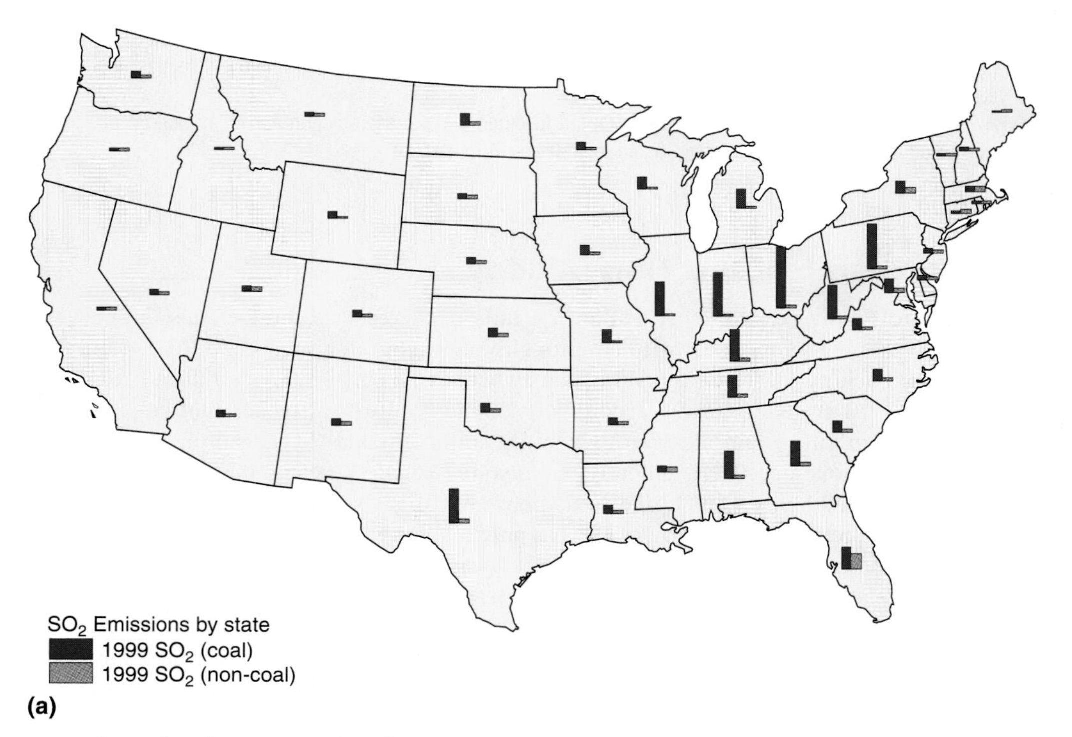

Source: http://www.epa.gov/airmarkets/emissions/score99/figureb1.html.

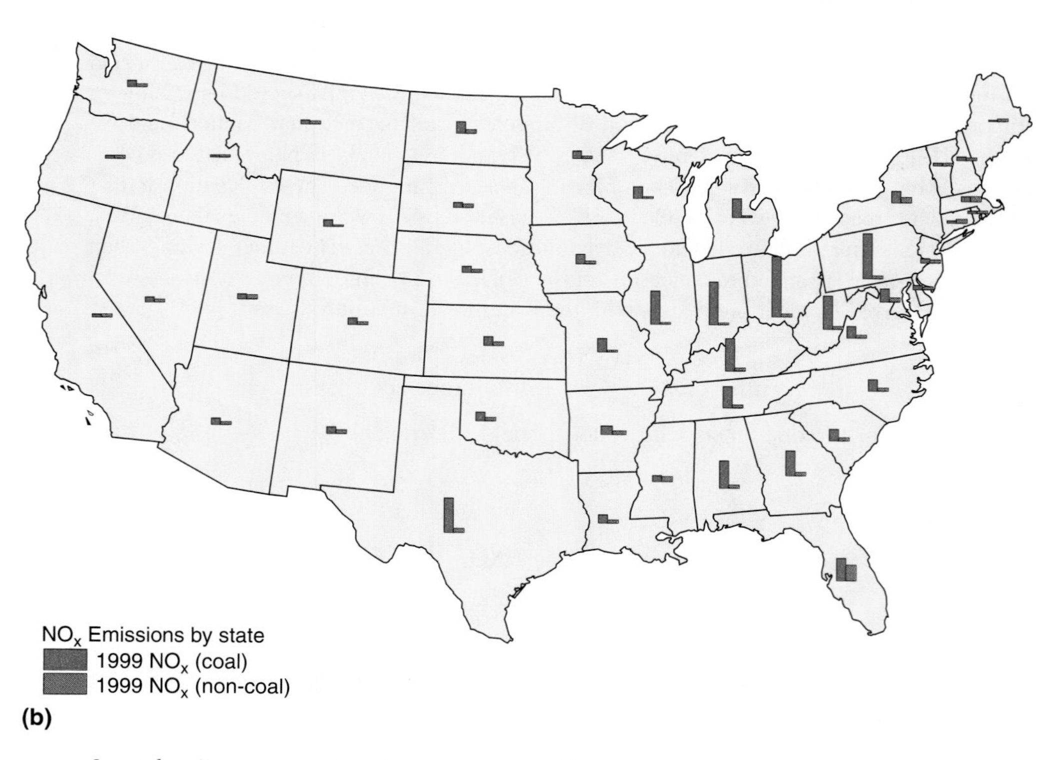

Source: http://www.epa.gov/airmarkets/emissions/score99/figureb2.html.

Figure 6.9

1999 emissions from coal- and non-coal-fired utilities: (a) SO_2 (b) NO_x.

$$4\ NO_2(g) + 2\ H_2O(l) + O_2(g) \longrightarrow 4\ HNO_3(aq) \quad [6.13]$$
nitrogen dioxide — nitric acid

$$HNO_3(aq) \longrightarrow H^+(aq) + NO_3^-(aq) \quad [6.14]$$
nitrate ion

Now that we see how oxides of sulfur and nitrogen contribute to acid rain formation, we need to get a closer look at how these oxides are formed and released into the atmosphere.

Figure 6.10
Burning sulfur produces SO_2 gas. Sulfur dioxide gas dissolved in water produces an acidic aqueous solution.

6.7 Sulfur Dioxide and the Combustion of Coal

Thus far, this chapter has established a relationship involving coal burning, atmospheric sulfur dioxide, and acid rain formation. Moreover, the fact that SO_2 and SO_3 react with water to yield acidic solutions is indisputable. What is not yet clear is why the combustion of coal should yield SO_2, the choking gas formed from burning sulfur (Figure 6.10). To answer this, we need to know something about the chemical nature of coal. At first glance, coal appears to be just a black solid, not very different from charcoal or black soot, both of which are essentially pure carbon. When carbon is burned, it forms carbon dioxide and liberates large amounts of heat (which of course is the reason for burning it).

$$C\ (\text{in coal}) + O_2(g) \longrightarrow CO_2(g) \quad [6.15]$$

As you learned in Chapter 4, coal is a complex substance. No two coal samples have exactly the same composition. Although coal is not a pure chemical compound, we can approximate its composition with the formula $C_{135}H_{96}O_9NS$. In addition to these five elements, coal also contains small amounts of silicon and various metal ions such as sodium, calcium, aluminum, nickel, copper, zinc, arsenic, lead, and mercury. When coal is burned, oxygen reacts with *all* the elements present to form oxides of those elements. Because carbon and hydrogen are the most plentiful, large quantities of gaseous CO_2 and H_2O are produced by the combustion. In addition, there is an unburned solid residue (ash) consisting of oxides of silicon, sodium, calcium, and the other trace elements just mentioned. But the sulfur is our primary interest right now. The combustion reaction of sulfur from coal with oxygen yields sulfur dioxide, a poisonous gas with an unmistakable choking odor.

• In ancient times sulfur was known as brimstone, thus the biblical admonition about "fire and brimstone."

$$S\ (\text{in coal}) + O_2(g) \longrightarrow SO_2(g) \quad [6.16]$$

You may be wondering why coal contains sulfur and why the amount varies so much. Coal formed 100–400 million years ago from decaying vegetation, often in swamps or peat bogs. Because sulfur is present in all living things, when the ancient plants decayed, some sulfur was left behind in the material that eventually became coal. However, most of the sulfur in coal originated from the sulfate ion (SO_4^{2-}) naturally present in seawater. Millions of years ago, bacteria on sea floors utilized the sulfate ion as an oxygen source, releasing the sulfide ion (S^{2-}). In turn, the sulfide ion became incorporated into the ancient rocks (including coal) that were in contact with seawater. In contrast, the freshwater peats that formed coal have a lower sulfur content.

• Sulfur's movement through the biosphere should remind you of the carbon cycle from Chapter 3.

While coals from different regions can vary considerably in their sulfur content, the combustion of almost all coals produces sulfur oxides. This fact is central to the acid rain story. In large coal-burning electrical power generating stations and industrial plants, the sulfur dioxide goes up the smokestack (unless control measures are used) along with the carbon dioxide, water vapor, and various metal oxides. Once in the atmosphere, SO_2 can react with more oxygen to form sulfur trioxide, SO_3.

$$2\ SO_2(g) + O_2(g) \longrightarrow 2\ SO_3(g) \qquad [6.17]$$

This reaction is fairly slow, but it is catalyzed (speeded up) by the presence of finely divided solid particles, such as the ash that goes up the stack along with the SO_2. Once SO_3 is formed, it reacts rapidly with any water vapor or water droplets in the atmosphere to form sulfuric acid (equation 6.11). There are also a variety of other agents and pathways for the conversion of sulfur dioxide into sulfuric acid. Of particular importance are tropospheric ozone, hydrogen peroxide, H_2O_2, and the hydroxyl radical ·OH, which is formed from ozone and water in the presence of sunlight. The reaction of SO_2 with ·OH accounts for 20–25% of the sulfuric acid in the atmosphere. The reaction goes faster in intense sunlight, and thus is more important in summer and at midday.

We can use a chemical calculation to better appreciate the vast quantities of SO_2 produced by coal-burning power plants. Such plants typically burn 1 million metric tons of coal a year, where 1 metric ton is equivalent to 1000 kg or 1×10^3 kg.

$$1 \times 10^6 \text{ metric tons coal/yr} = 1 \times 10^9 \text{ kg coal/yr} = 1 \times 10^{12} \text{ g coal/yr}$$

We will assume a low-sulfur coal that contains 2.0% sulfur; that is, 2.0 g sulfur per 100 g coal. First we can calculate the grams of sulfur released each year from one million metric tons (1×10^{12} g) of coal.

$$1 \times 10^{12} \text{ g coal/yr} \times 2.0 \text{ g S/100 g coal} = 2.0 \times 10^{10} \text{ g S/yr}$$

Next, we use the fact that one mole of sulfur reacts with oxygen to form one mole of SO_2 (equation 6.16). The molar mass of sulfur is 32.0 g, and the molar mass of SO_2 is 64.0 g, that is, 32.0 g + 2(16.0 g). Therefore, 32.0 g of sulfur burns to produce 64.0 g of SO_2.

- See Section 3.8 for a review of molar mass and moles.

$$2.0 \times 10^{10} \text{ g S} \times 64.0 \text{ g } SO_2/32.0 \text{ g S} = 4.0 \times 10^{10} \text{ g } SO_2$$

- Metric tons and short tons differ. Emissions data usually are in metric tons (1000 kg, 2200 pounds) or in short tons (2000 pounds). To add to the confusion, short tons are sometimes simply called tons.

This mass of SO_2 is equivalent to 40,000 metric tons or 88 million pounds of SO_2 per year. Power plants burning a higher sulfur coal emit more than twice this amount!

About two-thirds of the sulfur dioxide in the United States arise from the combustion of fossil fuels, chiefly coal, for the generation of electrical power. The coal burned in industrial processes contributes much of the remainder (Figure 6.11). While the large-scale production of metals such as nickel and copper contributes only a few percent to the total, huge quantities of SO_2 still can be generated in a particular region from metal production. Nickel and copper ores are typically sulfides, that is, compounds of the metal plus sulfur such as nickel sulfide. When heated to high temperatures in a smelter, the metal sulfides are decomposed and sulfur dioxide is released.

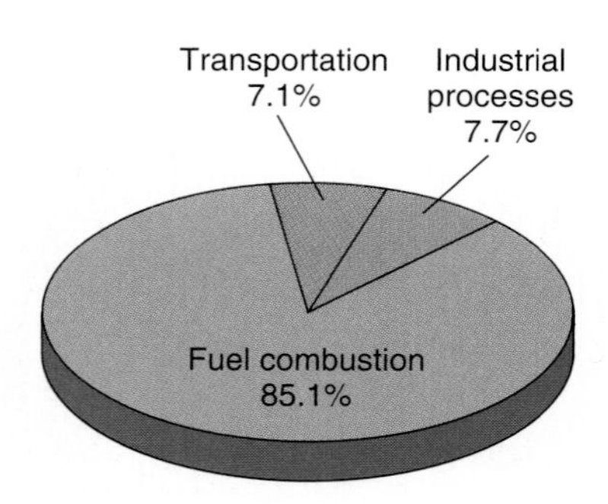

Figure 6.11
U.S. sulfur dioxide emission sources, 1998.
(Source: EPA/OAR.)

The world's largest smelter, in Sudbury, Ontario, converts nickel sulfide to nickel. The bleak, lifeless lunar landscape present in the immediate vicinity of the plant stands in mute testimony to earlier uncontrolled releases of SO_2. Today, after a major renovation in 1993, the two major smelters in the area have reduced their sulfur dioxide emissions substantially. Nonetheless, in 1997 over 250,000 metric tons of SO_2 still were released, some of it up a 1250-foot smokestack. The fact that this is the world's tallest smokestack—equal in height to the Empire State Building—simply means that the emissions were carried farther away from Sudbury by the prevailing winds. Lest we point any fingers, Canadians report that more than half of acid depositions in the eastern portion of their country originate elsewhere. The quantity of sulfur dioxide that drifts northward over the U.S. border into Canada is estimated to be 4 million tons per year.

6.12 Your Turn

a. Assume the composition of coal can be represented by $C_{135}H_{96}O_9NS$. Calculate the fraction and percent (by mass) of sulfur in the coal.
b. A power plant burns one million (1×10^6) tons of coal per year. Assuming the sulfur content calculated in **a,** calculate the number of tons of sulfur released per year.
c. Calculate the number of tons of SO_2 formed from this mass of sulfur.
d. Where does this SO_2 go, once it is released into the atmosphere?

Answers
a. 0.0168 or 1.68% **b.** 1.68×10^4 tons S (16,800 tons)

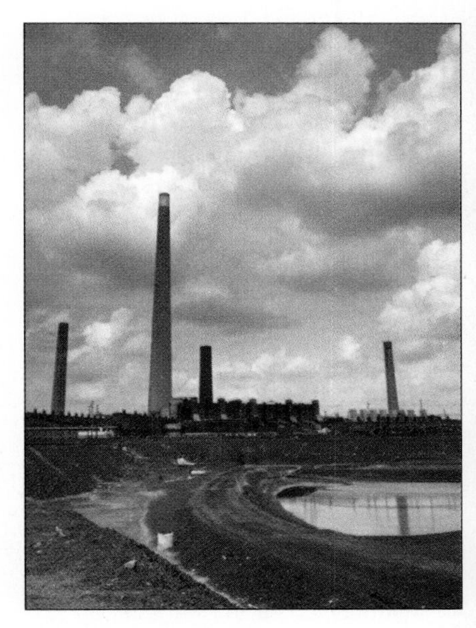

The 1250-foot smokestack in Sudbury, Ontario, the world's tallest.

6.8 Nitrogen Oxides and the Acidification of L.A.

The combustion of coal has been indicted as a major environmental offender, contributing sulfur dioxide to the atmosphere and to acid deposition. But SO_2 is not the only cause of acid precipitation, and another guilty party has been identified in California and other areas. The concentration of SO_2 in the smoggy air above the Los Angeles metropolitan area is relatively low, yet the pH of rain is still quite acidic. For example, in January 1982, fog near the Rose Bowl in Pasadena was found to have a pH of 2.5. Breathing it must have been like breathing a fine mist of vinegar. This level is at least 500 times more acidic than normal, unpolluted precipitation and 10 times more acidic than required to kill all fish in lakes. And, in December 1982, fog at Corona del Mar, on the coast south of Los Angeles, was more than 10 times more acidic, with a registered pH of 1.5. In both cases, something other than sulfur dioxide was involved.

That acidic "other" is emitted by the millions of cars that jam the Los Angeles freeways day and night. But it is not at all obvious why cars should contribute to acid precipitation. Cars run on gasoline, a mixture of hydrocarbons (Section 4.8) that burn to form mostly CO_2 and H_2O. Recall from Chapter 1 that this conversion is not always complete and some CO and unburned hydrocarbon fragments escape in the exhaust. Nevertheless, gasoline contains almost no sulfur, and therefore its combustion yields practically no SO_2. Consequently, we must look for another source of acidity.

Nitrogen oxides have already been identified as contributors to acid rain, but gasoline contains no nitrogen, either. Therefore, logic (and chemistry) assert that nitrogen oxides cannot be formed from burning gasoline. Literally, that is correct—after all, you cannot make something out of nothing. Remember, however, that nitrogen is present in air. In fact, 78% of air consists of N_2 molecules. These molecules are remarkably stable and do not readily undergo chemical reactions under ordinary conditions. Thankfully, that is why nitrogen remains unchanged as we breathe it in and out of our lungs. Nevertheless, nitrogen can and does react directly with a few elements *at high temperatures and pressures*. One of these elements is oxygen. All that is needed is sufficient energy in the form of high temperatures or an electric spark. Under these conditions, the two elements combine to form nitric oxide (nitrogen monoxide), NO.

$$\text{Energy} + N_2(g) + O_2(g) \rightarrow 2\ NO(g) \quad [6.18]$$

Like automobile engines, jet engines have temperatures high enough to produce NO from N_2 and O_2 in the air.

Because air is a mixture of nitrogen and oxygen, it is always a potential source for the production of nitric oxide. The energy necessary for the reaction can come from lightning bolts or the "lightning" inside an internal combustion engine. In such an engine, gasoline and air are drawn into the engine cylinders and compressed to a high pressure. Moreover, the high pressure means that the nitrogen and oxygen molecules are closer together and thus even more likely to react. Once ignited in the engine, the gasoline burns rapidly. The energy released powers the vehicle. But the unfortunate truth is that the energy also triggers reaction 6.18.

The reaction of N_2 with O_2 to form NO is not limited to lightning and the automobile engine. Chapter 2 mentioned the concern over NO production in jet aircraft engines.

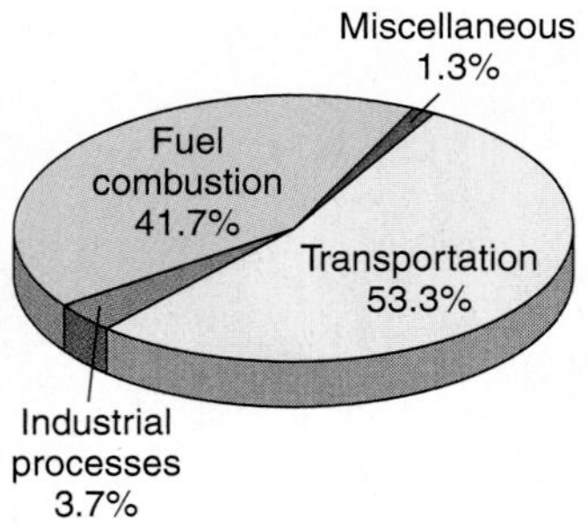

Miscellaneous 1.3%
Fuel combustion 41.7%
Transportation 53.3%
Industrial processes 3.7%

Figure 6.12
U.S. nitrogen oxide emission sources, 1998.

(Source: EPA Report 454/R-00-002, National Air Pollutant Emission Trends: 1900–1998, Figure 2-2.)

The same reaction occurs when air is heated to a very high temperature in the furnace of a coal-burning electrical power plant. Hence, such plants contribute vast amounts of *both* sulfur oxides and nitrogen oxides to acidify precipitation. On a national basis, stationary sources such as power plants release about a quarter of the nitrogen oxides (Figure 6.12). Mobile sources (such as motor vehicles, aircraft, trains, lawn mowers) account for about twice this. In urban environments, however, automobiles and trucks account for a far greater proportion of the atmospheric NO.

A green chemistry solution to reducing NO emissions and energy consumption has been introduced into U.S. glass manufacturing by Praxair Inc. of Tarrytown, NY. The award-winning innovative technology substitutes pure oxygen for air in the large furnaces used to make glass. Switching from air (78% nitrogen) to pure oxygen eliminates NO production and also saves fuel by more efficient burning. Glass manufacturers using the Praxair Oxy-Fuel technology save enough energy annually to meet the daily needs of one million Americans.

Unlike nitrogen, nitric oxide is very reactive. In Chapter 2, you read that NO can react with ozone in the upper atmosphere, thus destroying the O_3. Close to the surface of the Earth, it reacts primarily with O_2 to form nitrogen dioxide, NO_2.

$$2\ NO(g) + O_2(g) \longrightarrow 2\ NO_2(g) \qquad [6.19]$$

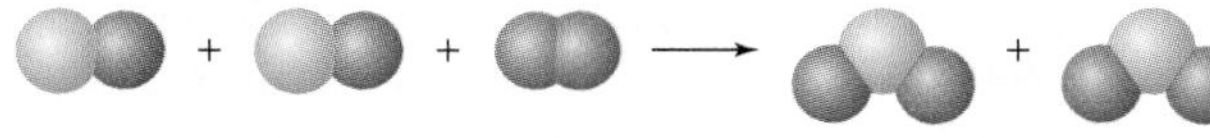

Several other oxides of nitrogen are formed from NO, with the most important being NO_2. It is a highly reactive, poisonous, red-brown gas with a nasty odor. For our purposes, its most significant reaction is the one that converts it to nitric acid, HNO_3. You saw one representation of that conversion in equation 6.13. Actually, a series of steps is involved in the chemistry that takes place in the urban atmosphere above Los Angeles, Phoenix, and other sunny metropolitan areas. Unraveling this complex web of reactions has proven to be a fascinating scientific detective story. Sunlight is required and volatile organic compounds, some released in the incomplete combustion of gasoline, are involved. An important intermediate is the hydroxyl radical, ·OH, which is formed in a reaction involving ozone, another common tropospheric pollutant. The hydroxyl radicals rapidly react with nitrogen dioxide to yield nitric acid.

$$NO_2(g) + \cdot OH(g) \longrightarrow HNO_3(l) \qquad [6.20]$$

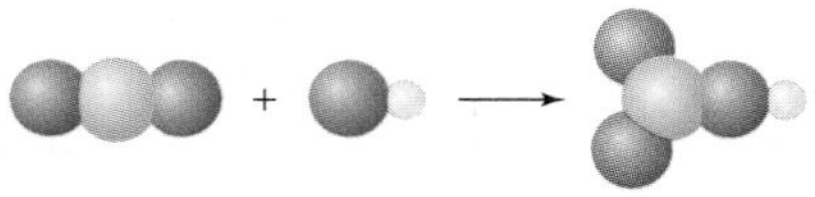

As you have already read (equation 6.14), HNO_3 dissociates completely in water to release H^+ and NO_3^- ions. The result is the alarmingly low pH values occasionally found in Los Angeles rain and fog.

6.9 SO_2 and NO_x: How Do They Stack Up?

• 1 ton ("short ton") = 2000 lbs = 0.9072 metric tons.

Now that we have identified SO_2 and NO_x as the two major contributors to acid precipitation, we should compare them in terms of the problems they pose. Currently, their annual U.S. anthropogenic emissions are of roughly equal magnitude, about 20 million tons for SO_2 and about 25 million tons for NO_x (Figures 6.13 and 6.14). As you saw earlier, about two-thirds of the sulfur dioxide emissions can be traced to large and stationary sources: coal-burning electric utilities (Figure 6.11). But these same utilities account for only about a quarter of the nitrogen oxides released. Small and mobile sources, the combustion engines that power various forms of transportation, emit the majority of the NO_x (Figure 6.12).

However, these pollutants were not always equal contributors. In years past, there was far less NO_x than SO_2 in our rain, fog, and snow. Current NO_x values are the result

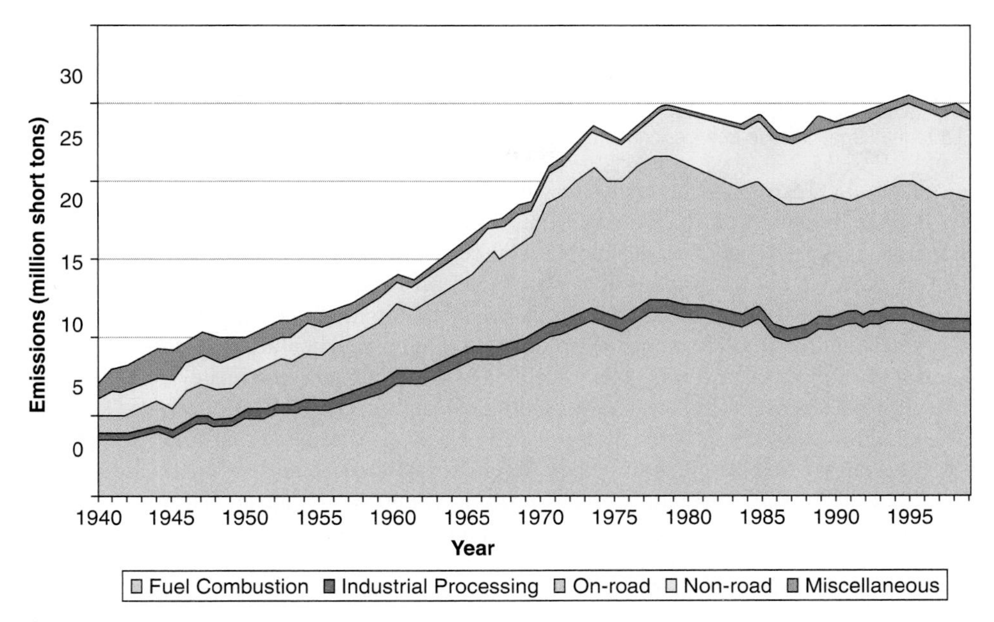

Figure 6.13

U.S. nitrogen oxide emissions 1940–1998.

(Source: EPA/OAR.)

of a relentless increase in emissions until about 1975, when the amounts leveled off. And unlike SO_2 emissions, today the nationwide NO_x emissions do not show the dramatic results of caps and controls. While SO_2 emissions have *decreased* about 40% since 1970, a tribute to the Clean Air Act and its amendments, NO_x emissions have actually *increased* slightly. In the final two sections of this chapter, we will look more closely at the costs, the control strategies, and the politics that accompany these changes.

- The chemistry of NO in the atmosphere is complicated. NO can destroy ozone, as seen in Chapter 2. But remember from Chapter 1 that NO can react with O_2 to form NO_2. In turn, NO_2 can react in sunlight to produce ozone.

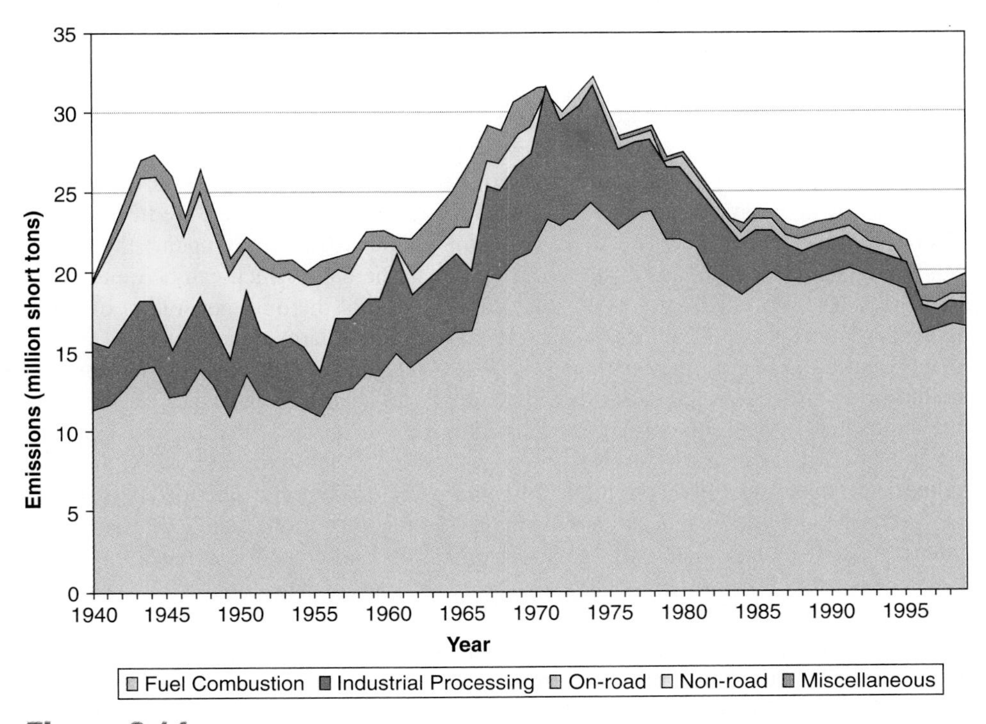

Figure 6.14

U.S. sulfur dioxide emissions 1940–1998.

(Source: EPA/OAR.)

6.13 Your Turn

Figure 6.13 shows five sources of NO_x emissions in the U.S. Which two did not change much between 1940 and 1995? Which three did? Similarly, in Figure 6.14 for SO_2, in this same time period, which sources led to the large increase in emissions in the 1970s?

How do U.S. emissions stack up against those of the rest of the world? We can better answer this question for SO_2 than for NO_x, as the latter originates in part from millions of small, unregulated, and mobile sources. In contrast, sulfur dioxide production can be estimated with the help of national data on fossil fuel consumption and on the smelting of metals such as copper. To do this, we start with the amount of fossil fuels (together with their sulfur content) produced in a country, then add in imports of fossil fuels and subtract out exports. Metal refining is a bit trickier to estimate, as the amount of sulfur released depends on the technologies used (which were not always known). Nonetheless, using these types of data, a 1999 analysis estimated that in 1990, approximately half of the world's sulfur dioxide emissions came from the combined smokestacks of the U.S., USSR, and China. European countries and Japan followed in turn.

But again, it is important to view the data over time. Not surprisingly, just as SO_2 emissions have dropped over the past decade in the U.S. the same is true in both western Europe and the former Soviet Union. These decreases occurred for different reasons—environmental regulations in Europe in contrast to economic depression in Russia. But such decreases have been more than offset by a massive increase in SO_2 emissions by the rapidly developing countries. For example, in 1970, the United States emitted about 30 million tons of sulfur dioxide and China emitted 10 million tons. In 1990, both countries released about 22 million tons of SO_2. With the start of the year 2000, however, China is emerging as the clear leader in sulfur dioxide emissions. Thus far, developing nations such as China have been unable to afford the pollution-reduction technologies or low sulfur fuels that have been adopted by the more wealthy nations. Nitrogen oxide emissions may pose an even more serious long-range problem. Although the technology exists, they have proven to be more difficult to control and appear to be increasing in most countries.

Table 6.1 presents a global view of SO_2 and NO_x emissions from both natural and human sources. On this worldwide scale, human activities release almost twice as much SO_2 as NO_x. Furthermore, if only fossil fuel combustion is considered, the mass of nitrogen oxides emitted per year is less than 40% of the mass of sulfur dioxide. According to *Vital Signs 94*, between 1980 and 1990, global SO_2 emissions from the burning of fossil fuels increased approximately 10% and NO_x emissions increased by 20%. Although the values in Table 6.1 are somewhat dated, research on global emissions is currently underway by an international network of scientists in over 30 countries, the International Global Atmospheric Project (IGAP). Forthcoming data will not only update the values, but also give better estimates of the uncertainties in the data, which can be quite large.

Table 6.1 also clearly points out that humans are not the only generators of sulfur oxides and nitrogen oxides. Oceans and volcanoes release large quantities of SO_2; soil, plants, and lightning are major sources of NO_x. In a typical year, natural emissions account for about 21% of the sulfur dioxide and 41% of the nitrogen oxides released into the atmosphere. Again, these data must be interpreted with care. Natural emissions are inherently variable and difficult to estimate. A 1997 publication from the IGAP network points out that tons of NO_x emitted by lightning vary widely by region (tending to be higher near the equator) and by month (higher during July in the north, January in the south). The NO_x emissions are subject to the updrafts and downdrafts of storms.

Occasionally, major geological events alter the pattern. The June 1991 eruption of Mount Pinatubo in the Philippines is a case in point. This eruption, the largest in a century, injected between 15 and 30 million tons of sulfur dioxide into the stratosphere. There the SO_2 reacted to form small droplets of sulfuric acid. For more than two years, much of this H_2SO_4 aerosol remained suspended in the atmosphere, reflecting and absorbing sunlight. The temporary drop in average global temperature observed in late 1991 and continued through 1992 has been attributed to the effects of the Mount Pinatubo eruption. Indeed, when the cooling effects of the Mount Pinatubo eruption are

Table 6.1

Estimated Global Emissions of Sulfur and Nitrogen Oxides (in millions of metric tons per year)

Source	SO_2*	NO_x**
Natural:		
Oceans	22	1
Soil and plants	2	43
Volcanoes	19	
Lightning		15
Subtotals	43	59
Anthropogenic:		
Fossil fuel combustion	142	55
Industry (mainly ore smelting)	13	
Biomass burning	5	30
Subtotals	160	85
TOTALS	203	144

Sources: *Data from Spiro et al., "Global Inventory of Sulfur Emissions with 1° × 1° Resolution" *Journal of Geophysical Research* 97, No. D5, 6023, 1992.

**United States Environmental Protection Agency, *Air Quality Criteria for Oxides of Nitrogen*, EPA/600/8–91/049aA.

included in the computer programs used to model global temperature changes, the predictions agree well with the observations, validating the models. There is also evidence that droplets and frozen crystals of H_2SO_4 formed as a result of the eruption provided many new microsites for chemical reactions leading to the destruction of stratospheric ozone. Quite obviously, the topics of this text are tightly interwoven.

Predicting SO_2 and NO_x emissions in the years ahead is perhaps the most difficult task, as estimates rely on a complex interplay of population and economic growth, energy sources, technological innovations, and government regulation. However, within the last few years there has renewed interest in creating emissions scenarios for NO_x and SO_2, as both these gases impact on global warming scenarios. The Intergovernmental Panel on Climate Change (IPCC), a scientific body open to members of World Meteorological Organization (WMO) and the United Nations Environment Programme (UNEP), was established in 1988 to assess the risks of human-induced climate change. In its 1999 *Special Report on Emissions Scenarios*, the IPCC pointed out the connection between sulfur dioxide emissions and climate change. The culprit is tiny aerosol particles of sulfates, formed in the atmosphere from sulfur dioxide, that can have a cooling effect by reflecting incoming sunlight. The IPCC Report states: "sulfates (primarily sulfur dioxides), . . . up until now have offset about half of human-caused global warming." On the other hand, the oxides of nitrogen (NO_x) are of interest because they are precursors to the formation of tropospheric ozone, which is a greenhouse gas with a warming effect.

6.10 The Effects of Acid Precipitation on Materials, Visibility, and Human Health

The evidence is persuasive that much of the rain, mist and snow in the United States is more acidic than normal, unpolluted precipitation. Fog, dew, and the bottom layers of clouds may even have a pH of 3.0 or lower. And there are clear indications that, on a regional basis, the acidity of precipitation has increased significantly since the Industrial Revolution. But does it really matter? To answer that fundamental question, we need to know something about the effects of acid deposition and how serious they really are. Clearly, these issues are central to the acid rain debate (Table 6.2).

In an effort to gain the information necessary to make informed decisions, during the 1980s the U.S. Congress funded a national research effort called the National Acid Precipitation Assessment Program (NAPAP). Over 2000 scientists were involved, with

Table 6.2

Effects of Acid Rain and Recovery Benefits

	Effects of Acid Rain	Recovery Benefits
Human health	Sulfur dioxide and nitrogen oxides in the air increase deaths from lung disorders (asthma and bronchitis) and impair the cardiovascular system.	Decrease visits to emergency room, hospital admissions, and deaths.
Surface waters	Acidic surface waters decrease the survivability of animal life in lakes and streams. In more severe instances, acidity eliminates some or all types of fish and other organisms.	Reduce the acidic levels of surface waters and restore animal life to the more severely damaged lakes and streams.
Forests	Acid deposition contributes to forest degradation by impairing trees' growth and increasing their susceptibility to winter injury, insect infestation, and drought. It also causes leaching and depletion of natural nutrients in forest soil.	Reduce stress on trees, thereby reducing the effects of winter injury, insect infestation, and drought, and reduce leaching of soil nutrients, thereby improving overall forest health.
Materials	Acid deposition contributes to the corrosion and deterioration of buildings, cultural objects, and cars, which decreases their value and increases costs of correcting and repairing damage.	Reduce the damage to buildings, cultural objects, and cars, and reduce the costs of correcting and repairing future damage.
Visibility	In the atmosphere, sulfur dioxide and nitrogen oxides form sulfate and nitrate particles, which impair visibility and affect the enjoyment of national parks and other scenic views.	Extend the distance and increase the clarity at which scenery can be viewed, thus reducing the haze.

Source: Adapted from Emission Trends and Effects in the Eastern U.S., United States General Accounting Office, Report to Congressional Requesters, March 2000.

In 1944

At present

Figure 6.15

Acid rain can damage limestone statuary. This statue of George Washington was first put outside in New York City in 1944. During the next 58 years, acid rain caused significant damage to the statue.

a total expenditure of $500 million. The project was completed in 1990 and the participating scientists prepared a 28-volume set of technical reports (NAPAP, *State of the Science and Technology*, 1991). Some of the material in the remainder of this chapter is drawn from the NAPAP reports plus other more recent documents prepared for Congress. We first will consider possible damaging effects of acid rain on materials, visibility, and human health, several of the items listed in Table 6.2.

- In March 2000, a report on acid rain to "congressional requesters" was prepared by the United States General Accounting Office, a nonpartisan research agency for Congress. Entitled *Emission Trends and Effects in the Eastern U.S.,* it is available on the Web.

Statues and monuments, such as those in the Gettysburg National Battlefield and New York City parks, are victims of irreparable pollution damage (Figure 6.15). Because they are made of limestone (a sedimentary rock composed mainly of calcium carbonate, $CaCO_3$), they slowly dissolve in the presence of H^+ ions.

$$CaCO_3(s) + 2\ H^+(aq) \rightarrow Ca^{2+}(aq) + CO_2(g) + H_2O(l) \qquad [6.21]$$

6.14 Your Turn

Some marbles may contain magnesium carbonate as well as calcium carbonate. Write a chemical equation analogous to equation 6.21 for the action of acid rain on magnesium carbonate.

Visitors to the Lincoln Memorial in Washington learn that huge stalactites growing in chambers beneath the Memorial are the result of acid rain eroding the marble (calcium carbonate or a mixture of calcium and magnesium carbonates). Some limestone tombstones are no longer legible. Other monuments and structures in the eastern United States are suffering similar fates. Worldwide, many priceless and irreplaceable marble and limestone statues and buildings are also being attacked by air-borne acids (Figure 6.16). The Parthenon in Greece, the Taj Mahal in India, and the Mayan ruins at Chichén Itzá show signs of acid erosion. Much of the acid deposition at these sites is due to the NO_x produced by traffic, including tour buses and numerous vehicles without emission control.

6.15 Consider This: Exposure to the Elements

Examine again the opening photo of the chapter as well as the one in Figure 6.16. While it may be tempting to blame all the damage on acid rain, other agents act as well. View the different types of deterioration for yourself by taking a photo tour of our nation's capitol, courtesy of the United States Geological

Figure 6.16

Acid rain knows no geographic or political boundaries. Acid rain has eroded Mayan ruins at Chichén Itzá, Mexico.

Survey. Locate the web site either by searching for "USGS," "acid rain" and "capitol," or use the direct link at the *Chemistry in Context* web site. What kinds of damage do the photos show? What factors promote damage by acid rain? What else has caused the buildings to deteriorate?

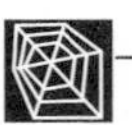

6.16 Consider This: Acid Rain Across the Globe

The issues and concerns of acid rain vary around the globe. Many countries in North America and Europe have web sites dealing with acid rain. Either locate one by searching or use the links provided at the *Chemistry in Context* web site. What are the concerns in the country you selected? Does part of the acid deposition originate outside the country you picked?

- Steel is an alloy containing mainly iron.

Another damaging effect of acidic rain is the corrosion of metals, particularly iron, which is undoubtedly the most important structural metal. Buildings, bridges, railroads, and vehicles of all kinds depend on iron and steel. Unfortunately, iron readily corrodes or rusts by undergoing a reaction with oxygen and water. The reaction requires hydrogen ions, but even pure water has sufficient H^+ concentration to promote slow rusting. In the presence of acid rain, the corrosion is greatly accelerated. The role of H^+ is evident in equation 6.22, which represents the first of a two-step process. Iron (Fe) reacts with oxygen and hydrogen ions to yield Fe^{2+} ions.

$$4\ Fe(s) + 2\ O_2(g) + 8\ H^+(aq) \longrightarrow 4\ Fe^{2+}(aq) + 4\ H_2O(l) \qquad [6.22]$$

Then Fe^{2+} reacts with more oxygen to produce iron oxide, the familiar reddish brown material we call rust.

$$4\ Fe^{2+}(aq) + O_2(g) + 4\ H_2O(l) \longrightarrow 2\ Fe_2O_3(s) + 8\ H^+(aq) \qquad [6.23]$$

The net result, rust formation, is simply the sum of equations 6.22 and 6.23:

$$4\ Fe(s) + 3\ O_2(g) \longrightarrow 2\ Fe_2O_3(s) \qquad [6.24]$$

- Automobile bumpers used to be coated with chromium metal (chrome plated) to protect against rusting. Reinforced plastic bumpers have largely replaced chrome-plated iron ones.

Because iron is inherently unstable when exposed to the natural environment, enormous sums of money are spent annually to protect exposed structural iron and steel in bridges, cars, and ships. Paint is the most common means of protection, although even paint degrades more rapidly when exposed to acidic rain and gases. Another means of protection is to coat the iron with a thin layer of a second metal such as chromium (Cr) or zinc (Zn). Iron coated with zinc is called *galvanized iron.* There is widespread evidence that galvanized iron corrodes more rapidly in the presence of acid rain. As a consequence, galvanized structures must be replaced more frequently than in the past.

6.17 Your Turn

Show that the sum of equations 6.22 and 6.23 is the same as equation 6.24.

Car and truck paint can be affected by acid deposition, whether generated by SO_2 or NO_2, leaving etched spots or pits in the paint. To prevent this, automobile manufacturers have begun to use acid-resistant paints on new vehicles, adding approximately $5 cost per car or truck (about $61 million per year for all new vehicles). It is a bitter irony that trucks and automobiles—the very icon of vehicle-worshipping citizens of Los Angeles and elsewhere—create an air pollutant (NO_x) whose acidic by-products can attack the gleaming finish that many vehicle owners work so hard to maintain.

Perhaps more obvious effects of acid deposition can often be observed by simply looking out of the window. Anyone living in the eastern half of the United States is familiar with the summer haze that usually clouds the landscape. (Ironically, you become more aware of it on the occasional really clear day when it does seem that you can see forever.) Travelers crossing the country by airplane can easily see the haze covering the east. And visitors to the Great Smoky Mountains National Park can view a prominent display of photographs showing reduced visibility in the mountains (Figure 6.17). Power plants in the Ohio Valley are identified as the primary cause. This haze in the eastern part of the country has become steadily worse for several decades. It consists primarily of microscopic aerosol particles containing a mixture of sulfuric acid, ammonium sulfate [$(NH_4)_2SO_4$], and ammonium hydrogen sulfate (NH_4HSO_4). The haze is most pronounced in summer when there is more sunlight to accelerate the photochemical reactions leading to sulfuric acid, and it is particularly evident when the air is stagnant. As a consequence of this haze, average visibility in the east is now 25 miles or less, and occasionally as low as 1 mile.

Volatile *natural* hydrocarbons emitted by trees also contribute to the haziness in the Great Smoky Mountains.

By contrast, visibility in the western states can be over 100 miles, but it is also affected by haze. Where you formerly might have been able to take in a 140 mile scenic view, the EPA estimates that, on the average, your visibility is now reduced 40 to 90 miles. National parks such as the Grand Canyon, Yellowstone, Glacier, Rocky Mountain, and Zion have all been affected. In the last few days of his presidency, Clinton signed a bill authorizing the EPA to issue regulations to help clear the skies in national parks

Visual range 20 miles

Visual range 100 miles

Figure 6.17

A hazy day and a clear day view from the Great Smoky Mountain National Park Look Rock Tower; photo taken with a digital camera.

and wilderness areas. The bill required hundreds of older power plants that emit vast quantities of SO_2, NO_x, and particulates to retrofit their operations with pollution controls. As of 2001, President Bush was moving forward with this plan.

6.18 Consider This: Hazy in Yellowstone?

What is the latest in the battle for clear skies? If you search for keywords such as "haze" and "national parks," you will be rewarded by a variety of news articles and press releases. Determine

a. the current status of EPA's efforts to reduce haze
b. the proposed timetable for the emissions cuts
c. any progress noted by industry or government in clearing the air.

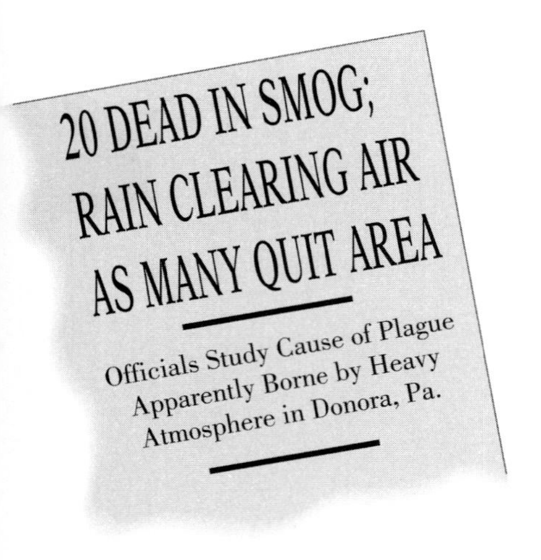
20 DEAD IN SMOG; RAIN CLEARING AIR AS MANY QUIT AREA

Officials Study Cause of Plague Apparently Borne by Heavy Atmosphere in Donora, Pa.

A news headline about the deaths in Donora, PA in 1948.

Inhaling sulfate and sulfuric acid aerosols can cause illness and even death. The acidic droplets are deposited directly in the lungs where they attack sensitive tissue. The elderly, the ill, and those with preexisting respiratory problems such as asthma, bronchitis, and emphysema are especially susceptible. One of the worst recorded instances of pollution-related respiratory illness occurred in London in December 1952. At that time, the English capital was still burning large quantities of sulfur-rich coal, much of it in home fireplaces. The deadly fog lasted five days and claimed approximately 4000 lives. A similar incident occurred in 1948, in Donora, PA, a steel mill town of 12,000 south of Pittsburgh. By noon, the skies had darkened as a choking mixture of fog and smoke became trapped in the region (Figure 6.18). An 81-year-old fireman who took oxygen door-to-door to the victims reported, "It may sound dramatic or exaggerated, but you could barely see." High concentrations of sulfuric acid soon caused widespread illness, 17 people died during the fog, and four later. To be sure, Donora and London were both extreme and unusual situations. But the U.S. EPA and the World Health Organization estimate that 625 million people are exposed to unhealthy levels of SO_2 released by burning fossil fuels.

Studies by the EPA have estimated that the reductions in SO_2 and associated acid aerosols pollution called for by the Clean Air Act Amendments of 1990 could result in saving 12 to 40 billion dollars in health care costs. The savings would come principally from reduced costs to treat pulmonary diseases such as asthma and bronchitis, and the decrease in premature deaths caused by them.

Although acidic fogs are more immediately hazardous to health than acid rain, concern is growing over the indirect effects of acid precipitation. The solubilities of certain toxic heavy metals, including lead, cadmium, and mercury, are significantly increased in the presence of acids. These elements are naturally present in the environment, but normally they are tightly bound in minerals that make up soil and rock. Dissolved in acidified water and conveyed to the public water supply, these metals can pose serious health threats. Elevated concentrations of heavy metals have already been discovered in major reservoirs in western Europe.

Figure 6.18
Donora, PA at noon during the deadly smog.

6.11 Damage to Lakes and Streams

Healthy lakes have a pH of 6.5 or slightly above. As the pH is lowered below 6.0, various species are affected (Figure 6.19). Only a few species survive below pH 5.0; and at pH 4.0, a lake is essentially dead. When acidic precipitation falls on surface waters, it seems reasonable to predict that they will become more acidic. Indeed, acidification of surface waters is another of the effects of acid rain listed in Table 6.2.

Numerous studies have reported the progressive acidification of lakes and rivers in certain geographic regions, along with reductions in fish populations. In southern Norway and Sweden, where the problem was first observed, one-fifth of the lakes no longer contain any fish, and half of the rivers have no brown trout. In southeastern Ontario, the average pH of lakes is now 5.0, well below the pH of 6.5 required for a healthy lake. In New England and the Adirondack Mountain region of northern New York, a puzzle has developed. Some of the lakes remain as acidic as ever, even though the rain and snow falling has become less acidic. Because of the significant reduction

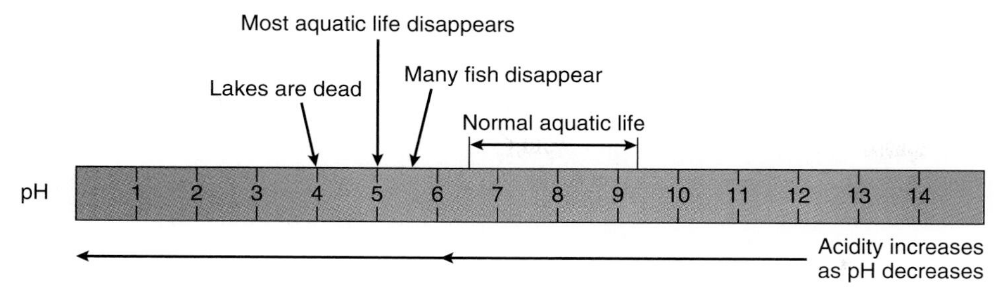

Figure 6.19
Aquatic life and pH.

of SO_2 emissions from power plants along the Ohio River valley since 1990, the acidity of precipitation measured along that area and the eastern United States, particularly in the Middle Atlantic region, has decreased over the past decade.

Going westward, we find that many areas of the Midwest have no problem with acidification of lakes or streams, even though the Midwest is supposed to be the major source of the acids in acid precipitation. This apparent paradox can be explained quite simply. When acidic precipitation falls on or runs off into a lake, the pH of the lake will drop (become more acidic) unless the acidic precipitation is neutralized or somehow utilized by the surrounding vegetation. In some regions, the surrounding soils contain bases that can neutralize the acid. The capacity of a lake to resist a decrease in pH is called its **acid neutralizing capacity (ANC).** The surface geology of much of the Midwest is mostly limestone ($CaCO_3$), which has a high acid neutralizing capacity, thus neutralizing the acid rain (equation 6.25).

$$CaCO_3(s) + 2\,H^+(aq) \longrightarrow Ca^{2+}(aq) + CO_2(g) + H_2O(l) \qquad [6.25]$$

But even more importantly, the lakes and streams have a relatively high concentration of calcium bicarbonate as a result of reaction of the limestone with carbon dioxide and water.

$$CaCO_3(s) + CO_2(g) + H_2O(l) \longrightarrow \underset{\text{calcium ion}}{Ca^{2+}(aq)} + \underset{\text{bicarbonate ion}}{2\,HCO_3^-(aq)} \qquad [6.26]$$

- Calcium carbonate (limestone) is not soluble in water; calcium bicarbonate is soluble.

The bicarbonate ion accepts an H^+ ion, thus neutralizing acids:

- Bicarbonate ion is also called hydrogen carbonate ion.

$$HCO_3^-(aq) + H^+(aq) \longrightarrow CO_2(g) + H_2O(l) \qquad [6.27]$$

- Reaction 6.27 occurs when sodium bicarbonate (baking soda) is taken to neutralize excess stomach acidity. The CO_2 released produces a burp!

Because the added acid is consumed by this reaction, the pH of the lake will remain more or less constant.

In contrast to the Midwest, many lakes in New England and northern New York (as well as in Norway and Sweden) are surrounded by granite, a hard, impervious, and much less reactive rock. Unless other local processes are at work, these lakes have very little acid neutralizing capacity. Consequently, they will most likely show a gradual acidification.

Experimental evidence indicates that fish populations most likely are affected through a chain of events that starts with acid rain and ends with the biological uptake of aluminum ions (Al^{3+}). Aluminum is the third most abundant element in the Earth's crust (after oxygen and silicon). Granite contains aluminum ions, and soil includes complex aluminum-silicate structures. Natural aluminum compounds have a very low solubility in water, but in the presence of acids, their solubilities increase dramatically. Thus, when the pH of a lake drops from, say, 6.0 to 5.0, the aluminum ion concentration in the lake may increase 1000-fold. When fish are exposed to a high concentration of aluminum ions, a thick mucus forms on their gills and the fish suffocate. Additionally, aluminum ions (Al^{3+}) react with water molecules to generate H^+ ions, increasing the acidity, which in turn dissolves more aluminum ions to further exacerbate the problem.

- A shift from pH 6.0 to 5.0 is a 10-fold *increase* in acidity.

$$Al^{3+}(aq) + H_2O(l) \longrightarrow H^+(aq) + [Al(OH)]^{2+}(aq) \qquad [6.28]$$

As it turns out, understanding the acidification of lakes is a good deal more complicated than simply measuring pH and acid neutralizing capacities. One level of complexity is added by seasonal and yearly variations. For example, in some years the heavy winter snowfalls persist into the spring and melt suddenly. As a result, the runoff may be far more acidic than usual, as it contains all the acidic deposits locked away in the winter snows. A surge of acidity may enter the waterways at just the time when fish are spawning or hatching from eggs and are more vulnerable. Thus, in some places, the damage done by acid rain is episodic; in others, it is chronic. Unfortunately, the episodic surges in acidity increase the number of lakes and streams that are susceptible to damage. In the Adirondacks, about 70% of the sensitive lakes are at risk for episodic acidification, in comparison to a far smaller percent that are chronically affected (19%). In the Appalachians, the episodically affected lakes (~30%) are seven times greater in number than the 4% chronically affected (Table 6.3).

The *chemistry* of the acidic species adds another level of complexity. As you might expect, the oxides of sulfur and nitrogen behave differently in the environment. For example, unlike sulfur dioxide, nitrogen oxides (in the form of the nitrates they deposit in surface waters) may be utilized as nutrients by the vegetation surrounding a watershed. The patterns of absorption depend on both the age of the vegetation (in general, younger, growing forests absorb more than older ones) and the time of year (plant growth does not take place in the winter). But this absorption seems to have its limits. Evidence suggests that, over time, regions may become overloaded, resulting in a situation called "nitrogen saturation."

When, if ever, will the lakes recover? An understanding of both the chemical behavior of pollutants and their seasonal variations can help us make better predictions. The good news is that the SO_2 emissions have been declining in recent years, and we have seen a corresponding decrease in the sulfate ion concentrations in the lakes of the Adirondacks. However, even though NO_x emissions have remained fairly constant, the amount of nitrates in the Adirondacks is increasing in more lakes than not. Thus, it appears that "nitrogen saturation" has occurred in the surrounding vegetation, with more of the acidity ending up in the lakes. It also is likely that the soil in the region of these lakes has lost some of its acid neutralizing capacity (ANC).

Recent predictions are fairly gloomy. A March 2000 report to Congress puts it bluntly, "The lakes in the Adirondack Mountains are taking longer to recover than lakes located elsewhere and are likely to recover less or not recover, without further reductions of acid deposition." An earlier EPA report factors in future nitrogen saturation to predict the health of these lakes. Consider Table 6.3, which offers two possible scenarios, one with nitrogen saturation occurring within 50 years and another within 100. Even with the implementation of the 1990 Clean Air Act Amendments, more lakes are likely to become acidic, both in the Adirondacks and elsewhere.

Table 6.3

Percentage of Waters in Three Environmentally Sensitive Areas Projected to Be Acidic in 2040 (with implementation of the 1990 amendments)

Watershed	% Acidic in 1984	% Acidic in 2040 if 50 Years Before Nitrogen Saturation Develops	% Acidic in 2040 if 100 Years Before Nitrogen Saturation Develops
Adirondack lakes	19	43	26
Mid-Appalachian streams	4	9	5
Southern Blue Ridge streams	0	4	0

Source: EPA, Acid Deposition Standard Feasibility Study: Report to Congress, Oct. 1995.

Nonetheless, there still is a bit of hopeful news. Evidence suggests that, with the proper treatment, acidified lakes can be reclaimed. Recent experiments with adding calcium hydroxide, $Ca(OH)_2$, to acidified lakes are encouraging. Calcium hydroxide (lime) is a base, and it reacts to neutralize acid. If large-scale liming were carried out and further input of acids halted, many lakes possibly could be restored to health.

Liming lakes in southwestern Sweden.

Similar neutralization reactions also take place in the atmosphere. Calcium hydroxide, calcium carbonate, and other basic compounds of calcium, magnesium, sodium, and potassium are common in the soil. They are also abundant in the fly ash released from industrial smokestacks. Particles of these alkalis are swept into the air where they can react with droplets of acid. Thus, efforts to reduce air pollution by reducing particulate emission and controlling dust can lead to a decrease in the acid neutralizing capacity of the atmosphere and, ironically, a drop in the pH of precipitation. In fact, this may explain why the recent decrease in SO_2 emissions in this country has not resulted in a decrease in the acidity (increase in the pH) of rain.

6.19 Your Turn

a. Write an equation for the reaction of slaked lime, $Ca(OH)_2$, with hydrochloric acid, HCl.

b. Write an equation for the reaction that takes place when "quick lime," CaO, is added to water to produce $Ca(OH)_2$ ("slaked lime").

6.12 The Mystery of the Unhealthy Forests

Of all the ravages attributed to acid precipitation, the most widely discussed has been damage to trees and forests. Pictures of dead and dying trees provoke strong emotional responses and the ire of many environmentalists. But the fact is that the effects of acid deposition on trees are less well understood than any other aspect of the acid rain story. Consequently, the topic is fraught with controversy.

At least in certain cases, the reality of forest decline seems indisputable. The phenomenon was first observed in the 1960s in what was then the German Federal Republic (West Germany), and extensive studies by German scientists have chronicled a steady decline. Fir and spruce trees at high elevations have been especially hard hit, first showing limp branches, then yellowing needles, and then loss of needles. This gradually spreads until the weakened trees finally are killed by drought, cold, insects, winds, or some combination of these. The decline was especially rapid from 1982 to 1985, when the number of damaged trees (defined as 10% needle loss) throughout West Germany increased from 8% to 52%.

Tree loss appeared to spread rapidly across western Europe during the 1980s, with damaged forests reported in Italy, France, the Netherlands, Sweden, Norway, and Britain. According to a 1993 study by the United Nations Economic Commission for Europe, almost one in four European trees has lost more than 25% of its leaves or needles. It is uncertain, however, when the reported damage actually began and how much is simply the result of better observing and reporting. For eastern Europe, where air pollution controls have been virtually nonexistent, information has become available only recently. Some reports claim that over 80% of the forests in eastern Germany, Poland, and the Czech Republic are damaged. In some areas of these countries, forests are said to have been totally destroyed by air pollution.

In North America, forest damage has been most dramatic in portions of the Appalachian Mountains. On Mount Mitchell and in the Great Smoky Mountains in North Carolina and Tennessee, you can now see the stark landscape of dead trees (Figure 6.20). In the Green Mountains of Vermont, half the red spruce trees on Camel's Hump died between 1965 and 1981. The sugar maple trees in New England and southeastern Canada, famed for autumn color and maple syrup, are claimed to be unhealthy, with some farmers reporting alarming decreases in maple syrup production.

Figure 6.20
Acid-rain damaged trees in the Great Smoky Mountains.

It is tempting to blame acidic precipitation for all these effects. After all, the acids rain down directly on the trees, and the effects are generally most pronounced in regions with highly acidic precipitation. But the story is not that simple. It is difficult to prove cause-and-effect relationships, especially when there are many effects and many possible causes. In some cases (such as the New England and Canadian maple trees), weather-related stresses are considered responsible for most of the damage. In other regions, the cause may be insect infestations (as in the loss of the fir trees in the southern Appalachians) or tropospheric ozone and other air pollutants (as in the damage to pines in the San Bernardino Mountains). The dead trees on Mount Mitchell and other parts of the southern Appalachians are primarily Fraser firs killed by an infestation of the balsam woolly adelgid, a tiny, hard-to-see pest introduced from Europe around 1900.

6.20 Your Turn

Ozone damages plants. Review the difference between tropospheric and stratospheric ozone. In this chapter, are we talking about tropospheric or stratospheric gases that contribute to acid deposition?

In mountainous regions losses are greatest at certain elevations, and only a few species of trees have shown complete die-back. Furthermore, careful field studies have shown that many of the effects were not as serious as had been believed. For example, although the famed Black Forest in Germany experienced a 10% needle loss, trees did not suffer the severe die-back widely reported in the press. The North American Sugar Maple Decline Project was initiated by the United States and Canada in 1987 to investigate alarming reports of unhealthy sugar maple trees. The project (now called the North American Maple Project) monitors tree health in 10 states and 4 provinces. Results to date show that over 90% of sugar maple trees in the study area are healthy, that the percentage of unhealthy trees has not changed significantly in the last decade, and that most of the damage can be traced to insects, disease, and drought.

Nevertheless, there is strong circumstantial evidence that acidic precipitation is at least a contributing factor to the declining health of trees. A careful study of tree growth rings in southern New Jersey pines showed a dramatic reduction in growth since 1965. This change, the greatest growth reduction in the 125-year record, correlates well with the acidity of nearby surface waters. Approximately 90% of the streams in the New

Figure 6.21
New trees growing in a formerly acid rain–affected area.

Jersey Pine Barrens area are acidic, the highest percentage in the country. This may be an example of the synergistic effects of acids, ozone, SO_2, and NO_x.

- If the total effect is greater than the sum of the individual effects, this is termed a synergistic effect or a synergism.

According to this theory, which is gaining acceptance, the ozone and nitrogen oxides attack the waxy coating on leaves, permitting hydrogen ions to deplete nutrients. Acidification of soil beneath the trees mobilizes metals (especially aluminum) that attack the tree roots, preventing absorption of nutrients and water. Simultaneously, potassium, calcium, magnesium, and other minerals essential for plant growth are leached out of the soil. These effects leave the trees susceptible to destruction by natural factors such as disease, insects, drought, or high winds. A further factor in high-elevation forests is that the mountains are often shrouded in fog, so-called cap clouds. As reported earlier in the chapter, cloud water often has a lower pH than even the most acidic rain, and thus the trees are perpetually exposed to an acidic mist. There is clear experimental evidence that acid deposition has contributed to the decline of red spruce at high elevations in the northern Appalachians by reducing the species' cold tolerance.

Whatever the complex causes and the extent of involvement of acid rain, the damage to forests is unfortunate and expensive. However, it is important to realize that even in the case of severe decline or total die-back, most of the damaged regions have shown surprisingly good growth of new trees (Figure 6.21). The new growth appears to be thick and healthy. Nature does indeed regenerate itself.

6.13 Costs and Control Strategies

The acid rain debate has turned increasingly to economic considerations. Policy makers want to know the costs of the damage already incurred, the costs of cleanup, and the projected costs of future abatement. They also want to know who might benefit and, of course, who will pay. Unfortunately, it is extremely difficult to assess accurately the costs of the damage now occurring as a result of acid precipitation. Not only are the cause-and-effect relationships unclear, but many assumptions must be made.

One such attempt was published in 1985 by Thomas Crocker, an economist, and James Regens, a natural resources scientist. They acknowledged that the only *conclusive* evidence for damage was to aquatic ecosystems. In spite of this caveat, Crocker and

Regens estimated that, for the eastern third of the United States, the maximum annual economic losses attributable to acid deposition exceeded $5 billion. Others have concluded that all forms of air pollution in aggregate cost the United States over $40 billion each year in health care and lost productivity. And another study estimated $30.4 billion were lost annually because of sulfur dioxide damage to European forests. These huge numbers take on a more personal perspective when you consider another study, which reported that acid deposition does nearly $300 million worth of damage to buildings in Chicago each year, a cost that corresponded to $45 per resident.

While efforts continue to gauge more accurately the current economic burden of acid rain, other studies are underway to evaluate strategies to control acid emissions and to estimate the costs of these measures. Most attention is being focused on ways to limit and eventually reduce the release of sulfur dioxide from coal-fired furnaces and nitrogen oxides from gasoline-fueled vehicles. The Acid Rain Program, established as part of the Clean Air Act Amendments of 1990, made reducing NO_x and SO_2 emissions a national priority.

For NO_x, the Acid Rain Program set a target of reducing the annual emissions by 2 million tons by 2000. Phase I of the NO_x program applied to about 170 coal-fired boilers that produce electricity, specifying an emission rate of either 0.50 or 0.45 pounds of NO_x per million Btu of heat input, depending on the type of boiler. Flexibility was built in, so that emission rates could be averaged over several units. Phase II began in 2000, tightening these emission standards and applying standards to still other types of boilers.

• Btu stands for British thermal unit.

In spite of these efforts, on a national scale the goal for NO_x emissions has not yet been achieved. According to the data released in 2000, nitrogen oxide emissions from all sources have *increased* slightly since 1990. While emissions by electric utilities (generating about a quarter of the NO_x) declined, NO_x emissions increased elsewhere, such as by the increasing number of trucks and automobiles on our highways. Reduction of nitrogen oxides from these vehicles is particularly challenging, because as sources they are small, individually owned, and by design, mobile. And there are more than 200 million motor vehicles in the United States (about 1 billion worldwide). Of these, the biggest contributors to NO_x pollution continue to be diesel engines, and these remain problematic for at least two reasons: Diesel vehicles currently are subject to fewer emission restrictions and the engines are built to last up to a million miles and are not replaced as quickly by ones with cleaner technology.

To reduce NO_x, a variety of techniques bearing a range of price tags are in use. It is chemically possible to reduce the NO emitted by cars and trucks by fitting them with catalytic converters and other emissions control devices. We already mentioned one of the functions of these catalysts: converting CO and unburned hydrocarbon fragments to CO_2. Other catalysts, typically in other parts of the catalytic converter, promote the reversal of the combination of nitrogen and oxygen that occurs in the engine at high temperatures. As the exhaust gases cool, there is a tendency for the NO to decompose into its constituent elements.

• See Figure 1.14 for a photo of an automobile catalytic converter.

$$2\ NO(g) \longrightarrow N_2(g) + O_2(g) \qquad [6.29]$$

Normally, this reaction proceeds slowly, but the appropriate catalyst can significantly increase its rate and thus decrease the amount of NO emitted. An alternative, of course, is to eliminate the internal combustion engine altogether. One possibility would be to convert to electric-powered vehicles, so-called zero emission vehicles.

• Electric-powered vehicles are discussed in Chapter 8.

6.21 Consider This: Electricity or Cars?

Your metropolitan area has been warned by the EPA to lower sulfur dioxide and nitrogen oxide emissions or it will lose substantial federal funding. One citizen's group calls for conservation as a means of drastically reducing electricity consumption. A second group advocates curtailing the use automobiles through a variety of means. Choose either of these or some other position to support. Research the position and present your findings to others. As a group, the goal is to come to a consensus of what actions should be taken.

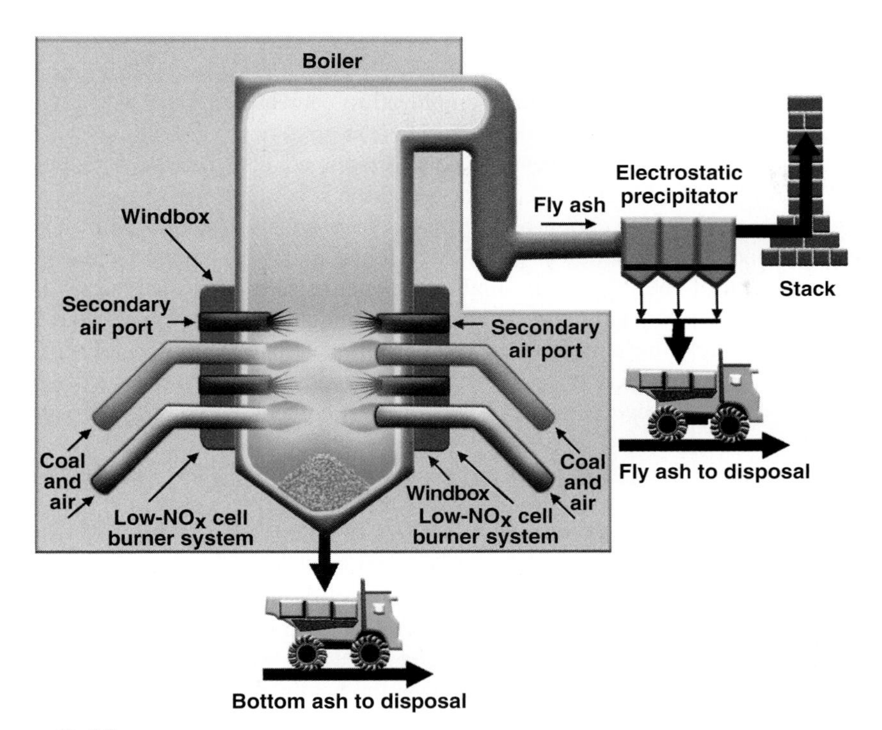

Figure 6.22

A low-NO_x burner being installed at an electrical utility plant.

(Source: US Dept. of Energy. Taken from Clean Coal Technology, a Report by the Assistant Secretary for Fossil Energy, November 1999. http://www.fe.doe.gov, p. 6.)

Coal-fired plants, another major source of NO_x emissions, demonstrate other new technologies. For example, the Clean Coal Technology (CCT) Demonstration Program has developed and installed low-NO_x burners on numerous coal-fired plants (Figure 6.22). These burners control the mixing of fuel and air during the combustion process to minimize NO_x production. Another CCT option is "reburning," where the combustion products are injected with additional fuel to strip away the oxygens from the NO_x. Both these new technologies are sufficiently complex that an artificial intelligence system may be used to optimize their performance. According to a 1999 report by the Department of Energy, the CCT Demonstration Program, has "built the foundation for meeting NO_x emissions well into the 21st century."

6.22 Consider This: Clean Coal Update

The CCT Program is funded both by government and industry, and seeks technologies that meet the needs of our environment. What is new in coal-cleaning technology? Look up one of the success stories on the Web and report the details. You might want to start with The Clean Coal Technology Compendium web site or use the links provided at the *Chemistry in Context* web site.

The Acid Rain Program also called for a 10-million-ton reduction of SO_2 emissions by the year 2000. Phase I, begun in 1995, required 263 mostly coal-burning boiler units at 110 electric utility power plants (located in 21 different states) to reduce their emissions. Phase II began in 2000 and further tightened the emissions on these plants and set further restrictions on power plants fired by natural gas and oil as well to encompass over 2000 boiler units. To date, the SO_2 emissions program has met with success. The fact that most anthropogenic SO_2 comes from a limited number of point sources (coal-burning power plants and factories) made the SO_2 problem easier to attack. As we already saw from Figure 6.14, great strides have occurred in reducing U.S. SO_2 emissions. According to the EPA, SO_2 from all sources in 1990 was 23.7 million tons and declined to 19.6 tons in 1998. However, there still were small increases in the years from 1996 to 1998.

Three major strategies have been employed to decrease SO_2 emissions: (1) switch to "clean coal" with lower sulfur content, (2) clean up the coal to remove the sulfur before use, and (3) use chemical means to neutralize the acidic sulfur dioxide in the power plant. We briefly consider the effectiveness and the cost of each of these.

Coal switching is an option because coals vary widely in their sulfur content and their heat content. Anthracite or "hard" coal, found mainly in Pennsylvania, yields the greatest amount of energy and the smallest amount of sulfur. But the anthracite supply is practically exhausted and more expensive. Bituminous or "soft" coal, abundant in the Midwest, has nearly the same heat content as anthracite but usually contains 3–5% sulfur. Western states have enormous deposits of low-sulfur sub-bituminous coal and lignite (brown coal); however, this coal has a low heat content and may contain up to 40% water.

Coal cleaning is relatively easy to do, and the technology is available. The coal is crushed to a fine powder and washed with water so that the heavier sulfur-containing minerals sink to the bottom. But the process removes only about half of the sulfur, and it is expensive—$500–1000 per ton of SO_2 eliminated.

An alternative to using coal switching or coal cleaning is to chemically remove the SO_2 during or after combustion in the power plant. The chief method for doing this is called *scrubbing.* The stack gases are passed through a wet slurry of powdered limestone, $CaCO_3$. Calcium carbonate neutralizes the acidic SO_2 to form calcium sulfate, $CaSO_4$.

- The carbon dioxide produced by equation 6.30 adds to the concentration of atmospheric CO_2.

$$2\ SO_2(g) + O_2(g) + 2\ CaCO_3(s) \longrightarrow 2\ CaSO_4(s) + 2\ CO_2(g) \qquad [6.30]$$

Limestone is cheap and readily available. Although the process is highly efficient, installing scrubbers is expensive, so that the cost of this method has been estimated at $400–600 per ton of SO_2 removed. Part of the expense is associated with the disposal of the $CaSO_4$ formed. We simply cannot avoid the law of conservation of matter. The sulfur must end up somewhere; either it goes up the stack as SO_2 or gets trapped as $CaSO_4$.

6.23 Consider This: Emissions Close to Home

Thanks to the EPA, you now can find the acid rain emissions data for the power plants in your state. Visit the EPA's web site for Clean Air Market Programs or use the link provided at the *Chemistry in Context* web site. Select a plant of your choice and report:

a. the name of the plant and the type(s) of fuel it burns
b. whether or not emissions controls are installed
c. the tons of SO_2 and NO_x emitted
d. the trend in emissions, by looking at previous years

The principal reason why compliance with the 1990 Clean Air Act Amendments regulations was achieved and even bettered was coal switching, in which high-sulfur coal was replaced by low-sulfur coal. By the early 1990s, the use of a new rail carrier and favorable railway tariffs made vast deposits of cheaper low-sulfur coal (even less than 1% S) in Montana and Wyoming available at costs lower than that for midwestern or eastern low-sulfur coal. In 1991, western low-sulfur coal averaged just $1.30 per million Btus; eastern low-sulfur coal was $1.60 to $1.70 per million Btus. High-sulfur eastern coal cost $1.35–1.55 per million Btus. Given this price advantage, it is not surprising that nearly 60% of SO_2 reduction came from switching to low-sulfur western coal rather than using more expensive alternatives, such as scrubbing.

But there are hidden costs in this conversion to low-sulfur coal. It ignores the social and economic impact on the states that produce high-sulfur coal. It has been estimated that since 1990, coal switching has caused a 30% decline in employment in the high-sulfur coal–mining regions of Pennsylvania, Kentucky, Illinois, Indiana, and Ohio, although half of the drop has been because of automation and other market factors.

Western states now produce nearly 33% of the coal mined in the United States, up from only 6% in 1970.

The shift to low-sulfur western coal has another side to it. Because the coal produces less heat per gram than eastern coals, power plants must burn more of it to generate the same amount of electricity, thus producing more CO_2 per unit of electricity generated. The increased carbon dioxide adds to the atmospheric burden and potential global warming. Mercury and other trace metals are more prevalent in western coal than other coals, thereby increasing the atmospheric concentration of these metals unless steps are taken to remove them before they go up the smokestacks (a costly proposition).

As the legislation that became the 1990 Clean Air Act Amendments moved through Congress, estimates were made about the costs that would be necessary for electrical utilities and other affected companies to comply with the proposed Act. The estimates varied widely (and wildly). The U.S. Congressional Office of Technology Assessment proposed a cost of between $3 and $4 billion per year. Estimates of $4–23 billion annually were made by the Electric Power Research Institute, an arm of the electrical utilities companies. In practice, the cost of acid rain controls has been much lower than estimated. Phase I costs to achieve compliance were actually less than 30% of the lowest estimated cost. In 1995, the first year for compliance, the annual cost for SO_2 reduction was $836 million, which works out to about $3 for each person in the United States.

Well in advance of the United States, Japan decided that SO_2 reduction was worth the cost. In 1968, the Japanese government issued strict SO_2 controls, and encouraged the use of low-sulfur fuels and desulfurization techniques. As a consequence, Japan's power plants cut their output of sulfur dioxide from almost 7 g/kWh in 1970 to less than 1 g/kWh in 1980. Quite obviously, the procedures work. Such dramatic decreases are related to the widespread use of scrubbers on Japanese power plants. In 1982, nearly 1200 scrubbers were in place, compared to about 200 in the United States. To be sure, Japan and the United States are different in many respects, including the fact that the U.S. electrical utilities chose coal switching rather than installing more expensive scrubbers. But the Japanese and United States experiences clearly indicate that the problems of reducing SO_2 emissions are political as well as technological.

6.14 The Politics of Acid Rain

It is not surprising that legislation to control acid rain has been the subject of intense political maneuvering. The neutralization of acid rain requires more than chemistry. For example, a few years ago the Indiana state legislature passed a bill requiring that a certain percentage of the coal burned in the state also had to be mined in Indiana. But the law was struck down after Wyoming coal producers filed suit against Indiana.

Nor is it surprising that the electrical power industry and the producers of high-sulfur coal, both strong and effective lobbies, found tactics to resist acid rain legislation and controls. For example, a clause in the Clean Air Act of 1970 exempted older dirty plants under a grandfather clause, presumably with the belief that these plants would close in the near future. But if any major improvements were made to these plants, they then would be subject to the same stricter rules as newer plants. To avoid the stricter standards, apparently the power industry made major improvements to the older plants, but billed the changes as "routine maintenance." To some (especially those on the receiving end of the pollutants), this appeared to be an abuse of the grandfather clause. But a clause in the Clean Air Act allows citizens to sue for violations. The attorney general of New York State (with help from the EPA) did just that, suing power plants in other states with the contention that much of the acidic pollution in New York was blowing in from places where controls were more lax. In November 2000, a large, polluting Virginia power company lost to New York and agreed to cut emissions from eight coal-burning plants by 70%. Other successful suits have followed.

As you read earlier in the chapter, the U.S. Congress entered into the controversy in 1990 by passing major new environmental legislation. The Clean Air Act Amendments were signed into law by President George H. W. Bush in November of that year. Although electrical utility companies initially opposed the new legislation, they soon realized the need to comply and did so mainly through coal switching. A unique feature of this 1990 legislation was setting up a system of "emissions trading," under which each company operates with a *permit* that specifies the maximum level of emissions that a company can legally release. This maximum was set based on factors such as prior fuel use and emission rates. Exceeding this maximum carries fines of up to $25,000 per day. Permits thus provide an environmental safety shield. In addition, companies are also assigned emission allowances that are set below the permit levels. Each *allowance* authorizes a unit to emit one ton of SO_2, either during the current year or any year thereafter.

The goal is for each company to achieve its individual allowance level. Some in fact perform better. A power plant with emissions below its allowance is assigned pollution credits, one credit per ton of SO_2. These credits can be sold to other power plants that have been unable to meet their emission allowances. There is thus a financial incentive for power producers to achieve significant reductions of acidic oxide emissions. On the other hand, the purchase of credits by those who cannot yet meet the more stringent standards allows them to continue operation, at or below the permit level, while the plant works to reduce emissions.

The first official trade of emissions credits under the provisions of the new law occurred in 1993. Since then, credits have been bought and sold in private transactions and at public auctions. There is even a commodity trading market in emission allowances on the Chicago Board of Trade. Prices have ranged between $400 (1993) to $68 (1996) per credit, nowhere near the $1000 per credit predicted by utility officials. Prices in 2000 were in the range of $150, which suggests that credits are not in as high demand as originally foreseen. This in turn indicates that progress has been made in overall reduction of SO_2 emissions largely because of coal switching. The patterns of allowance use have been complex. Some states cover essentially all their emissions with allowances originating within their state: Wisconsin, Missouri, Kansas, Michigan, Minnesota, and Utah. Other states use out-of-state allowances to cover up to a quarter of their emissions, with Illinois, Kentucky, and West Virginia being notable in this regard. Interestingly enough, a few allowances each year continue to be purchased by students, including those in programs of environmental law. These credits are in essence "retired" and are never used.

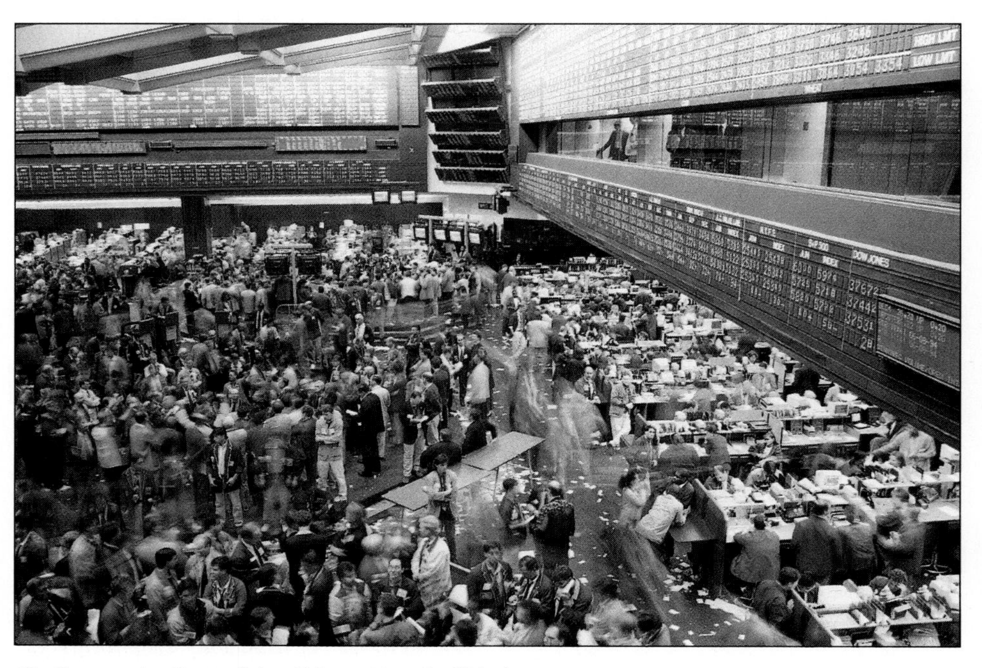

Trading on the floor of the Chicago Board of Trade.

An emissions trading program currently does not exist for NO_x. However, since 1999, a NO_x Budget Program has been in operation in the northeast. Similar to the SO_2 emissions trading program, an allowance can be bought, sold, or banked to emit one ton of NO_x. This program, however, is quite distinct from the Acid Rain reduction requirements and is aimed at alleviating tropospheric ozone, a secondary pollutant formed from NO_x in the presence of sunlight. Nonetheless, the forecast for acid rain may hold some sort of cap and trade program for NO_x, similar to the SO_2 program now in existence.

- NO_2 breaks down in the presence of sunlight to produce oxygen atoms that in turn react with oxygen molecules to produce ozone. This production of O_3 from O_2 and O is similar to that which you saw in Chapter 2, except that it occurs in the troposphere.

As these programs demonstrate, industry has been offered both flexibility and economic incentives to reduce its emissions. In turn, the public has been rewarded both with cleaner air and home electrical utility bills that have not risen significantly because of the Clean Air Act Amendments, even in eastern and midwestern states. Through a marriage of technology with economic forces, the average price charged by many large utilities for electricity has remained reasonably constant over the past decade.

6.24 Consider This: Up for Auction

The year 2002 marks the 10th annual auction for sulfur dioxide allowances conducted for the EPA by the Chicago Board of Trade (CBOT). How have allowances sales been going? You can learn more about emissions credits at the web sites of both the EPA and the CBOT. For example, you can find recent information about allowance auctions and price trends at the EPA's site entitled, "Acid Rain Program." Do some detective work on the Web and see if you can find out:

a. are the allowances more costly or less costly this year than last?
b. how many allowances were auctioned last year?
c. are most companies still achieving compliance without having to buy credits?

A more difficult Web research question that you might want to consider is which emissions credits, if any, are now auctioned for pollutants other than SO_2. Some starting points for your search are provided at the *Chemistry in Context* web site.

6.25 Consider This: Voting Record of Senators

Without consulting voting records or party affiliations, predict how senators from Kentucky, Wyoming, Texas, and Michigan might have voted on the Clean Air Act Amendments of 1990, and explain your reasoning. Then consult the *Congressional Record* to see how they actually did vote. If there is a difference between your expectation and the actual votes cast by these senators, offer some possible reasons for the difference.

CONCLUSION

If this chapter has taught anything, we hope it has been skepticism, prudence, and the recognition that complex problems cannot be solved by simple or simplistic strategies. "Acid rain" is not the dire plague once described by environmentalists and journalists. Nor is it a matter to be ignored. It is sufficiently serious that federal legislation, the Clean Air Act Amendments of 1990, has been enacted to reduce SO_2 and NO_x emissions, precursors to acid deposition. The Act already has helped to clean the air.

Current research indicates that acid deposition is a complex and tenacious problem, especially for certain watersheds and ecological niches. Any failure to acknowledge the intertwined relationships involving the combustion of coal and gasoline, the production of sulfur and nitrogen oxides, and the reduced pH of fog and precipitation is to deny some fundamental facts of chemistry. A knowledge of ecology and biological systems is needed as well, so that acid deposition can be understood in the context of entire ecosystems, a task that requires that experts from several disciplines work together.

One response that we as individuals and as a society might make to the problems of acid precipitation has hardly been mentioned in this chapter, yet it is potentially one of the most powerful. It is to conserve energy. Sulfur dioxide and nitrogen oxides are by-products of our voracious demand for energy—energy for electricity and energy for transportation. And, of course, carbon dioxide is an even more plentiful product. If our personal, national, and global appetite for fossil fuels continues to grow unchecked, our environment may well become a good deal warmer and a good deal more acidic, especially if developing countries fail to put into place environmental policies that restrict emissions from fossil fuel combustion. Moreover, the problem may be intensified as petroleum and low-sulfur coals are consumed and we become even more reliant on high-sulfur coal.

There are other sources of energy—nuclear fission, water and wind, renewable biomass, and the Sun itself. All of them are already being used, and their use will no doubt increase. We explore nuclear fission in the next chapter. But we conclude this chapter with the modest suggestion that, for a multitude of reasons, the conservation of energy by industry and collectively by individuals could have profoundly beneficial effects on our environment.

Chapter Summary

Having studied this chapter, you should be able to:

- Define and apply the definitions of acid, base, and neutralization (6.1–6.3)
- Use chemical equations to represent the dissociation of acids and bases (6.1–6.2)
- Describe solutions as acidic, basic, or neutral based on their pH or concentrations of H^+ and OH^- (6.3–6.4)
- Interpret pH values as being acidic, basic, or neutral (6.4)
- Describe acid rain (acid deposition) and factors causing it (6.5–6.8)
- Express the roles played by sulfur oxides and nitrogen oxides in causing acid rain, and describe regional variations (6.7, 6.8)
- Discuss the contributions of man-made emissions of pollutants to the atmosphere linked to acid rain and compare these with natural emissions (6.7–6.9)
- Summarize the uncertainties associated with implicating acid rain as the cause for certain environmental degradation; that is, destruction of forests, and death of lakes (6.10–6.12)
- Express the effects and economic impact of acid rain on materials and the environment in general (6.13)
- Discuss the nature of the 1990 Clean Air Act amendments, and the impact they have had on SO_2 emissions (6.13–6.14)
- Outline the various alternatives proposed to control acid rain, noting the cost-benefit considerations to be made for each (6.13–6.14)
- Explain why coal switching was the method of choice used by electrical power companies (6.13)
- Explain why acid rain control is such a politically sensitive issue (6.14)

Questions

Emphasizing Essentials

1. a. What properties are associated with acids?
   b. What structural feature characterizes an acid?
2. a. Rewrite equation 6.1 using Lewis structures.
   b. Rewrite equation 6.1 using sphere representations.
3. Write a chemical equation showing the release of one hydrogen ion from a molecule of each of these acids.
   a. HBr (hydrobromic acid)
   b. $H_3C_6H_5O_7$ (citric acid)
   c. $HC_2H_3O_2$ (acetic acid)
4. a. What properties are associated with bases?
   b. What structural feature characterizes a base?
5. a. Rewrite equation 6.4 using Lewis structures.
   b. Rewrite equation 6.4 using sphere representations.
6. Write a chemical equation showing the release of ions as each solid base dissolves in water.
   a. RbOH
   b. $Ba(OH)_2$
7. Write a balanced chemical equation for each neutralization reaction.
   a. potassium hydroxide, KOH, with nitric acid, HNO_3
   b. barium hydroxide, $Ba(OH)_2$, with hydrochloric acid, HCl
8. Classify each of these aqueous solutions as acidic, neutral, or basic.
   a. HI(*aq*)
   b. NaCl(*aq*)
   c. RbOH(*aq*)

9. Classify each of these aqueous solutions as acidic, neutral, or basic. Give the criterion for your choice.
   a. $H^+ = 1 \times 10^{-8}$ M
   b. $H^+ = 1 \times 10^{-2}$ M
   c. $H^+ = 5 \times 10^{-7}$ M

10. Formic acid, HCO_2H, is a natural component of rainwater. Write an equation showing how formic acid can contribute to the acidity of rainwater.

11. You just purchased a new car and are worried about whether its paint will be damaged by acid rain. Consult Figure 6.8 to find the data necessary to answer these questions.
    a. Which of these cities—Chicago, Atlanta, Seattle, or San Francisco—would have an atmosphere that is the most kind to your car? Why?
    b. How does the average pH of rain for each of those cities compare with the average pH of rain where you live?

12. Write a balanced chemical equation for the reaction shown in Figure 6.10b.

13. The text states that the reaction of SO_2 with an $\cdot OH$ free radical accounts for 20–25% of the sulfuric acid in the atmosphere.
    a. Write a balanced chemical equation for this reaction.
    b. Write the chemical equation in part **a** using sphere equations.
    c. What is the source of $\cdot OH$ free radical in the atmosphere?

14. Give the formulas for the acid anhydride of each of these acids.
    a. carbonic acid, H_2CO_3
    b. sulfurous acid, H_2SO_3

15. Assume that coal can be represented by the formula $C_{135}H_{96}O_9NS$.
    a. What is the percent of nitrogen by mass in coal?
    b. If 3 tons of coal are burned completely, what mass of nitrogen in NO will be produced? Assume that all of the nitrogen in the coal is converted to NO.
    c. Will the mass of nitrogen in NO calculated in part **b** be the *maximum* amount of nitrogen in NO produced in this combustion reaction? Why or why not?

16. Acid rain can damage marble statues and limestone building materials. Write a balanced chemical equation to represent this destruction.

17. a. What does the phrase "pH balanced" imply on the label of a shampoo bottle?
    b. Will the presence of the phrase "pH balanced" influence your decision to buy a particular shampoo? Why or why not?

18. Consider the information in this graph.

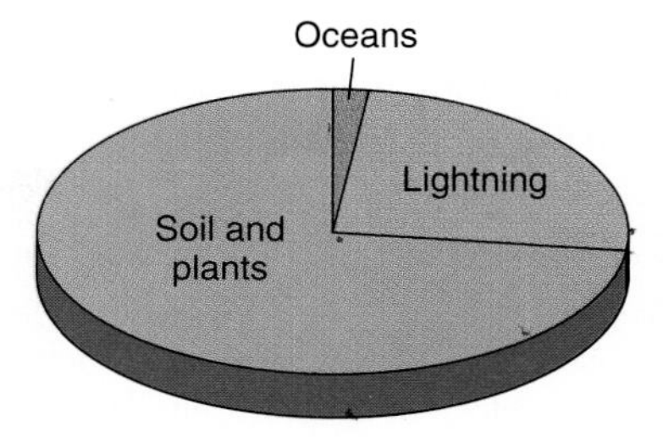

   a. Approximately what percent of natural global emissions of NO_x is contributed by each source?
   b. To check the approximations made from the graph, calculate the percent contribution by using the data in Table 6.1. Compare these results with your estimations.
   c. The graph and Table 6.1 both convey the same information about the natural sources of NO_x emissions. What do you see as the strengths of each method as a means of communicating that information? Give a reason to support your opinions.

19. Figure 6.11 offers information about the percentage of SO_2 emissions from fuel combustion, mainly for electrical power production, and the percentage of SO_2 produced by transportation. Figure 6.12 offers information about the percentage of NO_x from fuel combustion and the percentage of NO_x produced by transportation. Is the relative importance of fuel combustion and transportation the same for emissions of both SO_2 and NO_x? Why might they differ?

20. Given that almost equal *masses* of SO_2 and NO_x are produced by human activities in the United States, how does their production compare based on a *mole* basis? Assume that all the NO_x is produced as NO_2.

21. About 13% of all global emissions of SO_2 and 21% of all global emissions of NO_x come from the United States.
    a. Suggest reasons why the U.S. percentage of global emissions is greater for NO_x than for SO_2.
    b. How do the percentages of U.S. SO_2 and NO_x emissions compare with the percentage of the world population that lives in the United States?

22. Calculate the mass of $CaCO_3$ (in tons) necessary to react completely with 1.00 ton of SO_2 according to the reaction shown in equation 6.30.

23. Cost estimates of reducing SO_2 emissions in the United States have varied considerably. This is one such estimate, stated in terms of percent reduction of emissions.

Emission Reduction, %	Cost (billions of U.S. dollars)
40	1–2
50	2–4
70	5–6

a. Prepare a graph to represent the relationship between the percent reduction in emissions and the cost.

b. Comment on the prospect of achieving 100% reduction; that is, zero emissions.

24. A garden product called "Dolomite Lime" is composed of tiny chips of limestone that contain both calcium and magnesium carbonates. This product is "intended to help the gardener correct the pH of acid soils," as it is "a valuable source of calcium and magnesium."

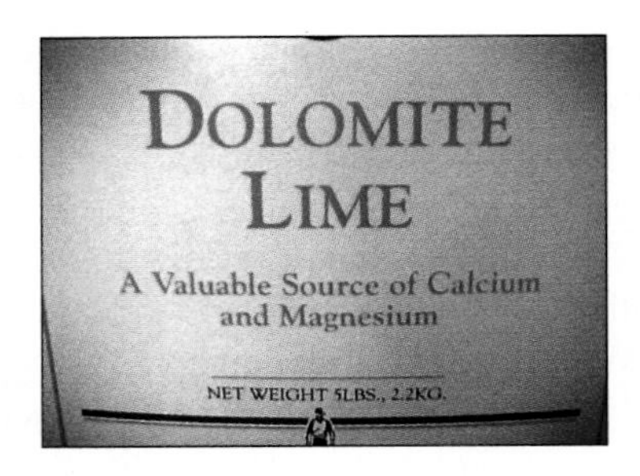

a. Write the chemical formulas for magnesium carbonate and calcium carbonate. In what form is the calcium in this compound, calcium ion or calcium metal?

b. Write a chemical equation that shows why limestone "corrects" the pH of acid soils.

c. Will the addition of Dolomite Lime to soils cause the pH to rise or fall?

d. Plants such as rhododendrons, azaleas, and camellias should not be given dolomite lime. Name a possible reason why.

25. a. The Clean Air Act has been discussed in this chapter, the Montreal Protocol in Chapter 2, and the Kyoto Accord in Chapter 3. What principal issue is each of these important pieces of legislation or international agreements trying to address?

b. Place each of these important legislative or international agreements, together with any significant amendments, on an appropriate time line. You may choose any format for the time line, but it should communicate the maximum amount of information.

Concentrating on Concepts

26. Judging by the taste, do you think there are more hydrogen ions in an equal volume of orange juice or in milk? Explain your reasoning.

27. The formula for acetic acid, the acid present in vinegar, is commonly written as $HC_2H_3O_2$. Many chemists write the formula as CH_3COOH.

a. Draw the Lewis structure for acetic acid.

b. Show that the two formulas both represent acetic acid.

c. What are the advantages and disadvantages of each formula?

d. How many hydrogen atoms can be released as hydrogen ions per acetic acid molecule? Explain.

28. Television and magazine advertisements remind us about the need for antacid tablets. A friend suggests that a good way to get rich quickly will be to market "antibase" tablets. Explain to your friend the purpose of an antacid tablet and offer some advice about the potential success of the "antibase" tablets.

29. In 6.7 Your Turn you listed the ions present in aqueous solutions of acids, bases, and common salts. Now add water, a molecular species, to this list.

a. List all molecular and ionic species in order of decreasing concentration in a 1.0 M aqueous solution of NaOH.

b. List all molecular and ionic species in order of decreasing concentration in a 1.0 M aqueous solution of HCl.

30. Which of these has the *lowest* concentration of hydrogen ions: 0.1 M HCl, 0.1 M NaOH, 0.1 M H_2SO_4, pure water? Explain your answer.

31. Explain why rain is naturally acidic, but not all rain is classified as "acid rain."

32. Do Figures 6.8 and 6.9 prove that SO_2 and NO_x emissions cause the pH of rain to drop (a *causal relationship*) or is there merely a *correlation* between the emissions and the pH drop? Explain your answer.

33. Ozone in the troposphere is an undesirable pollutant, while stratospheric ozone is beneficial. Does nitric oxide, NO, have a similar dual personality in these two atmospheric regions? Explain your thinking. *Hint:* Consult Chapter 2.

34. The mass of CO_2 emitted during combustion reactions is much greater than the mass of NO_x or SO_x, but there is less concern about the contributions of CO_2 to acid rain than from the other two oxides. Suggest two reasons for this apparent inconsistency.

35. The average pH of precipitation in New Hampshire or Vermont is relatively low, even though these states have low levels of vehicular traffic and virtually no industry that emits large quantities of air pollutants. How do you account for this low pH?

36. Consider these statements made at a public meeting discussing the issue of acid rain.

- Acid rain is a simple problem that can be solved by individuals.
- Acid rain is a somewhat straightforward problem that can be solved by industry alone.
- Acid rain is a somewhat complex problem but can be solved simply by banning all cars.
- Acid rain is a complex problem that cannot be solved by simplistic methods.

Explain which of these positions you support and which you do not, based on the information in this chapter and any other sources you consult. Cite your sources.

37. a. Efforts to control air pollution by limiting the emission of particulates and dust can sometimes con-

tribute to an increase in the acidity of rain. Offer a possible explanation for this observation.

b. In Chapter 2, stratospheric ice crystals in the Antarctic were involved in the cycle leading to the destruction of ozone. Is this effect related to the observations in part **a**? Why or why not?

38. a. Several strategies to reduce SO_2 emissions are described in the text. The most effective ones in the last 10 years have been coal switching and stack gas scrubbing. Prepare a list of the advantages and disadvantages associated with each of these methods.

b. Explain why coal cleaning has not been an effective strategy.

39. Discuss the validity of the statement, "Photochemical smog is a local problem, acid rain is a regional one, and the enhanced greenhouse effect is a global one." Describe the chemistry behind each of these air quality problems and explain why the problems affect different geographical areas.

Exploring Extensions

40. The text makes this statement. "By trying to use a technological fix for one problem, we inadvertently create another."

a. Explain how the problems associated with acid rain fit this statement.

b. Pick another example from any of the issues explored in Chapters 1–5. Briefly explain how your choice fits the statement as well.

41. Here are two substances that both contain OH in their chemical formulas. Explain why you cannot write an equation similar to equation 6.3 for either one.

a. $Al(OH)_3$ *Hint:* Consult a solubility table.

b. C_2H_5OH *Hint:* Consider bond energy and the nature of the bonds.

42. In 6.7 Your Turn, you listed ions present in aqueous solutions of acids, bases, and common salts. In question 29, you added molecular substances to the list. To quantify this list,

a. Calculate the molar concentration of all molecular and ionic species in a 1.0 M solution of NaOH.

b. Calculate the molar concentration of all molecular and ionic species in a 1.0 M solution of HCl.

43. A representative of the Electric Power Research Institute, making a presentation in a workshop to establish research priorities and criteria on factors that govern precipitation chemistry, made this statement.

"If whatever control strategy is hit upon is successful in cutting the acidity in half, an evil conspiracy of chemists will only allow the pH of precipitation to increase by 0.3."

As a Sceptical Chymist in attendance at this workshop, how would you respond to this statement? Explain the reasons for your response. *Hint:* See Appendix 3.

44. Equation 6.18 states that energy must be added to get N_2 and O_2 to react to form NO. A Sceptical Chymist wants to check this assertion and determine how much energy is required. Show the Sceptical Chymist how this can be done. *Hint:* Draw the Lewis structures for the reactants and products and then check Table 4.1 for bond energies.

45. The text describes a green chemistry solution to reducing NO emissions for glass manufacturers.

a. Identify the strategy.

b. Use the Web to research what other industries might use this green chemistry strategy. Write a report to summarize your findings.

46. Many things have been suggested to help reduce acid rain, and some examples of what an individual can do are given here. For each item, explain the connection between what you would or wouldn't do and the generation of acid rain:

a. Hang your laundry to dry it.

b. Walk, ride a bicycle, or take public transportation to work.

c. Run the dishwasher and washing machine only with full loads.

d. Add additional insulation on hot water heaters and hot water pipes.

e. Buy locally produced food and other items.

47. How do researchers determine whether the negative effects of acid deposition on aquatic life are a direct consequence of low pH or the result of Al^{3+} released from rocks and soil? Find at least one article that gives the details of such a study. In your own words write a summary of the experimental plan and its results.

48. One way to compare the acid neutralizing capacity (ANC) of different substances is to calculate the mass of the substance required to neutralize one mole of hydrogen ion, H^+.

a. Write a balanced equation for the reaction of $NaHCO_3$ with H^+, and use it to calculate the ANC for $NaHCO_3$.

b. Determine the cost to neutralize one mole of H^+ with $NaHCO_3$ if $NaHCO_3$ costs \$9.50 per kilogram.

49. Why are developing countries likely to emit an increasingly higher percentage of the global amount of SO_2? Pick a nation, research its current emissions of SO_2 and calculate its percentage of global emissions. Speculate on whether increasing emissions are likely to continue in the future and offer an explanation for your prediction.

50. Like diesel trucks, sport utility vehicles (SUVs) emit more than their share of pollutants. Do SUVs emit NO_x, SO_2, or both? What are the current proposals to clean up their emissions? Use the resources of the Web and/or an owner's manual to research this question.

51. Some local weather maps give forecasts for pollen, UV index, and air quality. Why do you suppose that no forecast for acid rain is provided?

7

The Fires of Nuclear Fission

Uranium fuel pellet to be used in a nuclear power reactor.

Nuclear fuel pellets are small and quite ordinary looking. They are made using a black, ceramic material that is cool to the touch. If you were to pick up a few and examine them, you would have no clue as to the extravagant amount of energy packed within them. But when 7–10 million fuel pellets are put together under the proper conditions in a nuclear reactor, these pellets become a highly concentrated energy source. Over 100 nuclear power reactors across our nation are fueled by pellets such as the one pictured.

Although each pellet contains a mere 4 grams of uranium oxide and is only slightly taller than a dime (Figure 7.1), one fuel pellet is roughly equivalent to the energy released when a ton of coal, a cord of wood, or 150 gallons of gasoline are burned. The current price of a pellet—under $3.00 each—is economically competitive with other fuel prices.

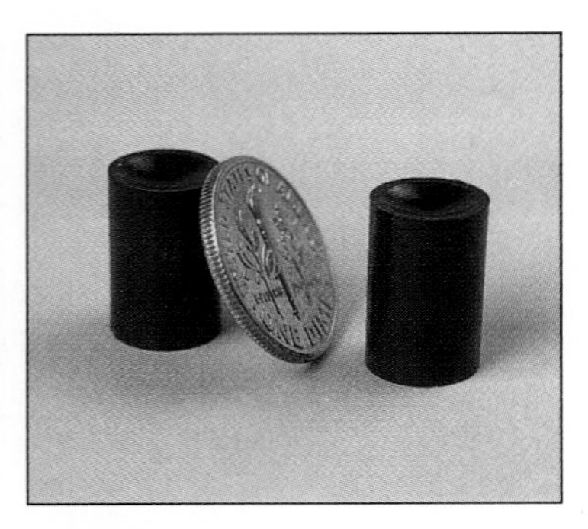

Figure 7.1
Nuclear fuel pellets.

The cost of the nuclear fuel is low compared to the cost of 150 gallons of gasoline, even if gas prices were to triple. However, once fuel pellets and the fuel rods that house them have been used to generate power in a nuclear reactor, they become highly radioactive. The cost for a fuel pellet does *not* include the price of handling it after it is "spent" and becomes nuclear waste. Similarly, the price of a gallon of gasoline does not include any costs to clean up the air pollutants released by automobile engines.

This talk about fuel pellets and nuclear reactors is getting a bit ahead of the story, but we hope it piqued your curiosity. As you study nuclear fission and radioactive decay, you will come to understand how fuel pellets are used in a nuclear power plant to generate electricity. It will make sense to you why the uranium used to make these pellets is called "enriched," leaving a by-product of "depleted" uranium. Similarly, it should become clear why the radiation emitted by the fuel pellets is low *before* they are used to produce energy, but very high afterwards. And finally, by knowing both the chemical and the nuclear processes involved, we hope you will understand why it is so difficult to find a solution that is suitable for everyone for handling the nuclear waste produced by the operation of a nuclear reactor. As has been the case in earlier chapters, your understanding will be based on a complex interplay of issues involving chemistry, energy, and societal issues.

Nuclear phenomena—probably no subject in all of physical science has been more likely in recent years to provoke an emotional response. The word "nuclear" carries a tremendous baggage of disturbing associations, including the bombing of Hiroshima and Nagasaki, radioactive fallout from bomb tests, radiation-induced cancer and birth defects, the risks of accidents and meltdowns (Three Mile Island and Chernobyl), the difficulties of disposing high-level radioactive wastes, and the ultimate threat of nuclear annihilation. And yet, people recognize that many benefits also spring from the nucleus—radiation therapy for the treatment of cancer, medical X-rays and nuclear diagnostic scans that can be done without anesthesia and surgery, and of course, the production of electricity by nuclear power plants. The applications of nuclear phenomena, harmful at one extreme and beneficial at the other, present us with a dilemma of risks and benefits. It is a double-edged sword of Damocles that hangs precariously over our heads, described succinctly by Dr. Hans Blix, former Director General of the International Atomic Energy Agency: "The dual challenge of the atom—to exploit in peace, not explode in war."

- In Greek legend, Damocles was courtier to a king, who made the courtier aware of the constant dangers associated with being king. At a banquet, he seated Damocles under a sword suspended by a single hair.

Certainly the largest, and the most controversial, nonmilitary application of nuclear energy is the generation of electricity by nuclear power plants. When it was first demonstrated that electricity could be obtained from the splitting of atoms, a new age appeared to be dawning. The first commercial nuclear power generating station in this country was completed in 1957 at Shippingport, Pennsylvania, along the Ohio River near Pittsburgh. This new source held the promise of unlimited, cheap electricity. During the early 1960s, proponents of nuclear power were plentiful, including the former head of the U.S. Atomic Energy Commission who suggested naively that electricity produced by this method would be so inexpensive that it would be inconsequential to even meter consumers' use of it. There would be plenty of electricity for everyone! In spite of the optimism, the prediction of costless electricity has not come true. Although the percentage of electricity generated by nuclear power has increased since that time, its critics are numerous and outspoken.

CHAPTER OVERVIEW

For over 50 years, critical issues have arisen over using nuclear power to generate electricity, ones that have raised both hopes and serious doubts. Citizens have asked and continue to ask important questions: How does nuclear fission produce energy? Is this energy produced safely? What are the safeguards against a meltdown? Can a nuclear power plant explode like an atom bomb? Is there a danger that nuclear fuel can be diverted to make nuclear weapons? What is radioactivity? How long will nuclear waste products remain radioactive and how will such wastes be disposed? What risks and benefits are associated with nuclear power? And finally, how crucial is nuclear power to our present and future energy requirements? In this chapter, we address these questions. In every instance, we combine scientific fundamentals, applications, technology, and societal implications. Moreover, we try to temper emotionalism with understanding to help readers weigh the risks and benefits of nuclear energy. We begin by examining the prospects for nuclear power in the coming years. But, before we start, we ask you to consider your own position regarding nuclear power by completing 7.1 Consider This.

7.1 Consider This: Personal Opinion of Nuclear Power

Record your answers to these questions and save them because you will be asked to revisit them at the end of the chapter in 7.32 Consider This.

a. Given a choice between electricity being generated by a nuclear power plant and a traditional coal-burning plant, which would you choose, and why?
b. Would you be more willing to live near a nuclear power plant or near a coal-burning power plant? Give reasons to support your answer.
c. If you already have a position about the use of nuclear power for generating electricity, under what circumstances, if any, would you be willing to change your position?

7.1 A Comeback for Nuclear Energy?

For many people, it is easy to mindlessly turn on a light switch, giving no thought to the source of the electricity and whether the energy will continue to be there. But for anybody who has lost electrical power for a period of time because of a storm or experienced repeated power blackouts, flipping a light switch may trigger a different set of memories. Will electricity be there when I turn on the switch? Is another power failure going to happen again soon? The answers to questions such as these lie in the choices that we make to produce our electrical energy now and in the years to come.

We have a high demand for electricity.

Suppose you brew some coffee. The odds are 1 in 5 that the electricity running your coffee pot came from a nuclear power plant. In 2001, about 20% of the electrical power in the United States was being produced by the 103 nuclear power units such as the one shown in Figure 7.2. However, *no* new nuclear plants have been licensed since 1978; a moratorium was placed on their construction after the Three Mile Island accident in 1979. Furthermore, nine nuclear plants ceased their operations, some of them before their licenses even expired (Table 7.1). They include what was once the nation's largest nuclear plant, the Zion nuclear power station on the shores of Lake Michigan. Reasons cited for plant closings included the competition of natural gas and the competitive pressures of energy deregulation. Therefore, when you brew your coffee next year or a decade from now, what will the source of electricity be?

7.2 Consider This: Nuclear Power State by State

Is a nuclear power plant operating in your state? If so, what percent of your state's electrical energy needs is furnished by nuclear power? Check the interactive map at the Nuclear Energy Institute. Either search

Figure 7.2
Watts Bar nuclear plant, Watts Bar, TN.

to locate it or use the direct link at the *Chemistry in Context* web site. In addition, identify a state where more than half of the energy is produced by nuclear power.

It is hard to predict how long the nuclear plants nearby you (or in neighboring states) are likely to continue to operate. In the late 1990s, the likely decommissioning of a dozen or so additional plants seemed a foregone conclusion, and the list in Table 7.1 was expected to grow. Some worried that so few nuclear plants would renew their licenses that the electrical power generated by them would drop from the current 20% to less than 10%. But, given the recent increased demands for electricity in the U.S., a renaissance in nuclear power is now being predicted. For example, the Oyster Creek plant in New Jersey was scheduled for shutdown in 2000, but the plant is still in operation, purchased by an international power company. Although it is too soon to tell, nuclear plants may be becoming a hot item on the energy market.

- The decommissioning (or shutdown) of a nuclear plant is a complex operation. All parts of the plant must be analyzed and removed according to strict hazardous waste criteria. We say more about nuclear waste in Sections 7.10 and 7.11.

The construction and continued operation of nuclear plants is not only a matter of energy supply and demand, but also a matter of public acceptance. Depending on your age, you may have little recollection of the controversy that surrounded some nuclear power plants when they were proposed or being constructed. People have been lining up on one side or the other of the nuclear fence for quite some time, such as those protesting the construction of the Seabrook nuclear power plant in New Hampshire. Figure 7.3 shows one of the many protests that took place at Seabrook.

Table 7.1

Nuclear Plant Closings Since 1990

Nuclear Plant	State	License Issued	Date shut down
Big Rock Point	Missouri	1964	1998
Millstone 1	Connecticut	1966	1998
Zion 1, Zion 2	Illinois	1973	1998
Maine Yankee	Maine	1973	1997
Haddam Neck	Connecticut	1974	1996
San Onofre 1	California	1967	1992
Trojan	Oregon	1975	1992
Yankee-Rowe	Maine	1963	1991

Source: Environmental Law and Policy Center, at http://www.elpc.org/energy/nuclear_closings.html.

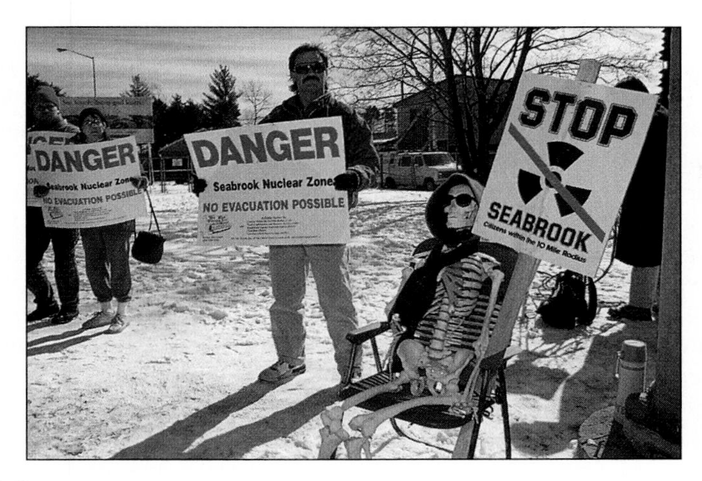

Figure 7.3
Protesting the construction of the Seabrook, New Hampshire nuclear power plant.

7.3 Consider This: *Boston Globe* Reporter

Imagine you are a reporter for the *Boston Globe* assigned to cover the Seabrook demonstration of 1977. You have appointments to interview those leading the protest and the officials of the Public Service Company of New Hampshire, the operators of the Seabrook nuclear plant. List the questions that you will ask to obtain the information you need to write an informative article. Make your questions as focused as possible, so you can gain the maximum amount of information in the time available.

As we next move into our explorations of nuclear fission and the energy it produces, consider these two statements.

> *"We are going to see a revival of nuclear energy; we don't really have an option."*
>
> **Dr. Gregory Choppin**
> Distinguished Professor of Chemistry
> Florida State University
> 2001 Spring American Chemical Society Meeting, San Diego, CA

> *"There has never been much middle ground for nuclear power—people either like it or they don't."*
>
> **Jeffrey W. Johnson**
> Senior Editor, Washington News Bureau
> *Chemical & Engineering News*, Oct 2, 2000

Although these individuals speak to different issues, both points of view need to be kept in mind as you read the sections that follow. Professor Gregory Choppin has been teaching and doing research in nuclear chemistry for decades (he was a codiscoverer of element number 101) and stands well among those who can speak to nuclear power as an option. Jeff Johnson, a staff writer for the American Chemical Society's weekly journal to its members, clearly captures the public sentiment about nuclear power. Nuclear energy *is* an option that stands poised for a revival. But the public is not of one opinion. In the section that follows, we will take a look at the process of nuclear fission, which is the first step in explaining both the sources of the controversy and the hopes for nuclear energy in the future.

7.2 How Does Fission Produce Energy?

The key to answering this question is probably the most famous equation in all of the natural sciences, $E = mc^2$. This equation dates from the early years of the 20th century and is one of the many contributions of Albert Einstein (1879–1955). The equation summarizes the equivalence of energy, E, and matter or mass, m. The symbol c represents the speed of light, 3.0×10^8 m/s, so c^2 is equal to 9.0×10^{16} m^2/s^2. The fact that this number is very large means that it should be possible to obtain a tremendous amount of energy, which we will soon consider, from a very small amount of matter, whether in a power plant or in a weapon.

For over 30 years, Einstein's equation was a curiosity. Scientists believed that it described the source of the Sun's energy, but as far as anyone knew, no one had ever observed on Earth a transformation of a substantial fraction of matter into energy. Then, in 1938, two German scientists, Otto Hahn (1879–1968) and Fritz Strassmann (1902–1980), discovered what appeared to be the element barium (Ba) among the products formed when uranium (U) was bombarded with neutrons. The observation was unexpected because barium has an atomic number of 56 and an atomic mass of about 137. Comparable values for uranium are 92 and 238, respectively. At first, the scientists were tempted to conclude that the element was radium (Ra, atomic number 88), which is a member of the same periodic family as barium. But Hahn and Strassmann were fine chemists, and the chemical evidence for barium was too compelling.

The German scientists were unsure of how barium could have been formed from uranium, so they sent a copy of their results to their colleague, Lise Meitner (1878–1968), for her opinion (Figure 7.4). Dr. Meitner had collaborated with Hahn and Strassmann on related research, but was forced to flee Germany in March 1938 because of the anti-Semitic policies of the Nazi government. When she received their letter, she was living in Sweden. She discussed the strange results with her physicist nephew, Otto Frisch (1904–1979), as the two of them took a walk in the snow. In a flash of insight, the explanation became clear: Under the influence of the bombarding neutrons, the uranium atoms were splitting into smaller atoms of lighter elements, such as barium. The nuclei of the heavy atoms were dividing, like biological cells undergoing **fission.**

That word from biology is applied to a physical phenomenon in the letter that Meitner and Frisch published on February 11, 1939, in the British journal *Nature.* In the letter, entitled "Disintegration of Uranium by Neutrons: A New Type of Nuclear Reaction," the authors state the following:

> *Hahn and Strassmann were forced to conclude that isotopes of barium are formed as a consequence of the bombardment of uranium with neutrons. At first sight, this result seems very hard to understand . . . On the basis, however, of present ideas about the behavior of heavy nuclei, an entirely different . . . picture of these new disintegration processes suggests itself . . . It seems therefore possible that the uranium nucleus . . . may, after neutron*

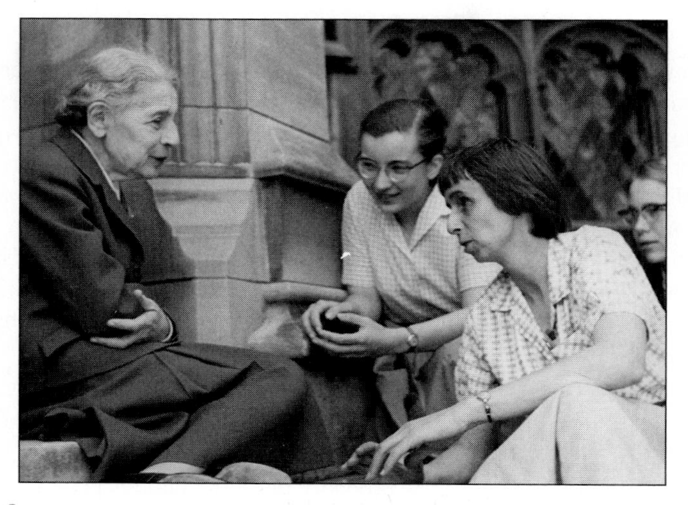

Figure 7.4
Lise Meitner (left) at Bryn Mawr College (1959).

capture, divide itself into two nuclei of roughly equal size . . . The whole "fission" process can thus be described in an essentially classical way.

The letter is just over a page long, but it would be difficult to think of a more important scientific communication. Its significance was recognized immediately. Niels Bohr (1885–1962), an eminent Danish physicist, learned of the news directly from Frisch and brought it to the United States on an ocean liner several days before its publication. Within a few weeks of Meitner and Frisch's letter in *Nature*, scientists in a dozen laboratories in various countries confirmed that the energy released by the fission of uranium atoms was that predicted by Einstein's equation. Lise Meitner's contributions to the discovery of nuclear fission have been honored by naming element 109 meitnerium. The other woman for whom an element is named is Marie Curie, an earlier nuclear pioneer (Section 7.7).

Energy is given off when an atom splits because the total mass of the products is slightly less than the total mass of the reactants. In spite of what you may have been taught, neither matter nor energy is *individually* conserved. Matter disappears and an equivalent quantity of energy appears. Alternately, one can view matter as a very concentrated form of energy, and nowhere is it more concentrated than in the atomic nucleus. Remember that an atom is mostly empty space. If an electron were to travel in a sphere that was half a mile in diameter, the nucleus would lie at the center of this sphere, the size of a baseball. Because almost the entire mass of an atom is associated with the nucleus, the density of the nucleus is incredibly high. Indeed, a pocket-sized matchbox full of atomic nuclei would weigh over 2.5 billion tons! Given the energy-mass equivalence of Einstein's equation, this means that the energy content of all nuclei is, relatively speaking, immense.

Only nuclei of certain elements undergo nuclear fission. Light atoms do not; rather, only heavier nuclei, such as those of uranium atoms, will split when struck with a neutron. Whether or not fission occurs depends on the relative number of protons and neutrons in a nucleus. For example, not all uranium atoms are fissionable in a typical nuclear power reactor. Which ones fission also depends on the particular isotope of uranium and the energy of the neutrons that cause fission. *All* uranium atoms contain 92 protons (accompanied by 92 electrons if the atom is electrically neutral), but uranium isotopes have different numbers of neutrons. Approximately 99.3% of uranium atoms contain 146 neutrons. The mass number of this isotope of uranium is 238, (92 protons plus 146 neutrons). The remaining uranium atoms in nature (about 0.7%) are a different isotope, one containing 143 neutrons; its mass number is 235, the sum of 92 protons and 143 neutrons. Thus, the difference between these two uranium isotopes—uranium-238 and uranium-235—is a mere three neutrons. In nuclear terms, however, this difference is significant. Under the conditions present in a nuclear power reactor, uranium-238 is *not* fissionable, yet uranium-235 is. Small differences in the nucleus can make large differences in *nuclear* behavior.

- Isotopes are atoms of the same element that differ in the number of neutrons. The mass number of an atom is the sum of the number of protons and neutrons. See Section 2.2.

7.4 Your Turn

A trace amount of uranium-234 is found in nature. How many protons are in an atom of uranium-234? How many neutrons?

Physicists and chemists conventionally write the mass number as a superscript preceding the elementary symbol. The atomic number (the number of protons in the nucleus) is written as a subscript. Hence, uranium-238 is represented as follows:

$$\begin{array}{r} \text{Mass number} = \text{number of protons} + \text{number of neutrons} \longrightarrow \\ \text{Atomic number} = \text{number of protons} \longrightarrow \end{array} {}^{238}_{92}\text{U}$$

A wide variety of possible products or fission fragments can be formed when the nucleus of an atom of U-235 is struck with a neutron. A typical reaction is shown here.

$${}^{1}_{0}\text{n} + {}^{235}_{92}\text{U} \longrightarrow [{}^{236}_{92}\text{U}] \longrightarrow {}^{141}_{56}\text{Ba} + {}^{92}_{36}\text{Kr} + 3\,{}^{1}_{0}\text{n} \qquad [7.1]$$

This nuclear equation makes use of the notation just introduced. Note that the subscript for a neutron (designated n) is 0, indicating zero charge; the superscript is 1 because the mass number of a neutron is one. In a balanced nuclear equation, *the sum of the subscripts on the left side equals that of the subscripts on the right side of the equation. Likewise, the sum of superscripts on each side of the equation is equal.* Coefficients in a nuclear equation, such as the 3 preceding the neutron symbol in the products, are treated the same way as in chemical equations: The coefficient multiplies the term following it. In the preceding case, the coefficient indicates three neutrons. Where no coefficient is given explicitly, a 1 is understood. We can check the correctness of nuclear equation 7.1 by applying these rules.

- Nuclear equations are similar to, but not the same, as conventional chemical equations.

	Left	**Right**
Superscripts:	$1 + 235 = 236$	$141 + 92 + (3\times1) = 236$
Subscripts:	$0 + 92 = 92$	$56 + 36 + (3\times0) = 92$

7.5 Your Turn

Find the relevant atomic numbers from the periodic table to help write the nuclear equations that occur when an atom of uranium-235 is struck by a neutron.

a. U-235 undergoes fission to form Ba-138, Kr-95, and neutrons.
b. U-235 undergoes fission to form an element with an atomic number of 52 and a mass number of 137, another element with an atomic number of 40 and a mass number of 97, and neutrons.

Answer
a. ${}^{1}_{0}\text{n} + {}^{235}_{92}\text{U} \longrightarrow {}^{138}_{56}\text{Ba} + {}^{95}_{36}\text{Kr} + 3\,{}^{1}_{0}\text{n}$

7.6 Your Turn

Strontium-90 (Sr-90) is a radioactive fission product that contaminated milk for some time after earlier atmospheric nuclear bomb tests. It can be formed from the fission of U-235 in a reaction that produces three neutrons and another element. Write the equation for this nuclear fission reaction that occurs when U-235 is bombarded by a neutron.

We mentioned earlier that energy is given off during fission because the mass of the products is slightly less than that of the reactants. However, from the nuclear equations we have just written it may appear that there is no mass change, because the mass numbers are equal on both sides. In fact, the actual mass does decrease slightly. To understand this, remember that the actual masses of the nuclei are not the mass numbers (the sum of the number of protons and neutrons); rather, they have measured values with many decimal places. For example, the atomic weight of an atom of U-238 is 238.0507847 atomic mass units. Were you to keep all seven decimal places and compare the masses on both sides of a nuclear fission equation, you would find that the products' mass is slightly less. As a consequence, the total potential energy of the product nuclei is less than that of the reactants, and this difference in potential energy corresponds to the energy released.

When atoms of U-235 split under neutron bombardment, about 0.1% or 1/1000th of the mass is lost and is replaced by an equivalent amount of energy. We can use this information and $E = mc^2$ to calculate just how much energy would be produced if 1.0 kilogram (2.2 pounds) of U-235 were to undergo nuclear fission. Since only 1/1000th of this mass is lost, the value for m that goes into the Einstein equation is $1.0 \text{ kg} \times 1/1000 = 1.0 \times 10^{-3}$ kg (or 1.0 g). As you already know, $c = 3.0 \times 10^8$ m/s. Substituting these values yields this result.

$$E = mc^2 = 1.0 \times 10^{-3} \text{ kg} \times (3.0 \times 10^8 \text{ m/s})^2$$

$$E = 1.0 \times 10^{-3} \text{ kg} \times 9.0 \times 10^{16} \text{ m}^2/\text{s}^2$$

Completing the calculation gives an energy answer in what appears to be unusual units.

$$E = 9.0 \times 10^{13} \text{ kg} \cdot \text{m}^2/\text{s}^2$$

The unit kg · m^2/s^2 is identical to a joule. Therefore, the energy released from the fission of an entire kilogram of uranium-235 is a whopping 9.0×10^{13} joules or 9.0×10^{10} kilojoules.

- As described in Section 4.1, joule (J) is a unit of energy: 1 J = 1 kg · m^2/s^2.

To put things into perspective, 9.0×10^{13} J is the amount of energy released by the explosion of 33,000 tons (33 kilotons) of TNT. Alternatively, you could get this much energy by burning 3300 tons of coal. This is enough energy to raise about 700,000 cars six miles into the sky or to turn 8.7 million gallons of water into steam. Yet, this massive amount of energy comes from the fission of a single kilogram of U-235, where only one gram (0.1% mass change) is actually transformed into energy.

All this energy is accessible because the fission of a uranium atom releases two or three neutrons; three neutrons are indicated in equation 7.1. Thus, there is a net production of neutrons. Each of these neutrons can in turn strike another U-235 nucleus, cause it to split, and release a few more neutrons. The result is a rapidly branching and spreading **chain reaction** (Figure 7.5) that can, under certain circumstances, sweep through a mass of fissionable uranium in a fraction of a second. Such a chain reaction will occur spontaneously if a critical mass, about 15 kg (33 pounds), of pure U-235 is brought together in one place. But as you will soon see, the uranium in a nuclear power plant is far from pure U-235.

7.7 Your Turn

At full capacity, the nuclear plant at Seabrook, New Hampshire, generates 1160 million joules of electrical energy per second. Calculate the total amount of electrical energy produced per day and the mass of U-235 actually converted to energy per day.

Hint: Start by calculating the quantity of energy generated not per second, but per day. Then use the equation $E = mc^2$ and solve for mass, m. Report your answer in grams.

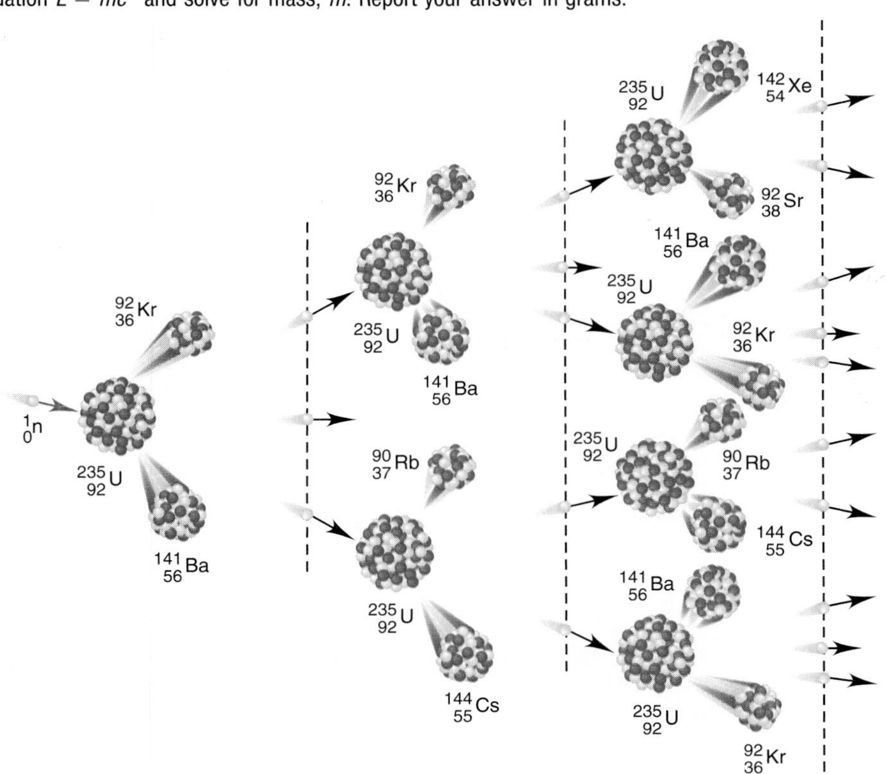

Figure 7.5
A chain reaction of the fissioning of U-235.

Figures Alive!

7.3 How Does a Nuclear Reactor Produce Electricity?

Chapter 4 described how a conventional power plant burns a fuel such as coal or oil to produce heat. The heat is then used to boil water, converting it into high-pressure steam that turns the blades of a turbine. The shaft of the spinning turbine is connected to large wire coils that rotate within a magnetic field, thus generating electrical energy. A nuclear power plant operates in much the same way, except that the water is heated not by fossil fuel, but by the energy released from the fission of nuclear "fuel" such as U-235. Like any power plant, a nuclear power plant is subject to the efficiency constraints imposed by the second law of thermodynamics. The theoretical efficiency for converting heat to work depends on the maximum and minimum temperatures between which the plant operates. This thermodynamic efficiency, typically 55–65%, is significantly reduced by other mechanical, thermal, and electrical inefficiencies.

• Section 4.13 contains information on operating efficiencies.

A nuclear power station consists of two parts: a nuclear reactor and a nonnuclear portion (Figure 7.6). The nuclear reactor is the hot heart of the power plant. The reactor, together with one or more steam generators and the primary cooling system, is housed in a special steel vessel within a separate reinforced concrete dome-shaped containment building. The nonnuclear portion contains the turbines that run the electrical generator. It also contains the secondary cooling system. In addition, the nonnuclear portion must be connected to some means of removing heat from the coolants. Accordingly, a nuclear power station will have one or more cooling towers and/or be located near a sizeable body of water.

The uranium fuel in the reactor core is in the form of uranium dioxide (UO_2) pellets, each about the size of a large pencil eraser, as shown earlier in Figure 7.1. These pellets are placed end to end in tubes made of a special metal alloy, which in turn are grouped into stainless-steel clad bundles (Figure 7.7). There are at least 200 pellets per tube. Although a fission reaction, once started, can sustain itself by a chain reaction (Figure 7.5), neutrons are needed to start the fission reaction (equation 7.1). One means of generating the required neutrons is to use a source that produces neutrons from a nuclear reaction (equation 7.2).

• Plutonium-238, Pu-238, is an isotope first made in 1940.

• Gamma radiation is also produced in the last step of equation 7.2.

$$^{238}_{94}\text{Pu} \longrightarrow {}^{4}_{2}\text{He} + {}^{234}_{92}\text{U} \quad \text{(alpha decay of Pu-238)}$$

$$^{4}_{2}\text{He} + {}^{9}_{4}\text{Be} \longrightarrow {}^{12}_{6}\text{C} + {}^{1}_{0}\text{n} \qquad [7.2]$$

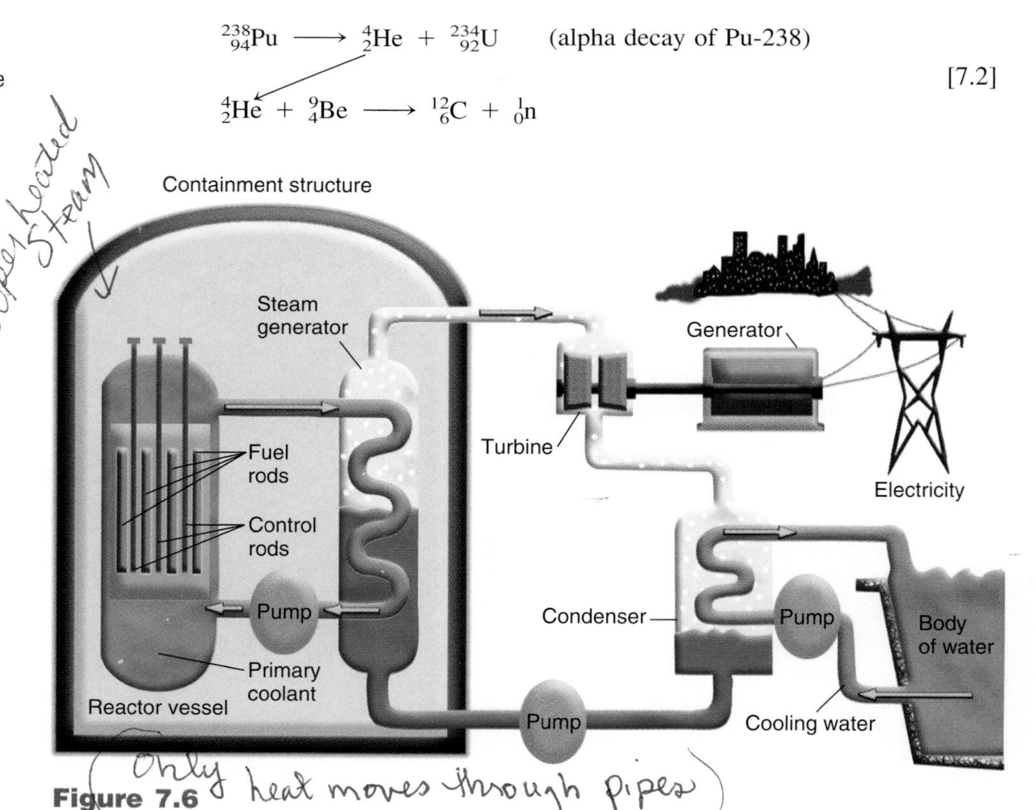

Figure 7.6
Diagram of a nuclear power plant.

Such a source contains beryllium-9 and a heavier element such as plutonium or americium. The heavier element releases alpha particles (helium-4 nuclei, as we will discuss in Section 7.7). When an alpha particle (helium-4 nucleus) strikes a beryllium atom, the two nuclei combine to form a carbon-12 nucleus, plus a neutron. The neutrons produced in this way initiate the nuclear fission of uranium-235 in the reactor core.

Once fission is initiated, the rate of fission and the amount of heat generated by the fission are controlled using a principle employed in the first controlled nuclear fission reaction, which took place at the University of Chicago in 1942. Rods composed primarily of the element cadmium, an excellent neutron absorber, are interspersed among the fuel elements. Modern control rods also contain silver and indium. As long as these rods are in place, the reaction cannot become self-sustaining because neutrons are absorbed by the control rods, preventing fission from becoming self-sustaining. When the rods are withdrawn, the reactor "goes critical"; that is, the fission chain reaction becomes self-sustaining. But the rods can be rapidly adjusted to halt the self-sustaining chain reaction in the event of an emergency (Figure 7.6).

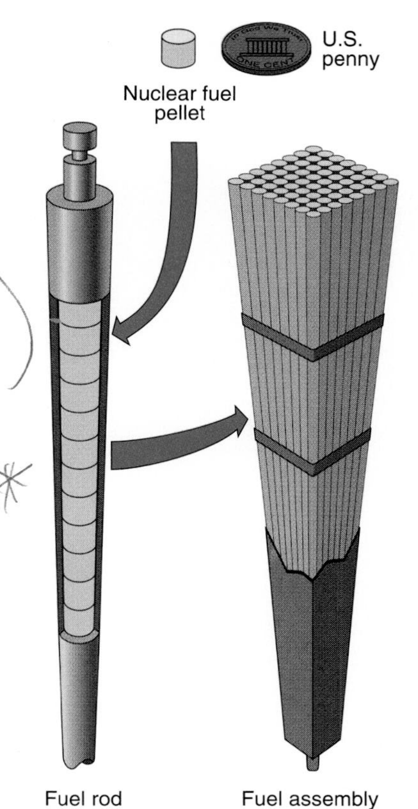

Figure 7.7

Fuel pellet, fuel rod, and fuel assembly making up the core of a nuclear reactor.

The fuel bundles and control rods are bathed in what is called the *primary coolant.* In the Callaway nuclear reactor shown in Figure 7.8, and in many others, the primary coolant is a water solution of boric acid, H_3BO_3. The boron atoms absorb neutrons and thus control the rate of fission and the temperature. The water solution also serves as a "moderator" for the reactor, slowing the speed of the neutrons and making them more effective in causing fission. Another major function of the primary coolant is to absorb the heat generated by the nuclear reaction. Because the primary coolant solution is at a pressure more than 150 times normal atmospheric pressure, it does not boil. It is heated far above its normal boiling point and circulates in a closed loop from the reaction vessel to the steam generators, and back again. This closed primary coolant loop thus forms the link between the nuclear reactor and the rest of the power plant (Figure 7.6).

The heat from the primary coolant is transferred to the water in the steam generators, sometimes called the secondary coolant. At the Callaway nuclear plant, over 30,000 gallons of water are converted to vapor each minute. The energy of this hot vapor turns the blades of a turbine, which is attached to an electrical generator. The water vapor is then cooled and condensed back into the liquid state, and returned to the steam generator to continue its heat transfer cycle. In many nuclear facilities the cooling is done using large cooling towers, which are commonly mistaken for the reactors. The reactor is actually housed in a relatively small dome-shaped building (Figure 7.8).

- Cooling towers are also used in coal-fired plants.

Figure 7.8

Cooling tower and containment building at the Union Electric Callaway (MO) nuclear power plant. The nuclear reactor is in the domed containment building.

7.8 Your Turn

Some days you can see a cloud coming out of the cooling tower, as shown in Figure 7.8. What causes the cloud? Does it contain any products of the nuclear fission reaction?

Nuclear power plants also use water from lakes, rivers, or the ocean to cool the condenser. For example, at the Seabrook nuclear power plant in New Hampshire, every minute 398,000 gallons of ocean water flow through a huge tunnel (19 feet in diameter and 3 miles long) bored through rock 100 feet beneath the floor of the ocean. A similar tunnel from the plant carries the water, now 22°C warmer, back to the ocean. Special nozzles distribute the hot water so that the observed temperature increase in the immediate area of the discharge is only about 2°C. The ocean water is two loops away from the fission reaction and its products. The primary coolant (water with boric acid) circulates through the reactor core inside the containment building. However, this boric acid solution is kept isolated in a closed circulating system, which makes the transfer of radioactivity to the secondary coolant water in the steam generator highly unlikely. Similarly, the ocean water does not come in direct contact with the secondary system, so the ocean water is well protected from radioactive contamination. It should be obvious that the electricity generated by a nuclear power plant is identical to the electricity generated by a fossil-fuel plant; the electricity is not radioactive, nor can it be.

7.9 The Sceptical Chymist: Watching the Watts

Consider the statistics quoted for the Seabrook plant. The energy generated in one day is said to be the equivalent of 10,000 tons of coal. Perform a calculation to determine if this value is consistent with the quoted power rating of 1160 megawatts (1160×10^6 J/s). Assume that burning coal releases 30 kJ/g or 30,000 J/g.

Hint: Start by rereading 7.7 Your Turn, or completing it if you did not do so earlier. Show that the quantity of energy per day would be equivalent to burning *fewer* than the 10,000 tons of coal quoted here. How big is this discrepancy and what important factor can account for such a discrepancy?

7.4 Could There Be Another Chernobyl? Safeguards Against a Meltdown

A 1979 film called *The China Syndrome* told the story of a near-disaster in a fictitious nuclear power plant. The heat-generating fission reaction almost got out of control. If such a thing were to happen, the intense heat might cause a "meltdown" of the uranium fuel and the reactor housing. Fancifully, the underlying rock might even melt "all the way to China." But in spite of various human and instrumental errors, the safety features of the system worked in the film and fictional disaster was averted. Seven years later on April 26, 1986, the engineers of the very real Chernobyl power plant in the Ukraine, then part of the Soviet Union, were far less fortunate (Figure 7.9). This plant consisted of four reactors, two built in the 1970s and two more in the 1980s, all near the town of Chernobyl (pop. 12,500). Water from the nearby Pripyat River was used to cool the reactors. Although the surrounding region was not heavily populated, nonetheless approximately 120,000 people lived within a 30 km radius.

- Chernobyl is the Russian spelling; Chornobyl is the more accurate Ukrainian usage.

Chernobyl stands as the world's worst nuclear power plant accident to date. What went wrong there? During an electrical power safety test at Chernobyl reactor 4, the flow of cooling water to the core was deliberately interrupted as part of the test. The temperature of the reactor rose rapidly. A combination of errors followed. An insufficient number of control rods had been left in the reactor (and couldn't be reinserted quickly enough), and the steam pressure was too low to provide coolant (due to both operator error and faulty design). An overwhelming power surge produced heat, rupturing the fuel elements and releasing hot reactor fuel particles that in turn exploded on contact with the coolant water. The reactor core was destroyed in seconds. The graphite used to slow neutrons in the reactor to fissionable speeds burned in the heat. When

- The explosion at Chernobyl was produced by combustion, a chemical reaction, not a nuclear process.

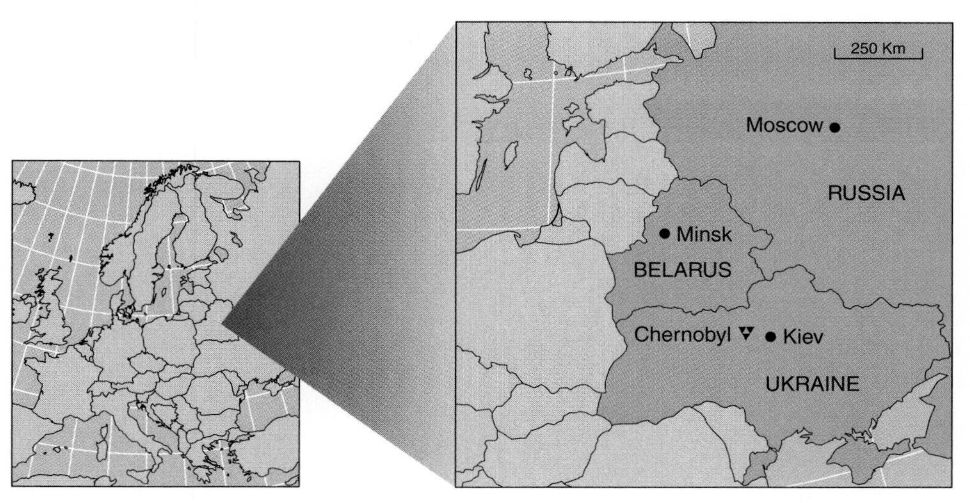

Figure 7.9
Chernobyl, in Ukraine of the former Soviet Union.

water was sprayed on the burning graphite, they reacted *chemically* to produce hydrogen gas, which exploded when it chemically reacted with oxygen in the air.

$$2\ H_2O(l) + C(s,\ \text{graphite}) \longrightarrow 2\ H_2(g) + CO_2(g)$$

$$2\ H_2(g) + O_2(g) \longrightarrow 2\ H_2O(g)$$

The explosion blasted off the 4000-ton steel plate covering the reactor (Figure 7.10). Although a "nuclear" explosion never occurred, the fire and explosions of hydrogen blew vast quantities of radioactive material out of the reactor core into the atmosphere.

Figure 7.10
The Chernobyl 4 reactor after the chemical explosion.

Fires began to burn in what remained of the building. In a short amount of time, the plant lay in ruins. The head of the crew on duty at the time of the accident has written: "It seemed as if the world was coming to an end . . . I could not believe my eyes; I saw the reactor ruined by the explosion. I was the first man in the world to see this. As a nuclear engineer I realized the consequences of what had happened. It was a nuclear hell. I was gripped with fear." (*Scientific American*, April 1996, p. 44.)

The disaster continued in the countryside for over a week. As the reactor burned, it continued to spew large quantities of radioactive fission products into the atmosphere for 10 days (Figure 7.11). The release of radioactivity was estimated on the order of about 100 of the atomic bombs dropped on Hiroshima and Nagasaki. People in nearby regions reported an odd, bitter, and metallic taste as they inhaled the invisible particles. The radioactive dust cut a swath across the Ukraine, Belarus, and up into Scandinavia. Nearly 150,000 people living within 60 kilometers of the power plant were permanently evacuated after the meltdown.

The human toll was immediate and continues to grow. Several people working at the plant were killed outright, and another 31 firefighters died in the cleanup process from acute radiation sickness, a topic we take up in a later section. An estimated 250 million people were exposed to levels of radiation that may ultimately shorten their lives. Included in this figure are 200,000 "liquidators," people who buried the most hazardous wastes and constructed a 10-story concrete structure ("the sarcophagus") to surround the failed reactor. As of 2001, more than 700 children in Belarus, a neighboring country, have been treated for thyroid cancer; most have survived. Presumably this illness is caused by radioactive iodine-131, one of the fission products blown from the reactor. As Dr. Akira Sugenoya, a Japanese physician who volunteered his expertise in Belarus to treat the children suffering from thyroid cancer, remarked "The last chapter of the terrible accident is far from written."

- The thyroid gland incorporates iodide ion to manufacture thyroxine, essential for growth and metabolism.

Given the demonstrable problems with their design, the four reactors at Chernobyl have been shut down. On Friday, December 15, 2000, the control rods slid into the core at unit 3, the last remaining reactor operating at Chernobyl, permanently shutting it

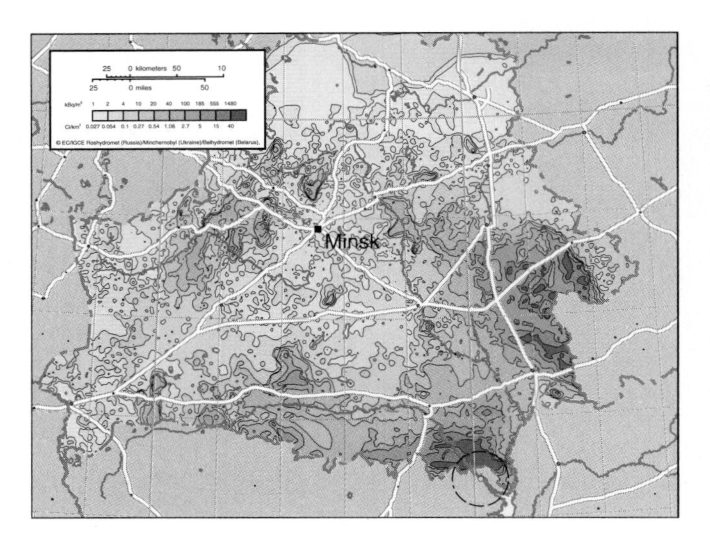

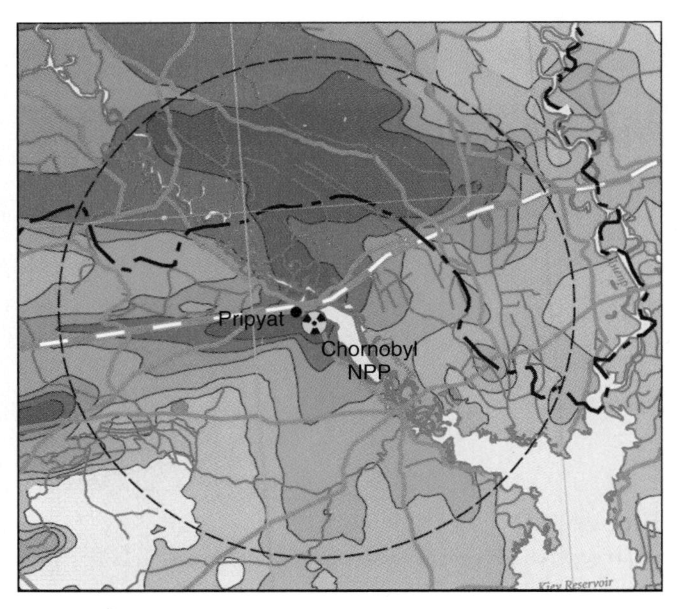

Figure 7.11

Radioactive fallout from the Chernobyl accident, in Belarus and in the vicinity of Chernobyl. Red areas range from 5 to over 40 curies per square kilometer. The curie, Ci, is a unit of radiation explained in Section 7.8.

Source: 20 April 2001 Science "Living in the Shadow of Chornobyl".

down. Ukranian President Leonid Kuchma reported "This decision came from our experience of suffering. We understand that Chernobyl is a danger for all of humanity and we forsake a part of our national interests for the sake of global safety."

7.10 Consider This: Nuclear Neighbors

The Chernobyl reactor site and the land near it continue to be among the most highly radioactive places on Earth. Ukranian President Kuchma is reported to have said, "We shall continue to bear this. This is our fate." Use the Web to find out what kinds of humanitarian aid continue to be offered to the victims of Chernobyl. What cleanup tasks are required? What are the medical needs? Are there any avenues to give assistance that you personally or as a class would like to pursue?

The Ukrainian government estimates that a replacement gas-burning power station and associated expenses will cost $4 billion. Thus far, Western countries have pledged $2.3 billion towards the project, and the United States is providing financial and technical assistance to establish an international nuclear safety and environmental research center near Chernobyl. But there are many hidden costs. A study by a Russian economist estimates the total cost of the Chernobyl meltdown at $358 billion, a figure that includes the expense of the cleanup and the loss of farm production. Present plans include encasing the ruined reactor in additional concrete and steel. No plans have been made to deal with the radioactive materials and dust inside.

7.11 Consider This: Chernobyl's Legacy

The Why? Files, a web site of the National Institute for Science Education, specializes in giving the science behind the news. Check out The Why? Files on the Web either by doing a search or by using the direct link from the *Chemistry in Context* web site. Find the news story about Chernobyl entitled "Radiation Reassessed." What happened at Chernobyl in 1986? Review the photos and read through the scenario. What are the latest reports about the survivor's exposure to low-level radiation?

This recounting of the solemn facts concerning the Chernobyl tragedy leads to an inevitable question: "Could it happen here?" America's closest brush with nuclear disaster occurred in March 1979, when the Three Mile Island power plant near Harrisburg,

Pennsylvania, lost coolant and only a partial meltdown occurred. There were no fatalities and no serious release of radiation. In spite of the initial failure, the system held and the damage was contained. Since then, refinements in design and safety have been made to existing reactors and those under construction. Nuclear engineers agree that no commercial nuclear reactors in the United States have the design defects that led to the Chernobyl catastrophe.

- Unlike at Chernobyl, an individual at the gates of the Three Mile Island plant at the time of the accident would have been exposed to less radiation in two weeks than that received from a single chest X-ray.

Consider, for example, the Seabrook nuclear power plant that has been hailed as an example of state-of-the-art engineering. The energetic heart of the station is the 400-ton reaction vessel with 44-foot high walls made of eight-inch carbon steel. It is surrounded by a reinforced concrete, dome-shaped containment building, a feature the Chernobyl plant did not have, but all reactors operating in the United States must have. As the name suggests, this structure is built to withstand accidents of natural or human origin and prevent the release of radioactive material. It is clearly visible in the photograph (Figure 7.12). The inner walls of the building are 4.5 feet thick and made of steel-reinforced concrete; the outer wall is 15 inches thick. Information supplied by North Atlantic Energy Service Corporation (NAESCO), the company that manages the Seabrook station, states that the containment building is constructed to withstand hurricanes, earthquakes, 360-mph winds, and the direct crash of a United States Air Force FB-111 bomber.

While it seems unlikely that a nuclear accident of the proportions of Chernobyl could occur in the U.S., the question still remains: Could a nuclear disaster happen in some other region of the world? There is that possibility, because such disasters result from the complex interplay of faulty plant design, human error, and political instability. All three of these must be minimized to keep a nuclear power plant operating safely. While the nuclear units in many parts of the world get high rankings on all three factors, this is not the case everywhere. For example, in July 2001 the German government urged the closing of a Czech nuclear power plant near the German border because of safety concerns. Several reactors in Russia have long histories of safety violations and raise similar concerns, both for operation and the handling of waste (the topic of Section 7.10). A plume of radioactive dust easily crosses international boundaries, and so the concerns of neighboring nations are justifiable.

Figure 7.12
Seabrook nuclear power plant. The dome is part of the containment building that houses the reactor.

7.12 Consider This: Nuclear Power in Russia

As this book goes to press, fewer than 30 reactors are operating in Russia, accounting for roughly 10% of the electricity produced. What is the current status of nuclear power in Russia? Use the resources of the Web to research this question. If possible, comment on any controversies over whether some reactors are now viewed as too dangerous to continue to operate.

7.5 Can a Nuclear Power Plant Undergo a Nuclear Explosion?

Many people ask this question. The devastation and destruction that atomic bombs brought to Hiroshima and Nagasaki are painfully etched in the memory of anyone who has even seen the pictures of those cities and their survivors. Therefore, it is reassuring that the answer to the question is an emphatic "No."

Obviously, nuclear power plants and nuclear bombs are constructed for different purposes. Although both derive energy from nuclear fission, the desired rate of reaction is quite different. A nuclear power plant requires a slow, controlled energy release; in a nuclear weapon, the release is rapid and uncontrolled. In both cases, the fuel is U-235 and the fission reaction is essentially the same, but there is one key difference. Commercial nuclear power plants typically operate using uranium that is 3–5% of the fissionable U-235 isotope and 97–95% U-238. This fuel is called "enriched uranium," which contains 3–5% U-235, greater than the less than the 1% U-235 found in naturally occurring uranium. Most of the neutrons given off during fission of U-235 are absorbed by U-238 atoms and elements such as cadmium and boron. As a consequence, the neutron stream cannot build up enough to establish an explosive chain reaction, such as that in a nuclear fission bomb. In contrast to the uranium used in nuclear power plants, atomic weapons use highly enriched uranium, that is enriched to uranium greater than 90% U-235.

- Naturally occurring uranium ores are about 99% U-238 and 0.7% U-235. See Section 7.2.

As we noted earlier, a spontaneously explosive nuclear chain reaction (such as in a nuclear bomb) will occur only if about 33 pounds of highly purified U-235 are quickly assembled in one place. Fortunately for our troubled world, it is not easy to prepare pure U-235. The separation of this fissionable isotope from nonfissionable U-238 is extremely difficult, because these two isotopes behave essentially the same in their chemical reactions. However, U-235 and U-238 do differ by mass (three neutrons) and this mass difference can be exploited to accomplish a separation. For example, on average, lighter gas molecules mover faster than heavier ones. So, at least in theory, gas molecules containing U-235 should diffuse slightly more rapidly than their analogs containing U-238. However, uranium ore is not a gas; it is a solid consisting mainly of UO_3 and UO_2. Most other uranium compounds are solids as well. But it turns out that UF_6, uranium hexafluoride ("hex"), is a solid at room temperature that readily vaporizes when heated to 56°C (about 135°F). To produce hex, the uranium ore is converted to UF_4, which is then reacted with fluorine.

- Since isotopes have essentially the same chemical reactivity, both U-235 and U-238 react identically with fluorine in equation 7.3.
- Note that equation 7.3 is a *chemical* reaction, not a nuclear reaction.

$$UF_4(g) + F_2(g) \longrightarrow UF_6(g) \quad [7.3]$$

One way to separate molecules of different masses is by gaseous diffusion, a process in which a gas is forced through a series of permeable membranes. Lighter gas molecules ($^{235}UF_6$) diffuse more rapidly through the membrane than heavier gas molecules ($^{238}UF_6$). On average, a $^{235}UF_6$ molecule diffuses through the membrane about 0.4% faster than a $^{238}UF_6$ molecule. If the diffusion is allowed to occur over and over through a long series of permeable barriers, significant separation of the fissionable and the nonfissionable isotopes can be achieved. For more than four decades, uranium isotopes were separated by gaseous diffusion at the Oak Ridge National Laboratory in Tennessee. Currently, one commercial plant at Paducah, Kentucky, carries out uranium enrichment (Figure 7.13); another recently closed at Portsmouth, Ohio. As of 2002, enrichment plants also are operating commercially in Canada, the U.K., France, and the former Soviet Union. Other separation methods, including centrifugation of UF_6 molecules, have been developed. Inspectors attempting to determine a nation's nuclear military capabilities often look for the apparatus necessary to concentrate U-235.

7.13 Your Turn

What is the percent mass difference in $^{238}UF_6$ and $^{235}UF_6$?

7.14 Consider This: Depleted Uranium

After fissionable U-235 is separated from uranium ore, "depleted uranium" is left behind. Over 700,000 metric tons of depleted UF_6 are estimated to be in storage at the Paducah site in Kentucky.

a. How is depleted UF_6 stored? Use the resources of the Web to research the issues and points of controversy. Your answer should clearly distinguish between the *chemical* and the *nuclear* properties of depleted UF_6.

b. Depleted uranium was used to construct anti-tank shells that have been used recently in several armed conflicts around the world. Explain why depleted uranium was used. Create a list of bullet points about the issues and points of controversy.

Figure 7.13
The transportation of "hex" at Paducah, Kentucky. A cylinder of enriched uranium product from the Paducah plant is shown being loaded into autoclave no. 2 of the Transfer and Shipping Facility. Cylinders are weighed and selectively sampled for purity and adherence to specifications.

7.6 Could Nuclear Fuel Be Diverted to Make Weapons?

Given the amount of processing that is required to produce highly enriched U-235 from reactor-grade fuel, such a diversion from peaceful to military uses is difficult and costly. A more likely fissionable material for clandestine weapons manufacturing is plutonium-239 (Pu-239), formed in a conventional reactor when a plentiful U-238 nucleus absorbs a neutron. The reaction subsequently emits 2 electrons as beta particles, ${}^{0}_{-1}e$, as shown in equation 7.4:

$${}^{1}_{0}n + {}^{238}_{92}U \longrightarrow {}^{239}_{94}Pu + 2\ {}^{0}_{-1}e \qquad [7.4]$$

This transformation was discovered early in 1940. The chemical and physical properties of plutonium were determined with an almost invisible sample of the element on the stage of a microscope. The chemical processes devised on such minute samples were scaled up a billionfold and used to extract plutonium from the spent fuel pellets from a reactor built on the Columbia River at Hanford, Washington. **Spent nuclear fuel** is the material remaining in fuel rods after they have been removed from a reactor. The reactor was called a **breeder reactor** because it was designed primarily to convert U-238 to fissionable Pu-239 by means of neutron absorption (equation 7.4). The plutonium was chemically separated from the uranium and used in the first fission test explosion on July 16, 1945, and in the bomb dropped on Nagasaki a little less than a month later.

• Beta particles are discussed later when we consider radioactivity.

Plutonium-239 can also be used to power nuclear reactors. Thus, a breeder reactor is one that creates both energy and new fuel (Pu-239) as it fissions the old fuel (U-235). This seems like a dream come true to an energy-hungry planet. France, the United Kingdom, Russia, Japan, and the United States have conducted research on breeder reactors that permit recovery of plutonium from spent fuel. However, such reactors represent another example of the very mixed blessings of modern technology. The problems are largely associated with the product, Pu-239. The radiation emitted by Pu-239 cannot penetrate the skin. Moreover, solid metallic plutonium is not easily absorbed into the body. But when plutonium is exposed to air, it reacts with oxygen to form plutonium oxide, PuO_2, a powdery compound. The PuO_2 dust can be easily dispersed and inhaled. Once it enters the body, plutonium is one of the most toxic elements known. A few micrograms (10^{-6} g) of PuO_2, lodged in the lungs, can induce lung cancer. Plutonium oxide can also slowly dissolve in the blood and be transported to other parts of the body, especially bone and liver, where its long-lived radioactivity can do serious damage.

• You will learn more about the radiation emitted by Pu-239 in 7.16 Your Turn.

Plutonium-239 poses an international problem because the plutonium produced in nuclear power reactors could possibly wind up in bombs. It has been widely speculated that the 1981 bombing of a nuclear facility in Iraq by war planes from Israel was done to prevent Iraq from being able to produce plutonium-containing nuclear weapons. If Iraq had succeeded in developing a nuclear arsenal, the outcome of Operation Desert Storm in 1991 might have been quite different. A more recent international crisis involved efforts to dissuade North Korea from building a reactor to produce plutonium. And, in 1998, United Nations weapons inspectors were denied access to sites in Iraq

A smuggled canister of military grade Pu-239 captured in Germany.

suspected of being nuclear weapons production facilities. Given the risks associated with Pu-239 and U-235, it is essential that the supplies and distribution of these isotopes be carefully monitored nationally and around the world. The United States for many years banned the reprocessing of commercial fuel elements. That ban was lifted in 1981, but no plutonium is currently being recovered from commercial reactors in this country. The price of uranium is currently so low that plutonium recovery is not competitive.

Safeguarding nuclear materials has taken on a new meaning since the end of the Cold War and the demise of the former Soviet Union. One part of the problem is the plutonium and highly enriched uranium in Russia's nuclear arsenal (about 20,000 warheads). At present, though, these warheads are stored with relatively good security. Furthermore, any thief would find it difficult to remove the plutonium and uranium from the warheads. In contrast, Russia's legacy from the Cold War, a stockpile of highly enriched uranium and plutonium (about 600 metric tons), is far more accessible and hence far more threatening to world security. The fissionable materials stored in labs, research centers, and shipyards across the former Soviet Union are far more vulnerable to theft and their monitors accepting bribes. These 600 tons of fissionable material translate into the capacity to construct approximately 40,000 new nuclear weapons.

The dangers of nuclear trafficking and the need for effective safeguards are now well recognized by the world community. Anita Nilsson, head of the Vienna-based International Atomic Energy Agency's Office of Physical Protection and Material Security, warned in 2001 that nuclear trafficking "has emerged as a real and dangerous threat." In May 2001, officials from over 70 countries met in Stockholm to turn international attention on nuclear smuggling. They reported 370 confirmed cases of international smuggling of radioactive materials since 1993, with 9% of these involving plutonium or highly enriched uranium. Researchers also expressed concern that the frequency of confirmed incidents was growing and detection of radioactive smuggled goods was difficult.

The topics of enriched uranium, spent reactor fuel, fissionable stockpiles, and nuclear waste all rest on an understanding of the topic of radioactivity. We now turn to this topic.

7.15 Consider This: The Reality of Reprocessing

The supply of uranium in the United States (and the rest of the world) is large, but not limitless. By failing to reprocess spent nuclear fuel, we are discarding a potential source of energy that some European countries are presently tapping. Is the current American practice justified? Why is it different from the European practice? List arguments on both sides of this issue and then take a stand.

7.7 What Is Radioactivity?

Radioactivity was discovered accidentally in 1896 by Antoine Henri Becquerel (1852–1908). The French physicist found that when a uranium-containing mineral sample was placed on a photographic plate that had been wrapped in black paper, the plate's light-sensitive emulsion darkened. It was as though the plate had been exposed to light. Becquerel immediately recognized that the mineral itself was emitting a powerful form of radiation that penetrated the light-proof paper. Further investigation by Marie Curie revealed that the rays were coming from the element uranium, a constituent of the mineral. In 1899, Marie Curie applied the term **radioactivity** to this spontaneous emission of radiation by certain elements. Subsequent research by Ernest Rutherford (1871–1937) in Canada and England led to the identification of two major types of radiation. Rutherford named them after the first two letters of the Greek alphabet, alpha (α) and beta (β).

Marie Sklowdowska Curie won two Nobel Prizes—one in chemistry, the other in physics—for her basic research on radioactive elements.

Alpha and beta radiation have strikingly different properties. A **beta (β) particle** is a high-speed electron emitted from a nucleus. A beta particle has a single negative electrical charge (-1) and only a tiny bit of mass, about 1/2000 that of a proton or a neutron. In contrast, an **alpha (α) particle** is positively charged ($+2$) and far heavier than an electron. An alpha particle consists of the nucleus of a helium atom—two protons and two neutrons. Since no electrons accompany the helium nucleus, it has a +2 charge. It was subsequently discovered that a third form of radiation, gamma (γ) radiation, frequently accompanies the emission of alpha or beta radiation. Unlike alpha and beta radiation, **gamma rays** do not consist of particles. Rather, they are made up of high-energy, short-wavelength photons of energy and are part of the electromagnetic spectrum, as are infrared (IR), visible, and ultraviolet (UV) light rays. Gamma rays have no charge and no mass. The properties of these three types of radiation are summarized in Table 7.2. Note that the term *radiation* is used in two different, but overlapping contexts. *Electromagnetic* radiation refers to all the different types of light, including gamma rays. *Nuclear* radiation (including gamma rays) refers to the radiation emitted by nuclei.

- Gamma rays were introduced as part of the electromagnetic spectrum in Chapter 2. They are similar in energy to X rays, and you will learn more about them in connection with food preservation in Chapter 11.

Whenever an alpha or beta particle is given off during radioactive decay, a remarkable transformation occurs: The emitting atom changes to an atom of a different element! For example, an atom of uranium-238 is converted into an atom of thorium-234 (Th-234) when the U-238 loses an alpha particle. Such a change might be understood by the ancient alchemists, who sought to transmute lead and other common metals into gold. But, according to modern chemistry, elements and atoms are supposed to be unchanging and unchangeable. Yet, there is ample experimental evidence that *whenever an atom emits an alpha particle, it changes into an atom of the element with an atomic number two less than the original.* Such a transformation can be represented with a nuclear equation, as in the case of alpha emission by uranium-238 to form thorium-234.

$$^{238}_{92}\text{U} \longrightarrow {}^{234}_{90}\text{Th} + {}^{4}_{2}\text{He} \qquad [7.5]$$

- The sum of the mass numbers on both sides of nuclear equation 7.5 are equal: 234 + 4 = 238. The sum of the atomic numbers is also equal: 90 + 2 = 92.

The species undergoing radioactive decay is called the **parent** (here U-238); the product is called the **daughter product** (here Th-234). For every alpha emission, the mass number of the daughter is 4 less than the mass number of the parent. The loss of 4 nuclear particles—2 protons and 2 neutrons—accounts for this change.

Sometimes radioactive decay yields a daughter product that still is radioactive. This is the case with the thorium-234 produced in equation 7.5, a daughter product that in turn undergoes beta emission to yield protactinium-234 (Pa-234).

$$^{234}_{90}\text{Th} \longrightarrow {}^{234}_{91}\text{Pa} + {}^{\;\,0}_{-1}\text{e} \qquad [7.6]$$

Just as with alpha emission, beta emission results in the transformation of one element into another. In this case, Th-234 with atomic number 90 decays to the element Pa-234 with an atomic number of 91, one greater than that of the parent. *The number of protons (the atomic number) increases by one when a beta particle is ejected from a nucleus.* This suggests that you can regard a neutron as consisting of a proton and an electron. The loss of an electron (a beta particle) converts a neutron to a proton.

$$^{1}_{0}\text{n} \longrightarrow {}^{1}_{1}\text{p} + {}^{\;\,0}_{-1}\text{e} \qquad [7.7]$$

The mass number (neutrons plus protons) in the nucleus remains constant during beta emission because a neutron changes into a proton. In nuclear equation 7.6, both the parent and the daughter have the same mass number 234.

Table 7.2

Radioactive Emissions

Type of Radiation	Symbol	Consists of	Charge	Changes to the parent nucleus when this type of radiation is emitted.
Alpha	$^{4}_{2}\text{He}$	2 Protons + 2 Neutrons	+2	The mass number decreases by 4 and the atomic number decreases by 2.
Beta	$^{\;\,0}_{-1}\text{e}$	Electron	−1	There is no change in mass number, and the atomic number increases by 1.
Gamma	$^{0}_{0}\gamma$	Photon of energy	0	There is no change in either the mass number or in the atomic number.

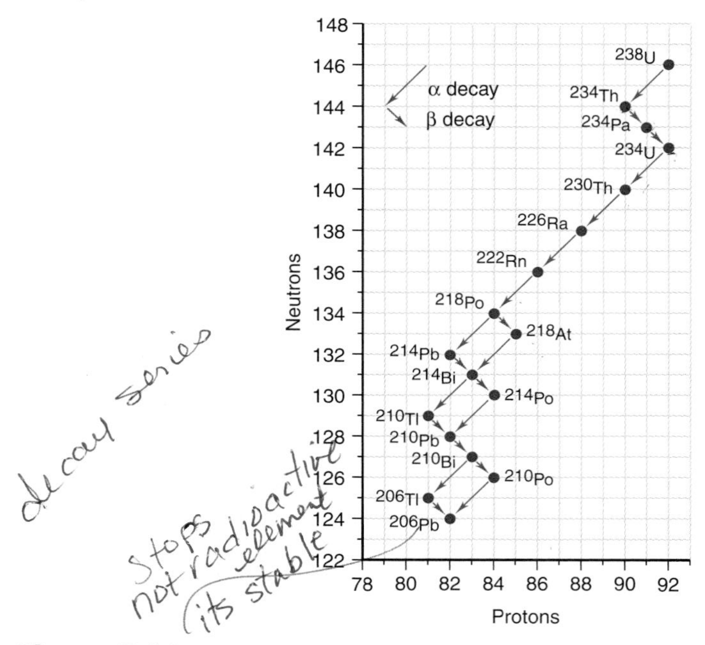

Figure 7.14
The U-238 decay series.

Most of the atoms that make up our planet are *not* radioactive. Whether an isotope is an alpha emitter, a beta emitter, or nonradioactive depends on the stability of the nucleus. Nuclear stability is related to the ratio of neutrons to protons in the particular isotope. By emitting alpha, beta, or other radiation, radioactive nuclei adjust this neutron-proton ratio until a stable neutron/proton ratio is achieved, and the nucleus is no longer radioactive. For example, the radioactive decay of U-238 and Th-234 (equations 7.5 and 7.6) are the first two steps in a long series. Naturally occurring U-238 decays through 14 steps until it reaches a stable isotope of lead, Pb-206 (Figure 7.14). For most elements, the most plentiful isotope is not radioactive. Most of the elements that make up our world have stable (nonradioactive) nuclei. However, *all* isotopes of *all* elements with an atomic number of 83 or greater are radioactive. This includes elements such as uranium, plutonium, radium, and radon, all of which we discuss later in this chapter.

7.16 Your Turn

a. Rubidium-86 (Rb-236) is produced in nuclear reactors and is a beta emitter. Identify its daughter product by name, symbol, atomic number, and mass number and write the nuclear equation for the radioactive decay of Rb-236.

b. Plutonium-239, a toxic isotope that causes lung cancer, is an alpha emitter. Identify its daughter product by name, symbol, atomic number, and mass number. Then write the nuclear equation.

c. Iodine-131 is an isotope used for the medical measurement of thyroid gland activity. I-131 is a beta emitter. Identify the daughter product produced by beta emission from I-131 by name, symbol, atomic number, and mass number. Then write a nuclear equation.

Answer

a. The daughter product is strontium-86, $^{86}_{38}\text{Sr}$. $^{86}_{37}\text{Rb} \longrightarrow {}^{86}_{38}\text{Sr} + {}^{0}_{-1}\text{e}$

7.8 What Hazards Are Associated with Radioactivity?

It may come as a surprise that the radiation exposure an American citizen receives from living near a nuclear power plant is about one-tenth the radiation he or she would get during a coast-to-coast trip on a commercial flight. Even under adverse circumstances, radiation exposure from nuclear power plants still can be low. If you had been at the gate of the Three Mile Island plant for the last week of March 1979 (the time of the accident), you would have been exposed to less radiation than from a single chest X-ray.

One nuclear accident stands as a notable exception—the explosion at Chernobyl discussed in Section 7.4

The evidence of the past makes clear that it would be a serious mistake to dismiss radioactivity as harmless. Unfortunately, some of the first scientists to study radioactivity, including Marie Curie (1867–1934), were not fully aware of the dangers inherent in the phenomenon. Madame Curie died of a form of leukemia that most likely was induced by her exposure to radiation. Often alpha and beta particles or gamma rays have sufficient energy to ionize the atoms and molecules they strike. As with bombardment by ultraviolet rays (Chapter 2), the resulting changes in molecular structure can have profound effects on living things. Rapidly growing cells are particularly susceptible to damage, a fact that has led to the use of radiation to treat some kinds of cancer. But bone marrow and white blood cells are also easily damaged, and anemia and susceptibility to infection are among the early symptoms of radiation sickness. Radiation-induced transformations of DNA can give rise to cancer or genetic mutation.

Today, considerable care is taken to shield medical, laboratory, and other workers from nuclear radiation. Protective shielding made of lead and other dense metals is used to absorb much of the radiation in the workplace. However, it must be recognized that it is impossible to be fully protected from exposure to radioactivity. The Earth itself, the building materials quarried or manufactured from it, our food, and even you and your best friends contain some naturally occurring radioactive atoms. Because of this, all these sources emit what is called *background radiation* (Figure 7.15). Note that the vast majority of background radiation (82%) is of natural origin. The major natural radiation source is radon, a radioactive gas released from rocks and minerals during natural decay of uranium-238 (Figure 7.14).

7.17 Your Turn

Radon-222 is an alpha emitter. Write the nuclear equation for the decay process. Would you expect the daughter to be radioactive?

- Radon was discussed in Section 1.14 as an indoor air pollutant.

Answer

$^{222}_{86}Rn \longrightarrow ^{218}_{84}Po + ^{4}_{2}He$; Polonium is radioactive, as are all elements with an atomic number of 83 or greater.

The amount of background radiation you receive depends on where you live, what type of residence you live in, the number of people you live with, and how close you get to them. The late Isaac Asimov, a prolific science writer, pointed out in one of his many books that a human contains approximately 3.0×10^{26} carbon atoms, of which 3.5×10^{14} are radioactive carbon-14 atoms. With each breath you inhale about three and a half million (3.5×10^{6}) C-14 atoms.

7.18 The Sceptical Chymist: Radioactive Carbon in Your Body

Assume that Isaac Asimov's figures are correct, and that 3.5×10^{14} of the 3.0×10^{26} carbon atoms in your body are radioactive. Calculate the fraction of carbon atoms that are radioactive carbon-14.

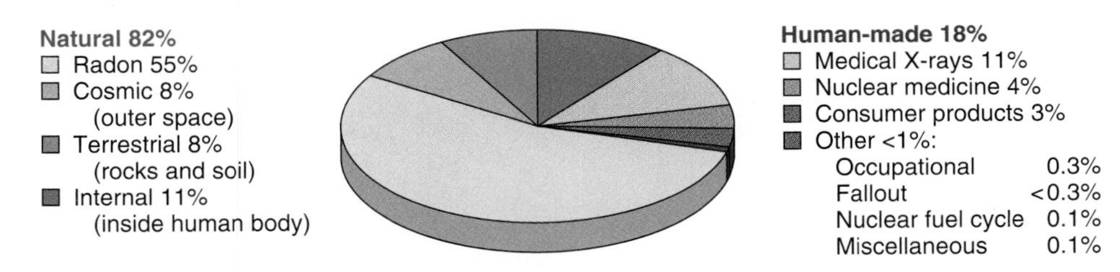

Figure 7.15

U.S. sources of background radiation. This chart shows that natural sources of radiation account for about 82% of all public exposure while human-made sources account for the remaining 18%.

(Source of data: National Council on Radiation Protection and Measurements (NCRP) Report No. 93, "Ionizing Radiation Exposure of the Population of the United States," 1987.)

To better evaluate radiation hazards, we need units to measure radioactivity and the damage it can cause. One simple way to measure radioactivity is in terms of the nuclear decay in a given time period. The curie (Ci), named after Marie Curie, is based on the number of decays per second from one gram of radium, a radioactive element that she isolated.

$$\text{One curie} = 1 \text{ Ci} = 3.7 \times 10^{10} \text{ disintegrations/second}$$

• The prefix pico means 1/1,000,000,000 or 1×10^{-9} and the prefix micro means 1/1,000,000 or 1×10^{-6}.

Here, disintegrations refer to alpha, beta, or gamma emissions. One curie is a large amount of radiation; typically amounts are measured in smaller units such as *pico*curies (pCi) or *micro*curies (μCi). For example, the amount of radon in a basement is measured in picocuries (7.22 Your Turn); laboratory radioactive samples are typically measured in terms of millicuries (mCi; 1 mCi = 10^{-3} Ci). A spill at a lab bench involving 100 millicuries would require special cleanup procedure, whereas one of 100 microcuries would not. To put these values in perspective, the explosion at Chernobyl spewed 100 to 200 million curies into the atmosphere, approximately 100 times the amount of radiation released by the atomic bombs that exploded on Nagasaki and Hiroshima. At the time of the accident, radiation near Chernobyl was measured at 5 to over 40 Ci per square kilometer (Figure 7.11).

The extent of biological damage caused by radiation depends, in part, on the total amount of energy absorbed. This energy is measured in a unit called the *rad*, short for radiation absorbed dose. Note that the curie and the rad measure radiation differently. The former focuses on the radioactive sample (how quickly it decaying), the latter on how much energy is being absorbed by the target. One **rad** is defined as the absorption of 0.01 joule of radiant energy per kilogram of tissue. But not all radiation is equally harmful to living organisms; some types are more destructive because of how often and how energetically they hit molecules. Therefore, to estimate the potential physiological damage, the number of rads is multiplied by a quality factor Q that is characteristic of the particular type of radiation. Highly damaging radiation, such as alpha particles and high-energy neutrons, has an Q of 10. Less damaging forms, including beta, gamma, and X rays, are assigned an Q of 1. When Q is multiplied by the number of rads, the product is called **rem,** short for "roentgen equivalent man".

$$\text{Number of rems} = Q \times (\text{number of rads})$$

Thus, a 10 rad dose of alpha radiation equals 100 rem. For beta and gamma radiation, a dose of 10 rads is the same in rems, 10 rems. The number of rems in a dose of radiation exposure is thus a measure of the power of the radiation to cause damage to human tissue. It would take approximately 10 times as many beta particles to do the same damage as a given amount of alpha particles. This is to be expected, because alpha particles deposit a greater amount of energy in the target tissue. The scientific community in the U.S., as well as most of the world, now uses the **sievert (Sv)** instead of the rem. The rem is a smaller unit than the sievert, and these units are related by a factor of 100.

$$1 \text{ Sv} = 100 \text{ rem}; \; 1 \text{ rem} = 0.0100 \text{ Sv}$$

The likely effects of a single dose of radiation at various levels are given in Table 7.3.

Table 7.3

Physiological Effects of a Single Dose of Radiation

Dose (rem)	Dose (Sv)	Likely Effect
0–25	0–0.25	No observable effect
25–50	0.25–0.5	White blood cell count decreases slightly
50–100	0.5–1	Significant drop in white blood cell count, lesions
100–200	1–2	Nausea, vomiting, loss of hair
200–500	2–5	Hemorrhaging, ulcers, possible death
>500	>5	Death

Because most doses of radiation are significantly less than one sievert (or one rem), smaller units such as **microsieverts** (μSv) and **millirems** (mrem) are employed.

1 microsievert = 1/1,000,000th of a sievert = 1×10^{-6} Sv = 1 μSv

1 millirem = 1/1000th of a rem = 1×10^{-3} rem = 1 mrem

- One microsievert is 0.1 millirem.

Table 7.4 uses microsieverts to report the radiation exposure associated with various activities and lifestyle factors. This information is the basis for estimating your personal annual radiation exposure in 7.19 Your Turn. Once you have completed this exercise, you can check Table 7.4 to see how your exposure compares with that of the average American.

7.19 Your Turn

Use Table 7.4 to estimate the approximate radiation dosage you receive each year.

Note that nearly 3000 μSv (82%) of the approximately 3600 μSv absorbed in a year by a typical resident of the United States comes from natural background sources—mostly radon, cosmic rays, soil, and rock—not from human-related sources. Of the 670 μSv of artificial radiation absorbed annually, about 600 are attributable to medical procedures such as diagnostic X-rays. As is evident from Table 7.4, the radiation emitted

Table 7.4

Your Annual Radiation Dose

Source of Radiation	μSv/yr
1. Location of your town or city	
a. Cosmic radiation at sea level (U.S. average 260 μSv*)	260 μSv
b. Additional dose if you are above sea level	______ μSv
1000 m (3300 ft) add 100 μSv	
2000 m (6600 ft) add 300 μSv	
3000 m (9900 ft) add 900 μSv	
2. House construction	
Building materials contain tiny amounts of radioisotopes.	
Brick, 700 μSv; wood, 300 μSv; concrete, 70 μSv	______ μSv
3. Ground	
Radiation from rocks and soil (U.S. average)	260 μSv
4. Food, water, and air (U.S. average)	400 μSv
5. Fallout from nuclear weapons testing (U.S. average)	40 μSv
6. Medical and dental X-rays	
a. Chest X-ray (100 μSv)	______ μSv
b. Gastrointestinal tract X-ray (5000 μSv)	______ μSv
c. Dental X-rays (100 μSv each visit)	______ μSv
d. Other X-rays (estimate)	______ μSv
7. Jet travel (exposure to cosmic radiation.)	
A 5-hr flight at 30,000 ft is 30 μSv	______ μSv
8. Other	
Live within 50 miles of a nuclear plant site, add 0.09 μSv	______ μSv
Live within 50 miles of a coal-fired power plant, add 0.3 μSv	______ μSv
Use a computer terminal, add 1 μSv	______ μSv
Go through X-ray check stations at airports, add 0.02 μSv	______ μSv
Smoke 1.5 packs of cigarettes a day, add ~13,000 μSv	______ μSv
Your Total Annual Dose of Radiation	______ μSv
Compare your annual dose to the U.S. annual average of 3,600 μSv.	

*Based on the "BEIR Report III," National Academy of Sciences, Committee on Biological Effects of Ionizing Radiation 1987. *The Effects on Populations of Exposure to Low Levels of Ionizing Radiation*. Washington, DC: National Academy of Sciences.

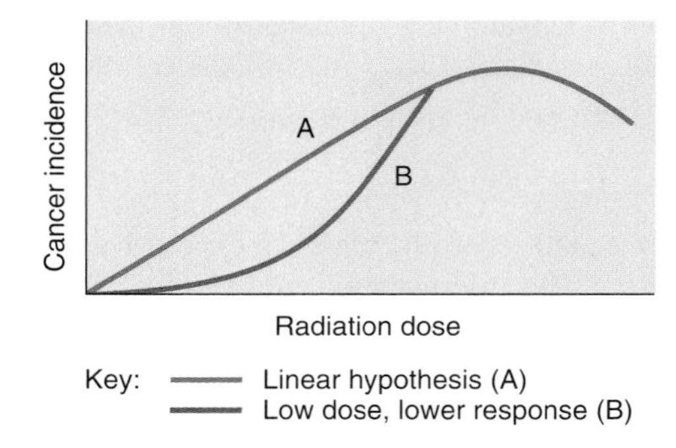

Figure 7.16

Dose response curves for radiation. The curves level off and then decrease as radiation doses get very high because more cells die than become cancerous.

by a properly operating nuclear power plant is negligible compared with normal background radiation, including the natural radiation of your own body. For example, about 0.01% of all the potassium ions (K^+) that are essential to your internal biochemistry are K-40, a radioactive isotope. These K-40 ions give off about 200 μSv per year, approximately 1000 times more radioactivity than that received as a result of living within 20 miles of a nuclear power plant. In fact, because bananas are rich in potassium, a steady diet of them contributes to your personal radioactivity.

To put things further in perspective, note that the immediate physiological effects of radiation exposure are generally not observable below a single dose of 0.25 sieverts (250,000 μSv; Table 7.3). This is nearly 70 times the average annual exposure. The more a given dose of radiation is spread out over time, the less harmful it appears to be. However, there is still uncertainty about the long-term effects of low doses of radiation. The assumption is usually made that there is no threshold below which no damage occurs. However, the effects of low doses are so small and the time span so great that scientists have not been able to make reliable measurements. Moreover, tests with animals are not always reliable because there is considerable species-to-species variation in the effect of radiation.

The issue then is how to extrapolate the known high-dose data to low doses. Two dose-response models are illustrated in Figure 7.16. The assumption of a linear relationship between the incidence of cancer and radiation dose is represented by curve A. In this model, doubling the radiation dose doubles the incidence of cancer, tripling it causes three times the number of cancers, and so on. This is the relationship currently used by federal agencies in setting exposure standards. Many scientists believe that the biological effect is relatively less at low levels of radiation because of the self-repairing mechanism of cells. This model is represented by curve B, which drops below the straight line of curve A.

7.20 Consider This: Radiation Dose Response

You have just read about two models for the dose-response curve for low level radiation (Figure 7.16). The linear hypothesis plot represents the more stringent hypothesis used by federal agencies in setting exposure limits. The low-dose, lower response curve is believed by many scientists to be closer to our true biological susceptibility.

The ramifications of adopting a specific model are both biological and economical. The more stringent linear dose model requires stricter limits on workers' acceptable radiation dose limit than the lower dose model. By using the linear model, are we being "better safe than sorry" or are we wasting a lot of money protecting ourselves from an emotional issue without looking at the science behind it?

As a nuclear medicine technician who must operate under the stricter federal limits for radiation safety at a hospital, write a letter to an interested friend giving your position on this issue and the reasons for it.

7.9 How Long Will Nuclear Waste Products Remain Radioactive?

Here is one area in which popular magazines, in spite of their frequent hyperbole, seldom exaggerate the problem. Some of the products formed in nuclear reactors do indeed have dangerously high levels of radioactivity for thousands of years. There is no way to speed up the radioactive decay process. The fact that most of us experience considerably more radioactivity from natural sources than from artificial ones should not lull us into a false sense of security. Nuclear waste disposal presents formidable problems because one radioactive isotope often generates others. As noted in the previous section, daughter products from decays act as parents for other radioactive emissions; for example, U-238 decays to Th-234, which in turn yields Pa-234 (Figure 7.14).

• See nuclear equations 7.4 and 7.5

A particularly significant consideration in the disposal of radioactive waste is the *rate* at which the level of radioactivity declines. Depending on the particular isotope, the decline can occur very rapidly over a short time or very slowly over a long period. The rate of decay is typically reported in terms of the **half-life,** the time required for the level of radioactivity to fall to one-half of its initial value. For example, plutonium-239, the alpha-emitting fissionable isotope formed in uranium reactors, has a half-life of 24,400 years. This means that it will take 24,400 years for the radiation intensity of a freshly generated sample of Pu-239 waste to drop to one-half its original value. At the end of a second half-life of another 24,400 years, the radiation will be one-fourth the original level. And in three half-lives (a total of 73,200 years), the level will be one-eighth of the original (Figure 7.17).

The half-life of a particular isotope is a constant, and is *independent* of the physical or chemical form in which the element is found. Moreover, the rate of radioactive decay is essentially unaltered by changes in temperature and pressure. But, when various radioisotopes are compared, their half-lives are found to range from millennia to milliseconds. For example, the half-life of uranium-238 (equation 7.5) is 4.5 billion years. (Coincidentally, this is approximately the age of the oldest rocks on Earth, a determination made by measuring their uranium content.) By contrast, other radioisotopes have much shorter half-lives.

Radioisotope	Half-life
Thorium-234	24 days
Plutonium-239	24,400 years
Plutonium-231	8.5 minutes
Polonium-214	0.00016 seconds

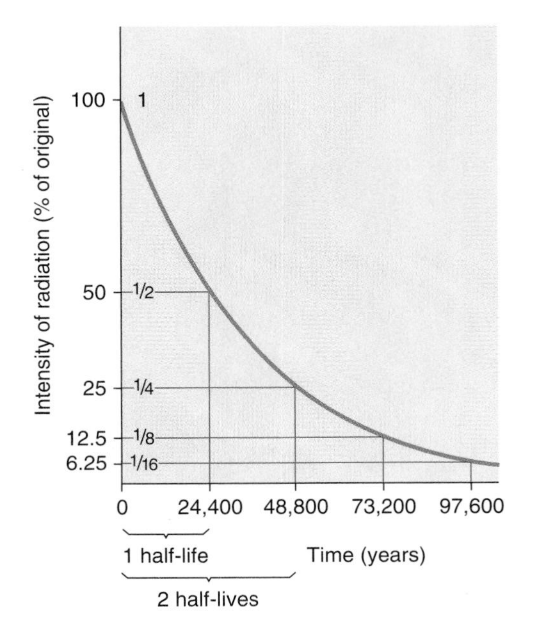

Figure 7.17
Radioactive decay of Pu-239.

7.21 Your Turn

Using the half-life of Pu-231 just given, what percent of a sample of Pu-231 would remain after about 25 minutes?

Answer

25 minutes is about three half-lives (8.5 minutes × 3). After one half-life, 50% of the sample has decayed and 50% remains. After two half-lives, 75% of the sample has decayed and 25% remains. And after three half-lives, 87.5% has decayed and 12.5% remains. Note that we could have asked this question as, "After 25.5 minutes (exactly three half lives), what percent remains?" However, when estimating radioactivity, it is helpful to be able to make a quick estimate.

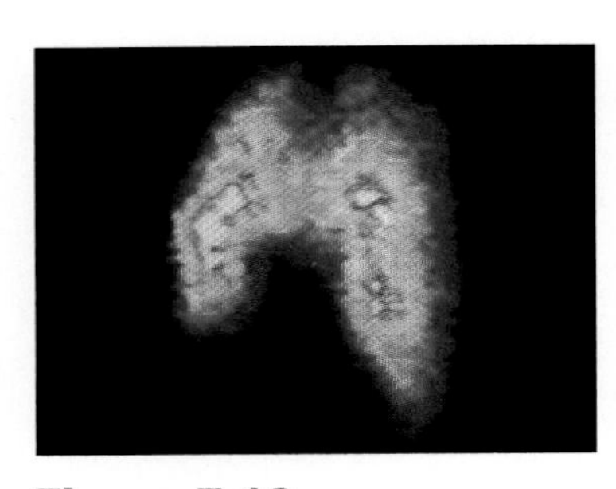

Figure 7.18

A thyroid image produced with I-131. The yellow and red regions show the area of the thyroid gland in which radioactive iodine has been concentrated.

Tritium (Hydrogen-3 or H-3) is sometimes formed in the primary coolant water of a nuclear reactor. Tritium is a beta emitter with a half-life of 12.3 years. Iodine-131, in the form of iodide ions, with a half-life of 8 days, is used to treat hyperthyroidism in persons with Graves' disease. In this procedure, the orally administered radioactive iodide ions concentrate in the overactive thyroid gland, fully or partially destroying it (Figure 7.18). In most patients, thyroxin, the iodine-containing hormone normally secreted by the thyroid, must then be supplemented with a synthetic substitute.

Strontium-90 (Sr-90) is a particularly dangerous isotope in the fallout from nuclear weapons testing. Strontium ions are chemically similar to calcium ions; both elements are in Group 2A of the periodic table. Hence, strontium (Sr^{2+}), like Ca^{2+}, concentrates in milk and bone. There, the radiation from the Sr-90 can pose a lifelong threat to an affected individual because of the 28.9-year half-life of Sr-90. Significantly, I-131 and Sr-90 are among the fission products produced in nuclear reactors, and there is concern that many persons living near Chernobyl were exposed to harmful levels of both isotopes. The incidence of thyroid cancer among children in the Chernobyl radioactive fallout area seems significantly higher than normal, and iodine-131 has been implicated (Figure 7.19).

7. 22 Your Turn

When people speak of radioactive iodine, they may be referring to iodine in the chemical form of an iodine atom, an iodine molecule, or an iodide ion. Write Lewis dot structures to distinguish among these three chemical species. In the previous paragraph, radioactive iodine is taken up by the thyroid gland in the form of the iodide ion.

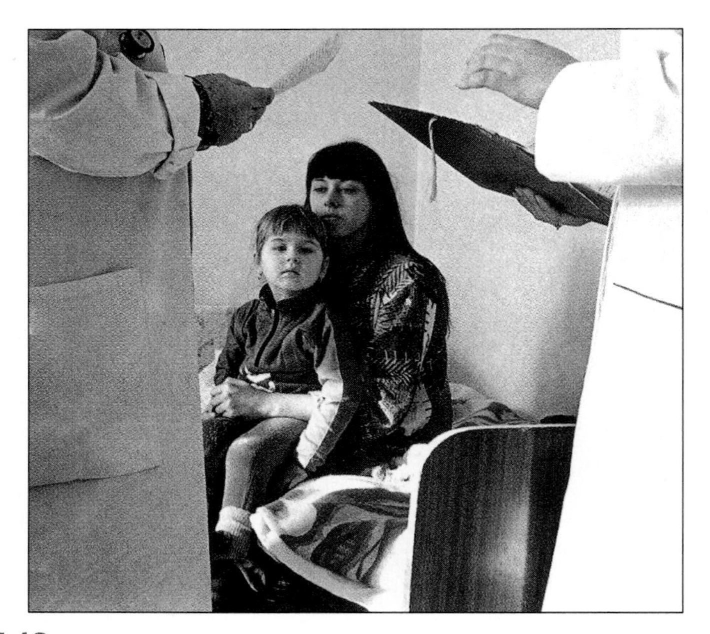

Figure 7.19

A child in Ukraine awaiting a thyroid examination at the Kiev Institute of Endocrinology.

Carbon-14 (C-14), with a half-life of 5730 years, is a beta emitter that decays to nitrogen-14 (N-14). This isotope of carbon is well known because it is often used to determine the age of the remains of once-living things or objects made from them. Atmospheric carbon dioxide contains a constant steady-state ratio of one radioactive carbon-14 atom for every 10^{12} atoms of nonradioactive carbon-12. Living plants and animals incorporate the isotopes in that same ratio. However, when the organism dies, exchange of CO_2 with the environment ceases. Thus, no new carbon is introduced to replace the C-14 converted to nitrogen-14 by beta decay. As a consequence, the concentration of C-14 decreases with time, dropping by half every 5730 years. In the 1950s, W. F. Libby first recognized that by experimentally measuring the C-14/C-12 ratio in a sample, the time at which the organism died could be estimated. Human remains and many human artifacts contain carbon, and fortunately, the rate of decay of C-14 is a convenient one for measuring human activities. Charcoal from prehistoric caves, ancient papyri, mummified human remains, and suspected art forgeries have all revealed their ages by this technique. The C-14 technique provides ages that agree to within 10% of those obtained from historical records, thus validating the legitimacy of the radiodating technique.

- Carbon-14 was used to establish the age of the Shroud of Turin.

7.23 Your Turn

Radon-222 is a radioactive gas produced by the decay of radium, which is naturally present in hard rocks. The half-life of Rn-222 is 3.8 days.

a. Where did the radium in rocks come from? *Hint:* See Figure 7.14.
b. Radon activity is usually measured in picocuries (pCi). Suppose that the radioactivity from Rn-222 in your basement was measured as 16 pCi (which would be high). If no additional radon entered the basement, how much time would pass before the radiation level fell to 0.50 pCi?
c. Why is it incorrect to assume that no more radon will enter your basement?

Hint: **b.** Note that in dropping from 16 pCi to 1 pCi, the radiation level is reduced by half four times: $16 \longrightarrow 8 \longrightarrow 4 \longrightarrow 2 \longrightarrow 1$. This corresponds to four half-lives, each 3.8 days.

7.10 How Will We Dispose of Waste from Nuclear Plants?

The experience of more than 50 years seems to suggest that the answer to this vitally important question is, "slowly and with difficulty." Whether that is the correct answer is another matter. In a June 1997 *Physics Today* article, John Ahearne, past Chairman of the U.S. Nuclear Regulatory Commission, reminds us that, ". . . Like death and taxes, radioactive waste is with us—it cannot be wished away. . ."

High-level nuclear waste (HLW) has high levels of radioactivity and, because of the long half-lives of the radioisotopes involved, requires essentially permanent isolation from the biosphere. This waste comes in a variety of chemical forms, including ones that are highly acidic or basic, and contains heavy metals that are toxic. Thus, HLW is sometimes labeled a "mixed waste" in that it is hazardous *both* because of the chemicals it contains *and* from their radioactivity. Furthermore, this waste also presents a security risk because it contains fissionable plutonium that could be extracted and used to construct nuclear weapons.

- Unlike uranium, all isotopes of plutonium can be used to construct nuclear weapons.

Federal statutes define HLW by its source, rather than by its chemical and nuclear characteristics. According to the regulations, *HLW consists of radioactive materials that result from the reprocessing of spent nuclear fuel*; for example, the waste created when fuel is reprocessed to produce plutonium for military uses or that from commercial nuclear power plants. HLW also includes several other highly radioactive materials that require permanent isolation. Approximately 99% of the volume of HLW in the United States originated in the nuclear weapons industry. In 1996, the U.S. Department of Energy reported that the accumulated high-energy radioactive wastes generated by the Department of Defense occupied a total volume of approximately 350,000 m^3 with a radioactivity of about 900 million curies. This volume corresponds to nine football fields covered to a depth of 30 feet. Military waste is in the form of solutions, suspensions, slurries, and salt cake stored in barrels, bins, and tanks.

Spent nuclear fuel (SNF), is regulated as HLW. Radioactive spent fuel is an unavoidable by-product of nuclear reactors. After about three or four years of use, the U-235 concentration in the fuel rods of a nuclear power reactor, initially at 3–5%, drops to the point where it is no longer effective in sustaining the fission process. Approximately one-fourth to one-third of the fuel rods are replaced annually, on a rotating schedule. Nevertheless, the spent fuel rods are still "hot"—both in temperature and radioactivity. They contain various isotopes of uranium plus plutonium-239 formed by the capture of neutrons by U-238, and a wide variety of fission products such as iodine-131, cesium-137, and strontium-90. Remotely controlled machinery, operated by workers protected by heavy shielding, removes the spent rods from the reactor and replaces them with new fuel rods. The spent rods are transferred to on-site deep pools where they are cooled by water containing a neutron absorber. As of January 1, 1997, over 34,000 metric tons of SNF has been discharged from the nation's commercial reactor sites. This figure does not include the spent waste from defense reactors or nuclear submarines, and is estimated to reach 52,000 metric tons by 2005. Most of this waste is currently under water in storage pools on site at the nuclear plant where it was used (Figure 7.20). For example, the storage facility at Seabrook is a 34-foot deep steel-lined concrete pool in a secure building. It has the capacity to hold up to 25 years' worth of nuclear waste.

- Depleted uranium (DU) was mentioned in 7.14 Consider This. Although officially defined by the Atomic Energy Act of 1954 as a "source material" and not a waste, the inventory of DU (mostly as UF_6) in the U.S. is now estimated to be over a billion pounds.

The on-site storage at nuclear power plants of high-level radioactive waste for 25 years is hardly ideal. John Ahearne points out that "Almost all of the [spent fuel] waste is currently being stored at the sites where it was generated, in facilities that were not built for long-term storage." The initial plans, begun in the 1950s and early 1960s, had been to reprocess the spent fuel to extract plutonium and uranium from it and recycle these elements as nuclear fuel to produce additional energy. Storage capacity for spent fuel rods on site was designed with such reprocessing in mind. However, only one of several planned reprocessing plants actually went into operation, and then only briefly (1967–75). Thus, reprocessing never was capable of keeping up with the rate of spent fuel production, about 2000 tons every year. In 1977, President Jimmy Carter, a nuclear engineer, declared a moratorium on commercial nuclear fuel reprocessing that continues to this day. Long-term geological storage of HLW, first proposed in 1957 by the National Academy of Sciences, became the alternative option. It is estimated that, by 2010, the country will face a total disposal problem of more than 100,000 tons of military and civilian high-level nuclear waste. The year is significant, because it is the earliest date when a permanent underground repository can possibly open. Given past and recent experience, 2010 seems an unrealistic goal.

- Each of the 103 U.S. commercial nuclear reactors produces about 20 tons of spent fuel annually.

Figure 7.20
Spent cooling rods from a nuclear power reactor in a cooling chamber.

Two feasible options for the storage of HLW are now under consideration: monitored storage on or near the surface, and storage in geological repositories deep underground. These differ in a key variable: *active management* (Figure 7.21). In surface storage, human societies over thousands of years must commit resources to maintain the integrity of the wastes. In geological repository storage, the wastes may be accessible and retrievable (although less easily) or sealed "forever," requiring minimal human vigilance. In a report published in 2000 by the National Academies of Science, the latter option of deep underground storage was favored, noting that it was not prudent to assume that future societies on Earth would be able to maintain surface storage facilities.

Yet, no long-term HLW storage facility of any type currently exists in the United States (or in any other country). The absence of a long-term repository is becoming a significant impediment to the use of nuclear power. In fact, in the 1970s, some states passed laws prohibiting the construction of any new nuclear power plants until the federal government demonstrated that radioactive wastes could be disposed of safely and permanently. The Department of Energy contracted with electrical utility companies to begin accepting spent fuel elements for underground long-term storage in 1998, but has still not met this deadline in 2002.

Progress in preparing a national underground disposal site for HLW continues to be painfully slow. A site must be found that is suitable for storage that will remain isolated from the ground water for tens of thousands of years. The concept is to carve out a chamber at least 1000 feet below ground, 1000 feet above the water table, in an appropriate rock formation. Salt, basalt, tuff, granite, and shale have all been considered. Salt domes, which are geological formations entirely of salt, are particularly attractive because they are very stable, extremely dry, and self-sealing if cracks should appear. Granite and basalt always contain cracks, but they have a great capacity to chemically absorb most wastes. HLW would be stored in such a chamber for at least 10,000 years, long enough for the high level of radioactivity to decrease significantly. It is estimated that the beta-emitting fission products need to be isolated for 300 to 500 years, about 10 half-lives for species such as Sr-90 and Cs-137. Uranium and heavier elements such as plutonium typically have much longer half-lives, but their radiation intensity is much lower. A sobering time line is shown in Table 7.5.

- After 10 half-lives, the radioactivity of a radioisotope has dropped essentially to background level.

Most plans call for encasing the spent fuel elements in ceramic or glass, packing the product in metal canisters, and burying them deep in the Earth in a designated repository. A method called vitrification has been developed to contain reprocessed defense wastes, including Pu-239, for future geological burial. The wastes are mixed with finely ground glass and melted to about 1150°C. The molten glass and wastes are then poured into stainless steel canisters, cooled, and capped for on-site storage until development

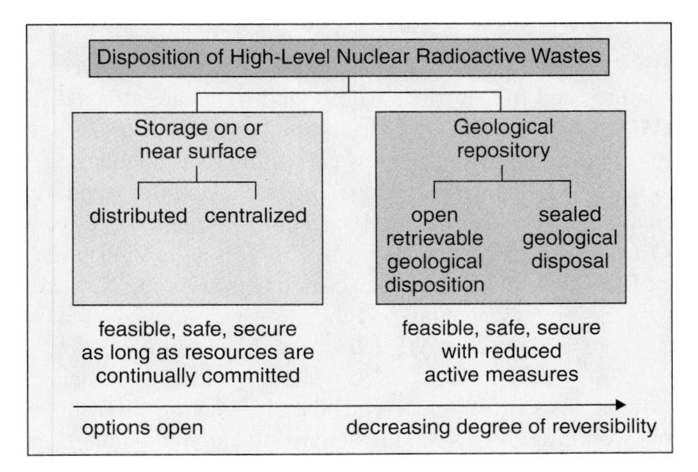

Figure 7.21

Methods of high-level nuclear waste deposition.

(Source: Disposition of High-Level Wastes and Spent Nuclear Fuel, National Academy Press, 2000.)

Table 7.5

Time line for HLW Underground Repository

Year	Event
2005	Construction on underground storage site begins.
2010	Waste storage begins.
2035	Loading ends.
2060	Waste packages retrievable until this time.
2316	Repository sealed by this year.
3000	Most dangerous radioactive substances have decayed to stable products. First waste package is assumed to fail because of manufacturing defects.
12010	End of regulatory period of 10,000 years. Radioactive exposure of farmers in nearby valley is predicted to be 0.007 μSv/yr, an insignificant amount.
102010	Waste packages predicted to begin failing from corrosion.
312010	Radioactive exposure for nearby farmers predicted to reach 250 μSv/yr, a dose that concerns regulators.
622010	Peak radioactive exposure for farmers predicted at 850 μSv/yr.

Source: Adapted from *New York Times*, Science Times, August 10, 1999, D4.

- In 2001, the EPA released standards for radiation exposure as part of a lengthy report about deep underground geological storage. The standards limit exposure by all pathways to 150 μSv/yr. These standards must be met before the repository can accept waste (and are not met by the estimated radiation releases in Table 7.5).

of a long-term underground repository (Figure 7.22). More than one million pounds of waste have been treated this way.

Figure 7.22
Encapsulating reprocessed HLW in glass canisters.

7.24 Consider This: Nuclear Waste Warning Markers

The Department of Energy recently asked 13 experts to design a system of markers to be installed near an underground nuclear waste repository in New Mexico, warning future generations of the existence of nuclear waste. The markers must last for at least 10,000 years (more than four times the age of the pyramids of Egypt), the end of the regulatory period. The message on the markers must be intelligible to Earthlings of the future. Try your hand at designing these warning markers, keeping in mind the changes that have occurred in *Homo sapiens* during the past 10,000 years and those that might occur in the next 10 millennia.

To establish deep geological storage of HLW, the federal government must deal with state legislatures and Indian tribes whose land rights are affected. The "not in my backyard" (NIMBY) syndrome has even led a number of states to adopt legislation prohibiting the disposal of high-level nuclear waste within their boundaries. The Mescalero Apaches of southern New Mexico may be unique in bucking this trend. In 1995 the tribe voted to implement a plan for storing high-level commercial radioactive waste on reservation lands. Over 30 utility companies are willing to pay a high price for temporary storage of spent fuel rods. But the plan has been opposed by the New Mexico state legislature and Congressional delegation, and the issue was unresolved as this book went to press in 2002. Southeastern New Mexico is also the location of a five-year waste isolation pilot project in which salt beds 2150 feet below ground are being tested as storage reservoirs. If the tests prove successful, the site may be used to store transuranic wastes, elements with lower levels of radioactivity and atomic numbers greater than that of uranium.

Currently, Yucca Mountain (Figure 7.23) in Nevada is the leading candidate for HLW storage by the Department of Energy. But it is not certain that the Yucca Mountain depository will ever become operational. In 1999, Warner North, senior vice president of Decision Focus Inc., summarized the situation: "The most formidable problems

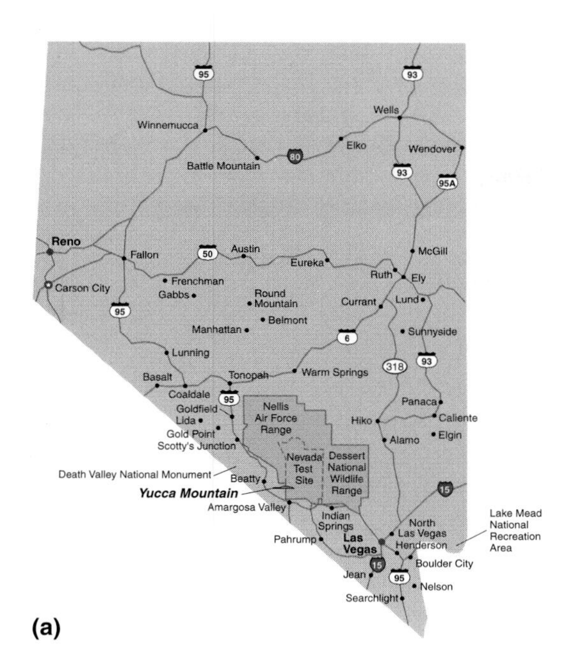

(a)

(b)

Figure 7.23
(**a**) Map of Yucca Mountain, NV and state of Nevada (Source of data: Department of Energy), (**b**) an aerial view of the proposed HLW repository at Yucca Mountain.

associated with using Yucca Mountain are political ones." The selection of Yucca Mountain as the site for study of underground burial of HLW required overcoming difficult political barriers as well as technical ones. The 1982 Nuclear Waste Policy Act projected that the initial geological repository would be constructed in time to receive high-level nuclear waste by 1998. In 1987, the Nuclear Waste Policy Amendments Act designated Yucca Mountain as the sole site to be studied as an underground long-term, high-level nuclear waste repository. To fulfill the requirements of the 1982 Act, the Department of Energy (DOE) contracted with nuclear utilities to begin accepting spent fuel beginning in 1998. To date, utility companies have paid over $14 billion to fund the development of a repository to do so. According to a recent court decision, that responsibility remains with DOE, even though no such facility is available now.

Meanwhile, tunnels are being dug 1400 feet beneath Yucca Mountain, Nevada (Figure 7.23) to determine its adequacy for deep, long-term storage of HLW. The Department of Energy has already spent an estimated $54 billion on the project. If completed, the site will be the largest radioactive storage facility in the world, with a capacity of 70,000 tons of spent fuel and 8,000 tons of high-level military waste. As shown in Table 7.5, roughly 25 years will be required just to transport the waste to the Nevada site, at a rate of 20 shipments per day.

Congress again got involved with HLW disposal by proposing legislation to create the Nuclear Waste Policy Act of 1997. Under this legislation, an *interim* above-ground storage site would be created for 40,000 tons of spent fuel at the Nevada Test Site, adjacent to the unfinished Yucca Mountain site. Spent fuel would be stored at the interim site, to be operational by January 2002, in metal canisters inside concrete bunkers until a permanent repository is ready. In discussing the bill, Nevada Representative Jim Gibbons complained that, "The people in support of this bill are the ones who have nuclear waste in their districts and want to get it out—get it from wherever it is into the state of Nevada." Nevada has no commercial nuclear reactors, and Nevada officials are concerned that a permanent site might never be approved, thereby leaving Nevada stuck with the temporary facility indefinitely.

Congressional opponents of the bill have dubbed it the "mobile Chernobyl" bill. They are concerned that the spent fuel high-level waste would be transported by rail and highway through 43 states within half a mile of 50 million Americans before it reached the proposed Nevada Test Site interim repository. Speaker of the House Dennis Hastert of Illinois, a supporter, declared that the bill "assures that another 15 years will not pass before the federal government lives up to its responsibility of accepting spent fuel." His

• In July, 2002, the Senate gave final approval for establishing an underground nuclear waste repository at Yucca Mountain. The site is projected to open in 2010.

position is not surprising given that Illinois has more commercial nuclear reactors (11) than any other state. Former President Clinton promised to veto the legislation, if passed by Congress. The veto was not necessary. In June 1998, unable to work out a satisfactory compromise between the House and Senate versions of the bill, Congress failed to take final action on the bill. Such controversial political maneuvering is sometimes described as the "NIMTO" phenomenon—Not In My Term in Office.

7.25 Consider This: Yucca Mountain

Use the resources of the Web to determine the current status of Yucca Mountain as a HLW long-term repository. Are the plans stalled or moving ahead to construct a long-term repository at this site? What other options are being considered? Be sure to cite your sources.

Some members of Congress have argued that a decision on a 10,000-year storage facility should be postponed and a 100-year storage facility should be developed. Congress (or its successor) may still be debating the issue 24,400 years from now, when the plutonium-239 completes its *first* half-life. Other disposal methods seem even less promising. Disposal in deep-sea clay sediments, under 3000–5000 m of water was investigated. Proposals to bury the radioactive waste under the Antarctic ice sheet or to rocket it into space have largely been discredited. But one thing is sure: Whatever disposal methods are ultimately adopted, they must be effective over an extremely long period of time.

7.26 Consider This: Storage of Nuclear Waste in Developing Countries

Some industrialized nations have proposed a novel method of foreign aid, with strings attached. The producers of nuclear power would ship their radioactive wastes to developing countries and pay them to store it. If you were the president of a developing country that was considering such an arrangement, what issues would you need to consider? Prepare a position paper in outline form listing the reasons for and against such an offer of foreign aid, and then reach a decision based on the relative importance of the factors in your list.

7.11 What Is Low-Level Radioactive Waste?

Not all nuclear waste is HLW. Nearly 90% of the volume of all nuclear waste is low-level waste. **Low-level radioactive waste (LLW** or LLRW) is described as waste contaminated with smaller quantities of radioactive materials than HLW, and specifically excludes spent nuclear fuel. LLW waste includes a wide range of materials. Some have radioactivity levels that can be quite low, such as laboratory clothing, gloves, and cleaning tools from medical procedures using radioisotopes, and from discarded smoke detectors. Other types of LLW have higher levels of radioactivity, such as the wastes generated by nuclear fuel fabrication facilities, the manufacture of radioactive pharmaceuticals, mining, and research facilities. Approximately twice the volume of LLW comes from military sources as from commercial and medical sources. It is estimated the United States will have 4.5 million cubic meters of low-level nuclear waste by the year 2030.

Since low-level nuclear waste is far less radioactive than high-level waste, it is disposed of differently. For example, certain types of LLW are put into sealed canisters and buried in lined trenches 10 m deep (Figure 7.24). Low-level military nuclear waste is disposed at federally owned sites maintained by the Department of Energy. Nonmilitary low-level waste disposal is the responsibility of the state where it is generated.

Although low-level waste poses significantly less danger than high-level waste, the NIMBY syndrome operates with low-level nuclear waste as well. The very idea of radioactive waste, even if low-level, is sufficient to generate considerable opposition to proposed sites for LLW. Congress expected that states, through the Low Level Radioactive Waste Policy Act of 1980, would form regional compacts by which compact members would send their low-level waste to a disposal site in one state. Such com-

Figure 7.24
Burial of low-level nuclear waste.

pacts have not been successful, including failed efforts in Illinois and in New York, each costing $55 million over eight years.

Currently, there are two commercial LLW disposal sites in operation, one at Barnwell, SC and another at Richland, Washington. About 70% of the waste shipped to the South Carolina site is from other states. The Washington state site accepts the other 30%, but only from a limited number of states, all nearby except Alaska and Hawaii. A third site in Clive, Utah, also accepts some commercial and Department of Energy low-level waste in addition to other types of waste. Thirty-five years ago, six commercial LLW sites were in operation. Two were closed in the 1970s, a third in 1986, and a fourth in 1992. In 2001, the National Academy of Sciences issued a report expressing concern, stating that efforts to open any new sites were "deadlocked" politically. There is an ongoing need to safely store the low-level radioactive wastes generated across the nation by hospitals, nuclear manufacturers, and research laboratories. Lack of access to such disposal sites could compromise the ability of such facilities.

If sufficient places cannot be found in a country for LLW disposal, why not send the waste to another country? This is not a rhetorical question; such transfer actually has been considered. For example, in 1997, the Taiwan Power Company (Taipower) began negotiations to ship low-level nuclear waste to the Democratic People's Republic of Korea (North Korea) for burial in abandoned coal mine shafts. Although the North Korean government, badly in need of money, agreed to accept an initial shipment of 60,000 barrels over two years, the deal did not go through. The agreement was for a cash settlement of about $1150 per barrel for a total payment of approximately $227 million U.S. dollars. In contrast, a proposed LLW waste site in the United States would be evaluated by the Nuclear Regulatory Commission, typically taking about eight years at a cost of $1 million or more, before taking final action.

7.12 Nuclear Power Worldwide

Globally, about 17% of the electricity produced and consumed globally is generated in about 440 nuclear power plants (and these figures are slowly rising). If this amount of energy were to be replaced, it would require the entire annual coal production of the United States or the former Soviet Union. Thus, there is already a relatively small but significant international reliance on nuclear energy. From Table 7.6, you can see that the United States has the largest number of nuclear reactors and is by far the largest generator of electricity from nuclear power. But, as of 2002, nearly one-third of the U.S. nuclear units are over 30 years old. Furthermore, as noted in the opening section of this chapter, no new reactors are currently under construction.

Table 7.6

Nuclear Power in Selected Countries

	Reactors		Electrical Power from Nuclear Reactors (MW)
	Operating June 2002	Under Construction	
Brazil	2	0	1,855
Canada	14	6	9,998
China	3	8	2,167
France	59	0	63,203
Germany	19	0	21,141
Hungary	4	0	1,755
India	14	2	2,548
Japan	53	4	43,505
Mexico	2	0	1,364
Russia	30	3	20,793
South Korea	16	4	12,970
Sweden	11	0	9,460
United Kingdom	33	0	12,528
United States	103	0	98,060

Source: World Nuclear Association, http://www.world-nuclear.org/info/reactors.htm.

The numbers of reactors and total power output, however, do not tell the complete story. A more interesting measure is the percentage of electrical power a country obtains from fission reactors, as graphed in Figure 7.25. On a percentage basis, France leads the world in nuclear power. As of 2001, it has 59 operational nuclear power plants that generate 76% of the electricity used in France. The Swiss also have a high nuclear dependency, producing 36% of their electricity with only five reactors. Most of the countries that generate over 40% of their electricity from nuclear power plants are in western Europe; Hungary and South Korea are exceptions. Reactors are now on order, or being planned, in North Korea, Egypt, Indonesia, and Iran.

Also noteworthy are the countries *not* included in Figure 7.25. Thus far, only industrialized nations have been able to afford major commercial development of nuclear fission. As of 2001, Mexico and Pakistan each had two operating reactors, and Romania had one operating reactor. India obtains less than 4% of its electricity from its 14 reactors. China had three operating nuclear power reactors, with eight more under construction and two more planned. Ironically, in spite of the fact that much of the world's uranium comes from Africa, most nations on that continent use no nuclear energy.

- Pakistan's and India's nuclear capabilities extend to military applications as well, as evidenced by the nuclear tests each conducted near the other's border in 1998. These tests raised the specter of nuclear weapons being used again after a period of 53 years without nuclear warfare.

7.27 Consider This: Worldwide Nuclear Power

a. From the data in Table 7.6, who are the top three producers of electricity from nuclear power in terms of megawatts of power? In general, how would you characterize these countries?

b. Name three countries missing from Table 7.6. In general, how would you characterize these countries?

c. How does this list of producers compare with the "Top 20" in carbon dioxide emissions found in 3.26 Consider This? Explain this relationship.

7.28 Consider This: Nuclear Neighborhood

The number of nuclear reactors in the United States changes. Some are being decommissioned and others are under construction. The Nuclear Regulatory Commission's (NRC) web site provides a map of all reactors now licensed to operate in the United States. How many reactors are there? Identify the three nuclear reactors nearest to where you live. You can find a direct link to the NRC at the *Chemistry in Context* web site.

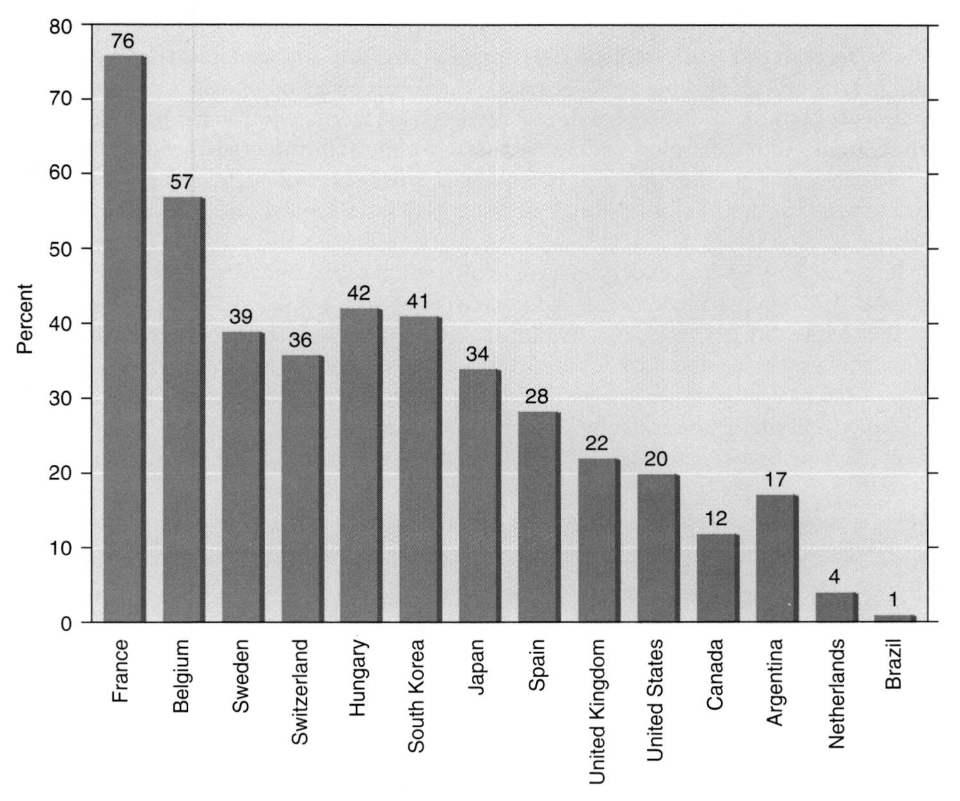

Figure 7.25
Percent of electrical power generated by nuclear power reactors in selected countries, 2001.
(Source: World Nuclear Association. Taken from www.world-nuclear/org/info/eractors.htm.)

7.13 Living with Nuclear Power: What Are the Risks and Benefits?

From the previous section, it is obvious that countries differ markedly in the extent to which they use nuclear power to generate electricity. What is *not* so obvious are the reasons for this variability. Some nations have fiscal problems so severe that it is difficult or impossible for them to fund the construction or expansion of nuclear power facilities. For others, an adequate supply of relatively cheap electricity is available from water power, fossil fuels, or other sources. Therefore, there is little need for them to use nuclear energy. Nuclear power provides a means for some countries to gain a greater independence from needing to import fossil fuels. In still other countries, such as France, a conscious choice has been made to use nuclear energy to produce the bulk of electrical energy to reduce dependency on imported fossil fuel. And, just the opposite conclusion has been reached by other nations. In Sweden, which currently obtains 39% of its electricity from fission, a referendum has called for the halt of nuclear power generation by 2010.

Regardless of whether a country contains many nuclear-powered electrical generators or only a few, associated risks and benefits must be weighed. Such risk-benefit analyses are never easy, though in a sense we do it every day. We commonly regard risk as the probability of being injured or losing something, but there are many types of risk. They can be voluntary, such as those associated with wind surfing or bungee jumping, or involuntary, such as inhaling someone else's cigarette smoke. When we drive a car we control the risks (at least to some extent), but we have no control over the increased risk of radiation exposure at high altitudes or of a commercial plane crash. Counterbalancing risks are benefits such as the improvement of health, increased personal comfort or satisfaction, saving money, or reducing fatalities. Everyday living inevitably involves risks and their related benefits: crossing a street, riding a motorcycle or in a car,

- The 19th century poet William Wordsworth spoke of technological risks and benefits as ". . . Weighing the mischief with the promised gain . . ." He was speaking, in this case, about the railroad, a technology new in his time.

cooking a meal or eating one, and even the simple act of getting up in the morning. Because there is some element of risk in everything we do, we almost automatically make judgments about what level of risk we consider acceptable. Most people do not intentionally put themselves at high risk, even when the potential benefit is also high, such as going into a burning building to save a child. On the other hand, there is an alarming increase in the number of people who expect "zero risk" in whatever they do or whatever surrounds them, although it is impossible to achieve; *there is no such thing as zero risk.*

In the case of nuclear energy, we are dealing with social benefits in relation to technological risks, but we must not make the mistake of only considering the risks and benefits that relate directly to fission. We must also weigh the risks associated with the alternatives—especially the coal-fired power plants that nuclear reactors are designed to replace. Recall, for example, the estimate in Chapter 4 that over 100,000 workers have been killed in American coal mines since 1900, many prior to the 1950s when higher safety standards were instituted. Table 7.7 summarizes the risk of fatalities from the an-

Table 7.7

Risks from Coal and Nuclear-Powered Electricity Generation

Hazard Type	Coal	Nuclear
Routine occupational hazard	Coal-mining accidents and black-lung disease constitute a uniquely high risk.	Risks from sources not involving radioactivity dominate.
Deaths	2.7	0.3 to 0.6
Routine population hazards	Air pollution produces relatively high, though uncertain, risk of respiratory injury. Significant transportation risks.	Low-level radioactive emissions are more benign than corresponding risks from coal. Significant transportation risks incompletely evaluated.
Deaths	1.2 to 50	0.03
Catastrophic hazards (excluding occupational)	Acute air pollution episodes with hundreds of deaths are not uncommon. Long-term climatic change, induced by CO_2, is conceivable.	Risks of reactor accidents are small compared to other quantified catastrophic risks. The problem lies in as yet unquantified risks for reactors and the remainder of the fuel cycle.
Deaths	0.5	0.04
General environmental degradation	Strip mining and acid runoff; acid rainfall with possible effect on nitrogen cycle, atmospheric ozone; eventual need for strip mining on a large scale.	Long-term contamination with radioactivity; eventual need for strip mining on a large scale.

Note: Deaths are the number expected per year for a 100-megawatt power plant. In all cases, 6000 man-days lost are assumed to equal one death.

(*Source:* Modified from *Perilous Progress: Managing the Hazards of Technology,* by Robert W. Kates, Ed., 1985, Westview Press, Boulder, Colorado.)

nual operation of a 100-megawatt power plant using either coal or nuclear power. The conclusion is that, at least for the hazards identified here, the risks associated with energy produced by nuclear power are considerably less than those of coal-burning plants.

Paradoxically, coal-fired power plants release more radioactivity on a daily basis than nuclear plants. In addition to C, H, O, N, and S, coal contains other elements, including uranium and thorium, as impurities. According to W. Alex Gabbard, a physicist at the Oak Ridge National Laboratory, trace quantities of uranium in coal can be as high as 10 ppm and the amount of thorium is usually more than twice that of uranium. Gabbard has estimated that in 1982, power plants in the United States burned a total of 616 million tons of coal and released 801 tons of uranium and 1971 tons of thorium into the environment. In fact, the quantity of uranium emitted by the coal-fired plants exceeded the mass of uranium consumed in nuclear plants. He predicted that if in the year 2040 the U.S. burns 2516 million tons of coal, over 145,000 tons of uranium will be released, noting that about 1000 tons of these will be uranium-235.

Coal-fired power plants also produce huge amounts of carbon dioxide, a waste product of fossil fuel combustion for which we currently have no large-scale remediation technology. This is part of the nuclear power/fossil fuel risk-benefit scenario. One 1000 megawatt coal-fired electric power plant releases about 4.5 million tons of CO_2 annually into the atmosphere, as well as generating 3.5 million cubic feet of waste ash each year, a substantial volume of solid waste to be handled. By comparison, a 1000 megawatt nuclear power reactor produces about 70 cubic ft of high-level waste per year.

7.29 Your Turn

a. If 1000 of the tons of uranium released in the year 2040 are U-235, what do the remaining tons of uranium consist of?

b. Why is thorium found together with uranium in coal?

As we have noted, nuclear energy carries tremendous emotional overtones, made of mystery, misunderstanding, and mushroom clouds. The risks of radiation, involuntary and uncontrolled, and the possibility of a major disaster, however remote, loom large in human consciousness. The accidents at Three Mile Island and Chernobyl, even though hardly equivalent, make the public wary. We have limited trust in technology and perhaps even less in people. We are apprehensive about human error in the design, construction, and management of nuclear power plants. After all, human errors and technicians' responses to them were the weak points in the prescribed safety procedures that caused the accidents at Three Mile Island and Chernobyl.

7.30 Consider This: Informed Citizens

a. What should you know to be an informed citizen about nuclear power plants? Make a list of questions (at least five) that would be important to ask about a specific nuclear reactor.

b. Check out the specifications of a particular reactor at the web site provided by the Nuclear Energy Institute (NEI). Choose any reactor in the country you wish. Does the information provided answer the questions you posed? Comment on what you would like to know that you were unable to find. If others in your class selected different reactors, you may wish to compare notes.

7.14 What Is the Future for Nuclear Power?

This final question is, in many ways, the most difficult one posed in this chapter. The answer remains uncertain. Early in the chapter, a provocative quote from Professor Gregory Choppin indicated that a revival of nuclear energy will occur because we have no other option. An accompanying statement by Jeff Johnson points out that the public remains divided on the issues of nuclear energy, as noted by others.

This question of the future of nuclear power does not stand in isolation from other international issues. At present, the world community is actively seeking ways to reduce greenhouse gases. Because heat from the fission of U-235 produces steam used to generate electricity, a nuclear power plant releases no carbon dioxide, a major greenhouse gas. Nuclear power also has been touted as a way to reduce acid rain, as nuclear fission releases no acidic oxides of sulfur and nitrogen. Recall that the Seabrook plant generates electricity at a rate of 1160 megawatts (1160 million joules per second) or 1×10^{14} joules per day. Approximately 10,000 tons of coal would have to be burned in a conventional power plant to generate this daily energy output. Burning this quantity of coal could easily release 300 tons of SO_2 and perhaps 100 tons of NO_x. Whether the risks associated with nuclear power outweigh those of global warming and acid rain is a difficult question for which there are no clear-cut answers, in spite of extended study and debate. Proponents line up on each side of the argument.

- Two tons of U-235 (40–60 tons of enriched uranium fuel) can fuel a 1000 megawatt–producing commercial nuclear reactor for about one and a half years. To produce this amount of electricity, a coal-fired plant would use the coal carried by a train with 200 coal cars each carrying 15 tons of coal *every day* for the same one and a half years.

In August 1988, during a summer of record heat, 15 U.S. senators co-sponsored a bill to fund research to combat the enhanced greenhouse effect by developing carbon dioxide-free energy sources, including safer and more cost-effective nuclear power plants of standardized design. Alan Crane, an energy policy specialist, appeared before a House subcommittee hearing and spoke of global warming and nuclear energy risks in these terms:

> *"There is significant, though not yet quantifiable, risk that the resulting climate changes will wreak devastating changes in agricultural production throughout the world, among other problems. Such changes could lead to the death of far more people and cause far greater environmental damage than any nuclear reactor accident, and appear to be considerably more likely."*

Others conclude that nuclear power can reduce global warming only slightly and suggest that it would be much less expensive to invest in enhanced energy efficiency. Bill Keepin and Gregory Kats of the Rocky Mountain Institute have estimated that to reduce carbon dioxide emissions significantly through the use of nuclear power would require the completion of a new nuclear plant every 2 days for the next 38 years. Oak Ridge National Laboratory staff members reported that before any massive replacement of fossil fuel with nuclear power could occur, several new techniques would have to be developed. These include commercial-scale recycling of nuclear fuels, breeder reactors to extend existing fuels, and possibly uranium recovery from seawater.

Hans Blix, former director general of the International Atomic Energy Agency, in describing the international dimensions of the problem, noted that developing nations will not likely build nuclear plants in the near future: "If nuclear power is to be relied on to alleviate our burdening of the atmosphere with carbon dioxide, it is therefore to the industrialized countries that we must first look. They are in the position to use these advanced technologies—and they are also the greatest emitters of carbon dioxide."

The original nuclear era, synonymous with the growth of nuclear power in the United States, began in the early 1960s and lasted until 1979. It fell victim to stabilized demand for electricity, which was brought on by enormous oil price hikes in 1975 and the Three Mile Island incident in 1979. As a consequence of that accident, the required number of nuclear plant personnel and their training requirements grew significantly. In addition, the mandatory retrofitting of existing nuclear facilities to enhance their safety added significantly to the cost of an already capital-intensive industry. As a result, the electrical power industry became understandably reluctant to invest further in nuclear facilities.

A second nuclear era may occur. Some experts believe that such a rebirth is possible through the development of smaller, more efficiently designed reactors in the 600 megawatt range, rather than the current 1000–1200 megawatt facilities. Unlike the many different designs used to build the current reactors, these new reactors would be of a standardized, easily replicated design, have a longer operational lifetime (60 years versus the current 30), and be demonstrably safer and more economical in operation. James Lake, past president of the American Nuclear Society and laboratory director at a DOE nuclear energy laboratory, says "The energy crisis has shined a spotlight on us [nuclear power] . . . There are 438 reactors worldwide, and people are thinking there should be 4,000 in the next 20 years." The Nuclear Regulatory Commission (NRC) certified three new nuclear reactor designs in the 1990s. Under new NRC rules, if a reactor with such a design were used, a company could bring a new nuclear power plant on line in just five to six years, rather than the eight to ten years needed to construct the plant for a yet-to-be-certified reactor.

But in all of this, the unsolved problem of the safe disposal of radioactive waste remains perhaps the greatest impediment. Nuclear engineers William Kastenberg and Luca Gratton conclude their *Physics Today* article (June 1997) with the sobering thought: "For a high-level waste depository of the type proposed for Yucca Mountain, it is clear that natural processes will eventually redistribute the waste materials. Present design efforts are directed toward ensuring that, at worst, the degraded waste configurations will eventually resemble stable, natural ore deposits, preferably for periods exceeding the lifetimes of the more hazardous radionuclides. Perhaps that's the best we can hope for."

7.31 Consider This: Risks and Benefits

We have examined nuclear fission as a source of electrical power in some detail. Now we ask you to list the risks and benefits associated with currently operating nuclear fission reactors. Using this list, take a stand on the question of the future use of nuclear fission-powered plants. Write an editorial for a local newspaper proposing your view of a viable 20-year national policy on the issues.

7.32 Consider This: Second Opinion Survey

Now that you are near the end of your study of nuclear power, return to the personal opinion survey of 7.1 Consider This and answer the questions one more time.

After completing the survey for a second time, compare your answers in the second survey with those from the first. Are there any striking differences in your opinions of nuclear power between the first and second surveys? If so, which of your opinions about nuclear power changed the most? What was responsible for this shift?

CONCLUSION

Nearly 45 years have passed since the first commercial nuclear power plant began producing electricity in the United States. The glittering promise of boundless, unmetered electricity, drawn from the nuclei of uranium atoms, has proved illusory. But the needs of our nation and our world for safe, abundant, and inexpensive energy are far greater today than they were in 1957. So scientists and engineers continue their atomic quest. Where the search will lead is uncertain, but it is clear that people and politics will have a major say in ultimately making the decision. Reason, together with a regard for those who will inhabit our planet in both the near and far future, must govern our actions.

Chapter Summary

Having studied this chapter, you should be able to:

- Tell how nuclear fission occurs (7.2)
- Write balanced nuclear equations for alpha and beta decay, and for nuclear fission (7.2)
- Use mathematical relationships to calculate the amount of energy produced by a fission reaction (7.2)
- Compare and contrast how electricity is produced by a conventional power plant with how it is produced by a nuclear power plant (7.3)
- Summarize the reasons why a nuclear power reactor cannot undergo a nuclear explosion (7.5)
- Develop a personal radiation dose inventory and describe the biological effects of nuclear radiations (7.8)
- Understand and apply the concept of half-life to the use of radioisotopes, radio-carbon dating techniques, and nuclear waste storage (7.9)
- Relate the issues surrounding the use of nuclear power in this country and abroad (7.12)
- Describe the issues associated with the production and storage of high-level nuclear waste, including spent fuel (7.10)
- Take an informed stand on the storage of high-level nuclear wastes (7.10)
- Read and hear news articles on nuclear power and nuclear waste issues with confidence in your ability to interpret the accuracy and efficacy of such reports (7.10-7.11)

- Summarize the nature of low-level nuclear waste and its storage (7.11)
- Report on the use of nuclear power for electricity generation globally and the reasons why several countries have very high percentages of electrical power production from nuclear reactors compared to the United States (7.12)
- Describe the risks and benefits of the use of nuclear power (7.13)
- Take an informed stand on the use of nuclear power for electricity production (7.14)
- Discuss why the 1950s and 60s promise of abundant and cheap nuclear energy was not realized in this country (7.14)
- Outline the factors that will allow or oppose the growth of nuclear energy in the next decade (7.14)

Questions

Emphasizing Essentials

1. $E = mc^2$ is one of the most famous equations of the 20th century. What does each of the letters in the equation represent?
2. What is the difference between the symbol N and symbols such as ^{14}N or ^{15}N?
3. a. How many protons does an atom of $^{239}_{94}Pu$ contain?

   b. How many neutrons does an atom of $^{90}_{38}Sr$ contain?

   c. How many protons and how many neutrons does an atom of $^{238}_{92}U$ contain?
4. For each of these nuclei, find the number of protons (from a periodic table) and the number of neutrons (from the atomic and mass numbers).

   a. Na-23 c. Cu-65

   b. Cl-37 d. Hg-200
5. C-12, Ca-40, Zn-64, and Sn-119 are stable isotopes that have an even number of protons. Based on this information, what generalization can you propose about the relative numbers of protons and neutrons in stable nuclei with an even number of protons as the atomic number increases?
6. a. Na-22 is a radioactive isotope, Na-23 is stable, and Na-24 is a radioactive isotope. Based on this information, what generalization can you propose about the relative number of protons and neutrons in stable nuclei? What is this relationship for nuclei with an odd number of protons?

   b. What additional information would be needed to draw a generalization about the relative number of protons and neutrons in stable nuclei with an odd number of protons as the atomic number increases?
7. In what ways does a nuclear equation differ from a chemical equation?
8. a. Boron, element number 5, exists in only two common stable isotopes, B-10 and B-11. Given that the periodic table lists the average atomic mass as 10.81, which isotope must be more abundant? Explain your reasoning.

   b. The average atomic mass of chlorine is 35.453. Does this mean that there are just two stable isotopes, one with atomic mass 35 and one with atomic mass 36? Explain your answer.
9. Show that for this nuclear equation, the sum of the subscripts on the left is equal to the sum of the subscripts on the right. Then show that the sum of the superscripts on the left is equal to the sum of the superscripts on the right.

   $$^{239}_{94}Pu + ^{4}_{2}He \longrightarrow [^{243}_{96}Cm] \longrightarrow ^{242}_{96}Cm + ^{1}_{0}n$$
10. Write equations to represent each of these nuclear reactions.

    a. Two H-2 nuclei join to form another nucleus and a neutron.

    b. U-238 is bombarded with N-14 to produce another nucleus and 5 neutrons.

    c. Pu-239 is bombarded with a neutron to form Ce-146, another nucleus, and 3 neutrons.
11. When 4.00 g of hydrogen nuclei fuse to form helium in the Sun, 0.0265 g of matter is converted into energy. Use Einstein's equation, $E = mc^2$, to calculate the energy equivalent of this change in mass.
12. Consider a reaction in which an isotope of hydrogen fuses with an isotope of helium to form a different isotope of helium and another isotope of hydrogen. The exact mass of a mole of each isotope is given below the isotope.

    $$\underset{2.01345\text{ g}}{^{2}_{1}H} + \underset{3.01493\text{ g}}{^{3}_{2}He} \longrightarrow [^{5}_{3}Li] \longrightarrow \underset{4.00150\text{ g}}{^{4}_{2}He} + \underset{1.00728\text{ g}}{^{1}_{1}H}$$

    a. What is the mass difference between a mole of the reactant and the product isotopes?

    b. How much energy (in joules) released in this reaction?
13. Einstein's equation, $E = mc^2$, also applies to chemical changes as well as to nuclear reactions. An important chemical change studied in Chapter 4 was the combustion of methane, which releases 50.1 kJ of energy for each gram of methane burned.

    a. What mass loss corresponds to the release of 50.1 kJ of energy?

b. To produce the same amount of energy, what is the ratio of the mass of methane burned in a chemical reaction to the mass loss converted into energy according to the equation $E = mc^2$?

c. Use your results in parts **a** and **b** to comment on why Einstein's equation, although correct for both chemical changes and nuclear changes, is usually only applied to nuclear changes.

14. This schematic diagram represents the reactor core of a nuclear power plant.

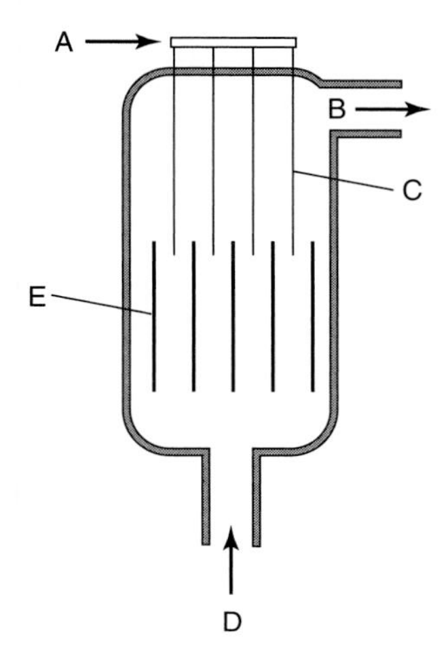

Match each letter with one of these terms.

fuel rods

cooling water into the core

cooling water out of the core

control rod assembly

control rods

15. Identify the segments of the nuclear power plant diagrammed in Figure 7.6 that are nuclear and those that are not nuclear. Briefly explain your choices.

16. One important distinction between the Chernobyl reactors and those in the U.S. is that those in Chernobyl used graphite as a moderator to slow neutrons whereas U.S. reactors use water. In terms of safety, give two reasons why water is a better choice.

17. One of the biggest challenges in preparing fuel for nuclear power plants is to separate the isotopes of uranium.

a. Which uranium isotopes occur naturally in uranium ore? Why is it necessary to separate these isotopes?

b. It is not possible to separate the isotopes of uranium by chemical means. Why not?

c. How are the uranium isotopes separated?

xtra

18. Write nuclear equations to represent each of these reactions.

a. U-235 and a neutron react to form Br-87, La-146, and more neutrons.

b. U-238 is bombarded with a nucleus to produce Fm-249 and 5 neutrons.

c. A neutron induces the fission of U-235 to form one nucleus with 56 protons, a second with a total of 94 neutrons and protons, and 2 neutrons.

19. Pu-239 is most hazardous when inhaled. Explain the reasons behind this observation.

20. What do α, ${}^{4}_{2}He$, and ${}^{4}_{2}He^{2+}$ represent? What is the difference among them?

21. Radioactivity involves the emission from a radioisotope of an alpha particle, a beta particle, a neutron, and/or a gamma ray. For each of these, what changes in the radioisotope do you expect after the emission?

a. a change in the mass number

b. a change in the atomic number

c. changes in both the atomic number and the mass number

d. changes in neither the atomic number nor the mass number

22. Write nuclear equations to represent each of these transformations.

a. I-131 (used in thyroid imaging) releases a beta particle.

b. U-238 (found in uranium ore) decays with the loss of an alpha particle.

c. Tl-206 emits a beta particle.

d. Mo-98 is bombarded with a neutron to produce Tc-99 (used widely in medical imaging) and another particle.

23. Given that the average U.S. citizen receives 3600 μSv of radiation exposure per year, use the data in Table 7.4 to calculate the percentage of radiation exposure the average U.S. citizen receives from each of these sources.

a. food, water, and air

b. a dental X-ray twice a year

c. the nuclear power industry

24. What percent of a radioactive isotope would remain after two half-lives, four half-lives, and six half-lives? What percent would have decayed after each time period?

25. Suppose somebody tells you that a radioisotope is "gone" after about seven half-lives. Critique this statement, explaining both why it could be a reasonable assumption and why it might not be.

26. What is the half-life of the radioisotope X in this graph?

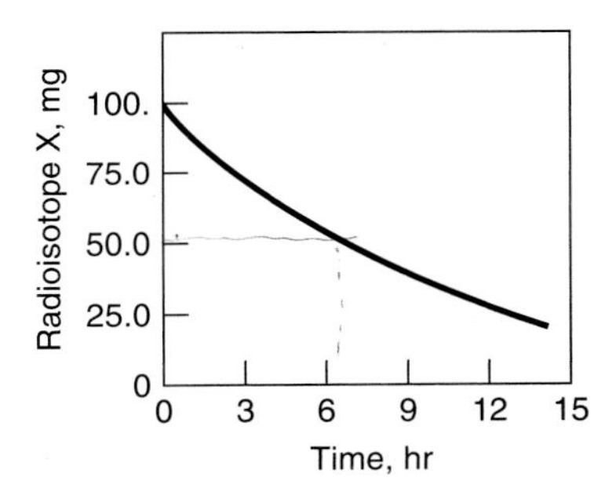

Concentrating on Concepts

27. In 7.1 Consider This, you were asked to answer several question about nuclear power. Extend this survey by asking the same questions of someone at least one generation older than you and someone still in high school. What similarities and differences did you find in their answers compared with your opinions?

28. a. Why were citizens' groups in Massachusetts concerned about construction of the Seabrook Nuclear Power Plant in New Hampshire? *Hint:* Consult a road atlas to find the location of Seabrook.

b. What aspects of the site chosen for the Seabrook plant were advantages in the minds of the designers and builders of the Seabrook plant, but were disadvantages in the minds of some protesting citizens' groups?

29. The Seabrook power plant at full capacity uses only a few pounds of uranium to generate 1160 megawatts of power, which is equivalent to 1.16×10^9 joules every second. To produce the same amount of energy would take about two million gallons of oil or about 10,000 tons of coal in a conventional power plant. What is the fundamental difference in the way that energy is produced in the Seabrook plant, compared with conventional power plants?

30. Considering that Einstein had proposed the equation $E = mc^2$ over 30 years earlier, why were Otto Hahn and Fritz Strassmann puzzled when, in 1938, they discovered the element barium among the products formed when uranium was bombarded with neutrons?

31. In a chemical reaction, it is often said that matter is conserved. Why is it incorrect to say that mass is conserved in a nuclear reaction?

32. If you look at nuclear equations in sources other than this textbook, you may find that the subscripts have been omitted. For example, you may see an equation for a fission reaction written this way.

$$^{235}U + {}^{1}n \longrightarrow [^{236}U] \longrightarrow {}^{87}Br + {}^{146}La + 3\ {}^{1}n$$

a. How do you know what the subscripts should be? Why can they be omitted?

b. Why are the superscripts *not* omitted?

33. More than two decades have passed since the incident at Three Mile Island. Use the Web to answer these questions.

a. Is Reactor 2, the site of the problem in 1979, back on line producing electricity?

b. How has the nuclear waste from the accident been treated?

34. What are the similarities between the incidents at Chernobyl and at Three-Mile Island? What are the differences?

35. Not all the neutrons in a nuclear reactor produce fission reactions. Name some other possibilities as to what might happen to these neutrons.

36. a. Is depleted uranium (DU) still radioactive? Explain your answer.

b. Is spent nuclear fuel (SNF) still radioactive? Explain your answer.

37. It is generally believed that terrorists would be more likely to construct a nuclear bomb using Pu-239 reclaimed from breeder reactors than using U-235. Use your knowledge of chemistry to offer reasons for this.

38. A web site describing an X-ray procedure reports, "Despite its negative connotations, people are exposed to more radiation on a daily basis than they may realize. For example, infrared radiation is released whenever there is extreme heat. The sun generates ultraviolet radiation, and a little exposure to it results in a tan. In addition, the body contains naturally radioactive elements." Critique this explanation.

39. What does the term "decommission" mean, as in decommissioning a nuclear power plant? What are the technical challenges involved? You might want to start by learning more about the decommissioning of the Yankee Rowe facility (Table 7.1). The resources of the Web can help you answer this question.

40. a. What are the characteristics of high-level radioactive waste? Name at least two processes that produce it.

b. Two options are being considered for the storage of HLW: monitored storage on the surface, and storage deep underground. Cite the advantages and disadvantages of both.

c. Explain how low-level waste (LLW) differs from HLW.

Exploring Extensions

41. Lise Meitner and Marie Curie were both pioneers in developing an understanding of atomic nuclei. You likely have heard of Marie Curie and her work, but may not have heard of Lise Meitner. How are these two women related in time and in their scientific work?

42. Gallium consists of just two stable isotopes, Ga-69 and Ga-71.

a. If the atomic mass of elemental gallium is 69.72, which isotope is present in a larger percentage?

b. If Ga-69 has a mass of 68.9257 and Ga-71 has a mass of 70.9249, what percentage of each isotope is present?

43. Alchemists in the Middle Ages dreamed of converting base metals, such as lead, into precious metals—gold and silver. Why did they never succeed? Has the situation changed since then? Explain your answer.

44. In question 12, the energy of a mole of H-2 joining with He-3 was calculated. What is the ratio of the energy released in this nuclear reaction to the energy released in the combustion of a mole of hydrogen gas? *Hint:* This value can be calculated using bond energies from Table 4.1.

45. Californium, element number 98, was first synthesized by bombarding an element with alpha particles. The products were californium-245 and a neutron. What was the target isotope used in this nuclear synthesis?

46. Consider this representation of a Geiger counter, a device commonly used to detect ionizing radiation. The probe contains a gas.

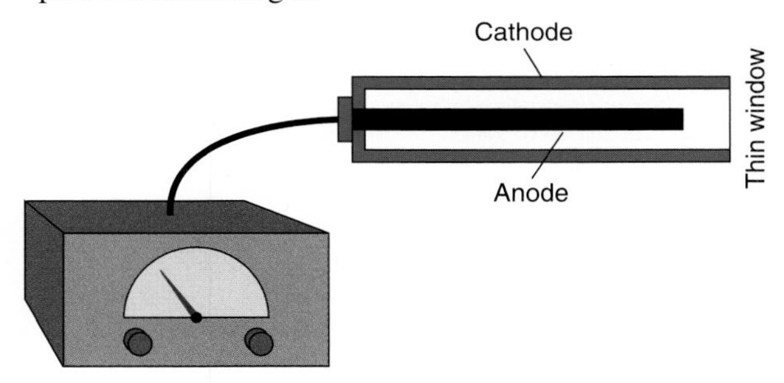

a. How does radiation enter the Geiger counter?

b. Why does this device only detect radiation that is capable of ionizing the gas contained in the probe?

c. What are other methods for detecting the presence of ionizing radiation?

47. A stockpile of approximately 50 metric tons of plutonium exists in the U.S. as a result of disassembling warheads from the nuclear arms race. What should the fate of this plutonium be? *Hint:* A search for "plutonium disposal" on the Web will bring up references. Try also including U.S. and DOE as search terms.

a. Some propose that the plutonium be sent to local nuclear power plants to "burn" as fissionable fuel. What are the advantages and disadvantages of such a course of action?

b. Others propose that it be stored permanently in a repository. Again, list the advantages and disadvantages.

48. Advertisements for Swiss Army watches stress their use of tritium. One ad states that the ". . . hands and numerals are illuminated by self-powered tritium gas, 10 times brighter than ordinary luminous dials. . . .". Another advertisement boasts that the ". . . tritium hands and markers glow brightly making checking your time a breeze, even at night. . . ." Evaluate these statements and, after doing some Web research, discuss what form the tritium is in these watches, and what its role is.

49. The amount of exposure from medical X-rays varies considerably, depending on the procedure. For example, dental X-rays may be taken as "bite-wings," full mouth, or panoramic, with each involving different amounts of radiation. Pick a particular type of medical X-ray and research the exposure to radiation involved. Report your data in both rems and sieverts (using mrem, milliSv or μSv as appropriate).

50. MRI, or magnetic resonance imaging, is a very important tool for some types of medical diagnoses.

a. What is the scientific basis for this technique?

b. What information can an MRI give a physician that cannot be obtained through direct examination of a patient?

c. This MRI method used to be called NMR, nuclear magnetic resonance. Why do you think the name was changed?

8

Energy from Electron Transfer

More than 65 million cell phone users depend on compact, lightweight, long-lasting, rechargeable batteries.

In two earlier chapters we considered two sources of energy that are used extensively in the United States and in most other countries to produce electrical energy. Chapter 4 emphasized the energy obtained from burning fossil fuels, and Chapter 7 focused on energy released by splitting fissionable atomic nuclei. Centralized power plants, whether fueled by coal or fission, distribute electricity regionally through vast power networks to offices, classrooms, and residences. But supplies of both fossil and fission fuels are limited, and when they are gone, they are gone forever. Moreover, these fuels also create a huge environmental cost. Combustion of coal and petroleum releases vast quantities of carbon dioxide, which contribute to global warming, and sulfur dioxide and nitrogen oxides, which give rise to acid precipitation. Nuclear fission is bedeviled by the as-yet unsolved problem of disposing of high-level nuclear wastes that will continue to emit dangerous radiation for thousands of years. The conclusion seems obvious. If our species is to continue to inhabit this planet, we must develop other sources of energy.

Our way of life is heavily dependent on both the availability of electricity on a large scale, and on the convenience of personal power sources, such as batteries. The portable electronics market's demand for compact, mobile, and lightweight energy sources has been a major factor in developing advanced battery technology. The common feature among all types of batteries is the transfer of electrons. The title of this chapter reflects the importance of this fundamental process.

8.1 Consider This: Batteries in Your Daily Life

To help understand your personal dependence on electron transfer in consumer products, think about the things you own or use that run on batteries.

a. Make a list of things you own or use that run on batteries. Indicate which items uses a battery as the main source of energy and which as a backup source.
b. What type of battery is used in each case?
c. Which batteries are rechargeable and which are not?
d. How do you dispose of batteries that are not rechargeable?

CHAPTER OVERVIEW

When we flip a light switch and nothing happens due to a power failure, what do we turn to for electricity? We typically use battery-powered flashlights and lamps. Other portable devices, such as cell phones, watches, and hearing aids, require the use of ever smaller and longer-lasting, reliable batteries. This chapter starts by discussing the electron transfer that takes place in different types of batteries. What is the difference between a rechargeable battery and one that must be discarded after it "runs down"? What are fuel cells, and how can they be used to produce electricity? Hydrogen is one of the fuels used in fuel cells. And so the chapter continues by considering hydrogen, a non-fossil fuel. After being obtained from a variety of possible sources, hydrogen can either be burned as a fuel or used to generate electricity in a fuel cell. The discussion of fuel cells leads to a related section on powering many modern devices, including the so-called zero emission electric vehicles (ZEVs). Although electric cars are now commercially available, questions about them remain: Why have them? How do they operate? How do they compare to gasoline-powered vehicles? Are electric vehicles competitive in the marketplace? Are people using them? We will address all these questions in this chapter, as well as discuss other electric vehicle technologies.

In the long run, the most promising alternate technology for generating electricity could well be photovoltaic cells, devices that convert the Sun's radiation directly into electricity. The principles governing semiconductors in photovoltaic cells, their operation, and their current and potential applications are discussed. The chapter ends with an assessment of future trends in developing energy sources from electron transfer.

8.1 Electrons, Cells, and Batteries

The famous inventor Thomas Edison was convinced in the late-19th century that batteries were an idea doomed to failure. Despite branding such devices as ". . . a sensation, a mechanism for swindling the public by stock companies, . . ." he went on to say that, although such batteries were scientifically all right, their commercial success would be ". . . as absolute a failure as one can imagine." He clearly was better at inventing than

• Edison's view might have been clouded by his ownership of a large, municipal electric company.

at judging the future success of batteries! Why are batteries important in our times? As you realize if you are wearing a battery-powered watch, carrying a cellular phone, or using a laptop computer, you depend on batteries every day.

A **battery** is a system for the direct conversion of chemical energy to electrical energy. Batteries are found everywhere in today's society because they are convenient, transportable sources of stored energy. Batteries are also big business in the U.S., with consumers buying \$3.5 billion worth of batteries in the year 2000 according to the Consumer Electronics Association. This market will continue to become even more important, including supplying power to run electric vehicles. Although the term battery is in common use, a standard flashlight "battery" is more correctly called a cell, or an electrochemical cell. A **galvanic cell** is a device that converts the energy released in a spontaneous chemical reaction into electrical energy. A collection of several galvanic cells wired together constitutes a true battery, such as the battery in your automobile. A galvanic cell is the opposite of an **electrolytic cell,** one in which electrical energy is converted to chemical energy.

- A galvanic cell converts chemical energy to electrical energy.

All galvanic cells produce electricity from reactions involving the transfer of electrons from one substance to another. The transfer of electrons involves two reactions, each known as a half-reaction. One of the half-reactions is **oxidation,** in which a reactant loses electrons. The other half-reaction is a **reduction,** involving the gain of electrons by some other reactant. Two half-reactions—one oxidation and the other reduction—are always involved.

- Oxidation = Loss of electrons; Reduction = Gain of electrons.

Let us start by looking at a simplified example of the reaction that takes place in a nickel-cadmium (NiCad) battery:

Oxidation: $Cd \longrightarrow Cd^{2+} + 2\,e^-$ [8.1]

Reduction: $2\,Ni^{3+} + 2\,e^- \longrightarrow 2\,Ni^{2+}$ [8.2]

Overall cell reaction: $2\,Ni^{3+} + Cd \longrightarrow 2\,Ni^{2+} + Cd^{2+}$ [8.3]

- The overall reaction is obtained by adding equations 8.1 and 8.2

In this case, two *electrons are given off,* or "lost," in the *half-reaction of oxidation* (equation 8.1). Two *electrons are gained* in the *half-reaction of reduction* (equation 8.2). This transfer of electrons through an external circuit is what creates the flow of electricity needed to drive a cordless razor or a power tool or countless other battery operated devices (Figure 8.1). The two half-reactions of the cell must be connected in such a way that the electrons released during the oxidation reaction are transferred to the reduction reaction (Figure 8.2). This is accomplished by using **electrodes,** electrical conductors placed in the cell as sites for chemical reaction. The electrode where oxidation takes place is called the **anode.** The electrons given up in the process flow from the anode through a wire to the cathode. At the **cathode,** the electrons are used in the reduction half-reaction. The electrical energy is the result of the spontaneous reaction that occurs in the cell. The difference in **voltage,** the difference in electrochemical potential between the two electrodes, is proportional to the energy evolved; the more voltage, the greater the energy. Voltage is expressed in units called volts (V). The greater the difference in potential between two electrodes, the higher the voltage. In the case of the NiCad battery, the maximum difference in electrochemical potential is experimentally measured as 1.46 volts.

- Anode: Electrode at which oxidation occurs.
  Cathode: Electrode at which reduction occurs.
- Electrochemical potential energy difference is measured in volts, a unit honoring Alessandro Volta.

Several different types of galvanic cells will be discussed in this chapter, and their electron transfer analyzed. To construct a commercially useful galvanic cell, more than just the electrochemical potential must be considered. Successful cells must be of reasonable cost, last a reasonable length of time, and be safe to use and discard or recharge. In some applications, the size and weight of the cell is of paramount importance. Solids, pastes, gels, or thick slurries act as electrolytes to carry ions between electrodes. Aqueous solutions are often too hazardous to use in commercial cells because of the potential for leakage. However, aqueous solutions are often used in simple laboratory cells. Electrons are transferred through the external circuit from anode to cathode, but positive and negative ions are often transported internally through a salt bridge to complete the circuit.

Figure 8.1

A portable power drill, such as this Black and Decker model, usually comes with two 9.6 V NiCad battery packs.

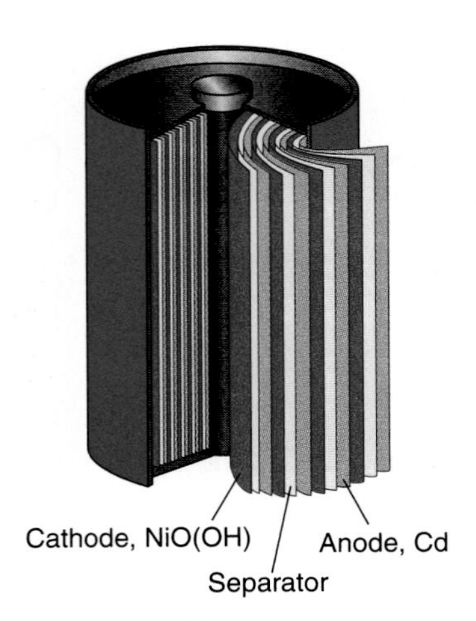

Figure 8.2

A galvanic cell in a NiCad battery in which cadmium metal is oxidized and Ni^{3+} is reduced to Ni^{2+}. In this cell, cadmium is oxidized at the anode, and Ni^{3+} is reduced at the cathode.

The complete reaction in a NiCad battery is actually just a bit more complicated than represented so far. The cell contains a water-based paste of either NaOH or KOH as an electrolyte. This changes the actual form of nickel and cadmium as the reaction proceeds. Cadmium metal, the anode, is oxidized to Cd^{2+} ions; simultaneously, Ni^{3+} ions (in hydrated NiO(OH)) on a nickel electrode are reduced to Ni^{2+} ions (in $Ni(OH)_2$).

Anode reaction (Oxidation): $Cd(s) + 2\ OH^-(aq) \longrightarrow Cd(OH)_2(s) + 2\ e^-$ [8.4]

Cathode reaction (Reduction): $2\ NiO(OH)(s) + 2\ H_2Oc + 2\ e^- \longrightarrow 2\ Ni(OH)_2(s) + 2\ OH^-(aq)$ [8.5]

Overall reaction: $Cd(s) + 2\ NiO(OH)(s) + 2\ H_2O(l) \longrightarrow 2\ Ni(OH)_2(s) + Cd(OH)_2(s)$ [8.6]

These three equations show exactly the same transfer of electrons shown in equations 8.1, 8.2 and 8.3, but now all the different states and chemical forms are indicated.

An important feature of the NiCad battery is that the battery is rechargeable. The starting materials that undergo oxidation and reduction are solids, as are the products. The solids formed by the forward (discharging) reaction cling to a stainless steel grid within the battery. Thus, they are still available and permit the reaction to be reversed during the recharging process. No gases are produced during either the recharging or the discharging, so the battery can be totally sealed. Transfer of electrons takes place during both the discharging and recharging processes, just in opposite directions. Equation 8.7 indicates this reversible process.

$$Cd(s) + 2\ NiO(OH)(s) + 2\ H_2O(l) \underset{\text{recharging}}{\overset{\text{discharging}}{\rightleftharpoons}} 2\ Ni(OH)_2(s) + Cd(OH)_2(s) \quad [8.7]$$

8.2 Your Turn

Consider this galvanic cell. A reddish coating of impure copper metal begins to appear on the surface of the copper cathode as the cell operates. This is the overall equation for the reaction.

$$Zn(s) + Cu^{2+}(aq) \longrightarrow Zn^{2+}(aq) + Cu(s)$$

1.10 V
e^-
e^-
Copper (cathode)
Zinc (anode)
1 M $Cu^{2+}(aq)$
1 M $Zn^{2+}(aq)$
Salt bridge to carry ions

a. Write the oxidation half-reaction taking place at the anode.
b. Write the reduction half-reaction taking place at the cathode.

8.3 Consider This: Cell Phone Batteries

Many high-end cell phones are equipped with lithium ion batteries. Use the resources of the Web to find out more about this type of battery by searching for "lithium battery chemistry."

a. Why is this battery suited for use in portable devices?
b. What materials form the anode and the cathode of a lithium ion battery?

c. What is the voltage of a lithium ion battery?

d. What other types of batteries are used in cell phones? What are their advantages and disadvantages compared to lithium ion batteries?

Figure 8.3

AAA to D cells all produce 1.54 volts, but the larger cells sustain larger currents.

Almost everyone has used an alkaline battery such as shown in Figure 8.3. How does the electron transfer work in this case? Figure 8.4 gives you an overview of the operation of a typical alkaline battery.

These are the half-reactions for the alkaline cell illustrated in Figure 8.4.

$$\text{Anode (oxidation):} \quad Zn(s) + 2\ OH^-(aq) \longrightarrow Zn(OH)_2(s) + 2\ e^- \quad [8.8]$$

$$\text{Cathode (reduction):} \quad 2\ MnO_2(s) + H_2O(l) + 2\ e^- \longrightarrow Mn_2O_3(s) + 2\ OH^-(aq) \quad [8.9]$$

The overall cell reaction is the sum of the two half-reactions.

$$Zn(s) + 2\ MnO_2(s) + H_2O(l) \longrightarrow Zn(OH)_2(s) + Mn_2O_3(s) \quad [8.10]$$

This cell produces 1.54 volts. The cell voltage depends primarily on which elements and compounds participate in the reaction. It does not depend on factors such as the overall size of the cell, the amount of material it contains, or the size of the electrodes. Thus all alkaline cells, from the tiny AAA size to the large D cells, give the same voltage, 1.54 V. On the other hand, the **current,** or rate of electron flow, does depend on the size of the cell: Larger cells generate larger currents. And, because power is obtained by multiplying the voltage by the current, larger cells are more powerful than smaller ones.

• Current is measured in amperes (amps) to honor the physicist André Ampère.

Many different galvanic cells have been developed for different specific purposes. Some of these cells are listed in Table 8.1, along with their voltages and an indication of whether they are rechargeable. Because mercury batteries can be made very small, they have been used widely in watches, camera equipment, hearing aids, calculators, and other devices that use transistors and integrated circuits that do not require large currents. Unfortunately, the toxicity of mercury (Chapter 5) makes the disposal of these cells a potential hazard. Burning trash may exacerbate and extend the problem by releasing mercury vapor into the atmosphere. The EPA estimated that as late as 1989, 88% of the 1.4 million pounds of mercury in urban trash came from non-rechargeable batteries. Development of safer battery alternatives and the need to recycle many types of batteries led to passage of the Mercury-Containing and Rechargeable Battery Management Act (The Battery Act) in 1996. The Act mandated the phaseout of mercury in batteries and represents a major step forward in the effort to facilitate recycling nickel-cadmium and certain small sealed lead-acid rechargeable batteries. The change in mercury use, including batteries, is shown in Figure 8.5.

• Lead-acid rechargeable batteries are discussed next.

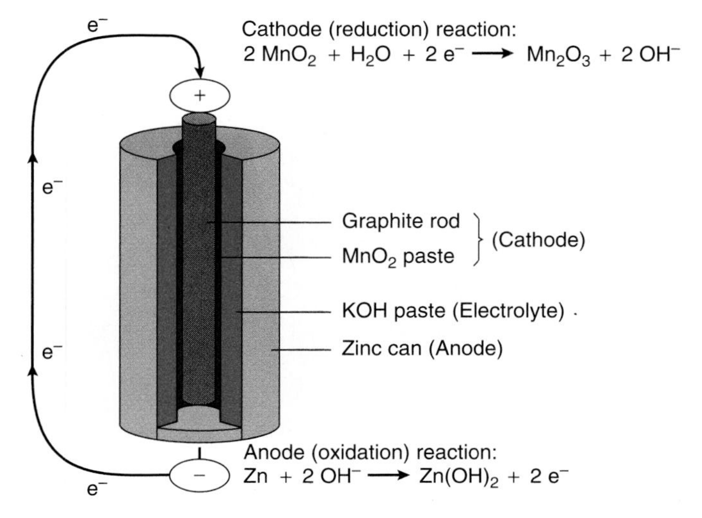

Figure 8.4

Diagram of an alkaline cell (battery).

Table 8.1

Some Common Galvanic Cells and Their Associated Voltages

Type	Voltage	Rechargeable?
Alkaline	1.54	No
Mercury	1.3	No
Lithium-iodine	2.8	No
Lithium ion	3.7	Yes
Lead storage	2.0	Yes
Nickel-cadmium	1.46	Yes

8.4 Consider This: The Federal Battery Act and You

Explore the provisions of The Battery Act by going to the EPA web site or using the link provided at the *Chemistry in Context* web site.

a. Why was it necessary at the federal level to regulate battery manufacture, use by consumers, and disposal of batteries?
b. What types of batteries are regulated by The Battery Act?
c. What are some potential hazards associated with the improper disposal of batteries?
d. What do you personally do with your "dead" batteries? Explain the options available to you in your community.

Lithium-iodine cells are so reliable and long-lived that they are used to power cardiac pacemakers. The lithium battery takes advantage of the low density of lithium metal to make a lightweight battery. A lithium-iodine pacemaker battery implanted in the chest can last as long as 10 years before it needs to be replaced. In fact, the widespread use of such pacemakers today has been due, in large part, to the improvements made in the bat-

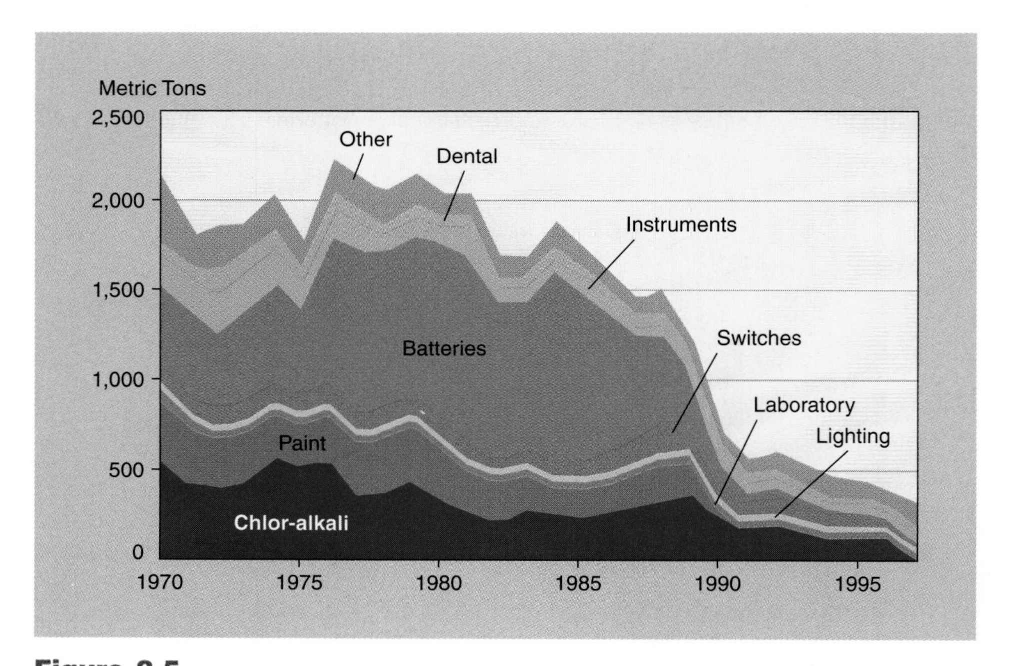

Figure 8.5

Mercury use for the manufacture of batteries has declined sharply in recent years.

Source: U.S. Geological Survey Minerals Yearbook. Taken from Chemical & Engineering News, *February 5, 2001, p. 22.*

teries used to power them, rather than in the pacemakers themselves. Persons with pacemakers are advised to avoid electromagnetic fields that can interfere with the operation of the device, such as holding a cell phone close to the location of the pacemaker.

Most batteries convert chemical energy into electrical energy with an efficiency of about 90%. This can be compared with the much lower efficiencies that typically characterize the conversion of heat to work (30–40%) in electricity-generating plants. However, it is important to remember that considerable energy is required to manufacture galvanic cells. Metals and minerals must be mined and processed, and the various components manufactured and assembled. Moreover, a battery has a finite life. Even rechargeable batteries will eventually fail and have to be replaced. Sooner or later, the chemical reaction in the battery will be complete, the voltage will drop below usable levels, and electrons will no longer flow. At this point, the battery is "dead" and ready for disposal.

The best-known rechargeable battery is the lead-acid storage battery. Lead-acid batteries are called **storage batteries** because they store electrical energy. Until very recently, such batteries were used in every automobile. The lead-acid storage battery is a true battery because it consists of six cells, each generating 2.0 V for a total of 12.0 V. The overall cell reaction is given by equation 8.11.

$$\underset{\text{lead}}{Pb(s)} + \underset{\text{lead dioxide}}{PbO_2(s)} + \underset{\text{sulfuric acid}}{2\,H_2SO_4(aq)} \underset{\text{recharging}}{\overset{\text{discharging}}{\rightleftarrows}} \underset{\text{lead sulfate}}{2\,PbSO_4(s)} + \underset{\text{water}}{2\,H_2O(l)} \qquad [8.11]$$

The anode is made of metallic lead and the cathode of lead dioxide, PbO_2. The electrolyte is a concentrated sulfuric acid solution (Figure 8.6). Although the weight of the lead and the corrosive properties of the acid are disadvantages, the lead-acid storage battery is dependable and long lasting. The key to its success is the fact that reaction 8.11 is reversible. As it spontaneously proceeds in the direction indicated by the arrow to the right, the reaction produces the energy necessary to power a car's starter, headlights, and various devices. But, as the reaction proceeds, the battery "discharges." The electrical demands of a modern car are so great that in a short time, most of the reactants would be converted to products, significantly reducing the voltage and the current. To counter this, the battery is attached to a generator, or alternator, turned by the engine. The alternator generates direct current electricity, which is run back through the battery. This input of energy reverses the reaction, represented by the arrow to the left in equation 8.11, and recharges the battery. In a high-quality lead-acid storage battery,

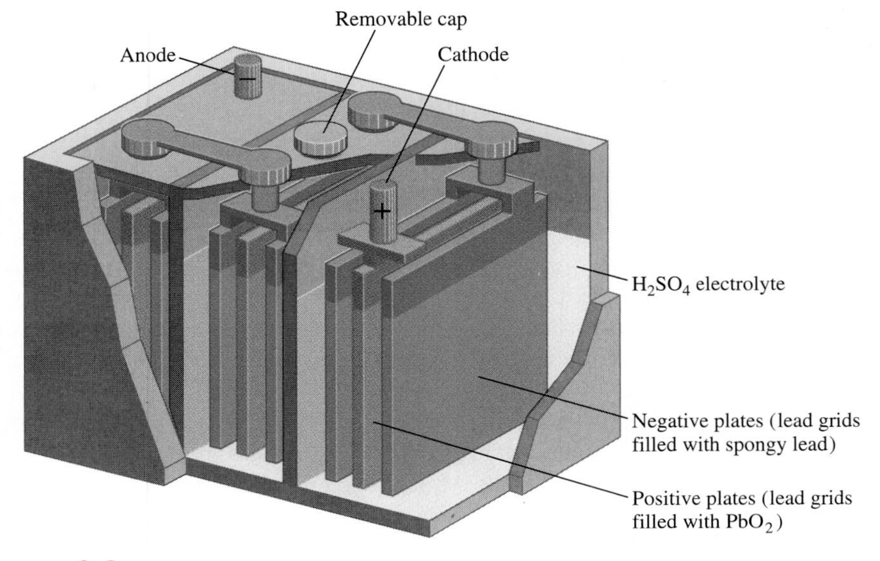

Figure 8.6

Cutaway view of a lead-acid storage battery.

this process of discharging and recharging can go on over a period of five or more years.

In environments where the emissions from internal combustion engines cannot be tolerated, sealed lead-acid storage batteries provide the only source of energy for locomotion. Thus, forklifts in warehouses, passenger carts in airport terminals, golf carts, and wheelchairs are typically powered by lead-acid storage batteries. Because of the dependability of such batteries, they are sometimes used in conjunction with wind turbine electrical generators. The generator charges the batteries when the wind is blowing, and the batteries provide electricity when the wind stops.

8.5 Your Turn

Examine equation 8.11 for the reaction that takes place in a lead-acid storage battery while it is discharging.

a. Which material is being oxidized?
b. Which material is being reduced?
c. Write the chemical equation for the *recharging* reaction. What material is being oxidized? Which is reduced?

Having discussed several types of rechargeable and nonrechargeable batteries, we now look at a different type of battery, a fuel cell, sometimes called a "flow" battery, because both fuel and oxidizer must constantly flow into the cell to continue the chemical reaction. After we consider this important type of battery, we can put into perspective new approaches to powering cars and other types of transportation vehicles by using fuel cells.

8.2 Fuel Cells

Suppose someone were to suggest that there was a way to combine H_2 and O_2 to form H_2O without the hazards of combustion. Suppose further that this individual claimed that the reaction could be carried out without allowing the hydrogen and oxygen to come in contact with each other. The Sceptical Chymist might well dismiss such assertions as sheer nonsense—an outright impossibility. And yet, sometimes what appears to be completely contrary to common sense can happen in the natural world. The operation of a fuel cell is a case in point. In a **fuel cell,** the chemical energy of a fuel is converted directly into electricity without burning the fuel. Sir William Grove, an English physicist, invented fuel cells in 1839. However, they remained a mere curiosity until the advent of the U.S. space program in the 1960s. Such devices are routinely used as sources of electrical energy in the space program. The space shuttle, for example, carries three sets of 32 cells fueled with hydrogen, but never "burns" it. Instead, the hydrogen is used in the fuel cell to produce electricity for the shuttle.

A fuel cell functions somewhat like a conventional flashlight battery. But, unlike batteries, fuel cells use a constant external supply of fuel (such as hydrogen) and oxidant (such as the oxygen in air). Therefore, they produce electricity as long as fuel and oxidant are provided and consequently do not need to be recharged. In a fuel cell, the chemical reaction is physically separated into two parts, each of which occurs in a separate region of the cell (Figure 8.7). Recently developed hydrogen fuel cells use a solid polymer to separate the reactants and act as the electrolyte. The polymer is a proton exchange membrane (PEM) permeable to H^+ ions, and coated on both sides with a platinum-based catalyst.

• Polymers are discussed in Chapter 9.

In a fuel cell, one of the reactions is always oxidation in which a reactant loses electrons. The other reaction is a reduction involving the gain of electrons by some other reactant. In a hydrogen fuel cell using a PEM, hydrogen gas (H_2) is the fuel used in conjunction with oxygen. The oxidation and reduction are represented by half-reactions in equations 8.12 and 8.13. As hydrogen passes through the membrane, it loses electrons to form H^+ ions, an oxidation half-reaction.

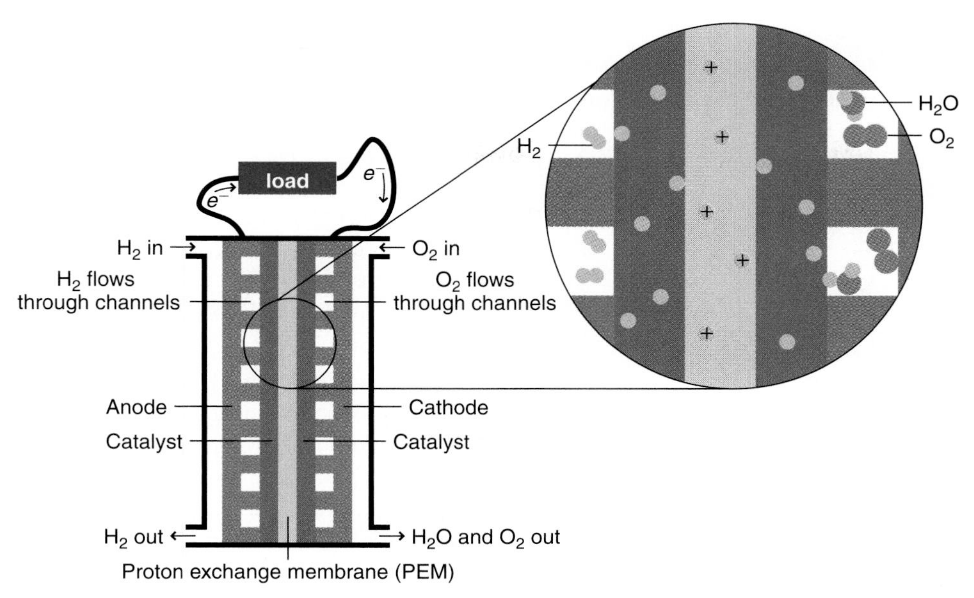

Figure 8.7

A PEM fuel cell. The anode reaction is $H_2 \longrightarrow 2\,H^+ + 2e^-$
The cathode reaction is $\frac{1}{2}\,O_2 + 2\,H^+ + 2\,e^- \longrightarrow H_2O$
The overall (net) reaction is $H_2 + \frac{1}{2}\,O_2 \longrightarrow H_2O$

 **Figures Alive!**

$$H_2(g) \longrightarrow 2\,H^+(aq) + 2\,e^- \qquad [8.12]$$

The H^+ ions flow through the proton exchange membrane to the other side, where they combine with oxygen (O_2) and electrons to form water in the reduction half-reaction.

$$\frac{1}{2}\,O_2(g) + 2\,H^+(aq) + 2\,e^- \longrightarrow H_2O(l) \qquad [8.13]$$

Thus, there is a transfer of electrons—from H_2 to O_2. An important thing to note again is that oxidation cannot occur alone; that would be rather like one hand clapping. Oxidation (electron loss) must always be paired with a reduction reaction (electron gain). The overall reaction (equation 8.14) combines the oxidation and the reduction half-reactions, taking into account gain and loss of electrons in the two half-reactions:

$$H_2(g) + \frac{1}{2}\,O_2(g) + 2\,H^+(aq) + 2\,e^- \longrightarrow 2\,H^+(aq) + 2\,e^- + H_2O(l) \qquad [8.14]$$

The two electrons lost in the oxidation half-reaction provide the two electrons gained in the reduction process. The two electrons and 2 H^+ that appear on each side of the arrow in equation 8.14 can be cancelled to give the net equation

$$H_2(g) + \frac{1}{2}\,O_2(g) \longrightarrow H_2O(l) \qquad [8.15]$$

- This same equation appears in Chapters 1, 4, and 5, an indication of its significance.
- In industry, fuel cells are sometimes called "flow batteries."

This is clearly the equation for the burning of hydrogen in oxygen. But in a fuel cell, it is "burning" without a flame and with relatively little heat. Water is the only product, besides a transfer of electrons, which is what we call electricity. Hence, fuel cells are a more environmentally friendly way to produce electricity than burning fossil fuel or fissioning uranium or plutonium atoms.

To produce electricity, the two half-reactions of the fuel cell must be connected in such a way that the electrons released during the oxidation reaction are *transferred* to the reduction reaction (Figure 8.7). As with other types of batteries, electrodes provide

Table 8.2

Comparison of Combustion with Fuel Cell Technology

Process	Fuel*	Oxidant	Products	Other Considerations
Combustion	H_2	O_2 from air	H_2O, heat, light, and noise energy	Rapid process, flame present, lower efficiency, most useful for producing heat
Fuel cell technology	H_2	O_2 from air	H_2O, electricity, some heat	Slower process, no flame, quiet, higher efficiency, most useful for generating electricity

*Other compounds containing chemically combined hydrogen, such as natural gas or alcohols, can be used as fuels. This results in other products, such as CO_2, being released.

the sites for chemical reaction. Oxidation occurs at the anode, and reduction takes place at the cathode. The electrical energy produced is the result of the spontaneous reaction that occurs in the fuel cell. Hydrogen and oxygen gases that have not been consumed can be recycled into the cell. The electrons flowing from the anode to the cathode of a fuel cell move through an external circuit to do work, which is the whole point of the device. On the space shuttle these electrons are used to illuminate bulbs, power small motors, and operate computers.

The net reaction represented by equation 8.14 releases 286 kJ of energy per mole of water formed. But instead of liberating most of this energy in the form of heat, the fuel cell converts 40–45% or more of it into electrical energy. This direct production of electricity eliminates the inefficiencies associated with using heat to do work to produce electricity. Internal combustion engines are only 20–30% efficient in deriving energy from fossil fuels. Moreover, a fuel cell does not "run down" or require recharging. It keeps functioning as long as the fuel (hydrogen, natural gas, or methanol) and oxygen are supplied. To generate useful amounts of electrical energy, fuel cells are stacked in layers, just as cells are combined to produce a more powerful battery. Figure 8.10 in the next section shows cells stacked in a fuel-cell powered automobile. A comparison of the combustion of hydrogen with oxygen and their chemical combination in a fuel cell is given in Table 8.2.

8.6 Consider This: Other Fuel Cell Technologies

Proton-exchange membrane fuel cells are not the only fuel cell technology being explored. For example, phosphoric acid is used as the electrolyte in many commercially operating fuel cell systems, and there are demonstration units for both molten carbonate and solid-oxide ceramic fuel cells. Use the resources of the Web to find out some additional information about the current status, advantages, and disadvantages of these alternate fuel cell technologies.

The U.S. space program accelerated the development of fuel cell technology, but there is also a good deal of interest in using fuel cells to generate power at stationary locations. The largest fuel cell assembly in the United States supplies electrical power to 1000 homes in Santa Clara, CA. The health maintenance organization giant Kaiser Permanente has installed fuel cell units to furnish electricity in three of its California hospitals. Ballard Generation Systems, in cooperation with partners in the U.S., Germany, Switzerland, and Japan, started a series of field trials of 250-kilowatt stationary power

Figure 8.8
This 250-kW stationary fuel cell power generator uses natural gas, rather than hydrogen, as the fuel.

generators in September 1999. These trials will continue through 2003 with the goal of obtaining performance data in real world operations. Figure 8.8 shows one of these units being used to provide high-quality power to the on-site power grid. The fuel cell power may also be used for standby or backup power.

Engineers and developers are now working on applications of fuel cell technology to provide portable power cleanly and quietly for remote, recreational, home, and emergency power markets. One day, fuel cells may be powering your lawn mower and even your laptop computer in remote locations; there is a prototype fuel cell to power a cell phone. Figure 8.9 shows one example of the lofty heights that can be reached while using fuel cell technology.

- A laptop computer on a hike seems out of place, unless the computer is being used to record and process research data.

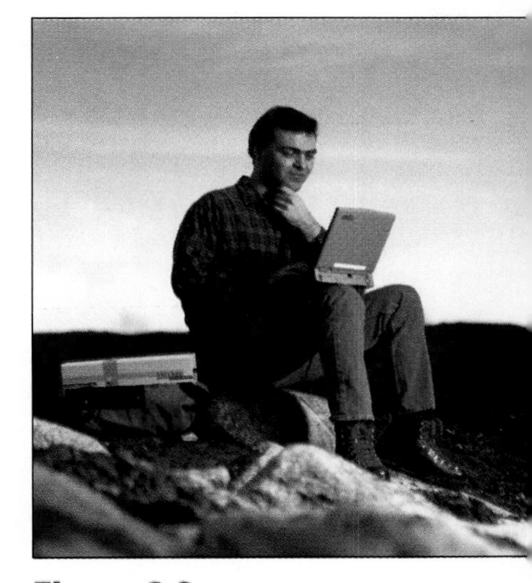

Figure 8.9
Portable fuel cell power. Note the small size of the fuel cell unit resting on the backpack behind the person.

8.7 Consider This: Fuel Cells in Place

a. What are some of the reasons that fuel cells are being used for generating power in stationary locations, but are not yet being actively considered for use in personal applications such as handheld calculators?
b. How do you expect applications of fuel cells to change in the future? Explain your reasoning.

8.8 Consider This: Military Electrons

Fuel cells find many applications in the military. Where and how? To find the military fuel cell demonstration site nearest you, check out the web site of the Department of Defense (DoD) on fuel cells. Check out several of the demonstration sites across the country to find out (a) the different uses of these fuel cells and (b) the cost savings reported.

8.9 Consider This: Batteries—Functions and Uses

The best match between technical capability and intended function determines the type of battery used for a specific application. Based on what you have learned to this point in Sections 8.1 and 8.2, consider each of these uses and discuss the criteria that will determine which battery is best for the intended use: (1) TV remote controls; (2) heart pacemakers; (3) deep-space probes; (4) cellular phones; (5) automobiles.

We turn next to how fuel cells, batteries, and combinations of energy sources dependent on the transfer of electrons can be used to power automobiles and other forms of transportation.

8.3 Alternate Energy Sources for Transportation

Because PEM fuel cells are compact, light, and do not require hazardous electrolytes such as potassium hydroxide used in previous generations of fuel cells, they are prime candidates for use in electric vehicles. Safe and efficient storage of hydrogen fuel has been a problem in developing fuel cell use for vehicles. Bulky storage containers have somewhat limited the vehicle's driving range. Nevertheless, as shown in Figure 8.10, hydrogen fuel can be used with stacked PEM fuel cells to power a car by electron transfer.

Another option is to use a fuel such as methanol (CH_3OH) as the source of hydrogen in PEM fuel cells. A device, called an "on-board methanol reformer" converts the hydrogen from methanol into H_2O. A different catalyst, an alloy made from four metals, replaces the platinum-based one used with hydrogen-only fuel cells, which cannot be used with methanol and water. At the anode of a methanol fuel cell, a 3% solution of methanol and water reacts to produce carbon dioxide, hydrogen ions, and electrons (equation 8.16).

$$H_2O(l) + CH_3OH(aq) \longrightarrow CO_2(g) + 6\,H^+(aq) + 6\,e^- \qquad [8.16]$$

In the cathode compartment, air or oxygen is blown into the cathode compartment where it reacts with the electrons and hydrogen ions to produce water.

$$\tfrac{3}{2}\,O_2(g) + 6\,H^+(aq) + 6\,e^- \longrightarrow 3\,H_2O(l) \qquad [8.17]$$

Equation 8.18 is the net reaction.

$$CH_3OH(aq) + \tfrac{3}{2}\,O_2(g) \longrightarrow CO_2(l) + 2\,H_2O(l) \qquad [8.18]$$

As with hydrogen fuel cells, the electricity produced by methanol fuel cells is used to power electric motors that provide the motor power for the vehicle. Note from equation 8.18 that electric vehicles powered by these fuel cells produce no nitrogen oxides, such as those produced from internal combustion engines. Also, the amount of CO_2 generated per unit of useful energy is lower than that emitted in the direct combustion of the fuel because the efficiency is higher. In addition, methanol is a renew-

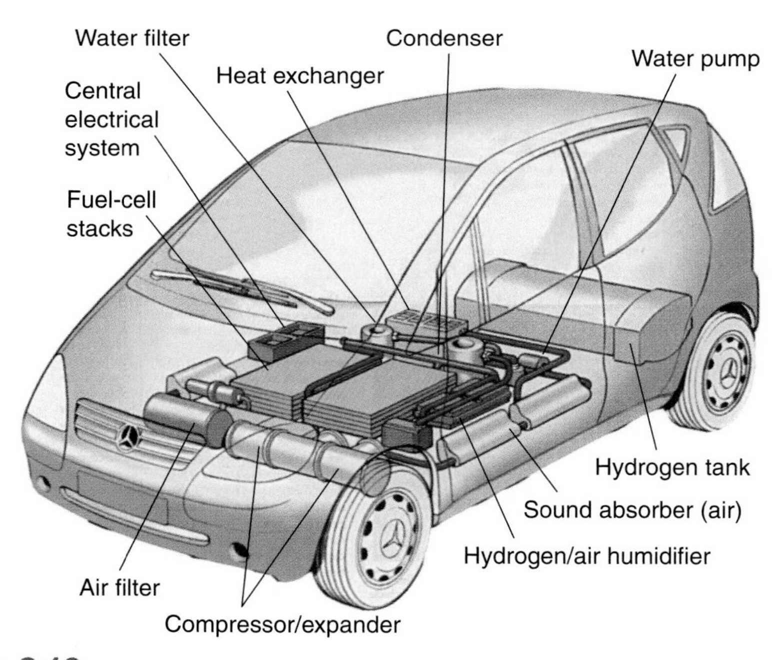

Figure 8.10

The electrochemical engine for vehicles.

Source: George Retseck Daimler Chrysler in July, 1999 issue of Scientific American.

able fuel, unlike gasoline. Furthermore, fuel cells have no moving parts, so the vehicles should require little repair and last longer. According to Dr. Halpert, Program Manager for Batteries and Fuel Cells at the Jet Propulsion Laboratory, "This [methanol] fuel cell may well become the power source of choice for energy-efficient, non-polluting electric vehicles."

8.10 Consider This: Fuel Cells in Your Future?

Many different types of fuel cells are under development. Another promising type is the propane-oxygen fuel cell. Propane is $C_3H_8(g)$. This is the equation for the chemical reaction that takes place in a propane-oxygen fuel cell.

$$C_3H_8(g) + 5\ O_2(g) \longrightarrow 3\ CO_2(g) + 4\ H_2O(l)$$

a. Identify the substance that undergoes oxidation and the substance that undergoes reduction in this reaction.
b. Unlike batteries, fuel cells do not store chemical energy. Explain the significance of this statement for the future of fuel cells.
c. What are some of the reasons why fuel cells have not been the energy source of choice in the past, but may become a viable choice in the future?

Electric vehicles powered by fuel cells are not a far-fetched idea. Buses carrying up to 60 passengers have been powered by PEM fuel cells in Chicago, IL and at Georgetown University in Washington, DC (Figure 8.11).

In 1997, Mercedes-Benz (now Daimler-Chrysler) unveiled NECAR 5 (New Electric Car), an electric vehicle that operates on methanol fuel cells (Figure 8.12). NECAR 5 averages about 25 miles per gallon of methanol and has a 250-mile range without refueling, close to that of more conventional vehicles. Dr. Ferdinand Panik with Daimler-Chrysler remarked "In the end, I believe the fuel cell can be done for the same price as the piston engine, or lower. And I believe it can let the owner travel 50% farther for the fuel used, with an engine that will be truly maintenance-free."

Whether Panik's prediction will be the case remains to be seen. Reliable predictions of the true costs of automobiles powered by fuel cells are difficult to obtain, but the prices of the cell components are declining. Also true is that the "big three" domestic automobile manufacturers each have put significant resources behind developing fuel cell-powered electric cars under the Department of Energy's Partnership for a New Generation of Vehicles. General Motors plans to have a fuel cell electric car ready for

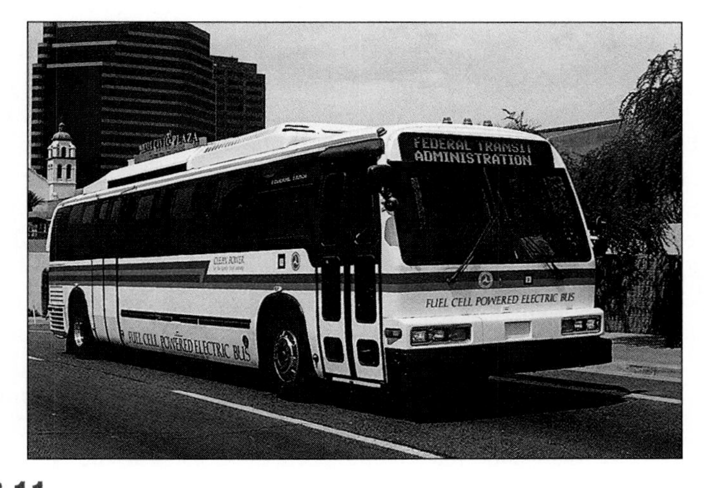

Figure 8.11
A methanol-based fuel cell bus at Georgetown University. The Federal Transit Administration funded the bus.

Figure 8.12
The NECAR 5 (New Electric Car) by Daimler-Benz (now Daimler-Chrysler) operates using methanol fuel cells.

production by 2004, Ford Motor Company had a test car ready in 2000, and Daimler-Chrysler had a prototype vehicle developed by that time as well. A fuel cell manufacturer, Ballard Power Systems, predicts that competitively priced, zero-emission vehicles will start to show up in automobile showrooms for the 2004-2005 model year. To make the prediction a reality, Ballard Power Systems has formed a worldwide alliance with Daimler-Chrysler and Ford Motor Company to become the largest commercial producer of fuel cell-powered electric drive trains and components for cars, trucks, and buses. The alliance is gearing up to hit their target of mass producing 100,000 fuel cell-powered electric vehicles by 2004-2005. "The beauty of fuel-cell vehicles is that they are pollution-free and energy efficient, and we can make the fuel right here in America," said Paul Lehman, a fuel cell researcher at Humboldt State University (CA). "In electric cars, fuel cells offer important advantages over batteries: They have greater range, and they take minutes to refuel—not hours to recharge."

Arthur D. Little, an energy-consulting firm, has announced the development of a prototype fuel cell that converts gasoline to hydrogen. In the prototype fuel cell, gasoline vapor is converted into hydrogen and carbon monoxide. The carbon monoxide, in contact with a special catalyst, is then reacted with steam to produce carbon dioxide and additional hydrogen. The hydrogen can then be used to make electricity, as in a conventional fuel cell. As a fuel to produce hydrogen for fuel cells, gasoline would have the advantage of using a pre-existing fuel distribution system, a feature not yet available for distributing hydrogen directly. Still in development, the gasoline fuel cell is not likely to be available commercially until at least 2003.

In contrast with automobile makers planning on fuel cells to produce pollution-free vehicles, General Motors decided to take a different approach. The first modern all-electric passenger car to be mass produced and available in showrooms around the country was GM's EV1. This zero emission electric vehicle (ZEV) started a transformation in the way in which the world considers alternatives to fossil fuels as energy sources. Twenty-six lead storage batteries, weighing a total of about 1100 pounds, are at the energetic heart of the 137-horsepower General Motors EV1 (Figure 8.13).

This new car was developed in response to legislation enacted in California and subsequently in New York and Massachusetts. The original goal was that by 2003, 10% of the new cars sold by major auto manufacturers in these states would meet "zero emission" standards. This goal was to be achieved through a series of intermediate steps. Each automobile corporation had a quota of zero emission cars that had to be sold in a given year. If the number actually sold was below that minimum, the company would have been fined $5000 for each car short of that goal. The regulations proposed for

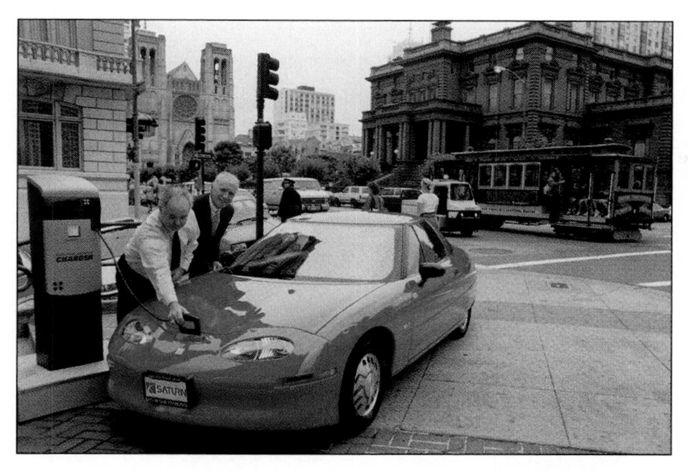

Figure 8.13
The EV1 electric car by General Motors at a charging station.

southern California were even more stringent. In an effort to remedy what has been called the worst air pollution in the United States, a tentative plan had been adopted that would require all cars in the Los Angeles basin to be converted to electric power or other clean fuel by 2007. Given current technology at the time these proposals were passed, regulators felt that the only way to achieve this goal was to rely on lead storage batteries.

California air regulators confirmed the lead-acid storage battery approach to meeting air quality goals in September 2000, and Governor Gray Davis signed a law establishing a state grant program to encourage the purchase or lease of zero emission vehicles. However, in December 2000, the regulators took action to ease these stringent requirements and to scale back the number of vehicles that would have to meet the 2003 and 2007 deadlines. Although recent court action has restored the original goal of having 10% of new cars sold by 2003 meet zero emission standards, there is a growing realization that emerging automobile battery technologies cannot provide the only answer. Fuel cell research, as well as new automobile battery technologies, may permit manufacturers to meet future deadlines, as will the development of hybrid technologies to be discussed in a subsequent section.

The good news is that the electric, battery-operated cars release into the atmosphere no carbon dioxide or carbon monoxide, no oxides of sulfur or nitrogen, and no unburned hydrocarbons or ozone. Such cars require no gasoline or other fuel. Unfortunately, zero emission on the road does not translate to a similar absence of pollution elsewhere. The batteries in an electric car must be recharged frequently, usually after only 90 miles of highway driving or 70 miles of city driving. And if the temperature drops to 20°F, battery efficiency decreases, and the driving range falls to about 25 miles. To re-energize the car, it must be plugged into a source of electricity for recharging, requiring at least three hours. Moreover, charging stations are currently few and far between, creating the electric car equivalent of "running out of gas" with no gas station nearby. However, 220-volt chargers to use at home are available at a cost of about $2000, which is built into the cost of the vehicle.

But recharging is only part of the problem. As you know, electrical power generation is notoriously inefficient; less than half of the energy released from burning fuel is converted into electricity (Section 4.13). Furthermore, power plants that burn fossil fuels for energy release sulfur dioxide, nitrogen oxides, and carbon dioxide. In fact, calculations indicate that the SO_2 and NO_x emitted from power plants generating the electricity to keep a fleet of battery-powered cars operational would exceed the amount of these two gases that would be released by the gasoline-powered cars that would have been replaced! The overall CO_2 emission does decline when electric cars are substituted for internal combustion automobiles, but by less than 50%.

8.11 Consider This: Look Ma, No Tailpipe!

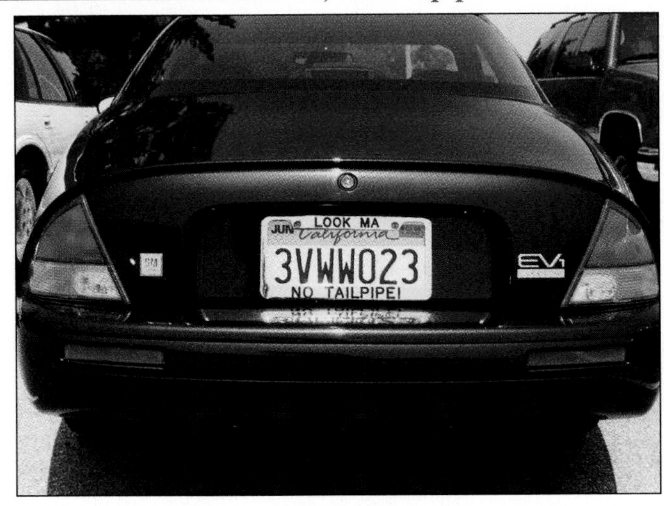

a. Why would a GM car dealer call attention to the missing tailpipe on an EV1 car?

b. What *else* is missing other than the tailpipe on an EV1 car? Why are these other components unnecessary?

An early and serious criticism of dependence on lead-acid storage batteries to power "pollution-free" electric cars was published in *Science* in May 1995 by Lester Lave, Chris Hendrickson, and Francis McMichael. These three Carnegie Mellon University professors argued that a switch to cars powered exclusively by lead-acid storage batteries would dramatically increase the amount of lead released into the environment. Their calculations include estimates of lead dispersed in mining, processing, and battery manufacture. They conclude that, with current technology, 1.34 grams of the toxic metal would be emitted per kilometer traveled by an electric car. This corresponds to 2.16 grams of lead per mile or 47.5 pounds in 10,000 miles, a typical annual mileage driven in the United States. Ironically, this amount of lead is 60 times that released over the same distance by a car burning leaded gasoline. To be sure, critics have questioned some of the assumptions made by Lave, Hendrickson, and McMichael and argued that their conclusions greatly exaggerate the problem. Moreover, power plants, lead mines, lead refineries, and battery factories are point, not mobile, sources of pollution. This makes it easier to control emissions from them than from the millions of cars that clog the California and other freeways and spew their emissions over thousands of square miles. But experience should alert us to the fact that it is dangerous to focus on one part of a problem without considering broad systemic environmental impact, lest the cure be worse than the ailment it sets out to remedy.

It is unlikely that electric cars powered by lead storage batteries alone will become a major, long-term answer to the problem of automotive air pollution. At best they appear to be a temporary solution. But the development of lighter, more efficient, more environmentally benign batteries would be a major step forward. Among the designs being used is a nickel-metal hydride cell that would double the range of an electric car and cut recharging time to 15 minutes. Unlike lead-acid batteries that must be replaced after 25,000–50,000 miles, these batteries last the life of the vehicle. Honda uses nickel-metal hydride batteries in its EV+ four-seater vehicle. Other possible battery combinations include nickel-cadmium, sodium-nickel-chloride, sodium-sulfur, zinc-air, aluminum-air, and lithium polymer. The U.S. Advanced Battery Consortium (USABC), a joint venture of major U.S. automakers and the Department of Energy, continues to sponsor intensive research aimed at developing advanced battery systems for motor vehicles, including development of nickel-metal hydride batteries. USABC Chairman Robert L. Davis said that USABC-sponsored research "has identified lithium-based batteries as the long-range solution to competitive electric vehicles and has worked successfully to demonstrate technical design feasibility for these technologies."

8.12 Consider This: Nickel versus Lithium-based Batteries

Are lithium-based batteries better than nickel-metal hydride ones? Use the Web to find details about these two types of batteries. Then, write a brief summary of your findings and give your conclusion as to which battery would be more suitable for use in an electric vehicle.

Concerns about energy supplies and environmental effects have driven the development of electric vehicles. However, a compromise is needed so that consumers can enjoy the convenience and range of a gasoline-powered car combined with the environmental advantages of an electric vehicle powered by batteries. Toyota and several other auto manufacturers have developed such a vehicle, called a **hybrid car.** The Toyota Prius, a five-passenger hybrid about the size of a Toyota Corolla, was first available in Japan in 1998 (Figure 8.14).

With a 1.5-liter gasoline engine sitting side-by-side with nickel-metal hydride batteries, an electric motor, and an electric generator, the Prius does not need to be recharged. It consumes only about half the gasoline, emits 50% less carbon dioxide and far less nitrogen oxides than a conventional car, while delivering 45 miles per gallon of gasoline around town and 52 out on the highway. The electric motor draws power from the batteries to get the car moving or to power it at low speeds. Using a process called regenerative braking, the kinetic energy of the car is transferred to the generator, which charges the batteries during deceleration and braking. The gasoline engine assists the electric motor during normal driving, with the batteries boosting power when extra acceleration is needed. The attractive $18,000 price tag has been set artificially low to stimulate sales in Japan, which have been brisk enough for demand to outpace supplies. The Prius became available in the United States in 2000. In the same year, Honda introduced a slightly smaller hybrid passenger car, the Insight, which gets 70–80 miles per gallon overall. GM is entering the field of hybrids with a five-passenger hybrid car, the Precept, designed to achieve 80 miles per gallon.

8.13 Consider This: Hybrid Cars

Visit web sites for at least two different companies to find the latest information on their hybrid vehicles. Search for "hybrid car" or "advanced vehicles" to find the latest information. The *Chemistry in Context* web site has links to several sites describing hybrid vehicles.

a. What does the manufacturer say about them?
b. What information can you find about test driving, leasing, or buying hybrid cars?
c. What do the owners or users of these cars say about each type of hybrid car?
d. Are hybrid cars available in your region of the U.S.? If so, at what location?

(a)

(b)

Figure 8.14
(a) The Toyota Prius hybrid automobile.
(b) The "engine" of the Prius.

8.14 Consider This: Batteries, Fuel Cells, and Hybrids

The text describes two potential sources of electrical power for cars and other motor vehicles: rechargeable batteries and fuel cells. Both use chemical reactions to generate electricity but differ in the details. Hybrid systems also are under development. Demonstrate your knowledge of the chemical principles involved by explaining the similarities and differences in the operation of rechargeable batteries and fuel cells to a classmate. Then explain the advantages that a hybrid car could bring to the consumer.

Automobile industry leaders agree that there will be no mass market for alternative energy vehicles—battery-powered, hybrid cars or ones using fuel cells—unless the performance and price of such vehicles match those of conventional models. Jack Smith, former CEO of General Motors, said: "People are too practical. There's a certain element who will fall in love with new technology, but technology won't survive unless it's cost effective." The common strategy for companies now is to provide more than one option for fuel-efficient, low emission vehicles, then allow consumers to choose the best option for their personal transportation needs.

8.15 Consider This: Advanced Vehicles—Visit the Showroom!

Many of the major manufacturers are advertising their versions of advanced vehicles, those that use something other than a conventional gasoline-driven internal combustion engine. Are they available in your area? If so, visit the showroom and talk with a sales agent. (If not, visit an on-line salesroom.) What features would convince consumers to buy these cars? Design either a poster or a radio or television announcement that would help market an advanced vehicle.

Much progress has been in the development of smaller and lighter batteries and the production of hybrid cars. The planned introduction of fuel cell cars by most major automakers by 2004 requires that we take a more careful look at an essential element, hydrogen. Where better to get hydrogen than from the abundant water supplies that are widely distributed around the world?

8.4 Splitting Water: Fact or Fantasy?

In Jules Verne's 1874 novel *Mysterious Island*, a shipwrecked engineer speculates about the energy resource that will be used when the world's coal supply has been used up. "Water," the engineer declares, "I believe that water will one day be employed as fuel, that hydrogen and oxygen which constitute it, used singly or together, will furnish an inexhaustible source of heat and light."

Is this simply science fiction, or is it energetically and economically feasible to break water into its component elements? Can hydrogen really serve as a useful fuel? And what does this have to do with advanced vehicles and light from the sun? To answer these questions and assess the credibility of the claim by Verne's engineer, a Sceptical Chymist needs to examine the energetics of the reaction represented by equation 8.19.

$$2\ H_2(g) + O_2(g) \longrightarrow 2\ H_2O(l) \qquad [8.19]$$

Experiment shows that the reaction, as written, gives off 572 kJ when two moles of liquid water are formed from the combination of two moles of hydrogen and one mole of oxygen. It follows that the burning of one mole of H_2 will yield one mole of H_2O and $\frac{1}{2}(572)$ or 286 kJ of energy. This indicates a highly exothermic reaction, represented by equation 8.20.

$$H_2(g) + \tfrac{1}{2}\ O_2(g) \longrightarrow H_2O(l) + 286\ kJ \qquad [8.20]$$

8.16 Sceptical Chymist: Checking with Bond Energy

Use the values in Table 4.1 for bond energies to check the energy released by the reaction in equation 8.20. To be convincing, you will need to clearly show your reasoning.

Because energy is *evolved*, the energy change in this combustion reaction is −286 kJ for each mole of H_2 burned to form liquid water. This is equivalent to releasing 143 kJ per gram of H. In comparison, the heat of combustion of coal is 30 kJ/g, octane (a major component in gasoline) is 46 kJ/g, and methane (natural gas) is 54 kJ/g when the products of combustion are $CO_2(g)$ and $H_2O(l)$. Clearly, hydrogen has the potential of being a powerful energy source. In fact, on a per gram basis, hydrogen has the highest heat of combustion of any known substance. The extraordinary energy per gram of hydrogen when it burns raises a tantalizing prospect—the practical use of hydrogen as a fuel to power motor vehicles that would produce only water vapor. No pollutants such as the carbon monoxide and nitrogen oxides would form, unlike what happens when burning fossil fuels. Simple as this might seem, there are challenges to overcome before burning hydrogen as fuel in automobiles becomes common practice.

A fundamental question to be answered is: How could we obtain a sufficient supply of hydrogen if it were to be used for fueling motor vehicles? On the one hand, things look promising because hydrogen is the most plentiful element in the universe. Over 93% of all atoms are hydrogen atoms. Although hydrogen is not nearly that abundant on Earth, there is still an immense supply of the element. But essentially all of it is tied up in chemical compounds. Hydrogen is too reactive to exist for long in its diatomic form, H_2, in the presence of the other elements and compounds that make up the atmosphere and the Earth's crust. Therefore, to obtain hydrogen for use as a fuel, it is necessary to extract hydrogen from hydrogen-containing compounds, and this requires energy.

8.17 Your Turn

Calculate the number of moles and grams of H_2 that would have to be burned to yield an American's daily energy share of 260,000 kcal (1 kcal = 4.18 kJ).

Answer

3800 mole, 7600 g

8.18 Your Turn

You have just learned that the formation of one mole of liquid water from hydrogen and oxygen releases 286 kJ of energy. In contrast, the direct formation of one mole of gaseous water from hydrogen and oxygen releases 242 kJ of energy. Offer a possible explanation for this observed difference in energy.

The traditional laboratory method of generating hydrogen gas involves the action of certain acids on certain metals. Perhaps the most common combination is sulfuric acid and zinc:

$$H_2SO_4(aq) + Zn(s) \longrightarrow ZnSO_4(aq) + H_2(g) \qquad [8.21]$$

Although convenient as a small-scale source of hydrogen, the reaction of a metal with an acid is far too expensive to scale up for industrial applications. Therefore, we turn back to the most abundant earthly source of hydrogen—water.

Because the formation of one mole of water from hydrogen and oxygen releases 286 kJ (equation 8.15), an identical quantity of energy must be absorbed to reverse the reaction to produce hydrogen. Figure 8.15 summarizes the processes.

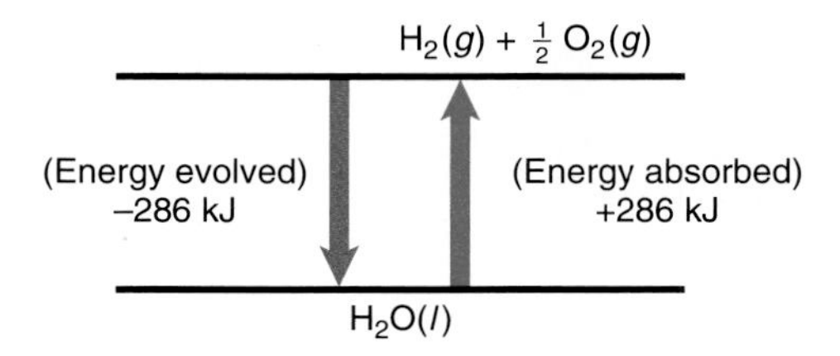

Figure 8.15

Energy differences in the hydrogen-oxygen-water system.

All that is needed to bring about this reaction is a source of 286 kJ. The most convenient method of decomposing water into hydrogen and oxygen is by **electrolysis,** the passage through water of a direct current of electricity of sufficient voltage to decompose it into H_2 and O_2 (Figure 8.16). When water is electrolyzed, the volume of hydrogen generated is twice that of oxygen. This suggests that a water molecule contains twice as many hydrogen atoms as oxygen atoms, testimony to the formula H_2O.

$$286 \text{ kJ} + H_2O(l) \longrightarrow H_2(g) + \tfrac{1}{2}\, O_2(g) \qquad [8.22]$$

Of course, the question remains: How will the electricity for large-scale electrolysis be generated? Most electricity in the U.S. is produced by burning fossil fuels in conventional power plants. If we only had to deal with the first law of thermodynamics, the best we possibly could do would be to burn an amount of fossil fuel equal in energy content to the hydrogen produced in electrolysis. But recall from Chapter 4 that we must also deal with the consequences of the second law of thermodynamics. Because of the inherent and inescapable inefficiency associated with transforming heat into work, the maximum possible efficiency of an electrical power plant is 63%. When we add the additional energy losses caused by friction, incomplete heat transfer, and transmission over power lines, it would require at least twice as much energy to produce the hydrogen than we could obtain from its combustion. This is comparable to buying eggs for 10 cents each and selling them for five; it's no way to do business. Furthermore, most methods of generating electricity have a variety of negative environmental effects. At one time it was thought that the "cheap" extra electricity from nuclear fission could be used to produce hydrogen to fuel the economy, but that energy utopia has hardly been realized. It should be apparent, therefore, that electricity generated from fossil fuels or nuclear fission does not offer a practical way to split water to produce hydrogen for use as a fuel.

A second possibility is to use heat energy to decompose water. One process used to produce commercial hydrogen does, in fact, use heat. You have already encountered it, in Chapter 4, in the discussion of substitutes for fossil fuels. Hot steam is passed over coke (essentially pure carbon) at 800°C.

$$131 \text{ kJ} + H_2O(g) + C(s) \longrightarrow H_2(g) + CO(g) \qquad [8.23]$$

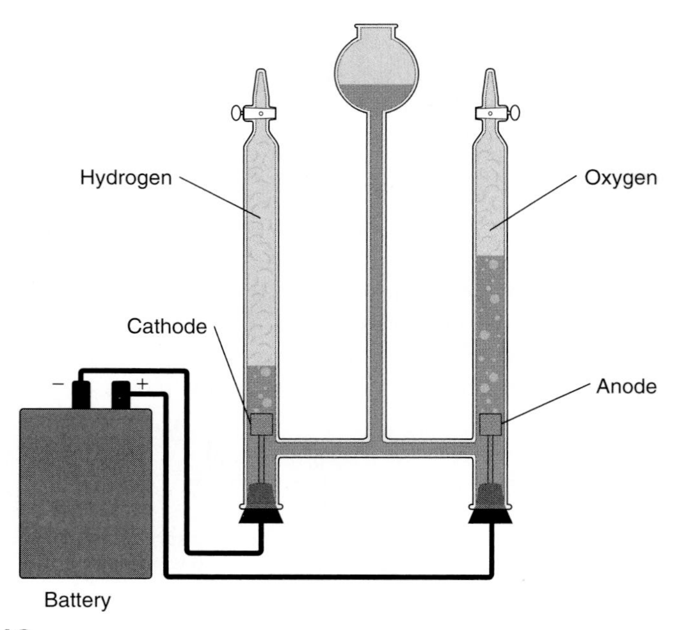

Figure 8.16

Electrolysis of water.

The mixture of hydrogen and carbon monoxide produced can be burned directly. It can serve as the starting material for the synthesis of hydrocarbon fuels and other important compounds, or the hydrogen can be separated from the mixture and used as needed. Reaction 8.23 is being studied in an effort to find catalysts, including biological catalysts, that will make it possible to carry out the reaction at lower temperatures.

Simply heating water to decompose it thermally into H_2 and O_2 is not commercially promising. To obtain reasonable yields of hydrogen and oxygen, temperatures of over 5000°C would be required. The attainment of such temperatures is not only extremely difficult, but it would also consume enormous amounts of energy—at least as much as would be released when the hydrogen was burned. Thus, we have again reached a point where we would be investing a great deal of time, effort, money, and energy to generate a quantity of hydrogen that would, at best, return only as much energy as we invested and, in practice, a good deal less.

Methane, CH_4, the major component of natural gas, is currently the chief source of hydrogen. The hydrogen is formed from the endothermic reaction of methane with steam.

$$165 \text{ kJ} + CH_4(g) + 2\ H_2O(g) \longrightarrow 4\ H_2(g) + CO_2(g) \qquad [8.24]$$

Researchers continue to find ways to increase the efficiency of this method of producing hydrogen.

8.19 Your Turn

Use the values in Table 4.1 for bond energies to check the energy required by the reaction in equation 8.23. Clearly show your reasoning.

8.5 The Hydrogen Economy

If and when we succeed in developing economically feasible methods for producing hydrogen cheaply and in large quantities, we will still face significant problems in storing and transporting it. Although H_2 has a high energy content per gram, it occupies a very large volume—about 12 liters (a bit over 12 quarts) per gram at normal atmospheric pressure and room temperature. If H_2 is to be stored and transported in its gaseous state, large, heavy-walled metal cylinders will be required, thus eliminating much of the advantage of the favorable energy/mass ratio. To save space, gases are typically converted to the liquid state under high pressure, as in the case of "bottled gas" or "liquid propane." But hydrogen must be cooled to −253°C before it liquefies. This means that keeping it in liquid form requires low temperatures and high costs.

A number of attempts have been made to circumvent these problems by storing H_2 in other forms. One of these, proposed recently, involves absorbing gaseous H_2 into a solid, such as activated carbon, which can hold a great deal of H_2 at only moderate pressures. The carbon can be heated as needed to release the H_2 for combustion. A different approach involves reacting the H_2 with certain metals to produce compounds called hydrides, which are relatively stable solids that have a reasonably high storage capacity. For example, when 10 liters (slightly over 10 quarts) of H_2 gas at 25°C and 1 atmosphere of pressure react with lithium metal, the resulting LiH occupies a mere 4.3 mL, or slightly less than a teaspoon.

$$Li(s) + \tfrac{1}{2}\ H_2(g) \longrightarrow LiH(s) \qquad [8.25]$$

fairly stable.

When such hydrides react with H_2O, they produce H_2, which can then be burned in the usual fashion.

$$LiH(s) + H_2O(l) \longrightarrow H_2(g) + LiOH(aq) \qquad [8.26]$$

Prototype vehicles that operate on this principle have been built and used in a number of locations. Clearly, such approaches would greatly improve the safety and convenience of handling H_2, and perhaps be the decisive factor in determining the extent of its

Figure 8.17

A battery-powered electric car being charged from photovoltaic cells at the Sacramento Municipal Utilities District. The bank of photovoltaic (solar) cells is in the panels mounted on the columns.

acceptance as a fuel. As we discussed earlier in this chapter, a hydrogen-fueled car would produce only water vapor and none of the carbon monoxide or nitrogen oxides emitted from using gasoline-fueled internal combustion engines.

Even if we manage to solve all of the production, storage, and transport problems just identified, we must consider how best to extract energy from our hydrogen. The most obvious way would be to burn it in power plants, vehicles, and homes. A stream of pure hydrogen burns smoothly, quietly, and safely in air, delivering 143 kJ per gram and forming only nonpolluting water as an end product. But when hydrogen is mixed directly with oxygen, a spark can be sufficient to produce a devastating explosion, limiting the usefulness of hydrogen as a direct fuel. However, the promise of using hydrogen in fuel cells is creating renewed interest in the hydrogen economy.

Meeting the energy needs of our modern world will not be possible with only one technology. In the long run, pollution-free transportation will benefit from a mix of improved fuel cells, new battery technologies, and changes in public attitudes towards utilization and conservation of energy. However, one more source must be considered, one that may well play an increasingly important role in our energy mix. Figure 8.17 presents an intriguing possibility. It is a photograph of a battery-powered car being charged outside the offices of the Sacramento (CA) Municipal Utilities District. If you look closely, you will note that the car is "plugged into" the Sun via a bank of photovoltaic cells. In the next section you will see this is yet another example of electron transfer.

8.20 Consider This: Iceland's Hydrogen Economy

The small country of Iceland is taking bold steps to be the first to cut its ties to fossil fuels. The strategy, announced in February 1999, is to demonstrate that it can produce, store, and distribute hydrogen as a means to power both public and private transportation.

a. Who are the partners in this joint venture?
b. What will be the first tangible outcome of this venture?
c. Do you think the lessons to be learned in Iceland will be relevant for the U.S.? Explain your reasoning.

8.6 Photovoltaics: Plugging in the Sun

The Earth receives more energy from the Sun each day than is consumed by our planet's nearly six billion inhabitants in 27 years. But, currently, less than 0.5% of the power generated in the United States comes directly from the Sun. Electricity supplies about 35% of U.S. energy needs, two-thirds of it used in residential and commercial buildings.

Therefore, it would be to our advantage if solar radiation could be directly converted into electricity, without the intermediary of hydrogen or some other fuel. This is the function of **photovoltaic cells,** also called *solar cells*. Such devices have already demonstrated their practical utility for both large- and small-scale electrical generation such as for satellites, highway signs, street lighting, navigational buoys, automobile recharging stations, and remote residences. They presently may represent the best hope for capturing and using solar energy.

- A photovoltaic cell converts radiant energy to electrical energy.

Electricity involves a stream of electrons, flowing from a region of higher electrical potential (voltage) to one of lower voltage. For a photovoltaic cell to generate electricity, light must induce such a flow in the cell. This flow depends on the interaction of matter and photons of radiant energy—a topic already treated in considerable detail in Chapters 2 and 3. In those chapters, we pointed out that the portion of the Sun's radiation reaching the Earth's surface is mainly in the visible and infrared regions of the spectrum, with a maximum intensity near a wavelength of 500 nm. Light of this wavelength is green and has an energy of about 4×10^{-19} J per photon, corresponding to 240 kJ per mole of photons. A photovoltaic cell must be made of atoms or molecules that will release electrons when struck by radiation of approximately this same wavelength.

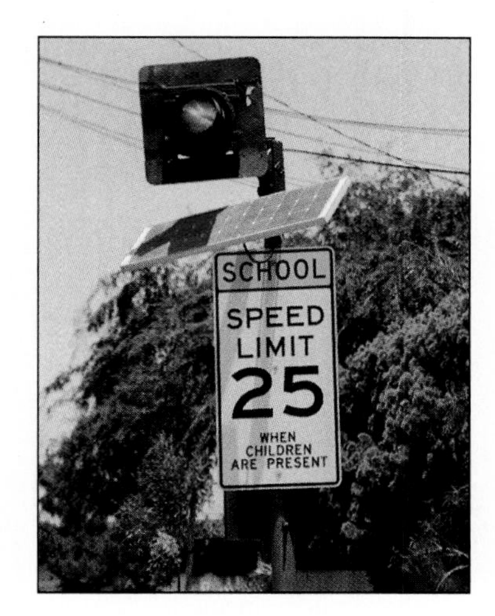

Photovoltaic (solar) cells are used to power this traffic signal.

Among substances that can be used for photovoltaic cells is a class of materials known as **semiconductors**—materials that do not normally conduct electricity well but do so under certain conditions. One of the first semiconductors identified was the element silicon, which you may recognize as being used in computers, radios, pocket calculators, and digital watches. Many of the businesses that developed the use of silicon were clustered in "Silicon Valley," which took its nickname from the element. A crystal of silicon consists of an array of silicon atoms, each bonded to four others by means of shared pairs of electrons, as represented in Figure 8.18a. These shared electrons are normally fixed in bonds and are unable to move about through the crystal. Consequently, silicon is not a very good electrical conductor under ordinary circumstances. However, if an electron absorbs sufficient energy, it can be excited and released from its bonding position, as indicated in Figure 8.18b. Once freed, the electron can move throughout the crystal lattice, thus making the silicon an electrical conductor. The energy required for silicon to release an electron from a bond is 1.8×10^{-19} J per photon, which is equivalent to radiation with a wavelength of 1100 nm. Visible light has a wavelength range of 400–700 nm. Recall that the shorter the wavelength of radiation, the greater the energy per photon. Therefore, photons of visible sunlight have more than enough energy to excite electrons in silicon semiconductors. Indeed, handheld calculators are now powered with solar cells, thus eliminating the added expense of buying batteries.

The fabrication of photovoltaic cells is not without some major problems. Although the starting material from which silicon is extracted (common sand—silicon dioxide) is

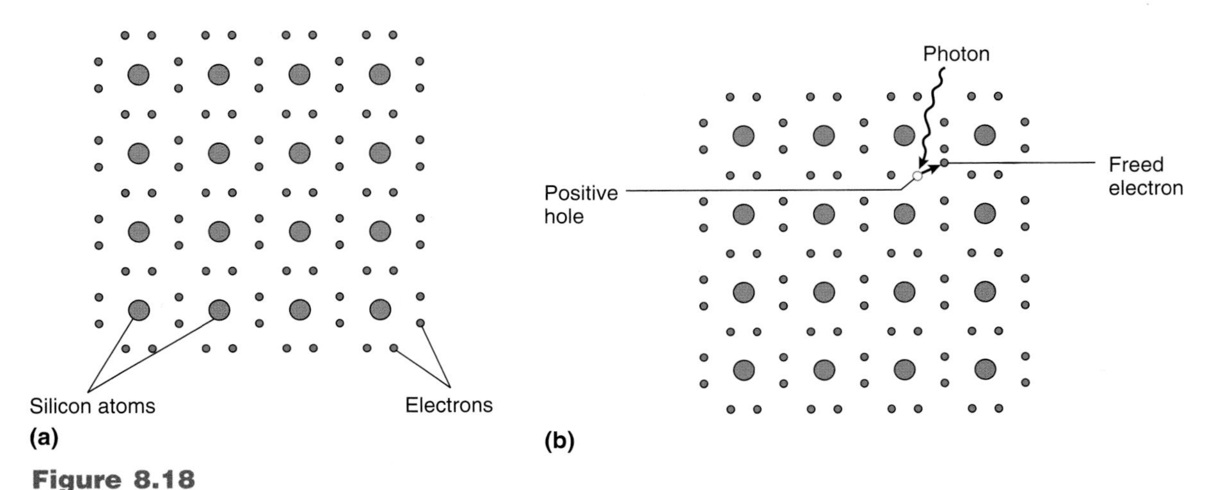

Figure 8.18

(a) Schematic of bonding in silicon.

(b) Photon-induced release of a bonding electron in a silicon semiconductor.

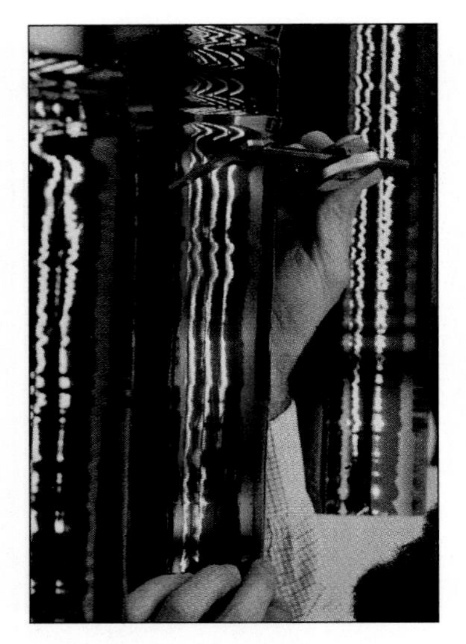

The individual is holding an ultrapure silicon ingot.

cheap and abundant, purifying silicon to the appropriate level (99.999% purity) is fairly expensive. A second complication is that the direct conversion of sunlight into electricity is not very efficient. A photovoltaic cell could, in principle, transform into electricity up to 44% of the radiant energy to which it is sensitive. But more than a third of this (16% of the total) is lost to internal cell processes. This leaves a theoretical efficiency limit of 28%. In practice, efficiencies between 10% and 17% have been achieved.

In Chapter 4, we lamented the 63% maximum efficiency of converting heat to work in a conventional power plant. It might seem that we should be even more distressed at the lower limits that can be achieved by photovoltaics. But remember that the Sun is an essentially unlimited energy source, and there is a good deal of empty space on the planet that is well suited for large arrays of photovoltaic cells and for little else. Moreover, the fact that solar energy is free of many of the environmental problems associated with burning fossil fuels or with nuclear fission adds impetus to research and development of solar cells.

The first use of solar cells was to provide electricity in NASA spacecraft, where cost seemed to not be a major concern and the intensity of radiation is so high that the low efficiency is not a serious limitation. But, because most commercial applications must be more cost conscious, a great deal of effort is being directed to increasing the efficiency of solar cells and lowering manufacturing costs. One promising innovation is replacing crystalline silicon with the non-crystalline form of the element. Photons are more efficiently absorbed by the less highly ordered Si atoms, a phenomenon that permits reducing the thickness of the silicon semiconductor to 1/60th of its former value. The cost of materials is thus significantly reduced. More common is the "doping" of silicon with other elements. This process consists of intentionally introducing about 1 ppm of elements such as gallium (Ga) or arsenic (As) into the silicon. These two elements and others from the same periodic families are used because their atoms differ from silicon by a single outer electron. Silicon has 4 electrons in its outer energy level, gallium has 3, and arsenic has 5. Thus, when an atom of As is introduced in place of Si in the silicon lattice, an extra electron is added. The replacement of an Si atom with a Ga atom means that the crystal is now one electron "short."

- Ga is in Group 3A; Si is in Group 4A; As is in Group 5A

The extra electrons in arsenic-doped silicon are not confined to bonds between atoms. Rather, electrons move easily through the lattice, increasing the electrical conductivity of the material over that of pure silicon. Silicon doped in this manner is called an **n-type semiconductor** because its electrical conductivity is due to negative carriers—electrons. On the other hand, for each silicon atom replaced with a gallium ion, an electronic vacancy or "hole" is introduced into what is normally a 2-electron bond. When an electron moves into this hole, a new hole appears where the mobile electron formerly was located. Therefore, holes and electrons move in opposite directions. Because holes can be regarded as positive carriers of electricity, gallium-doped silicon is called a **p-type semiconductor.** Figure 8.19 illustrates both n- and p-type semicon-

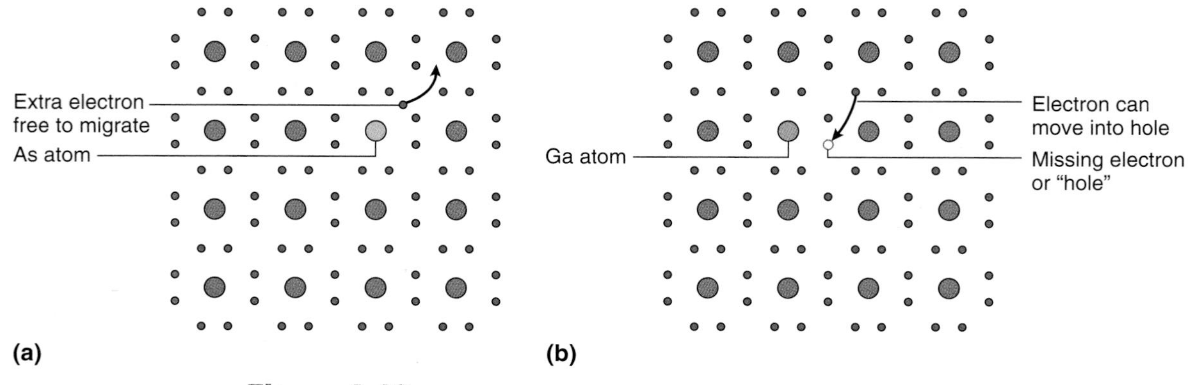

Figure 8.19

(a) An arsenic-doped n-type silicon semiconductor. (b) A gallium-doped p-type silicon semiconductor.

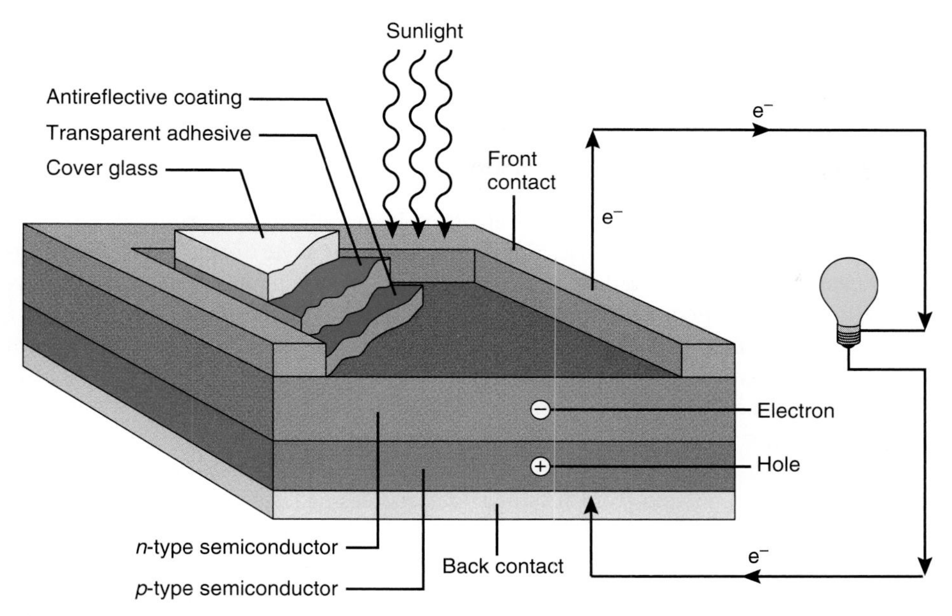

Figure 8.20
Schematic diagram of a photovoltaic (solar) cell.

ductors. Both types of doping increase the conductivity of the silicon because less energy is needed to get extra electrons or holes moving. This means that photons of low energy (light of longer wavelength) can induce electron release and transport in doped crystals.

"Sandwiches" of n- and p-type semiconductors are used in transistors and many of the other miniaturized electronic devices that have revolutionized communication and computing. Similar structures are central to the direct conversion of sunlight to electricity. A photovoltaic cell typically includes sheets of n- and p-type silicon in close contact (Figure 8.20). The n-type semiconductor is rich in electrons and the p-type is rich in positive holes. When they are placed in contact, there is a tendency for the electrons to diffuse from the n-region into the p-region and for the positive holes to move from the p-region to the n-region. This generates a voltage or potential difference at the junction between the semiconductors. The voltage difference accelerates the electrons released when sunlight strikes the doped silicon. If the two layers are connected by a wire or another type of conductor, electrons flow through the external circuit from the n-semiconductor, where their concentration is higher, to the p-semiconductor, where it is lower. The result is a direct current of electricity that can be intercepted to do essentially all the things electricity does. As long as the cell is exposed to light, the current will continue to flow, powered only by radiation.

In addition to making an effort to improve the performance of silicon semiconductors by doping, scientists have been searching for other substances that exhibit the same or better semiconductor properties. Among promising substitutes are germanium, an element found in the same family of the periodic table as silicon, and compounds in which the elements have the same number of outer electrons as Sn or Ge. Included in this latter list are gallium arsenide (GaAs), indium arsenide (InAs), cadmium selenide (CdSe), and cadmium telluride (CdTe). Some of these new semiconductors have enhanced the efficiency of radiation-to-electricity conversion and made possible photovoltaic cells that are responsive to certain regions of the spectrum. Siemens Solar Industries recently developed a copper indium selenide semiconducting thin film that has significant advantages over amorphous silicon and cadmium telluride.

8.21 Your Turn

Using the periodic table as a guide, show why the electron arrangement and bonding in GaAs and CdSe would be very similar to that in pure Ge.

Some researchers have taken a different approach to improving semiconductors. A power company and the commercial arm of the University of New South Wales, Australia, formed a research consortium called Pacific Solar. Its main goal has been to develop a multilayer solar cell into a viable commercial product. By creating up to 10 very thin layers of alternating p-type and n-type doped silicon, each electron has only a short distance to travel to reach the next p-n junction. This lowers the internal resistance within the cell and raises the efficiency of the cells to at least 15% at an increasingly competitive price. Figure 8.21 gives a sense of just how thin these layers actually are. The project has been able to greatly reduce processing costs because smaller quantities of low-grade silicon are required and the process has become highly automated.

As a result of these and other advances in photovoltaic technology, the cost of producing electricity in this manner has dropped dramatically from $3 per kilowatt-hour in 1974 to 28 cents per kilowatt-hour in 1998. Although electricity produced by this technology is still not competitive with that produced from fossil fuels, the long-range prospects for photovoltaic solar energy are encouraging. Its cost is decreasing while the cost of electricity generated from fossil fuels is increasing. But the increasing cost of clean fuel, the limited supply, and the added expense of pollution controls is driving the cost higher. Given this situation and the expected improvements in the performance and decreases in the cost of solar cells, electricity from photovoltaic systems could become competitive with that from fossil fuels by early in the 21st century. Photovoltaic power has grown at an average rate of 16% per year since 1990 (Figure 8.22). "World solar markets are growing at 10 times the rate of the oil industry, which has expanded only 1.4% per year since 1990," wrote Christopher Flavin and Molly O'Meara of the Worldwatch Institute. Most major energy companies, such as Amoco and British Petroleum have invested heavily in the solar business, now valued at about $1 billion a year. Even so, solar power still represents only about 1% of global power supplies.

- Using solar cells rather than batteries in navigational buoys saves the U.S. Coast Guard an estimated $6 million annually through reduced maintenance and repair.

The largest solar installation in the United States, located in Carrisa Plains, CA, was built by ARCO Solar, Inc. and Pacific Gas and Electric Company. It generates about seven megawatts at peak power. Although this generating capacity is small compared to that of fossil fuel, nuclear, and hydroelectric plants, much larger solar installations are expected in the future. A 200-megawatt plant, which could provide household power for 300,000 people, could be erected on a square mile of land at relatively modest capital expense and minimal maintenance. At currently attainable levels of operating efficiency, all the electricity needs of the United States could be supplied by a photovoltaic generating station covering an area of 85 miles by 85 miles, roughly the area of New Jersey.

Photovoltaic technology is already in use in a number of countries. From small, remote villages in developing countries to upscale suburbs in Japan and the United States, 500,000 homeowners worldwide use solar cells to generate their own electricity. Aided by generous tax credits, over 23,000 homes in Japan have rooftop solar units, installed to overcome the high cost of electricity. In sunny Sacramento, CA, the Municipal Utility District has constructed an array of 1600 photovoltaic cells that produces two megawatts (2×10^6 watts) of electricity, sufficient to serve 660 homes (Figure 8.23). An additional 420 homes have rooftop photovoltaic systems. The homeowners sell back excess elec-

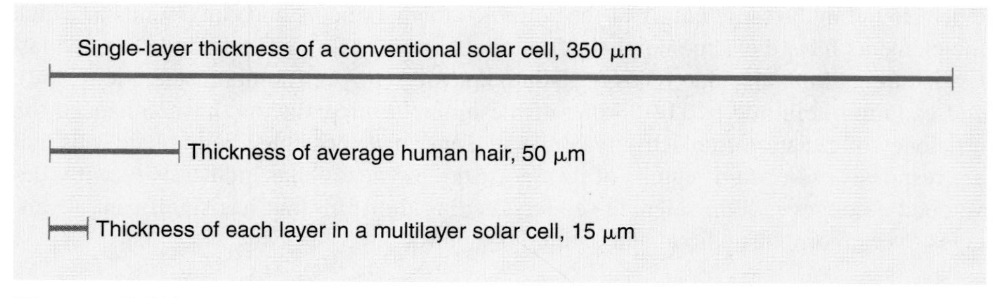

Figure 8.21

Relative thicknesses of layers in single and multilayer solar cells.
Note: $1\ \mu m = 1 \times 10^{-6}$ m.

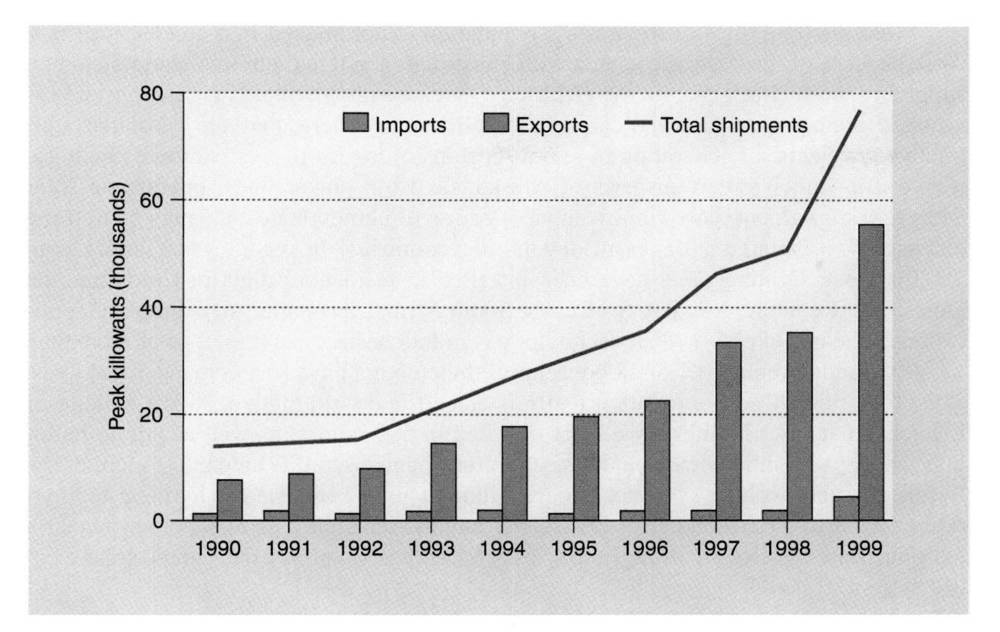

Figure 8.22

The growth of world photovoltaics. Import and export shipments of photovoltaic cells, 1990–1999.

(Source: www.eia.doe.gov/fuelrenewable.html.)

tricity to the utility company. In a touch of irony, the photovoltaic site is adjacent to a now-closed nuclear power plant. The Sacramento Municipal Utility District voted to close its nuclear reactors in favor of using photovoltaics and other "cleaner" energy technologies.

The European Union and the United States are partners in the Million Roofs program, an initiative to install a million rooftop photovoltaic systems by 2010 (Figure 8.24). The U.S. program proposes a 15% tax credit and funds to subsidize partnerships among builders, utilities, and local governments. Already supported by private companies as well as local and state governments, it remains to be seen whether the program will receive adequate funds from Congress to achieve the goal. A study in western Germany pointed out that half the electricity needs of that region could be met using rooftop solar cells.

Figure 8.24

Rooftop photovoltaic cells as part of the Million Roofs program.

Figure 8.23

This array of 1600 photovoltaic cells in Sacramento, CA produces 2 megawatts (2 million watts) of electricity.

Figure 8.25
Solar (photovoltaic) cells bring electricity to remote areas.

More that one-third of the Earth's population is not hooked into an electrical network because of the costs associated with constructing and maintaining equipment, and supplying the fuel to generate the electricity. Because photovoltaic installations are essentially maintenance free and can be used almost anywhere, they are particularly attractive for electrical generation in remote regions of the Earth. For example, the highway traffic lights in certain parts of Alaska, far from power lines, operate on solar energy. A similar, but more significant application of photovoltaic cells may be to bring electricity to isolated villages in developing countries. In recent years, more than 200,000 solar lighting units have been installed in residential units in Colombia, the Dominican Republic, Mexico, Sri Lanka, South Africa, and India. An exemplary application of the use of photovoltaic technology is in Indonesia, an archipelago of more than 13,000 islands. About 70% of all households there do not have access to electrical lines. Therefore, installing photovoltaic cells is a legitimate alternative. In the village of Leback, solar electric units have been installed in 500 homes, as well as public buildings, shops, 11 public television units, streetlights, and a satellite antenna (Figure 8.25). Before the photovoltaic systems, Leback villagers used kerosene for lighting and batteries for radios. Kerosene costs \$6–12 per month, depending on availability. Under a loan-purchase agreement, villagers pay only \$4.25 per month for their home solar electric system.

- Energy costs are high in remote areas, such as in Alaska, or in developing nations. Thus, electricity from photovoltaics is economically competitive in these areas.

8.22 Consider This: A "Million Roofs" Perspective

Many success stories are associated with the Million Roofs initiative. Use the resources of the Web to investigate a project other than those discussed in the text.

a. What is the name of the project you picked and where is it located?
b. Who are the partners in this project?
c. What is the major purpose for which the energy provided by photovoltaics will be used?
d. What potential sociological problems could be created by bringing electricity to the people in the project you studied? Explain.

The transfer of electrons taking place in solar cells can even power cars. Sunrayce USA has developed into an 11-day, 2,300-mile race sponsored by the U.S. Department of Energy (DOE), its National Renewable Energy Laboratory (NREL), Terion Communications, and several other private corporations. It has become a popular activity for engineering students. Student teams design, build, test, and drive cars powered by photovoltaic cells and battery packs (Figure 8.26). In spite of their vanguard designs, the cars are not ready to be put on the roads for everyday use. Australia's SunRace 21.C is an equally demanding 10-day, 2300 km race for solar and electric vehicles. The 2001 race went from Adelaide through Melbourne and Canberra, finishing in front of the famous Opera House in Sydney. The Ultimate Solar Car Challenge in 2002 was a 4000-km transcontinental race from Perth on the west coast to Sydney on the east coast. This race celebrated the 20th anniversary celebration of the world's first solar car journey. Both countries regularly hold solar bike racing events and the U.S. has an intercollegiate solar boat competition called Solar Splash. In addition, an airplane powered only by battery-charging amorphous (non-crystalline) silicon solar cells has flown more than 2400 miles in less than 120 hours.

8.23 Your Turn

a. Which solar vehicle race is longer—the U.S. Sunrayce 2001 (2300 miles) or the Australian SunRace 21.C (2300 km)? Explain.
b. Which solar vehicle race is longer—the U.S. Sunrayce 2001 (2300 miles) or the Australian Ultimate Solar Car Challenge (4000 km)? Again, explain.

Earlier in the chapter, we explored the potential importance of hydrogen as a fuel for the generation of electricity. Recently, researchers at the National Renewable Energy Laboratory in Golden, Colorado, developed a photovoltaic cell that in one step gener-

Figure 8.26

The Solar Phantom V, built by students at Rose-Hulman Institute of Technology, Terre Haute (Indiana), is shown driving along Route 66 from Albuquerque to Gallup, NM, in preparation for the 2001 Sunrayce. The 2001 race for solar-powered vehicles ran from Chicago to Claremont, California.

ates sufficient electricity to extract hydrogen by decomposing water (Figure 8.27). The cell uses paired gallium-based semiconductors to create the electricity to decompose the water. The efficiency of the process, 12.5%, is nearly double that of the previous record. In that case, the process took two steps in separate devices. The first step generated the electricity; the second step used the electricity to split water into hydrogen and oxygen.

Although prodigious amounts of sunshine hit the Earth daily, the rays do not strike any specific spot on our planet for 24 hours a day, 365 days a year. This means that the electricity generated by photovoltaic cells during the day must be stored for use at night. Given current technology, electrical storage requires batteries, electrochemical devices afflicted with the disadvantages discussed in Section 8.3. Nevertheless, the direct conversion of sunlight to electricity has many advantages. In addition to freeing us from our dependence on fossil fuels, an economy based on solar electricity would decrease the environmental damage that frequently occurs when these fuels are extracted or transported. Further, it would help to lower the levels of atmospheric pollutants such as sulfur oxides and nitrogen oxides, and it would also help avert the dangers of global warming by decreasing the amount of carbon dioxide released into the atmosphere. It is likely that fossil fuels will continue to be the preferred form of energy for certain applications. On balance, however, the future looks sunny for solar-based energy. Indeed, the only better source of energy might be if some modern Prometheus could steal a part of the Sun and bring it down to Earth.

- Prometheus was a Greek mythic figure who stole fire from the gods.

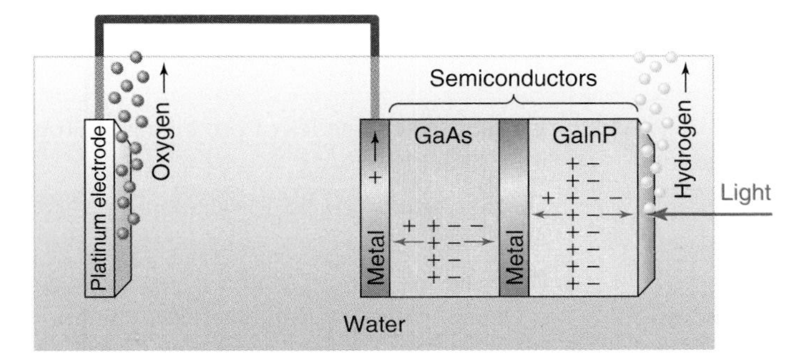

Figure 8.27

A solar semiconductor device that splits water into hydrogen and oxygen. GaInP is made of the elements gallium, indium, and phosphorus.

CONCLUSION

Fossil fuels, the Sun's ancient investments on Earth, are fast disappearing and we must seek alternatives for tomorrow. For that future, we are looking to use many different forms of electron transfer. Cells and batteries provide the means for transferring electrons, and these electrons can be captured for useful energy. Fuel cells are one of the most successful new strategies, and may become a major energy source for future transportation needs. With the flexibility in fuel they can use, fuel cells are developing very rapidly in the near term. Of course, for some purposes, conventional fuels are more convenient than electricity. Perhaps advances in research and changes in global economies will make it fiscally and energetically feasible to use solar radiation to extract hydrogen from water or some other hydrogen source. The hydrogen can either be burned directly as a clean fuel or combined with oxygen in a fuel cell that generates electricity rather than heat. Automobile engineers and designers are actively creating prototypes that use fuel cells, batteries, and hydrogen as energy sources for motor vehicles. Many of these energy systems, once thought to be only experimental, have moved beyond the prototype stage to being mass produced. It is fortunate that what seems to be one of the most promising methods for capturing and transforming solar energy is also the most direct. Photovoltaic cells convert sunlight directly into electricity. There is no need for inefficient intermediate steps in which fuels are synthesized and burned, and the resulting heat energy is transformed first into mechanical and then into electrical energy.

But the laws of thermodynamics and human nature are such that these transformations will not occur spontaneously. Energy alternatives cannot be developed without hard work and the investment of intellect, time, and money. Yet, in the United States, the amount of effort and money devoted to research on new energy sources appears to be directly proportional to the cost of oil. When international crises drive up the price of petroleum or when regional energy shortages occur, there is a sudden flurry of official interest in energy conservation and the development of alternate technologies. When oil supplies are plentiful and prices are low at the gasoline pump, few seem to care about preparing for the time when fossil fuels will be depleted or they become much too polluting or too expensive to burn. Someday, maybe teams of chemists, physicists, and engineers will even succeed in simulating the Sun by controlling the Sun's nuclear reactions here on Earth as an affordable energy source, not a mere laboratory curiosity. But, until then, we need to establish national and personal priorities, and act on them. We have been the beneficiaries of a bountiful nature, but in turn, we have an obligation to assure energy sources for unborn generations.

Chapter Summary

Having studied this chapter, you should be able to:

- Discuss the principles governing the transfer of electrons in galvanic cells, including the processes of oxidation and reduction (8.1)
- Describe the design, operation, applications, and advantages of several different types of batteries (8.1)
- Describe the design, operation, applications, and advantages of fuel cells (8.2)
- Describe the advantages and disadvantages of hydrogen as a fuel (8.2–8.5)
- Compare and contrast the principles, advantages, and challenges of producing and using fuel-cell powered, battery-powered, and hybrid vehicles (8.3)
- Explain the energetics of producing hydrogen and using it as a fuel (8.4)
- Discuss issues related to developing a hydrogen economy (8.6)
- Describe the principles governing the operation of photovoltaic (solar) cells and their current and potential uses (8.6)
- Express informed opinions about the future development of all types of electron transfer technology for producing electrical energy on personal, regional, national, and global scales (8.1–8.5)

Questions

Emphasizing Essentials

1. a. What is the definition of oxidation?

   b. What is the definition of reduction?

   c. Why is it necessary for these processes to take place together?

2. Which half-reactions represent oxidation and which reduction? Explain your reasoning.

   a. $Fe \rightarrow Fe^{2+} + 2\ e^-$

   b. $Ni^{4+} + 2\ e^- \rightarrow Ni^{2+}$

   c. $2\ H_2O + 2\ e^- \rightarrow H_2 + 2\ OH^-$

3. Consider this galvanic cell. A coating of impure silver metal begins to appear on the surface of the silver electrode as the cell discharges.

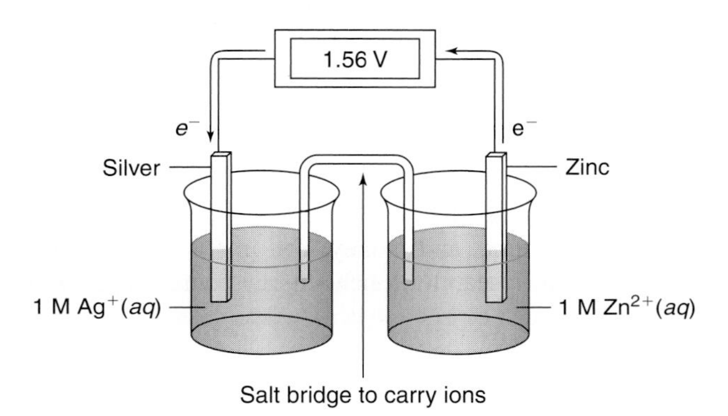

   a. Identify the anode in this cell.

   b. Write the oxidation half-reaction.

   c. Identify the cathode.

   d. Write the reduction half-reaction.

4. Explain the difference between a rechargeable battery and one that must be discarded. Contrast a NiCad battery with an alkaline battery in your explanation.

5. Is there a difference between a galvanic cell and an electrochemical battery? Explain, giving examples to support your answer.

6. In the lithium-iodine cell, Li is oxidized to Li^+; I_2 is reduced to $2\ I^-$.

   a. Write equations for the two half-reactions that take place in this cell, labeling the oxidation half-reaction and the reduction half-reaction.

   b. Write an equation for the overall reaction in this cell.

   c. Identify the half-reaction that occurs at the anode and the half-reaction that occurs at the cathode.

7. Two common units associated with electricity are volts and amps. What does each unit measure?

8. Explain the significance of the title of this chapter, "Energy from Electron Transfer."

9. Is the voltage from a tiny AAA-size alkaline cell the same as that from a large D alkaline cell? Why or why not?

10. The mercury battery is used extensively in medicine and the electronics industries. Its overall reaction can be represented by this equation.

$$HgO(s) + Zn(s) \rightarrow ZnO(s) + Hg(l)$$

   a. Write the oxidation half-reaction.

   b. Write the reduction half-reaction.

11. Figure 8.5 shows how the use of mercury has changed since 1970. What was the most common use in 1970? 1980? 1990? What are some reasons for the observed changes in mercury use?

12. These are the *unbalanced* equations for the half-reactions in a lead storage battery. These half-reactions do not show the electrons either lost or gained.

$$Pb(s) + SO_4^{2-}(aq) \rightarrow PbSO_4(s)$$

$$PbO_2(s) + 4\ H^+(aq) + SO_4^{2-}(aq) \rightarrow PbSO_4(s) + 2\ H_2O(l)$$

   a. Balance both equations with respect to charge by adding electrons on either side of the equations, as needed.

   b. Which half-reaction represents oxidation and which represents reduction?

   c. One of the electrodes is made of lead, the other of lead dioxide. Which is the anode and which is the cathode?

13. a. What is the function of the electrolyte in a galvanic cell?

   b. What is the electrolyte in an alkaline cell?

   c. What is the electrolyte in a lead-acid storage battery?

14. What is the difference between a storage battery and a fuel cell?

15. Is the conversion of oxygen gas, $O_2(g)$, to $H_2O(g)$ in a fuel cell an example of oxidation or reduction? Use electron loss or gain to support your answer.

16. Consider this diagram of a hydrogen-oxygen fuel cell used in earlier outer space missions.

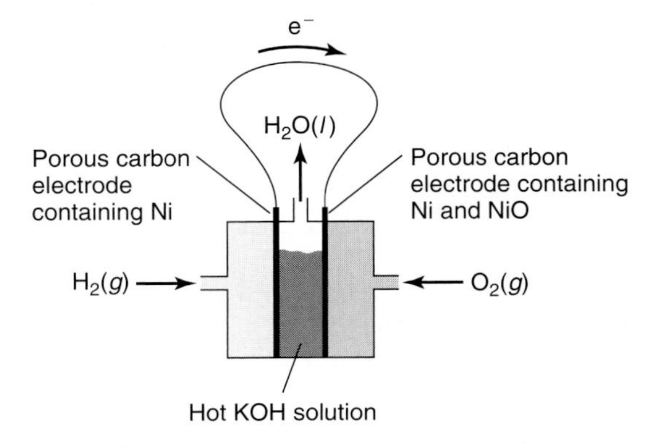

   a. How does the reaction between hydrogen and oxygen in a fuel cell differ from the combustion of hydrogen with oxygen?

b. Write the half-reaction that takes place at the anode in this fuel cell.

c. Write the half-reaction that takes place at the cathode in this fuel cell.

17. What is a PEM fuel cell and how does it differ from the fuel cell represented in question 16?

18. In addition to the studies on hydrogen fuel cells, experiments are being done on fuel cells using methane as a fuel. Balance the given oxidation and reduction half-reactions and write the overall equation for a methane-based fuel cell.

Oxidation: ___ CH_4 + ___ OH^- → ___ CO_2 + ___ H_2O + ___ e^-

Reduction: ___ O_2 + ___ H_2O + ___ e^- → ___ OH^-

19. This *unbalanced* equation represents the last step in the production of pure silicon for use in solar cells.

___ $Mg(s)$ + ___ $SiCl_4(l)$ → ___ $MgCl_2(l)$ + ___ $Si(s)$

a. How many electrons are transferred per atom of pure silicon formed?

b. Is the production of pure silicon an oxidation or a reduction reaction? Why do you think so?

20. What is meant by the term "hybrid car"?

21. a. How are equations 8.19 and 8.20 the same and how are they different?

b. How will the energy released in the reaction shown in equation 8.19 compare to the energy released in the reaction represented by equation 8.20? Explain your reasoning.

22. Given that 286 kJ of energy are released per mole of H_2 burned, how much energy will be released when 375 kg of H_2 are used?

23. a. Use bond energies from Table 4.1 to calculate the energy released when one mole of hydrogen burns.

b. Compare your result from part a with the stated value of 286 kJ. Account for any difference.

24. The symbol • represents an electron and the symbol ○ represents a silicon atom. Does this diagram represent a gallium-doped p-type silicon semiconductor, or does it represent an arsenic-doped n-type silicon semiconductor? Explain your answer.

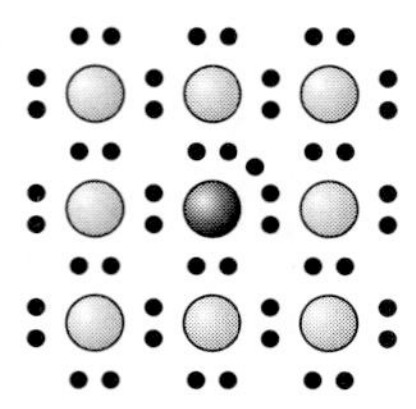

25. a. What is the Million Roofs program?

b. Who are the partners in this program?

c. What does this program hope to achieve?

Concentrating on Concepts

26. Assuming that electric cars are available in your area, what questions would you ask the car dealer before deciding to get this type of car? Which questions do you consider most important? Offer reasons for your choices.

27. a. What is meant by the phrase "the hydrogen economy"?

b. Even if methods for producing hydrogen cheaply and in large quantities become available, what problems still remain for the hydrogen economy?

28. Although hydrogen gas can be produced by the electrolysis of water, this reaction is usually not carried out on a large scale. Suggest a reason for this fact.

29. Consider this diagram of two water molecules in the liquid state.

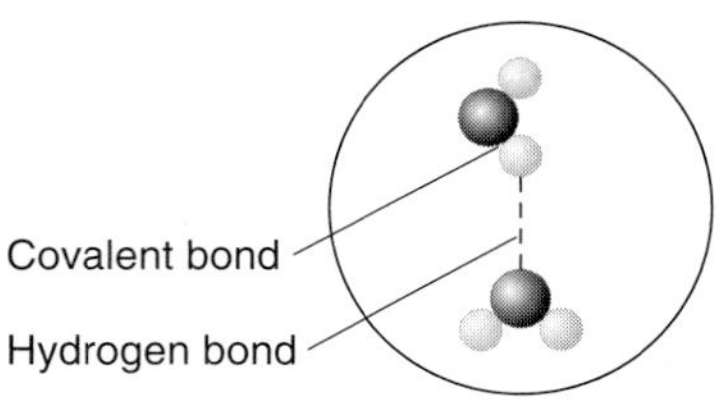

a. What bonds are broken when water boils? Are these intermolecular or intramolecular bonds? (See Chapter 5.)

b. What bonds are broken when water is electrolyzed? Are these intermolecular or intramolecular bonds?

30. You have seen several examples of redox reactions in this chapter and identified the exchange of electrons taking place. Now examine these equations and decide which are redox reactions and which are not. Give the reasons for your decisions.

Equation 1 $Zn(s) + 2\ MnO_2(s) + H_2O(l) \rightarrow Zn(OH)_2(s) + Mn_2O_3(s)$

Equation 2 $HCl(aq) + NaOH(aq) \rightarrow NaCl(aq) + H_2O(l)$

Equation 3 $CH_4(g) + 2\ O_2(g) \rightarrow CO_2(g) + 2\ H_2O(g)$

31. The text describes a prototype fuel cell that converts gasoline to hydrogen and carbon monoxide. The carbon monoxide, in contact with a special catalyst, then reacts with steam to produce carbon dioxide and additional hydrogen.

a. Write a set of reactions that describe this prototype fuel cell, using octane (C_8H_{18}) to represent the hydrocarbons found in gasoline.

b. When is this fuel cell expected to be commercially available?

c. Speculate as to the future economic success of this prototype fuel cell.

32. Fuel cells were invented in 1839, but never developed into practical devices for producing electrical energy until the advent of the U.S. space program in the 1960s.

What advantages did fuel cells have over other power sources that led to widespread utilization of fuel cells in the space program?

33. Hydrogen, H_2, and methane, CH_4, can each be used with oxygen in a fuel cell. Hydrogen and methane also can be burned directly. Which has a higher heat content when burned, 1.00 g of H_2 or 1.00 g of CH_4? *Hint:* Write the balanced chemical equation for each reaction and use the bond energies in Table 4.1 to help answer this question.

34. The battery of a cell phone discharges when the phone is in use. A manufacturer, while testing a new "power boost" system, reported these data.

Time, min	Voltage, V
0.00	6.56
1.00	6.31
2.00	6.24
3.00	6.18
4.00	6.12
5.00	6.07
6.35	6.03
8.35	6.00
11.05	5.90
13.50	5.80
16.00	5.70
16.50	5.60

a. Prepare a graph of these data.

b. The manufacturer's goal was to retain 90% of its initial voltage after 15 minutes of continuous use. Has that goal been achieved? Justify your answer using your graph.

35. Why are electric cars powered by lead-acid storage batteries alone considered only a short-term solution to the problem of air pollution emissions from automobiles? Outline your reasons.

36. 5.6×10^{21} kJ of energy come to Earth from the Sun every year. Why can't this energy be used to meet all of our energy needs?

37. The cost of electricity generated by solar thermal power plants currently is greater than that of electricity produced by burning fossil fuels. Given this economic fact, suggest some strategies that might be used to promote the use of environmentally cleaner electricity.

38. Prepare a list of the environmental costs and benefits associated with hybrid vehicles. Compare that list with the environmental costs and benefits of vehicles powered by gasoline. On balance, which energy source do you favor, and why?

39. William C. Ford, Jr., the chief executive officer of Ford Motor Company, is quoted as saying that going "totally green" with zero-emission vehicles will be a real challenge. Regular drivers won't buy high-tech clean cars, Ford admits, until the industry has a "no-trade-off" vehicle widely available. What do you think he means by a "no-trade-off" vehicle? Do you think he is justified in this opinion?

40. What are some of the current applications of photovoltaic cells *other* than the production of electricity in remote areas?

Exploring Extensions

41. A laboratory method to prepare hydrogen gas is by reacting metallic sodium with water, as shown in this equation.

$$2\ Na(s) + 2\ H_2O(l) \rightarrow H_2(g) + 2\ NaOH(aq)$$

a. Calculate the grams of sodium needed to produce 1.0 mole of hydrogen gas.

b. Calculate the grams of sodium needed to produce sufficient hydrogen to meet an American's daily energy requirement of 1.1×10^6 kJ.

c. If the price of sodium is $94 per kilogram, what is the cost of producing 1.0 mole of hydrogen? Assume the cost of water is negligible.

42. a. There are several advantages and disadvantages of using hydrogen as a fuel. Set up parallel lists that give the advantages and the disadvantages of using hydrogen as the fuel for transportation and for producing electricity.

b. Do you advocate the use of hydrogen as a fuel for transportation or for the production of electricity? Explain your position by writing a short article to your student newspaper.

43. The aluminum-air battery is being explored for potential use in automobiles. In this battery, aluminum metal undergoes oxidation to Al^{3+} and forms $Al(OH)_3$. Oxygen, O_2, from the air undergoes reduction to OH^- ions.

a. Write equations for the oxidation and reduction half-reactions. Use H_2O as needed to balance the number of hydrogen atoms present, and add electrons as needed to balance the charge.

b. Add the half-reactions to obtain the equation for the overall reaction in this cell.

c. Specify which half-reaction occurs at the anode and which occurs at the cathode in the battery.

d. What are the potential benefits of the widespread use of the aluminum-air battery? What are some of the limitations? Write a brief summary of your findings.

e. What is the current state of development of this battery? Is it in use in any vehicles at the present time? What is its projected future use?

44. An iron-based "super battery" is a promising alternative for delivering more power with fewer potential environmental effects than alkaline batteries. Find out how the super-iron battery is designed and its state of commercial acceptance.

45. The text discusses projections made by Ballard Power Systems, in worldwide alliances with Daimler-Chrysler and Ford Motor Company, to become the largest commercial producer of fuel cell-powered electric drive trains and components for cars, trucks, and buses. The alliance plans to produce 100,000 fuel cell-powered electric vehicles by 2004–2005. What is the progress toward that goal? See if you can find whether there have been any revisions to their predictions. Write a brief report explaining their progress, gearing your report towards an investor who wants to know the potential future growth of the company.

46. Figure 8.23 shows an array of photovoltaic cells installed by the Municipal Utility District in Sacramento, CA. Where else in the United States or in the world is there a comparable array? Use the Web to learn of other large-scale photovoltaic cell installations. What factors help to influence this approach, one that uses a centralized array rather than using individual rooftop solar units?

47. There has been little publicity about the Million Roofs program. Design a poster to explain to the general public what this program is all about and enlist its support.

48. Consider three artificial sources of light—a candle, a battery-powered flashlight, and an electric light bulb. Provide this information for each source.

a. The origin of the light.

b. The immediate source of the energy that appears as light.

c. The original source of the energy that appears as light. (*Hint:* Trace this back stepwise as far as possible.)

d. The end-products and by-products of using each of the light sources.

e. The environmental costs associated with each light source.

f. The advantages and disadvantages of each light source.

49. The text refers to fossil fuels as the ". . . Sun's ancient investment on Earth." How would you interpret this statement to a friend who is not enrolled in this course?

50. If electric cars become the choice of consumers in the near future, the popular saying "fill 'er up" will have to be replaced by "plug 'er in." Design a one-minute TV public service announcement that conveys what the new phrase means and why the former phrase no longer applies. Present this announcement to your class either live or on film and gather feedback on its effectiveness.

9

The World of Plastics and Polymers

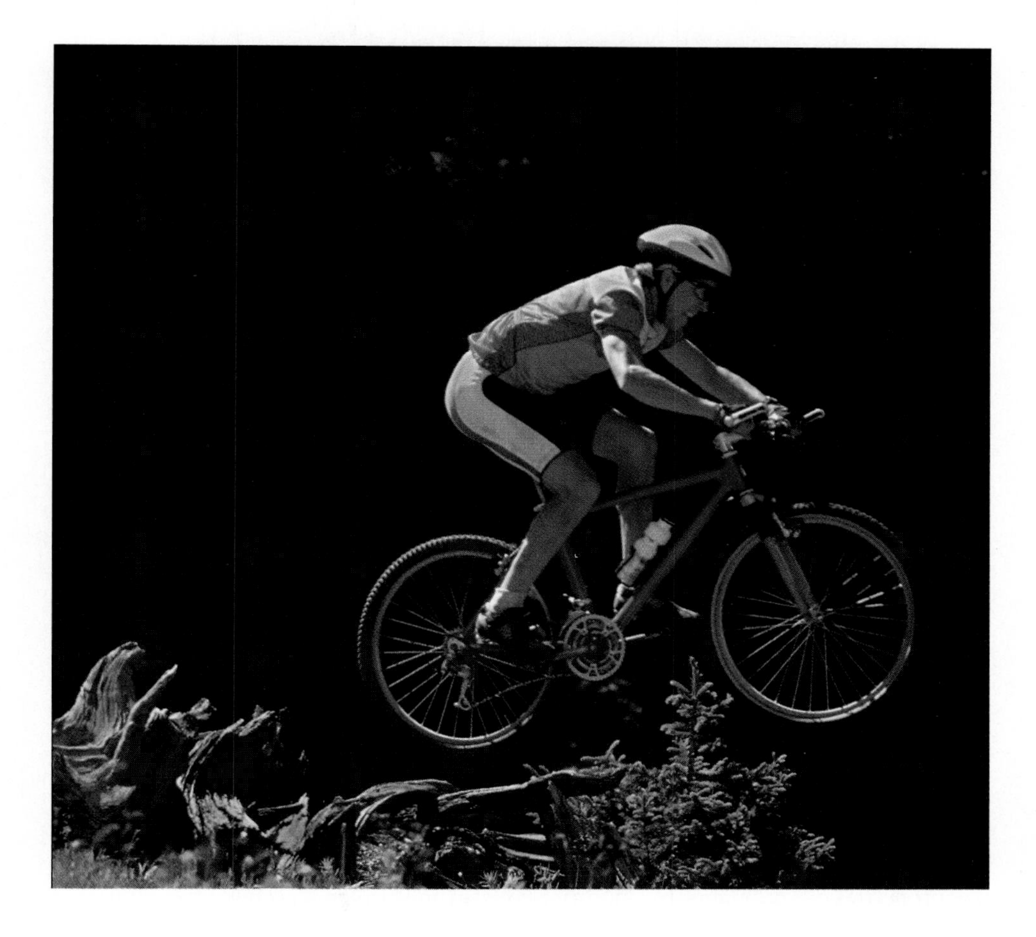

Mountain biking makes extensive use of polymers, both synthetic and natural ones.

"Hey, I'm out of here this weekend. I sure need a break from studying and exams. I've got a date with this mountain bike before the cold weather really sets in. But I don't even care if it's cold or rainy. Check out the new biking gear I just bought. I've got Spandura tights that are 18 times tougher than my older nylon/Lycra ones. Do you like my C-Tech™ polyester microfiber jersey and these new Porelle Drys socks that will keep me warm? In case I need it, I have my Thinsulate-lined Gore-Tex jacket that will keep me dry. And I can stash the rest of my gear in my lightweight nylon backpack with polyurethane padded straps for comfort. Like I said, I'm outta here. It will feel great to get away from all that synthetic, artificial stuff and commune with Ma Nature for a change. Catch ya later."

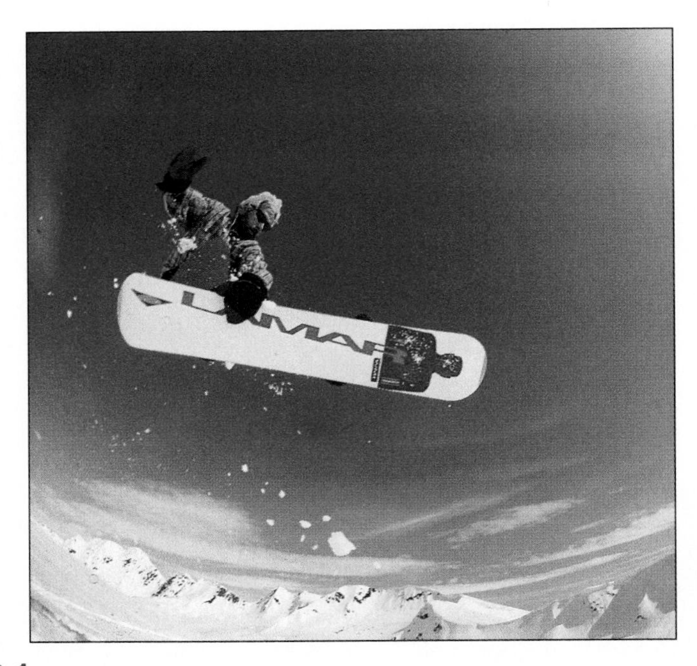

Figure 9.1
An example of using polymers to play.

9.1 Consider This: Skiing Gear Analysis

Plastics, polymers, and other synthetic materials have revolutionized sports equipment in recent years. Such technology has created lighter, stronger, and more responsive materials to match recreational needs. This is especially true with biking, skiing, hiking, and camping. Look a little more closely at the snowboarder shown in Figure 9.1. Identify the parts of his equipment and outer clothing that are created by chemists, and describe the properties of these materials that make them well-suited for their intended uses.

CHAPTER OVERVIEW

We begin the chapter by stating the obvious: Plastics are all around us. After considering a few examples, we relate plastics with polymers—materials whose molecules consist of long chains of atoms. Polymers can be natural or synthetic. Here we are more concerned with the latter, particularly the six polymers we normally encounter, the "Big Six," and their impact on the American economy, lifestyle, and leisure. Chemists alter the properties of polymers by modifications in molecular structure and choice of monomers, the building blocks of polymers. To explain the chemistry of how polymers are made, we consider polymerization by addition and condensation reactions. A short, but significant, aside addresses proteins, an important class of natural polymers, and nylon, a related synthetic polymer. Because polyethylene is the simplest and most widely used plastic, we look at it in some detail. Its composition, structure, production, properties, modifications, and uses are all considered.

The phenomenal success of plastics and their widespread distribution has not been without cost. Therefore, the final third of the chapter is devoted to the problems associated with what to do with plastic articles after we are done using them. The issue of plastic versus paper supermarket bags provides a focus and a brief case study. We examine disposal options including incineration, biodegradation, reuse, recycling, and source reduction. Not surprisingly, we find that there are no easy solutions to the problems posed by plastics, these useful and ubiquitous materials.

9.1 Polymers: The Long, Long Chain

It should be obvious that our biking student missed the point. You cannot get away from synthetic polymers—especially not when biking, hiking, or camping; or anywhere else, for that matter. At this moment you are probably wearing or carrying at least a half dozen different materials that did not exist 70 years ago, perhaps not even 10 years ago. Your shoes alone may consist of six or more different kinds of plastic: in the sole, the trim, the foam padding, the sock liner, the upper, the laces, and even the lace tips. Very likely at least some of your clothing contains synthetic fibers. The pen you are taking notes with is made largely of plastic. Your calculator has a plastic case and so does your cell phone and computer. And the CDs and DVDs you play are made of plastic, as are their containers. Several kinds of plastics are also essential components of your car, which contains nearly 400 pounds of them.

We apply the term "plastic" to a wide range of materials with an equally broad range of properties and applications. According to a standard dictionary definition, *plastic* is an adjective meaning "capable of being molded" or a noun referring to something that is capable of being molded. More specifically, the *Merriam-Webster Collegiate Dictionary*, 10th edition mentions "any of numerous organic synthetic or processed materials that are mostly thermoplastic or thermosetting polymers of high molecular weight and that can be molded, cast, extruded, drawn, or laminated into objects, films, or filaments." You are familiar with the common or brand names of many different plastics: rayon, nylon, Lycra, polyurethane, Teflon, Saran, Styrofoam, and Formica to list only a few. In this text we will reserve the term plastics for such synthetic substances—all are polymers and all are creations of polymer chemists and chemical engineers. What natural and synthetic polymers have in common is evident at the molecular level.

9.2 Consider This: How Much Plastic Do You Throw Away?

Keep a journal of all the plastic and plastic-coated products you throw away (not recycle) in one week. Include plastic packaging from food and other products that you purchase. Keep the journal handy because you will be asked to review it in a later activity (9.33 Consider This).

All plastics are large molecules made up of long chains of atoms covalently bonded together. Like a linked strand of paper clips, the molecular chain in a plastic consists of subunits that are repeated many times. This basic subunit is called a **monomer** (from *mono* meaning "one" and *meros* meaning "unit"). Many monomers join together to form a long molecular strand called a **polymer** (*poly* means "many"). These polymer molecules can be very long indeed. Sometimes they involve thousands of atoms, and molecular masses can reach over a million. No wonder that polymers are sometimes referred to as **macromolecules.**

Although this chapter will focus primarily on synthetic polymers, it is important to note that chemists did not invent polymers. All plastics are polymers, but not all polymers are plastics. Polymeric materials were here long before chemists, and chemists are also made up of many polymers. Natural polymers are found in plants as well as in animals: wood, wool, cotton, starch, rubber, skin, hair—even some minerals, such as asbestos and quartz—are polymers. Polymeric molecules give strength to an oak tree, delicacy to a spider's web, softness to goose down, and flexibility to a blade of grass. Much of the motivation for the synthesis of plastics has been a desire to reproduce such properties in artificial materials. Indeed, many synthetic polymers were originally created as substitutes for expensive or rare naturally occurring materials, or to improve on natural polymers.

To some people, the word "plastic" may carry the connotation "cheap" or "tacky." But the fact remains that synthetic polymers have revolutionized modern life. Few advocates of natural materials would be willing to give up sports shoes, synthetic rubber tires, sporting gear, and the dozens of other plastic objects that have become an accepted part of today's lifestyle. As chemists developed new polymers, the variety of properties and uses expanded dramatically. For example, plastics have become increasingly important in automobile manufacturing because the plastics are considerably less dense than structural metals, making cars more fuel efficient. Plastic packaging reduces weight, eliminates breakage, and helps save fuel during shipping; plastic grocery bags now use 30% less material than they did in 1994. Construction materials of plastic have replaced wood in some applications, and plastic pipes substitute effectively for ones made from lead, iron, copper, and tile. As the introductory episode indicates, recreation has been revolutionized by the introduction of synthetic polymers. Football is played on artificial turf by players wearing plastic helmets, padding, and pants. Tennis balls, tennis rackets, and strings are all made from synthetic polymers. In-line skates are made almost entirely of synthetic polymers. Carbon fibers embedded in plastic resins provide the strength and flexibility required in trail bikes, fishing rods, golf-club shafts, and sailboat hulls and sails. Ice skaters and hockey players skate without ice on rinks of Teflon or high-density polyethylene. Most modern canoes are made of Kevlar, ABS, or polyethylene synthetic polymers, not birchbark, wood, or aluminum. Professional baseball, that bastion of conservatism and tradition, still clings to natural polymers in the form of wooden bats, leather gloves, and a ball made of cowhide covering woolen yarn and a cork center. But even here, cotton/polyester blend uniforms have replaced the hot, scratchy wool ones worn by Babe Ruth, Joe DiMaggio, and even Hank Aaron early in his career.

9.3 Consider This: Polymers All Around Us

Information regarding the impact of plastics on our lives is available from the American Plastics Council at the *Chemistry in Context* web site. From the web site, select two areas where plastics are used. Identify the plastics, their properties, and uses.

9.4 Consider This: Sporty Polymers

Choose a favorite outdoor activity—on the land, water, or even in the air. Do a Web search to find a company that manufactures or sells equipment for this activity. Which polymers can you find mentioned in this web site? Make a table of their names, the equipment they are used in, and any desirable properties cited as advertising points. *Hint:* Many polymer names start with "poly," such as polyester or polypropylene. Other polymers have trade names, such as Gore-Tex, Orlon, or Styrofoam. Still other polymers are coatings and resins, and may be mentioned as epoxides or acrylics.

Plastics manufacturing has a considerable economic impact. For example, plastics manufactured in 2000 had a value of over $40 billion and were used in products valued at nearly $100 billion. In fact, plastics production has eclipsed that of metals. Since 1976, the United States has manufactured a larger volume of synthetic polymers than the volume of steel, copper, and aluminum combined. Six polymeric materials account for about 66% of all the plastics used in the United States. Much of this chapter will deal with the "Big Six," to which we now turn.

9.5 Consider This: Personal Polymers

A Teflon ear bone, fallopian tube, or heart valve? A Gore-Tex facial implant or hernia repair? Some polymers are biocompatible and now used to replace or repair body parts. List four properties that would be desirable for polymers used *within* the human body.

Other polymers may be used outside your body—but in close contact with it. For example, no surgeon is needed for you to use your contact lenses—you insert, remove, clean, and store them yourself. From which plastics are contact lenses made? What properties are desirable in these plastics? Either a call to an optometrist or a search on the World Wide Web may provide some answers. For the latter, a search using "contact lenses" and "plastics" will provide a useful start.

9.2 The "Big Six" of Polymers: Theme and Variations

A good way to begin your study of polymers is by collecting various types of plastic and making some observations.

9.6 Consider This: Plastics You Use

Gather up a variety of plastic items from your residence or your residence hall—plastic bags, soda bottles, whatever happens to be at hand. Make a list of the objects and note the properties of the plastics. Include color, transparency, flexibility, elasticity, hardness, and other properties that could be used to classify and identify the plastics. Try to draw conclusions about which objects are made from the same material. *Hint:* Table 9.1 will be of some help.

You have no doubt discovered from the 9.6 Consider This activity (or from prior experience) that plastics exhibit a wide range of properties. We can illustrate this with a few objects that might well be found in your room or residence. Your backpack is water resistant, lightweight, and sturdy, chosen from a myriad of styles and colors, possibly even your school's colors. The plastic that makes up most soft-drink and water bottles is transparent and has moderate hardness and flexibility. The ubiquitous styrofoam cup is white, opaque, light, soft, and easily deformed and torn, but it is an excellent heat insulator. A CD "jewel case" is hard, brittle, transparent, and almost glasslike in its transparency. The CD inside the case is very strong, yet delicate; it can be scratched easily. Videotape is shiny, flexible, and durable. A plastic milk bottle is somewhat opaque or at least translucent. Although it can be deformed, it is not as soft and flexible as many plastics. Finally, in our brief survey, a typical grocery plastic fruit bag is light, transparent, flexible, and easily stretched. But it is sometimes difficult to correlate these properties unambiguously to the chemical composition of the plastics. If you were trying to sort for recycling the items described in this paragraph, you might find it hard to do so. In fact, it may be surprising that objects as different as a foam coffee cup and a CD case are made of the same plastic—polystyrene. The videotape and soda bottle are composed primarily of polyethylene terephthalate, and the plastic bag and the milk bottle are both polyethylene.

Today, more than 60,000 plastics are known. Most have been developed for special purposes, ranging from frying pan coatings to resins for restoring antiques. Yet the bulk of the plastics we regularly encounter are the "Big Six"—**low-density polyethylene** (LDPE), **high-density polyethylene** (HDPE), **polypropylene** (PP), **polystyrene** (PS), **polyvinyl chloride** (PVC), **and polyethylene terephthalate** (PET or PETE). All are ultimately derived from petroleum. Over 30 million tons of these six polymers are made annually in the United States.

Table 9.1 summarizes information about the "Big Six." Six monomers are used to make six different polymers. *The difference among the polymers is the monomer used to form the polymer.* Ethylene, vinyl chloride, styrene, and propylene monomer molecules are similar in that they each contain two carbon atoms connected by a double bond, C=C. In ethylene, two hydrogen atoms are attached to each of the double-bonded carbon atoms, $H_2C{=}CH_2$ (the theme). But, in vinyl chloride, propylene, and styrene molecules, one of the hydrogen atoms has been replaced with something else (the variations). In the case of vinyl chloride, hydrogen is replaced by a chlorine atom. A methyl group, $—CH_3$, substitutes for a hydrogen atom in propylene. In styrene the substituent is a phenyl group, $—C_6H_5$, which consists of six carbon atoms, each at the corner of a hexagon. Five of the six carbon atoms are bonded to hydrogen atoms; the sixth carbon atom is linked to some other atom, which in the case of styrene is the C atom in CH of $CH_2{=}CH$.

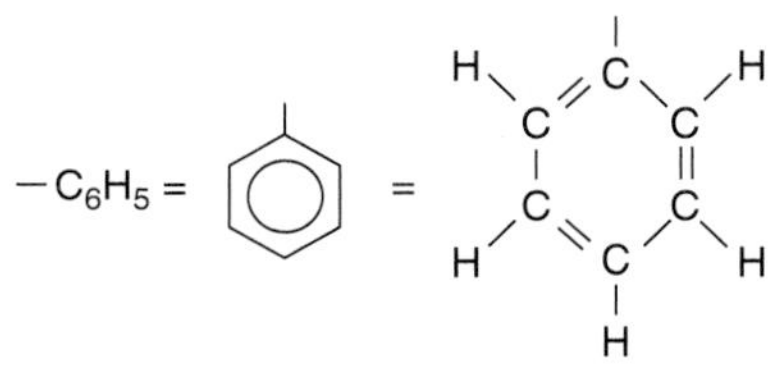

$-C_6H_5 =$

• The phenyl group is one of the most common structures in organic chemistry.

These replacements in the monomers create variety in them and consequently in the polymers formed from them. Moreover, the substituents give the chemist greater latitude in designing plastics for particular uses. As we shall see in the next section, the first five of the Big Six are made by addition polymerization from monomers containing C=C

Table 9.1

The Big Six (Including Identifying Code of the Polymers)

Polymer	Monomer	Properties of Polymer	Uses of Polymer
Polyethylene (LDPE) 4 LDPE	Ethylene $H_2C{=}CH_2$	Opaque, soft, flexible, impermeable to water vapor, unreactive toward acids and bases, absorbs oils and softens, melts at 100°–120°C, does not become brittle until –100°C, oxidizes on exposure to sunlight, subject to cracking if stressed in presence of many polar compounds	Plastic bags, toys, electrical insulation
Polyethylene (HDPE) 2 HDPE	Ethylene $H_2C{=}CH_2$	Similar to LDPE, more opaque, denser, mechanically tougher, more crystalline and rigid	Milk and water jugs, gasoline tanks, cups
Polyvinyl chloride 3 V	Vinyl chloride $H_2C{=}CHCl$	Rigid, thermoplastic, impervious to oils and most organic materials, transparent, high impact strength	Plumbing pipe, garden hoses, "bubble" package wrap
Polystyrene 6 PS	Styrene $H_2C{=}CH(C_6H_5)$	Glassy, sparkling clarity, rigid, brittle, easily fabricated, upper temperature use 90°C, soluble in many organic materials	Styrofoam insulation, inexpensive furniture, drinking glasses
Polypropylene 5 PP	Propylene $H_2C{=}CH(CH_3)$	Opaque, high melting point (160°–170°C), high tensile strength and rigidity, lowest density commercial plastic, impermeable to liquids and gases, smooth surface with high luster	Battery cases, indoor-outdoor carpeting, bottle caps, auto trim
Polyethylene terephthalate 1 PETE	Ethylene glycol $HOCH_2CH_2OH$; Terephthalic acid $HOOC{-}C_6H_4{-}COOH$	Transparent, high impact strength, impervious to acid and atmospheric gases, not subject to stretching, most costly of the six	Clothing, soft-drink bottles, audio- and videotapes, film backing

- PET or PETE is used to abbreviate polyethylene terephthalate.

double bonds. The sixth polymer, polyethylene terephthalate, is a special case among the "Big Six"; it is made by a condensation polymerization process using two monomers. We will return to polyethylene terephthalate, but only after spending more time with the other members of this sextet.

Table 9.1 also lists some of the more important properties of these six polymers. They are all **thermoplastic;** that is, they can be melted and shaped, and all tend to be flexible. Three of them, low- and high-density polyethylene and polypropylene, have both crystalline and amorphous regions. Because of their structural regularity, the crystalline regions impart toughness and resistance to mechanical abrasion, making polypropylene and high-density polyethylene opaque. The amorphous regions do not have a regular arrangement of molecules. The molecular chains are more random, thus promoting flexibility. The other three polymers—polyethylene terephthalate, polystyrene, and polyvinyl chloride—are not crystalline. Their molecular chains are bonded together tightly but more or less randomly. The range of properties among different polymers means that they are differently suited for specific applications. The next Consider This gives you an opportunity to match polymers with their properties and uses.

9.7 Consider This: Uses of the "Big Six" Polymers

For each of these uses, specify the desirable properties of a plastic and, using the information in Table 9.1, suggest the most suitable polymer or polymers.

a. a bread bag
b. a soft-drink bottle
c. "bubble" packaging around a toy
d. bottle caps
e. outdoor lawn furniture
f. a drinking water bottle

Whatever use is made of them, the six major plastics also generally have small amounts of other materials added to them. Because all six are colorless, coloring agents are often introduced. Plasticizers, substances that improve the flexibility of the polymer, are commonly added, as are a variety of other substances that enhance the performance and durability of the plastic. Indeed, the smell associated with certain plastics (and interiors of new cars) is sometimes due to vaporizing plasticizers.

9.3 Addition Polymerization: Adding up the Monomers

We already noted that ethylene, vinyl chloride, styrene, and propylene monomer molecules each contain a carbon-carbon double bond, three hydrogen atoms, and one variable atom or group attached to these two carbon atoms, $H_2C{=}CH{-}R$, where R is a hydrogen atom, a chlorine atom, or a group of atoms (Table 9.1). In ethylene, $CH_2{=}CH_2$, there is an additional hydrogen atom on the second carbon. As seen in Table 9.1, vinyl chloride, styrene, and propylene monomer molecules contain a different atom or group of atoms rather than a hydrogen atom as in ethylene. All these monomers form polymers by a process called **addition polymerization.** In addition polymers, the monomers simply add to the growing polymer chain in such a way that the product contains all the atoms of the starting material. No other products are formed, and no atoms are eliminated. Thus, ethylene molecules become bonded together to form polyethylene. The two carbon atoms in ethylene are linked with a double bond that is capable of reacting with another ethylene molecule. In the process, the C=C double bonds are converted to C—C single bonds, and the polymer contains only C—C single bonds with hydrogen atoms attached to the carbon atoms. The production of polyethylene begins with the joining of two ethylene molecules to form a four-carbon chain (equation 9.1).

$$2\ H_2C{=}CH_2 \longrightarrow {-}CH_2{-}CH_2{-}CH_2{-}CH_2{-} \qquad [9.1]$$

An additional ethylene molecule joins by adding to the growing chain (equation 9.2).

$${-}CH_2{-}CH_2{-}CH_2{-}CH_2{-} + H_2C{=}CH_2 \longrightarrow {-}CH_2{-}CH_2{-}CH_2{-}CH_2{-}CH_2{-}CH_2{-} \qquad [9.2]$$

By continuing to add ethylene molecules, the chain grows rapidly to form a long polymer containing n monomeric units (equation 9.3).

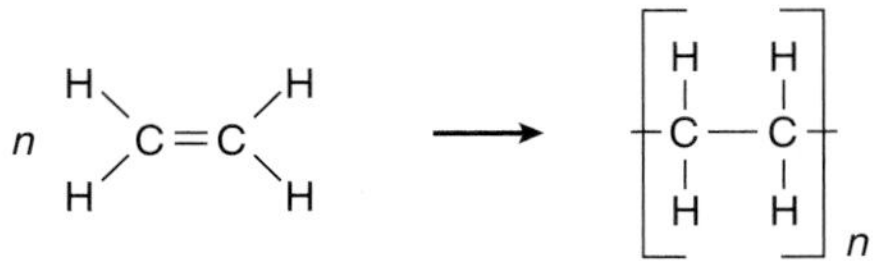

$$n\ \mathrm{H_2C{=}CH_2} \longrightarrow \left[\mathrm{-CH_2-CH_2-}\right]_n \qquad [9.3]$$

The numerical value of n and hence the length of the chain varies with the reaction conditions. Often n is in the hundreds or thousands, the properties of the polymer varying with the length of the polymer chain. Moreover, n can vary because a typical synthetic polymer is a mixture of individual polymer molecules of varying length and mass. Molecular masses of polyethylene are generally between 10,000 and 100,000. In every case, however, the carbon atoms are attached to each other by single bonds, and the hydrogen atoms are bonded to the carbon atoms. Thus, a polyethylene molecule is a macromolecular version of a hydrocarbon molecule, such as those in petroleum.

9.8 Your Turn

Write the structural formula of a polyethylene chain containing eight ethylene units.

Ethylene and polyethylene are made up only of carbon and hydrogen atoms. But the fact that a vinyl chloride molecule, CH_2CHCl, contains a chlorine atom introduces an opportunity for variability in the structure of polyvinyl chloride (PVC).

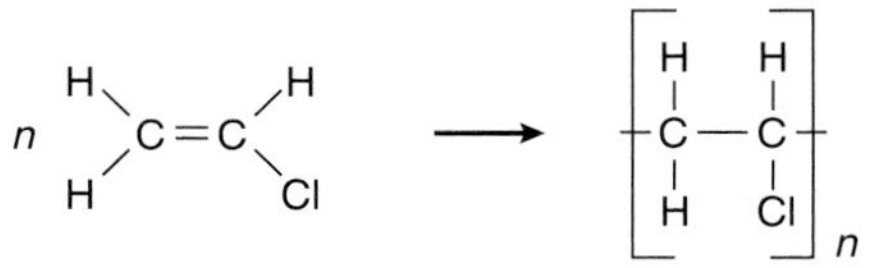

$$n\ \mathrm{H_2C{=}CHCl} \longrightarrow \left[\mathrm{-CH_2-CHCl-}\right]_n \qquad [9.4]$$

The chlorine atom creates an asymmetry in the molecule. Let us arbitrarily think of the carbon atom bearing two hydrogens (CH_2) as the "head" of a vinyl chloride molecule and the chlorinated carbon atom (CHCl) as its "tail." (We could just as easily have made the reverse assignments.) Because of the chlorine atom, when vinyl chloride molecules add to each other to form polyvinyl chloride, the molecules can be oriented in three possible arrangements: alternating head-to-head and tail-to-tail; repeating head to tail; and a random distribution of heads and tails. Figure 9.2 should help make this molecular interplay more evident.

In a head-to-head/tail-to-tail arrangement of PVC, chlorine atoms are on adjacent carbons. In the head-to-tail structure, chlorine atoms are on alternate carbons. And in the

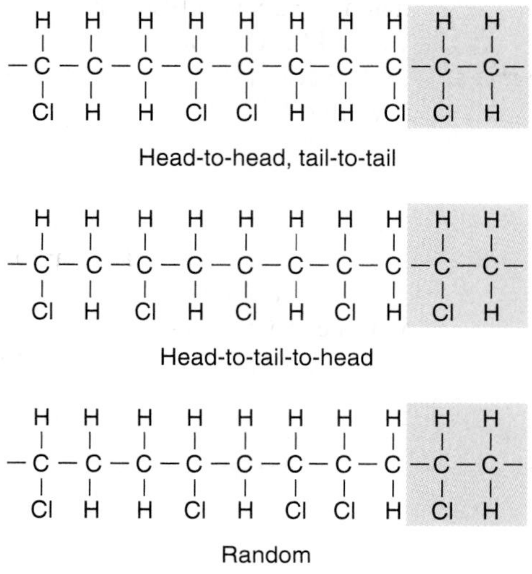

$$-\mathrm{CHCl-CH_2-CH_2-CHCl-CHCl-CH_2-CH_2-CHCl-CHCl-CH_2-}$$

Head-to-head, tail-to-tail

$$-\mathrm{CHCl-CH_2-CHCl-CH_2-CHCl-CH_2-CHCl-CH_2-CHCl-CH_2-}$$

Head-to-tail-to-head

$$-\mathrm{CHCl-CH_2-CH_2-CHCl-CH_2-CHCl-CHCl-CH_2-CHCl-CH_2-}$$

Random

Figure 9.2

Three possible arrangements of monomer units in PVC.

random polymer, an irregular mixture of the previous two types occurs. In each case, the properties are somewhat different. *Controlling monomer orientation within the chain is one method used to influence polymer properties.* The head-to-tail arrangement is the usual product for polyvinyl chloride. Depending on the orientation of monomer units in the chain, PVC can be stiff or flexible. Stiff PVC finds use in drain and sewer pipes, credit cards, house siding, toys, furniture, and various automobile parts. The flexible version is familiar in wall coverings, upholstery, shower curtains, garden hoses, insulation for electrical wiring, and packaging films.

Polypropylene (PP) is also formed by addition polymerization using propylene monomers (see Table 9.1). A particularly useful form of polypropylene has the monomeric units bonded in a head-to-tail fashion. This regularity imparts a high degree of crystallinity and makes the polymer strong, tough, and able to withstand high temperatures. Its uses reflect these properties. Polypropylene is found in indoor-outdoor carpeting, videocassette cases, and cold weather underwear. Strength and chemical resistance make polypropylene a good choice for applications in which structural ruggedness is required, such as in indoor-outdoor carpeting.

The familiar white foam hot beverage cup is the most common example of polystyrene (PS). The styrene monomer has the —C_6H_5 ring in place of a hydrogen atom on one of the double-bonded (C=C) carbons (equation 9.5). Typically the symbols for the carbon and hydrogen atoms in the phenyl group are omitted, and the entire —C_6H_5 structure is represented by a hexagon. The structure on the right of equation 9.5 indicates that bonding in the —C_6H_5 ring can be viewed as consisting of three carbon-carbon single bonds and three carbon-carbon double bonds, alternating around the hexagon. This arrangement of electrons conforms to the octet rule. Experiment shows that the electrons are in fact uniformly distributed around the ring, with all the carbon-carbon bonds of equal strength and length. This uniformity is implied by the circle within the hexagon. Under appropriate catalytic conditions, styrene polymerizes to polystyrene, usually with the head-to-tail arrangement. The by-now-familiar type of addition polymerization equation applies; here n equals about 5000.

$$n\ \mathrm{H_2C{=}CH(C_6H_5)} \longrightarrow \left[\mathrm{-CH_2-CH(C_6H_5)-}\right]_n \qquad [9.5]$$

Styrene Polystyrene

A —C_6H_5 ring in polystyrene.

We noted earlier that the hard, brittle, transparent "jewel" cases for CDs are chemically almost identical to light, white, opaque foam coffee cups; both are polystyrene. Such styrofoam cups are made by expansion molding. Polystyrene beads containing 4–7% of a low-boiling liquid are placed in a mold and heated using steam or hot air. The heat causes the liquid to vaporize and the expansion of the gas also expands the polymer. The expanded particles are fused together into the shape determined by the mold. Because it contains so many bubbles, this plastic foam is not only light, but it is also an excellent thermal insulator. Chlorofluorocarbons were used as foaming agents, but concern over the involvement of CFCs in the destruction of stratospheric ozone (Chapter 2) led to their replacement in 1990. Gaseous pentane (C_5H_{12}) and carbon dioxide are now frequently used for this purpose. The hard, transparent version of polystyrene is made by molding the melted polymer without the foaming agent. It is used to fabricate wall tile, window moldings, and radio and television cabinets, in addition to CD cases.

- Technically speaking, the term "Styrofoam" is a registered trademark of the Dow Chemical Company for its distinctive blue housing insulation product. But it has become common to use styrofoam for containers, such as beverage cups, made from expanded polystyrene beads.

The Dow Chemical Company developed a new process that uses pure carbon dioxide as a blowing (foaming) agent to produce styrofoam for packaging material. Using the Dow 100% CO_2 technology eliminates the use of three and a half million pounds of CFC-12 (Section 2.12) or HCFC-22 (Section 2.16) as blowing agents. The CO_2 used in the process is a by-product from existing commercial and natural sources, such as ammonia plants and natural gas wells, thus it does not contribute additional CO_2 to global warming. Because it is nonflammable, carbon dioxide is preferred as a blowing agent over pentane, which is flammable.

9.9 Consider This: Polypropylene—A Tough Plastic

Polypropylene, one of the "Big Six," is used to construct a number of items in which toughness counts. It may not be as familiar to you as polyethylene and PET, because many polypropylene items are not marked with a recycling symbol (and are not collected at curbside). Search for "polypropylene" on the Web to identify half a dozen specific items manufactured from polypropylene.

9.10 Your Turn

Write the structural formula of a polystyrene chain containing eight styrene units arranged in the head-to-tail arrangement. See Table 9.1 for the structural formula of styrene. Why do you think this arrangement is favored rather than the head-to-head arrangement?

9.11 Your Turn

Write the structural formula of a polypropylene chain containing eight units arranged in a head-to-head, tail-to-tail arrangement. See Table 9.1 for the structural formula of propylene.

9.12 Your Turn

The monomer used to form Teflon is tetrafluoroethylene, CF_2CF_2.

a. Write the structural formula of a polytetrafluoroethylene chain containing eight tetrafluoroethylene units.
b. Why is a head-to-tail arrangement not possible for this polymer?

- Roy Plunkett, a DuPont chemist, accidentally discovered Teflon while experimenting with gaseous tetrafluoroethylene. Plunkett was curious enough about the solid that formed accidentally to study it sufficiently to realize that he had made a previously unknown polymer. This exemplifies Louis Pasteur's maxim that "Chance favors only the prepared mind."

9.4 Polyethylene: A Closer Look at the Most Common Plastic

We use polyethylene to exemplify some additional strategies for modifying polymer chains to create differing properties of the polymer. Polyethylene has a wide variety of uses, which suggests a similarly wide range of properties for this single polymer. Yet, as we have seen in the previous section, all polyethylene is made from the same starting material—ethylene, CH_2CH_2. How can this one material form polymers that can be used in so many different ways?

- Ethylene is also known as ethene.
- Curiously, ethylene acts as a natural plant hormone that ripens fruit.

Ethylene is a compound extracted from petroleum (Section 4.8). At ordinary temperatures and pressures, ethylene is a gas. However, in the 1930s, it was discovered that by using a special catalyst to initiate the reaction, individual ethylene molecules could be made to bond to each other to form a polymer. How ethylene polymerizes involves what happens to its electrons. The reaction is initiated by a catalyst that is a free radical with one unpaired electron. In Figure 9.3, a representation of the polymerization of polyethylene, the free radical is represented by R· (the dot indicates an unpaired electron). The radical reacts readily with a CH_2CH_2 molecule. One of the two bonds be-

- Free radicals are very reactive species, as you may recall from the role of ·Cl atoms in stratospheric ozone destruction (Section 2.9).

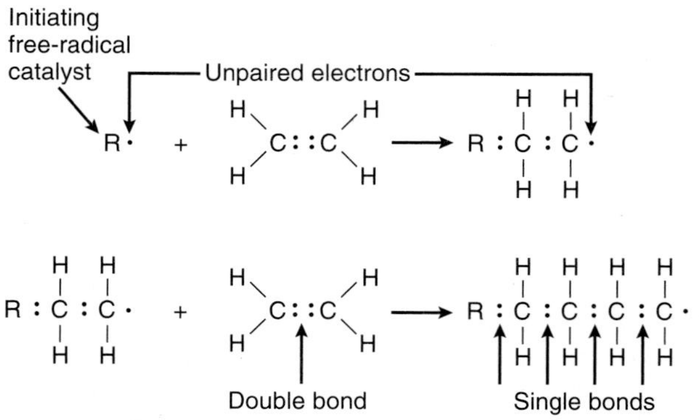

Initiating free-radical catalyst → R· + H₂C::CH₂ ⟶ R:CH₂:CH₂· (Unpaired electrons)

R:CH₂:CH₂· + H₂C::CH₂ (Double bond) ⟶ R:CH₂:CH₂:CH₂:CH₂· (Single bonds)

- The tetrahedral arrangement around each carbon atom makes the carbon atoms in polyethylene align in a zig-zag arrangement rather than a straight chain.

Figure 9.3
The polymerization of ethylene.

 **Figures Alive!**

tween the carbon atoms in ethylene breaks, and one of the electrons from that bond pairs with the unpaired electron of the radical to form a covalent bond (Figure 9.3). The new molecule formed, $RCH_2CH_2\cdot$, is also a free radical because it carries an unpaired electron left over from the broken carbon-carbon bond. Therefore, $RCH_2CH_2\cdot$ can react with another ethylene molecule that bonds to the carbon atom with the unpaired electron at the reactive, growing end of the polymer. This process is repeated many times over in many chains at the same time. Occasionally, the free radical ends of two polymers interact to form a bond and stop the chain growth. The result of all this chemistry is that gaseous ethylene is converted to solid polyethylene.

Many of the properties of polyethylene are related to the presence of these long chains of polymer molecules. Relatively speaking, they are very long indeed. If a polyethylene molecule were as wide as a piece of spaghetti, the molecular chain could be as much as half a mile long. To continue the analogy, in the polyethylene used to make plastic bags, these chains are arranged somewhat like spaghetti on a plate. The strands are jumbled up and not very well aligned, although there are quasi-crystalline regions where the molecular chains are parallel. Moreover, the polyethylene chains, like spaghetti strands, are not bonded to each other. Evidence of this molecular arrangement can be obtained by doing a little experiment. Cut a strip from a heavy-duty transparent polyethylene bag, grab the two ends of the strip, and pull. A fairly strong pull is required to start the plastic stretching, but once it begins, less force is needed to keep it going. The length of the plastic strip increases dramatically as the width and thickness decrease (Figure 9.4a). A little shoulder forms on the wider part of the strip and a narrow neck almost seems to flow from it in a process called "necking." Unlike the stretching of a rubber band, the necking effect is not reversible and eventually the plastic thins to the point where it tears.

Figure 9.4b is a representation of the necking of polyethylene from a molecular point of view. As the strip narrows and necks down, the previously mixed-up molecular chains move. They shift, slide, and align parallel to each other and the direction of the pulling force. In some plastics, such stretching or "cold drawing" is carried out as part of the manufacturing process to alter the three-dimensional arrangement of the chains in the solid to obtain ordered polymer chains. Of course, as the force and stretching continue, the polymer eventually reaches a point at which the strands can no longer realign, and the plastic breaks. Paper, a natural polymeric material, tears when pulled because the strands (fibers) in paper are rigidly held in place and are not free to slip like the long molecules in polyethylene.

9.13 Consider This: Necking Polyethylene

Necking polyethylene changes the properties of polyethylene.

a. Does necking affect the number of monomer units, *n*, in the average polymer?
b. Does necking affect the bonding between the monomer units within the polymer? Explain your reasoning.

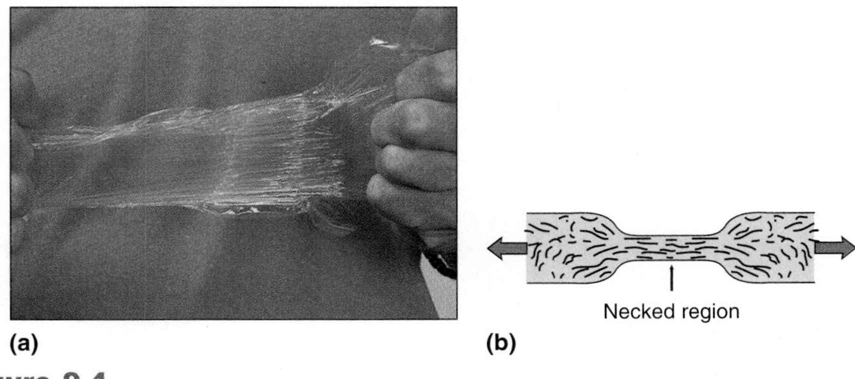

Figure 9.4
(a) A plastic bag stretched until it "necks." (b) Molecular rearrangement as polyethylene is stretched.

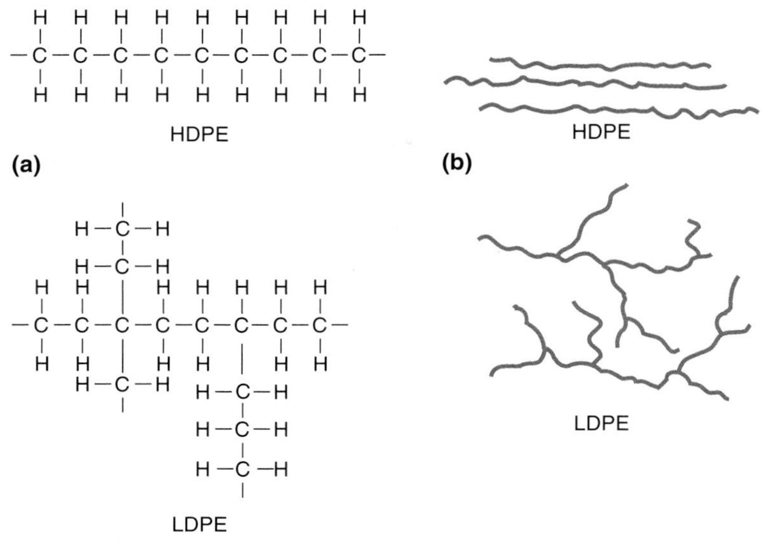

Figure 9.5

(a) Detail of bonding in high-density (linear) polyethylene and low-density (branched) polyethylene. (b) Representation of high-density (linear) polyethylene and low-density (branched) polyethylene.

Another strategy to control the molecular structure and physical properties of polymers is to regulate the branching of the polymer chain. This approach is used to produce the two general types of polyethylene. The version found in plastic bags, such as for supermarket fruits and vegetables, is low-density polyethylene (LDPE), which is soft, stretchy, transparent, and not very strong. This low-density form was the first type of polyethylene to be manufactured. Study of its structure reveals that the molecules consist of about 500 monomeric units and that the central polymer chain has many side branches, like the limbs radiating from a central tree trunk (Figure 9.5).

About 20 years after the discovery of LDPE, chemists were able to adjust reaction conditions to prevent branching and make another form of polyethylene called high-density polyethylene (HDPE). In their Nobel Prize–winning research, Karl Ziegler and Giulio Natta developed new catalysts that enabled them to make linear (unbranched) polyethylene chains consisting of about 10,000 monomer units. Having no side branches, these long chains can be arranged parallel to one another (Figure 9.5). The structure of HDPE is thus more like a regular crystal than the irregular tangle of the polymer chains in LDPE. The more highly ordered structure of HDPE gives it greater density, rigidity, strength, and a higher melting point than LDPE. Furthermore, the high-density form is opaque; the low-density form tends to be transparent.

The differences in properties of high- and low-density polyethylene give rise to different applications. HDPE is used to make toys, gasoline tanks, radio and television cabinets, and heavy-duty pipes. One new use of HDPE has been spurred by the AIDS epidemic. A surgeon who breaks her or his skin during an operation on an HIV-positive patient runs the risk of acquiring the human immunodeficiency virus through contact with the patient's blood. Allied-Signal Corporation has produced a linear polyethylene fiber called Spectra that can be fabricated into liners for surgical gloves. Spectra gloves are said to have 15 times the cut resistance of medium-weight leather work gloves, but they are so thin that a surgeon can retain a keen sense of touch. A sharp scalpel can be drawn across the glove with no damage to the fabric or the hand inside. Such strength is in marked contrast to the properties of the LDPE common plastic grocery bag.

9.14 Consider This: Shopping for Polymers

Macrogalleria is an excellent place to find information about polymers. At Level One of the site is a "shopping mall" featuring a host of different polymers. Use the web site to find at least six products made of LDPE and HDPE other than those mentioned in the textbook. Make your selections from two different "stores" at the "mall" and find two types of uses. Identify the polymer for each use.

It would be a mistake, though, to conclude that polyethylene is restricted to the extremes represented by highly branched or strictly linear forms. By modifying the extent and location of branching in LDPE, its properties can be varied from soft and wax-like (coatings on paper milk cartons) to stretchy (plastic food wrap). Because of its linearity and close packing, HDPE is sufficiently rigid to be used for plastic milk bottles. Unfortunately, consumers are sometimes unaware of the consequences of such structural tinkering. For example, the higher melting point of HDPE (130°C) permits plasticware made from it to be washed in automatic dishwashers. But objects made of LDPE, with a melting point of 120°C, melt in dishwashers.

- The plastic is melted by the high temperature of the dishwasher's heating element, not the hot water.

Finally, we should note that one of the first and most important uses of polyethylene was a consequence of the fact that it is a good electrical insulator. During World War II, polyethylene was used by the Allies as insulation to coat electrical cables in aircraft radar installations. Sir Robert Watt, who discovered radar, described polyethylene's critical importance in these words.

> *The availability of polythene [polyethylene] transformed the design, production, installation, and maintenance problems of airborne radar from the almost insoluble to the comfortably manageable . . . A whole range of aerial and feeder designs otherwise unattainable was made possible, a whole crop of intolerable air maintenance problems was removed. And so polythene played an indispensable part in the long series of victories in the air, on the sea, and on land, which were made possible by radar.*[1]

9.15 Consider This: High- and Low-Density Polyethylene

Use the structures of HDPE and LDPE found in Figure 9.5 to explain why the density of HDPE is greater than that of LDPE.

9.16 Consider This: Yet Another Type of Polyethylene

In addition to LDPE and HDPE, polyethylene is manufactured in other forms. Use the Web to find out an alternative form and its properties and uses.

9.5 Condensation Polymers: Bonding by Elimination

Unlike the other five polymers of the Big Six described earlier, polyethylene terephthalate (PET or PETE) is not formed by an addition reaction. Rather, it is produced via a condensation reaction. Many polymers are formed by condensation reactions. Natural ones formed this way include cellulose, starch, wool, silk, and proteins; synthetics are nylon, Dacron, Kevlar, ABS, and Lexan.

- ABS is a condensation polymer composed of linked *a*crylonitrile, *b*utadiene, and *s*tyrene monomer units.

In **condensation polymerization,** monomer units join by eliminating (splitting out) a small molecule, often water. Thus, a condensation polymerization has two products—the polymer itself plus the small molecules split out during the polymer's formation. Polyethylene terephthalate is compounded of two monomers—ethylene glycol and terephthalic acid—hence, it is called a copolymer. Ethylene glycol, $HOCH_2CH_2OH$, the chief ingredient in automobile antifreeze, is a dialcohol; it has an —OH on each end of the molecule. The —OH is a functional group found in all alcohols, and this group is responsible for the chemical and physical properties that characterize alcohols. **Functional groups** are distinctive arrangements of groups of atoms that impart characteristic chemical properties to the molecules that contain them. A molecule of

[1]Quoted by J.C. Swallow in "The History of Polythene" from *Polythene—The Technology and Uses of Ethylene Polymers* (2d ed.) A. Renfrew (ed) London: Iliffe and Sons, 1960.

- —OH is the alcohol functional group; —COOH is the functional group of an organic acid. Other functional groups are given in Table 10.2 and Section 10.3. See Section 4.10 about alcohols in fuels.

terephthalic acid, HOOCC$_6$H$_4$COOH, has a —COOH group at each end, the functional group found in all organic acids. Because terephthalic acid has two organic acid groups per molecule, it is a diacid. The six carbon atoms of C_6H_4 group are arranged in the same hexagonal ring you just encountered in styrene.

A key point about the condensation polymerization reaction of ethylene glycol and terephthalic acid to form polyethylene terephthalate is that each time a monomer reacts, the —OH of an acid group and the H from an alcohol group react to form a water molecule, which is given off by the polymerization reaction as polyethylene terephthalate forms. The remaining portions of the alcohol and the acid join by forming an **ester** linkage, another class of compounds. The condensation polymerization reaction of ethylene glycol with terephthalic acid is represented by equation 9.6; the ester linkage is enclosed in a shaded box.

- In equation 9.6, the benzene ring is represented by a hexagon.

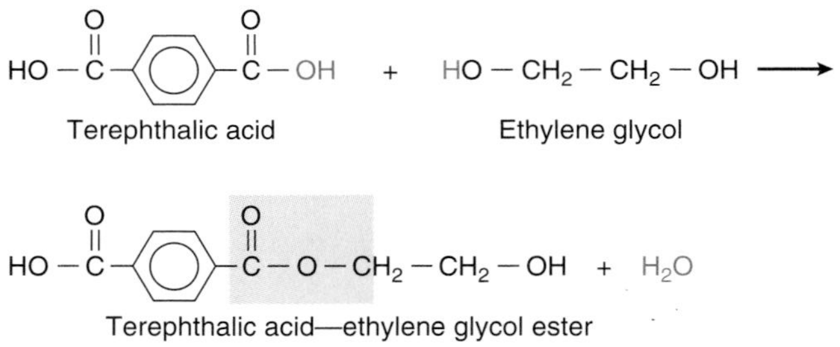

$$\underset{\text{Terephthalic acid}}{HO-\overset{O}{\overset{\|}{C}}-C_6H_4-\overset{O}{\overset{\|}{C}}-OH} + \underset{\text{Ethylene glycol}}{HO-CH_2-CH_2-OH} \longrightarrow \underset{\text{Terephthalic acid—ethylene glycol ester}}{HO-\overset{O}{\overset{\|}{C}}-C_6H_4-\boxed{\overset{O}{\overset{\|}{C}}-O}-CH_2-CH_2-OH} + H_2O \qquad [9.6]$$

Note that the ester produced in equation 9.6 still has a —COOH group on one end and an —OH on the other. The acid group can react with an alcohol group of another ethylene glycol molecule; likewise, the alcohol group of the growing polymer can react with an acid group of another terephthalic acid molecule. This process, represented in Figure 9.6, occurs many times over to yield a long polymeric chain of polyethylene terephthalate.

Polyethylene terephthalate (PET) is classified as a **polyester** because it contains many ester linkages. Since their introduction, polyester fibers have found many uses in fabrics and clothing. The polymer is perhaps most familiar under the trade name Dacron. This polyester is frequently mixed with cotton, wool, or other natural polymers, but it has many other uses. Indeed, about five million pounds of PET are produced annually in the United States. Narrow, thin-film ribbons of it (under the trade name Mylar) are coated with metal oxides and magnetized to make audiotapes and videotapes. Dacron tubing is used surgically to replace damaged blood vessels, and artificial hearts contain parts made of PET. Photographic and X-ray film are made from PET, and containers are made from it for medical supplies to be sterilized by irradiation. The most common use for this plastic is in soft-drink bottles because PET is semirigid, colorless, and gas-tight.

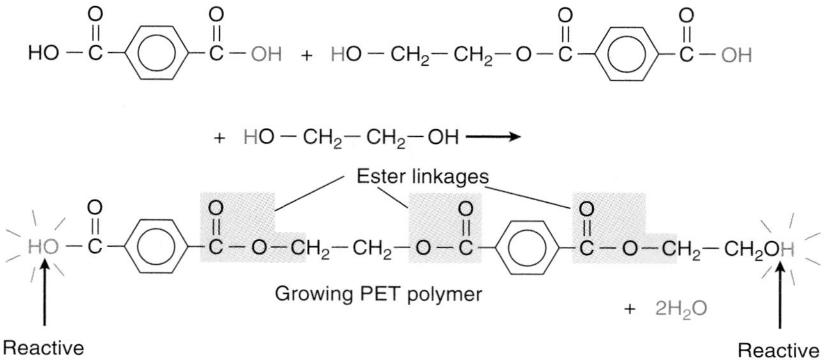

$$HO-\overset{O}{\overset{\|}{C}}-C_6H_4-\overset{O}{\overset{\|}{C}}-OH + HO-CH_2-CH_2-O-\overset{O}{\overset{\|}{C}}-C_6H_4-\overset{O}{\overset{\|}{C}}-OH + HO-CH_2-CH_2-OH \longrightarrow$$

$$\underset{\text{Reactive}}{HO}-\overset{O}{\overset{\|}{C}}-C_6H_4-\boxed{\overset{O}{\overset{\|}{C}}-O}-CH_2-CH_2-\boxed{O-\overset{O}{\overset{\|}{C}}}-C_6H_4-\boxed{\overset{O}{\overset{\|}{C}}-O}-CH_2-CH_2\underset{\text{Reactive}}{OH} + 2H_2O$$

Ester linkages

Growing PET polymer

Figure 9.6
A growing PET polymer chain.

9.17 Consider This: Can All Acids and Alcohols Form Polyesters?

You have seen that terephthalic acid and ethylene glycol are capable of forming a polyester. Consider these structures for another organic acid, acetic acid, and another alcohol, ethyl alcohol (ethanol):

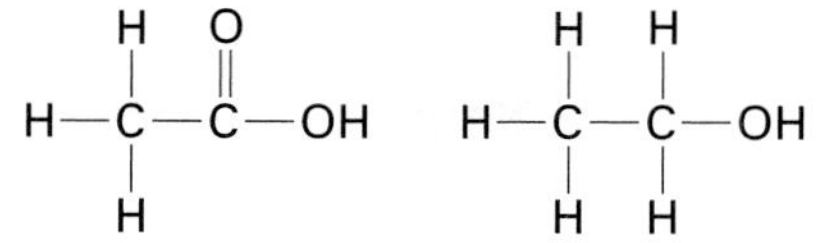

$$\underset{\text{Acetic acid}}{CH_3-C(=O)-OH} \qquad \underset{\text{Ethyl alcohol}}{CH_3-CH_2-OH}$$

a. Can these two substances form an ester?
b. Can they form a polyester? Explain your reasoning.

9.18 Consider This: New Combinations

Polyethylene naphthalate (PEN) is a polymer widely used in bar code labels. In both PET and PEN, the alcohol monomer is ethylene glycol, but the organic acid monomers differ slightly. The structural formula of the organic acid monomer used to produce PEN is

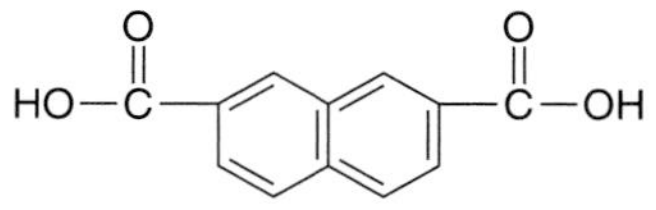

$$HO-C(=O)-C_{10}H_6-C(=O)-OH$$

Naphthalic acid

Draw the structural formula to show the formation of the PEN polymer after two units of naphthalic acid and two units of ethylene glycol have polymerized.

9.6 Polyamides: Natural and Nylon

No discussion of condensation polymerization can be complete without including proteins, one of the most important classes of natural polymers, and nylon, the synthetic substitute that brilliantly duplicates some of the properties of the natural material.

A wide variety of these biological macromolecules make up our skin, hair, muscle, and enzymes. All proteins are **polyamides,** which are polymers of amino acids. The word "amino" refers to the —NH_2 functional group, which is found in a class of compounds called **amines.** Molecules of *amino acids* contain amine groups (—NH_2) as well as acid groups (—COOH). A general formula for an amino acid follows. The amine and acid groups are attached to the same carbon atom. In addition, a hydrogen atom and another group (represented by an R) are bonded to that carbon as well.

• An amine group: —NH_2

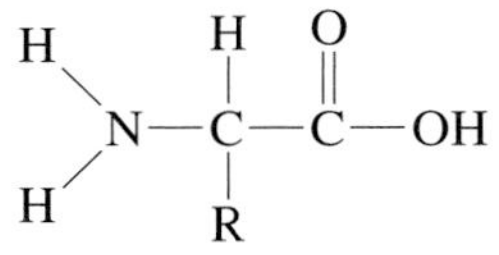

$$H_2N-CH(R)-C(=O)-OH$$

The 20 amino acids found in most proteins differ in the identity of their R groups. In some amino acids, R consists of carbon and hydrogen atoms, as in alanine, where R is a methyl group, —CH_3. In others, R also includes oxygen, nitrogen, or sulfur atoms. Some R groups are acidic and others are basic.

• See Sections 11.6 and 12.4 for more about amino acids and proteins.

Chapters 11 and 12 include a good deal of additional information about amino acids, and proteins, the polymers formed from them. At present, we focus on some fundamentals. The amino acids are the monomers for protein polymerization. The crucial point in protein formation is the fact that the —COOH group of one amino acid reacts with the —NH_2 group of another in a condensation reaction analogous to polyester formation. In the protein-forming reaction, an OH from the acid group of one amino acid reacts with an H from the —NH_2 group of another amino acid to form an H_2O molecule, which is eliminated. A **peptide bond** forms between the remaining portions of the two amino acids. The reaction is represented by equation 9.7, where the peptide bond is enclosed in a box. One amino acid contains the substituent R_1, the other R_2.

- Reaction 9.7 is similar to the neutralization of an acid by a base (see Section 6.3).

$$H_2N-CH(R_1)-C(=O)-OH + HN(H)-CH(R_2)-C(=O)-OH \longrightarrow H_2N-CH(R_1)-\boxed{C(=O)-N(H)}-CH(R_2)-C(=O)-OH + H_2O \quad [9.7]$$

In the sophisticated chemical factories called biological cells, this condensation reaction is repeated many times over to form long polymeric protein chains. Given the fact that there are 20 different naturally occurring amino acid building blocks, a great variety of proteins can be synthesized. Some proteins are made up of hundreds of amino acids, whereas others, like the hormone oxytocin, contain only eight.

Chemists are often well advised to attempt to replicate the chemistry of nature. What we have learned about the structures of natural polymers informs our syntheses of synthetic polymers. For example, in the 1930s, a brilliant chemist working for the DuPont Company set out to do just that. Wallace Carothers (1896–1937) was studying a variety of polymerization reactions, including the formation of peptide bonds (Figure 9.7). Instead of using amino acids, Carothers tried combining adipic acid, $HOOC(CH_2)_4COOH$, and hexamethylene diamine, $H_2N(CH_2)_6NH_2$ (also known as 1,6-diaminohexane). Note from equation 9.8 that a molecule of adipic acid has an acid group on both ends and the hexamethylene diamine molecule has an amine group on each end. As in the case of protein synthesis, the acid and amine groups react to eliminate water and form peptide bonds. But in this instance, the polymer consisted of alternating adipic acid and hexamethylene diamine monomer units.

$$\underset{\text{Adipic acid}}{HO-C(=O)-(CH_2)_4-C(=O)-OH} + \underset{\text{Hexamethylene diamine}}{H_2N-(CH_2)_6-NH_2} \longrightarrow \underset{\text{Part of a nylon molecule}}{HO-C(=O)-(CH_2)_4-C(=O)-N(H)-(CH_2)_6-NH_2} + H_2O \quad [9.8]$$

Reactive (at the HO– end) Reactive (at the –NH_2 end)

DuPont executives decided the new polymer had promise, especially after company scientists learned to draw it into thin filaments. These filaments were strong and smooth and very much like the protein spun by silkworms. Therefore, "Nylon" was first introduced to the world as a substitute for silk. The world greeted it with bare legs and open pocketbooks. Four million pairs of nylon stockings were sold in New York City on May 15, 1940, the first day that they became available (Figure 9.8). But, in spite of consumer passion for "nylons," the civilian supply soon dried up, as the polymer was diverted from hosiery to parachutes, ropes, clothing, and hundreds of other wartime uses. By the time World War II ended in 1945, nylon had repeatedly demonstrated that it was superior to silk in strength, stability, and rot resistance. Today this polymer, in its many modifications, continues to find wide applications in clothing, sportswear, camping equipment, the workroom, the kitchen, and the laboratory.

Figure 9.7
Wallace Carothers, the inventor of nylon.

9.19 Your Turn

Locate the bond analogous to a peptide bond in the segment of a nylon molecule shown in equation 9.8.

9.20 Your Turn

Kevlar is a condensation polymer used to make bulletproof vests. It is made from terephthalic acid (see equation 9.6) and phenylenediamine (structure shown below).

$$H_2N-C_6H_4-NH_2$$

Use the structural formulas of terephthalic acid and phenylenediamine to draw a segment of a Kevlar molecule containing three units of terephthalic acid and three phenylenediamine units.

Figure 9.8

Women eagerly lined up to buy nylon stockings in 1940, when they were first available commercially.

9.7 Plastics: Where From and Where To?

Given the constraints of the law of conservation of matter, it is obvious that we should pay close attention to the raw materials incorporated into plastics and the disposal or recycling of this matter after its use. You have already read that petroleum is the source of the monomers used to make most synthetic polymers (Section 4.9). Crude oil is a mixture of many compounds refined into various fractions on the basis of boiling point and molar mass. Figure 9.9 graphically depicts those fractions and their primary uses. Not surprisingly, given our discussion in Chapter 4, the great majority of petroleum is burned as fuel. Only 3% of it is reserved as a chemical feedstock (reactants) for manufacturing polymers and other chemicals, including new medicines.

This 3% is essential to current methods of manufacturing the polymers that have reshaped modern life. But the planet's supply of petroleum is limited and non-renewable, a fact that creates a serious dilemma. You already know from previous chapters (4 and 6) that petroleum is not an ideal energy source. Burning it releases carbon dioxide that can contribute to global warming, and unburned carbon fragments and other compounds

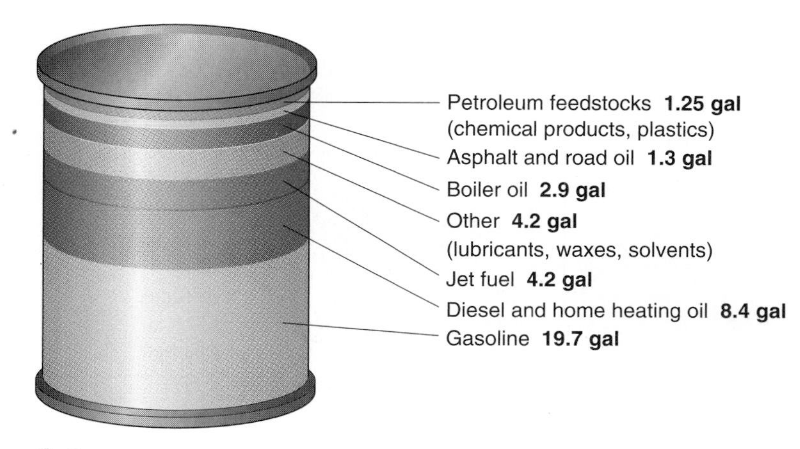

Figure 9.9

Fractions and their uses from a barrel of crude oil (42 gal.).

that give rise to smog and air pollution. But, compared to coal, petroleum is quite clean and convenient. Therefore, we return to a question posed earlier: To burn or not to burn? Should petroleum be burned, or should more of it be diverted for use as a raw material for the synthesis of materials that cannot now be made from any other source? What are the risks and benefits—the economic and social trade-offs—in using oil as a source of energy or as a source of synthetic products? If a greater share of petroleum is not reserved for uses other than fuel, the age of plastics may be very short.

It is important to note that chemistry may rescue society from this dilemma. In principle, at least, polymers can be made from any carbon-containing starting material. Crude oil is simply the most convenient and the most economical. But it might also prove possible to convert renewable biological materials such as wood, cotton fibers, straw, starch, and sugar into new polymers. After all, chemists at the Arthur D. Little Company once actually made a silk purse out of a sow's ear. But to transform biomass into synthetic polymers, new methods and new technologies would have to be developed. The cost of the research and the manufacturing would likely be substantial. Moreover, it would be essential to estimate the supply of the starting materials and the demand for the finished products. There are implications for land use, crop productivity, the environment, and no doubt much more.

There seems to be a good deal more concern about where plastics go than where they come from. Much of the plastic we use eventually ends up in a landfill, along with lots of other types of municipal and domestic solid wastes, in the usual "out of sight, out of mind" approach. As a nation, we daily discard enough trash to fill two Superdomes. Of course that's just a ballpark figure, but it corresponds to about 3 tons of trash annually per family. The EPA estimates that about 73% of all municipal solid waste is put into landfills. Of the remainder, 13% is recycled and 14% is incinerated. Figure 9.10 provides information about the contents of a typical landfill given in terms of percent by volume. It is the volume of the buried materials, not the mass, that causes landfills to reach their capacity.

You will note from Figure 9.10 that only 8% of municipal solid waste is plastic, about 30% of which is used in packaging. But the largest percentage of municipal solid waste (40%) is paper and paper products. This raises a question that has been in the news: Which constitutes the lesser environmental burden, paper or plastic? The section that follows is a real-life glimpse into this controversy.

9.21 Consider This: All Those Disposable Diapers

True or false? Disposable diapers are the main items in municipal landfills. Check out the link at the *Chemistry in Context* web site.

9.22 Consider This: The Polymer Age: From Start to Finish

It has been stated that "there seems to be a good deal more concern about where plastics go than where they come from."

a. What are some of the reasons why this is true within the United States?
b. How do you expect this perception to change within your lifetime? Explain your reasoning.

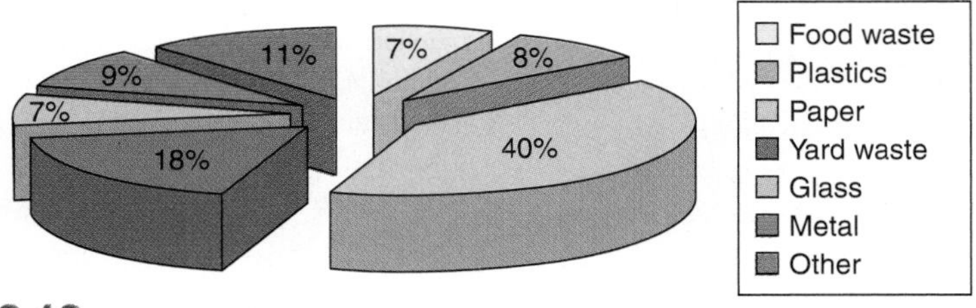

Figure 9.10

What's in your garbage? Composition of municipal solid waste.

(Source of data: U.S. Environmental Protection Agency, 1997.)

9.8 Paper or Plastic? The Battle Rages

The rather melodramatic title of this section appeared as a headline in the *Syracuse Herald-Journal.* It introduced a heated exchange of opinions, forcefully expressed by readers. At issue was whether grocery bags should be made of paper or plastic. The actual letters that follow are selected from many more, and reprinted here to illustrate this controversy in the context of a supermarket.

> *In these environmentally conscious times I am often appalled at the number of people who choose plastic bags over paper ones at the supermarket. They must be aware that plastic bags are neither biodegradable nor as easily recycled as the paper variety.*
>
> *It's common knowledge that plastic bags are not biodegradable, and although some forms of plastic are recyclable, no recycling center in the local area takes plastic bags. What most consumers fail to realize is that the production and processing of plastic involves a great amount of highly toxic chemicals.*
>
> *Improper land disposal of hazardous wastes, emissions of toxic chemicals into the air, and discharges of toxic industrial effluents into waterways as a result of plastic production seriously threaten the public health and the environment.*[2]

The local supermarket took the opposite position and argued that, by using plastic bags, it was acting in an environmentally responsible manner. As evidence, it printed the following message on its grocery bags.

> *Thank you for using plastic bags. If all of the Wegmans shoppers using plastic bags last year had insisted on paper, they would have increased the amount of solid waste by over eight million pounds and taken up nearly seven times more space in landfills.*
>
> *1000 plastic bags equal 17 lbs and 1219 cubic inches*
>
> *1000 paper bags equal 122 lbs and 8085 cubic inches*[3] [See Figure 9.11.]

A second letter to the editor, based on information such as that just cited, provides the perspective of a consumer and an employee.

> *As an employee of Wegmans Food markets you may determine that my opinion is biased and it certainly is . . . Clearly the plastic bags take up less space than paper . . . In our backroom of the store, an entire pallet of paper bags takes up as much space as the plastic, however, there are only approximately 10,000 paper bags on a pallet compared to approximately 30,000 plastic sacks . . . [T]he cost certainly is a benefit. Plastic bags cost 1.5 cents whereas paper ones cost 3 cents. If all customers would insist on plastic bags, the savings would certainly be passed on to the consumer.*
>
> *Finally, I would like to address the landfill issue. As mentioned on our plastic bags, the use of paper sacks fills landfills much faster than the use of plastic bags. Plastic bags take up considerably less space than paper . . . While environmental groups claim that paper bags degrade at a rapid rate, they are simply misleading the public.*[4]

There is much in these letters to engage the Sceptical Chymist, but we may not yet have enough information to pass critical judgment on the many complex issues involved. Indeed, we may not achieve such knowledge and wisdom within the limitations of this chapter. Nevertheless, in the next section we press on in our efforts to become better informed.

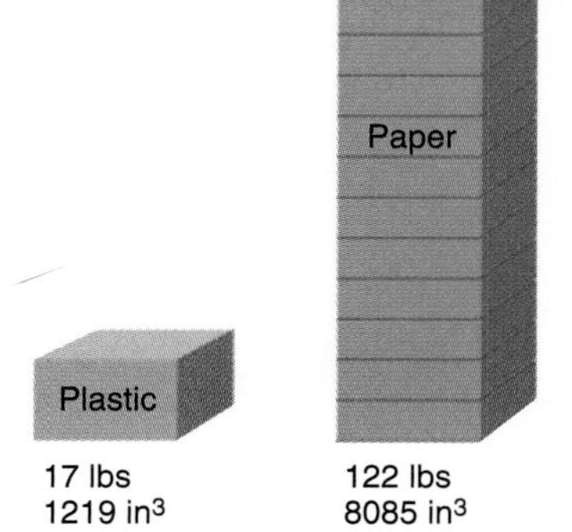

Figure 9.11
A scaled representation of volumes occupied by 1000 plastic bags and 1000 paper bags.

[2,3,4] *Syracuse Herald Journal,* Syracuse N.Y.

9.23 The Sceptical Chymist: Analyzing Letters to the Editor

Analyze the two letters with their opposing views. Pay particular attention to the initial assumptions, the evidence cited, the logic used, and the conclusions drawn.

a. Which makes the more compelling case and why?
b. On the basis of these two letters only, which position would you support?

9.24 Consider This: Plastic versus Paper—Where Do You Stand?

When you are in a supermarket, which do you usually request, plastic or paper grocery bags? List the advantages and disadvantages of plastic versus paper bags, and decide which is preferable. (Your preference may change depending on the products to be bagged.) Then prepare a letter to the editor of your local newspaper clearly stating your position and explaining your reasons for it.

9.9 Dealing with Plastics

Every year, about 100 billion pounds of plastic are produced in the United States—a bit more than 350 pounds for every woman, man, and child. Most of this ultimately finds its way into landfills. Given this huge quantity, there is little consolation in the fact that landfills contain considerably more paper than plastic. The reduction of the amount of plastic going into landfills remains a high priority. Five strategies suggest themselves: *incineration, biodegradation, reuse, recycling*, and *source reduction*. In the paragraphs that follow, we will examine each of these approaches and attempt to weigh the relative merits.

Because the Big Six and most other polymers are composed primarily of carbon and hydrogen, *incineration* is an excellent way to dispose of used plastics. Indeed, a recent study in Germany led to the conclusion that burning waste plastic does less damage to the environment than any other method of disposal. The chief products of combustion are carbon dioxide, water, and a good deal of energy. In fact, pound for pound, plastics have a higher energy content than coal. Plastics account for about 7% of the weight of municipal solid waste, but approximately 30% of its energy content. The German study found that the greater the percentage of plastic in the refuse burned in a garbage incinerator, the more efficient the burning, the greater the quantity of energy released, and the lower the emission of airborne pollutants. By contrast, recycling polymers requires energy, and if the waste plastic is dirty or of low quality, more energy is needed to recycle it than to produce a comparable quantity of new virgin plastic. It has been estimated that incineration can decrease the volume of plastic headed for landfills by as much as 90%.

But incineration of plastics is not without some drawbacks. The repeated message of Chapters 1–4, that burning does not destroy matter, applies here as well. Effluent gases produced by combustion may be "out of sight," but they had best not be "out of mind." Burning plastics produces CO_2, which potentially contributes to global warming. Of special concern in incineration are chlorine-containing polymers such as polyvinyl chloride, which release hydrogen chloride during combustion. Because HCl dissolves in water to form hydrochloric acid, such smokestack exhaust could make a serious contribution to acid rain. Moreover, some plastic products have inks containing heavy metals such as lead and cadmium. These toxic elements concentrate in the ash left after incineration and thus contribute to a secondary disposal problem. Burning plastics also converts the carbon they contain into carbon dioxide, just as if petroleum, the raw material from which plastics are made, had been burned directly. Thus, burning plastics converts their carbon into a by-product that is not easily converted directly into other useful materials. On balance, however, if carefully monitored and controlled, incineration can lead to a large reduction in plastic waste, generate much-needed energy, and have little negative impact on the environment.

9.25 Your Turn

When polypropylene burns completely, it produces just carbon dioxide and water. Write a balanced chemical equation for the combustion of polypropylene. Assume an average chain length of 2500 monomer units.

Hint: Check the molecular formula of propylene in Table 9.1.

Another strategy for disposing of plastic wastes is to enlist bacteria to do the job—in other words, to employ *biodegradation*. The problem is that bacteria and fungi do not find most plastics very appetizing. These microorganisms lack the enzymes necessary to break down plastics. However, because bacteria and fungi evolved in our natural environment, they possess enzymes to break down naturally occurring polymers into simpler molecules. Indeed, many strains of bacteria use cellulose from plants or proteins from plants and animals as their primary energy sources. You have already encountered several instances of such processes in this text. In Chapter 3 you read about the release of methane by belching cattle. Actually, the methane is produced when bacteria decompose cellulose in the cow's rumen. In the same chapter, we also mentioned that methane is generated by natural decomposition of organic material in landfills.

Scientists are now engineering biodegradability into some synthetic polymers. Certain bonds or groups are introduced into the molecules to make them susceptible to fungal or bacterial attack, or to decomposition by moisture. Recently, research scientists at DuPont have developed a biodegradable polymer called Biomax, which decomposes in about eight weeks in a landfill. This new polymer is a close chemical relative of PET. Biomax uses other monomers in conjunction with those conventionally used to prepare PET (ethylene glycol and terephthalic acid). When polymerized, these co-monomers create sites in the polymer chains that are susceptible to degradation by water. Once the moisture does its job of breaking the polymer into smaller chains, naturally occurring microorganisms feed off the smaller chains, converting them to CO_2 and water. Biomax could be used in a variety of applications such as lawn bags, bottles, liners of disposable diapers, disposable eating utensils, and cups.

9.26 Consider This: Will Biomax Work?

a. Identify some appropriate applications of Biomax, other than those just given.
b. Identify some applications of Biomax that would be inappropriate.

Achieving biodegradability in polymers raises some concerns, as an EPA report cautions.

> *Before the application of these technologies can be promoted, the uncertainties surrounding degradable plastics must be addressed. First, the effect of different environmental settings on the performance (e.g., degradation rate) of degradables is not well understood. Second, the environmental products or residues of degrading plastics and the environmental impact of degradables on plastic recycling is unclear.*[5]

Part of the difficulty is that even natural polymers do not decompose as completely in landfills as was suggested earlier in this chapter. Modern waste disposal facilities are covered and lined to prevent leaching of waste and waste by-products into the surrounding ground. Landfill linings and coverings create anaerobic (oxygen-free) conditions that impede bacterial and fungal action. As a result, many supposedly biodegradable substances decompose slowly or not at all when buried. Excavation of old landfills have found 37-year-old newspapers that are still readable and five-year-old hot dogs that, while hardly edible, are at least recognizable (Figure 9.12).

9.27 Consider This: Landfill Liners

A Web search for "landfills" can bring up high quality information about garbage. If you browse through these sites, you will see that a fair amount of controversy surrounds the plastic materials that can be used to construct the liner. Search for "landfill liners" and check out the plastics involved.

a. Which polymers are now used for liners?
b. What are their drawbacks?
c. Are there new polymers that offer more desirable properties? Cite the author and the URL of the web sites.

---

[5]From *EPA Report to Congress: Methods to Manage and Control Plastic Wastes*, February, 1990.

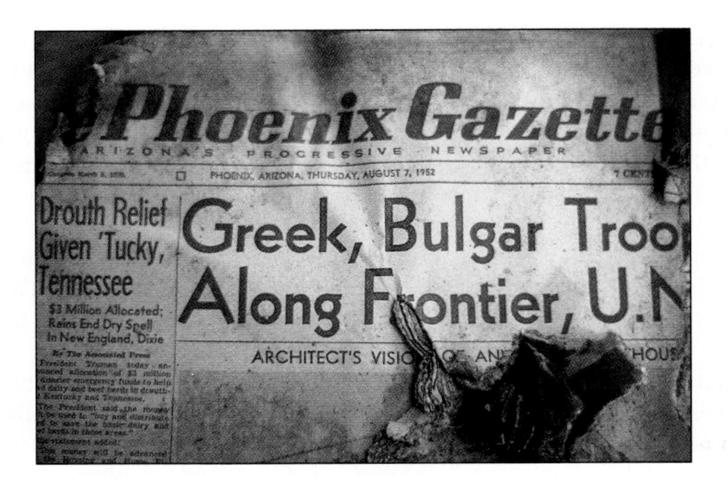

Figure 9.12
Some buried wastes can remain intact for a long time.

Reusing a plastic product is one way to divert it from a landfill. Although not all plastics are directly reusable, many are. Plastic bottles can be reused by cleaning them and filling them with the same substance (milk, water, shampoo, etc.), or used in other ways. A recent study reported that 80% of Americans reuse plastic products, such as food storage containers and refillable bottles.

In the United States, nearly 50% of certain plastic parts from damaged or discarded cars are repaired and reused. In a bold move to foster reuse on a large scale, new cars in Germany must be designed and built so that when the life of the car is over, its plastic parts can be removed easily and used to build other automobiles, thus creating automobiles that renew themselves. Through reusing rather discarding, far less petroleum is needed to make new plastics in the first place. The savings in this case can be substantial considering that the use of polymers in automotive applications has risen from an average of approximately 60 pounds per vehicle in 1970 to more than 360 pounds today. Estimates show that in five years, the average plastics content per modern vehicle will be nearly 200 kg (440 lb).

9.28 Consider This: Plastics in an Automobile

Use the Web engine to find the kinds of plastics in an automobile and the uses of them. In particular, find this information for

a. the exterior of the car
b. the interior of the car
c. the fuel system
d. the engine

Given the problems associated with landfill disposal of natural and synthetic polymers, attention has logically turned to *recycling* both. Although recycling plastics does not literally dispose of them as does incineration or biodegradation, it helps to reduce the amount of new plastic entering the waste stream. Nearly 20,000 communities, representing 78% of the nation's population, provide either curbside or drop-off collection of plastics. Through these efforts, more than one billion pounds of plastic packaging, about 2% of the total produced in the United States, were recycled. Although this represents a 12% increase over 1992, this country still lags far behind Germany and other developed nations in the percentage of plastics recycled. Germany has a particularly aggressive program, that, in 1994, resulted in recycling 52% of the plastics used in packaging. That same year, the Germans collected 10.9 billion pounds of packaging materials, including glass, paper, and tin cans, and recycled 10.1 billion pounds of it.

- 200 million milk jugs joined end-to-end would form a chain long enough to reach nearly twice around the Earth.

In the United States, more than 200 million milk and water jugs have been recycled to convert their high-density polyethylene into a fiber. The fiber is then made directly into TYVEK, a material used in such broad applications as sports clothing, durable mailing envelopes, and insulating wrap for new buildings.

9.29 Consider This: How Effective Is Recycling?

Go to the American Plastics Council web site to find the latest about plastics recycling.

a. What plastics make up the bulk of recycled plastics?
b. What percent of PET (PETE) bottles are now being recycled?
c. The American Plastics Council has initiated an All Plastic Bottle Collection Program. Use information from this program to list the major opportunities and the major obstacles present for a recycling program that collects all types of plastic bottles, not just those made of PET and HDPE.

Increasing quantities of "post-consumer" plastics are being used in the United States. The demand for recycled "Big Six" polymers grew by 32% between 1988 and 1993, and it is expected to grow. In 1998, about 24% of all plastic bottles made in the United States were recycled (more than 1.45 billion pounds). Polyethylene terephthalate soft drink bottles are particularly easy to melt and reuse. About 710 million pounds of PET plastic containers were recycled in the United States in 1998. Much of recycled PET was converted into polyester fabrics, including carpeting, T-shirts, the popular "fleece" used for jackets and pullovers, and the fabric uppers in jogging shoes (Figure 9.13). Five recycled 2-liter bottles can be converted into a T-shirt or the insulation for a ski jacket; it takes just about 450 such bottles to make polyester carpeting for a 9×12 foot room.

The chemist Nathaniel Wyeth used his creativity to develop the plastic soda bottle in contrast with his artist brother Andrew Wyeth, who expresses his creativity on canvas. Nathaniel Wyeth had an even bigger vision than T-shirts or rugs from recycled PET. In *ChemMatters* magazine (October 1994) he said: "One of my dreams is that we're going to be able to melt the returned bottles down, mix them with reinforcing fibers, and make car bodies out of them. Then, once the car has served its purpose, rather than put it in the junk pile, melt the car down and make bottles out of it."

Although such technology is possible, its use depends on economic factors as well. These factors have helped increase the demand for recycled polyethylene terephthalate and other polymers. Cleaned, recycled PET sells for 28–30 cents per pound (1998 prices). By contrast, virgin PET sells for about \$1.50 per pound. Not surprisingly, the laws of supply and demand work here as they do throughout the economy. When virgin PET prices rise, recycled PET suppliers raise their prices in response to the market.

Another major recycling initiative involves national supermarket chains recycling their HDPE grocery bags. In fact, if you read the labels on plastic materials, you will increasingly find them made of a mixture of virgin and recycled (post-consumer) plastics. Some of the use of post-consumer plastics is now mandated by law. For example, since 1995, all HDPE packaging used in California has been required to contain 25% recycled material.

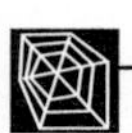

9.30 Consider This: Polymers in Your Computer

Twenty years ago, recycling personal computers was not a concern because there weren't enough of them around to matter. These days, given the number of monitors, keyboards, and "mice" in circulation, there is good reason to keep these out of the landfill.

a. What polymers are used in making your computer?
b. Is it possible to recycle the polymers in your computer? Use the Web to find out.
c. What is being done currently to recycle plastics from computers? You might want to search for "computers" and "recycling" to get started. Which plastic-containing computer supplies and accessories should have recycling programs?

Figure 9.13
T-shirts and activewear made from recycled PET.

However, simply collecting plastics to be recycled is not enough. For recycling to be successful and self-sustaining, a number of factors must be coordinated. They involve not only science and technology, but economics and sometimes politics as well. True

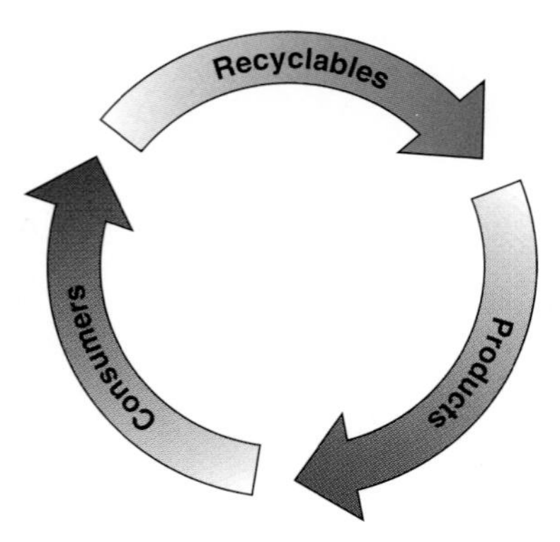

Figure 9.14

True recycling involves a never-ending loop.

Reprinted by permission of NAPCOR (National Association for PET Container Resourses).

• Density is discussed in Section 5.8.

recycling involves a closed loop (Figure 9.14) in which plastics are collected and sorted, then converted into products that consumers buy, use, and later recycle. First of all, there must be a dependable *supply* of used plastic, consistently available at designated locations. This creates the formidable task of *collecting* the discarded objects, along with the associated job of *sorting and separating* the various polymers. The codes that appear on plastic objects (Table 9.1) are provided to help facilitate this process. But, because of the large volume of material to be sorted, automated sorting methods are being developed. The plastics industry has spent more than $1 billion nationally on recycling research and developing environmentally responsible and sustainable plastic recycling programs.

9.31 Your Turn

Plastics vary in density. When placed in a liquid, a substance, such as a plastic, floats if its density is less than that of the liquid, and the substance sinks if it is more dense than the liquid.

The densities (g/mL) of four common plastics are: high-density polyethylene 0.95–0.97; polyethylene terephthalate 1.38–1.39; polypropylene 0.90–0.91; and polyvinyl chloride 1.18–1.65. The six liquids and their densities are

Liquid	Density (g/mL)
Methanol	0.79
An ethanol/water mixture	0.92
A different ethanol/water mixture	0.94
Water	1.0
Saturated solution of $MgCl_2$	1.34
Saturated solution of $ZnCl_2$	2.01

Using the densities of the liquids and of the plastics given, develop a procedure by which the individual plastics can be separated from a mixture containing pieces of all four plastics.

• Automated sorting methods are needed to accommodate the huge volume of plastic waste, especially in urban areas. An estimated 40 million plastic bottles are discarded *daily* in metropolitan New York City.

Once the entropy and disorder have been overcome by sorting, the *reprocessing* is relatively simple. Almost any polymer that is not extensively cross-linked can be melted. If the supply of waste is homogeneous, that is, if it contains only one type of plastic, the molten polymer can be used directly in the *manufacturing* of new products. Alternatively, it can be solidified, pelletized, and stored for future use. However, when mixtures of various polymers are melted, the product tends to be darkly colored with varying properties, depending on the nature of the mixture. Although this reprocessed material does not have outstanding working properties, it is good enough for general lower grade uses such as parking lot bumpers, disposable plastic flower pots, and cheap plastic lumber. Such mixed material is obviously not as valuable as the pure, homogeneous recycled polymer. Hence, the importance of sorting plastics. For similar reasons, manufacturers prefer to use only a single polymer in a product to avoid having to separate the various polymers.

A variant on this method of reprocessing plastics is to decompose the polymers into simpler molecules, in some cases back to the actual original monomers. Fanciful as Nathaniel Wyeth's bottles-to-cars-to-bottles idea might seem, it is now a reality. DuPont chemists won a Presidential Green Chemistry Challenge Award for improving on Wyeth's dream. They developed a proprietary process for treating post-consumer PET that, like pulling beads apart from a necklace, unlocks (depolymerizes) the polymer back to its original monomers. These can then be reused to make new PET for other products. Mary Johnson, a DuPont employee, says "Because these monomers retain their original properties, they can be reused over and over again in any first-quality application. A popcorn bag can become an overhead transparency, then a polyester peanut butter jar, then a snack food wrapper, then a roll of polyester film, then a popcorn bag again."

• The patented DuPont process is called Petretec Polyester Regeneration Technology.

The essential final step in recycling plastics is *marketing*. But a company or a city would be well advised to determine or create the demand for recycled polymers before completing all the other steps. Without a product and buyers, recycling programs are doomed to fail. In fact, recycling laws in a number of cities have not been implemented and enforced because one of the links in this polymeric chain of supply, collecting, sorting, processing, manufacturing, and marketing is missing. Without all these, the system will not work, unless it is heavily subsidized. Most municipalities have been unwilling to provide the necessary funds. As this fourth edition of *Chemistry in Context* went to press, the market for post-consumer plastics was strong and prices were high enough to justify significant recycling activity.

9.32 Consider This: Responsible Consumerism

Unless consumers buy products made from recycled plastics, there will be little financial incentive to produce such products, and plastics recycling will fail. Use the Web to find at least half a dozen products from recycled plastics, other than those mentioned in the textbook. Choose products that show the range of types of plastics and their use. Identify the plastic(s) used in each product.

The remaining option for dealing with plastics, *source reduction*—less waste in the first place—appears to be simplest and most direct. Simply decrease the quantity of plastics produced and used. The advantages are many—resources would be conserved, pollution would be reduced, and potentially toxic materials would be minimized. Examples of source reduction are the improvements made in plastic technologies that have reduced the amounts of plastic needed to make high-volume products; the 2-liter soda bottle now uses 25% less plastic than when it was introduced in 1975, and the 1-gallon milk jug weighs 30% less than a decade ago.

To reduce the amounts of materials, solvents, and labor, engineers at Chrysler Corporation have developed the Plymouth Pronto Spyder (Figure 9.15). What sets the Plymouth Pronto Spyder apart is that its body is made entirely of low-cost PET. This innovation could cut production expenses enough for the Spyder to cost only half the price of a regular car. Using just six molded PET body pieces that can be glued together eliminates using the usual 80 or so pieces and does away with expensive surface painting, stamping, and forming operations. Paint is molded directly into the PET, coloring the plastic all the way through, thus eliminating conventional surface painting, a costly and time-consuming step in auto production.

The Baja is a new lightweight, all-terrain two-door SUV, the world's second all-plastic body and chassis vehicle, featuring thermoformed body panels and a composite chassis (Figure 9.16). It is intended for off-road recreational use in the West Indies, Central and South America, and the Pacific Rim countries, where its composite body

Figure 9.15
The Plymouth Pronto Spyder.

Figure 9.16
The Baja SUV.

and chassis resist the sand and seawater. The lightweight plastics make the Baja a fuel-efficient vehicle and also eliminate the need to paint, saving hundreds of thousands of dollars in production costs.

Yet source reduction, a seemingly innocuous option, is far more complicated than it appears. The problem is that something else is generally used to replace the plastic, and this substitution can be fraught with hidden pitfalls. In making choices between alternative materials, the decisions must be informed by the source and nature of chemical feedstocks (reactants), the method of manufacturing, waste products produced during manufacturing and their disposal, and many other factors. Energy costs as well as economic costs must be taken into account. How much energy must be expended in the entire life cycle of a product from raw material to final disposal?

Obviously, the identification and proper weighing of all of the possible variables is a complex and difficult task. But, when the job is done properly, one sometimes discovers that attempts to reduce the amount of plastic waste by substitution for the plastic may actually increase the overall amount of waste and the associated negative environmental impact. For example, two pounds of plastic are enough to make containers able to hold about eight gallons of juice, soft drinks, or water. To hold that same amount of beverage would require three pounds of aluminum, eight pounds of steel, or 27 pounds of glass.

As another, more detailed example, consider the replacement of a plastic cup with one made of paper. Each occupies about the same volume in a landfill, where both will probably remain undecomposed for a long time. The "conventional wisdom" of public opinion has it that paper cups are more environmentally friendly than styrofoam ones. But a detailed analysis of the issue made by Martin Hocking of the University of British Columbia counters that position (*Science* Vol 251, 1 February, 1991, pp 504–50). Hocking did a cradle-to-grave type of life cycle analysis in which he considered all aspects of the production and disposal of the two types of cups (Table 9.2). Hocking's conclusion is that paper cups are not as environmentally friendly as commonly thought. The paper cups require more raw material and consume as much petroleum as that used to produce the styrofoam cups. The harsh nature of the chemicals used, the large volume of water required, and the nature of effluents generated into the air and water in paper making are far greater than those affiliated with producing polystyrene cups. Both types of cups are not very biodegradable in sealed landfills, so paper offers no significant advantage there. The styrofoam cups are easier to reuse and recycle, and about as easy to incinerate as paper ones.

9.33 Consider This: Reuse and Alternatives

To complete 9.2 Consider This, you kept a journal of all the plastic and plastic-coated products you discarded in a week. Review that list and indicate ways you could reuse those discarded products or suggest alternatives for the plastic in the product.

Of course, the best method of source reduction is not to replace plastics, but do without them or their substitutes whenever possible. One correspondent in the great plastic versus paper battle said it well:

> *There is a danger in this grocery bag controversy of losing sight of issues of greater importance. One of these is the matter of legitimate, responsible use of resources. Plastics are made from one of the most precious resources, one which cannot be renewed or replaced. In many respects we should regard it as more precious than gold or diamond. There are products essential to human health and well-being which can be made only from petroleum. There are also non-essential, wasteful uses of this priceless commodity. Where did we get the idea that it is our right to waste millions of barrels of oil each year exceeding the speed limit? Who said we're justified in manufacturing and using plastic items like shopping bags, burger boxes, and disposable*

Table 9.2

Paper versus Styrofoam Cups

Item	Paper Cup	Plastic Cup
Per cup		
Raw materials		
Wood and bark, g	33	0
Petroleum, g	4.1	3.2
Finished mass, g	10	11.5
Cost	2.5 times that of plastic	1
Per million grams of material		
Utilities		
Steam, kg	9,000–12,000	5,000
Power, 10^9 J	3.5	0.4–0.6
Cooling water, m^3	50	154
Water effluent		
Volume, m^3	50–190	0.5–2
Suspended solids, kg	35–60	Trace
Air emissions, kg	7–22	35–50
Recycling potential		
Primary user	Possible	Easy
After use	Low	High
Ultimate disposal		
Heat recovery (million J/kg)	20	40
Mass in landfill, g	10.1	1.5
Biodegradable	Slowly, if at all	No

diapers which are instant garbage? How did we get hooked on the consumer habits that are destroying not only a level of comfort we take for granted, but the very air and water we need to survive?[6]

There is no single, best solution to the problems posed by plastic waste and solid waste in general. Incineration, biodegradation, reuse, recycling, and source reduction all provide benefits and all have associated costs. Therefore, it is likely that the most effective response will be the development of an integrated waste management system that will employ all four of these strategies. The goal of such an integrated system would be to match the methods to the composition of the waste stream, thus optimizing efficiency, conserving energy and material, and minimizing cost and environmental damage.

CONCLUSION

The letter quoted in the last section goes well beyond a choice of plastic or paper shopping bags. Once more we come to an issue of lifestyle. Over the past nearly 70 years, chemists have created an amazing array of polymers and plastics—new materials that have made our lives more comfortable and more convenient. Many of these plastics represent a significant improvement over the natural polymers they replace. Furthermore, many products we take for granted today would be impossible without synthetic polymers and plastics. There would be no CDs or DVDs, no cell phones, no videotape, no kidney dialysis apparatus, and no heart-lung machines or artificial hearts. We

[6] *Syracuse Herald Journal,* Syracuse, N.Y.

have become dependent on plastics, and it would be difficult if not impossible to abandon their use. The chemical industry has given consumers what they want. But there now appears to be rather more of it than we would like or perhaps can deal with responsibly—mountains of soft drink bottles and miles of plastic bags. We must learn to cope with this glut of stuff while saving matter and energy for tomorrow. To create a new world of plastics and polymers will require the intelligence and efforts of policy planners, legislators, economists, manufacturers, consumers, and above all, chemists.

Chapter Summary

Having studied this chapter, you should be able to:

- Understand the nature of plastics and polymers, their typical properties and molecular structures (9.1)
- Describe typical uses for the "Big Six" polymers (9.2)
- Understand the molecular mechanism of addition polymerization (9.3) and condensation polymerization (9.5)
- Recognize the chemical composition and molecular structure of the "Big Six" polymers:

  Low-density polyethylene (LDPE) and high-density polyethylene (HDPE) (9.4)
  Polyvinyl chloride (PVC) (9.3)
  Polystyrene (PS) (9.3)
  Polypropylene (PP) (9.3)
  Polyethylene terephthalate (PET) (9.5)
- Tell how amino acids and proteins are related chemically (9.6)
- Use structural formulas to write the chemical equation for the synthesis of nylon (9.6)
- Identify sources of materials for manufacturing plastics (9.7)
- Compare the relative costs and benefits of plastic and paper grocery bags (9.8)
- Relate the technical, economic, and political issues in methods for disposing of waste plastic: incineration, biodegradation, reuse, recycling, and source reduction (9.9)

Questions

Emphasizing Essentials

1. This chapter deals with plastics and polymers. Do these terms mean the same thing?
2. Why are polymers sometimes referred to as macromolecules?
3. Some people object to synthetic polymers, saying that polymers are not "natural." Give examples of two natural polymers.
4. Survey 10 friends and ask what descriptive adjectives they associate with the noun "plastic." Share your responses with others in the class to see if others have found the same associations. Do your friends consider plastics to be "cheap" or "tacky"?
5. What is the major reason plastics are substituted for metal in automobiles?
6. Consider these data.

Year	U.S. population, millions	Billions of pounds of plastics produced in U.S.
1977	220	34
1997	269	89
1999	280	100

   a. How many pounds of plastic were produced per person in the United States in 1999?
   b. How many pounds of plastic were produced per person in the United States in 1977?
   c. What is the percent change in the total number of pounds of plastic produced per year between 1977 and 1999?
   d. What is the percent change in the number of pounds of plastic produced per person per year between 1977 and 1999?
7. Do you expect the heat of combustion of polyethylene, as reported in kJ/g, to be more similar to that of hydrogen, coal, or octane, C_8H_{18}? Explain your prediction.
8. Equations 9.1 and 9.2 show the polymerization of two ethylene monomers to form a small segment of polyethylene. Use the bond energies of Table 4.1 to calculate the energy change during the reaction in equation 9.1. Is the reaction endothermic or exothermic?
9. How will your result from question 8 differ if, rather than using ethylene as the monomer, tetrafluoroethylene is used as the monomer, forming a small segment of the Teflon polymer, polytetrafluoroethylene? This is the structure of the monomer.

$$F_2C{=}CF_2$$

10. a. Determine the number of CH_2CH_2 monomeric units, n, in one molecule of polyethylene with a molar mass of 40,000 g.

b. What is the total number of carbon atoms in this polyethylene?

11. Explain the role of a free radical in the polymerization of ethylene.

12. This is a representation of a small segment of polyethylene. The hydrogen atoms are omitted for the sake of clarity.

—C—C—C—C—

Does this representation predict that the carbon-to-carbon-to-carbon bond angles in polyethylene are all 180°? Explain your reasoning.

13. This is the usual two-dimensional representation for the formation of polyvinyl chloride polymer from vinyl chloride. As this process takes place at the molecular level, how does the approximate Cl—C—H bond angle change?

$$n\ H_2C{=}CHCl \longrightarrow \left[\!-CH_2-CHCl-\!\right]_n$$

14. Describe how each of these strategies would be expected to affect the properties of polyethylene and give an atomic/molecular level explanation for each effect.

a. increasing the length of the polymer chain

b. aligning the polymer chains with one another

c. increasing the degree of branching in the polymer chain

15. Both bottles are made of polyethylene. How do the two bottles differ at the molecular level?

16. The manufacturers of some plastic household containers say that it is fine to place the item in the dishwasher, particularly on the top shelf. Others do not claim that their products are dishwasher safe.

a. Why is it recommended that dishwasher-safe plastics be put only on the top shelf?

b. Assume that you have lost the information sheet that originally accompanied the dishwasher in your kitchen. Also, assume that you want to use the dishwasher as much as possible, rather than washing plastic containers by hand. What general properties of the plastic containers should help to guide you in avoiding problems in the dishwasher?

17. Do all the Big Six have a common structural feature (see Table 9.1)? If so, identify that feature. If not, identify which of the six do share a common structural feature.

18. Table 9.1 lists features of the "Big Six" polymers.

a. Which of the "Big Six" are the most flexible?

b. Which of the "Big Six" do not have crystalline regions?

c. Which of the "Big Six" are soluble in organic materials?

d. Which is the most costly of the "Big Six"?

19. **a.** The structure for the styrene monomer is given in Table 9.1. Rewrite this shorthand structure showing all of the atoms.

b. What is the molecular formula for styrene?

c. What is the molar mass of a polystyrene molecule consisting of 5000 monomer units?

20. Vinyl chloride monomers can join in several different orientations to form polyvinyl chloride. Several different arrangements are shown in Figure 9.2. Which of those arrangements is shown here?

—C(Cl)(H)—C(H)(H)—C(H)(H)—C(Cl)(H)—C(Cl)(H)—C(H)(H)—

21. Butadiene, $H_2C{=}CH{-}CH{=}CH_2$, is polymerized to make buna rubber. Write an equation representing this process. Is this an example of addition or condensation polymerization?

22. The Dow Chemical Company has developed a new process that uses CO_2 as the blowing or foaming agent to produce Styrofoam packaging material. What compound does CO_2 likely replace in the process, and why is this substitution environmentally beneficial?

23. Kevlar is a type of nylon called an *aramid* that contains aromatic rings. Because of its great mechanical strength, Kevlar is used in radial tires. The two monomers for producing Kevlar are these.

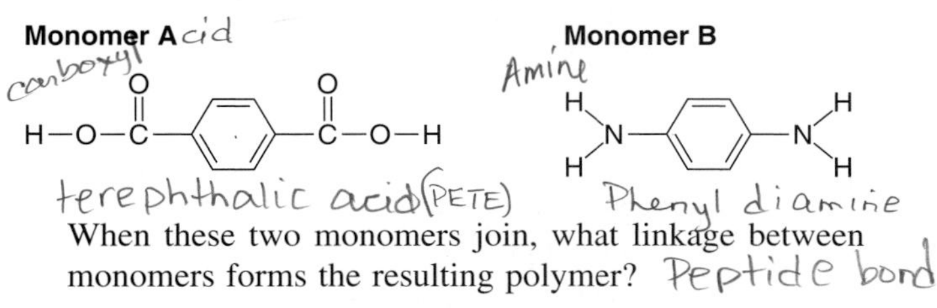

When these two monomers join, what linkage between monomers forms the resulting polymer?

24. Polyacrylonitrile is a polymer made from the monomer acrylonitrile, CH_2CHCN.

a. Draw a Lewis structure of this monomer.

b. Polyacrylonitrile is used in making Acrilan fibers

used widely in rugs and upholstery fabric. What danger do rugs or upholstery made of this polymer create in the case of house fires?

25. Section 9.7 has this heading: "Plastics: Where From and Where To?" Answer both questions posed in this heading, concentrating on the major source and major means of disposal.

Concentrating on Concepts

26. What are the relationships among these terms?
Natural, synthetic, polymers, nylon, protein
Show these relationships by writing an outline with enough information to show how the terms are related.

27. You were asked in 9.2 Consider This to keep a journal of all the plastic and plastic-coated products you *throw away* in one week. Now consider all of the plastic items that you *recycle* in one week. Are there any from your first list of items thrown away that could be on your second list? Why or why not?

28. Celluloid was the first commercial plastic, developed in response to the need to replace ivory for billiard balls and piano keys.
    a. Speculate on what specific properties were required for celluloid that allowed it to be a suitable substitute for ivory in these products.
    b. Bakelite was an early plastic. Was it also developed in response to a specific need? Explain your reasoning.

29. This graph shows U.S. production of plastics from 1977–1999.

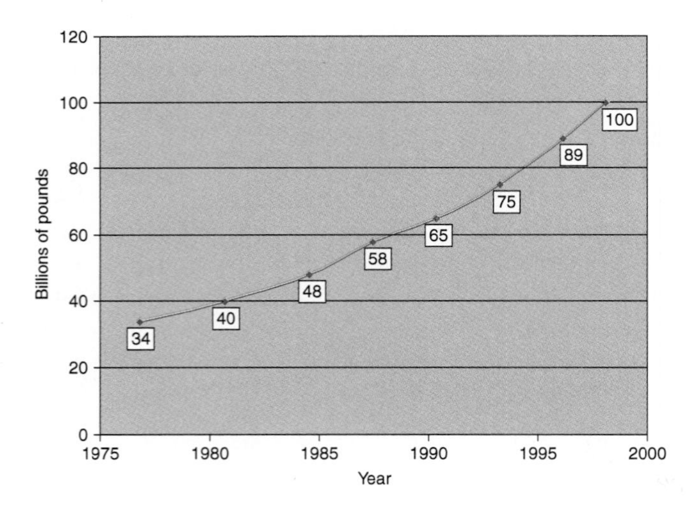

    a. Is there a uniformly spaced scale representing billions of pounds of plastics along the y-axis? If so, what does each division represent?
    b. Is there a uniformly spaced scale of years along the x-axis? If so, what does each division represent?
    c. How many years were required for plastics production in 1977 to double?
    d. Redraw the graph as a bar graph showing the relationship of the year to the pounds of plastics produced. Discuss whether the bar graph is easier to use than the line graph to establish the doubling time.

30. The properties of a plastic are a consequence of more than just its chemical composition. What are some of the other features of polymer chains that have an influence on the properties of the polymer formed?

31. Consider the polymerization of a thousand ethylene monomers to form a large segment of polyethylene.

$$1000\ H_2C{=}CH_2 \rightarrow \text{+}CH_2CH_2\text{+}_{1000}$$

    a. Calculate the energy change during this reaction. (Use Table 4.1 of bond energies.)
    b. Should heat be supplied or must heat be removed from the polymerization vessel to carry out this reaction? Explain.

32. Catalysts are used to help control the average molar mass of polyethylene, an important strategy to control polymer chain length. During World War II, low-pressure polyethylene production used varying mixtures of triethyl aluminum, $Al(C_2H_5)_3$, and titanium tetrachloride, $TiCl_4$, as a catalyst. Here are some data showing how the molar ratio of the two components of the catalyst affects the average molar mass of the polymer produced.

Moles $Al(C_2H_5)_3$	Moles $TiCl_4$	Average Molar Mass of Polymer, g
12	1	272,000
6	1	292,000
3	1	298,000
1	1	284,000
0.63	1	160,000
0.53	1	40,000
0.50	1	21,000
0.20	1	31,000

    a. Prepare a graph to show how the molar mass of the polymer varies with the mole ratio of $Al(C_2H_5)_3/TiCl_4$.
    b. What conclusion can be drawn about the relationship between the molar mass of the polymer and the mole ratio of $Al(C_2H_5)_3/TiCl_4$?
    c. Use the graph to predict the molar mass of the polymer if an 8:1 ratio of $Al(C_2H_5)_3$ to $TiCl_4$ were used.
    d. What ratio of $Al(C_2H_5)_3$ to $TiCl_4$ would be used to produce a polymer with a molar mass of 200,000?
    e. Can this graph be used to predict the molar mass of a polymer if either pure $Al(C_2H_5)_3$ or pure $TiCl_4$ were used as the catalyst? Explain.

33. When you try to stretch a piece of plastic bag, the length of the piece of plastic being pulled increases dramatically and the thickness decreases. Does the same thing happen when you pull on a piece of paper? Why or why not? Explain on a molecular level.

34. Consider Spectra, Allied-Signal Corporation's HDPE fiber, used as liners for surgical gloves. Although the Spectra liner has a very high resistance to being cut, the polymer allows a surgeon to maintain a delicate sense of touch. The interesting thing is that Spectra is *linear* HDPE, which is usually associated with being rigid and not very flexible.

a. Suggest a reason why branched LDPE cannot be used in this application.

b. Offer a molecular-level reason for why linear HDPE is successful in this application.

35. One limitation of the "Big Six" is the relatively low temperatures at which they melt, 90–170°C (see Table 9.1). Suggest ways to raise the upper temperature limits while maintaining the other desirable properties of these substances.

36. All the "Big Six" polymers are insoluble in water, but some of them dissolve or at least soften in hydrocarbons or in chlorinated hydrocarbons (Table 9.1). Use your knowledge of molecular structure and solubility concepts to explain this behavior.

37. Vinyl chloride monomers can join in several different orientations to form polyvinyl chloride (see Figure 9.2). Do these two structures represent the same possible arrangement?

$$-CHCl-CH_2-CHCl-CH_2-CHCl-CH_2- \quad \text{and} \quad -CHCl-CH_2-CHCl-CH_2-CHCl-CH_2-$$

Explain your answer by identifying the orientation in each arrangement.

38. What structural features must a monomer possess to undergo addition polymerization? Explain, giving an example.

39. What structural features must a monomer possess to undergo condensation polymerization? Explain, giving an example.

40. Plastics are widely used in packaging. Check the recycling code on 10 containers to identify the plastic used for each of them (see Table 9.1). How many of these containers are made from addition polymers? From condensation polymers?

Exploring Extensions

41. "Buckminsterfullerene," a form of elemental carbon, is shaped somewhat like a soccer ball. Each carbon is located at a corner where two six-membered rings and one five-membered ring come together. Locate a structural drawing of buckminsterfullerene on the Web and find a corner of the drawing that fits this description.

42. Dr. Richard Smalley, Dr. Harry Kroto, and Dr. Robert Curl, Jr. won the 1996 Nobel Prize in chemistry. Research the connections among their work and write a short report to describe why they won the Nobel Prize.

43. What is the difference in the material used in "hard" and "soft" contact lenses? How do the differences in properties affect the ease of wearing of contact lenses?

44. Two terms that have been added to the vocabulary of plastics are "virgin plastic" and "post-consumer waste" plastics. What do these terms imply about the plastics being used?

45. Free radical peroxides promote the polymerization of ethylene into polyethylene. They also play a key role in tropospheric smog formation. Use the Web to learn more about how the peroxides promote ethylene polymerization and how peroxides are involved with photochemical smog formation in the troposphere. Write a brief report comparing the types of peroxides important to each of these cases. Give references for the Web information.

46. Synthetic rubber is usually formed through addition polymerization. An important exception is silicone rubber, which is made by the condensation polymerization of dimethylsilanediol. This is a representation of the reaction.

$$n\ HO-Si(CH_3)_2-OH \longrightarrow \left[-O-Si(CH_3)_2-O-\right]_n + n\,H_2O$$

a. Predict some of the properties of this polymer. Explain the basis for your predictions.

b. Silly Putty is a popular form of silicone rubber. What are some of the properties of Silly Putty?

47. Who first synthesized Kevlar? What was the background and academic training of these scientists? Was the potential for using this polymer in radial tires immediately understood? What are other applications of Kevlar? Write a short report on the results of your findings, giving references either to books or Web information.

48. This is the structural formula for Dacron, a condensation polyester:

$$\left[-O-CH_2-CH_2-O-\overset{O}{\overset{\|}{C}}-C_6H_4-\overset{O}{\overset{\|}{C}}-\right]_n$$

Dacron is formed by the reaction of an alcohol with an —OH group at each end of its molecule and a dicarboxylic acid, one with —COOH groups at opposite ends of the acid molecule. Write the structural formulas for the alcohol and the acid monomers used to produce Dacron.

49. Cotton, rubber, silk, and wool are natural polymers. Consult other sources to identify the monomer unit in each of these polymers; specify which are addition polymers and which are condensation polymers.

50. How does your college's or university's community dispose of plastics? Of all the strategies for disposing plastics described in this chapter, which are used? How are the alternatives presented to the people in the community? Find out the current practices in your community, and then offer some suggestions for improving them.

10

Manipulating Molecules and Designing Drugs

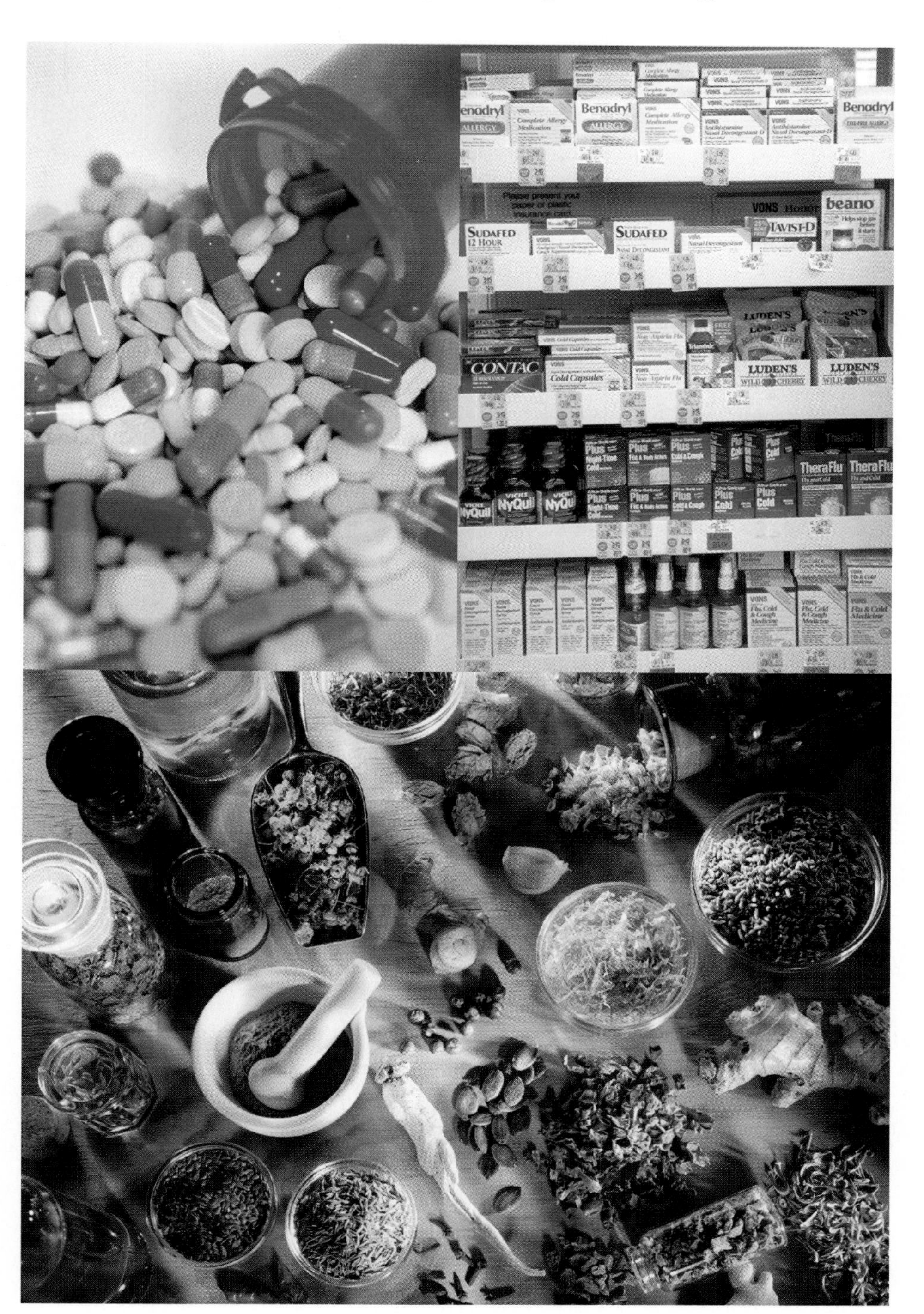

Medications range from prescription drugs . . .

. . . to generic medicines . . .

. . . to herbal alternatives.

Since ancient times, people by chance have discovered natural remedies that heal injuries and cure illnesses. More recently, chemists have designed, synthesized, and characterized a vast array of prescription and over-the-counter drugs in response to advances in medical knowledge. In discussing such drugs, we will consider these questions:

- From where do the ideas and resources to develop drugs come?
- What is the process by which pharmaceuticals make it to market?
- Why can't all medicines be generic and over-the-counter?
- What are the merits and pitfalls of herbal medicines?

CHAPTER OVERVIEW

Pharmaceuticals (drugs) are *substances that prevent, moderate, or cure illnesses.* We begin the chapter with a drug that has probably been used by all readers of this book. Its discovery and development is an introduction into how a new drug often comes into existence. All the drugs mentioned (indeed, most of the drugs used today) contain the element carbon. Therefore, we embark on a brief excursion into organic chemistry, the realm of carbon compounds. The principles governing the structure of organic molecules are those we have already applied to other molecules in previous chapters. The molecular structures of organic compounds offer the opportunity to explore new features such as isomers and functional groups. Both are of great importance in linking molecular structure to drug function. Examples of characteristic drug activity are provided by a group used to control pain, but a more general treatment of the topic introduces the concept of the fit between the drug and the biochemical site at which it acts. Sometimes, activity depends on chirality, the subtle property of three-dimensional structure that makes some molecules related like right and left hands.

Discussion then shifts to steroids, one of the most interesting and important families of biologically active compounds. Members of the steroid family that are treated in some depth are cholesterol, sex hormones, contraceptives, aborting agents, and anabolic steroids. These compounds, or at least their uses, are familiar to almost everyone because they have often been steeped in controversy. By setting the steroids in their social context we explore a number of these issues. We turn next to drug testing and approval, the lengthy and demanding process that is necessary to obtain consent to sell, distribute, and use a new drug. Here again, we will find that risks and benefits are present, whether using the original or generic forms of drugs or whether they are prescription or over-the-counter medications. The chapter ends with a section outlining some of the chemical issues of alternative medicines such as herbal remedies.

10.1 A Classic Wonder Drug

In the fourth century B.C., Hippocrates, perhaps the most famous physician of all time, described a "tea" made by boiling willow bark in water. The concoction was said to be effective against fevers. Over the centuries, that folk remedy, common to many different cultures, ultimately led to the synthesis of a true "wonder drug"—one that has aided millions of people.

One of the first systematic investigators of willow bark (Figure 10.1) was Edmund Stone, an English clergyman. His report to the Royal Society (1763) set the stage for a series of further chemical and medical investigations. Chemists were subsequently able to isolate small amounts of yellow, needle-shaped crystals of a substance from the willow bark extract. Because the tree species was *Salix alba*, this new substance was named *sali*cin. Experiments showed that salicin could be chemically separated into two compounds. Clinical tests provided evidence that only one of these components reduced

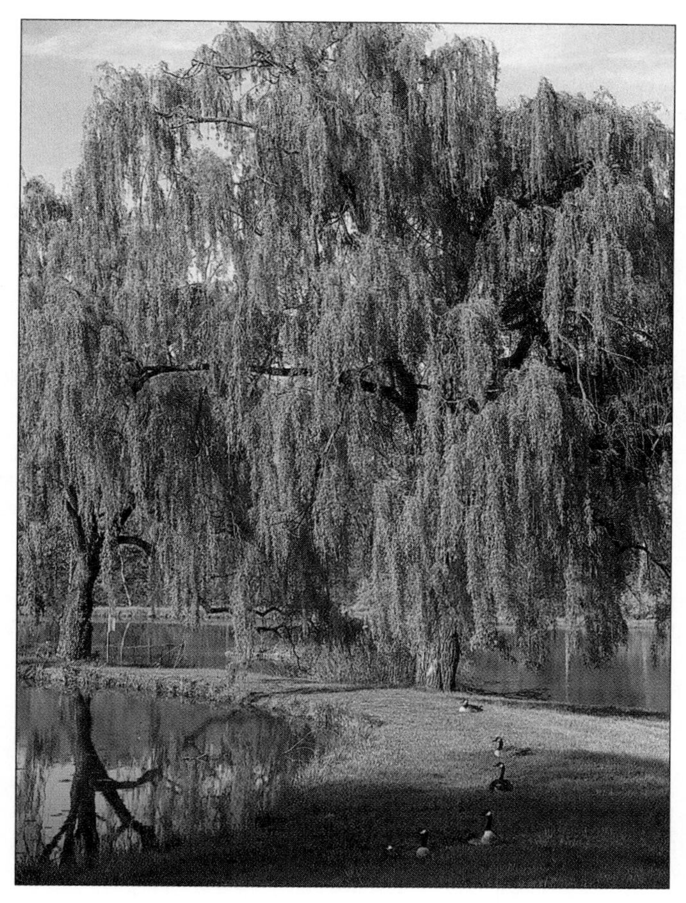

Figure 10.1
The willow tree *Salix alba*, source of a miracle drug.

fevers and inflammation. It was also demonstrated that the active component was converted to an acid in the body. Unfortunately, the clinical testing revealed some troubling side effects. The active component not only had a very unpleasant taste, but its acidity also led to acute stomach irritation in some individuals.

The active acid was used as a treatment for pain, fever, and inflammation. But because of its serious side effects, chemists set out to modify the structure of the active acid to form a related compound that still would be effective, but without the undesirable taste or stomach distress. The first modification attempt took a very simple approach. The acid was neutralized with a base, either sodium hydroxide or calcium hydroxide, to form a salt of the acid. It turned out that the resulting salts had fewer side effects than the parent compound. Based on this finding, chemists correctly concluded that the acidic part of the molecule was responsible for the undesirable properties. Consequently, the next step was to seek a structural modification that would lessen the acidity of the compound without destroying its medicinal effectiveness.

One of the chemists working on the problem was Felix Hoffmann, an employee of a major German chemical firm. Hoffmann's motivation was more than just scientific curiosity or assigned work. His father regularly took the acidic compound as treatment for arthritis. It worked, but he suffered nausea. The younger Hoffmann succeeded in converting the original compound into a different substance, a solid that reverted back to the active acid once it was in the body. This molecular modification greatly reduced nausea and other adverse reactions; a new drug had been discovered (1898).

Extensive hospital testing of Hoffmann's compound began along with simultaneous preparation for its large-scale manufacture by a well-known pharmaceutical company. The new drug itself could not be patented because it was already described in the chemical literature. However, the company hoped to recoup its investment by patenting the manufacturing process. Clinical trials showed the drug to be nonaddicting and relatively nontoxic. Its toxicity by ingestion is classed as low, but 20–30 g ingested at one time may be lethal. At the suggested dose of 325–650 mg (0.325–0.650 g) every four hours, it is a remarkably effective antipyretic (fever reducing), analgesic (anti-pain), and anti-inflammatory agent. Data from clinical tests uncovered the side effects noted in Table 10.1. The drug was also found to increase blood clotting time and to cause at least some small, almost always medically insignificant, amounts of stomach bleeding in about 70% of users.

10.1 Consider This: Miracle Drug

In the United States, the final step for approval of a drug is the submission of all of its clinical test results to the Food and Drug Administration (FDA) for a license to market the product.

a. If you were an FDA panel member presented with the information in Table 10.1, Side Effects of the "Wonder Drug," would you vote to approve this drug that treats pain, fever, and inflammation?
b. If approved, should this drug be released as an over-the-counter drug or would you restrict its availability by making it a prescription drug? Write a one-page report stating and defending your position.

Perhaps you have already guessed the identity of the miracle drug related to willow bark tea. Its chemical name 2-(acetyloxy)-benzoic acid or (more commonly) acetylsalicylic acid, may not help much. But the power of advertising is such that, had we revealed that the firm that originally marketed the drug was the Bayer division of I. G. Farben, we would have let the tablet out of the bottle. The compound in question is the world's most widely used drug. A century after its discovery, Americans annually consume nearly 80 billion tablets of this miracle medicine. You know it as aspirin.

Admittedly, we have compressed the time somewhat. Most of the development, testing, and design of aspirin occurred in the 18th and 19th centuries. Stone's letter to the Royal Society was written in 1763, and Felix Hoffmann's modification of salicylic acid to yield aspirin was done in 1898. Furthermore, the clinical testing of aspirin was somewhat less systematic than our account implies. But the basic facts and the steps that led

Table 10.1

Side Effects of the "Wonder Drug"

Symptoms	Frequency	Severity[a]
Drowsiness	rare	4
Rash, hives, itch	rare	3
Diminished vision	rare	3
Ringing in the ears	common	5
Nausea, vomiting, abdominal pain	common	2
Heartburn	common	4
Black or bloody vomit	rare	1
Black stool	rare	2
Blood in the urine	rare	1
Jaundice	rare	3
Anaphylaxis (severe allergic reaction)	rare	1
Unexplained fever	rare	2
Shortness of breath	rare	3

[a]The severity scale ranges from 1—life threatening, seek emergency treatment immediately to 5—continue the medication and tell the physician at the next visit.

Source: Data from H.W. Griffith, *The Complete Guide to Prescription and Non-Prescription Drugs*, 1983, HP Books, Tucson, Arizona.

to aspirin's full development are essentially correct. We must also add one more very important fact. Aspirin did not have to receive drug approval before being put on the market; no such certifying process was in place at that time. Had such approval based on clinical test results been necessary, it is quite likely that aspirin might only be available on a prescription basis.

10.2 Consider This: What Should a Drug Be Like?

Make a list of the properties you would like a drug to have. Then compare your list with those of your classmates. Note similarities and differences in the lists.

a. Are there items that are listed that should not be?
b. After further consideration, are there items that were missing from the original list that should be there?

10.2 Counting to Four: Introduction to Organic Chemistry

One of the properties of the element carbon is forming molecules, especially ones containing bonds linking multiple carbon atoms. As a result, the great majority of known chemical compounds contain carbon atoms. This element is so widely distributed throughout nature that the largest subdiscipline of chemistry, **organic chemistry,** is devoted to the study of carbon compounds. The name "organic" is historical and suggests a biological origin for the substances under investigation, but this is not necessarily true in every instance. In practice, most organic chemists confine themselves to compounds in which carbon is combined with a relatively small number of other elements—hydrogen, oxygen, nitrogen, sulfur, chlorine, phosphorus, and bromine. Even with this restriction, there are over 12 million organic compounds out of 14 million total known compounds. The chemical behavior (i.e., properties and reactivity) of organic molecules enables us to organize them into a relatively small number of categories. As a result, in this chapter we concentrate on only a few organic compounds and stress their important role within the functions of living things. The molecular structure of aspirin, an organic acid, gives it chemical and physiological properties, and so its molecular shape will prove to be of special significance.

To identify a specific organic compound from among the myriad of possibilities, the compound must be named. Chemists use a formal set of nomenclature rules established by an international committee so each of the 14 million compounds can be uniquely named. However, many of these compounds have been known for a long time by their common names, such as alcohol, sugar, and morphine. When a headache strikes, even chemists do not call out for 2-(acetyloxy)-benzoic acid; they simply say "Give me some aspirin!" Likewise, prescriptions specify penicillin-N rather than 6[(5-amino-5-carboxy-1-oxopentyl)amino]-3,3-dimethyl-7-oxopentyl-4-thia-1-azabicyclo[3.2.0] heptane-2-carboxylic acid. Mouthfuls like this are the cause of great merriment to those who like to satirize chemists, but they are important and unambiguous to those who know the system. You can rest easy because in this chapter we will use common names in almost all cases.

The incredible variety of organic compounds exists because of the remarkable ability of carbon atoms to bond in multiple ways with other carbon atoms or with atoms of other elements. To better understand such possibilities, we need a few basic rules for bonding in organic molecules. The most fundamental generalization is one you used as early as Chapter 2—the octet rule: *Each bonded carbon atom shares eight electrons, an octet.* A pair of shared electrons is a covalent bond, and each carbon atom's eight shared electrons give rise to four covalent bonds. The most common ways those four bonds to carbon occur in organic structures are as (a) four single bonds, (b) two single bonds and one double bond, (c) one single bond and one triple bond, or (d) two double bonds. These arrangements are illustrated in Figure 10.2. Other elements exhibit different bonding behavior. A hydrogen atom is always attached to a molecule by a single covalent bond. An oxygen atom in a molecule typically has two pairs of bonding electrons,

• The octet rule is discussed in Section 2.3.

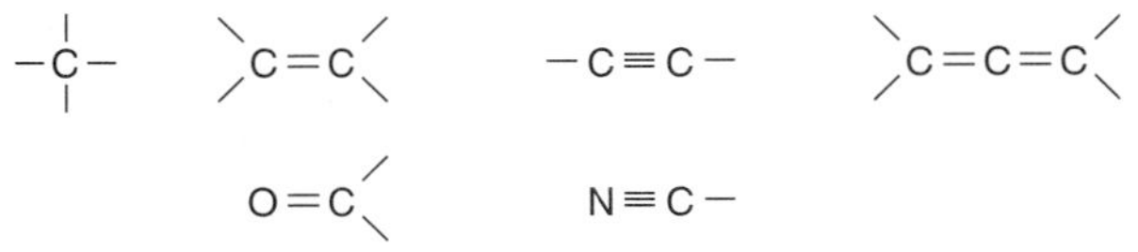

$-\overset{|}{\underset{|}{C}}-$ (a) $\quad >C=C<$, $O=C<$ (b) $\quad -C\equiv C-$, $N\equiv C-$ (c) $\quad >C=C=C<$ (d)

Figure 10.2
Representation of some carbon bonding possibilities.

either in the form of two single bonds or one double bond. A nitrogen atom shares in three pairs of bonding electrons and hence can form three single bonds, one triple bond, or one single and one double bond.

Molecular formulas, such as C_4H_{10}, indicate the kinds and numbers of atoms present in a molecule, but do not show how the atoms are arranged. To get that higher level of detail, structural formulas are used. These representations show the atoms and their arrangement with respect to each other in a molecule. In the case of C_4H_{10} (butane, a hydrocarbon used in cigarette lighters and camp stoves), a structural formula can be written as follows.

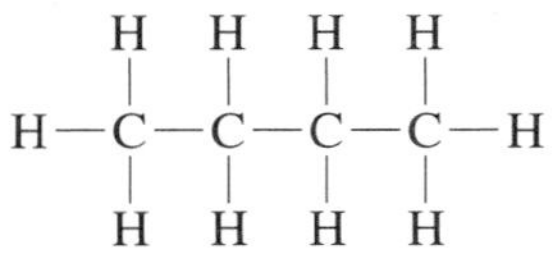

```
    H   H   H   H
    |   |   |   |
H — C — C — C — C — H
    |   |   |   |
    H   H   H   H
```

Note that, in this representation, the bonding and position of each atom relative to all others are specified. But a drawback to writing structural formulas, at least in a textbook, is that they take up considerable space. Instead, modified or condensed structural formulas can be used to convey the same information and are easily typeset into a single line. In these, carbon-to-hydrogen bonds are not drawn out explicitly, but simply understood to be single bonds. These are condensed structural formulas for C_4H_{10}.

$$CH_3—CH_2—CH_2—CH_3 \text{ or } CH_3CH_2CH_2CH_3$$

These representations imply that the first carbon on the left is bonded to three hydrogens and to the carbon atom directly to its right. The second carbon from the left is attached to two hydrogens and to the carbons on its left and right. Thus, the carbons are bonded directly to other carbon atoms in a straight chain. The hydrogen atoms do not intervene in the chain. Rather, two or three are attached to each carbon atom, depending upon its position in the molecule.

• Structural isomers are described in Section 4.8.

One reason why there are so many different organic compounds is because the same number and kinds of atoms can be arranged in unique ways to form a variety of arrangements called isomers. **Isomers** are compounds with the same chemical formula (same number and kinds of atoms) but different molecular structures and properties. You already encountered isomers in the discussion of octane, C_8H_{18} in Chapter 4. Here we illustrate the idea with C_4H_{10}. One way the atoms can be arranged is the linear isomer called butane, shown previously. However, another arrangement is possible in which the four carbon atoms are not in a "straight" line. This other isomer, known as isobutane, is represented by the structural formula at the left. The linear butane (at the right) is also shown for comparison.

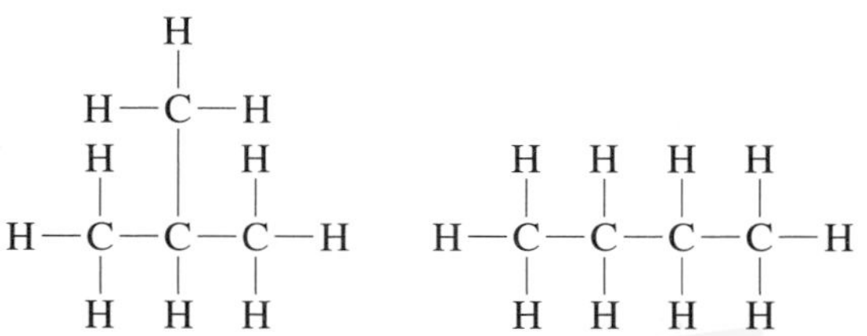

```
        H
        |
    H — C — H
    H   |   H                    H   H   H   H
    |   |   |                    |   |   |   |
H — C — C — C — H            H — C — C — C — C — H
    |   |   |                    |   |   |   |
    H   H   H                    H   H   H   H
    Isobutane                       Butane
```

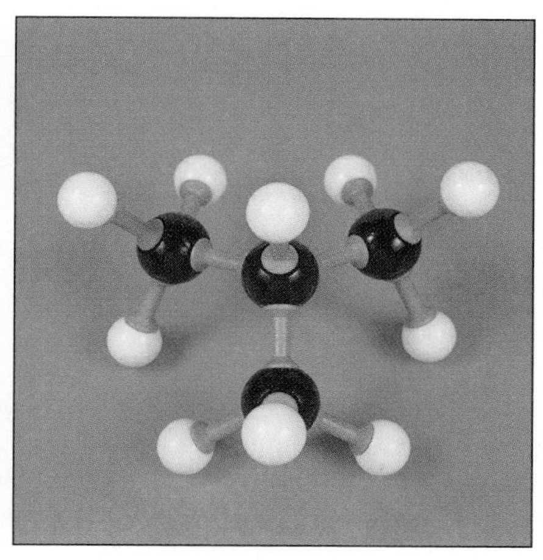

Figure 10.3

Photograph of molecular models of the two isomers of C_4H_{10}: butane (left) and isobutane (right). The carbon atoms are the black spheres.

Note that the central carbon in isobutane has three carbons connected to it and all of the other carbons have one other carbon (and three hydrogens) connected to them. Rotating this representation or moving it doesn't change the connectivity of the atoms. The number and kinds of atoms in these two isomers are the same; the connectivity of their atoms is different.

Just like its linear isomer butane, the isobutane formula can also be written in condensed form.

$$\begin{array}{ccccc} & & CH_3 & & \\ & & | & & \\ CH_3 & — & CH & — & CH_3 \end{array} \quad \text{or} \quad CH_3CH(CH_3)CH_3$$

In the condensed formula, the parentheses around the CH_3 indicate that its carbon is attached to the carbon to its left. Note from the structural formula that the $—CH_3$ attached to the CH carbon atom introduces a "branch" into the molecule. Molecular models of both isomers of C_4H_{10} are pictured in Figure 10.3.

Only two structural isomers of C_4H_{10} exist. It might be tempting to draw another representation (isobutane view 1). However, a bit of inspection should reveal that this is simply the previous structure for isobutane (isobutane view 2) written upside down, and not a different compound. What is important is which atoms are connected to which others, not their orientation in the image. As the number of atoms in a hydrocarbon increases, so do the number of possible isomers. Thus, there are 18 isomers of C_8H_{18} and 75 isomers of $C_{10}H_{22}$.

```
                                  H
                                  |
                              H — C — H
    H   H   H                 H   |   H
    |   |   |                 |   |   |
H — C — C — C — H         H — C — C — C — H
    |   |   |                 |   |   |
    H   |   H                 H   H   H
    H — C — H
        |
        H
  Isobutane view 1          Isobutane view 2
```

10.3 Your Turn

The formula C_5H_{12} represents three different compounds, all structural isomers of pentane. Draw structural formulas for these isomers.

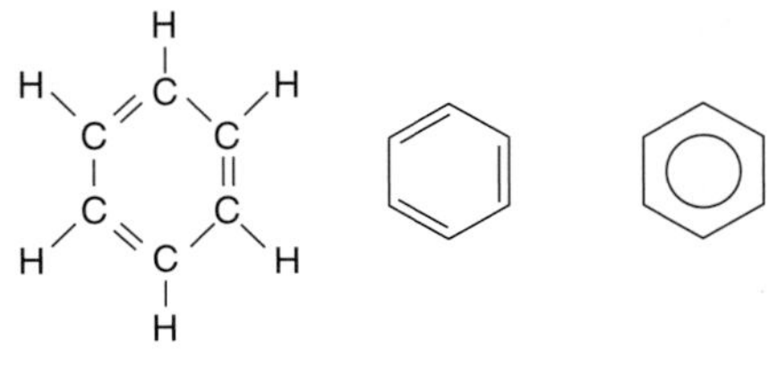

Figure 10.4
Representations of benzene, C_6H_6.

Carbon atoms are not limited to being bonded in a linear or branched fashion. In many molecules, including aspirin, carbon atoms are also arranged in a ring. Such rings most commonly contain five or six carbon atoms. In aspirin, the carbon atoms are joined in a six-membered hexagonal ring called a benzene ring after the compound with the formula C_6H_6. The Lewis octet rule applied to C_6H_6 predicts alternating single and double bonds between adjacent carbon atoms. The complete representation of this structure appears on the left in Figure 10.4. But the benzene ring is so widely distributed in organic molecules that symbols for individual carbon and hydrogen atoms are generally not written. Instead, the C_6H_6 molecule is written as a hexagon with a ring of electrons. One carbon atom with one attached hydrogen atom is assumed to occupy each corner of the hexagon. The representation in the center of Figure 10.4 indicates single and double bonds connecting the carbon atoms, but this picture is somewhat misleading. Experiment shows all carbon-carbon bonds in benzene to be identical. In strength and in length, these bonds are somewhere between single bonds and double bonds. This means that the electrons connecting the carbon atoms must be uniformly distributed around the ring. The circle within the hexagon in the drawing on the right is an effort to convey this idea. This same hexagonal structure is found in the —C_6H_5 phenyl group that is part of many molecules, including styrene and polystyrene (Chapter 9).

• The uniform distribution of electrons in the benzene ring is an example of resonance (Section 2.3).

10.3 Functional Groups

Fortunately, the plethora of organic compounds is somewhat simplified by the existence of a relatively small number of functional groups that appear with considerable frequency. **Functional groups** are distinctive arrangements of groups of atoms that impart characteristic physical and chemical properties to the molecules that contain them. Indeed, these groups are so important that we often focus our formulas on them and symbolize the remainder of the molecule with an R. The R is generally assumed to include at least one carbon atom connected to the functional group, but it can be practically anything. You already encountered some functional groups in Chapter 9. The generic formula for an alcohol is ROH, as in methanol, CH_3OH, (an alcohol derived from degradation of wood) and ethanol, CH_3CH_2OH (alcohol derived from fermentation of grains and sugar). The presence of the —OH group attached to carbon makes the compound an alcohol.

• Section 9.5 discussed alcohols in polymerization reactions.

Similarly, acidic properties are conferred by a carboxylic acid group,

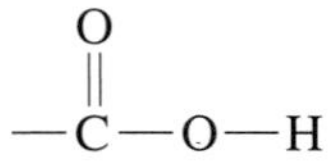

$$-\overset{\overset{\displaystyle O}{\|}}{C}-O-H$$

commonly written as —COOH or —CO_2H. In aqueous solution, an H^+ ion (a proton) is transferred from the —COOH group to an H_2O molecule to form a hydronium ion, H_3O^+. We represent an organic acid with the general formula RCOOH. In acetic acid, CH_3COOH, the acid in vinegar, R is —CH_3, a methyl group. Table 10.2 lists eight of the most important functional groups found in drugs and other organic compounds. Each functional group is characteristic of an important class of compounds. The table presents the general, generic formula of the class. In every case, the functional group is highlighted in color. In addition, Table 10.2 includes the formula and molecular structure of an example of each compound class.

The presence and properties of functional groups are responsible for the action of all drugs. Aspirin has three such subunits, boxed and numbered in Figure 10.5. You will

Table 10.2

Some Important Organic Functional Groups

Group	Generic Formula	Structural Formula	Examples: Name	Condensed Structural Formula
alcohol	R—OH	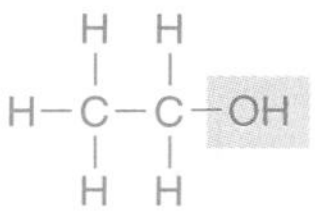 H—C(H)(H)—C(H)(H)—OH	ethanol (ethyl alcohol)	CH_3CH_2—OH
ether	R—O—R′	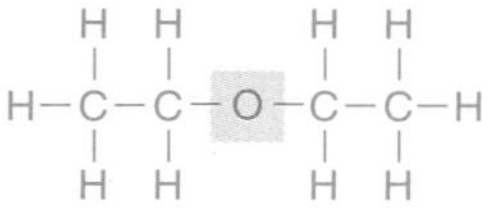 H—C(H)(H)—C(H)(H)—O—C(H)(H)—C(H)(H)—H	diethyl ether	CH_3CH_2—O—CH_2CH_3
aldehyde	R—C(=O)—H	C_6H_5—C(=O)—H	benzaldehyde	C_6H_5—C(=O)—H
ketone	R—C(=O)—R′	H—C(H)(H)—C(=O)—C(H)(H)—H	acetone	CH_3—C(=O)—CH_3
carboxylic acid	R—C(=O)—OH	H—C(H)(H)—C(=O)—OH	acetic acid	CH_3—C(=O)—OH
ester	R—C(=O)—OR	H—C(H)(H)—C(=O)—O—C(H)(H)—C(H)(H)—H	ethyl acetate	CH_3—C(=O)—OCH_2CH_3
amine	R—NH_2	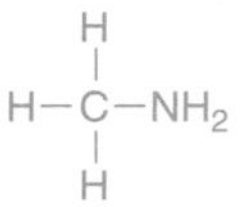 H—C(H)(H)—NH_2	methyl amine	CH_3—NH_2
amide	R—C(=O)—NH_2	H—C(H)(H)—C(H)(H)—C(=O)—NH_2	propanamide	CH_3CH_2—C(=O)—NH_2

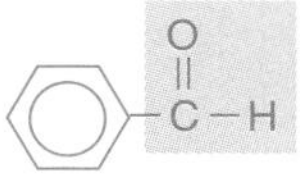

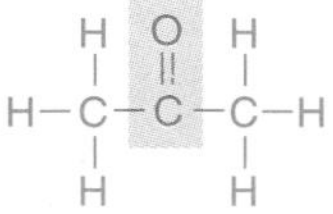

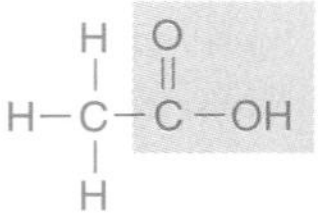

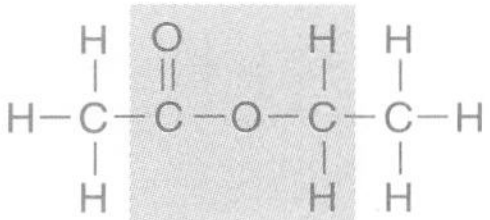

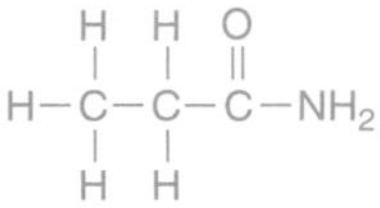

recognize that box 1 encloses a benzene ring. Its presence makes aspirin soluble in fatty compounds that are important cell membrane components. The other two portions, boxes 2 and 3, are responsible for the drug activity. You have just been reminded that the —COOH group indicates an organic acid (box 2). The remaining functional group (box 3) is an ester. An ester is formed by the reaction of an acid and an alcohol when a water molecule is eliminated in the process.

Felix Hoffmann prepared aspirin by modifying the structure of salicylic acid. But note that he did not modify the carboxylic acid group on the molecule. Salicylic acid also contains a special alcoholic —OH group, and it was this part of the molecule that Hoffmann reacted with acetic acid via equation 10.1.

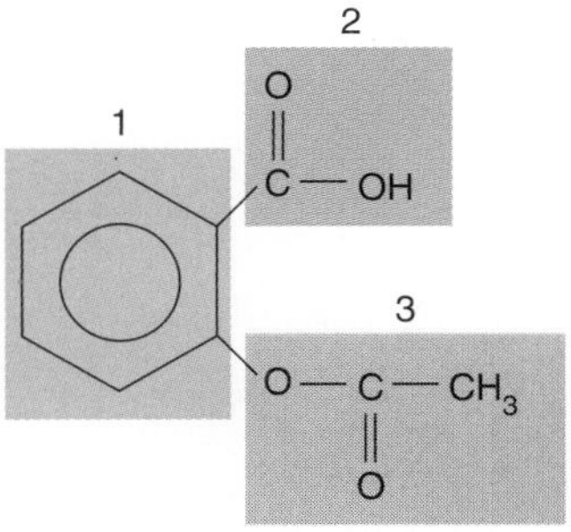

1 2 3
C—OH, O
O—C—CH₃, O

Figure 10.5
Structural formula of aspirin

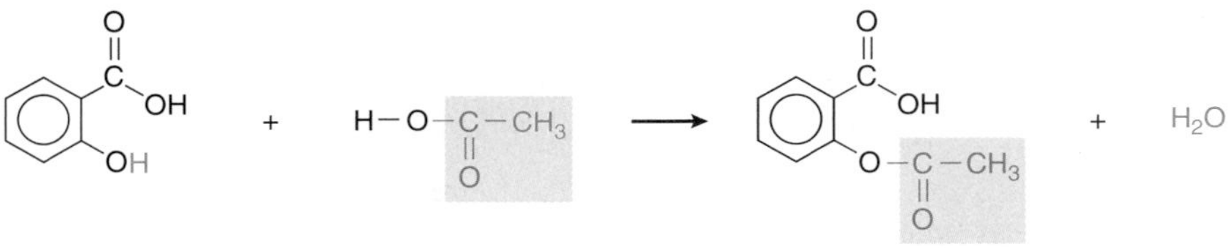

$$\text{C}_6\text{H}_4(\text{OH})\text{—C(=O)—OH} + \text{H—O—C(=O)—CH}_3 \longrightarrow \text{C}_6\text{H}_4(\text{O—C(=O)—CH}_3)\text{—C(=O)—OH} + \text{H}_2\text{O} \quad [10.1]$$

• Formation of an ester is shown for condensation polymers (Section 9.5, equation 9.6).

The product was an ester of acetic acid and salicylic acid, which accounts for one of aspirin's names—acetylsalicylic acid.

Because aspirin retains the —COOH group (box 2, Figure 10.5) of the original salicylic acid, it still has some of the undesirable acidic properties of the parent compound. However, the presence of the ester group (box 3, Figure 10.5) reduces the strength of the acid group and makes the compound more palatable and less irritating to the stomach lining. Once aspirin is ingested and reaches the site of its action, reaction 10.1 is reversed. The ester splits into acetic acid and salicylic acid, and the latter compound exerts its *antipyretic* (fever-reducing) and *analgesic* (pain-reducing) properties.

Functional groups can play a role in the solubility of a compound, an important consideration in the uptake, rate of reaction, and residence time of drugs in the body. The general solubility rule, "like dissolves like," applies in the body as well as in the test tube. When that rule was introduced in Chapter 5, a distinction was made between polar and nonpolar molecules. A polar molecule has a nonsymmetrical distribution of electric charge. This means that a negative charge builds up on some part (or parts) of the molecule, while other regions of the molecule bear a positive charge. Water is an excellent example of a polar molecule. Relatively speaking, the oxygen atom is slightly negatively charged and the hydrogen atoms are slightly positive. Because the molecule is bent, it has a nonsymmetrical charge distribution. Functional groups containing oxygen and nitrogen atoms (for example, —OH, —COOH, and —NH_2) usually increase the polarity of a molecule. This in turn enhances its solubility in a polar substance such as water.

By contrast, compounds whose molecules do not contain such atoms, but consist primarily or exclusively of carbon and hydrogen atoms, are typically nonpolar. A hydrocarbon such as octane, C_8H_{18}, is a good example; this compound is insoluble in water. However, it does dissolve in nonpolar solvents that are structurally similar to it. For the same reason, drugs with significant nonpolar character tend to accumulate in cell membranes and fatty tissues, which are themselves largely hydrocarbon and nonpolar.

Drugs with similar physiological properties often have similar molecular structures, including some of the same functional groups. Of the approximately 40 alternatives to aspirin that have been produced, ibuprofen and acetaminophen are the most familiar. Figure 10.6 gives the structural formulas of the three leading analgesics. All are based on a benzene ring with two substituents, but the substituents differ in detail. In 10.4 Your Turn you have an opportunity to identify the structural similarities and differences of these compounds.

10.4 Your Turn

Look at the structural formulas in Figure 10.6. Identify the structural features and functional groups that aspirin, ibuprofen, and acetaminophen have in common.

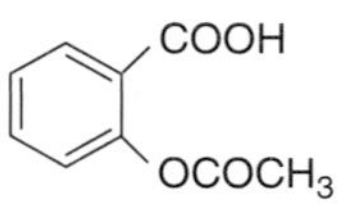

COOH / $OCOCH_3$

Aspirin

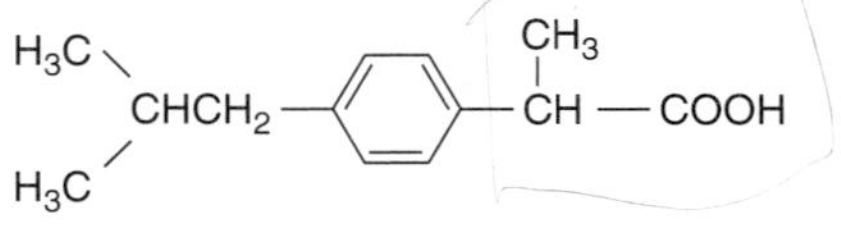

H_3C / H_3C > $CHCH_2$—⟨benzene⟩—CH(CH_3)—COOH

Ibuprofen (Advil®)

CH_3CONH—⟨benzene⟩—OH

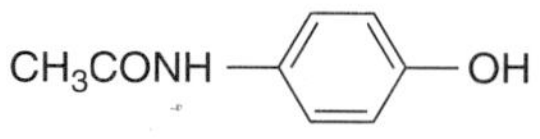

Acetaminophen (Tylenol®)

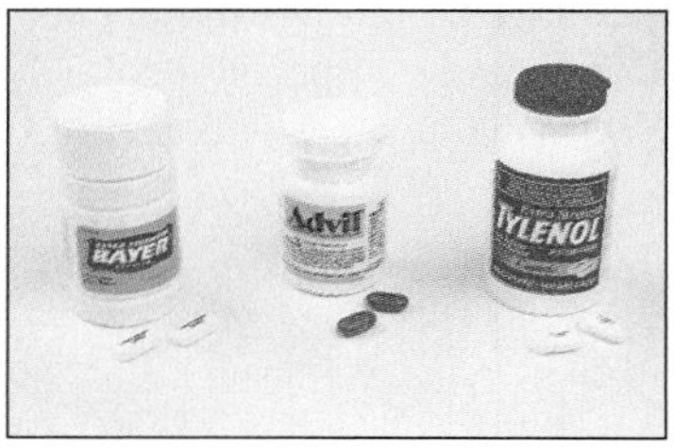

BAYER Advil TYLENOL

Figure 10.6

Structural formulas and samples of some common analgesics.

The current commercial method for producing ibuprofen is a stunning application of green chemistry. Previous methods of ibuprofen production required six steps, used large amounts of solvents, and generated significant quantities of wastes. By using a catalyst that also serves as a solvent, BHC Company, a 1997 Presidential Green Chemistry Challenge Award winner, makes ibuprofen in just three steps with a minimum of solvents and waste. In the BHC process, virtually all the reactants are converted to ibuprofen or another usable by-product; any unreacted starting materials are recovered and recycled. Nearly eight million pounds of ibuprofen, enough to make 18 billion 200-mg pills, are produced annually in Bishop, Texas at the BHC facility, built specifically for the commercial production of the drug.

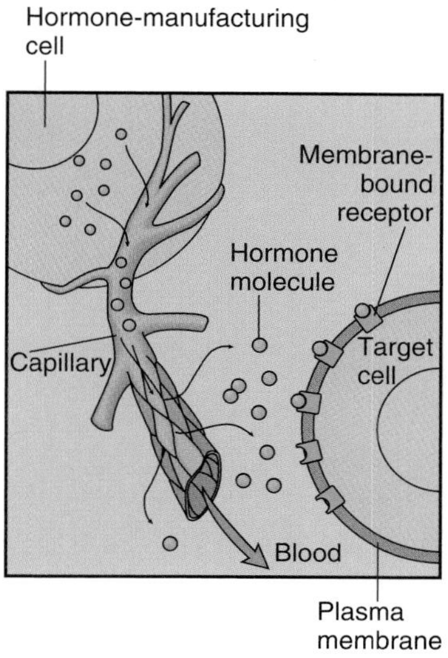

Figure 10.7

Chemical communication in the body. Hormone molecules travel from the cell where they are made, through the bloodstream, to the target cell.

10.4 How Aspirin Works: Relating Molecular Structure to Activity

That a drug like aspirin elicits certain physiological responses requires that some aspect of its structure and properties make its presence felt by the body. To understand the action of aspirin it is necessary to know something about the body's chemical communication system. We normally think of internal communication as consisting of electrical impulses traveling along a network of nerves. This is certainly true for the system that triggers movement, breathing, heartbeats, and reflex actions. Most of the body's messages, however, are conveyed not by electrical impulses, but through chemical processes. In fact, your very first communication with your mother was a chemical signal saying "I'm here; better get your body ready for me." It is much more efficient to release chemical messengers into the bloodstream, which then circulates them to appropriate body cells, than to "hardwire" each individual cell with nerve endings.

These chemical messengers are called **hormones** and they are produced by the body's endocrine glands. Figure 10.7 is a representation of such chemical communication. Hormones encompass a wide range of functions and a similarly wide range of chemical composition and structure. Thyroxine, an iodine-containing amino acid, is one of the simpler ones, but is essential for regulating metabolism. The chemical breakdown of "blood sugar" (glucose) requires insulin. This hormone, a small protein of only 51 amino acids, is secreted by the pancreas. Persons who suffer from diabetes are often required to take daily injections of insulin. Yet another well-known hormone is adrenaline (epinephrine), a small molecule that prepares the body to "fight or flee" in the face of danger. And the hormonal messages that are so compelling in adolescents are carried by steroids, a sexy set of molecules that we will visit in a few pages.

- Section 12.5 describes the use of genetic engineering to obtain human insulin from bacteria.

Aspirin and other drugs that are physiologically active, but not anti-infectious agents, are almost always involved in altering the chemical communication system of the body. A significant problem is that this system is very complex, allowing many compounds to be used to send more than one message simultaneously. The wide range of aspirin's therapeutic properties, as well as its side effects, are clear evidence that the drug is involved in several chemical communication systems. It works in the brain to reduce fever, it relieves inflammation in muscles and joints, and it apparently decreases the chances of stroke and heart attack. It may even lessen the likelihood of colon, stomach, and rectal cancer.

In large measure, the versatility of aspirin and similar *n*o*n*steroidal *a*nti-*i*nflammatory *d*rugs (NSAIDs) is related to their remarkable ability to block the actions of other molecules. Research on the activity of aspirin indicates that one of its modes of action involves blocking cyclooxygenase (COX) enzymes. These enzymes, like all others, are biochemical **catalysts**. They are proteins that influence the rate of a chemical reaction. Most enzymes speed up reactions and channel them so that only one product (or a set of related products) is formed. In the case of cyclooxygenases, the reaction is the synthesis of a series of hormone-like compounds called **prostaglandins.** Prostaglandins cause a variety of effects. They produce fever and swelling, increase sensitivity of pain receptors, inhibit blood vessel dilation, regulate the production of acid and mucus in the stomach, and assist kidney functions. By preventing prostaglandin production, aspirin

- You encountered catalysts in several other contexts, including automobile emissions control (Section 1.11), petroleum refining (Section 4.5), and addition polymerization (Section 9.3).

reduces fever and swelling. It also suppresses pain receptors and so functions as a painkiller. Because the benzene ring conveys high fat solubility, aspirin is also taken up into cell membranes. In certain specialized cells, the drug blocks the transmission of chemical signals that trigger inflammation. This process also appears to be related to aspirin's effectiveness as a pain reliever.

The NSAIDs exhibit these same properties in varying degrees. For example, because acetaminophen blocks cyclooxygenase (COX) enzymes, but does not affect the specialized cells, it reduces fever but has little anti-inflammatory action. On the other hand, ibuprofen is a better enzyme blocker and specialized cell inhibitor. Consequently, ibuprofen is both a better pain reliever and fever reducer than aspirin. Ibuprofen has fewer functional groups than aspirin, which may be the reason why ibuprofen has fewer side effects. With fewer functional groups, ibuprofen is less polar and more lipid soluble than aspirin. Its anti-inflammatory activity is five to 50 times that of aspirin.

All these structurally related anti-inflammatory drugs appear to affect the way cell membranes respond to stimuli. Research has shown that this is yet another possible mode of action for aspirin and its chemical relatives. On the other hand, aspirin is unique among these three compounds in its ability to inhibit blood clotting. This property has led to the suggestion that low regular doses of aspirin can help prevent strokes or heart attacks. Of course, these anticoagulation characteristics also mean that aspirin is not the painkiller of choice for surgical patients or those suffering from ulcers. That is why "more hospitals use Tylenol." You already read that some people experience stomach irritation when they take aspirin. Another drawback of the drug is that in rare cases in the presence of certain viruses, it can trigger a sometimes fatal response known as Reye's syndrome, particularly in children under the age of 15.

Recently, scientists have been able to better understand how aspirin, ibuprofen, and acetaminophen affect the two cyclooxygenase (COX) enzymes. These drugs block one of the enzymes, COX-2, which makes prostaglandins associated with inflammation, pain, and fever, thereby reducing these symptoms. But the drugs also inhibit the other enzyme in the pair, COX-1, which primarily makes hormones that maintain proper kidney function and keep the stomach lining intact. Thus, the drugs are not sufficiently selective to affect COX-2 without shutting down COX-1 as well. By determining the crystal structure of COX-2 in 1992, researchers were then guided in making nearly a dozen new candidate drugs that block COX-2 alone, thereby creating new medicines that have been termed "superaspirins." Regarding the future of superaspirins, medicinal chemist Phil Portoghese predicts that "Eventually all those [NSAID] molecules on the market will become dinosaurs." A compound that inhibits both enzymes, but preferentially acts on COX-2, is now available in Europe. One of the other new COX-2 inhibitors involves some clever molecular modification. While retaining the acetyl group, as in aspirin, the acid group is replaced with a long molecular side chain that includes a sulfur atom and seven carbon atoms, two of which have a triple bond between them (Figure 10.8).

A few final comments about aspirin seem appropriate. Because it is a specific chemical compound, aspirin is aspirin—acetylsalicylic acid, regardless of its manufacturer. Indeed, about 70% of all acetylsalicylic acid produced in the U.S. is made by a single manufacturer. But, although all aspirin molecules are identical, not all aspirin tablets are the same. The commercial products are mixtures of various components, including inert fillers and bonding agents that hold the tablet together. Buffered aspirin tablets also include weak bases that counteract the natural acidity of the aspirin. Some coated

• Triple bonds between carbon atoms, C≡C, are unusual in natural compounds.

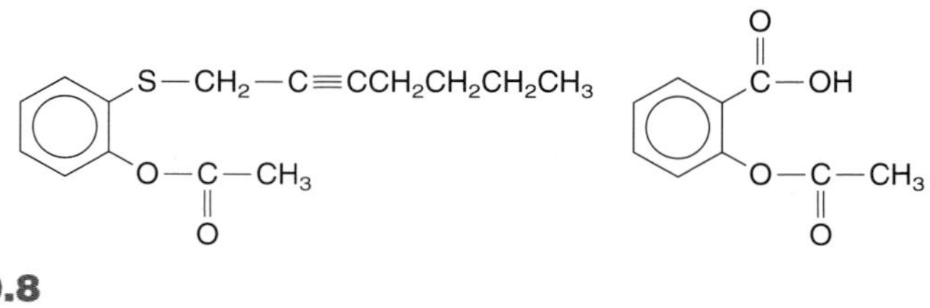

S—CH$_2$—C≡CCH$_2$CH$_2$CH$_2$CH$_3$
O—C(=O)—CH$_3$

C(=O)—OH
O—C(=O)—CH$_3$

Figure 10.8
A new COX-2 inhibitor (left) compared to aspirin (right).

aspirins keep the tablet intact until it leaves the stomach and enters the intestine. These differences in formulation can influence the rate of uptake of the drug and hence, how fast it acts, and the extent of stomach irritation. Furthermore, although standards for quality control are high, it is conceivable that individual lots of aspirin may vary slightly in purity. Aspirin also decomposes with time, and the smell of vinegar can signify that such a process has begun. Fortunately, none of this poses a significant threat to health, and the benefits of aspirin far outweigh the risks for the great majority of people.

10.5 Consider This: Aspirin in a Large Bottle

A friend who suffers from heart disease has been told by the doctor to take one aspirin tablet a day. To save money, your friend often buys the large 300-tablet bottle of aspirin. You, on the other hand, rarely take aspirin, but cannot pass up a good bargain. You also buy the large bottle.

a. Why is the "giant economy size" bottle of aspirin not as good a deal for you as it is for your friend?
b. What chemical evidence supports your opinion?

10.5 Modern Drug Design: Optimizing Function Through Structural Changes

The evolution of "willow bark tea" to aspirin and further modifications to this painkiller's structure to enhance its beneficial effects and decrease its side effects represent stages in historical drug design. Penicillin is another example of a miracle drug whose origin lies in "natural" sources. Its story includes an accidental discovery by the British bacteriologist Alexander Fleming in 1928. Fleming's curiosity was aroused by the chance observation that in a container of bacterial colonies, the area contaminated by the mold *Penicillium notatum* was free of bacteria (Figure 10.9). He correctly concluded that the mold gave off a substance that inhibited bacterial growth, and he named this biologically active material penicillin.

A careful reconstruction has indicated that a series of critical, but fortuitous events, had to occur for the discovery to be made. Spores from the mold, part of an experiment in a nearby lab, drifted into Fleming's laboratory and accidentally contaminated some Petri dishes containing *Staphylococcus* (bacteria) growing on a nutrient medium. Then came a series of chance incidents involving poor laboratory housekeeping, a vacation, and the effects of weather. Fleming fortunately noticed the dishes in which the *Staphylococcus* had been killed. Using former experience, he correctly interpreted the phenomenon, recognizing that the unknown substance being given off by the *Penicillium* was a potential antibacterial agent for the treatment of infection. "The story of penicillin," Fleming wrote, "has a certain romance in it and helps to illustrate the amount of chance, or fortune, of fate, or destiny, call it what you will, in anybody's career." But of course, the discovery would not have happened without Fleming's powers of observation and insight in response to the unexpected. The episode admirably illustrates the often misquoted maxim of the great French scientist, Louis Pasteur: "In the fields of observation, chance favors only the prepared mind." Most versions of this famous aphorism neglect the "only." It was *only* because Fleming's mind was prepared that he was able to capitalize on this chain of unlikely events.

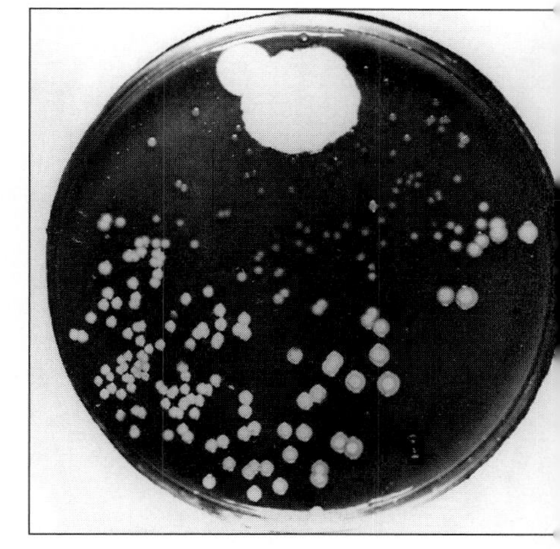

Figure 10.9
A Petri dish showing the antibacterial properties of penicillin. The large white area at the 12 o'clock position is the mold *Penicillium notatum*; the smaller white spots are areas of bacterial growth. The reduced growth of bacteria in the vicinity of the mold is evidence that a compound produced by the mold inhibits bacteria. That compound is penicillin.

Taking penicillin from the Petri dish to the pharmacy was not easy. The first step was a systematic effort to isolate the active agent produced by *Penicillium notatum*. Once identified, the substance had to be separated, purified, and concentrated by new, sophisticated techniques. Also, the efficacy of penicillin in treating humans had to be demonstrated. World War II gave increased impetus to this research and to the development of new methods for preparing large quantities of penicillin. Because the scientists were successful in doing so, thousands of lives were saved during the war, and millions since then.

The discovery of penicillin may have been serendipitous, but the next several "generations" of antibiotics in this class have involved systematic and careful research. Small changes are designed into a drug and its efficacy examined and tested. Just as in the case of NSAIDs, beneficial effects are optimized while side effects are decreased. More than 10 different penicillins are currently in clinical use: penicillin G (the original discovered by Fleming and the form that causes an allergic reaction in about 20% of the population), ampicillin, oxacillin, cloxacillin, penicillin O, and amoxicillin (the pink, bubble gum-flavored concoction you might have been given when you were a child). Bacteria develop resistance to penicillin by secreting an enzyme that destroys the penicillin structure before it can act. The newer penicillins differ in the organisms against which they are most effective and in the drug's susceptibility to the enzyme. Closely related to the penicillins are the cephalosporins (like Keflex®) that are particularly effective against resistant strains of bacteria. Careful research on structural modifications has led to other effective medicines like cyclosporin, a major anti–tissue rejection drug that has revolutionized organ transplant surgery.

So how do chemists know what structural features are important to a drug's function? The modern approach to chemotherapy and drug design probably began early in the 20th century with Paul Ehrlich's search for an arsenic compound that would cure syphilis without doing serious damage to the patient. His quest was for a "magic bullet," a drug that would affect only the diseased site and nothing else. He systematically varied the structure of many arsenic compounds, simultaneously testing each new compound for activity and toxicity on experimental animals. He finally achieved success with Salvarsan 606, so named because it was the six hundred sixth compound investigated. Since then, medicinal chemists have adopted Ehrlich's strategy of carefully relating chemical structure and drug activity.

Drugs can be broadly classified into two groups: those that produce a physiological response in the body and those that inhibit the growth of substances that cause infections. You already learned that aspirin falls in the first group. So do synthetic hormones and psychologically active drugs. These compounds typically initiate or block a chemical action that generates a cellular response, such as a nerve impulse or the synthesis of a protein. Antibiotics exemplify drugs that prevent the reproduction of foreign invaders. They do so by inhibiting an essential chemical process in the infecting organism. Thus, they are particularly effective against bacteria.

Although drugs vary in their versatility, many of them act only against particular diseases or infections. This specificity is consistent with the relationship that exists between the chemical structure of a drug and its therapeutic properties. Both the general shape of the molecule and the nature and location of its functional groups are important factors in determining its physiological efficacy. This correlation between form and function can be explained in terms of the interaction between biologically important molecules. Although many of these molecules are very large, consisting of hundreds of atoms, each molecule often contains a relatively small **active site** or **receptor site** that is of crucial importance in the biochemical function of the molecule. A drug is often designed to either initiate or inhibit this function by interacting with the receptor site.

- The lock-and-key analogy was first proposed in 1894 by Emil Fischer, a famous biochemist.

An example is provided by a receptor site that controls whether a cell membrane is permeable to certain chemicals. In effect, such a site acts as a lock on a cellular door. The key to this lock may be a hormone or drug molecule. The drug or hormone bonds to the receptor site, opening or closing a channel through the cell membrane. Whether the channel is open or closed can significantly influence the chemistry that occurs in the cell. In fact, under some circumstances, the cell may be killed, which may or may not be beneficial to the organism.

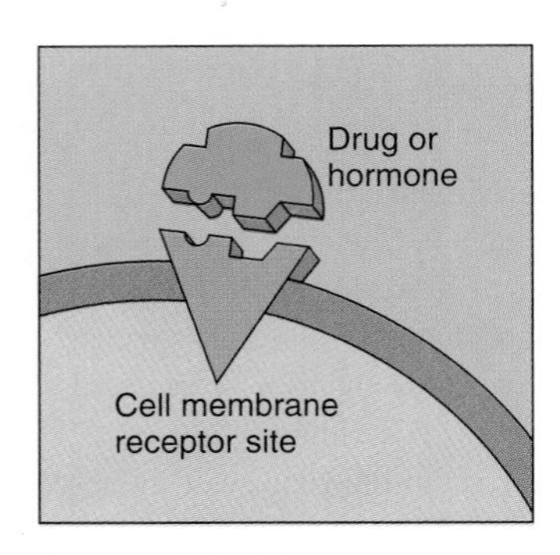

Figure 10.10
Lock-and-key model of biological interactions.

This lock-and-key analogy is often used to describe the interaction of drugs and receptor sites. Just as specific keys fit only specific locks, a molecular match between a drug and its receptor site is required for physiological function. The process is illustrated in Figure 10.10. If a perfect lock-and-key match were required in the body, it would mean that each of the millions of physiological functions would have a unique receptor site and a specific molecular segment to fit it. Simple logic suggests that such rigid demands would not promote cellular efficiency. Consequently, the lock-and-key model, although a good starting point that works in a limited number of cases, must be modified.

Using another analogy, a receptor site is like a size-9 right footprint in the sand. Only one foot will fit it exactly, and many feet (all left feet and all right feet much larger or smaller than size 9) will not fit adequately. But many other right feet can fit into the print reasonably well. So it is with receptor sites and the molecules or functional groups that bind to them. Some active sites can accommodate a variety of molecules including drugs. Indeed, the way most drugs function is by replacing a normal protein, hormone, or other substrate in the invading organism. The presence of the drug molecule thus prevents the enzyme, cell membrane, or other biological unit from carrying out its required chemistry. As a result, the growth of an invading bacterium is inhibited, or the synthesis of a particular molecule is turned off (Figure 10.11).

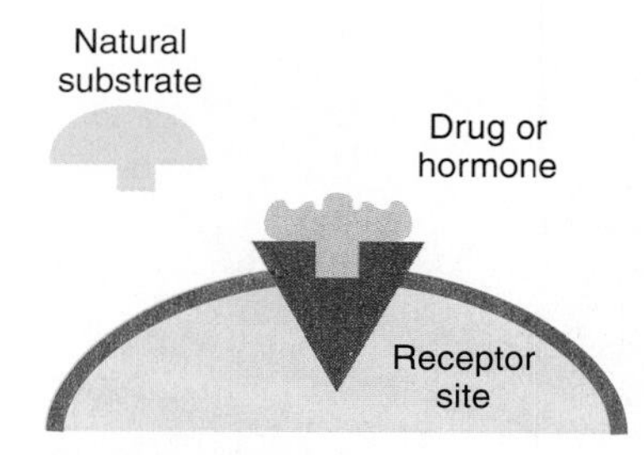

Figure 10.11
The substrate inhibition model of enzyme activity. A drug (gray) occupies the active site of an enzyme, thus blocking the normal substrate (light blue) and the normal activity that it elicits.

Generally speaking, the drug that best fits the receptor site has the highest therapeutic activity. In some cases, however, a drug molecule does not need to fit the receptor site particularly well. The bonding of functional groups of the drug to the receptor site may even alter the shape of the drug, the site, or both. Often what counts is for the drug to have functional groups of the proper polarity in the right places. Therefore, one important strategy in designing drugs is to determine the specific part of the molecule that gives the compound its activity. Medicinal chemists then synthesize a molecule having that specific active portion, but with a much simpler, nonactive remainder. These researchers custom design the molecule to meet the requirements of the receptor site. In effect, they design feet to fit footprints.

10.6 Consider This: 3-D Drugs

See for yourself how drug molecules appear in three dimensions by visiting the Three-Dimensional Drug Structure Data Bank at the National Institutes of Health (NIH). A direct link is provided by the *Chemistry in Context* web site, or you can locate the site by searching for "Center for Molecular Modeling" and/or "NIH." Note: To view the molecules you will have to download and install a browser "plug-in" called Chime. Directions for using Chime are provided at the *Chemistry in Context* web site.

a. Select several drugs and examine their three-dimensional structure. How do these computer representations differ from the structural formulas of drugs shown in this chapter?
b. What are the advantages of the computer representations over two-dimensional drawings? What are their limitations compared to "real" molecules? Are there any disadvantages?

An outstanding example of this approach is provided by opiate drugs such as morphine. Morphine, a very complex molecule, is difficult to synthesize. However, the particular portion of the molecule responsible for opiate activity has been identified and is highlighted in Figure 10.12. The flat benzene ring fits into a corresponding flat area of the receptor, and the nitrogen atom binds the drug molecule to the site. Incorporating this particular portion into other less complex molecules, such as demerol, creates opiate activity. Demerol is an non-addictive treatment for severe pain.

The discovery that only certain functional groups are responsible for the therapeutic properties of pharmaceutical molecules has been an important breakthrough. Sophisticated computer graphics are now used to model potential drugs and receptor sites. Thanks to these representations, with their three-dimensional character, medicinal

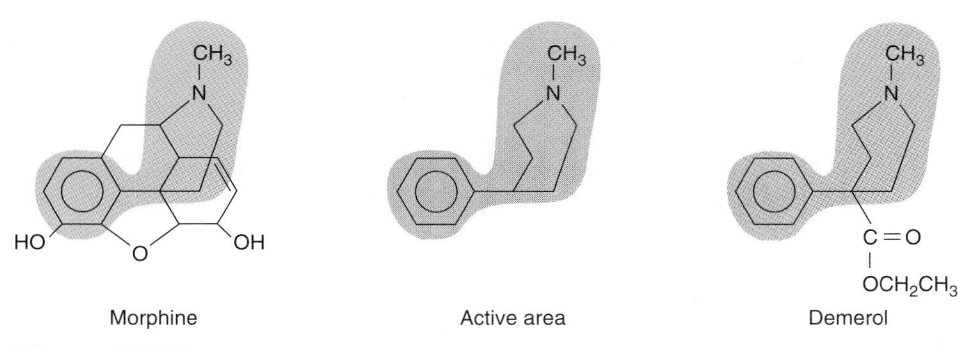

Figure 10.12
Molecular structures of morphine and demerol. The highlighted "active areas" are the portion of the molecules that interact with the receptor.

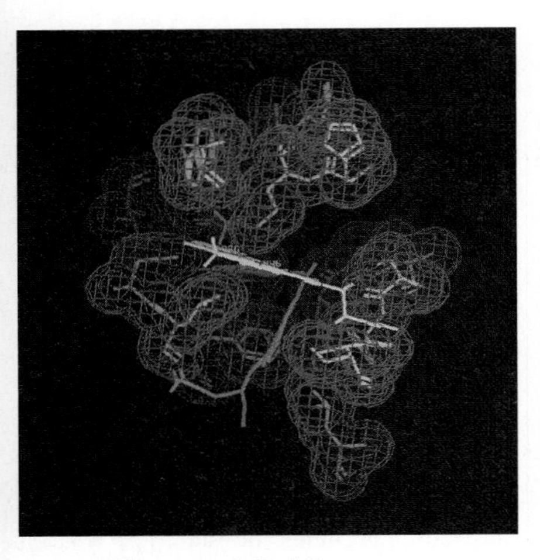

Figure 10.13
Computer modeling of the arthritis drug methotrexate (pink) as it binds at a protein receptor site (light blue structural components covered by the blue grid lines). A search of a combinatorial library found many molecules, including one represented in yellow, that would be expected to bind at the same receptor site.

chemists can "see" how drugs interact with a receptor site. Computers can then be used to search for compounds that have structures similar to that of an active drug. Chemists can also modify structures in computer models and visualize how the new compounds will function.

Such techniques help to minimize the time it takes to prepare a so-called lead compound, one that shows high promise for becoming an approved drug. Combinatorial chemistry is a recent development that accelerates the creation of lead compounds. Combinatorial chemistry uses the fact that organic molecules contain functional portions, that is, "pieces" of a molecule responsible for the chemical property of the molecule, and nonfunctional portions. For example, an 8×12 array of small wells can be used to examine 96 combinations of structural features for their potential as drugs. Each row in one direction contains 8 functional group variations in a target molecule; each row in the perpendicular direction contains 12 different functional groups or variations in another portion of the same target molecule. In other words, in a given row, one part of the molecule remains unchanged while the second site is changed. Each of the 96 wells is examined to see whether any of these structural variants show the sought-after chemical or biomedical activity. If activity is observed in a well, those wells are analyzed for their chemical composition.

The process can be repeated several times, each time seeking to determine when reactions have formed potential drugs containing diverse functional groups. Unpromising reactions can be screened out quickly. From the continuing candidates, the company can develop a so-called library of molecular diversity for a huge array of synthesized compounds, any of which might become a lead compound in the search for a drug. Used in conjunction with computers, combinatorial chemistry can minimize the trial-and-error aspects and expense, thus speeding up drug design and development (Figure 10.13). Using traditional methods, a medicinal chemist could prepare perhaps four lead compounds per month at an estimated cost of $7000 each. With combinatorial chemical methods, the chemist can prepare nearly 3300 compounds in that same time for only about $12 each.

10.7 Consider This: Orphan Drugs

Aspirin and other drugs have a large and profitable market around the world. However, development and marketing of essential drugs needed by only a small number of people suffering from rare diseases can be an enormous economic drain on a pharmaceutical company. If the pharmaceutical companies decide not to make and market these "orphan drugs" because of their low economic return, how will people who need these drugs obtain them? Should the government step in and require successful drug companies to contribute a percentage of their profits to a fund for research, development, and production of these "orphan drugs?" Take a position on this issue and outline the reasons to support your position.

10.6 Right- and Left-Handed Molecules: Structures in Three Dimensions

Drug design is further complicated when drug-receptor interaction involves a common, but subtle phenomenon called optical isomerism or chirality. **Chiral** or **optical isomers** have the same chemical formula, but they differ in their molecular structure and their interaction with polarized light, hence the name. Chirality most frequently arises when four *different* atoms or groups of atoms are attached to a carbon atom. A compound having such a carbon atom can exist in two different molecular forms that are non-superimposable mirror images of each other. These are **chiral isomers,** also called **optical isomers.**

- Polarized light waves move in a single plane; nonpolarized light waves move in many planes.

Non-superimposable mirror images should be familiar to you. You carry two of them around with you all the time—your hands. If you hold them with the palms up you can recognize them as being mirror images. For example, the thumb is on the left side of the left hand and on the right side of the right hand. Your left hand looks like the reflection of your right hand in a mirror. But your two hands are not identical. Figure

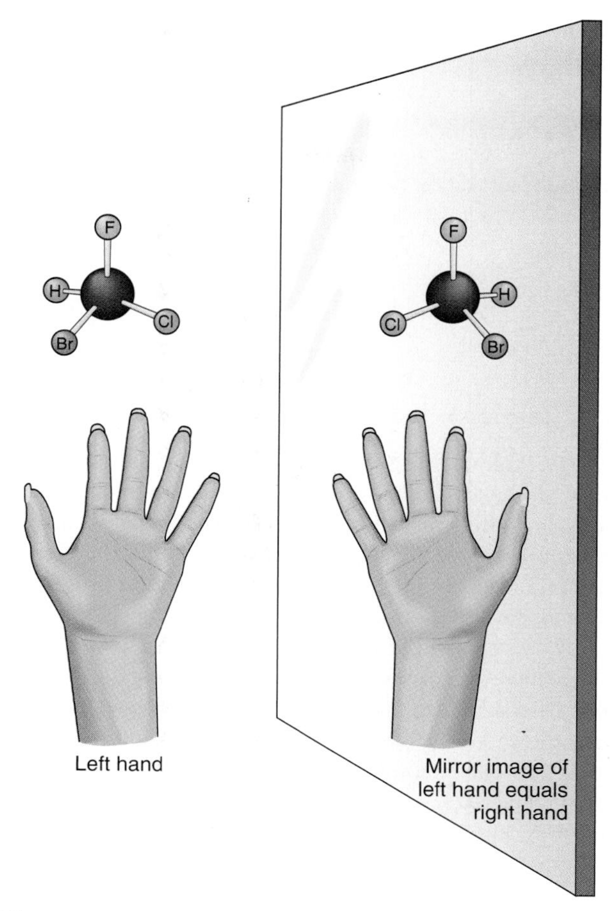

Figure 10.14

Mirror images of molecules and hands. In the molecule CHClFBr, each bond connects a different atom to carbon.

10.14 illustrates this relationship for both hands and molecules. Note that the four atoms or groups of atoms bonded to the central carbon atom are in a tetrahedral arrangement. *The positions of these four atoms correspond to the corners of a three-dimensional figure with equal triangular faces (like the pyramids of Egypt).* The "handedness" of these molecules gives rise to the term chirality, from the Greek word for hand.

It turns out that many biologically important molecules, including sugars and amino acids, exhibit chirality. This is significant because, although most chemical and physical properties of a pair of optical isomers are very nearly identical, their biological behavior can be profoundly different. Generally, the explanation for this difference is related to the necessity of a good molecular fit between a molecule and its receptor site. (Maybe Lewis Carroll's Alice had some inkling of this when, in *Through the Looking Glass*, she remarked to her cat, "Perhaps looking-glass milk isn't good to drink.")

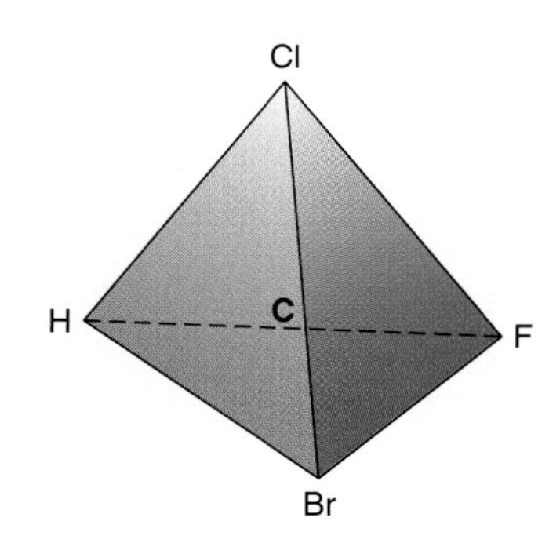

A tetrahedron has four equilateral triangular faces.

You can illustrate this relationship between chirality and biological activity by taking things in your own hands. Your right hand fits only a right-handed glove, not a left-handed one. Similarly, a right-handed drug molecule fits only a receptor site that complements and accommodates it. Any drug containing a carbon atom with four different atoms or groups attached to it will exist in chiral isomers, only one of which usually fits into a particular asymmetrical receptor site (Figure 10.15).

The extreme molecular specificity created by chirality makes the medicinal chemist's job more complex. A drug molecule must include the appropriate functional groups, and these groups must be arranged in the biologically active configuration. Often the "right" and "left" optical isomers are produced simultaneously. Such a situation results in a **racemic mixture,** consisting of equal amounts of each optical isomer. But frequently

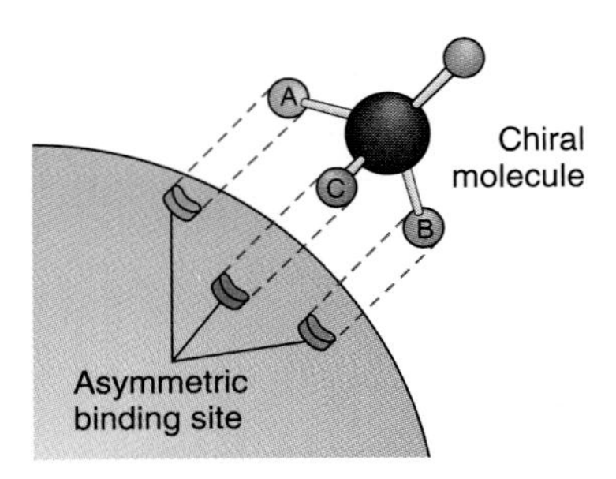

Figure 10.15
A chiral molecule binding to an asymmetric site.

only one optical isomer is pharmaceutically active. For example, many opiate drugs exist as optical isomers, only one of which may have opiate activity. Levomethorphan, the left-handed (levo) isomer of methorphan, is an addictive opiate. On the other hand, its right-handed (dextro) mirror image is a nonaddictive cough suppressant. This permits the use of dextromethorphan in many over-the-counter cough remedies, but the right-handed (dextro) isomer must either be synthesized in pure dextro form or separated from a mixture with its levo isomer.

HN OCH₃ Levomethorphan | NH CH₃O Dextromethorphan

Many other drugs exhibit chirality and are active only in one of the isomeric forms. This is true for some antibiotics and hormones, and for certain drugs used to treat a wide range of conditions, including inflammation, cardiovascular disease, central nervous system disorders, cancer, high cholesterol levels, and attention deficit disorder. Among the widely used chiral drugs are ibuprofen, the anti-rejection drug cyclosporin used in organ transplants, and the lipid-reducing drug Lipitor. Ibuprofen is sold as a racemic mixture of D- and L-isomers (Figure 10.16). L-ibuprofen is a pain reliever whereas D-ibuprofen is not. However, in the body the D-form is converted to the L-isomer. Therefore, it is likely that someone taking ibuprofen is just as well off taking the racemic mixture rather than the more expensive L-ibuprofen. On the other hand, Naproxen, a common pain reliever, is one example of many in which one isomer is preferred, even required. One form of Naproxen relieves pain; the other causes liver damage.

- The vitamin E sold in stores is generally a racemic mixture of D- and L-isomers. The D-isomer is the physiologically active one, which can be purchased in pure form at a significantly higher price.

Consequently, drug companies have active research programs designed to create chirally "pure" drugs, those having only the beneficial isomer of a drug in pure form. In fact, 40% of all dosage-form drugs sold in 2000 were of single enantiomers, rather than as racemic mixtures. Although making the proper, single isomer might seem like an exercise of interest only to chemists, it is big business. Worldwide sales of chirally pure drugs were $150 billion in 2002 and predicted to reach $200 billion by 2008, a very good return for knowing how to produce molecules of the correct chirality.

- William Knowles, Barry Sharpless, and Ryoji Noyori shared the 2001 Nobel Prize for their research that developed new catalytic methods used to synthesize chiral drugs.

10.8 Your Turn

Carefully examine the structural formula for ibuprofen given in Figure 10.6. Which is the chiral carbon atom?

D-ibuprofen | L-ibuprofen

Figure 10.16
D- and L-ibuprofen.

10.7 Steroids: Cholesterol, Sex Hormones, and More

Certain cellular components, contraceptives, muscle-mass enhancers, abortive agents; what do they have chemically in common? They are all **steroids,** a family of compounds that arguably best illustrates the relationship of form and function. Certainly no other

Table 10.3

Steroid Functions

Function	Examples
Regulation of secondary sexual characteristics	Estradiol (an estrogen); testosterone (an androgen)
Reproduction and control of the reproductive cycle	Progesterone and the gestagens
Regulation of metabolism	Cortisol; cortisone derivatives
Digestion of fat	Cholic acid; bile salts
Cell membrane component	Cholesterol

group of chemicals is more controversial than steroids because their uses range from contraception to vanity promoters. The naturally occurring members of this ubiquitous group of substances include structural cell components, metabolic regulators, and the hormones responsible for secondary sexual characteristics and reproduction. Among the synthetic steroids are drugs for birth control, abortion, and bodybuilding.

In spite of their tremendous range of physiological functions represented in Table 10.3, all steroids are built on the same molecular skeleton. Thus, these compounds also provide a marvelous example of the economy with which living systems use and reuse certain fundamental structural units for many different purposes. The body synthesizes the many large molecules necessary for life by combining smaller molecular fragments. Once such a process is established, the same fundamental biochemical reactions are used to incorporate these molecular fragments into a variety of complex compounds. This wonderfully efficient process is rather like having a standardized house plan that can be reproduced readily—a unit that gains individuality by changes in the types of windows and doors or by the interior decorations.

The common characteristic of steroids is a molecular framework (nucleus) consisting of 17 carbon atoms arranged in four rings—"three rooms and a garage" if you like (six-membered rings are the "rooms"; the five-membered ring the "garage"). This steroid nucleus is illustrated.

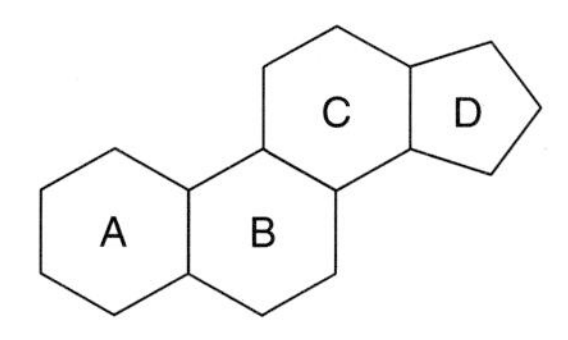

Recall that in such a representation, carbon atoms are assumed to occupy the vertices of the rings, but they are not explicitly drawn. The three six-membered carbon rings of the steroid nucleus are designated A, B, and C, and the five-membered ring is designated D. Although the steroid nucleus is drawn flat, it has a three-dimensional shape. The dozens of natural and synthetic steroids are all variations on this theme. They differ only slightly in structural detail, but can differ profoundly in physiological function. Extra carbon atoms and/or functional groups at critical positions on the rings are responsible for this variation.

A shorthand system is used to represent the molecular structures of the steroids. This system concentrates on the backbone of connected carbon atoms. One carbon atom is assumed to occupy each vertex of the structure. A line protruding from a molecular structure also signifies a carbon atom (actually a —CH_3 group), unless the symbol for another element is attached to it. Hydrogen atoms, which also are not indicated, are bonded to the carbon atoms as is necessary to satisfy the octet rule. This notation follows here for estradiol, a female sex hormone. The figure on the left includes all the atoms in the molecule; the one on the right gives the skeletal representation.

Estradiol

This same system is used in Figure 10.17 to represent the molecular structures of six vitally important steroids. The boxes enclose the regions in which structural variations occur. Careful examination of the figure indicates that some very subtle molecular differences can result in profoundly altered properties. For example, the only differences between a molecule of estradiol and one of testosterone are associated with ring A. The female sex hormone has no localized double bonds in the ring and an attached —OH group; the male hormone has one double bond in the A ring, an =O in place of the —OH, and a $—CH_3$ group (represented by the vertical line where the A and B rings come together). It is, of course, naive to suggest that the differences between men and women are all due to a carbon atom and a few hydrogen atoms, but it is tempting.

In this chapter we concentrate on only a small number of the many steroid compounds. We begin with cholesterol, the most abundant steroid in the body and probably the best known. The average-sized adult has about half a pound of cholesterol in his or her body. Cholesterol is a starting point for the production of steroid-related hormones and a major component of cell membranes. Because their shape is relatively long, flat, and rigid, cholesterol molecules help to enhance the firmness of cell membranes. Although cholesterol is essential for human life, there are concerns that too much of the compound in the blood can lead to the buildup of plaque, fatty deposits in the blood ves-

Female sex hormone
Estradiol

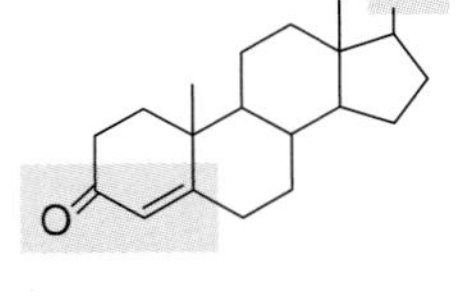

Male sex hormone
Testosterone

Metabolic regulator
Cortisone

Pregnancy hormone
Progesterone

Bile salt for fat digestion
Cholic Acid

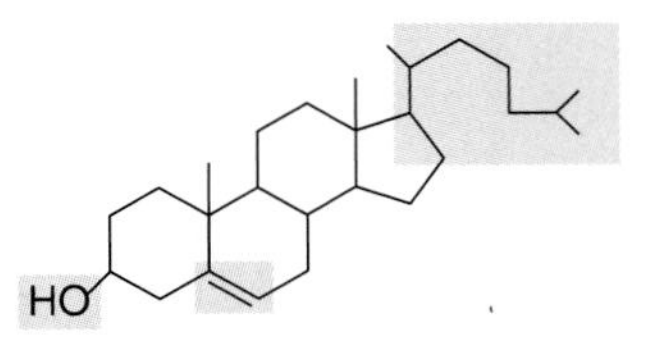

Cell membrane component
Cholesterol

Figure 10.17

Molecular structures of some important steroids.

Figures Alive!

sels. This plaque restricts blood flow and can lead to a stroke or heart attack. Therefore, people are advised to regulate their dietary intake of cholesterol, which is found in milk, butter, cheese, egg yolks, and other foods rich in animal fats. But one must keep in mind that some "cholesterol-free" foods can nevertheless contribute to the buildup of cholesterol in the body, where it is synthesized from fatty acids of animal or vegetable origin. A diet rich in "saturated fats" (those without double bonds between carbon atoms) is particularly likely to lead to elevated serum cholesterol.

- More information about dietary cholesterol appears in Section 11.5.

10.9 Your Turn

Carefully examine the structural formulas given in Figure 10.17. Identify the similarities in the structures of each of these pairs.

a. estradiol and testosterone
b. estradiol and progesterone
c. cholic acids and cholesterol

The role of cholesterol in the body is relatively passive, but steroid hormones are involved in a tremendous range of physiologically vital processes, including such popular pastimes as digestion and reproduction. Because of the importance of these functions, medicinal chemists have, over the past 50 years, synthesized many derivatives of naturally occurring steroid hormones. These drugs, developed to mimic or inhibit the activities of the hormones in the body, have been variously described as "miracle drugs," "killer compounds," or "sleazy therapeutic agents." Perhaps more than any other type of pharmaceutical, steroid-related drugs are involved with social and ethical issues. These issues include birth control, abortion, diet, bodybuilding, drug abuse, and drug testing. We begin by looking at drugs related to sex hormones.

10.8 "The Pill"

Sex hormones are the chemical agents that determine the secondary sex characteristics of individuals. Female sex hormones are classified as **estrogens;** male sex hormones as **androgens.** All males have a low concentration of female sex hormones, and there are low levels of male sex hormones in all females. However, androgens predominate in males and estrogens in females.

Because of their importance, androgens and estrogens were the first steroidal hormones studied in great detail. When this work was just beginning in the 1930s, techniques for determining molecular structure were in their infancy. A sample of several milligrams of the pure substance was required—much more than is needed today. Because sex hormones occur only in very small quantities, Herculean efforts were required to obtain sufficient amounts for the early chemical studies. For example, one ton of bull testicles was processed to yield just 5 mg of testosterone, and four tons of pig ovaries provided only 12 mg of estrone, a precursor of estradiol. Fortunately, improved technology and instrumentation allow modern chemists to determine molecular structures with samples weighing only a fraction of a milligram. After the molecular structures of the sex hormones were determined, work could proceed on the synthesis of drugs of similar structure. These efforts ultimately led to the creation of "the Pill"—the oral contraceptive that has had such a profound effect on modern society.

- A precursor is a molecule that can be converted directly to a different molecule.

As with aspirin, birth control drugs came about through molecular modifications, in this case, changing substituents selectively on the steroid nucleus. Interestingly, the initial motivation of the research that ultimately led to oral contraceptives was the enhancement of fertility in women who found it difficult to conceive. When fertilization occurs, the hormone progesterone is released, carrying a number of chemical messages. Some of these messages help prepare the uterus for the implantation of the embryo. Others block the release of pituitary hormones that stimulate ovulation. The reason for this is clear. Ovulation during pregnancy could lead to very serious complications. Gregory Pincus and John Rock injected progesterone into patients to block ovulation

Prepackaged contraceptive pills.

and stimulate body changes related to pregnancy. Their hope was that when the therapy was discontinued, a kind of rebound would occur and ovulation would be stimulated. Such a response, now known as the "Rock rebound," does, in fact, take place and fertility increases.

Unfortunately, progesterone was expensive and not very effective when administered orally. It also caused some serious side effects in a small percentage of patients. Therefore, chemists working in a number of pharmaceutical firms set out to develop a synthetic analog for progesterone that could be taken orally, would reversibly suppress ovulation, and would have few side effects. The ultimate goal of these efforts soon became the inhibition of fertility, not its enhancement. In the mid-1950s, Frank Colton, a chemist at G. D. Searle, synthesized norethynodrel. The molecular structure of this compound (Figure 10.18) shows some subtle but significant differences from progesterone, notably the replacement of —$COCH_3$ on the D ring with —OH and —C≡CH. As a consequence of these changes, the norethynodrel molecule is tightly held on a receptor site, which prevents its rapid breakdown by the liver and permits its oral administration. Norethynodrel became the active ingredient in Enovid, the first commercially available oral conceptive, which was approved for sale in 1960.

The drug's availability had an immediate and substantial impact. In 1962, 1.2 million women in the United States used the "Pill," then containing 150 micrograms (μg) of estrogen and approximately 10 mg of progestin, a close chemical relative of progesterone. Since the development of the original birth control pill, further molecular modifications (many of them minor) have led to decreased dosage and minimized side effects. Currently, over 16 million users ingest one of about 50 commercial varieties of oral contraceptives with just 20–35 μg of estrogen and approximately 1 mg of progestin or the "mini-pill" that just contains the progestin. About 8 out of 10 women in the U.S. have used the birth control pill at some point in their lives, and 4 out of 10 American women between the ages of 18 to 29 currently take the Pill. The current dosages are evidence of how just scant amounts of the hormones act as major chemical messengers with profound effects. Recent advances include an implant (Norplant®) comprised of six small plastic rods surgically placed under the skin of the upper arm for up to five years, during which time the rods slowly release progestin into the body. Another development is "the shot" (Depo-Provera®), a highly-effective progestin injection given by a physician every three months, although infertility may last up to a year.

The mechanism for the action of steroid-based contraceptives is diagrammed in the simple schematic of Figure 10.19. In effect, the drug "fools" the female reproductive system by mimicking the action of progesterone in true pregnancy. Birth control steroids, being progesterone-like molecules, send a chemical message that is similar to the message carried by progesterone. Because pregnancy is simulated, ovulation is inhibited. In effect, the message this time is not "Hey, Mom, I'm here!" but rather "Hey, you think I'm here, but I'm not!"

In the fall of 2000, a University of Edinburgh study reported that researchers found a drug, desogestrel, a synthetic hormone, that suppresses daily sperm production while maintaining normal levels of testosterone in the body. Some minor side effects occur, including mood swings, weight gain, and increased appetite, some of the same side ef-

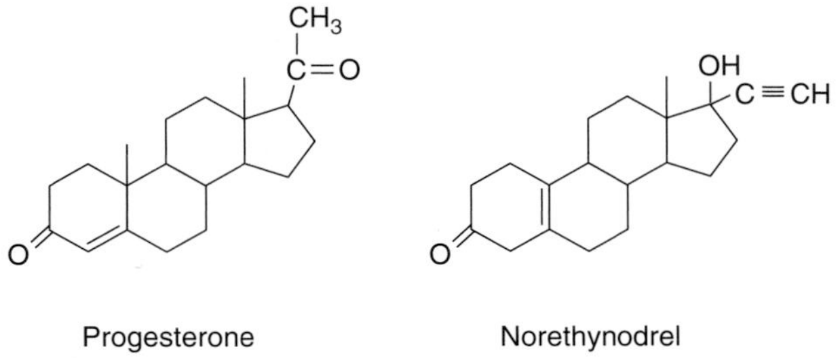

CH_3
C=O
O
Progesterone

OH
C≡CH
O
Norethynodrel

Figure 10.18

Molecular structures of progesterone and norethynodrel.

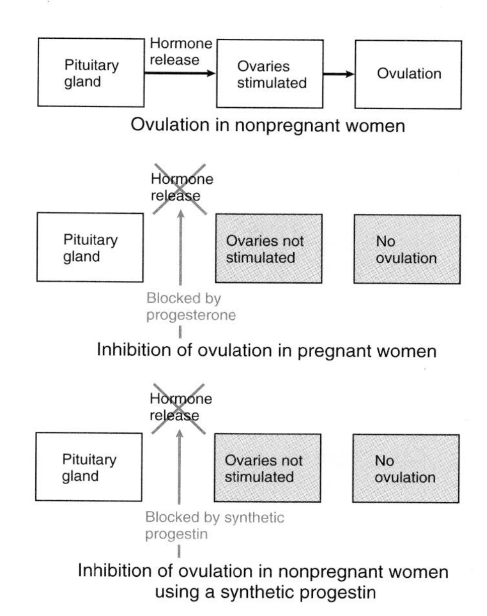

Figure 10.19

Action of a steroid contraceptive.

From *Drugs and the Human Body with Implications for Society,* by Liska, Ken. ©1994. Reprinted by permission of Prentice-Hall, Inc., Upper Saddle River, NJ.

fects felt by women who take contraceptive pills. After discontinuing the dose of desogestrel, sperm concentrations in all of the men in the study returned to prestudy levels within 16 weeks. A pill or implantable male contraceptive is expected by 2005. Perhaps gender equity in contraception will soon be achieved!

10.10 Consider This: Oral Contraceptives

The advent of the first commercially available birth control pill in 1960 was coincident with a cultural and sexual revolution. Direct control of fertility was put into the hands of women, the sex most directly affected by the birth of children. With this new control came added responsibility. Suggest some of the scientific reasons why chemical control of reproduction was developed first for female rather than for male reproductive systems. Suggest some cultural reasons why it might be so.

10.9 Emergency Contraception and the "Abortion Pill"

Some controversy still surrounds the synthetic steroidal hormones that control fertility by inhibiting ovulation. But a far more controversial approach to birth control are drugs taken after ovulation has occurred. Such drugs fall into two categories. The first are Emergency Contraceptive Pills (ECPs, or "the morning-after pills"), actually a large dose of ordinary oral contraceptives taken after intercourse. They function in the same way as other hormonal methods, by suppressing ovulation and making the uterus inhospitable to a fertilized ovum. ECPs were first used in the 1960s for rape victims, but the FDA has approved their emergency use when a woman has had noncontraceptive intercourse within the previous 72 hours. ECPs can only be obtained from a physician.

The second category came into legal use in the U.S. in September 2000, when the Food and Drug Administration approved RU-486 for marketing under the name Mifeprex. Since its discovery was announced in France in 1982, RU-486, otherwise known as "the French abortion pill," has been used successfully by over 400,000 women in France, Sweden, the United Kingdom, and China. RU-486 was extensively tested and found to be 96% reliable. Moreover, it appears to have a relatively low incidence of serious side effects. Mifeprex, which is actually two pills (mifepristone and misoprostol) taken under a doctor's supervision over two days, has proven to be very effective in ending pregnancies up to the 50th day following conception.

- China manufactures its own version of RU-486, a copy of the Rousell-Uclaf product first synthesized in France.

Roussel-Uclaf, which holds the drug's patent, has sold RU-486 in Europe since 1988. Although abortion is legal in the United States, it is certainly not universally accepted by the American public and the medical community. Because of threats of boycotts and protests, Roussel-Uclaf did not introduce RU-486 into this country until the company had arranged to donate the rights to the drug to the Population Council, a nonprofit research organization. The Council began clinical testing of RU-486 in October 1994, administering it at a dozen clinics and hospitals in various parts of the country. In March 1996, after completion of its clinical trials, the Population Council requested approval of RU-486 from the FDA. The FDA final approval took almost 4 years because of the enormous pressure from both political and medical organizations.

RU-486 and progesterone have very similar molecular structures. Thus, RU–486 is an **antagonist** for progesterone—it occupies the progesterone binding site, but shows no activity. Consequently, progesterone is not released, and no pregnancy protein production signal is sent. Because progesterone activity is essential for implantation of the embryo in uterine cells, the developing embryo is spontaneously aborted.

RU–486

An alternative method for drug-induced abortion has been developed in the United States that could make the potential availability of RU-486 in the United States a moot point. In August 1995, gynecologists announced that two widely available prescription drugs used in combination had been shown to have results very similar to those associated with mifepristone for nonsurgical abortions no more than seven weeks after conception. The medical protocol is much like that followed with RU-486, but the initial drug administered is methotrexate. Unlike RU-486, methotrexate is not a progesterone mimic. Rather, it works as an abortifacient by blocking a B vitamin called folic acid required for normal cell growth and division. Hence, methotrexate inhibits the development of the embryo and placenta. As with RU-486, the expulsion of the fetus is induced by misoprostol.

What makes the use of methotrexate and misoprostol for this purpose particularly interesting is the fact that each had been already approved by the FDA, although for different uses. Methotrexate has been used for some time, in large doses to treat some cancers and in small doses for rheumatoid arthritis and psoriasis. Misoprostol is prescribed to protect the stomach linings of people who require daily doses of powerful anti-inflammatory drugs or anti-ulcer medications. Once a drug has FDA approval, a licensed physician can use it for any purpose, including "off-label" uses not originally specified or intended. Thus, no exhaustive approval process was required before these drugs could be employed in combination as abortive agents.

10.11 Consider This: "Off-Label" Drugs

Additional information about "off-label" uses of drugs and many other subjects appear on the FDA web site. Drugs such as methotrexate and misoprostol can be prescribed for "off-label" uses (also called "unapproved," "unlabeled," or "extra-label" uses).

a. In general, does prescribing medications to be used in this way strike you as a reasonable thing for physicians to do? Explain your reasoning. Compare your opinion with what the FDA thinks.

b. Use the search engine provided at the site and enter "off label." From the lengthy list of citations, pick two that interest you and summarize their contents.

c. Explain how what you learned from this exercise strengthened or modified your opinion of "off-label" uses.

10.10 Anabolic Steroids: What Price Glory?

Androstenedione—is it or is it not a steroid? Attention was drawn to this controversy by the 1998 major league baseball home run race and the declared use of "andro" by the home run champion, Mark McGwire. To answer this question, we need a closer look at **anabolic** ("building-up") **steroids.** Like birth control drugs, anabolic steroids are controversial. Moreover, like some of their contraceptive chemical cousins, anabolic steroids also were created for quite a different purpose than their ultimate use. These steroids were developed initially to help patients suffering from wasting illnesses to regain muscle tissue. Ironically, their use has now become perverted by the strong who seek to become even stronger.

- In 2002, a former most valuable player alleged that 50% of Major League Baseball players use anabolic steroids. Major League Baseball does not test for them.

It has long been known that testosterone promotes muscle growth as well as the development of male secondary sexual characteristics. Drug companies sought to pursue this avenue to produce a testosterone-like drug that would stimulate muscle growth in debilitated patients, such as those recovering from long-term illness. The intent was to modify the testosterone molecule in such a way that its analog would have the desired effects on muscle development without serious negative side effects. Anabolic steroids were the result.

This research, as any involving sex hormones, required the use of a suitable animal model to evaluate the effectiveness and safety of the drugs. Ethical considerations and public opinion preclude the use of human subjects for testing in such unpredictable circumstances. Therefore, castrated rats were used to test anabolic steroids. Various trials compared prostate gland weight with the weight of an isolated abdominal muscle. The idea was to develop a drug that increased muscle mass (an anabolic effect) without increasing prostate mass (an undesirable side effect). Side effects such as this are said to be androgenic. They result from changes in the level of sex hormones, and they often accentuate female characteristics in males (enlarged breasts) and male characteristics in females (deep voices, beards.)

Several drug companies eventually succeeded in greatly reducing the androgenic effects of synthetic steroids, while not affecting their desired anabolic effects. The structural formulas of the two most potent anabolic steroids, norethandrolone and ethylestrenol, are given in Figure 10.20, along with that of testosterone for comparison. Note their very close similarities. In other synthetic anabolic steroids, the molecular shape has been altered by adding substituents to the A ring. These substituents interfere with the fit of the molecule on the androgen activity receptor, but they do not impair anabolic activity. Two examples are stanozol and oxymetholone.

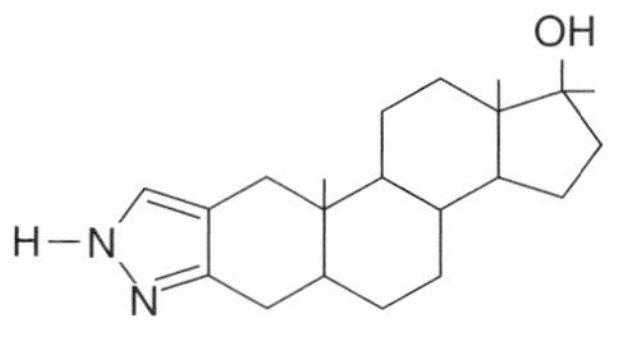

OH
H–N
N
Stanozol

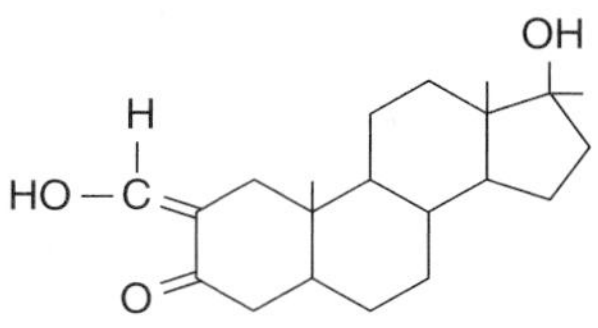

OH
H
HO–C
O
Oxymetholone

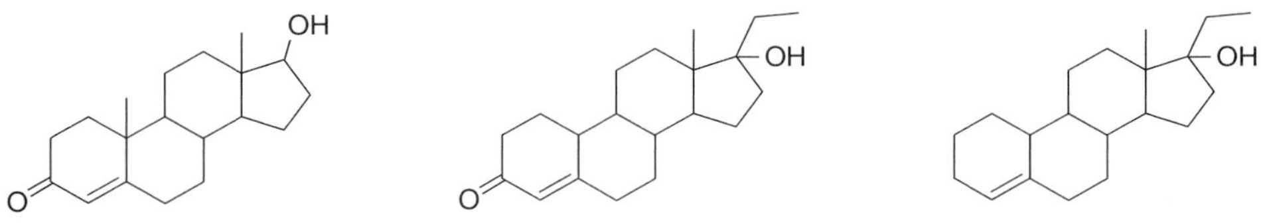

OH
O
Testosterone

OH
O
Norethandrolone

OH
Ethylestrenol

Figure 10.20

Molecular structures of testosterone and two anabolic steroids.

Research efforts to properly balance anabolic and androgenic effects in synthetic anabolic steroids have not been completely successful. Unfortunately, this serious drawback has not precluded the widespread use of these drugs. A vast new market has sprung up on the world's playing fields and in athletic training facilities, even though the drugs can be obtained legally only by prescription. Because anabolic steroids increase muscle mass, they appeal to some athletes who compete in strength-related sports such as football, weight lifting, certain track and field events, even baseball. The desire to win has led athletes to take anabolic steroids to gain a purported competitive edge. And the problem seems to be pervasive. It has been estimated that half of recent Olympic athletes, in sports ranging from weight lifting to figure skating, have used anabolic steroids at some time in their careers. Annual sales of illegal steroids to athletes are estimated to be in excess of $400 million—in spite of the fact that legitimate experts in exercise physiology and related fields hold opposing opinions about the merits of whether steroid use enhances athletic performance.

On the other hand, it is well established that using large doses of anabolic steroids over time can cause a variety of undesirable side effects. In males, these include shrinking testes, difficulty in urination, impotence, fluid retention, baldness, high blood pressure, and heart attack (in short, the symptoms of aging). Women suffer from masculinization in which female secondary characteristics are lost. Both sexes show increased aggressiveness, unpredictable periods of violent mood changes, and other behavior disorders, the so-called "roid rage." Heavy users often compound the problem of drug abuse by using more than one anabolic steroid at a time—sometimes in untested combinations. Many abusers seem to be convinced that "more is better." For example, steroids that are prescribed in legitimate therapeutic doses of a few milligrams have been taken in doses 20 times larger, even though such a large overdose may be lethal.

One of the most challenging competitions in sports is that between athletes who use performance-enhancing illegal drugs and the chemists who test for them. Very sophisticated chemical separation and analytical techniques have been developed to detect banned substances in blood and urine samples at the parts-per-billion level. Detection of synthetic anabolic steroids is difficult because they are generally used in relatively small amounts. But the real detection problem arises because they are chemically very similar to compounds that occur normally in the body. The success of synthetic chemists now becomes the analytical chemists' burden.

Athletes who use anabolic steroids illegally have resorted to many strategies to avoid detection. These range from simple substitution of a "clean" urine sample for their own, to the rather extreme practice of draining the bladder and then using a catheter to fill the bladder with a sample of "pure" urine just before the drug test. Methods for steroid testing must take into consideration the fact that the fat-soluble steroids take some time to completely clear the body. Because of this time lag, elaborate schemes have been developed for tapering off illegal drugs to get below allowable limits just before competitions. Not surprisingly, some coaches and athletes with very little previous curiosity of medicinal chemistry now make it a significant area of interest—at least the part that applies to steroids.

Returning to the question posed at the beginning of this section: Is androstenedione a steroid? It was developed in the 1970s in East Germany to enhance the Olympic performances of its athletes. "Andro," as it is called, is sold in the United States as a nutritional supplement, not a steroid. Therefore, it can be purchased "over the counter" without a prescription. Its purveyors tout its effectiveness, as one advertisement says: "Think of one compound that will let you put on 10 pounds of muscle in three weeks while adding 20 pounds to your bench press. Now try and come up with a compound that will do that legally . . . One customer gained 18 pounds of muscle in six weeks! $29.17 for 100 capsules (100 mg androstenedione). Typical doses are one capsule three to six times daily." Although gains in muscle mass are debatable and may vary with "andro" use, what is not in question is that androstenedione increases testosterone levels. Classified as a steroid, although technically not an anabolic steroid, androstenedione (Figure 10.21) is a very close chemical relative of testosterone. Given the similarities, once ingested, "andro" is converted directly to testosterone, which is an anabolic steroid.

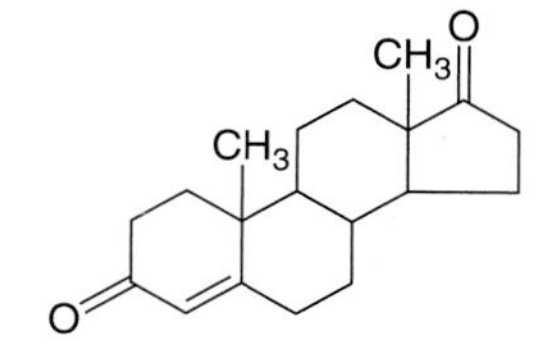

Figure 10.21
Structural formula of androstenedione.

10.12 Your Turn

Carefully examine the structural formula for androstenedione given in Figure 10.21 and the structural formula for testosterone given in Figure 10.20. Identify the similarities in these two structures.

10.13 Consider This: Olympic Advantage

In a survey of world-class athletes, 50% said they would take a drug that would enable them to win an Olympic gold medal, even though it would probably cause their death within 10 years. Assume you are an editor of the sports page for your college newspaper. Write an editorial that will help your readers understand the potential risks and benefits that are associated with taking performance-enhancing drugs.

It is well established that testosterone stimulates the creation of more protein, which replaces and rebuilds muscle fibers broken down by exercise. This allows athletes to exercise harder, yet recover more quickly. Dr. Linn Goldberg, head of Health Promotion and Sports Medicine at the Oregon Health Science University stated in the September 8, 1998 *New York Times* that "When androstenedione is converted to testosterone, it is no different than taking anabolic steroids. It's an attempt to cheat to improve your performance, by unnaturally boosting your testosterone levels."

Although allowed by Major League Baseball, the National Hockey League, and the National Basketball Association, "andro" is banned by the National Football League, the National Collegiate Athletic Association (NCAA), and the International Olympic Committee. The Association of Professional Team Physicians has issued a position statement recommending that androstenedione be removed from over-the-counter status and be banned from all competitive sports. Yet some trainers and coaches insist that no compelling research evidence exists that the side effects from androstenedione are as adverse as those from excessive testosterone use. Patrick Arnold, the person who introduced androstenedione commercially into the United States, contends that using it is "a very safe way of performance enhancement, as safe as drinking a cup of coffee." Countering this is Dr. Gary Wadler, a steroid authority, who cautions: "It would be foolish to delude anyone that this is an innocuous substance."

10.14 Consider This: Buying "Andro"

As this edition went to press, "andro" could be purchased via the Web. Is this still possible? To find out, search for "androstenedione" and see what turns up.

a. Is "andro" still sold over the counter (via the Web)?
b. What do web sites claim about "andro"?
c. What does this steroid cost individuals both in dollars and in health effects? As always cite both the sponsor and URL of the web sites that you visit.

10.15 Consider This: FDA in Handcuffs?

Writing about "andro," Pamela Zurer, a columnist for *Chemical & Engineering News* (September 28, 1998, page 37) comments: "I lay the blame squarely on Congress, which handcuffed FDA with the Dietary Supplement Health & Education Act (DSHEA) of 1994. Until some regulatory body is given the authority to require the safety . . . of so-called nutritional supplements . . . we're going to see a lot more of these controversies. I hope not too many people ruin their health in the meantime." It may take a bit of detective work, but you can find the details of the DSHEA on the Web.

a. See for yourself—did this really handcuff the FDA? Give the rationale for your choice.
b. Whose needs was the DSHEA intended to meet?

10.11 Drug Testing and Approval

Many drugs and medicines were once available without a prescription. Before the Food and Drug Administration regulations existed, virtually anything could be sold as a cure. In fact alcohol, cocaine, and opium were included in some early products without notification to users. The Food, Drug, and Cosmetic Act of 1938, and its amended versions in 1951 and 1962, defined **prescription drugs** as ones that could be habit forming, toxic, or unsafe for use except under medical supervision. Virtually everything else is available for sale over the counter.

Manufacturing prescription drugs is done on a colossal scale to meet patients' demands. In 2000, about three billion prescriptions were filled in the United States, accounting for spending of about $132 billion. But the pathway for a new drug from a laboratory to a pharmacy shelf is long and complicated. All proposed new drugs, whether extracted from natural materials or synthesized in the laboratory, are subjected to exacting series of tests before they obtain FDA approval. Current law requires evidence that the drugs are safe as well as effective before such approval is granted. The steps for approval are summarized in Figure 10.22.

From discovery to approval, the development of a new drug takes, on average, nearly 12 years and about $500 million—over three times the cost of a decade ago. The expenses are principally for the various stages of drug testing, probably the most complicated and thorough premarketing process ever developed for any product. Although the number of pills getting through the funnel of Figure 10.22 gets progressively smaller with time, the diagram does not begin to convey the high rejection rate of proposed drugs. Currently, the odds of getting a candidate drug from identification to approval are 1 in 10,000. For every 10,000 trial compounds that begin the process, 20 make it to the level of animal studies, half that many get clearance for use in clinical testing with humans, and finally one gets FDA approval.

Examples already encountered in this chapter have suggested the long process of chemical hide-and-seek that often precedes the identification of a compound as possibly having therapeutic properties. Once the promising candidates have been identified, they are subject to *in vitro* studies, those carried out in laboratory flasks. Simultaneously, a wide range of activity is undertaken by the pharmaceutical company. Chemists and chemical engineers investigate whether the compound can be produced in large volume with consistent quality control. Chemists and pharmacists together carry out studies of the most effective way to formulate the drug for administration—as capsules, pills, injection, syrup, or perhaps something more unusual such as a nasal spray, skin patch, or

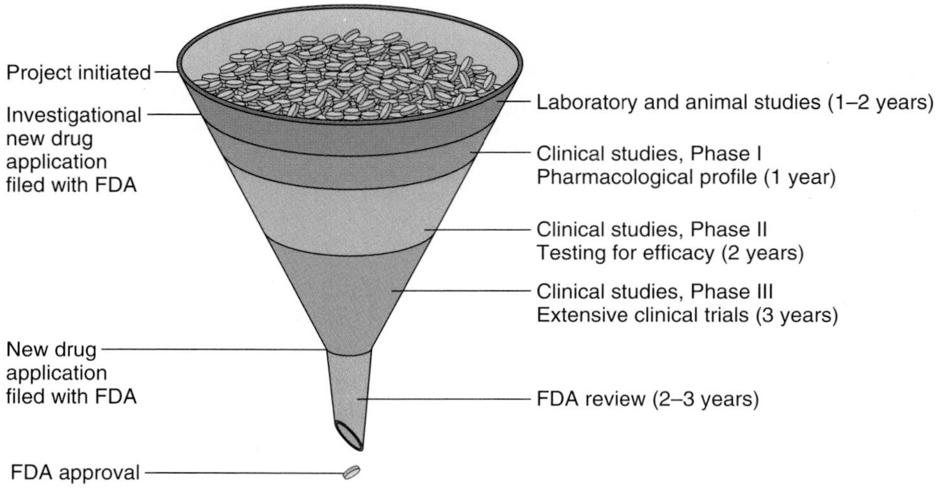

• *In vitro* literally means "in glass."

Figure 10.22

Schematic of the drug approval process in the United States.

implant. Chemical stability and shelf life are evaluated. Economists, accountants, patent attorneys, and market analysts conduct research on the likelihood of deriving a profit from the product. A fair, responsible price must be established that allows the corporation to recapture the extensive development costs while keeping the drug affordable.

Only a small fraction of compounds survive this scrutiny to move on to animal testing. Such *in vivo* ("in life") tests are designed to determine the drug's efficacy, safety, dosage, and side effects. It is typically at this stage that pharmacologists determine the drug's mode of action, how it is metabolized, and its rate of absorption and excretion. The tests are carefully controlled, requiring the collection of very specific kinds of data. For example, drugs are evaluated for their short- and long-term effects on particular organs (such as the liver or kidneys) and on more general systems (such as the nervous or reproductive system). Perhaps the most controversial toxicity testing involves the determination of the **lethal dose-50 (LD_{50})**, the minimum dose that kills 50% of the test animals.

10.16 Consider This: Animals and Drug Testing

Animal rights groups often target the LD_{50} standard as an example of callous indifference to animal welfare. Other groups argue that standards such as LD_{50} are necessary to ensure drug safety and effectiveness. Take a position on the issue and write a letter to editor of your school newspaper defending your position.

Results of animal tests must be submitted to the FDA for evaluation before permission is granted to proceed to the next stage—clinical testing of the drug on humans. In addition, approval must be obtained from local agencies and authorities such as a hospital's ethics panel or medical board. The FDA must establish whether the drug is effective *and* safe before it can be sold to the public. What needs to go on the label regarding use, side effects, warnings, etc., must also be determined. Typically, clinical studies involve the three phases identified in Figure 10.22: Phase I. Developing a pharmacological profile, Phase II. Testing the efficacy of the drug, and Phase III. Carrying out the actual clinical tests. Most of the safety tests of Phase I are done with healthy volunteers, who are given single and repeat doses of the drug in various amounts. It is also at this stage that researchers look for interactions with other drugs. Double-blind placebo tests are administered to small patient groups in Phase II to test the drug's effectiveness on patients having the condition that the drug is designed to affect. In this protocol, neither the patient nor the physician knows which patients are receiving the drug and which are receiving a placebo, an inactive imitation that looks like the "real thing." Such tests are designed to eliminate bias from the interpretation of the results. Long-term toxicity studies are also initiated during Phase II. The clinical trials are expanded in Phase III, while manufacturing processes are scaled up and tests are carried out on the stability of the drug. The entire process often requires six years or more. In the early 1980s an average of 30 clinical trials were done on drugs that were ultimately approved; that number has risen to over 70, the trials have become more complex, and the number of patients treated per trial (about 4000) has more than doubled over the past 20 years.

Large-scale clinical trials are desirable because a large pool will more likely include a wide range of subjects. Variety is important because the drug in question may have markedly different effects on the young and the old; men and women; pregnant or lactating women; infants, nursing infants, and unborn infants; and persons suffering from diabetes, poor circulation, kidney problems, high blood pressure, heart conditions, and a host of other maladies.

Once clinical trials have been completed successfully—typically by only 10 drugs out of an original pool of 10,000 compounds—the test data are submitted to the FDA as part of a new drug application. This document can easily exceed 3500 pages. On review, the Agency may require the repetition of experiments or the inclusion of new ones, thus adding years to the approval process. Of the drugs submitted to clinical testing, only about one in 10 is finally approved.

10.17 Consider This: Double-Blind Testing

Double-blind protocols have other uses than testing for the effectivenss of a drug. For example, physicians may use a double-blind test for diagnosing food allergies. In such tests, the physician administers a series of foods and placebos in disguised form. The test substances are labeled in code known only to a third person. Why do you think double-blind tests for the effectiveness of a drug or for establishing a food allergy are necessary? Compared to the single-blind tests, in which only the patient is unaware of the drug or food being administered, how do double-blind tests effect the reliability of the information gained?

Once it receives the FDA's imprimatur, a drug can be sold in the United States. Nevertheless, it still remains under scrutiny, monitored through reports from physicians. Drugs are removed from the market if serious problems occur. Some side effects show up only when large numbers of users are involved. A new drug application, for example, typically includes safety data on several hundred to several thousand patients. An adverse event occurring in 1 in 15,000 or even 1 in 1,000 users, could be missed in clinical trials, but it could pose a serious safety problem when the drug is used by many times that number of patients. Such an example is Seldane, a second-generation antihistamine for treating seasonal allergies without causing drowsiness. A small, but statistically significant number of patients who took Seldane along with certain antibiotics or antifungal medicines developed abnormal heart rhythms. In 1998, the FDA removed Seldane from the approved list.

A number of other recalls were for drugs on which the public had come to rely. Some of these are noted in Table 10.4. The spate of withdrawls has encouraged some to question FDA's procedures.

The lengthy process for drug testing and approval is not without controversy. Most people probably favor thorough screening of any proposed drug, but the price of such protection is high. The most obvious costs are monetary. Bringing a new drug to market is incredibly expensive, and the number of new drugs being developed has risen even as research and development (R&D) costs have risen (Figure 10.23). Of course much of the expense is passed on to the consumer. For example, in a hospital a single dose of Aggrastat, a medication that dramatically increases the likelihood of surviving a heart attack, costs $1100. Such prices have driven the costs of medical care and medical insurance to astronomical levels. One issue in the debate over health care reform is who will pay for the research and development that ultimately leads to new medication.

10.18 Consider This: Who Should Pay?

Who should pay for Aggrastat, a life-prolonging drug—individuals, medical insurance, pharmaceutical companies, or the government? Take a stand and defend your position.

Table 10.4

Recently Recalled Prescription Drugs

Drug	Use/Treatment	When Removed	Reason
Pondimin	Weight loss	1997	Heart valve abnormalities
Redux	Weight loss	1997	Heart valve abnormalities
Posicor	Blood pressure, chest pains	1998	Dangerous reactions with 25 other medications
Duract	Pain reliever	1998	Severe liver damage
Hismanal	Antiallergies	1999	Cardiac arrest
Propulsid	Heartburn	2000	Cardiac arrest
Raxar	Antibiotic	1999	Liver toxicity
Rezulin	Adult diabetes	2000	Liver failure
Lotronex	Irritable bowel syndrome	2001	Severe bleeding of the colon
Baycol	Lipid lowering	2001	Severe muscle damage

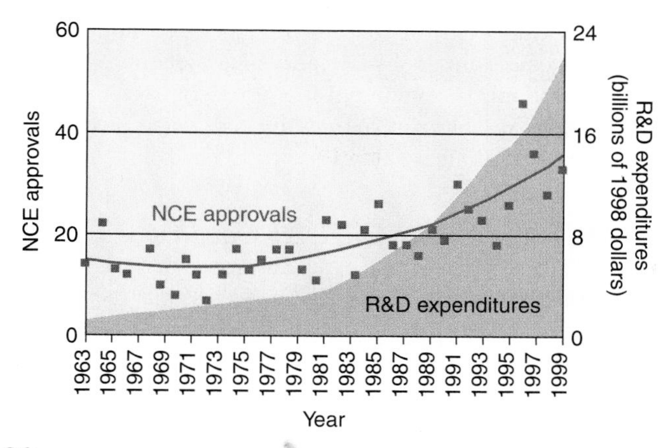

Figure 10.23

Drug development statistics: new certified approvals (NCE or new chemical entity) and development costs (R&D expenditures).

Source: Tufts Center for the Study of Drug Development, Impact Report, Vol. 2, December 2000.

But more than money is at stake. In some cases, the costs of the protracted drug approval process may be human lives. When a patient is suffering from an almost certainly fatal disease such as some cancers, the risk/benefit equation changes. When there is nothing to lose, people are willing to take great risks, including imperfectly tested drugs. Some mortally ill patients have smuggled drugs from countries where the approval process is less stringent than in the United States. Some have grasped at the straw of largely unproven remedies. And, within the system, some advocates have urged that the FDA approval process be short-circuited to permit the use of experimental drugs on patients who have no other options.

Within the limits of its legal responsibilities to balance benefits with risks, the FDA has responded appropriately. Ten years ago, an FDA review for a new drug required nearly three years whereas today it is less than a year. A new "fast-track" system has been instituted for priority drugs—those that address life-threatening ailments or new drug therapies for conditions that had no such therapies. The fast-track policy promises to have priority drugs, if found to be acceptable, approved within six months of application. Action on non-priority drugs is to be taken within 10 months, down from the initial target of 12 months.

People suffering from rare diseases may not be able to purchase appropriate medication at any price because it may not exist. There is a significant financial disincentive for a pharmaceutical company to invest heavily in developing "orphan drugs" that will be used by only a small fraction of the population.

10.19 Consider This: A Lottery for Life?

In 1996, a lottery was held to select 2000 patients with advanced AIDS to receive an experimental drug called ritonavir. In other tests, similar drugs had proved effective in almost completely suppressing the HIV virus in most subjects. Therefore, thousands of people suffering from the disease signed up for what might be a chance at prolonged life. Responses to this unconventional means of distributing a limited supply of the drug were mixed. Ben Chang, a San Francisco AIDS activist, supported it: "I think everybody agrees the lottery is not the best way of doing this, [but] it's the only fair and equitable way of distributing what little of the drug is available." In contrast, Bob Chapman, an AIDS victim and volunteer, said: "I'm opposed to these lottery things. They are playing with people with terminal illness and putting people in competition with each other. It's medicine hitting a new low." List the ethical arguments on both sides of the issue, and indicate which position you favor.

To such considerations, one must add the objections some have to standard test protocols. Animal rights advocates are highly critical of the use of any animal subjects in drug screening. The sacrificing of test animals in establishing LD_{50} values is especially controversial. For others, the generally accepted methods of human testing are at issue. The argument is that because the drug may have some benefit and will probably do no significant harm, it is unethical to withhold it from a control population. Some terminally ill AIDS patients, by mixing and sharing test drugs and placebos, have refused to cooperate with double-blind clinical studies.

10.20 Consider This: The High Price of Drugs

Several other issues arise from prescription drug manufacturing and sale outside of the United States. In some cases American prescription drugs are sold in foreign countries, often at prices that are much lower than those in the U.S. As a result, the press has reported about senior citizens who travel to Canada or Mexico to buy their prescription drugs at reduced prices. The situation with several successful medicines used in AIDS patients poses another dilemma. Brazil has decided to disregard patents on the anti-AIDs drug Viracept and allow its manufacture in government laboratories. Compile a list of arguments supporting each side of these actions. Which position do you support?

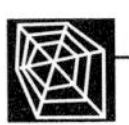

10.21 Consider This: Safety and Standards on the Fast Track

The credibility of the FDA's fast-track drug testing program is called into doubt when 11 drugs released under the program were withdrawn within a period of 5 years (1997–2001). For example, some would argue that the recall of the pain killer Duract or heartburn medicine Propulsid show that adequate testing had not taken place on these drugs. Others could counter that standards of testing have not changed, only the speed with which data are evaluated, and that no tests can produce data that are 100% reliable in humans. Search the FDA and other web sites to gather information about Duract or Propulsid. Use what you learn to reach a decision as to whether you think fast-track drug testing is overall a benefit or a risk to the health of Americans.

10.12 Brand-Name or Generic Prescriptions?

A customer gives a prescription to a pharmacist and is asked "Brand name or generic?" This scenario is played out daily in thousands of cases across the country. How is the person to decide? For millions of Americans, the cheaper generic version can mean the difference between getting the necessary medication and not being able to afford it, although not all approved drugs are available in generic form.

The two forms can be differentiated rather simply. A **pioneer drug** is the first version of a drug that is marketed under a brand name, such as Valium for the anti-anxiety drug. The **generic version** is a drug that is equivalent to the pioneer drug, but cannot be marketed until the patent protection on the pioneer drug has run out after 20 years. The lower-priced drug is commonly marketed under its generic name, in this case Diazepam instead of Valium. The 20-year patent protection on the pioneer drug begins when it is patented, not when first put on the market. In cases requiring a long preapproval time, the actual marketing period can be relatively short, even less than six years. In such a situation, a drug company has very little time to recapture its research and development costs before a generic competitor can be manufactured. Almost 80% of generic drugs are produced by brand-name firms (Figure 10.24). Like pioneer drugs, generic drugs must also be approved by the FDA.

In 1984, Congress passed the Drug Price Competition and Patent Restoration Act, which greatly expanded the number of drugs eligible for generic status. This Act eliminated the need for generics to duplicate the efficacy and safety testing done on counterpart pioneer drugs. Doing so saves drug manufacturers considerable time and money. The FDA also issued specific guidelines for a generic drug's comparability to the pioneer drug. By FDA mandate, the generic and pioneer versions must be bioequivalent in dosage form, safety, strength, route of administration, quality, performance characteristics, and intended use. In other words, it must deliver the same amount of active ingredient into a patient's bloodstream at the same rate.

Health insurance companies and the FDA suggest that policyholders choose generic rather than brand-name drugs when possible, for obvious economic reasons. According

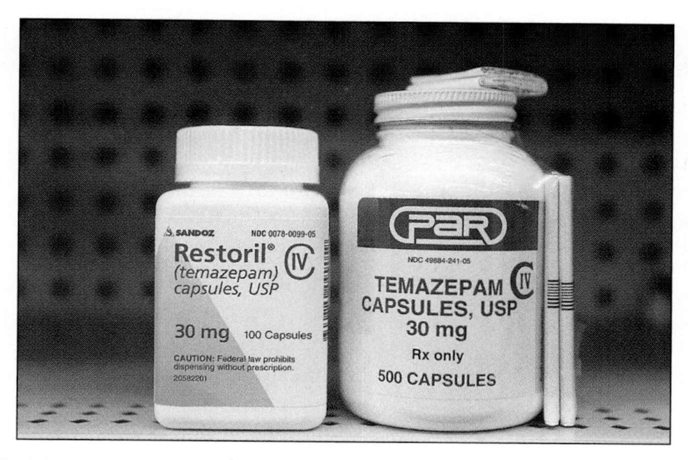

Figure 10.24
A brand name drug (Restoril) and its generic counterpart (Temazepam).

to the Congressional Budget Office, generic drugs saved consumers an estimated $8 to $10 billion at retail pharmacies in 2000. Additional billions of dollars are saved when hospitals use generics. The concern for health care costs, along with the graying of baby boomers, will likely accelerate the use of generics, as will patents that expire on additional brand-name drugs, making their generic versions possible.

10.13 Over-the-Counter Drugs

Over-the-counter (OTC) drugs are ones that allow people to relieve many annoying symptoms and cure some ailments without the need to see a physician. In accordance with laws and regulations prevailing in this country, drugs including OTC drugs are subjected to an intensive, extensive, and expensive screening process before they can be approved for sale and public use. Just as in the case of prescription drugs, the ultimate question to be answered with over-the-counter drugs is, "Is the drug safe to use?" The answer to this question for an OTC drug depends on whether a consumer is using it properly. Therefore, the FDA and pharmaceutical manufacturers must try to balance OTC safety and efficacy. Table 10.5 contains several of the major categories of OTC products and their chief components.

The over-the-counter pain relievers and NSAIDs and their mode of action have been described earlier. Their side effects include increased stomach bleeding (aspirin), gastrointestinal upset (aspirin, ibuprofen), aggravating asthma (aspirin), kidney (acetaminophen) or liver (acetaminophen) damage at high dosage or with chronic use.

More than 100 viruses are responsible for the misery from the common cold. A whole host of cold remedies, most with multiple components, are designed to help the sufferer. Antihistamines relieve the running nose and sneezing, but cause drowsiness and often lightheadedness. Because they induce drowsiness, it is not surprising that approved sleep aids are all antihistamines. However, children often experience insomnia and hyperactivity after taking them. Decongestants reduce swelling when viruses invade the mucous membranes but the adverse effects include nervousness and insomnia. Nasal sprays relieve the swollen nasal tissues, but their use beyond a three-day limit often leads to a physical dependency.

Coughing is an natural way to rid the lungs of excess secretions. Expectorants make the phlegm easier to cough up, while suppressants provide relief and restful sleep. The presence of both in most cough remedy preparations seems senseless. Codeine and dextromethorphan have equally good cough-suppressing potential but the former has a reputation for being habit forming, a property that limits its over-the-counter availability in some states.

Diet aids contain phenylpropanolamine or benzocaine, a topical painkiller, that apparently deadens the sensitivity of the tastebuds. Phenylpropanolamine, also a

Table 10.5

Examples of Over-the-Counter Drugs and Their Components

- Anagesics and Anti-inflammatory Drugs (NSAIDs)
  - Aspirin
  - Ibuprofen, Ketoprofen, Naproxen (Advil, Aleve)
  - Acetaminophen (Tylenol)
- Antacids and Indigestion Aids
  - Neutralizing Agents (Tums, Rolaids)
    - Aluminum, Magnesium salts
    - Calcium Salts
    - Calcium Carbonate
    - Sodium Bicarbonate
  - Agents to Block Acid Formation
    - Tagamet (Cimetidine)
    - Pepcid (Famotidine)
    - Axid (Nizatidine)
- Diet Aids
    - Phenylpropanolamine (recently banned)
    - Benzocaine
- Motion Sickness Drugs
    - Marzine or Neo-Devomit (Cyclizine)
    - Travelin or Vomex (Dimenhydrinate)
    - Benodin or Benylin (Diphenhydramine)
- Cough Remedies
  - Expectorant
    - Guaifenensin
  - Cough suppressants
    - Codeine (only in some states)
    - Dextromethorphan
- Cold Remedies
  - Antihistamines
    - Brompheniramine
    - Chlorpheniramine
    - Diphenhydramine
  - Decongestants
    - Pseudoephedrine
    - Phenylpropanolamine
    - Phenylephrine
- Sleep Aids
  - Benodin or Benylin (Diphenhydramine)
  - Unisom (Doxylamine)

decongestant, has an unimpressive effect in weight loss and has an efficacy limited to 3–4 months. In the fall of 2000, the U.S. Food and Drug Administration started advising people to stop taking over-the-counter cold medicines or appetite suppressants that contain phenylpropanolamine, citing a possible risk of stroke. The regulatory agency also asked drug manufacturers to discontinue use of the ingredient in other products.

Heartburn, indigestion, and "acid" stomach are targets of antacids and related drugs. Antacids are basic compounds containing aluminum, magnesium, and calcium salts that neutralize excess stomach acid. The popularity of the calcium-containing alternatives has grown with the promotion of the need for younger and older adults to maintain a regular supply of dietary calcium to prevent degradation of bones (via osteoporosis). A second type of antacid uses a completely different mode of activity. Histamine-2 blockers prevent the formation of excess acid. We may all be familiar with the television commercials that point out the immediate activity of the compounds that neutralize excess stomach acid versus the long-term relief that come from those that block acid production. The latter group contains brands like Axid and Pepcid, which represent drugs that went from creation in the prescription stage to OTC status in about 20 years.

The self-care revolution of the last several decades has encouraged the availability of safe and effective OTC drugs, and has provided additional pressure for the reclassification of many prescription drugs to OTC status. The third-generation allergy medicines Allegra, Claritin, and Zyrtec are under current consideration for such reclassification. Additional pressure to have the switch occur comes from the health insurance industry. Changing widely used prescription drugs to over-the-counter status greatly diminishes the insurance companies' share of payments.

10.22 Consider This: Losing Weight the Phen-Fen Way

More than 20 years ago, fenfluramine (fen), an appetite suppressant, and phentermine (phen), an amphetamine-like drug, were each FDA approved for short-term use as diet aids, although they were not very effective. In the 1980s, the drugs were taken together in a single study of 121 obese patients, resulting in an average weight loss of 30 pounds. In the 1990s, physicians began prescribing the combination for weight loss, even though long-term use of the drugs had not been approved. Eventually an estimated six million Americans, mostly women, took the combination drug. In 1995, after a one-year study, the combination use was approved by the FDA. However, the FDA approval was removed in 1997 because of evidence of serious heart valve abnormalities among phen-fen users.

a. The physician who conducted the original combination study said "I figured, gee whiz, these drugs have been on the market for 10, 12 years. Everything must be known about them" (*New York Times*, 9/23/97). With the advantage of hindsight, what is your response to his statement?
b. Assume that you were among the many who considered taking phen-fen prior to 1997. What questions would you have asked your physician at that time regarding the medication?
c. Physicians are free to prescribe licensed drugs however they see fit. Make a list of questions you have about such a policy.

10.14 Herbal Medicine

In concluding our discussion about drugs and medicines, we draw our attention back to the discovery of aspirin, the classic wonder drug. People still seek other ingredients from nature to help them live longer and better lives. Herbal remedies and folk medicines abound in most cultures. That this should be so is not surprising or astounding; nature is a very good chemist. Some (but not all) compounds found in plants and simple organisms are likely to have positive physiological effects in humans. Table 10.6 lists herbs and plants and their reported potential reduction of effects.

The Herb Research Foundation reports studies involving over 2000 patients in 23 clinical studies that have consistently found that a preparation of St. John's wort, a plant, is just as effective against mild to moderate depression as standard antidepressant drugs. In April 2001, however, a study published in the *Journal of the American Medical Association* reported that St. John's wort is ineffective against severe depression. The herb worked no better than the placebo in over 200 adults diagnosed with severe depression. The study was funded partly by the National Institutes of Mental Health and partly by Pfizer Incorporated, which makes Zoloft, the most commonly prescribed antidepression drug in the United States, with $2.14 billion is sales in the year 2000. Dr. Richard Skelton from Vanderbilt University, a co-author of the new study, recommended further studies on the use of the herb for mild depression, saying "I would like to see people with mild depression studied, and see if it works in those folks. If it works, that would be great." He recommended against using the herbal medicine until further studies are done.

In the spring of 2001, representatives from the American Society of Anesthesiologists reported concerns that patients undergoing surgery may risk unexpected bleeding when they take certain herbs within two weeks before surgery. To this point, no

Table 10.6

Common Herbs and Their Potential Decrease in Effects

Herb or Plant	Potential Decrease in Effects
Valerian, passion flower	Anxiety
Licorice, wild cherry bark, thyme	Coughs
Echinacea, garlic, goldenseal root	Colds, flu
St. John's wort	Depression
Chamomile, peppermint, ginger	Nausea, digestion problems
Valerian, passion flower, hops, lemon balm	Insomnia
Ginkgo biloba	Memory loss
Valerian, passion flower, Kava Kava, Siberian ginseng	Stress, tension

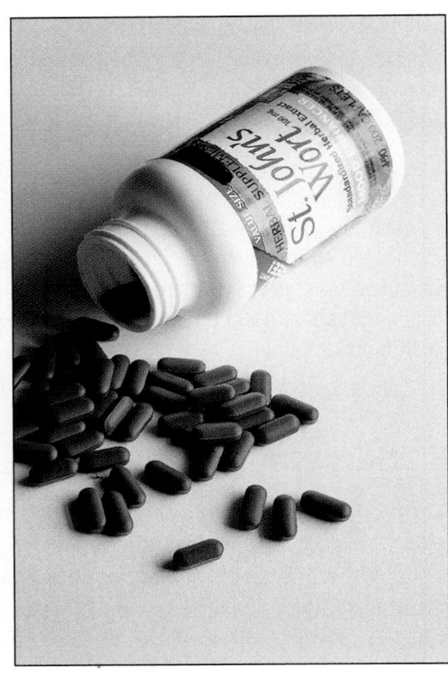

St. John's Wort plant (Hypericum) and an extract in tablet form.

scientific studies linking the bleeding and a specific herb have been published. According to Dr. John Neeldt, president of the American Society of Anesthesiologists, the familiar question "Are you taking any medications?" should be augmented with, "Are you taking any herbal remedies?"

These examples point to several issues with herbal medicines. Psychiatrists report that many of their depression patients have tried St. John's wort before coming for medical help. One estimate suggests that over a million people in the U.S. alone have tried it or are using St. John's wort. Irrespective of the accuracy of the estimate, large numbers of Americans are apparently using herbal medicines in amounts and under circumstances with little or no professional medical supervision. The manufacture of the herbal remedies is essentially unregulated. Unlike prescription or over-the-counter drugs, no regulatory group assesses purity of preparations or concentrations of active ingredients, sets amounts or delivery protocols, or studies interactions among herbal medicines or between them and the traditional medicines.

10.23 Consider This: A Female Aphrodisiac

In 2001, Niagra, a dietary supplement that is a caffeinated, nonalcoholic energy drink made with several South American herbs touted itself as an aphrodisiac that boosts the female sex drive. Its similar name, Niagra, prompted Pfizer, the manufacturer of Viagra, to sue the maker of the dietary supplement. Make a list of the arguments that the producers of Viagra and Niagara might make during a court appearance.

CONCLUSION

Molecular modifications by chemists have created a vast new pharmacopeia of wonder drugs that have significantly increased the number and quality of our days. Thanks to penicillin, sulfa drugs, and more recent antibiotics, the great majority of bacterial infections are easily controlled. Once dreaded killers such as typhoid, cholera, and pneumonia have been largely eliminated—at least in wealthy, industrialized societies. Synthetic steroids, used for birth control to bodybuilding, have transformed society. And the humble aspirin tablet has been supplanted by newer NSAIDs, which likely will be replaced by a new generation of innovatively designed "superaspirins."

But no drug can be completely safe and almost any drug can be misused. These issues become the focus when the FDA is asked to change the status of a drug from prescription to over the counter. Taking a medication is a conscious choice between the benefits derived from the drug and the risks associated with its side effects and limits of safety. Because most drugs have very wide, carefully established margins of safety, their benefits far outweigh their risks for the general population. For some drugs, however, the trade-off between effectiveness and safety involves a different balance. A drug with severe side effects may be the only treatment available for a life-threatening disease. Someone suffering from AIDS or advanced, inoperable cancer understandably has a different perspective on drug risks and benefits than a person with a severe cold. And the impersonal anonymity of averages takes on new meaning at the bedside of a loved one. That some countries like Brazil have ignored patents on anti-AIDS drugs is the first evidence of decisions based on the moral dilemmas. Herbal and alternative medicines raise new questions. Who is responsible for defining their efficacy, monitoring their purity, and developing contraindications to their use with other drugs? When chemistry is applied to medicine, science must be guided by morality, and reason must be tempered with compassion.

Chapter Summary

Having studied this chapter, you should be able to:

- Describe the discovery, development, and physiological properties of aspirin (10.1)
- Understand bonding in carbon-containing (organic) compounds (10.2)
- Apply the concept of isomerism to organic compounds (10.2)
- Recognize functional groups and the classes of organic compounds that contain them (10.3)
- Relate the molecular structure of aspirin to other analgesics (10.4)
- Understand the mode of action of aspirin and other analgesics (10.4)
- Describe the lock-and-key mechanism of drug action (10.5)
- Understand differences in molecular structure between chiral (optical) isomers (10.6)
- Recognize the structure of the steroid nucleus (10.7)
- Consider the chief functions of steroids, and some specific examples: sex hormones (testosterone, progesterone), metabolism regulators (cortisone), cell-membrane components (cholesterol) (10.7)
- Understand how birth control pills inhibit ovulation (10.8)
- Understand the mechanisms by which RU-486 and methotrexate induce abortion (10.9)
- Compare uses and abuses of anabolic steroids (10.10)
- Discuss the ethical issues in the use of steroids for birth control, abortion, and muscle building (10.8–10.10)
- Discuss the procedure for drug testing and approval and the associated benefits and costs (10.11)
- Understand the similarities and differences between brand name and generic drugs (10.12)
- Identify some of the over-the-counter drug categories and their uses (10.13)
- Describe some of the potential benefits and risks of herbal medicine (10.14)

Questions

Emphasizing Essentials

1. a. What is the intended effect of an antipyretic drug?
   b. What is the intended effect of an analgesic drug?
   c. What is the intended effect of an anti-inflammatory drug?
   d. Can any single drug exhibit all of these effects?
2. There are many subdisciplines within the field of chemistry.
   a. What do organic chemists study?
   b. How does this differ from what biochemists study?
3. Write condensed structural formulas for the three isomers of pentane assigned in 10.3 Your Turn.
4. Write the structural formula of each different isomer of hexane, C_6H_{14}. *Hint:* Be sure the bonding really is different, not just a different paper-and-pencil representation in two dimensions of the same structure.
5. Consider the isomers of butane. How many different isomers could be formed by replacing a single hydrogen atom with an —OH group; that is, how many different alcohols have the formula C_4H_9OH? Write the structural formula for each possible isomer.

6. For each compound, identify each functional group present and name the class of compounds to which it belongs.

a. $CH_3{-}O{-}CH_3$

b. $CH_3CH_2{-}C(=O){-}OH$

c. $CH_3CH_2{-}C(=O){-}CH_3$

d. $CH_3CH_2{-}C(=O){-}NH_2$

e. $CH_3CH_2{-}C(=O){-}OCH_3$

7. Which of these classes of compounds has a compound that contains only one carbon atom?

a. alcohol
b. aldehyde
c. carboxylic acid
d. ester
e. ether
f. ketone

Write a structural formula for each of these one-carbon compounds, and explain why the other classes of compounds are limited to compounds containing more than one carbon atom.

8. Some organic compounds exist in isomeric forms that are members of different classes of compounds. For example, in some cases the compound is an alcohol or an ether. For each of these, identify the class of compound represented by the formula as written.

a. $CH_3CH_2{-}OH$

b. $CH_3CH_2{-}C(=O){-}H$

c. $CH_3CH_2{-}C(=O){-}OCH_3$

Then, write the formula of an isomer with the same composition that is a member of a *different* class of compound. Identify the new class.

9. Histamine causes runny noses, red eyes, and other symptoms in hay fever sufferers. Here is the structural formula of histamine.

$H_2N{-}CH_2{-}CH_2{-}$(imidazole ring, N–H)

a. What is the molecular formula of this compound?

b. Identify the amine functional group in histamine.

c. Which part (or parts) of the molecule do you think make the compound water soluble?

10. Figure 10.6 shows a somewhat condensed structural formula for acetaminophen, the active ingredient in Tylenol.

a. Write the complete structural formula for acetaminophen, showing all atoms and all bonds.

b. What is the molecular formula for this compound?

c. Children's Tylenol is a flavored aqueous solution of acetaminophen. Predict what part or parts of the molecule make acetaminophen dissolve in water.

11. Identify the functional groups in each of these medications.

a. PABA (an ingredient in sunscreen)

$H_2N{-}C_6H_4{-}C(=O){-}OH$

b. Barbital (a sedative)

(ring: H–N–C(=O)–C(CH_2CH_3)(CH_2CH_3)–C(=O)–N–H, with O=C closing the ring)

c. Penicillin-G

$CH_3{-}C_6H_4{-}C(=O){-}NH{-}CH{-}CH{-}S{-}C(CH_3)_2{-}CH(COOH){-}N{-}C(=O)$ (β-lactam ring)

12. Ibuprofen is relatively insoluble in water but readily soluble in most organic solvents. Explain this solubility behavior based on its structural formula found in Figure 10.6.

13. This is the structural formula for methamphetamine, a stimulant.

$C_6H_5{-}CH_2{-}CH(CH_3){-}NH_2$ (as drawn: N bonded to two H)

Judging from its structural formula, do you expect it to be more soluble in fats or in aqueous solutions? Why?

14. Interpret this sentence by giving the meaning of each acronym and explaining the effect being stated. "NSAIDs have an effect on COX enzymes."

15. Compare the physiological effects of aspirin with those of acetaminophen and ibuprofen. Relate differences to the nature of each compound at the molecular and cellular levels.

16. Would aspirin be more active if it were to interact with prostaglandins directly, rather than by blocking the activity of COX enzymes? Explain your reasoning.

17. What are "superaspirins"? How do they differ from regular aspirin and other NSAIDs?

18. Identify the functional groups in morphine and demerol, using the structural formulas found in Figure 10.12. Can these molecules be assigned to a particular class of compound? Why or why not?

19. Sulfanilamide is the simplest of antibiotics known as sulfa drugs. It appears to act against bacteria by replacing *para*-aminobenzoic acid, an essential nutrient for bacteria, with sulfanilamide. Use these structural formulas to explain why this substitution is likely to occur.

$H_2N-C_6H_4-S(=O)_2-NH_2$ $H_2N-C_6H_4-C(=O)-OH$

Sulfanilamide   *Para*-aminobenzoic acid

20. Which of these compounds can exist in chiral forms?

a. NH_2 / $H_3C-C-CH_3$ / OH

b. NH_2 / $H_3C-C-COOH$ / $C\equiv N$

c. OH / $H_3C-C-COOH$ / CH_3

21. Which of these compounds can exist in chiral forms?

a. OH / $H_5C_2-C-CH_3$ / H

b. NH_2 / $H_3C-C-COOH$ / CH_2SH

c. H / H_5C_6-C-H / NH_2

22. Use the structural formulas in Figure 10.17 to answer these questions.

a. Identify the type and number of functional groups in cortisone.

b. Suggest a reason why cholic acid is more soluble in water than cholesterol is.

23. Molecules as diverse as cholesterol, sex hormones, and cortisone all contain common structural elements. Write the structural formula that represents the common structural elements.

24. In 10.12 Your Turn, you compared the structural formulas of androstenedione and testosterone.

a. What is the argument for banning the use of "andro" by athletes, which is the current policy of the International Olympic Committee?

b. What is the argument for *not* banning the use of "andro" by athletes, which is the current policy of Major League Baseball?

Concentrating on Concepts

25. The text states that some remedies based on the medications of earlier cultures contain chemicals that have been verified to be effective against disease, others are ineffective but harmless, and still others are potentially harmful. Describe how it might be determined into which of these three categories a recently discovered substance fits.

26. Draw structural formulas and determine the number and type of bonds (single, double, or triple) used by each carbon atom in these molecules.

a. H_3CCN (acetonitrile, used to make a type of plastic)

b. $H_2NC(O)NH_2$ (urea, an important fertilizer)

c. C_6H_5COOH (benzoic acid, used as a food preservative)

27. Carbon usually forms four covalent bonds, nitrogen usually forms three bonds, oxygen usually forms two bonds, and hydrogen can only form one bond. Use this information to write structural formulas for each of these compounds.

a. A compound that contains one carbon atom, one nitrogen atom, and as many hydrogen atoms as needed.

b. A compound that contains one carbon atom, one oxygen atom, and as many hydrogen atoms as needed.

28. In 10.3 Your Turn, students are asked to draw the structural formulas for each of the three isomers of pentane,

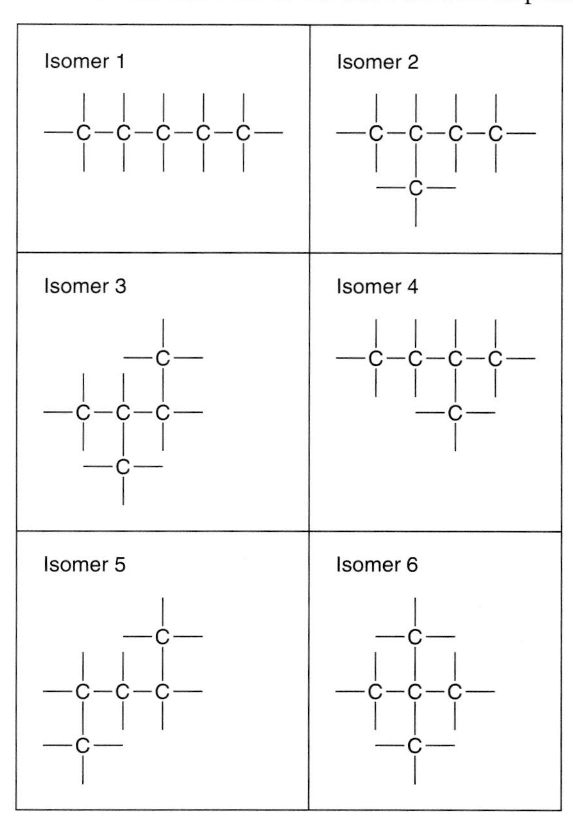

C_5H_{12}. One student submitted this set of isomers, with a note saying that six isomers had been found. Help this student to see why some of the answers are incorrect.

29. Styrene, $C_6H_5CH{=}CH_2$, is another compound that contains a ring similar to benzene, C_6H_6. One hydrogen has been replaced by the side chain —$CH{=}CH_2$. Write a set of structural formulas for styrene to show that this molecule, like benzene, also has resonance structures.

30. If aspirin is a specific compound, what justifies the claims for the superiority of one brand of aspirin tablets over another?

31. Consider Figure 10.7, which shows a schematic representation of chemical communication within the body. Write a paragraph explaining what this figure means to you in helping to explain chemical communication.

32. Consider this statement. "Drugs can be broadly classed into two groups: those that produce a physiological response in the body and those that inhibit the growth of substances that cause infections." In which class does each of these drugs fall?

a. aspirin **c.** antibiotics

b. superaspirin **d.** hormones

33. Consider the structure of morphine in Figure 10.12. Codeine, another strong analgesic with narcotic action, has a very similar structure. The only difference is that the —OH group attached to the six-member ring is replaced by an —OCH_3 group.

a. Draw the structural formula for codeine and label the functional groups present.

b. The analgesic action of codeine is only about 20% as effective as morphine. However, codeine is less addictive than morphine. Is this enough evidence to conclude that replacement of —OH groups with —OCH_3 groups in this class of drugs will always change the properties in this way? Why or why not?

34. Dopamine is found naturally in the brain. The drug L-dopa is found to be effective against the tremors and muscular rigidity associated with Parkinson's disease. Identify the chiral carbon in L-dopa, and comment on why L-dopa is effective whereas D-dopa is not.

L-dopa

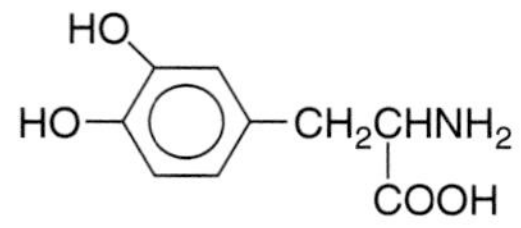

HO
HO—⟨benzene ring⟩—CH_2CHNH_2
COOH

35. Vitamin E is often sold as a racemic mixture of the D- and L-isomers.

a. Which is the more physiologically active isomer?

b. How does the cost of the racemic mixture compare with the price of the pure, physiologically active isomer?

36. Consider the fact that L-methorphan is an addictive opiate, but D-methorphan is safe enough to be sold in many over-the-counter cough remedies. Explain how this is possible from a molecular point of view.

37. Figure 10.19 diagrams the action of a steroid contraceptive. Study this diagram and then explain it in your own words to an interested friend.

38. Why are many projects to isolate or synthesize new drugs started in this country, but few actually receive FDA approval for general use?

Exploring Extensions

39. One avenue for successful drug discovery has been to use the initial drug as a prototype for the development of other similar compounds, called analogs. The text states that cyclosporin, a major anti–tissue rejection drug used in organ transplant surgery, is an example of a drug discovered in this way. Research the discovery of this drug to verify this statement. Write a brief report describing your findings, giving your references.

40. Dorothy Crowfoot Hodgkin first determined the structure of a naturally occurring penicillin compound. What was her background that prepared her to make this discovery? Write a short report on the results of your findings, giving your references.

41. Before the cyclic structure of benzene was determined (Figure 10.4), there was a great deal of controversy about how the atoms in this compound could be arranged.

a. Count the number of outer electrons that are available for C_6H_6, and then draw the structural formula for a possible linear isomer.

b. Give the condensed formula for the possible structure.

c. Compare your structure with those drawn by your classmates. Are they all the same? Why or why not?

42. Antihistamines are widely used drugs for treating allergies caused by reactions to histamine compounds. This class of drug competes with histamine, occupying receptor sites on cells normally occupied by histamine. Here is the structure for a particular antihistamine.

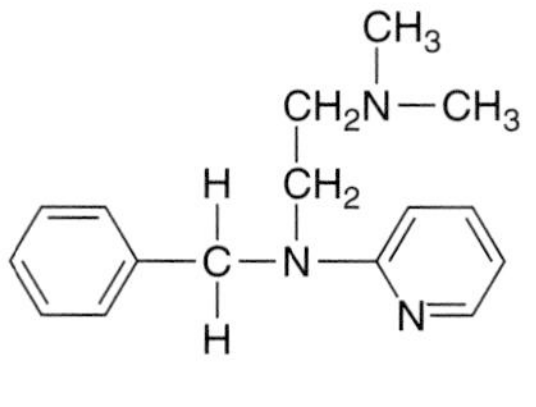

CH_3
$CH_2N—CH_3$
H CH_2
C—N
H N

a. What is the molecular formula for this compound?

b. What similarities do you see between this structure and that of histamine (shown in question 9) that would allow the antihistamine to compete with histamine?

43. Over the next few years, the FDA may consider deregulating more than a dozen drugs, nearly as many as have already been approved for over-the-counter sales during the past decade. The products that have led this trend have been the widely advertised drugs for heartburn.

a. What questions need to be answered before a drug is deregulated?

b. Will these questions change if you are considering this need from the viewpoint of the FDA, a pharmaceutical company, or as a consumer?

44. Find out more about the new process for the manufacture of ibuprofen that won the 1997 Presidential Green Chemistry Challenge Award. How does this process differ from the earlier process for manufacturing ibuprofen? Write a brief report on your research, giving your references.

45. Testosterone and estrone were first isolated from animal tissue. One ton of bulls' testicles were needed to obtain 5 mg of testosterone and four tons of pigs' ovaries were processed to yield 12 mg of estrone.

a. Assuming complete isolation of the hormones was achieved, calculate the mass percentage of each steroid in the original tissue.

b. Explain why the calculated result very likely is incorrect.

46. Herbal remedies are prominently displayed in supermarkets, drug stores, and discount stores.

a. What influences your decision whether to buy one of these remedies?

b. Choose one such remedy and carefully examine its label for as much information as possible about the active ingredients, any inert ingredients, any anticipated side effects, the suggested dosage, and the cost per dose.

c. How confident are you that the safety and efficacy of these remedies is assured? Explain your answer.

47. Habitrol was the most successful smoking-cessation prescription until the introduction of Zyban in 1997, which now has over 50% of the market. How are these two drug therapy approaches different? What is the current market share of each drug?

48. The drug approval laws of other nations are not the same as those of the United States. Choose a country and find out how its drug-approval process works compared to the process in the United States. Construct relative time lines that reveal what steps must be taken and approximately the length of time each step may require. Also, find out if specific drugs are available in that country that are not available in the United States and comment on what factors may be influencing the policies of each country.

49. Over-the-counter drugs allow consumers to treat a myriad of symptoms and ailments. An advantage is that the user can purchase and administer the treatment without effort or the expense of consulting a physician. To provide a wider margin of safety for these circumstances, the OTC versions of drugs are often administered in lower doses. Examine how general a situation this appears to be by looking at information for prescription and OTC versions of painkillers like Motrin and heartburn treatments like Axid and Pepcid. Report on your findings.

50. Herbal or alternative medicines are not regulated in the same way as prescription or OTC medicines. In particular, the issues of concern are identification and quantification of the active ingredient, quality control in manufacture, and side effects when the herbal remedy is used in conjunction with another alternative or prescription medicine. Look for evidence from herbal supplement manufacturers that address these issues, and write a report documenting your findings, giving your references.

11

Nutrition: Food for Thought

Fruits and vegetables are part of a balanced diet.

"Your choice of diet can influence your long-term health prospects more than any other action you might take."

Dr. C. Everett Koop (1988)
Former U.S. Surgeon General

If Dr. Koop, the Federal official who was charged with assuring the health of the American people feels that strongly, it might be a good idea for you to take a careful look at what you eat.

You are hungry and in a hurry, so you decide to eat at a fast-food restaurant. You get a cheeseburger, a large order of French fries, and a small chocolate shake. In doing so, you made some dietary decisions. What are the nutritional components of this meal? To what extent does the meal satisfy the Percent Daily Values? You get an opportunity to assess all this in 11.1 Consider This.

- 1 Dietary Calorie = 1 Calorie = 1 kcal = 1000 calories

11.1 Consider This: A Closer Look at Your Food

The composition of the food in your fast-food meal is given here. The Percent of Daily Values given there are calculated based on a 2000-Calorie diet. A Percent Daily Value of 100% means that you have met the daily recommendations for that substance. Your Percent Daily Values may be higher or lower depending on your calorie needs, as noted in this table.

	Cheeseburger	French fries	Shake*
Calories	330	540	360
Calories from fat	130	230	80
Total fat, g	14	26	9
% Daily Value	22	40	14
Saturated fat, g	6	4.5	6
% Daily Value	31	23	30
Cholesterol, mg	45	0	40
% Daily Value	15	0	14
Sodium, mg	830	350	250
% Daily Value	35	15	11
Carbohydrates, g	36	68	60
% Daily Value	12	23	20
Sugars, g	7	0	54
Protein, g	15	8	11
Vitamin A (% Daily Value)	6	< 2	6
Vitamin C (% Daily Value)	4	15	2
Calcium (% Daily Value)	25	2	35
Iron (% Daily Value)	15	8	4

* Not made with whole milk.

Recommended Daily Values for 2000 Cal and 2500 Cal Diets

	2000 Calories	2500 Calories
Total fat	Less than 65 g	80 g
Saturated fat	Less than 20 g	25 g
Cholesterol	Less than 300 mg	300 mg
Sodium	Less than 2400 mg	2400 mg
Total carbohydrate	300 g	375 g

Calories per gram: Fat = 9; Carbohydrate = 4; Protein = 4

a. Does the total for the three foods in any category exceed the Percent Daily Value? If so, which categories do (or come close)?

b. If the total in part **a.** is close to or exceeds the Percent Daily Value, what does that mean in terms of a daily diet?

c. There are 26 g of fat and 230 Calories from fat in the French fries. How well do these values agree with the ratio that there are 9 Calories per gram of fat?

d. Health care specialists recommend that no more than 30% of total Calories should come from fat in the daily diet. Calculate the percent of Calories in this meal that came from fat. Is it within the 30% guideline if you are on a 2000-Calorie diet? A 2500-Calorie diet?
e. What terms in the table are new to you? What terms have you seen, but actually do not know what they mean? Keep a list of these terms and refer to it as you study the chapter.

You can gain valuable information and insight about the food you eat by doing a bit of nutritional research, such as that in 11.1 Consider This. This activity included nutritional information and some questions regarding it to help focus your study. In addition, you can inspect the labels on the food packages or you can search the Web for nutritional information. The law requires that processed and packaged foods list on the package ingredients in order of decreasing concentration. Fast-food vendors distribute nutritional information either in pamphlets or on their web sites. If you completed 11.1 Consider This, you learned quite a bit about some fast foods and the substances they contain. This chapter is devoted to several aspects associated with nutrition and the food we eat.

CHAPTER OVERVIEW

In this chapter, we approach nutrition from both a personal perspective and a molecular structure point of view. We begin with an examination of what we eat and why we eat. This leads to a general consideration of the three main classes of food components: carbohydrates, fats, and proteins. Sources of these macronutrients are considered and we examine the current dietary recommendations for these three groups. What follows is a series of sections that investigate carbohydrates, fats, and proteins in considerably greater detail. In every case, we devote particular attention to molecular structure and its relationship to properties and functions. And we also include individual and public health issues. Sugars, starches, and other carbohydrates are discussed, and lactose intolerance is treated briefly. Our study of fats includes information about saturated and unsaturated fats and oils, dietary cholesterol and heart disease. The two sections on proteins treat such topics as amino acids, phenylketonuria, and the importance of a complete diet, one with adequate essential amino acids.

Because food is the source of the energy that powers our bodies and our brains, we next consider the caloric content of various foods, the recommended food energy intake for men and women of various ages and weights, and the energy expenditures associated with a number of activities. But calories are not enough to ensure a balanced diet, which also requires the correct amounts of a wide array of vitamins and minerals. Therefore, we devote individual sections to the roles of a few of these micronutrients and the hazards of an insufficient or an excessive supply of them. The chapter concludes with a consideration of how food supplies are preserved, including the use of gamma radiation, a controversial technology in the eyes of some critics. The chapter contains a wide range of information, but our goals are straightforward: (1) to help you see the connection between chemistry and nutrition by applying chemical principles in the context of the composition and reactions of foodstuffs; (2) to provide information that you can use in making daily choices about personal nutrition and health; and (3) to analyze a number of nutrition-related controversies that have appeared in the popular press.

11.1 You Are What You Eat

The title of this section states the obvious. During your lifetime, you will eat foods that collectively weigh about 700 times your body weight. Some of this prodigious mass of food will consist of delicious culinary creations elegantly prepared to celebrate some memorable social occasion. Some, no doubt, will be junk food gobbled on the run. But whatever the circumstances or the cuisine, all of us eat because food provides the four fundamental types of materials required to keep our bodies functioning. *These materials are water, energy sources, raw materials, and metabolic regulators.*

11.2 The Sceptical Chymist: A Lifetime of Food

During a lifetime, you will eat a truly prodigious amount of food, estimated to be about 700 times your adult body weight. This statement is itself quite a prodigious assertion. Do calculations to check that the statement is in the ballpark. State all of your assumptions clearly.

Hint: Start by assuming a lifespan of approximately 78 years and that your present weight is your adult weight. Estimate the weight of food eaten daily at present, and use these data to project your lifetime consumption of food.

- Many of these important properties of water are discussed in Chapter 5.

Water serves as both a reactant and a product in metabolic reactions, as a coolant and thermal regulator, and as a solvent for the countless substances that are essential for life. Our bodies are over 60% water, but H_2O cannot be burned in the body or elsewhere. Therefore, we need food as a source of energy to power processes as diverse as muscle action, brain and nerve impulses, and the movement of molecules and ions in suitable ways at appropriate times and places. We also eat because raw materials are needed for the syntheses of new bone, blood, enzymes, muscles, hair, and for the replacement and repair of cellular materials. And finally, food supplies chemical regulators such as enzymes and hormones that control the biochemical reactions associated with metabolism and all other vital processes.

Eating properly is a matter of consuming the proper foods, not simply a case of eating sufficient amounts of food. It is possible to consume food regularly, even to the point of being overweight, and still be malnourished. The meaning of this commonly used term is important. **Malnutrition** is caused by a diet lacking in the proper mix of nutrients, even though the energy content of the food eaten may be adequate. Malnutrition contrasts with **undernourishment.** The daily caloric intake of people who are undernourished is insufficient to meet their metabolic needs. According to the Food and Agricultural Organization of the United Nations (FAO) 2000 report, more than 800 million people worldwide were undernourished during 1998–1999, 99% of them in underdeveloped countries. Although the number of undernourished people in 37 developing countries fell by 100 million, principally in India and China, another 59 developing nations suffered an increase of 60 million individuals. Translating these sterile statistics into terms of human misery means that nearly one in six people in the developing countries is undernourished, evidence of the magnitude and tenacity of hunger in the world (Figure 11.1).

Contrast this dire accounting with the dietary circumstances of most people in the United States. Although there are malnourished and undernourished people in this country, hunger is typically not a consideration in a nation where nearly 55% of the population is up to 40% over their ideal weights, each consuming nearly 1000 excess dietary Calories (kilocalories) per day. One result of such excess is that about 50% of adult women and 25% of adult men annually attempt to practice girth control by dieting, generally with very limited long-term success. No wonder that Dr. Koop said "Americans are zany about food and diet. No other country gorges itself on junk food the way we do, and no other country has as many 'experts' on health diets. We have become more concerned about what we should not eat than what we should eat."

For many foods, what we eat is conveyed by food labels, which prominently display the content of carbohydrates, fats, and proteins. These are the macronutrients that provide essentially all of the energy and most of the raw material for repair and synthesis. Sodium and potassium ions (not the metals) are present in much lower concentrations, but these ions of metallic elements are essential for the proper electrolyte balance in the body. A number of other minerals and an alphabet soup of vitamins are listed in terms of the percent of recommended daily requirements supplied by a single serving of the product. It should be self-evident that all these substances, whether naturally occurring or added during processing, are chemicals. Unfortunately, this fundamental

- Vitamins and minerals are discussed in Sections 11.10 and 11.11.

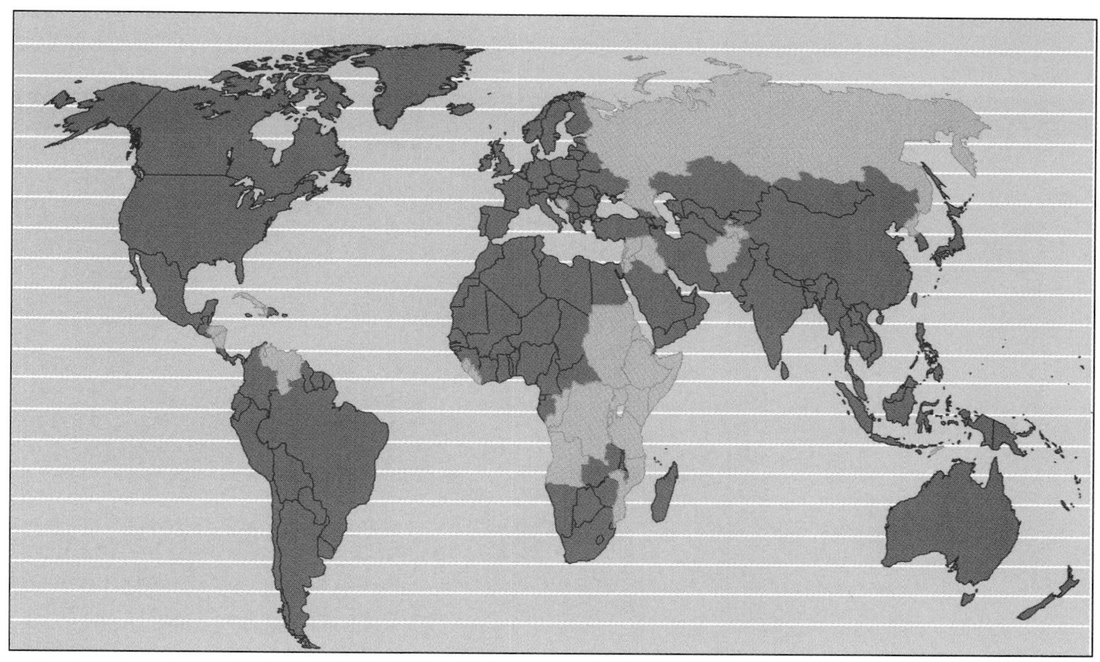

Figure 11.1

Countries (in salmon color) experiencing food supply shortfalls and requiring exceptional assistance.

(Source: The State of Food and Agriculture, 2000. Rome. Food and Agriculture Organization of the United Nations, 2000. Taken from http://www/fao.org/docrep/x4400e/x4400e04/htm.)

fact is apparently lost on those who pursue the impossible dream of a "chemical free" diet. *All* food is inescapably and intrinsically chemical, even that claimed to be "organic" or "natural."

Figure 11.2 indicates the percentages of water, carbohydrates, proteins, and fats in six familiar foods. The pie charts reveal that for this particular selection of foods, the variation in composition is considerable. But in every case, these four components account for almost all of the matter present. The percentage of water ranges from a high of 90% in 2% milk to a low of 2% in peanut butter. Peanut butter beats out steak in percent of protein and also leads these six foods in fat content. Chocolate chip cookies have the highest percentage of carbohydrate because of high sugar and starch concentrations.

Now compare Figure 11.2 with Figure 11.3, which presents similar data for the human body. It is not surprising that the composition of the human body is roughly similar to the composition of the stuff we stuff into it. We are wetter and fatter than bread, and contain more protein than milk; we are more like steak than like chocolate chip cookies. From the data of Figure 11.3, we can calculate that a 150-pound human being consists of 90 pounds of water (150 lb $\times$ 60 lb water/100 lb body) and 30 lb of fat. The remaining 30 lb is almost all composed of various proteins and carbohydrates plus the calcium and phosphorus in the bones. The other minerals and the vitamins together total less than 1 lb. This indicates that a little bit of each of them goes a long way, a point discussed in Section 11.11.

Of the nearly 90 naturally occurring elements, 11 make up over 99% of the mass of your body. Figure 11.4 is a periodic table outline highlighting these elements: hydrogen, carbon, nitrogen, oxygen, phosphorus, sulfur, chlorine, sodium, magnesium, potassium, and calcium. The first seven are nonmetals, the latter four are metals. Hydrogen, carbon, nitrogen, and oxygen are the "building-block" elements used to construct body cells and tissues; the other seven elements are the macronutrients. Table 11.1 (page 441) lists the mass percentages (grams of element/100 g body weight) of these 11 elements in the human body and gives their relative atomic abundances. Oxygen is the most abundant element when measured in grams per 100 g body weight, but hydrogen is the most plentiful element in terms of actual number of atoms present.

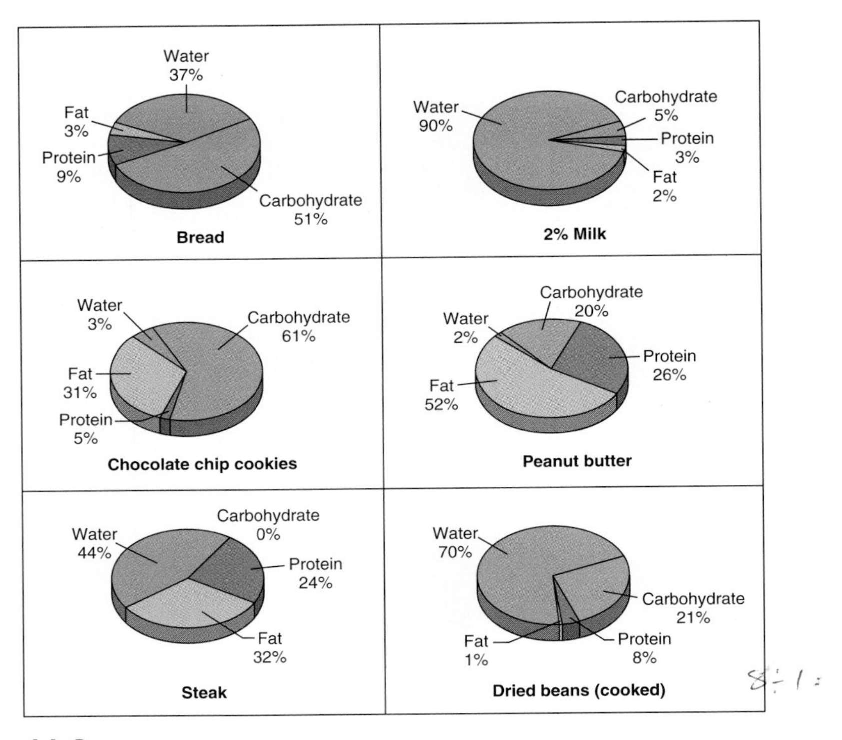

Figure 11.2

Percentage of water, carbohydrates, proteins, and fats in several foods.

(Source: The State of Food and Agriculture, 1990. Rome. Food and Agriculture Organization of the United Nations, 1991.)

11.3 Your Turn

Table 11.1 (page 441) gives both the mass of the major elements per 100 grams of body weight and the relative abundance in the number of atoms per million atoms in the body.

a. Why is oxygen the most abundant element when measured in grams per 100 g of body weight, but only the second most abundant when measured in terms of relative abundance per million atoms in the body?

b. How are these two values related?

Hint: Consider the atomic masses of each element.

The preponderance of oxygen, carbon, hydrogen, and nitrogen is fully consistent with the composition of the major chemical components of the human body. Hydrogen and oxygen are, of course, the elements of water. Moreover, along with carbon atoms, they constitute all carbohydrates and all fats. Finally, these three elements plus nitrogen are found in all proteins. Thus, nature uses very simple units—aggregates of oxygen, carbon, hydrogen, and nitrogen atoms—in a myriad of elegantly functional combinations to produce the major constituents of a healthy body and a healthful diet.

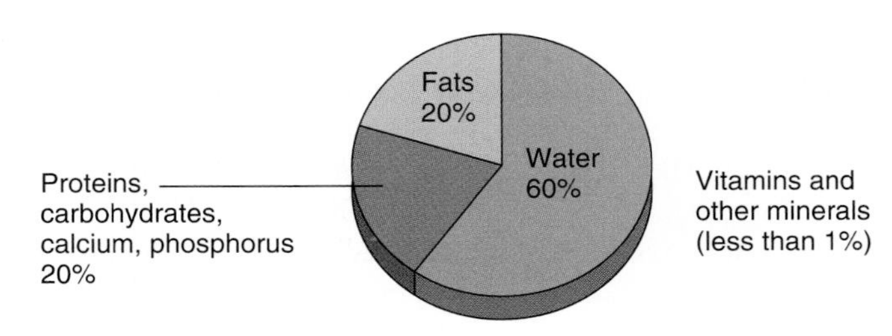

Figure 11.3

Composition of the human body.

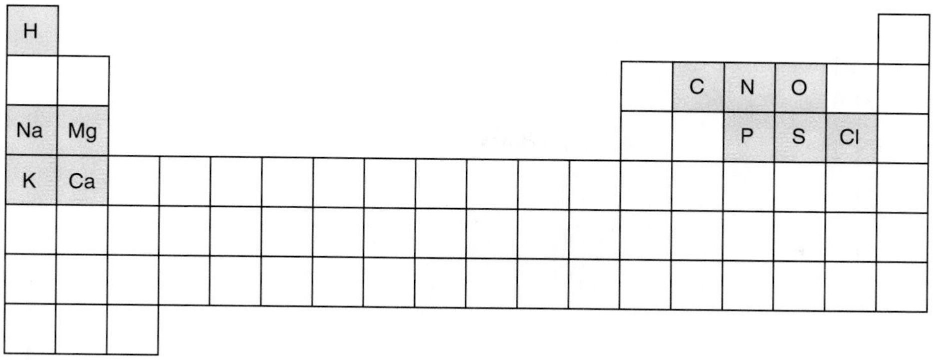

Figure 11.4
Periodic table highlighting the eleven major elements of the human body.

The current recommendations for a healthful diet, approved in 1991 by the U.S. Department of Agriculture, are represented by the food pyramid of Figure 11.5. The pyramid incorporates the traditional basic four food groups of the 1958 pie chart, but with different emphases. In particular, the pyramid calls for eating habits that increase the proportions of foods at the base of the pyramid and decrease those near or at the top. We are now urged to eat proportionally more bread, cereal, rice, and pasta, and more fruits and vegetables. Simultaneously, we are urged to reduce the percentage of fats, oils, and sweets in our daily diet.

11.4 Consider This: The Food Pyramid

Politics entered the picture when the U.S. Department of Agriculture (USDA) initially released the food pyramid (Figure 11.5). The meat and dairy industries pressured the USDA to delay public dissemination of the pyramid. Look at the 1958 basic-four pie chart and the new food pyramid. Although both charts list approximately the same number of servings for meat and dairy products, the number of some of the servings suggested on the pyramid is smaller than those on the pie chart. In addition to the smaller serving sizes, the pyramid includes a category not found on the pie chart, namely "fats, oils, and sweets." The category of "Fruits and vegetables" found on the pie chart is separated into two categories, "Vegetables" and "Fruits," on the pyramid.

Table 11.1

Major Elements of the Human Body

Element	Symbol	Grams/100 g Body Weight	Relative Abundance in Atoms/Million Atoms in the Body
Oxygen	O	64.6	255,000
Carbon	C	18.0	94,500
Hydrogen	H	10.0	630,000
Nitrogen	N	3.1	13,500
Calcium	Ca	1.9	3,100
Phosphorus	P	1.1	2,200
Chlorine	Cl	0.40	570
Potassium	K	0.36	580
Sulfur	S	0.25	490
Sodium	Na	0.11	300
Magnesium	Mg	0.03	130

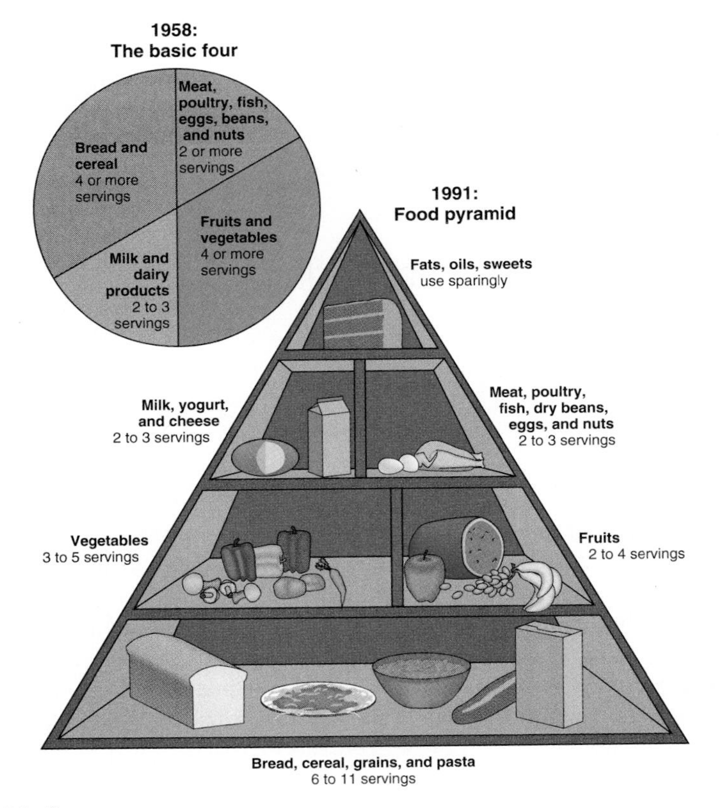

Figure 11.5

The basic four food groups and the food pyramid.

(Source: United States Dept. of Agriculture.)

Some nutritionists claim that the USDA, which is obligated to both promote and regulate agricultural products, has a conflict of interest between this responsibility and its obligation to promote the health of American citizens. Imagine that you were an administrator of the American National Cattlemen's Beef Association or the American National Cattlewomen, Inc.

a. How would you reconcile your concern for your members' livelihood, which depends on the quantity and price of the beef sold, and your interest in preserving their health?
b. Would you support or oppose the USDA's food pyramid, which encourages eating more grains and less red meat? Give reasons for your answer.
c. Do you think the USDA should be the agency to make this determination? Draft a letter to a friend stating and defending your position on this issue.

In 1998 an expert panel at the National Institutes of Health (NIH) developed new guidelines for body weight based on a body mass index (BMI). Under the new guidelines, people with a BMI of less than 25.0 have a healthy weight. Those whose BMI is 25.0 to 29.9 are defined as overweight; adults with a BMI of 30 or greater are classified as obese. The BMI is defined as a person's weight in kilograms (1 kg = 2.2 lb) divided by her or his height in meters squared, that is, kg/m^2 (1 m = 39.37 in.; 1 m^2 = 1.55×10^3 in.^{2}). For example, a person who is 5 ft 4 in. should weigh no more than 145 lb to have a BMI below 25:

$$145 \text{ lb} \times 1 \text{ kg}/2.2 \text{ lb} = 65.9 \text{ kg};$$
$$5 \text{ ft } 4 \text{ in.} = 64 \text{ in.}; (64 \text{ in.})^2 = 4.09 \times 10^3 \text{ in.}^2;$$
$$4.09 \times 10^3 \text{ in.}^2 \times 1 \text{ m}^2/(1.55 \times 10^3 \text{ in.}^2) = 2.64 \text{ m}^2$$
$$\text{BMI} = 65.9 \text{ kg}/2.64 \text{ m}^2 = 24.9 \text{ kg/m}^2$$

The new guidelines apply best to average people; exceptions are possible. For example, slightly muscled folks with small bones could have too much body fat and yet have a BMI below 25. On the other hand, persons who are naturally large-boned and heavily muscled may not have excess body fat, yet can have BMIs over 25. The NIH guidelines have a BMI lower limit of 18.5.

11.5 Your Turn

a. Calculate your BMI.
b. Explain the BMI to another person and calculate that person's BMI.

Although the food pyramid guidelines recommend roughly the same number of servings of meat and dairy products as the 1958 scheme, the newer dietary guidelines call for smaller portions. Because of concerns over dietary fat and cholesterol, the consumption of red meat, butter, and whole milk has decreased nationally. There is still some controversy surrounding the ideal balance of the macronutrients and the best sources of these compounds, but there is no question that a healthful diet requires carbohydrates, fats, and proteins. We therefore turn to a consideration of each of these macronutrients—their chemical composition, molecular structure, properties, and sources.

11.6 Consider This: Another Food Pyramid

The Mediterranean diet is favored by some health professionals in preference to the USDA Food Pyramid. The diet is prevalent in countries bordering the Mediterranean Sea, especially those along the eastern and southern shores.

a. Use a search engine to find out the details of such a diet and compare it with the Food Pyramid.
b. What are the major differences in the diets?
c. Find out why the Mediterranean diet is favored by some health professionals.

11.2 Carbohydrates—Sweet and Starchy

The best known dietary carbohydrates are sugars and starch. **Carbohydrates** are compounds containing carbon, hydrogen, and oxygen, the last two elements in the same 2:1 atomic ratio as found in water. Glucose, for example, has the formula $C_6H_{12}O_6$. This composition gives rise to the name, "carbohydrate," which implies "carbon plus water." But the hydrogen and oxygen atoms are not bonded together to form water molecules. Rather, carbohydrate molecules are built of rings containing carbon atoms and an oxygen atom. The hydrogen atoms and —OH groups are attached to the carbon atoms. This general arrangement provides many opportunities for differences in molecular structure. Thus, there are 32 distinct isomers (including chiral optical isomers) with the formula $C_6H_{12}O_6$. The isomers differ slightly in their properties, including intensity of sweetness.

• Chirality and optical isomerism are discussed in Section 10.6.

The simplest sugars are *mono*saccharides or "single sugars" such as fructose and glucose, both $C_6H_{12}O_6$. Figure 11.6 indicates that molecules of these compounds include rings consisting of four or five carbon atoms and one oxygen atom. The best way to consider these two-dimensional representations of a three-dimensional structure is to imagine that the ring is perpendicular to the plane of the paper, with the bold-print edges facing you. The H atoms and —OH groups are thus either above or below the plane of the ring. This results in two forms of glucose—alpha (α) and beta (β). In *alpha* glucose, the —OH group on carbon 1 is on the opposite side of the ring from the —CH_2OH group attached to carbon 5. As shown in Figure 11.6, the *beta* glucose form has the two groups on the *same* side of the ring. This is also the case for *beta* fructose, with the —OH on carbon 2 and the —CH_2OH group at carbon 5 being on the same side of the ring.

Ordinary table sugar, sucrose, is an example of a *di*saccharide, a "double sugar" formed by joining two monosaccharide units. In a sucrose molecule, an alpha glucose and a beta fructose unit are connected by a C—O—C linkage created when an H and an OH are split out from the monosaccharides to form a water molecule. This reaction and the structure of the $C_{12}H_{22}O_{11}$ sucrose molecule are also shown in Figure 11.6.

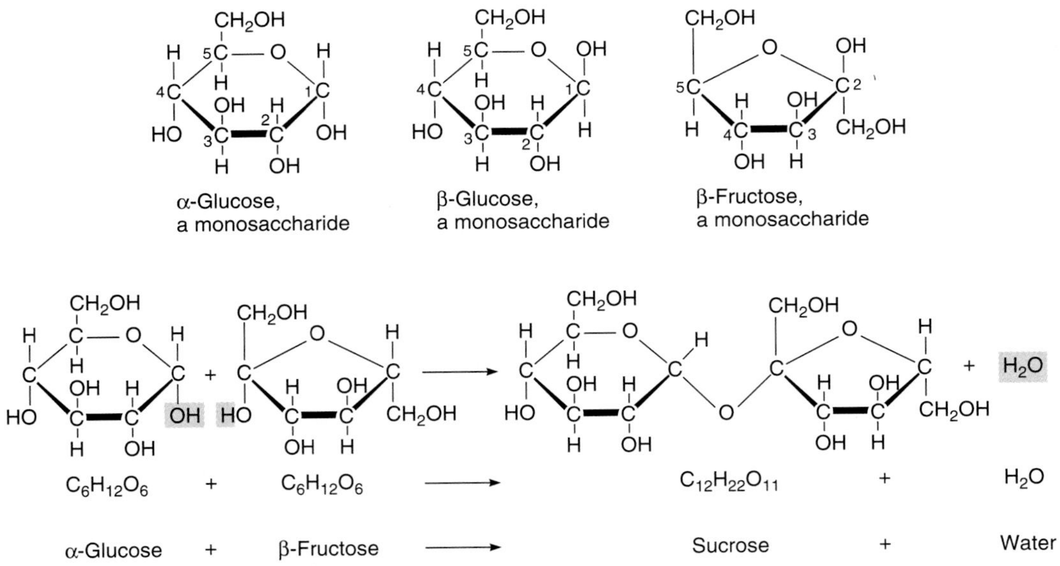

α-Glucose, a monosaccharide

β-Glucose, a monosaccharide

β-Fructose, a monosaccharide

$C_6H_{12}O_6$ + $C_6H_{12}O_6$ ⟶ $C_{12}H_{22}O_{11}$ + H_2O

α-Glucose + β-Fructose ⟶ Sucrose + Water

Figure 11.6

Molecular structures of some sugars.

- Disaccharides and polysaccharides are formed by the condensation polymerization of monosaccharides.

The linking of monosaccharide molecules by this reaction is by no means restricted to the formation of disaccharides. Some of the most common and abundant carbohydrates are *poly*saccharides, polymers made up of thousands of glucose units. As the name implies, these macromolecules consist of "many sugar units." Polysaccharides form when monsaccharide monomer units join into a chain through splitting out a water molecule each time a monomer unit is incorporated into the chain. This is a condensation polymerization reaction such as that also described in Section 9.5. Water produced in this way is used for other reactions in the body.

Starch, glycogen, and cellulose are three familiar examples of polysaccharides, sometimes called complex carbohydrates. A healthful diet derives more of its carbohydrates from starch than from simple sugars. Enzymes in our saliva initiate the process of breaking down the long polysaccharide chains into glucose molecules, an important first step in metabolism. But the body also synthesizes glucose from a variety of precursors, including other sugars. Some of the glucose molecules are polymerized to form another polysaccharide called glycogen, whose molecular structure is similar to that of starch. But the chains of glucose units in glycogen are longer and more branched than those in starch. Glycogen is vitally important because it is a storehouse of energy in molecular form. It accumulates in muscle and especially in the liver, where it is available as a quick source of internal energy.

11.7 Your Turn

Sustained chewing of an unsweetened, unsalted cracker results in a sweet taste. What is the molecular explanation for this phenomenon?

Like starch, cellulose is a polymer of glucose, but these two polysaccharides behave very differently in the body. Humans are able to digest starch by breaking it down into individual glucose units; on the other hand, we cannot digest cellulose. Consequently, we depend on starchy foods such as potatoes or pasta as carbohydrate sources rather than literally devouring toothpicks or textbooks. The reason for this is a subtle difference in how the glucose units are joined in starch and in cellulose. In the alpha (α) linkage in starch, the bonds connecting the glucose units have an angular orientation, whereas the beta (β) linkage between glucose units in cellulose is at a different angle.

α linkage

β linkage

Our enzymes, and the enzymes of many mammals, are unable to catalyze the breaking of beta linkages in cellulose. Consequently, we can't dine on grass, textbooks, or trees. However, cows, goats, sheep, and other ruminants manage to break down cellulose with a little help. Their digestive tracts contain bacteria that decompose cellulose into glucose monomers. The animals' own metabolic systems then take over. Similarly, the fact that termites contain cellulose-hungry bacteria means that wooden structures are sometimes at risk. And, as you have already read in Chapter 3, the methane produced by these bacteria may be contributing to global warming.

• See Section 3.10 for more about methane and other greenhouse gases.

Lactose intolerance, a common metabolic anomaly, is somewhat related to the difference in digestibility between starch and cellulose. But "anomaly" is hardly the correct term for a condition that is shared by about 80% of the world's population. Although most Northern Europeans, Scandinavians, and people of similar ethnic background eat milk, cheese, and ice cream with no ill effects, they are the exception. The great lactose-intolerant majority have difficulty digesting dairy products. Consumption of these foods is often followed by diarrhea and excess gas. The symptoms result from the inability to break down lactose (milk sugar) into its component monosaccharides, glucose and galactose. The linkage between the two monosaccharides in lactose is a beta form, similar to that in cellulose. People who are lactose intolerant have a lack of or a low concentration of lactase, the enzyme that catalyzes the breaking of this bond. In such individuals, the intact lactose is instead fermented by intestinal bacteria. This process generates carbon dioxide and hydrogen gases, and lactic acid, the principal cause of the diarrhea. Given milk's importance for growing bones and teeth, it is significant that infants of all ethnic groups generally produce sufficient lactase to digest a milk-rich diet. But, as we age, this production decreases. By adulthood, most people in the world do not have enough of the enzyme to accommodate a diet heavy in dairy products.

11.8 Your Turn

Speculate why the slight difference in molecular structure between starch and cellulose is enough to make the latter polysaccharide indigestible to human beings.

Hint: Section 10.5 can be helpful.

11. 9 Consider This: Lactose Intolerance—A Closer Look

Here are three questions on which to try your chemical detective skills using the Web by searching for "lactose intolerance." Reading the labels in your cupboard or at a nearby store can also help you milk out some information.

a. Over-the-counter digestive aids allow you to increase your intake of dairy products. How do these work? What are their advantages and disadvantages?

b. Even with digestive aids, you may risk not getting enough calcium, an essential mineral that you will learn more about in Section 11.11. What other foods can you eat to obtain enough calcium in your diet?

c. Sometimes lactose turns up in foods in which you least expect it, such as bread. Although lactose may not be listed on the label, you will see ingredients such as whey, milk products, nonfat dry milk, or dry milk solids—all of which contain lactose. Find three other non-dairy foods that you may have to watch out for if you are lactose intolerant.

11.3 The Fat Family

Everyone knows from personal experience the properties of fats. They are greasy, slippery, soft, low-melting solids that are not soluble in water. Butter, cheese, cream, whole milk, and certain meats are loaded with them. But margarine and some shortenings are evidence that fats can also be of vegetable origin. Oils, such as those obtained from olives or corn, exhibit many of the properties of animal-based fats, but in liquid form. The fact that these properties are also shared by petroleum-based oils and greases suggests a chemical and structural similarity. But there are some important differences. You are well aware that petroleum is made up almost exclusively of hydrocarbons. In the molecules of these compounds, carbon atoms are bonded to each other (often in chains) and to hydrogen atoms. Hydrocarbon molecules are nonpolar, and hence they do not mix well with water or other polar substances.

- Hydrocarbons in petroleum were discussed in Section 4.8. Reasons why nonpolar substances do not dissolve in water were examined in Section 5.12.

The inference that edible fats and oils must also be nonpolar is fully justified. The molecules of fats of animal and vegetable origin also include long hydrocarbon chains. But biological fats are a bit more structurally complex than their petroleum-based chemical cousins. Of particular significance is the fact that edible fats and oils always contain some oxygen. Most of these compounds are classified chemically as triglycerides. **Fats** are triglycerides that are solid at room temperature, whereas **oils** are obviously liquid under these same conditions. Whether solid or liquid, triglycerides are a major portion of a broader class of compounds called **lipids.** Cholesterol and other steroids (see Chapter 10) are also classified as lipids, and molecules of some complex compounds, such as lipoproteins, contain fatty segments.

To a chemist, the term 'triglyceride' reflects the composition and the formation of fats. A **triglyceride** is formally defined as an ester of three fatty acid molecules and one glycerol molecule. The formation of a triglyceride can be represented by a word equation:

$$\text{3 Fatty acids} + \text{Glycerol} \longrightarrow \text{Triglyceride} + \text{3 Water} \qquad [11.1]$$

In this process, as in the formation of polysaccharides, smaller units join to form more complex molecules by splitting out water molecules. The presence of three fatty acid units in the final product molecule makes it a *tri*glyceride. If you studied Chapter 9, you already encountered esterification in the formation of polyesters from acids and alcohols (Section 9.5). To see the connection, it is necessary to consider the molecular structures of fatty acids and glycerol.

Naturally occurring fatty acid molecules are characterized by two structural features: a long hydrocarbon chain generally containing an even number of carbon atoms (typically 12 to 24) including a carboxylic acid group (—COOH) at the end of the chain. This functional group is what puts the *acid* in fatty acid because —COOH can release a hydrogen ion (H^+). The long hydrocarbon chains, on the other hand, give fats most of their characteristic properties. Stearic acid, $C_{17}H_{35}COOH$, is a fatty acid found in animal fats; its structural formula and condensed chemical formula are given below.

$$CH_3CH_2CH_2CH_2CH_2CH_2CH_2CH_2CH_2CH_2CH_2CH_2CH_2CH_2CH_2CH_2CH_2-\overset{\overset{\large O}{\|}}{C}-OH$$

Stearic acid, a fatty acid

$$CH_3(CH_2)_{16}\overset{\overset{\large O}{\|}}{C}-OH$$

Condensed chemical formula of stearic acid

Glycerol, or glycerine as it is commonly called, is a sticky, viscous liquid that is sometimes added to soaps and hand lotions. The following structural formula indicates that a molecule of this compound includes three —OH groups, which classifies it as an alcohol.

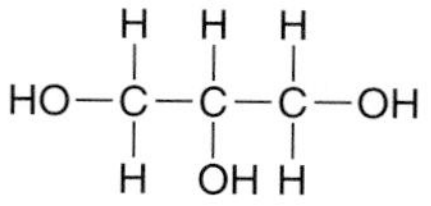

$$HO-\underset{H}{\overset{H}{C}}-\underset{OH}{\overset{H}{C}}-\underset{H}{\overset{H}{C}}-OH$$

Glycerol

In fatty acids and in glycerol we have the acid and alcohol functional groups, respectively, required to form an ester (Section 9.5). In fact, each of the three —OH groups in a glycerol molecule can react with a fatty acid molecule. Thus, equation 11.2 represents the combination of three stearic acid molecules with a glycerol molecule to form a triester or triglyceride. This is the process involved in the formation of most animal and vegetable fats and oils. Variety is built in by having up to three different fatty acids incorporated into the same triglyceride rather than just one, stearic acid, as in the example given.

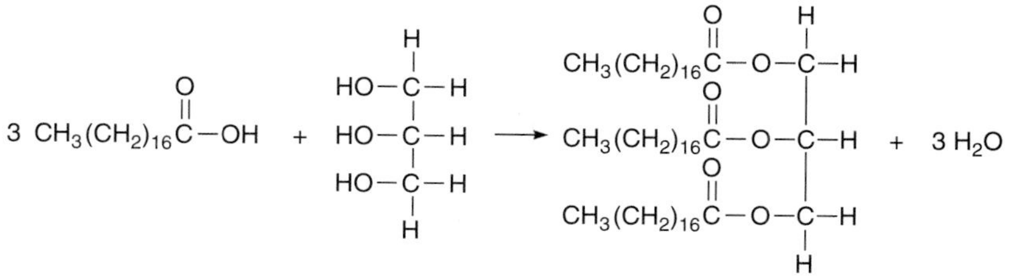

$$3\ CH_3(CH_2)_{16}\overset{O}{\overset{\|}{C}}-OH + \text{glycerol } (HO-CH_2-CH(OH)-CH_2-OH) \longrightarrow \text{triglyceride } (CH_3(CH_2)_{16}\overset{O}{\overset{\|}{C}}-O-CH_2,\ CH_3(CH_2)_{16}\overset{O}{\overset{\|}{C}}-O-CH,\ CH_3(CH_2)_{16}\overset{O}{\overset{\|}{C}}-O-CH_2) + 3\ H_2O \quad [11.2]$$

- Organic acid + Alcohol ⟶ Water + Ester
- This group is characteristic of an ester:

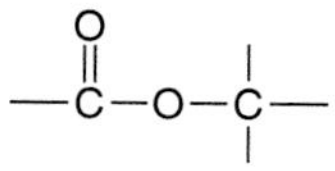

$$-\overset{O}{\overset{\|}{C}}-O-\overset{|}{\underset{|}{C}}-$$

11.4 Saturated and Unsaturated Fats and Oils

Animal and vegetable fats and oils exhibit considerable variety. A chief reason for this diversity is that not all fatty acids are identical. As we already noted, they vary in the number of carbon atoms and hence the length of the hydrocarbon chain. Additionally, fatty acids can contain one or more C=C double bonds and can differ where these double bonds are located in the molecule. If the hydrocarbon chain contains only C—C single bonds between the carbon atoms and no C=C double bonds, the fatty acid is said to be **saturated.** This is the case with stearic acid. If, however, the molecule contains one or more C=C double bonds between carbon atoms, the fatty acid is **unsaturated.** Oleic acid, with one C=C double bond per molecule, is classified as **monounsaturated.** Those fatty acids containing more than one C=C double bond per molecule are called **polyunsaturated.** Linoleic acid, which contains two C=C double bonds per molecule, and linolenic acid with three C=C double bonds per molecule, are polyunsaturated. Note that each of these three different unsaturated fatty acids contains the same number of carbon atoms, 18.

$$CH_3(CH_2)_{16}COOH$$

Stearic acid, a saturated fatty acid

$$CH_3-(CH_2)_7-CH=CH-(CH_2)_7-COOH$$

Oleic acid, a monounsaturated fatty acid

$$CH_3-(CH_2)_4-CH=CH-CH_2-CH=CH-(CH_2)_7-COOH$$

Linoleic acid, a polyunsaturated fatty acid

$$CH_3-CH_2-CH=CH-CH_2-CH=CH-CH_2-CH=CH-(CH_2)_7-COOH$$

Linolenic acid, a polyunsaturated fatty acid

The overwhelming majority of fatty acids in the body, almost 95%, are transported and stored in the form of triglycerides. The three fatty acids in a single triglyceride molecule can all be identical, two can be the same and the third can be different, or all three can be different. Moreover, the fatty acids in a triglyceride molecule can exhibit

Table 11.2

Melting Points of Some Fatty Acids

Name	Carbon Atoms per Molecule	Melting Point, °C
Saturated fatty acids		
Capric	10	32
Lauric	12	44
Myristic	14	54
Palmitic	16	63
Stearic	18	70
Unsaturated fatty acids		
Oleic (1 double bond/molecule)	18	16
Linoleic (2 double bonds/molecule)	18	5
Linolenic (3 double bonds/molecule)	18	11

• Normal body temperature is 37°C; room temperature is approximately 20°C.

varying degrees of unsaturation. The fatty acids a given fat or oil contains govern its extent of unsaturation.

The physical properties of fats also depend on their fatty acid content. Table 11.2 indicates that within a given family of fatty acids, for example, those that are saturated, melting points increase as the number of carbon atoms per molecule (and the molecular mass) increase. On the other hand, in a series of fatty acids with a similar number of carbon atoms, increasing the number of C=C double bonds decreases the melting point. Thus, when the melting points of the 18-carbon fatty acids are compared, saturated stearic acid (no C=C double bonds per molecule) is found to melt at 70°C, oleic acid (one C=C double bond per molecule) melts at 16° C, and linoleic acid (two C=C double bonds per molecule) melts at 5°C. These trends are carried over to the triglycerides containing the fatty acids. This explains why fats rich in saturated fatty acids are solids at room or body temperature, whereas highly unsaturated ones are liquids.

• Stearic acid is a solid at body temperature, whereas oleic and linoleic acids are liquids.

Figure 11.7 gives evidence of this generalization. The bar graphs present the composition of various dietary fats and oils in terms of saturated, polyunsaturated, and monounsaturated components. Typically, these naturally occurring lipids are mixtures of various triglycerides. In general, solid or semisolid animal fats, such as lard and beef tallow, are high in saturated fats. In contrast, olive, safflower, and other vegetable oils consist mostly of unsaturated triglycerides. However, the figure reveals that there are some surprising differences in the composition of oils. For example, palm and coconut oil contain much more saturated fat than corn and canola oil. Ironically, the coconut oil used in some non-dairy creamers is 92% saturated fat, far more than the percentage found in the cream it replaces. In fact, coconut oil contains more saturated fats than pure butterfat. Concern over the high degree of saturation in coconut and palm oil accounts for the statement sometimes printed on food labels: "Contains no tropical oils."

• Coconut oil has been used for making popcorn at movie concession stands and for cooking French fries at fast-food establishments.

11.10 Your Turn

Using Figure 11.7 and information from this section, identify the predominant fatty acids likely to be present in each of these.

a. canola oil
b. olive oil
c. lard

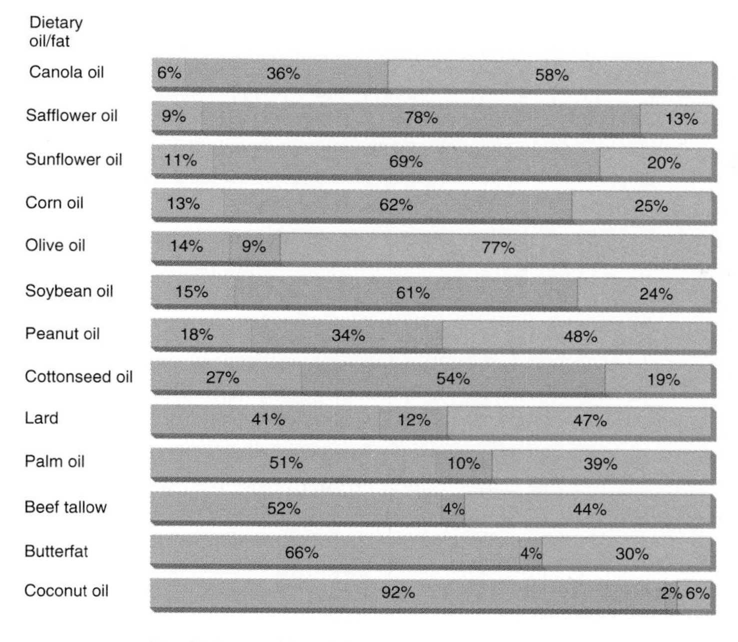

Figure 11.7

Saturated and unsaturated fats.

(Source: Food Technology, April 1989.)

11.11 Consider This: Are Unsaturated Fatty Acids Better for You Than Saturated Ones?

Unsaturated fatty acids can be either in a *cis* or *trans* orientation. There is controversy about the role of *trans* fatty acids in the diet. Use the Web to answer these questions. Write a news article about your findings.

a. What is the structural difference between *cis* and *trans* fatty acids?
b. What is the controversy regarding dietary *trans* fatty acids?
c. Have studies been completed that resolve the controversy? If so, cite them.

Some food labels also reveal that not all the fats and oils in our diet are consumed in their naturally occurring molecular forms. Unless you eat "natural" peanut butter, the jar on your shelf probably is labeled something like, "oil modified by partial hydrogenation." Peanuts ground to make peanut butter always release a quantity of peanut oil. This oil separates on standing and must be stirred back into the solid before use. The oil extracted from peanuts is rich in mono- and polyunsaturated fats, and thus is a liquid. It can be treated chemically and converted into a semisolid that does not separate from peanut butter. This is done by reacting the oil with hydrogen gas over a metallic catalyst. The hydrogen adds to the double bonds in the oil, converting some, but not all, of the $CH{=}CH$ bonds into CH_2CH_2 bonds, such as with the linoleic acid in peanut oil.

$$CH_3(CH_2)_4{-}CH{=}CH{-}CH_2{-}CH{=}CH{-}(CH_2)_7COOH + H_2 \longrightarrow$$
$$CH_3(CH_2)_4{-}CH_2{-}CH_2{-}CH_2{-}CH{=}CH{-}(CH_2)_7COOH$$

As a result of this partial hydrogenation, the number of double bonds in the lipid decreases, it becomes less unsaturated, and it is transformed from an oil into a semisolid fat. The extent of hydrogenation can be carefully controlled to yield products of desired unsaturation and resultant melting point, softness, and spreadability. Such customized fats and oils are in many products, including margarines, cookies, and candy bars. But "natural food" advocates can take some comfort in the realization that these triglycerides are not really artificial. Their degree of saturation may be chemically altered, but the fats formed through partial hydrogenation certainly also exist somewhere in nature in some other food.

In our preoccupation with dietary fat, it is important to realize that fats often enhance our enjoyment of food. They improve "mouth feel" and intensify certain flavors. Of prime significance, however, is the fact that fats are essential for life. They are the most concentrated source of energy in the body (Section 11.9), and they provide insulation that retains body heat and cushions internal organs. Moreover, triglycerides and other lipids, including cholesterol, are the primary components of cell membranes and nerve sheaths. Although "fathead" is hardly a compliment, in fact our brains are rich in lipids. These various functions require a variety of triglycerides incorporating a wide range of fatty acids—saturated, monounsaturated, and polyunsaturated. Fortunately, our bodies can synthesize almost all of the necessary fatty acids from the starting materials provided by a normal diet. The exceptions are linoleic and linolenic acids. These two essential fatty acids must be obtained directly from the foods we eat; our body cannot produce them. Generally this does not create a problem because linoleic and linolenic acids are found in many foods including plant oils, fish, and leafy vegetables.

• Cholesterol is discussed in Section 10.7.

11.12 Consider This: Spreadables and Fat Content

Here are the fat contents for Crisco (a partially hydrogenated shortening) and three soft, butter substitutes.

	Crisco	Brummel & Brown	I Can't Believe It's Not Butter	Benecol
Total fat	12 g	10 g	10 g	9 g
Saturated	3 g	2 g	2 g	1 g
Monounsaturated	6 g	3 g	2.5 g	4 g
Polyunsaturated	3 g	2 g	4.5 g	3 g

The three butter substitutes have less fat than butter, as advertised. All the butter substitutes are partially hydrogenated. Notice that the sum of the saturated, monounsaturated, and polyunsaturated fats does not equal the mass of total fat in the three products. The difference is the amount of *trans* fats, so called "stealth" fats. Thus, Brummel & Brown contains 10 g − 7 g = 3 g of *trans* fats per serving.

• *Trans* in this context refers to fats that have been transformed by partial hydogenation. This is different from *trans* in *cis-trans* isomerism.

a. Calculate the grams of *trans* fats in each of the other two butter substitutes. Of the three butter substitutes, which one has the highest percentage of *trans* fats?

b. Conduct a mini-survey of butter and margarine products in a supermarket in your area. List the total fat content and the mass of polyunsaturated, saturated, monounsaturated, and *trans* fats for examples of each of these five products: butter stick; regular margarine stick; regular margarine tub; "light" margarine stick; and "light" margarine tub.

11.5 Controversial Cholesterol

Although dietary fat is an essential part of a balanced diet, the fact remains that many Americans consume too much of it, and too much of the wrong kind. Fats provide about 40% of the calories in the average American diet. Health care specialists recommend that this value should be 30% or less. Much of the concern and controversy regarding cardiac health problems is focused on cholesterol, one of the steroids introduced in Section 10.7.

Cholesterol has drawn heavy media attention, but not always with total accuracy or objectivity. At issue is the connection between cholesterol in the blood (serum cholesterol) and cardiovascular disease. Over two decades, several national reports appeared and made conflicting recommendations. In 1980, the Food and Nutrition Board of the National Academy of Science issued "Toward Healthful Diets." This report cited the inconsistency of data collected in medical studies done up to that time. A direct causal connection between dietary cholesterol and atherosclerosis, the thickening of arterial walls, had not been unambiguously established. Therefore, the report concluded that healthy adults did not need to reduce dietary cholesterol. However, by the end of the 1980s, this position was reversed, based on the pooled data from a wide variety of new and continuing studies. These studies led to the conclusion that high serum cholesterol levels *appear* to predict the potential for a stroke or heart attack. Although an irrefutable direct linkage has not yet been demonstrated, the data seem compelling. Waxy deposits of excess cholesterol (plaque) cause arteries to narrow and harden, which may elevate blood pressure and increase the risk of heart disease (Figure 11.8).

Today there is general agreement that elevated blood cholesterol levels are associated with atherosclerosis, although there is not quite consensus on what constitutes dangerously high concentrations. Many medical researchers and the American Heart Association (AHA) consider values equal to or greater than 240 mg cholesterol per 100 mL of blood as the critical point for medical intervention. Cholesterol concentrations of 200 to 239 mg/100 mL are considered borderline high. According to the AHA, individuals with a total cholesterol of 240 mg/100 mL, in general, have twice the risk of heart attack than those who have a cholesterol level of 200 mg/100 mL.

One obvious response to elevated serum cholesterol is to restrict consumption of cholesterol. The American Heart Association recommends a maximum intake of 300 mg of cholesterol per day. This means cutting back on animal fats, which are rich in the compound. Included are fatty red meats, cream, butter, and cheese. Egg yolks are particularly rich in cholesterol, each yolk containing on average a whopping 213 mg. In contrast, egg whites contain no cholesterol, nor do fruits, vegetables, or vegetable oils.

11.13 Consider This: Cholesterol Content of Various Foods

The text mentions that an egg yolk contains significant amounts of cholesterol, whereas foods that are not derived directly from animals such as fruits and vegetables have no cholesterol. Use the Web to find the cholesterol content of various foods. List two foods that are high in cholesterol and two foods (not fruits or vegetables) that have much lower cholesterol levels.

But restricting dietary cholesterol is only one part of a two-part situation. It is possible that even a strict vegetarian with negligible cholesterol intake might have elevated serum cholesterol. This is because most of the body's cholesterol does not come from

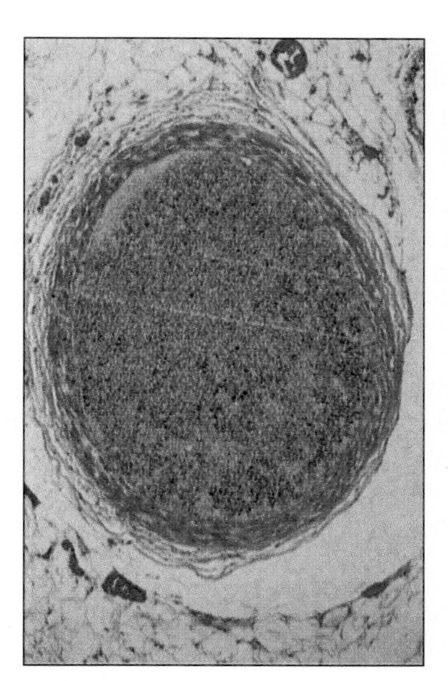

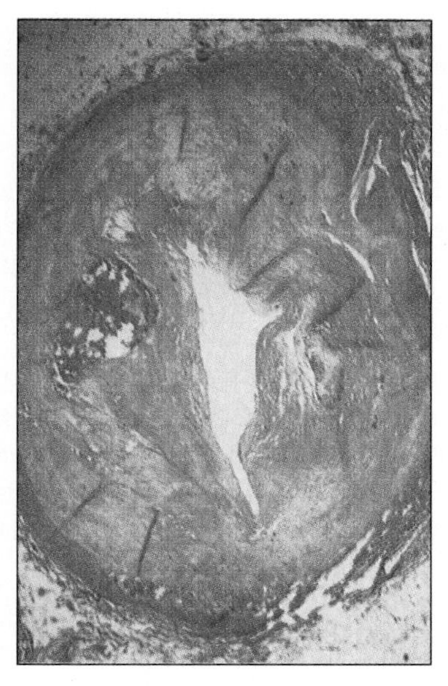

Figure 11.8
Cross sections of a healthy artery (left) and an artery clogged with atherosclerotic plaque (right).

the diet directly, but is synthesized by the body. About one gram of cholesterol is synthesized daily in the liver to maintain the minimum concentration required for use in cell membranes and to produce estrogen, testosterone, and other steroid hormones. The liver produces cholesterol principally from dietary saturated fats. Consequently, a high intake of saturated fats can result in a high concentration of cholesterol. Although cutting down on cholesterol consumption is an important step in lowering serum cholesterol, reducing the amount of saturated fats in the diet may be even more significant. The AHA recommends that only 8–10% of total Calories should come from such fats, and to especially limit the intake of those with certain numbers of carbon atoms per molecule: 12 (lauric acid), 14 (myristic acid), and 16 (palmitic acid).

11.14 Your Turn

Do calculations to determine whether the meal eaten in 11.1 Consider This meets the guideline that only 8–10% of total calories should come from saturated fats.

Diet, however, is only one of several factors influencing cholesterol synthesis. Perhaps the most important factor is genetic. This may explain why some people seem to eat fatty foods without suffering from heart disease, while others who carefully watch their diets are afflicted with it. Because we are not in complete control of our genes (at least not yet), physicians urge us to do what we can to lower our serum cholesterol. Reducing dietary cholesterol and fatty acids, exercising regularly, decreasing weight and stress, and eating certain types of dietary fiber appear to be important for maintaining good health.

It is possible that some of these habits also serve to lower the concentration of **low density lipoprotein (LDL)** and increase the concentration of **high density lipoprotein (HDL).** These compounds combine with cholesterol and triglycerides and carry them through the bloodstream, thus preventing the buildup of plaque in the arteries. The HDLs are the "good" lipoproteins, more effective in transporting cholesterol than LDLs. The AHA recommends a concentration of greater than 35 mg HDL/100 mL of blood and an LDL value of less than 130 mg/100 mL. It appears that people with high values for the LDL/HDL concentration ratio are particularly susceptible to heart disease. No doubt other discoveries linking diet and heart disease will continue to be made. But it is important to remember that the most effective way to reduce the risk of heart disease has to do with inhaling as well as ingesting. Dr. Richard Peto, an Oxford University epidemiologist writing in the *Harvard Medical School Health Letter*, December 1989, minces no words: "You can't offer eternal life to old people. But what you can do is to avoid death in middle age. At the moment, about a third of all Americans die in middle age, and that isn't necessary. About half of those premature deaths could be avoided if people took smoking, blood cholesterol, and blood pressure more seriously."

11.15 Consider This: Current AHA Recommendations

Have the American Heart Association's recommended values for cholesterol, HDL, and LDL changed from those reported in this section? Check the current recommendations at the AHA web site. If there are any differences, explain why.

11.16 Sceptical Chymist: Olean, A "Fake" Fat

Americans eat about 22 pounds of salty snack foods per capita every year. Olean, a nonfattening, nonmetabolizable fat developed by the Procter & Gamble Company, was approved by the FDA in 1996 for use in salty snack foods such as potato chips and tortilla chips. In spite of having FDA approval, Olean remains controversial, with supporters and detractors. Use the Web to locate the Olean web site as well as other sites that present contrasting viewpoints.

a. How does Olean work? Why is it not digested?
b. What are the drawbacks to using Olean in snack foods?
c. Why is the use of Olean in this way controversial?
d. Present accounts expressing opposing viewpoints regarding the use of Olean.
e. What is your decision about using products containing Olean? Explain your reasoning.

The campaign to lower blood cholesterol is not without cautions. For example, when researchers worldwide investigated blood cholesterol concentrations in relation to all premature deaths, not just those from heart disease, they discovered a very interesting result. As seen in Figure 11.9, the graphs of the results were a slightly U-shaped curve for men and a flat, declining line for women. In other words, men were at greater risk at both high and low cholesterol levels, whereas women showed no such signs, even at high cholesterol levels. Men with cholesterol levels greater than 240 mg/100 mL tended to die prematurely of heart disease while men with cholesterol levels lower than 160 mg/100 mL died prematurely from cancer and from respiratory and digestive diseases. Thus, although low-fat diets might prevent heart disease, they may increase susceptibility to other conditions. In response to these findings, the American College of Physicians now suggests that "cholesterol reduction is certainly worthwhile for those at high, short-term risk of coronary heart disease but of 'much smaller or . . . uncertain' benefit for everyone else." (*Science* Vol. 291 30 March 2001 p 3543).

11.17 The Sceptical Chymist: Dietary Fat

As you just read, the role of diet and cholesterol in heart disease and other health conditions is controversial. Ron Krauss, Chair of the American Heart Association Dietary Committee, suggests that it is "scientifically naïve to expect that a single dietary regime can be beneficial for everybody: The 'goodness' or 'badness' of anything as complex as dietary fat and its subtypes will ultimately depend on the context of the individual."

a. Why does Krauss think that it is "scientifically naïve to expect that a single dietary regime can be beneficial for everybody"?

b. Explain his sentence "The 'goodness' or 'badness' of anything as complex as dietary fat and its subtypes will ultimately depend on the context of the individual."

11.6 Proteins: First Among Equals

The word "protein" derives from *protos*, Greek for "first." The name is misleading. Life depends on the interaction of thousands of chemicals, and to assign primary importance to any single compound or class of compounds is naive. Nevertheless, proteins are an essential part of every living cell. They are also major components in hair, skin, and muscle; and they transport oxygen, nutrients, and minerals through the bloodstream. Many of the hormones that act as chemical messengers are proteins, as are all the enzymes that catalyze the chemistry of life.

Proteins are polyamides or polypeptides, polymers made up of amino acid monomers. The great majority of proteins are made from various combinations among 20 different naturally occurring amino acids. Molecules of all these amino acids share

• Amino acids and proteins were mentioned briefly in Section 9.6.

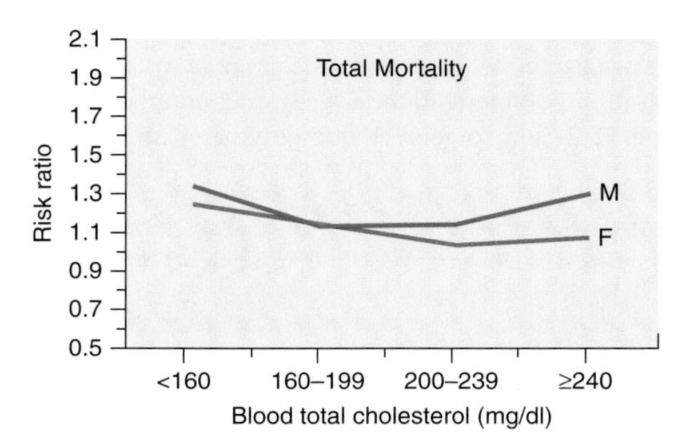

Figure 11.9

Risk of death and blood cholesterol levels. (Source: *Science,* Vol. 291, 30 March 2001.)

a common structural pattern. Four chemical species are attached to a carbon atom: (1) an acid group, —COOH; (2) an amine group, —NH_2; (3) a hydrogen atom, —H; and (4) a side chain designated R in the following structure.

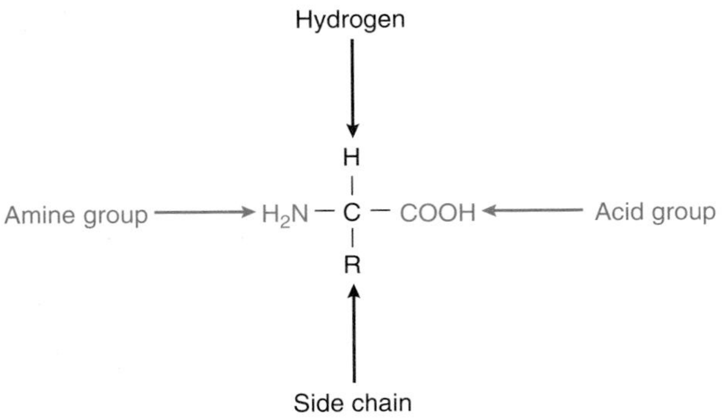

Hydrogen

Amine group → $H_2N-\underset{\large R}{\overset{\large H}{C}}-COOH$ ← Acid group

Side chain

Variations in the R side-chain group differentiate the individual amino acids. In glycine, the simplest amino acid, the R is a hydrogen atom. In alanine, R is a —CH_3 group; in aspartic acid (found in asparagus), it is —CH_2COOH; and in phenylalanine, it is a group with the formula —$CH_2(C_6H_5)$. Here, C_6H_5 designates the hexagonal phenyl ring first introduced in Chapter 9. Note the structural relationship between alanine and phenylalanine.

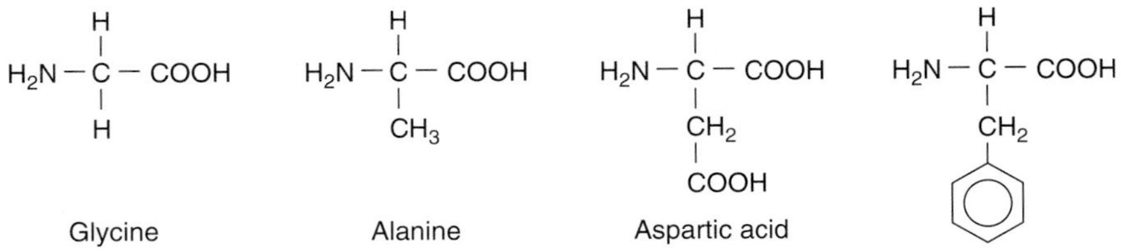

$H_2N-CH(H)-COOH$ Glycine

$H_2N-CH(CH_3)-COOH$ Alanine

$H_2N-CH(CH_2COOH)-COOH$ Aspartic acid

$H_2N-CH(CH_2C_6H_5)-COOH$ Phenylalanine

Two of the 20 naturally occurring amino acids have R groups that bear a second acidic —COOH functional group, three have R groups containing amine groups, and two others contain sulfur atoms. Because all amino acids except glycine involve four different units bonded to a central carbon atom, they all exhibit chirality and optical isomerism. All the naturally occurring amino acids that are incorporated into proteins are in the left-handed isomeric form.

- See Section 10.6 for a discussion of chirality.

The combination of amino acids to form proteins depends on the presence of the two characteristic functional groups that give this family of compounds its name—the amine group and the acid group. Equation 11.3 represents the reaction of glycine with alanine to form a dipeptide, a compound composed of the two amino acids. Here, the acidic —COOH group of a glycine molecule reacts with the —NH_2 group of alanine, and an H_2O molecule is eliminated. In the process, the two amino acids become linked by a peptide bond (indicated in the box). Once incorporated into the peptide chain, the amino acids are known as amino acid residues. The reaction in equation 11.3 is another example of condensation polymerization, already encountered in the formation of polysaccharides (Section 11.2) and some synthetic polymers (Section 9.5).

$$H_2N-CH_2-C(=O)-OH + H_2N-CH(CH_3)-C(=O)-OH \longrightarrow H_2N-CH_2-C(=O)-NH-CH(CH_3)-C(=O)-OH + H_2O \quad [11.3]$$

Glycine + Alanine ⟶ Dipeptide + Water

- Equation 11.3 is equivalent to equation 9.7.

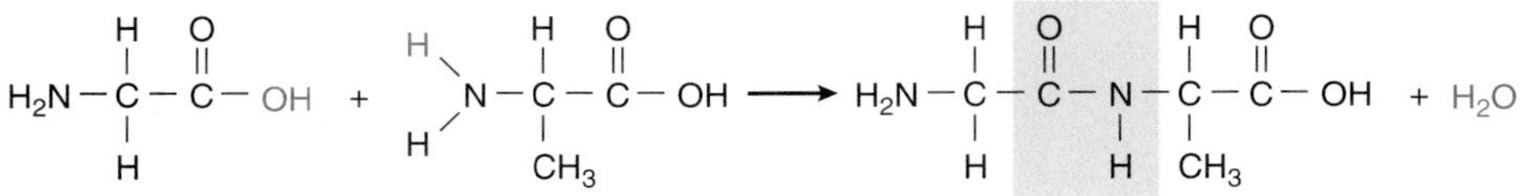

Because each amino acid bears an amine group and an acid group, there are two ways the amino acids can join. Hence, two dipeptides are possible. We illustrate the options with simple block diagrams for the amino acids. The first case is that of equation 11.3: Glycine acts as the acid and alanine as the amine.

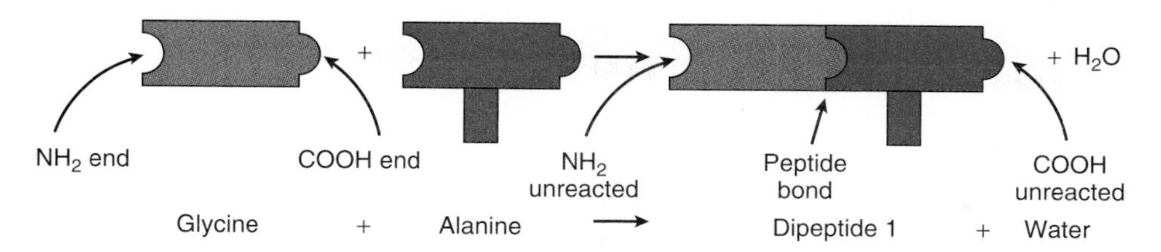

In the second case, the amino acids reverse roles; alanine provides the —COOH and glycine the —NH_2 for the reaction.

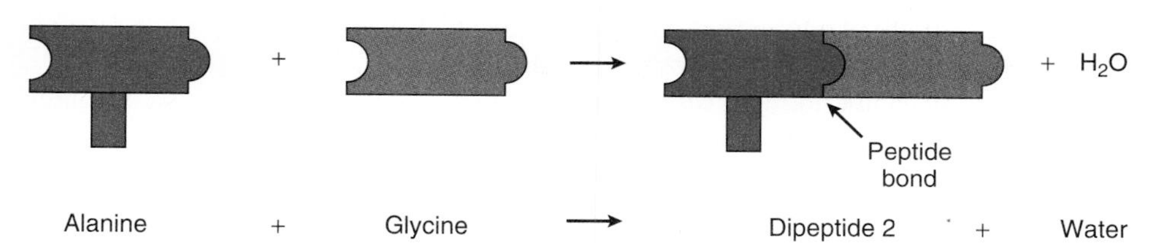

Examination of the molecular structures of the two dipeptides indicates that they are not identical. In dipeptide 1, the unreacted amine group is on the glycine residue and the unreacted acid group is on the alanine residue; in dipeptide 2, the —NH_2 is on the alanine residue and the —COOH on the glycine residue.

The point of all this is that the order of amino acid residues in a peptide makes a difference in its properties. The particular protein formed depends not only on what amino acids are present, but also on their sequence in the protein chain. Assembling the correct amino acid sequence to make a particular protein is like putting letters in a word; if they are in a different order, a completely new meaning results. Thus, a tripeptide consisting of three different amino acids is like a three-letter word containing the letters *a*, *e*, and *t*. There are six possible combinations of these letters. Three of them—*ate, eat,* and *tea*—form recognizable English words; the other three—*aet, eta*, and *tae*—do not. Similarly, some sequences of amino acids may be biological nonsense.

- Putting the amino acids of a protein into their proper order is like assembling a train correctly by placing each car in the right sequence.

Still restricting ourselves to three-letter words and only the letters *a*, *e*, and *t*, but allowing the duplication of letters, we can make perfectly good words such as *tee* and *tat*, and lots of meaningless combinations such as *aaa* and *tte*. There are, in fact, a total of 27 possibilities, including the six identified earlier. Just as many words use letters more than once, most proteins contain specific amino acids incorporated more than once. More information about the structure and synthesis of proteins is included in Chapter 12.

11.18 Your Turn

You can see from the block diagrams in this section that one glycine (Gly) and one alanine (Ala) molecule can combine to form two dipeptides: GlyAla and AlaGly. If one permits multiple use of each of the two amino acids, two other dipeptides are possible: GlyGly and AlaAla. Thus, a total of four different dipeptides can be made from two amino acids if each amino acid can be used more than once. Eight different tripeptides can be made from supplies of two different amino acids, assuming that each amino acid can be used once, twice, three times, or not at all. Use the symbols Gly and Ala to write down representations of the amino acid sequence in all eight of these tripeptides.

Hint: Start with GlyGlyGly.

11.19 Consider This: 3-D Amino Acids

Structural features of amino acids are more readily apparent if you look at their three-dimensional representations. At the *Chemistry in Context* web site, you can view and rotate the molecular structures for several different amino acids.

a. How is the three-dimensional structure of glycine different from the two-dimensional structure shown in your text?

b. Glycine is the simplest amino acid. It contains only two functional groups (—NH_2 and —COOH) and only the elements C, H, O, and N. Browse through the Chime collection of amino acids and then describe two ways in which their structures are more complex than glycine's.

c. In leucine, what four different groups are bonded to a central carbon atom? Is this molecule optically active? Explain your answer.

11.7 Enough Protein: The Complete Story

Dietary protein requirements are usually expressed in terms of grams of protein per kilogram of body weight per day, which vary with age, size, and energy demand. Infants require 1.8 g protein/kg body mass/day, middle school children about 1.0 g/kg/day, and adults 0.8 g/kg/day. Therefore, a 20-lb (9-kg) child needs 16 g of protein daily to provide the raw materials for body growth and development. A 165-lb (75-kg) adult requires 60 g each day to maintain proper physiological function.

The body does not normally store a reserve supply of protein, so foods containing these nutrients must be eaten every day. As the principal source of nitrogen for the body, proteins are constantly being broken down and reconstructed. A healthy adult on a balanced diet is in **nitrogen balance,** excreting as much nitrogen (primarily as urea in the urine) as she or he ingests. Growing children, pregnant women, and persons recovering from long-term debilitating illness have a positive nitrogen balance. This means that they consume more nitrogen than they excrete because they are using the element to synthesize additional protein. A negative nitrogen balance exists when more protein is being decomposed than is being made. This occurs in starvation, when the energy needs of the body are unmet from the diet, and muscle is metabolized to maintain physiological functions. In effect, the body feeds on itself.

Another cause of a negative nitrogen balance may be a diet that does not include enough of the essential amino acids. Of the 20 natural amino acids that make up our proteins, we can synthesize 11 from simpler molecules, but nine must be ingested directly. If any of the nine essential amino acids identified in Table 11.3 are missing from the diet, many important proteins cannot be produced in the body in sufficient quantity. The result can be severe malnutrition.

Good nutrition thus requires protein in sufficient quantity and suitable quality. Beef, fish, poultry, and other meats contain all the essential amino acids in approximately the same proportions found in the human body. Therefore, meat is termed a complete protein. However, most people of the world depend on grains and other vegetable crops rather than meat as their major sources of protein. If such a diet is not sufficiently diversified, some essential amino acids may be lacking. For example, Mexican and Latin American diets are rich in corn and corn products, a protein source that is *incomplete* because corn is low in tryptophan, an essential amino acid. A person may eat enough corn to meet the total protein requirement, but still be malnourished because of insufficient tryptophan.

Fortunately for millions of vegetarians, a reliance on vegetable protein does not necessarily doom one to malnutrition. The trick is to apply a principle nutritionists call *complementarity*, combining foods that complement each others' essential amino acid content so that the total diet provides a complete supply of amino acids. You do this, likely unknowingly, every time you eat a peanut butter sandwich. Bread is deficient in lysine and isoleucine, but peanut butter supplies these amino acids. On the other hand, peanut

Table 11.3

The Essential Amino Acids

Histidine	Lysine	Threonine
Isoleucine	Methionine	Tryptophan
Leucine	Phenylalanine	Valine

butter is low in methionine, a compound provided by bread. The traditional diets of many countries meet protein requirements through nutritional complementarity. In Latin America, beans are used to complement corn tortillas; soy foods are eaten with rice in parts of Southeast Asia and Japan. People in the Middle East combine bulgur wheat with chickpeas or eat hummus, a sauce of sesame seeds, and chickpeas, with pita bread. In India, lentils and yogurt are eaten with unleavened bread.

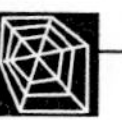

11.20 Consider This: Vegetarian Complementarity

Use the Web to find at least two additional examples of complementarity.

a. What essential amino acids are involved in the combination?

b. Do these combinations involve common foods of that country (such as peanut butter and bread in the U.S.)?

Livestock, especially beef cattle, also benefit from complementarity. They are fed a variety of grains with a complete set of amino acids to incorporate ultimately into steaks and hamburger. However, the second law of thermodynamics applies to beef cattle as well as to everything else. There is a loss of efficiency with each step of energy transfer, whether in electrical power plants or in cells during metabolism. Cattle are notoriously inefficient in converting the energy in their feed into meat on the hoof. It takes about seven pounds of grain to produce one pound of beef. Put into human terms, the 1.75 pounds of grain used to produce a "quarter pounder" can provide two days of food for someone on a vegetarian diet. Other animals are more efficient than cattle in converting grain to meat. Hogs require six pounds of grain per pound of meat, turkeys need four, and chickens even less, only three pounds. It is obviously much more efficient to get food energy directly from grains, rather than through secondary or tertiary sources further along the food chain. On the other hand, one should keep in mind that pasture land used to graze cattle is often unsuitable for growing crops. Moreover, much of the food consumed by animals would be indigestible or unpalatable to humans.

A postscript to the protein story is provided by the unusual case of aspartame, a sweet dipeptide. Because of the great American preoccupation and battle with excess Calories and excess pounds, artificial sweeteners have become a billion dollar business. Gram for gram, these compounds are much sweeter than sugar, but they have little if any nutritive value. Hence, they are nonfattening. The principal use (75%) of artifical sweeteners is in soft drinks. Currently the most widely used artificial sweetener is aspartame, the principal ingredient in NutraSweet and Equal. Somewhat surprisingly, the compound is related to proteins. Aspartame is a dipeptide made from aspartic acid and a slightly modified phenylalanine. The molecular structure of aspartame is given in the following diagram.

- Americans drink an annual average of about 53 gallons of soft drinks per person, up nearly 50% from the amount in 1985.

O O CH_2
HO—C—CH_2 C C N CH C O CH_3
H_3N (+) H H O

Alone, neither of the two amino acids in aspartame tastes sweet. Yet, the compound that results from their chemical combination is about 200 times sweeter than sucrose. The fact that sucrose and aspartame are chemically and structurally very different invites speculation about the molecular features that convey sweetness. But this digression will be long enough without taking up the issue of sweetness. For whatever reason, aspartame is sufficiently sweet to be used by millions of people worldwide. A few cases of adverse side effects have been attributed to aspartame, but exhaustive reviews have failed to show an unequivocal and direct connection between the symptoms and the sweetener. For the vast majority of consumers, aspartame is a safe alternative to sugar. There is, however, one group of people who definitely should not use aspartame. The

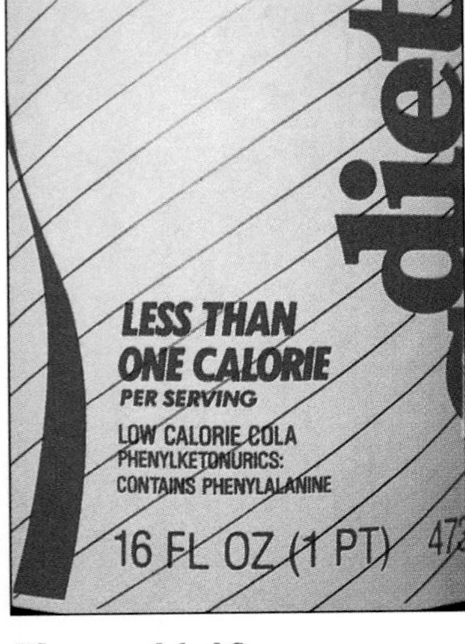

Figure 11.10
Warning: "Phenylketonurics: Contains Phenylalanine" on a diet beverage can.

warning on packets of artificial sweeteners and products containing aspartame is explicit: "Phenylketonurics: Contains Phenylalanine" (Figure 11.10).

This is a case where one person's meat is another person's poison. Phenylalanine is an essential amino acid converted in the body to tyrosine, another amino acid. Individuals with phenylketonuria, a genetically transmitted disease, lack the enzyme that catalyzes this transformation. Consequently, the conversion of dietary phenylalanine to tyrosine is blocked and the phenylalanine concentration rises. To compensate for the elevated phenylalanine, the body converts it to phenylpyruvic acid, excreting large quantities of this acid in the urine. Phenylpyruvic acid is termed a "keto" acid because of its molecular structure; hence, the disease is known as phenyl*keto*nuria or PKU. People with the disease are called phenylketonurics.

Excess phenylpyruvic acid causes severe mental retardation. Therefore, the urine of newborn babies is tested for this compound, using special test paper placed in the diaper. Infants diagnosed with PKU must be placed on a diet severely limited in phenylalanine. This means avoiding excess phenylalanine from milk, meats, and other sources rich in protein. Commercial food products are available for such diets, their composition adjusted to the age of the user. Because phenylalanine is an essential amino acid, a minimum amount of it must still be available, even in phenylketonurics. Supplemental tyrosine may also be needed to compensate for the absence of the normal conversion of phenylalanine to tyrosine. A phenylalanine-restricted diet is recommended for phenylketonurics at least through adolescence. Adult phenylketonurics must also limit their phenylalanine intake, and hence curtail their use of aspartame.

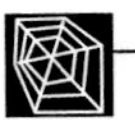

11.21 Consider This: Searching for Sweetness

Aspartame received FDA approval in 1981, but the search for artificial sweeteners has continued. Use the Web to find out if any artificial sweeteners have received FDA approval since 1981.

a. Identify the artificial sweetener(s) and its (their) intended uses.
b. Describe the chemical composition of the sweetener.

11.22 Consider This: How Sweet They Are!

Consider this information about the sugar content of different food products.

Food Product	Sugar	Calories	Serving Size
Altoids, Peppermint	2 g	10	3 pieces (2 g)
Ginger Snaps	9 g	120	4 cookies (28 g)
Critic's Choice Tomato Ketchup	3 g	15	1 tbsp (13 g)
Delmonte Pineapple Cup	13 g	50	Individual cup (113 g)
Dr Pepper Soft Drink	40 g	150	1.5 cups
French Vanilla Coffee Mate	5 g	40	1 tbsp (15 mL)
Hostess Twinkies	14 g	150	1.5 oz
LifeSavers, WintOgreen	15 g	60	4 mints (16 g)
Tropicana Home Style Orange Juice	22 g	110	8 oz (1 cup)
Snickers Bar	29 g	200	2.1 oz
Sunkist Orange Soda	52 g	190	1.5 cups
Wheatables Crackers	4 g	130	13 crackers (29 g)

a. Examine this list of 12 food products. Which item has the highest ratio of grams of sugar to the number of Calories (g sugar/Cal) in one serving?
b. Does the sugar content of any of these foods surprise you? Explain your response.
c. Do you expect that the specific sugars in Dr Pepper are the same sugars found in Sunkist orange soda? In cranberry juice? In the pineapple cup? Why or why not?
d. There are 16 g of total carbohydrates listed on the label for WintOgreen Lifesavers, of which 15 g are sugars. What type of compounds do you think account for the other 1 g of carbohydrates?

11.8 Energy: The Driving Force

All the energy needed to run the complex chemical, mechanical, and electrical system called the human body comes from carbohydrates, fats, and proteins. This energy initially arrives on Earth in the form of sunlight, which is absorbed by green plants during photosynthesis. Under the influence of a catalyst called chlorophyll, carbon dioxide and water are combined to form glucose. In the process, the Sun's energy is stored in chemical bonds of the sugar.

- If you need a refresher on chemical energetics, see Section 4.4.

$$\text{Energy (from sunshine)} + 6\ CO_2 + 6\ H_2O \longrightarrow C_6H_{12}O_6 + 6\ O_2$$

During metabolism, the photosynthetic process is reversed, the food is converted into simpler substances and the stored energy is released.

$$C_6H_{12}O_6 + 6\ O_2 \longrightarrow 6\ CO_2 + 6\ H_2O + \text{Energy (from metabolism)}$$

The breaking of chemical bonds in glucose and oxygen molecules requires the absorption of energy. But, more energy is released in the exothermic reactions in which carbon dioxide and water are formed. Thus, there is a net release of energy. This energy balance is schematically represented in Figure 11.11.

Energy provided by food we eat is used to drive the chemical reactions that constitute the processes of life. The most obvious example of an energy-requiring process is muscular motion, including the beating of the heart. But most of the energy released by metabolism goes to maintain differences in ionic concentrations across cell membranes. The natural tendency is for diffusion to move substances from regions of higher concentration to those of lower concentration. Energy is required to prevent this from happening. The proper concentration differences that are essential for nerve action and other physiological functions are maintained at great energetic expense. In short, spontaneous reactions furnish the energy to non-spontaneous reactions so that they can occur. As an analogy, consider an automobile storage battery. The battery produces electrical energy spontaneously because of chemical reactions in the battery. These spontaneous processes provide energy that can be used to permit non-spontaneous processes to take place, for example, starting the car or making the headlights and horn work.

- Each heartbeat uses one joule of energy.

In addition to having a supply of sufficient energy, the body must have some way of regulating the rate at which the energy is released. Without such control, wild temperature fluctuations and high inefficiency could result. Again, the automobile provides an analogy. Dropping a lighted match into the fuel tank would burn all the gasoline (and the car as well). This is a drastic but not particularly effective way to

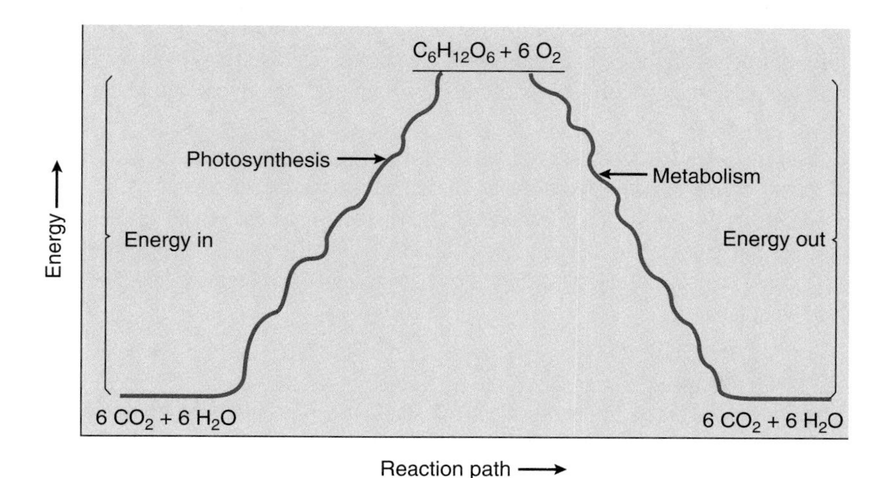

Figure 11.11
Energy from photosynthesis and metabolism.

From Chemistry for Health-Related Sciences *by Sears/Stanitski, © 1976. Reprinted by permission of Prentice-Hall, Inc., Upper Saddle River, NJ.*

move a car. Under normal operating conditions, just enough fuel is delivered to the ignition system to supply the automobile with the energy it needs without raising the temperature of the car and its occupants beyond reason. In this way, by releasing a little energy at a time, the efficiency of the process is enhanced. So it is with the body. The conversion of foods ultimately into carbon dioxide and water occurs over many small steps, each one involving enzymes, enzyme regulators, and hormones. As a result, energy is released gradually, as needed, and body temperature is maintained within normal limits.

- 1 dietary Calorie = 1 Calorie = 1 kcal = 1000 calories.
- Energy released: Carbohydrates 4 Cal/g; Fats 9 Cal/g.
- Oxygenated fuels were discussed in Section 4.10.

The chief sources of this energy in a well-balanced diet are carbohydrates and fats. When metabolized, *carbohydrates* provide about 4 kcal/g, *fats* release about 9 kcal/g. A **kilocalorie** (kcal) is identical to a dietary Calorie, written with a capital C. Package labels and nutritional tables that state how much energy we get from our meals typically use Calories, as in "one chocolate chip cookie provides 50 Calories." We will do the same. Keep in mind that whatever units are used, on a gram-for-gram basis, fats provide about 2.5 times as much energy as carbohydrates.

The reason for this dramatic difference is implicit in the chemical composition of these two types of material. Compare the formula of a fatty acid, lauric acid, $C_{12}H_{24}O_2$, with that of sucrose (table sugar), $C_{12}H_{22}O_{11}$. Both compounds have the same number of carbon atoms per molecule and very nearly the same number of hydrogen atoms. When the fatty acid or the sugar burns, its carbon and hydrogen atoms combine with added oxygen to form CO_2 and H_2O, respectively. But more oxygen is required to burn a gram of lauric acid, $C_{12}H_{24}O_2$, than a gram of sucrose, $C_{12}H_{22}O_{11}$. This is evident from the equations for the two reactions.

$$C_{12}H_{24}O_2 + 17\ O_2 \longrightarrow 12\ CO_2 + 12\ H_2O \qquad [11.4]$$

$$C_{12}H_{22}O_{11} + 12\ O_2 \longrightarrow 12\ CO_2 + 11\ H_2O \qquad [11.5]$$

In the jargon of chemistry, the sugar is already more "oxygenated" or more "oxidized" than the fatty acid. There are more CH bonds in the fatty acid to "burn" to CO_2 and H_2O and release more energy than from sucrose. Hence, sucrose is chemically and energetically "closer" to the end products of CO_2 and H_2O and needs less oxygen to form them. Therefore, when one gram of sucrose is burned, 3.8 Calories are released, compared to 8.8 Calories per gram of lauric acid.

The fact that fats are such concentrated energy sources means that it is easy to get an unhealthy percentage of our daily Calories from fats. The problem is nicely illustrated by 11.23 Sceptical Chymist and 11.24 Your Turn. Nutritionists and the American Heart Association advise that no more than 30% of your caloric intake should come from fats, and 55–60% should be derived from carbohydrates, especially polysaccharides. The remainder, 10% or less, should be contributed by protein. Although proteins, like carbohydrates, yield about 4 Calories per gram, they are not a major energy source, but rather a store of molecular parts for building skin, muscles, tendons, ligaments, blood, and enzymes.

11.23 The Sceptical Chymist: Low-Fat Cheese

A popular brand of low-fat shredded cheddar cheese advertises that it provides 1.5 g of fat with 15 Calories from total fat per serving. There are 50 Calories per serving and of the total fat, 1.0 g is saturated fat. A serving is defined as 1/4 cup or 28 g. Is this a "low-fat" cheese? Defend your decision with some calculations. Remember that the dietary recommendation is that no more than 30% of Calories should come from fat.

11.24 Your Turn

Bagel chips are an alternative to potato chips. The label on one popular brand of bagel chips lists 130 Calories per serving, with 35 Calories from fat. The total fat is listed as 4 g per serving, with 1 g being saturated fat, 1 g being polyunsaturated fat, and 2 g being monounsaturated fat.

a. What percentage of daily value of calories based on a 2000-Calorie diet is provided by one servings of these chips?

b. Potato chips have 150 Calories in a 28 g serving, 10 g of which is fat. Are bagel chips a more healthful alternative to potato chips in your opinion? Give reasons for your answer.

11.9 Energy: How Much is Needed?

The current unhealthy level of overeating and obesity in the United States, and the occasional reports of anorexia and bulimia prompt the question that serves as the title of this section. The answer is vague: "It depends." The number of Calories your diet should supply each day depends on your level of exercise or activity, the state of your health, your sex, age, body size, and a few other factors. You can probably identify yourself in Table 11.4, which summarizes the daily food energy intakes that have been recommended for Americans.

Some generalizations can be drawn from Table 11.4. Most men require more Calories per day than women do, but the difference is not totally due to differences in body weight. The column showing the recommended number of Calories per kilogram of body weight indicates that the magnitude of this indicator decreases with age. Growing children need a proportionally large energy intake to fuel their high level of activity and provide raw material for building muscle and bone. Therefore, children are particularly susceptible to undernourishment and malnutrition. Mortality rates among infants and young children are disproportionately high in famine-stricken countries. In many sub-Saharan Africa countries, the average energy intake is below 500 Calories daily, far below the required minimum for children or adults.

The minimum amount of energy required daily, the **basal metabolism rate (BMR),** is the amount necessary to support basic body functions—to keep the heart beating, the lungs inhaling and exhaling, the brain active, the blood circulating, all major organs working, and body temperature maintained at 37°C. This corresponds to approximately one Calorie per kilogram (2.2 lb) of body weight per hour, although it varies with size and age. The BMR is experimentally determined in a resting state, and the quantity of energy used in digestion is eliminated by having the subject fast for 12 hours before the measurement is made. To put this on a personalized basis, consider a 20-year old female weighing 55 kg (121 lb). If her body has a minimum requirement of 1 Cal/(kg/hr), her daily basal metabolism rate will be 1 Cal/(kg/hr) × 55 kg × 24 hr/day or about 1300 Cal/day. According to Table 11.4, the recommended daily energy intake for a woman of this age and weight is 2200 Cal. This means that 59% of the energy derived from this food goes just to keep her body systems going.

- Your basal metabolism rate is approximately 1 Cal/kg body mass per hour.

- $\frac{1300\ \text{Cal}}{2200\ \text{Cal}} \times 100\% = 59\%$

Table 11.4

Recommended Daily Energy Intake (United States)

Age (yrs)	Avg. weight (kg)	Avg. weight (lb)	Avg. height (in.)	Avg. Cal/kg	Avg. Cal/day
0.5–1.0	9	20	28	98	850
4–6	20	44	44	90	1800
7–10	28	62	52	70	2000
Males					
15–18	66	145	69	45	3000
19–24	72	160	70	40	2900
25–50	79	174	70	37	2900
51+	77	170	68	30	2300
Females					
15–18	55	120	64	40	2200
19–24	58	128	65	38	2200
25–50	63	138	64	36	2200
51+	65	143	63	30	1900

Table 11.5

Energy Expenditure (Cal/min) for Various Activities in Relation to Body Mass (lbs)

Activity	120 lbs (Cal/min)	150 lbs (Cal/min)	180 lbs (Cal/min)	200 lbs (Cal/min)
Aerobics (vigorous)	7	9	11	12
Basketball (vigorous)	10	13	15	17
Bicycling (11 mph)	6	7	9	10
Golf (carrying clubs)	5	6	7	8
Jogging (10 min/mile)	9	11	14	15
Rollerblading (12 mi/hr)	10	12	13	14
Running (7 min/mile)	12	15	18	20
Studying	1.3	1.7	1.9	2.0
Swimming (fast)	9	11	13	14
Volleyball	5	6	7	8
Walking (20 min/mile)	3	4	5	6

Where the rest of it goes depends on what she does. The law of conservation of energy decrees that the energy must go somewhere. If she "burns off" the extra Calories in exercise and activity, none will be stored as added fat and glycogen. But, if the excess energy is not expended, it will accumulate in chemical form. Putting it more crassly, "those who indulge, bulge," unless they work and play hard.

Some indication of how hard and how long we have to work or play to use up dietary Calories is given in Tables 11.5 and 11.6. The former reports the energy expenditures for various activities as a function of body weight. Table 11.6 quantifies exercise in readily recognizable units such as hamburgers, potato chips, and beers.

11.25 Your Turn

First calculate your BMR. Then select from Table 11.5 the activities you do in a typical day. Calculate the supplemental energy you need for these activities. Then add your BMR and your supplemental energy needs to determine the total number of Calories you require per day. How does this result compare with the recommended energy intake for your age and sex (see Table 11.4)?

Table 11.6

How Much Exercise Must I Do If I Eat This Cookie? Calories and Minutes of Exercise for a 150-Pound Person

Food	Calories	Walk (min)	Run (min)
Apple	125	31	8
Beer (regular) 8 oz	100	25	7
Chocolate chip cookie	50	12	3
Hamburger	350	88	23
Ice cream, 4 oz	175	44	12
Pizza, cheese, 1 slice	180	44	12
Potato chips, 1 oz	108	27	7

Table 11.7

Average Dietary Energy Supplies (DES) in Calories Per Person Per Day

Region	1979–81	1986–88	1997–99
Developed countries			
Averages	3300	3389	3220
North America	3487	3626	3300
Western Europe	3371	3445	3140
Developing countries			
Averages	2317	2352	2627
Africa	2148	2119	2415
Latin America	2675	2732	2770
Near East	2794	2914	Not available
Far East	2185	2220	2620
World (averages)	2587	2671	Not available

Source: Data from *The State of Food and Agriculture*, United States, New York, Rome: 1990 and 2000; *Food and Agriculture Organization of the United Nations*, United Nations, New York, 1991 and 1999.

11.26 Your Turn

A 150-lb person consumes a meal consisting of two hamburgers, 3 oz of potato chips, 8 oz of ice cream, and a 12-oz beer. Calculate the number of Calories in the meal and the number of minutes the person would have to run to "work off" the meal.

Answer

1524 Cal, 102 min running or 139 min jogging (from Table 11.6).

In many parts of the world, a major nutritional problem is not how to get rid of excess Calories, but how to get enough of them. This dietary discrepancy is evident from Table 11.7, which reports the average Dietary Energy Supplies (DES) in Calories per person per day for different global regions. Although the DES has generally increased since 1979, in each of the three time intervals listed, the average DES for developed nations was significantly greater than that for developing countries. In 1986–88, the average daily individual energy intake in the developed countries was 44% higher than that in less industrially advanced nations. DES values for North America are conspicuously high—well above the recommended daily energy intake. Another way to compare the data of Table 11.7 is by calculating the DES in various parts of the world as a percentage of the North American level. Again for 1997–99, the DES for Africa was only 73% that for North America, the Far East was 79% the North American value, and Latin America was 84%.

11.10 Vitamins: The Other Essentials

Your daily diet should supply an adequate number of Calories, but Calories alone are not enough. You have already read about the essential fatty acids and amino acids that must be ingested for good health, and you are well aware that a balanced diet must also provide certain vitamins and minerals. Unfortunately, many popular foods that are high in sugars and fats lack these essential micronutrients. It is thus possible that a person can be overfed with excess Calories but malnourished through a diet lacking adequate vitamins and minerals.

A detailed understanding of the role of vitamins and minerals is of relatively recent origin. Over the ages, humans learned that if certain foods were lacking, illness often resulted, but the correlation between diet and health was often accidental and anecdotal. More systematic studies began early in this century, with the discovery of "Vitamine B_1" (thiamine). The particular designation, B_1, was the label on the test tube in which the sample was collected. The general term "vitamin" was chosen because the compound, which is *vit*al for life, is chemically classed as an *amine*. The final "e" disappeared with the discovery that not all vitamins are amines. Today, **vitamins** are defined by their properties: they are essential in the diet, although required in very small amounts; they all are organic molecules with a wide range of physiological functions; and they generally are not used as a source of energy, although some of them help break down macronutrients.

- The relationship between molecular structure and solubility was discussed in Section 5.12.
- Hydrogen bonds were discussed in Section 5.8.

Vitamins are often classified on the basis of solubilities; they either dissolve in water or in fat. Vitamins A, D, E, and K dissolve in fat but not in water, because they are nonpolar molecules. For example, the molecular structure of vitamin A, which follows, is based almost exclusively on carbon and hydrogen atoms. Thus, it is similar to the hydrocarbons derived from petroleum. Vitamins that are not fat soluble are soluble in water because their polar molecules contain several —OH groups, which form hydrogen bonds with water molecules. Vitamin C is a case in point.

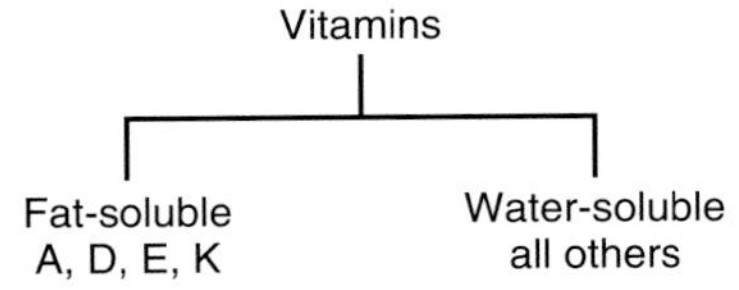

Vitamins
Fat-soluble A, D, E, K
Water-soluble all others

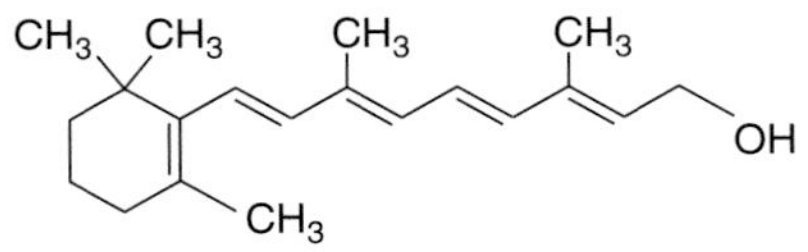

CH_3 CH_3 CH_3 CH_3 OH CH_3

Vitamin A, a fat-soluble vitamin

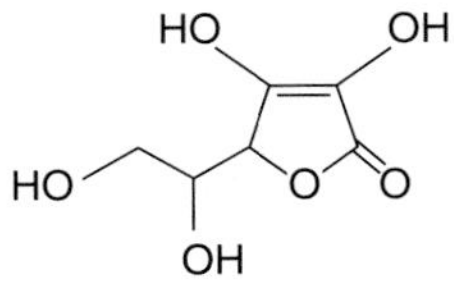

HO OH HO O O OH

Vitamin C, a water-soluble vitamin

11.27 Your Turn

Examine the molecular structures given for vitamin A and vitamin C. Use solubility and molecular structure relationships described in Section 5.12 under the general principle that "like dissolves like" to explain why vitamin A is fat soluble, not water soluble. Explain why vitamin C has the opposite solubility behavior.

These solubility differences among vitamins have significant implications for nutrition and health. Because of their fat solubility, vitamins A, D, E, and K are stored in cells rich in lipids, where they are available on biological demand. This means that the fat-soluble vitamins need not be taken daily. It also means that these vitamins can build up to toxic levels if taken far in excess of normal requirements. For example, high doses of vitamin A can result in fatigue, headache, dizziness, blurred vision, dry skin, nausea, and liver damage. Vitamin D toxicity occurs at just four to five times its Recommended Daily Allowance (RDA), making vitamin D the most toxic vitamin. Such high levels of the vitamin are reached using vitamin supplements, not through a normal diet. Cardiac and kidney damage can result. Water-soluble vitamins, by contrast, are not generally stored; any unused excess is excreted in urine. Thus, they must be consumed frequently and in small doses. Unfortunately, when taken in very large doses, water-soluble vitamins can also accumulate until they reach toxic levels, although such cases are rare. For example, there are reports that vitamin B_6, taken at 10 to 30 times the recommended dose per day for extended periods, results in nerve damage, including paralysis. Even higher doses of vitamin B_6 supplements, up to 1000 times the recommended dosage, have been consumed to alleviate the symptoms of pre-menstrual syndrome (PMS), again causing abnormal neurological symptoms. For most people, a balanced diet should provide all the necessary vitamins and minerals in appropriate amounts, making vitamin supplements unnecessary.

Even a brief review of various essential vitamins and minerals is well beyond the scope of this book, but a few observations might be of interest. For example, niacin or nicotinic acid illustrates the way in which vitamins, especially members of the B family, act as coenzymes. **Coenzymes** are generally small molecules that work in conjunction with enzymes to enhance the enzymes' activity. **Niacin** plays an essential role in energy transfer during glucose and fat metabolism. The synthesis of niacin in the body requires the essential amino acid tryptophan. Thus, a diet deficient in tryptophan may lead to niacin deficiency. Such a deficiency causes pellagra, a condition involving a darkening and flaking of the skin, as well as behavioral aberrations.

Vitamin C (ascorbic acid) must also be supplied in the diet, typically via citrus fruits and green vegetables. An insufficient supply of the vitamin leads to scurvy, a disease in which collagen, an important structural protein, is broken down. The link between citrus fruits and scurvy was discovered more than 200 years ago when it was found that feeding British sailors limes or lime juice on long sea voyages prevented the disease. Ascorbic acid is also required for the uptake, use, and storage of iron, important in the prevention of anemia. The claims that high doses of vitamin C can prevent colds and ward off certain cancers remain largely unsubstantiated.

- This practice also led to British sailors being called "limeys."

The last vitamin in this brief overview is **vitamin E,** important in the maintenance of cell membranes and as protection against high concentrations of oxygen, such as those that occur in the lungs. Vitamin E is so widely distributed in foods that it is difficult to create a diet deficient in it, although people who eat very little fat may need supplements. Vitamin E deficiency in humans has been linked with nocturnal cramping in the calves and fibrocystic breast disease.

11.28 Consider This: Megadosing Vitamin C

In the past, there have been claims that megadoses of vitamin C prevent the common cold and may be effective against certain types of cancers. Use the Web to determine whether other evidence to the contrary has been discovery about the efficacy of megadoses of vitamin C.

a. What dose of vitamin C constitutes a megadose?
b. Where was the research conducted?
c. What were the results of the research?

11.11 Minerals: Macro and Micro

An adequate supply of a number of **minerals** (ions or inorganic compounds) is also essential for good health. Table 11.1 lists calcium, phosphorus, chlorine, potassium, sulfur, sodium, and magnesium among the major elements in the body. These seven **macrominerals,** although not nearly as abundant as oxygen, carbon, hydrogen, or nitrogen, are nevertheless necessary for life. The adult RDAs for these macrominerals typically range from one to two grams. The body requires lesser amounts of iron, copper, zinc, and fluorine—the so-called **microminerals.** Trace minerals, including iodine, selenium, vanadium, chromium, manganese, cobalt, nickel, molybdenum, and tin are usually measured in micrograms (1×10^{-6} g). Arsenic, cadmium, and even lead, which are generally classified as toxic, are needed in very small amounts. Although the total amount of trace elements in the body is only about 25–30 grams, their slight amounts belie the disproportionate importance they have in good health.

- Macrominerals are needed dietarily per day in amounts greater than 100 mg (0.100 g) or are present in the body in amounts greater than 0.01% of body weight.

The essential minerals are highlighted in the periodic table outline that appears as Figure 11.12. The metallic elements exist in the body as cations, for example, Ca^{2+} (calcium), Mg^{2+} (magnesium), K^{+} (potassium), and Na^{+} (sodium). The nonmetals typically are present as anions, thus chlorine is found as Cl^{-} and phosphorus appears in the phosphate ion, PO_4^{3-}. The physiological functions of minerals are widely diverse.

Calcium is the most abundant mineral in the body. Along with phosphorus and smaller amounts of fluorine, it is a major constituent of bones and teeth. Blood clotting, muscle contraction, and transmission of nerve impulses also require Ca^{2+} ions.

Sodium is also essential for life, but not in the relatively excessive amounts supplied by the diets of most Americans. Physicians recommend a maximum of 1.2 grams (1200 mg) of sodium (Na^{+}) per day. This corresponds to 3 grams of salt (NaCl) and is twice the estimated minimum requirement. Most Americans exceed the recommended daily sodium intake, sometimes by three- to fourfold. The major culprits are processed

- The average U.S. daily diet contains about nine grams of NaCl (3.6 g Na^{+}). Some common dietary sources of sodium (mg Na^{+} per serving): processed macaroni and cheese (1086); canned chicken noodle soup (1100); baked ham (950); hot dog (640); potato chips (200); white bread slice (114). Recall that 1 g = 1000 mg.

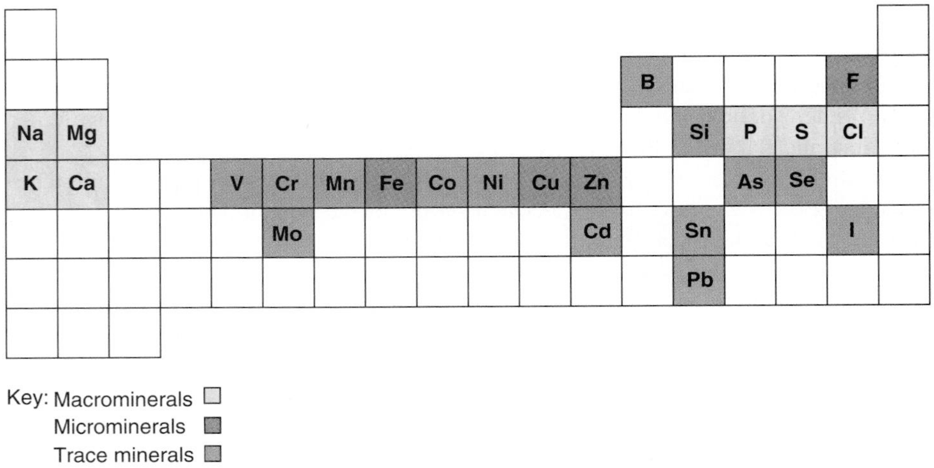

B F
Na Mg Si P S Cl
K Ca V Cr Mn Fe Co Ni Cu Zn As Se
Mo Cd Sn I
Pb

Key: Macrominerals
Microminerals
Trace minerals

Figure 11.12

Periodic table indicating macrominerals, microminerals, and some trace minerals necessary for human life. Boron, silicon, arsenic, lead, and cadmium are essential in animals and likely essential in humans.

- The Latin word for *salt* is *sal.* Salt was so highly valued in Roman times that soldiers were paid in *sal*, thereby forming the root for the modern word *salary.*

foods and fast foods, which are very heavily salted for flavor. The major concern with excess dietary sodium is its correlation with high blood pressure (hypertension).

Oranges, bananas, tomatoes, and potatoes help supply the recommended daily requirement of 2 grams of **potassium,** a mineral that is essential for the transmission of nerve impulses and intracellular enzyme activity. Potassium and sodium are elements in the first column of the periodic table. Neutral atoms of the two elements each have one electron in the outer level. These electrons are readily lost to form K^+ and Na^+ ions, the forms found in the human body. Because sodium and potassium ions have similar chemical properties, their physiological functions are also closely related. In intracellular fluid (the liquid within cells), the concentration of potassium ions is considerably greater than that of sodium ions. The reverse situation holds in the lymph and blood serum outside the cells. There the concentration of potassium ions is low and that of sodium ions is high. We have already noted the large amount of energy that must be expended to properly maintain these essential concentration gradients. The relative concentrations of K^+ and Na^+ are especially important for the rhythmic beating of the heart. Individuals who take diuretics to control high blood pressure commonly also take potassium supplements to replace potassium excreted in the urine. However, such supplements should be taken only under a physician's directions because of the potential danger that they could dramatically alter the potassium/sodium balance and lead to cardiac complications.

In most instances, the microminerals and trace elements have very specific biological functions and may be incorporated in only a relatively small number of biomolecules. **Iron,** for example, is an essential part of hemoglobin, the protein that transports oxygen in the blood, and of myoglobin, which is used for temporary oxygen storage in muscle. There are four Fe^{2+} ions in a hemoglobin molecule, and each binds reversibly to an O_2 molecule. Insufficient iron in the diet causes iron-deficiency anemia, a condition in which the red blood cells are low in hemoglobin and correspondingly can carry less oxygen. Symptoms include fatigue, listlessness, and decreased resistance to infection. Iron-deficiency anemia is a major problem in developed as well as developing nations. Iron deficiencies have been estimated as high as 20% in the United States, particularly among post-puberty women. On the other hand, too much iron in the diet can cause gastrointestinal distress and contribute to cirrhosis of the liver. Children have been fatally poisoned by ingesting iron supplement tablets. To be utilized by the body, iron must be absorbed as Fe^{2+} ions, not in the Fe^{3+} form or simply as elemental iron. Therefore, it is a bit surprising that some iron-fortified cereals contain metallic iron dust that can be removed by a magnet from a slurry of the cereal and water. Foods naturally rich in iron include liver and spinach.

Iodine is another element with a specific biological function. Most of the body's iodine is concentrated in the thyroid gland, where it is incorporated into thyroxine, a hormone that regulates basal metabolism rate. Excess thyroxine is associated with hyperthyroidism or Graves disease, in which basal metabolism is accelerated to an unhealthy level, rather like a racing engine. On the other hand, a thyroxine deficiency, sometimes caused by insufficient dietary iodine, slows metabolism and results in tiredness and listlessness. Both hyper- and hypothyroidism can lead to goiter, an enlargement of the thyroid gland. One way to help prevent goiter is by consuming adequate amounts of iodine. Seafood is a rich source of the element, but it is also provided by iodized salt, normal sodium chloride to which 0.02% of potassium iodide (KI) has been added. The tendency of the thyroid gland to concentrate iodine is key to the use of radioactive iodine-131 as a treatment for an overactive thyroid and to the risks of accidental exposure to this isotope.

11.29 Consider This: Getting Well Using Radioactive Iodine

Hyperthyroid individuals suffer from an overactive thyroid and accelerated metabolism. Use the Web to find out how radioactive iodine-131 is used to treat hyperthyroidism.

a. What role does I-131 play in the treatment?
b. Isn't radioactivity bad for the patient? Explain.
c. What further treatment does the patient need after the I-131 is used?

It is not enough that a society have an adequate food supply to furnish the necessary vitamins, minerals, and macronutrients. To be useful, the food also must be kept uncontaminated until it is consumed. In the next section, we consider food preservation methods, including food irradiation.

11.12 Food Preservation

Over the centuries, people have used a variety of methods to preserve foods. Heavily salting foods or storing them in concentrated sugar syrups were the traditional methods used before modern refrigeration. These two methods create a salt or sugar concentration very much greater than that in any contaminating microorganisms such as bacteria, yeasts, and molds. At such high concentrations of salt or sugar, osmosis causes water to leave the cells of the microorganisms, killing them by rupturing their cell membranes. Heat is also used to kill microorganisms, as in home canning or pasteurization. Modern refrigeration retards, but does not ultimately prevent, spoilage.

- Osmosis was discussed in Section 5.17.
- Refrigeration lowers the temperature so that the rates of reactions in offending microorganisms in foods are slowed, thus retarding spoilage.

To supplement these methods, substances are added to foods to reduce spoilage and extend useful shelf life. As increasing numbers of consumers turn to packaged foods for meals rather than cooking "from scratch," adequate shelf life of packaged products assumes greater importance. Such products carry warning labels like "Best if used by (date)" or ones that give a specific expiration date.

Anti-oxidants are one type of such additives, compounds that prevent packaged, processed foods from becoming rancid due to oxidation of oil or fats, which form harmful free radicals. If you examine the label of any such processed foods (dry cereals, potato chips and other so-called junk food snacks), you are likely to see the letters BHT or BHA. These stand for *b*utylated *h*ydroxy*t*oluene and *b*utylated *h*ydroxy*a*nisole, respectively, the two most common anti-oxidant food additives.

BHT BHA

Anti-oxidants such as BHT and BHA act by preventing the buildup of free radicals, which are molecular fragments formed when fats and oils react with oxygen from the air in the food package.

$$\text{Fat (or oil)} + \text{oxygen} \longrightarrow \text{Free radicals} + \text{other products}$$

A free radical has an unpaired electron (designated by a dot), which makes the species highly reactive. BHT, BHA, and other anti-oxidants scavenge the unpaired electron from the free radical to form a stable radical species. This prevents further oxidation of the fat, thus preventing rancidity.

- Free radicals were discussed in Sections 2.11 and 9.4 in relation to ozone depletion and polymerization, respectively.

A modern method of food preservation uses radiation. Food, of course, is irradiated in a number of circumstances, but not usually to preserve it. It's a matter of what part of the electromagnetic spectrum is used (Section 2.4). It should be rather obvious that radiation is essential for food production; visible light from the Sun drives photosynthesis to give us fruits and vegetables. We use longer-wavelength IR radiation from a stove, or even longer-wavelength microwaves to cook food or warm up leftovers. Irradiating foods to preserve them is an entirely different case because it uses short-wavelength, high-energy gamma radiation to kill microorganisms. Such radiation is *ionizing* radiation, in contrast to that from visible, infrared, or microwave radiation, which are non-ionizing.

Classified as a food additive by Congress in 1958, food irradiation was approved by the FDA in 1963. It has been used to preserve food for astronauts to take with them as they travel in space. The method is used in more than 30 countries and is especially prevalent in Europe, Mexico, and Canada. It has the enthusiastic endorsement of the Food and Agricultural Organization of the United Nations. Irradiated foods even have their own international logo (Figure 11.13). Yet irradiated foods are controversial, especially in the United States. Why the controversy?

The irradiation procedure is relatively simple in the 160 such facilities worldwide where it is done. The material to be irradiated is placed on a conveyer belt that moves past a tight beam of high-energy gamma radiation generated by a cobalt-60 or cesium-137 source. The source and the irradiation facility are enclosed and shielded so that extraneous radiation does not escape. Over 40 different foods have been approved internationally for preservation by irradiation. Yet, only a small number of irradiated foods have been approved by the United States, including potatoes and strawberries for domestic consumption, and fish, shrimp, and grapefruit for export (Figure 11.14). In 1999, the USDA announced its allowance of irradiation of red meat—beef, pork, and lamb—as a way to reduce or eliminate disease-causing organisms. Irradiation of poultry was approved in 1992.

Those opposed to food irradiation question whether irradiated foods are safe to eat. The effectiveness as well as the need for this technology are questioned, including the proliferation of radioactive material for a possibly unneeded commercial application. Critics also express concern that food irradiation will be used to cover up improper food handling processes. Even so, critics and proponents agree that irradiating foods to preserve them does not make them radioactive. But the irradiated foods do have the normal background radiation that all foods naturally possess (Section 7.8).

The most serious charge brought by critics concerns the formation of possible "unique radiolytic products" (URPs) generated by gamma radiation breaking chemical bonds. Recall from Section 2.4 that, with its short wavelength, gamma radiation has sufficient energy to break chemical bonds and create free radicals or ions. Gamma radiation has much more energy than microwave, infrared, visible, or even ultraviolet radiation. Because most foods contain a high percentage of water, the gamma radiation is absorbed by water to form irradiation products in extremely small amounts, which then react with other components of food to form stable products. It must be kept in mind that cooking food also causes chemical changes in the food, changes that are many times greater than those from gamma irradiation. Nearly five decades of research suggests that the by-products of irradiation are the same chemical substances formed by conventional cooking or other preservation methods. Studies based on animal feeding research have repeatedly demonstrated no toxic effects from irradiated foods. The World Health

Figure 11.13
The international label for irradiated food.

Organization, the Food and Agricultural Organization of the United Nations, and the U.S. FDA all have concluded that food irradiation is safe when proper procedures and practices are used.

The need for food preservation is serious and should be kept in perspective. Worldwide, food spoilage and contamination is a significant problem, claiming up to 50% of a food crop in some parts of the world, including many developing nations. Closer to home, we are not immune to such contamination in our food supply. Outbreaks of food poisoning in the United States occur periodically from chicken tainted with bacteria, commonly *Salmonella*, due to inadequate treatment in processing plants. Food contaminated with this organism has been linked to about 4000 deaths in the United States alone. The symptoms of food poisoning—abdominal pain, diarrhea, nausea, and vomiting—mimic those of short-term flu. Therefore, food poisoning is often mistaken as being caused by flu. It is estimated that almost half of the raw chicken sold in the United States is contaminated with *Salmonella*. With proper handling and cooking, even chicken that is contaminated with bacteria can be prepared so as to prevent illness. Irradiation of chicken meat would lower the threat of accidental poisoning by *Salmonella*.

- Refrigerated chicken has a shelf life of three days; after gamma irradiation, chicken can have a three-week refrigerated shelf life.

Food irradiation can be viewed in terms of risk versus benefit: Does the benefit outweigh the risk? Some would respond that irradiating strawberries to keep them fresh for a few days longer is a misuse or misapplication of a suspect technology. People are not likely to become ill or die from eating strawberries that are a bit past their peak. On the other hand, trichinosis, a serious disease, can occur by eating pork contaminated with the *Trichinella spiralis* parasite. Low-level gamma irradiation of pork kills the parasite, making the pork safe to eat. Although irradiated beef, pork, and chicken have been approved by the FDA, firms that process these meats are wary that consumers will not buy these irradiated products. This is in spite of the fact that in countries where humans have consumed irradiated foods for years, including poultry and seafood, no adverse effects have been observed. Apparently the U.S. chicken processing companies feel that the risks (and costs) do not outweigh the benefits (at least to them).

Figure 11.14
Strawberries preserved by irradiation.

11.30 Consider This: Food Irradiation...Thanks or No Thanks?

Food irradiation remains controversial. The Foundation for Food Irradiation Education, at its web site, suggested that the Web provides "a unique opportunity to communicate the facts about food irradiation to journalists, educators, food company executives and the general public." Indeed, the Web can link a host of constituents with differing viewpoints on a topic such as food irradiation. Use the Web to prepare a position paper on whether food should be irradiated. The paper can be written from the standpoint of a food company executive, a manufacturer of irradiation equipment, a government official, or a consumer activist. Be sure to cite your sources. Later, you may wish to join with others to stage a class debate about the issues involved.

11.13 Feeding A Hungry World

Even when food preservation methods are available, sufficient food must still be produced to feed the population. To meet growing populations, global grain and cereal production has almost doubled over the past quarter century, and supplies of vegetables, fruits, milk, meat, and fish have also increased. It is estimated that at current production levels there is enough food to provide for 6.1 billion people—the projected world population at the turn of the 21st century. Nevertheless, many in the world go hungry; over 800 million of our fellow human beings are undernourished, mostly women and children. Approximately 24,000 people die daily of undernourishment. Clearly, the world's food supply is not equally distributed among its inhabitants. Piles of corn and wheat rot in the American Midwest or on docks around the world, while half a billion men, women, and children go to bed hungry.

Economic, political, and social, as well as agricultural reasons account for this inequity. Some are geographic fates—certain areas of the world simply do not have enough arable land and adequate soil to produce sufficient food for their people. This is especially true in areas in which population growth is exploding. In some areas, prolonged droughts reduce crop yields, and episodic floods in other locales wash away crops. The fertilizers needed to supplement mineral-depleted soils may not be available because of costs, and beasts of burden, to say nothing of tractors, may be too expensive for farmers in some regions.

A nation that cannot feed itself must import food, which means that it must have something to sell. Consequently, economics and international trade dictate the nutrition equation. Under these conditions, some countries face the difficult choice between supporting domestic agriculture to achieve self-sufficiency or investing in manufacturing goods for export to establish a favorable trade balance. And, civil strife in some countries, created by political and military actions, blocks the flow of food and other agricultural products.

The disparity in food production is particularly great between developed and developing countries. The developed nations, including the United States and Canada, supply over 50% of the world's food, but have only 20% of its population. Food production per capita is one of the most meaningful ways of looking at the data, because it corrects for differences in population. In general, overall per capita food production has increased more rapidly in developed countries over the past two decades than it has in most developing countries, even though the latter started at much lower levels. One mitigating factor is the higher rate of population growth in developing regions. In many developing countries, food production has generally not kept pace with growing populations. The exception has been Asia, where the rate of increase in per capita food production has been greater than that even in the developed world. In stark contrast, sub-Saharan Africa has experienced a long-term, continuing decline in per capita food production. One major reason for the increase in Asian crop yields has been greater use of fertilizers and pesticides. The application of both has often been criticized as being harmful to the environment, but the fact remains that millions have been saved from starvation, thanks to fertilizers and pesticides.

Urea $(NH_2)_2CO$, is a major fertilizer used worldwide. It provides nitrogen to soil by decomposing to ammonia and carbon dioxide when acted on by urease, an enzyme in soils. The ammonia is then taken up by plants.

$$(NH_2)_2CO + H_2O \xrightarrow{\text{urease}} 2\ NH_3 + CO_2 \qquad [11.6]$$

However, the efficiency of urea as a fertilizer is typically reduced because of the direct loss of ammonia through evaporation, in excess of 30%, before it can be taken up by plant roots. To overcome this inefficiency, the IMC-Agrico Company, a 1997 entrant in the Presidential Green Chemistry Challenge Awards program, developed AGROTRAIN, a formulation containing a compound that is converted into a urease inhibitor. Spread on a field, the AGROTRAIN-linked product produces the urease inhibitor, which reduces the rate at which urease decomposes urea so that ammonia is released more slowly and efficiently. The higher efficiency is important, especially in no-till applications, an environmentally friendly approach where there is little or no disturbance of topsoil. This method reduces soil erosion and requires much less energy for application of the fertilizer.

An even more striking, and generally less controversial, contribution to world agriculture has been the Green Revolution. A fundamental component of this enterprise has been the development of high-yield grains, principally wheat, rice, and corn, which were genetically modified to grow best in particular regions. These new varieties mature faster, permitting more harvests per year, so that the same amount of cultivated land can produce more crops. Since 1960, the Green Revolution has helped world grain harvests to more than double. Billions of people in India, Asia, and Africa have benefited from the practice. But the Green Revolution is neither a panacea nor the ultimate answer. In spite of its successes, the Green Revolution is not universally applicable, and has not been without costs. It works best in areas where water for irrigation is abundant; where money is available for supplemental fertilizers such as ammonia, urea, or nitrates; and where technological understanding and application exist.

Researchers estimate that, within 20 years, global demand for the world's three most important crops—rice, maize (a type of corn), and wheat—will have increased by 40%, simply to keep pace with global food requirements. Genetic engineering and other applications of biotechnology now hold out promise for a second Green Revolution to meet such demands. In the next chapter, we turn to more closely examine genetic engineering—its methods, accomplishments, and limitations.

CONCLUSION

This chapter began at a fast-food restaurant and continued by considering why we eat many different kinds of foods. Even though our individual tastes vary, our biological needs are much the same. We need carbohydrates as our primary energy source; fats for cell membranes, synthesis, and lubrication; proteins to build muscle and create the enzymes that catalyze the wonderful chemistry of life; and vitamins and minerals to help make that chemistry happen. Nutrition, like water quality, is a global issue that affects the health of all human beings, regardless of where they live. People with too much to eat, like most Americans, seem preoccupied with food, although generally with too little regard for what they eat. The hungry and the starving think of little else beyond how to feed themselves. Chemistry is only part of the solution to one of the great challenges of our time—how to meet all individual dietary needs, regardless of region or wealth.

Chapter Summary

Having studied this chapter, you should be able to:

- Recognize the frequency and regional occurrence of malnutrition and undernourishment (11.1)
- Understand the physiological functions of food (11.1)
- Describe the distribution of water, carbohydrates, fats, and proteins in the human body and some typical foods (11.1)
- Identify the major chemical elements found in the human body (11.1)
- Recognize and use the chemical composition and molecular structure of carbohydrates (11.2)
- Differentiate among the structures and properties of sugars, starch, and cellulose (11.2)
- Describe the symptoms and cause of lactose intolerance (11.2)
- Recognize and use the chemical composition and molecular structure of fats and oils or triglycerides (11.3)
- Identify sources of saturated and unsaturated fats and their significance in the diet (11.4)
- Differentiate among saturated, monounsaturated, and polyunsaturated fatty acids and fats (11.4)
- Discuss sources of cholesterol and its significance in the diet (11.5)
- Give the general molecular structure of amino acids (11.6)
- Identify and use the chemical composition and molecular structure of proteins (11.6)
- Discuss the importance of essential amino acids and their dietary significance (11.7)
- Describe the symptoms and cause of phenylketonuria (11.7)
- Explain carbohydrates, fats, and proteins as energy sources (11.8)
- Discuss typical recommended daily energy intakes (11.9)
- Relate energy expenditures in various activities (11.9)
- Differentiate among international variations in dietary energy supplies (11.9)
- Identify and use basal metabolism rate (BMR) (11.9)
- Discuss the effects of selected vitamins on human health (11.10)
- Differentiate chemically between fat-soluble and water-soluble vitamins (11.10)
- Describe the effects of selected minerals on human health (11.11)
- Discuss the necessity of macrominerals, microminerals, and trace minerals for human health (11.11)
- Discuss various methods of food preservation, including the advantages and disadvantages of food irradiation (11.12)
- Describe various strategies for feeding the world's growing population (11.13)

Questions

Emphasizing Essentials

1. Food provides four fundamental types of materials to keep our bodies functioning. What are those types of materials?
2. Is it possible for a person to be malnourished even when eating a sufficient number of Calories every day to meet metabolic needs? Explain.
3. Use the information in Figure 11.1 to answer these questions.
   a. Which areas of the world are experiencing food supply shortfalls and require exceptional assistance?
   b. Give reasons why these areas do not include North America.
4. a. What are macronutrients and what role do they play in keeping us healthy?
   b. Name the three major classes of macronutrients.
5. Consider this chart.

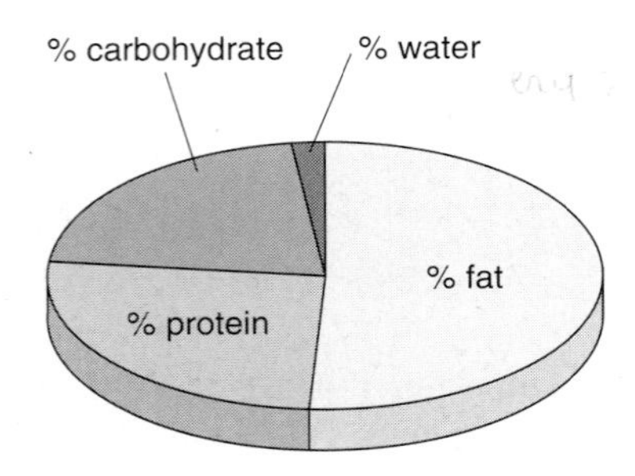

   Based on the relative percentages of protein, carbohydrate, water, and fat given, is this graph more likely a representation of steak, peanut butter, or chocolate chip cookies? Justify your choice based on the relative percentages of the components shown.
6. Answer each of these questions about the common foods shown in Figure 11.2.
   a. Identify the top three foods that are good sources of carbohydrates and arrange them in order of decreasing percentage of carbohydrates.
   b. Identify the top three foods that are good sources of protein and arrange them in order of decreasing percentage of protein.
   c. Which of these foods should be avoided if you are controlling dietary intake of fat? Identify the top three and arrange them in order of decreasing percentage of fat.
7. Answer each of these questions about the common foods shown in Figure 11.2.
   a. Which food has the highest protein-to-fat ratio? Calculate that ratio.
   b. Which food has the highest fat-to-protein ratio? Calculate that ratio.
8. An 18-oz steak is the manager's special at a local restaurant. Use the information in Figure 11.2 to calculate the number of ounces of protein, of fat, and of water that the customer eating this entire steak would consume.
9. Water is not considered as a macronutrient, but it clearly is essential in maintaining health. What are some of the roles that water plays in our bodies? *Hint:* You may want to refer to Chapter 5.
10. Examine the data in Table 11.1 and explain why hydrogen ranks first in atomic abundance in the human body, but third behind oxygen and carbon in terms of mass percent.
11. Use the information in Table 11.1 to answer these questions.
    a. What is the ratio of the relative abundance of potassium to sodium in the human body?
    b. What is the ratio of grams of potassium to grams of sodium in the human body?
    c. Are these elements included in the composition of the human body shown in Figure 11.3? Why or why not?
12. a. Consider the composition of the human body shown in Figure 11.3. What are the principal elements that make up water, proteins, carbohydrates, and fats? Which elements are in common among these major components of the human body?
    b. Compare your answers to part **a** with the relative abundance of the elements in the body given in Table 11.1. Is there a correlation? Explain the correlation between your lists and the relative abundances of the elements in the body.
13. This figure is a schematic diagram of the "food pyramid." What foods are in each section of the pyramid, and how many servings of each should you consume daily?

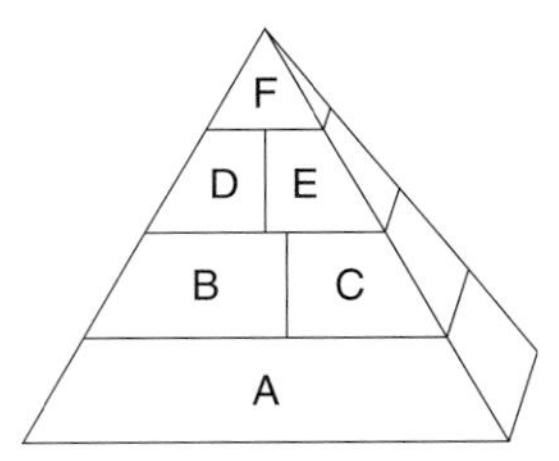

14. A large piece of sausage pizza would contain several food groups. Identify each group and name the part of the pizza that is responsible for representing that particular food group.
15. Fructose, $C_6H_{12}O_6$, is a carbohydrate.
    a. Rewrite the formula for fructose to emphasize the original meaning of the term "carbohydrate."
    b. Write a structural formula for one of the isomers of fructose.
    c. Do you expect the different isomers of fructose to all have the same sweetness? Explain why or why not.

16. Fructose and glucose both have the formula, $C_6H_{12}O_6$. How do their structural formulas differ?

17. State what is meant by each term, and give an example of a substance that fits that term.

  a. monosaccharide **c.** polysaccharide

  b. disaccharide

18. What problems can arise from regularly consuming excess dietary servings of carbohydrates?

19. Use the lock-and-key model discussed in Section 10.5 to offer a possible explanation why individuals who suffer from lactose intolerance can digest other sugars such as sucrose and maltose, but not lactose.

20. a. What are the similarities between fats and oils?

  b. What are the differences between fats and oils?

21. From the entries in Figure 11.7, identify the fat or oil with the highest percentage of each type of fat.

  a. polyunsaturated fat **c.** total unsaturated fat

  b. monounsaturated fat **d.** saturated fat

22. The label of a popular brand of soft margarine lists "partially hydrogenated" soybean oil as an ingredient. What does "partially hydrogenated" mean? Why does the label not simply say soybean oil, rather than partially hydrogenated soybean oil?

23. The text describes substitutes that have been developed for fat (Olean) and sugar (NutraSweet). Why have there not been attempts to develop a comparable substitute for protein?

24. What is the nutritional significance of the elements shaded on this periodic table?

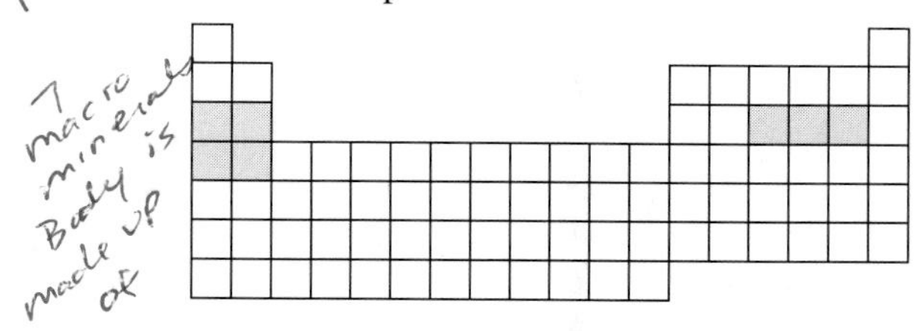

25. Why is it safer to take large doses of vitamin C than it is to take large doses of vitamin D?

Concentrating on Concepts

26. Explain to a friend why it is impossible to go on a highly advertised "all organic, chemical-free" diet.

27. Your friend wants to cut food costs and has learned that peanut butter is a good protein source. What additional information should your friend consider before making the decision to make peanut butter the major dietary protein source? *Hint:* There is relevant information in Figure 11.2.

28. **a.** What percentage of the total number of elements in the periodic table are utilized by the human body to produce proteins, carbohydrates, and fats?

  b. What relationship is there between the type of bonds these elements can form and what makes them so prevalent in the human body?

29. When the USDA made the decision to change dietary recommendations, it also changed the way the information was visually displayed to consumers. Rather than using a restructured pie chart, the food pyramid was introduced. Is the food pyramid a better communication tool for dietary recommendations than the pie chart used previously? Why or why not?

30. According to one USDA study, nearly 40% of the food that the average American eats each day consists of milk or dairy products. Would such a diet be possible and still meet the guidelines of the food pyramid?

31. In people who exhibit lactose intolerance, the enzyme lactase that normally catalyzes the lactose breakdown, is either missing or is present at levels too low to support normal enzymatic activity. How does this inability to break down lactose parallel our ability to metabolize starch, but not cellulose?

32. For each statement, indicate whether it is always true, may be true, or cannot be true. Justify your answers by explaining your reasoning.

  a. Plant oils are lower in saturated fat than are animal fats.

  b. Lard is more healthful than butterfat.

  c. There is no need to include fats in our diets because our bodies can manufacture fats from other substances we eat.

33. Experimental evidence suggests that some physiological effects of saturated fats, compared to unsaturated fats, may be caused by differences in folding or wrapping of the molecules. The hydrocarbon chains in saturated fatty acids can fold or wrap more tightly than those of unsaturated or polyunsaturated fatty acids.

  a. Explain why saturated fatty acid molecules are able to fold more tightly than molecules of unsaturated or polyunsaturated fatty acids. *Hint:* If you have a model set available, make suitable molecular models to help you see the effect single or double bonds can have on the ease of folding.

  b. Explain why the extent of molecular folding influences the melting points of stearic, oleic, linoleic, and linolenic acids. See Table 11.2 for melting point values.

34. Some people prefer to use non-dairy creamer rather than real cream or milk. Some, but not all non-dairy creamers, use coconut oil derivatives to replace the butterfat in cream. Should a person trying to reduce dietary saturated fats by using non-dairy creamer use non-dairy creamers such as these? Why or why not?

35. The reaction of free radicals and oxidizing agents with unsaturated and polyunsaturated fats in the body has been suggested as a cause of premature aging. What is the chemical basis for this assertion? *Hint:* You might find it helpful to consider the mechanism of addition polymerization in Chapter 9.

36. Why is it more difficult for a person to control her or his cholesterol level than to control her or his fat intake? What steps are effective in minimizing cholesterol in the blood?

37. **a.** Which are the "good" lipoproteins—LDL or HDL?

b. What function do the "good" lipoproteins perform?

c. Does a person with an LDL reading of 100 mg/100 mL and an HDL reading of 150 mg/100 mL meet current guidelines from the AHA?

38. Consider the structure for riboflavin, one of the B vitamins found in leafy green vegetables, milk, and eggs.

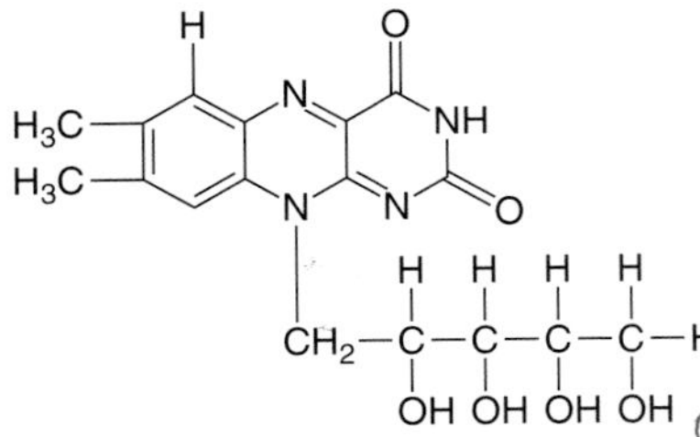

Why is it safer to take large doses of vitamin B than it is to take large doses of vitamin D?

39. American diets depend heavily on bread and other wheat products. A slice of whole wheat bread (36 g) contains approximately 1.5 g of fat (with 0 g saturated fat), 17 g of carbohydrate (with about 1 g of sugar), and 3 g of protein.

a. Calculate the total Calorie content in a slice of this bread.

b. Calculate the percent Calories from fat.

c. Do you consider bread a highly nutritious food? Explain your reasoning.

40. What is your opinion about food preservation by irradiation? Are there some cases in which you feel irradiation is justified as a way to ensure better quality food to the consumer? Explain your position and be prepared to defend it.

Exploring Extensions

41. How has the proportion of undernourished people in different areas of the world changed over the past 30 years? How have the *total numbers* of undernourished people changed during that time? Focus on any one region of the world and find the necessary information to speak to these two points. Then devise a visual way to represent these data.

42. Compare these two pie charts for the percentage of macronutrients in soybeans and wheat.

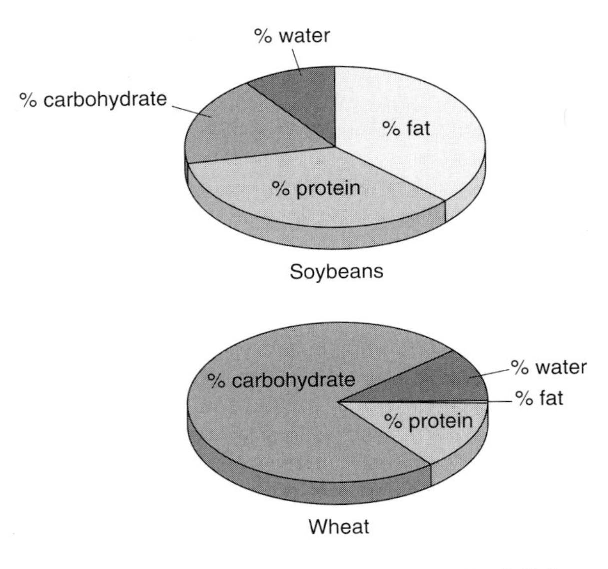

a. Use these charts to help explain why the World Health Organization has helped develop several soy-based, rather than wheat-based, food products for distribution in parts of the world where protein deficiency is a major problem.

b. Suggest some cultural reasons why soy might be preferable to wheat for some areas of the world.

43. The Sceptical Chymist finds the statement that the composition of the human body is ". . . roughly similar to the stuff we stuff into it" an idea hard to believe, but is willing to try to justify this statement, at least for the macronutrients. Compare the information found in Figures 11.2 and 11.3. Does it give you adequate information to decide whether the ". . . roughly similar to the stuff we stuff into it" statement is reasonable, assuming you eat only the foods shown in Figure 11.2? Why or why not?

44. How does the elemental composition of the human body compare to the elemental composition of the Earth's crust? How does the human body's composition compare to the elemental composition of the universe? Table 11.1 gives the values for the human body. Research the composition of the Earth's crust and that of the universe, listing your references. Then comment on the comparative values for the first five elements listed in order of mass abundance in each of the three circumstances—the human body, the Earth's crust, and the universe.

45. The food guide pyramid gives a range of servings for each food group. What factors do you think determine the number of servings you should eat?

46. The food guide pyramid gives a range of servings for each food group. To use this information, the consumer must know what constitutes a reasonable serving size. Investigate what constitutes reasonable serving sizes for one of the food groups, and then prepare a poster with your results to share with others who are investigating the reasonable serving sizes for other food groups. Were you surprised by any of the serving sizes? Which ones?

47. Not everyone considers milk nature's "perfect food." Compare and contrast the viewpoints of the dairy industry with activist coalitions that work against the dairy industry. What are some of the specific benefits attributed to milk, and what are some of the reasons that milk has been called "nature's not-so perfect food"?

48. Here is the label information from a popular brand of canned chicken noodle soup.

Serving Size: 1/2 cup (4 oz; 120 g)

Servings per container: about 2.5

Amount per serving

Calories 75	Calories from Fat 25
	%Daily Value
Total Fat 2.5 g	4%
Saturated Fat 1.5 g	8%
Cholesterol 20 mg	7%
Sodium 970 mg	40%
Total Carbohydrates 9 g	3%
Dietary Fiber 1 g	4%
Sugars 1 g	
Protein 4 g	
Vitamin A	15%
Vitamin C	2%
Calcium	2%
Iron	4%

a. Analyze this information to see if the soup conforms to dietary recommendations of the AHA.

b. Is the serving size recommended on the label adequate? Explain.

c. What effect would changing the serving size have on your answer to part a?

49. Examine this figure, which gives the world grain harvests from 1966 to 1997.

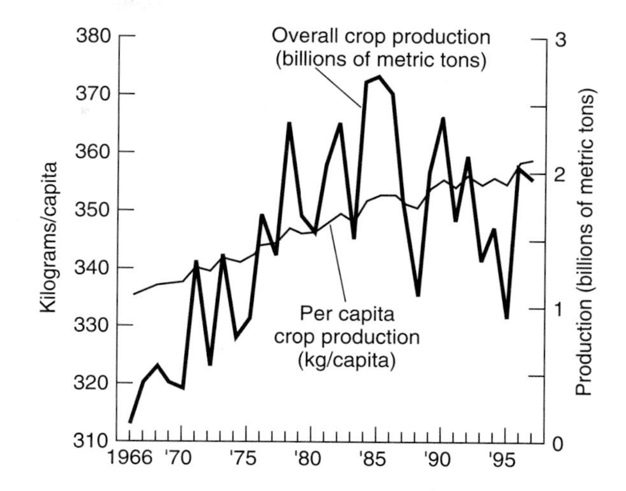

a. Write a paragraph summarizing the information displayed by the graph.

b. Has this information changed since the text version of the graph? Explain.

50. Every month, a certain consumer-advocate organization presents an "Unnatural Living Award" to a person, product, or institution that demonstrates an unnatural ability to provide an unnatural product to the American people. What are the criteria by which you would make your nomination for this award? What do you consider would be a good candidate to receive this award? Explain your reasons for suggesting this candidate.

12

Genetic Engineering and the Chemistry of Heredity

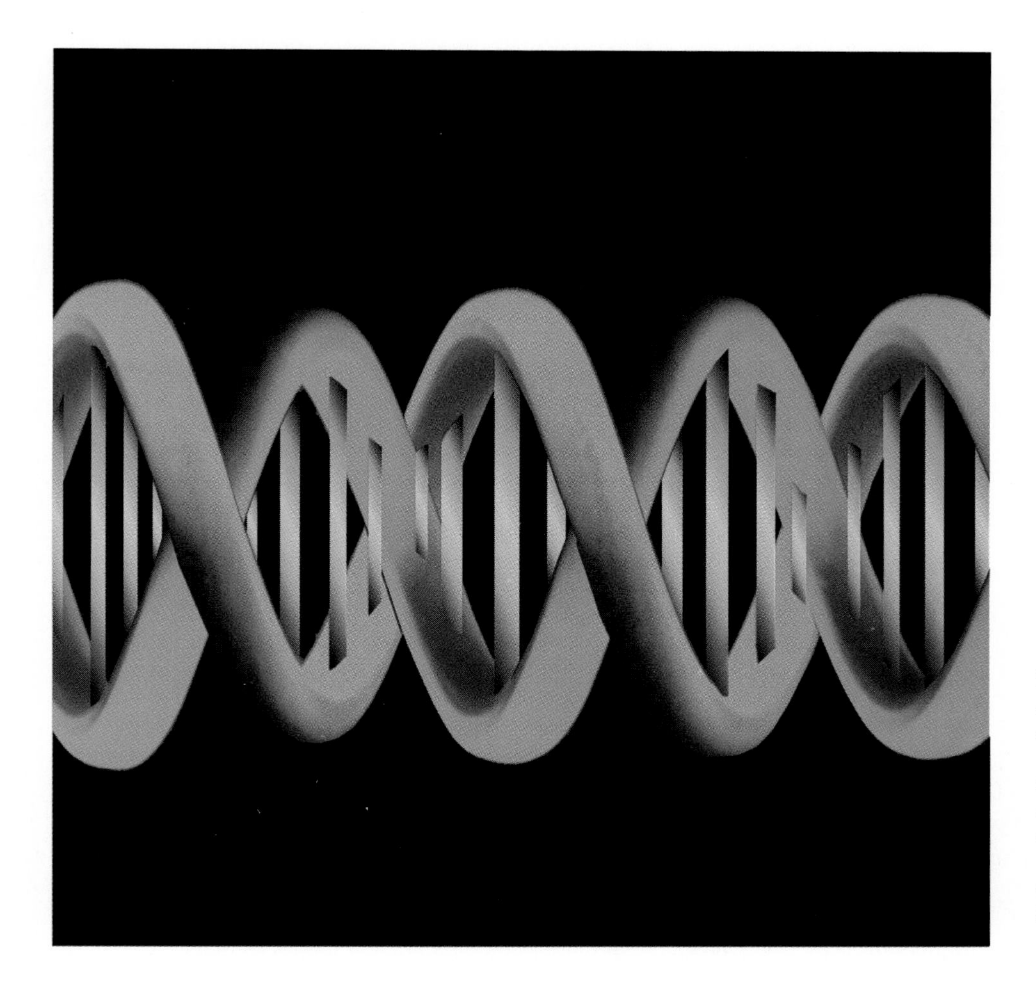

This is a stylized illustration of DNA (deoxyribonucleic acid), the so-called "molecule of life." The DNA molecule is a double-stranded helix of deoxyribose and phosphate that forms the "spirals" in blue and green, and the nitrogen-containing bases that form the yellow "rungs."

New Concerns Rise on Keeping Track of Modified Corn

Efforts to trace shipments of a bioengineered corn unapproved for human consumption have raised concern among food and grain industry officials that the corn—which has already been discovered in two brands of grocery products—may have made its way more widely into production channels for the nation's food supply . . . There is no evidence that the corn causes health problems in humans, but the discoveries have led to nationwide recalls of two brands of store-bought taco shells, a move that was extended yesterday to a larger group of brands and products.

Headline and excerpts from October 14, 2000 *New York Times*

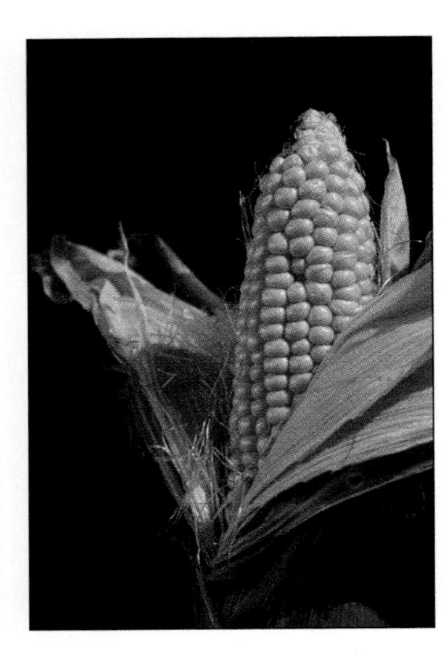

It might take weeks to figure out how the insect-killing trait in genetically altered StarLink corn migrated into a variety of corn that was not supposed to be genetically modified, according to the Garst Seed Company, the producer of the corn . . . Garst said it did not know whether the transfer came during pollination as the seed was being grown or through mixing of seeds after harvest.

November 24, 2000 *New York Times*

The Agriculture Department agreed to tighten testing requirements for corn bound for Japan to prevent shipments of grain contaminated with a biotech variety not approved for food use. Last month, Japanese inspectors found traces of genetically engineered StarLink corn in grain that had tested negative for the biotech variety before it left the United States . . . Starlink was never approved for human consumption because of unanswered questions about its potential to cause allergic reactions.

February 23, 2001 *New York Times*

What does the term "bioengineered corn" mean? Why is there such a hubbub about StarLink® corn? Biotechnology critics cite StarLink corn, as the first example of so-called "Frankenfoods," products that contain genetically modified traits introduced and spread in an unauthorized or unpredictable manner. Biotechnology includes a molecular-based understanding of plants, animals, and humans; the improvement to them; and the continued propagation of the species. It should be noted that advances in agriculture and biotechnology, in general, have been around almost since the beginning of time. In the words of former U.S. Agriculture Secretary, Dan Glickman, ". . . It's cavemen saving seeds of a high-yielding plant. It's Gregor Mendel, the father of genetics, cross-pollinating his garden peas. It's a diabetic's insulin, and the enzymes in your yogurt . . . Without exception, the biotech products on our shelves have proven safe." Now, StarLink corn gives us the opportunity to examine some of those biotechnology practices, and the safety of the products that have been produced.

The quotations that start the chapter are all about StarLink corn, a genetically modified variety developed by Aventis CropScience. StarLink corn contains a gene for the production of a protein that is toxic to corn borers and related insects. This biotechnological modification eliminates or greatly diminishes the need for corn farmers to use pesticides. StarLink has no known deleterious health effects on humans. Proteins structurally similar to the one in StarLink that is toxic to corn pests are already widely used in agriculture. In 1997, the Environmental Protection Agency decided that the genetically modified corn should be kept out of human food until it could be shown that the protein did not cause food allergies. Farmers are caught in a potential conflict—eliminating or reducing personal and environmental contact with pesticides through the use of StarLink, or facing a boycott of their products by some countries. The battle has been particularly intense in Europe, where genetically modified (GM) crops have been banned since April 1998. Anti-biotech sentiment has spread beyond Europe. In Japan there is a crackdown on GM food, and a Brazilian state has declared itself as a "transgenic-free zone." In the United States, the controversy has been building also, including reports of "eco-vandals" attempting to destroy GM crops, and the Greenpeace organization using "biological pollution" to describe genetically modified food. In many cases, the fear and misunderstanding by critics come from a limited or shallow knowledge of fundamentals and details behind genetically engineered foods. Thus, this chapter begins with a description of the chemical basis for heredity and continues with a description of the fundamentals of genetic engineering and issues related to it.

CHAPTER OVERVIEW

During the past 50 years, biology has been transformed by the application of chemical knowledge and methodology. In this chapter we first examine our understanding of how genetic information is transmitted and used. We begin with an introduction to the molecular basis of heredity—deoxyribonucleic acid (DNA) and its component chemical parts. These parts—four nitrogen-containing bases, a sugar (deoxyribose), and phosphate group—are combined into a double helix. That structure and the story of its remarkable discovery are recounted next. Thanks to superb chemical cryptographers, DNA has been decoded. Indeed, the entire molecular code for humans recently has been described for the first time. The genetic code directs the synthesis of proteins and determines the sequence of their constituent amino acids in all living organisms.

The remainder of the chapter addresses some of the many applications of genetic engineering. First you will encounter the recombinant DNA techniques that have made it possible to use bacteria to produce proteins such as human insulin and growth hormones. A whole generation of new drugs and vaccines has also been created through molecular manipulation. New methods of diagnosing diseases are presented next, followed by a discussion of gene therapy, in which normal genes are introduced into patients lacking them. It is now possible to take a unique genetic fingerprint of any one of us. Such information is valuable in solving crimes, identifying human remains, and constructing genetic family trees. Some very old DNA has already been isolated and studied, raising the possibility of a Dino-Disneyland. Some unusual inter-species genetic combinations also surface and are the basis for introducing important new characteristics into agricultural products like StarLink corn. The transfer of genetic material from one species to another gives rise to the "Frankenfood frenzy." We visit the technology that made the cloning of a sheep named Dolly several years ago and issues pertaining to that achievement. Then we devote some attention to the recently-completed, massive project to map all the genes in the human species. The chapter ends with a hope and a warning as we look to a future filled with the benefits and risks of cloning and genetic engineering.

12.1 The Chemistry of Heredity

The human body is the world's most complicated chemical "factory." Thousands of chemical reactions involving an even greater number of chemicals occur each second. Some compounds are decomposed and others are synthesized; energy is released, transformed, and used; and chemical signals are transferred and processed. But, in spite of the dazzling complexity of these processes, the last half-century has seen a phenomenal increase in our knowledge of the chemistry of life. Indeed, in many respects, biology has become a chemical science. Today, much biological research is focused on molecules, not cells or organisms. This "molecular" research has led to an understanding of the very basis of life itself.

The practical manifestations of this intellectual achievement are manifold. In 1900, average life expectancy at birth in the United States was under 50 years; today it is almost 77 years (Figure 12.1). Reasons for this dramatic increase are many: better nutrition, improved sanitation, advances in public health, more accurate medical diagnoses, new medical procedures, and numerous new medicines, drugs, and vaccines. Chemistry has contributed to all these innovations, and it is an integral part of the latest revolution in health care—biotechnology and molecular engineering. There seems little doubt that genetic engineering will profoundly affect human life in the 21st century.

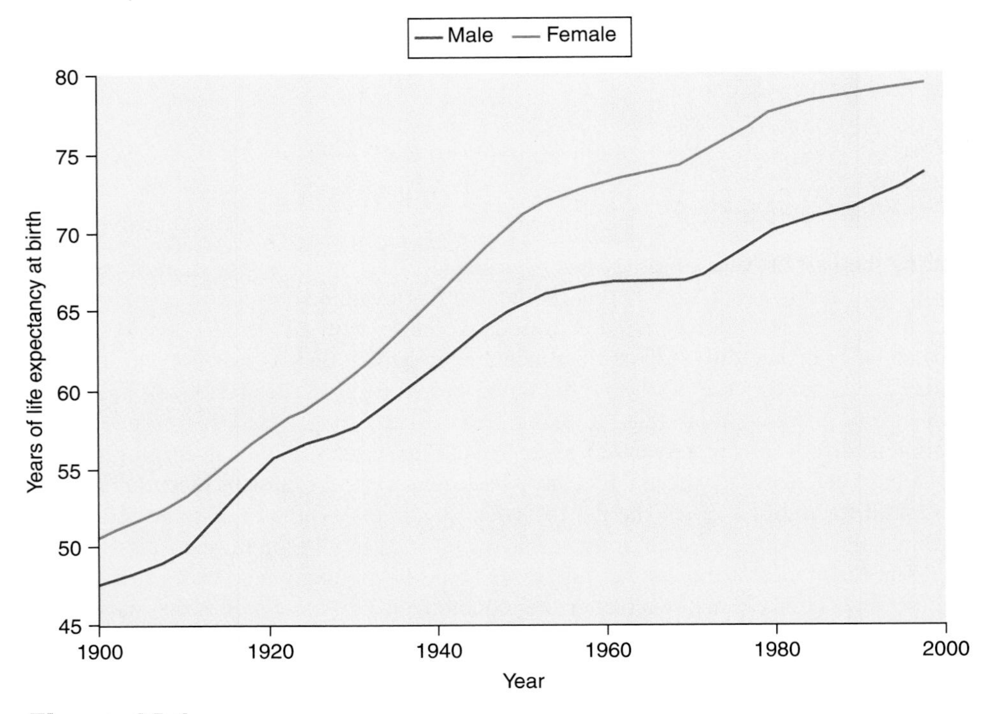

Figure 12.1

Average life expectancy at birth in the U.S. since 1900.

Source: Center for Disease Control and Prevention, National Center for Health Statistics, National Vital Statistics System, ***48,*** *#18, 2001.*

12.1 Your Turn

a. Use Figure 12.1 to determine the percent increase in average human life expectancy at birth for women and for men in the U.S. between 1900 and 1950; Between 1950 and 1998.
b. Speculate as to why the average life expectancy of women and men differ.

Some of the most significant advances in biochemistry have involved our rapidly growing knowledge of the molecular basis of heredity. To put this into a personal perspective, you contain about 10 million million (10×10^{12}) cells that have a nucleus. Each of these cell nuclei contains a complete set of the genetic instructions that make you what you are—at least biologically. The genetic instruction is organized in 23 pairs of **chromosomes** and approximately 100,000 genes, each of which conveys one or more hereditary traits. This is the **human genome,** the totality of human hereditary information in molecular form.

• Human red blood cells do not have nuclei.

Your special template of life is written in a molecular code on a tightly coiled thread, one invisible to the unaided eye. This thread is **deoxyribonucleic acid (DNA),** the molecule that carries genetic information in all species. Unraveled, the DNA in *each* of your cells is about two meters (roughly two yards) long. If all of the DNA in all 10 million million of your cells were placed end to end, the resulting ribbon would stretch from here to the Sun and back more than 60 times! But, you will soon discover that this astronomical figure is far from the most astounding feature of this amazing molecule.

12.2 The Sceptical Chymist: Stretching DNA

Sometimes authors get carried away with their rhetoric. Check the correctness of the claim that the DNA in an adult human being would stretch from the Earth to the Sun over 60 times. It is approximately 93 million miles from the Earth to the Sun. The other necessary information is in the preceding paragraphs and the unit conversion factors are in Appendix 1.

The molecular structure of deoxyribonucleic acid dictates the way DNA encodes genetic information. A strand of DNA consists of fundamental chemical units, repeated thousands of times. Each of the units is comprised of three parts: **nitrogen-containing bases,** the sugar **deoxyribose,** and **phosphate groups.** All are illustrated in Figure 12.2. Two of the bases, adenine (symbolized A) and guanine (G), are built on a six-membered ring fused to a five-membered ring. Carbon and nitrogen atoms make up the rings. Cytosine (C) and thymine (T) each contain six-membered molecular rings consisting of four carbon atoms and two nitrogen atoms. These four compounds are bases because they react with water to produce basic solutions. H^+ ions (protons) are transferred from H_2O molecules to nitrogen atoms of the nitrogen-containing bases, creating OH^- ions in solution, forming a basic (alkaline) solution.

- H^+ and OH^- ions are discussed in Sections 6.1 and 6.2. The double arrow indicates that reactants are not completely converted to products.

$$H_2O(l) + \text{N-base } (aq) \rightleftharpoons {}^+\text{HN-base } (aq) + OH^- \ (aq) \qquad [12.1]$$

Deoxyribose is a monosaccharide (a "single" sugar) with the formula $C_5H_{10}O_4$. Figure 12.2 reveals that the deoxyribose molecule is a five-membered ring formed by four carbon atoms and one oxygen atom. The phosphate group can be represented as $PO_4{}^{3-}$, but, depending on the pH, an $H^+\cdot$ ion can be attached to one or more of the

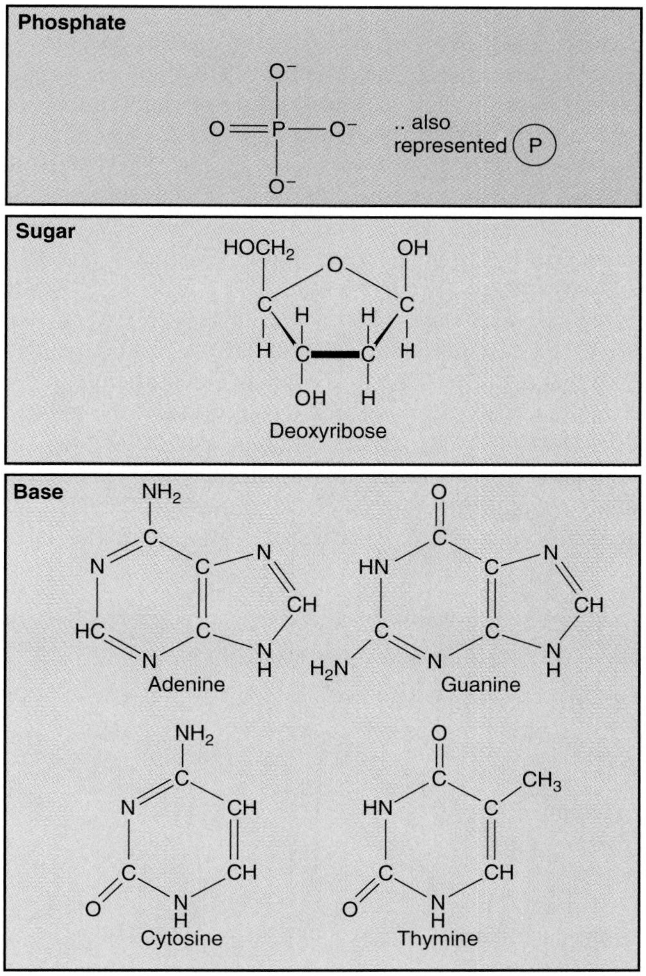

Figure 12.2

The components of DNA. Deoxyribose means that a hydroxyl or —OH group in ribose has been replaced by a hydrogen atom H, to form deoxyribose. That replacement or substitution occurs at the ring carbon shown with two hydrogen atoms in the sugar structure.

O atoms. The completely protonated form of phosphate is H_3PO_4, phosphoric acid. It is the ionizable hydrogen atoms on the phosphate groups that make nucleic acids acidic.

A **nucleotide** is a combination of a base, a deoxyribose molecule, and a phosphate group. Figure 12.3 indicates how these units are linked in a nucleotide called adenosine phosphate. A covalent bond exists between one of the ring nitrogen atoms of the adenine molecule and one of the ring carbons in deoxyribose. Another covalent bond connects the deoxyribose sugar molecule to the phosphate group. The other three bases form similar nucleotides of deoxyribose, each comprised of phosphate and a base joined covalently to the deoxyribose sugar. Although there are other possible molecular sites for linking the base, sugar, and phosphate units, the arrangement pictured in Figure 12.3 is found in DNA.

• Sugars are discussed in Section 11.2.

A typical DNA molecule consists of thousands of nucleotides covalently bonded in a long chain. Consequently, a segment of DNA may have a molecular mass in the millions. The phosphate groups link the individual nucleotides. Note that in Figure 12.3, one —OH group on the deoxyribose ring remains unreacted. The phosphate group of another nucleotide can react with this —OH group, forming and eliminating an H_2O molecule, and connecting the two nucleotides, as shown in Figure 12.4 for four nucleotides linked in this manner to form a segment of DNA. This alternating chain of deoxyribose-phosphate-deoxyribose-phosphate units runs the length of the nucleic acid molecule, like the vertical rails of a ladder. Attached to each of the deoxyribose rings is one of the four possible bases.

DNA can be represented as a polymer (see Chapter 9) of nucleotide monomers or repeating units. The schematic shows the sugar-phosphate-sugar-phosphate chain with different bases attached to each of the sugars, deoxyribose.

The specific bases and their sequence in a strand of DNA turn out to have great significance. Some of the early clues to the structure of DNA and the mechanism by which it conveys genetic information came as a result of the research of Erwin Chargaff in the 1940s and 1950s. Chargaff and his co-workers were able to determine the percentage of the four bases present in DNA from a variety of species. They found that the relative amounts of the bases in a DNA sample are identical for all members of the same species. Moreover, these percentages are independent of the age, nutritional state, or environment of the organism studied. For example, according to Chargaff's data, the DNA from all members of our species *Homo sapiens* contains 31.0% adenine, 31.5% thymine, 19.1% guanine, and 18.4% cytosine. Table 12.1 also contains such findings for other species. Humans, fruit flies, and bacteria do not seem to have very much in common, and it is perhaps reassuring that the mix of the four bases is quite different in the three species. But it turns out that the more closely related the species are, the more similar the base composition of the DNA. This observation certainly suggests that the base composition of the nucleic acid must have something to do with inherited characteristics.

A more careful examination of Table 12.1 discloses that the DNA of *Homo sapiens* and the DNA of *Escherichia coli* (*E. coli*), a species of bacteria that inhabits the human

Table 12.1

The Base Compositions of DNA for Various Species

Species	Adenine	Thymine	Guanine	Cytosine
Homo sapiens (human)	31.0	31.5	19.1	18.4
Drosophila melanogaster (fruit fly)	27.3	27.6	22.5	22.5
Zea mays (corn)	25.6	25.3	24.5	24.6
Neurospora crasa (mold)	23.0	23.3	27.1	26.6
Escherichia coli (bacterium)	24.6	24.3	25.5	25.6
Bacillus subtilis (bacterium)	28.4	29.0	21.0	21.6

Note that the percentages of adenine and thymine are consistently similar, as are the percentages of cytosine and guanine.

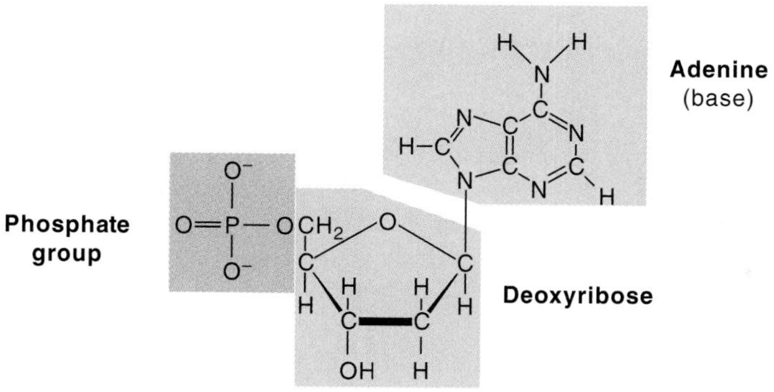

Figure 12.3

The molecular structure of a nucleotide, adenine phosphate. Adenine, a nitrogen base, is highlighted in green, the deoxyribose molecule is highlighted in blue, and the phosphate group is highlighted in pink.

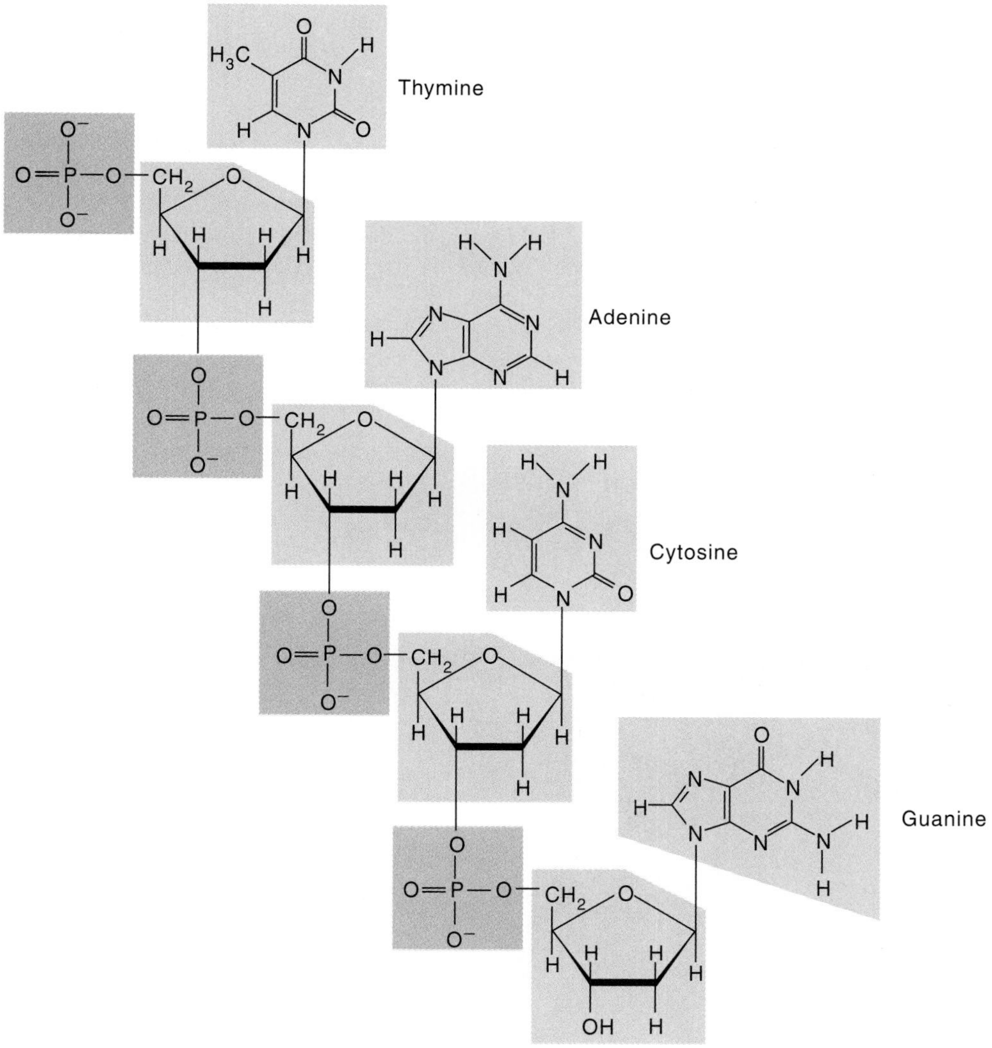

Figure 12.4

The molecular structure of a segment of DNA. A phosphate group connects one deoxyribose to an adjacent one. Each of the four different bases—thymine (T), adenine (A), cytosine (C), and guanine (G)—is attached to a deoxyribose sugar.

intestine, exhibit a very important common characteristic. They obey the same compositional regularity, now called **Chargaff's rules.** In every species, the percent of adenine almost exactly equals the percent of thymine. Similarly, the percent of guanine is essentially identical to the percent of cytosine. Put more simply: A = T and G = C. Such a correlation can hardly be coincidental; it must be based on biochemical form and function at the molecular level. As soon as Chargaff's rules were announced, the conclusion seemed obvious: The nitrogen-containing DNA bases somehow come in pairs. Adenine always appears to be associated with thymine, and guanine is consistently matched with cytosine.

12.2 The Double Helix of DNA

- X rays have short wavelengths and high energy.

What was not so obvious was how the paired bases were part of the overall molecular structure of DNA. Therefore, scientists set out to determine the way in which nucleotides were incorporated into the DNA molecule. The most fruitful experimental strategy was X-ray diffraction, a technique that had been known since early in the 20th century. In **X-ray diffraction,** a beam of X rays is directed at a crystal. The X rays strike the atoms in the crystal, interact with their electrons, and bounce off the atoms. Stated a bit more precisely, the X rays are diffracted, or scattered, by the atoms. The crucial point is that the X rays are only scattered at certain angles, which are related to the distance between atoms (Figure 12.5).

Rosalind Franklin, whose work contributed significantly to elucidating the DNA structure.

In the instruments used during the 1950s, the scattered X-ray beams struck and exposed photographic film. The resulting spots correspond to the angles of diffraction, and they represent a two-dimensional map of a three-dimensional structure. Figure 12.5 is such a map; it is the X-ray diffraction pattern of a DNA fiber obtained in late 1952 by the British crystallographer Rosalind Franklin. To the uninitiated, the photograph does not appear to contain much useful information, but the correct interpretation of these spots led to the determination of the structure of DNA.

The scientists responsible for this revelation were James D. Watson, a 24-year-old American, and Francis H. C. Crick, a loquacious and supremely self-confident Cambridge

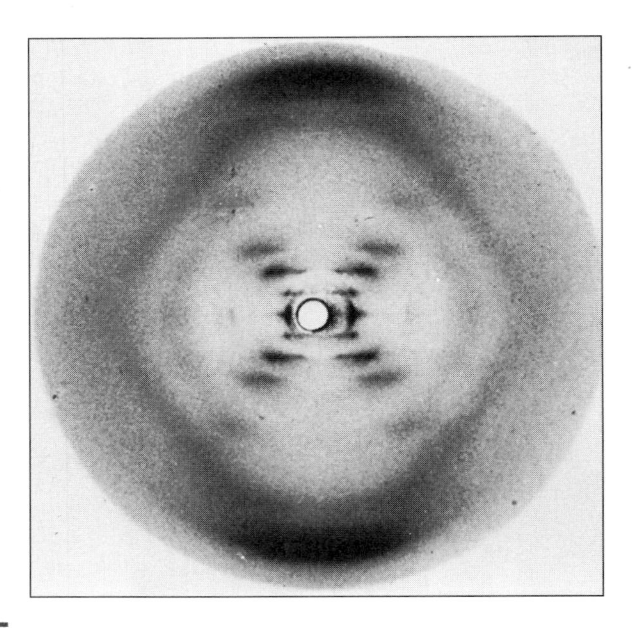

Figure 12.5
Photograph of the X-ray diffraction pattern of a fiber of DNA, obtained in 1952 by Rosalind Franklin. The X-ray pattern suggests a helical structure, and the spacing of features at the top and bottom correspond to a molecular spacing of 0.34 nanometers. Note: This is *not* a photograph of the molecular structure (see text).

James Watson and Francis H. C. Crick, codiscoverers of the double-helical structure of DNA.

University biophysicist. Watson and Crick concluded that the X-shaped pattern in Franklin's diffraction photograph was consistent with a repeating helical arrangement of atoms, similar to a loosely coiled spring. Moreover, the spacing of the large smudges at the top and bottom of the figure were evidence of a regular repeat distance of 0.34 nanometers (1 nm $= 1 \times 10^{-9}$ m) within a DNA molecule.

With these clues, Crick and Watson set out to combine the molecular pieces. After a variety of attempts, they finally came up with a structure that agreed with the data. A major breakthrough was the recognition that the adenine and thymine portions of the molecule fit together almost perfectly, like pieces in a jigsaw puzzle. Moreover, these two bases can be linked by two hydrogen bonds (Figure 12.6). Similarly, cytosine and guanine align by forming three hydrogen bonds. Adenine and thymine are said to be **complementary bases,** as are cytosine and guanine. This base pairing is the molecular basis underlying Chargaff's rules: A pairs with T and C pairs with G.

In the model of DNA developed by Watson and Crick, the hydrogen bonds between the complementary bases help hold together two strands of a **double helix.** Perhaps an

- Hydrogen bonds are not covalent chemical bonds, but relatively weak interactions between molecules like water or DNA bases. See Section 5.9.

Thymine
Adenine
To deoxyribose and DNA chain
To deoxyribose and DNA chain

Cytosine
Guanine
To deoxyribose and DNA chain
To deoxyribose and DNA chain

Figure 12.6

Base pairing of adenine with thymine and cytosine with guanine in DNA. Chemical bonds are drawn as solid lines, while the intermolecular attraction due to hydrogen bonds are shown as dashed lines.

even better metaphor is a spiral staircase. The steps of this molecular staircase are the bases, always paired A with T and C with G. One member of each base pair belongs to one strand of a helix, the other to the matching, complementary helical strand, thus creating a double helix. Recall that the bases are connected to the deoxyribose rings, which in turn are linked by phosphate groups. Thus, the deoxyribose and phosphate units are, in effect, the stair rails to which the steps are attached. The Anglo-American team concluded that the base pairs are parallel to each other, perpendicular to the axis of the DNA fiber, and separated by 0.34 nm, the repeat distance calculated from the diffraction pattern. In addition, Franklin's results also suggested another repeat distance of 3.4 nm. Watson and Crick took this to be the length of a complete helical turn consisting of 10 base pairs.

12.3 Your Turn

Identify the base sequences that are complementary to each of these sequences.

a. ATACCTGC
b. GATCCTA

Answers
a. TATGGACG **b.** CTAGGAT

12.4 Your Turn

The distance between bases in a molecule of DNA is 0.34 nm.

a. Calculate the length (in centimeters) of the shortest human chromosome, which consists of 50,000,000 base pairs.
b. Mark off that length on your paper. If this is the length of the unstretched DNA molecule, what does this imply about the organization of DNA in the chromosome?

Answers

a. $\frac{0.34 \text{ nm}}{\text{pair}} \times \frac{1 \text{ m}}{10^9 \text{ nm}} \times \frac{10^2 \text{ cm}}{1 \text{ m}} \times \frac{50{,}000{,}000 \text{ pairs}}{\text{chromosome}} = \frac{1.7 \text{ cm}}{\text{chromosome}}$

b. This is a small distance, about two-thirds of an inch. The only way that this many base pairs could fit into such a small space would be to have them tightly packed in a spiral.

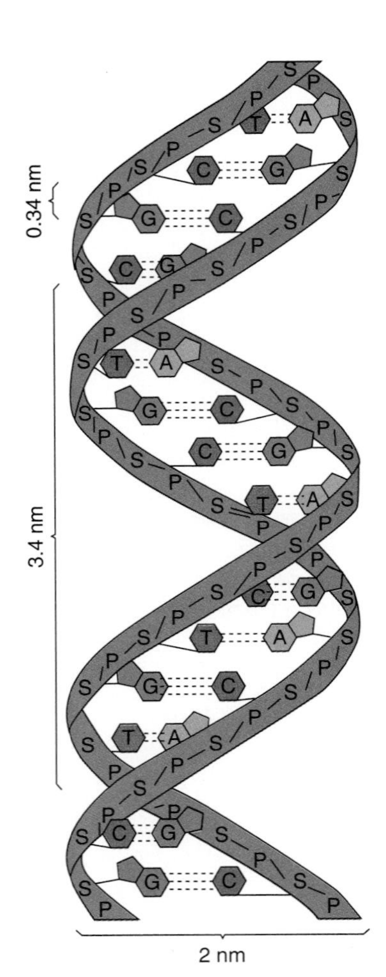

Figure 12.7
This is a molecular representation of DNA.
Key: P = phosphate group
S = the sugar, deoxyribose
A = adenine
T = thymine
C = cytosine
G = guanine

The overall structure of DNA is represented by Figure 12.7. In this simplified drawing, the alternating sugar (S) and phosphate (P) groups are represented by two twisting ribbons. The four bases are attached to this backbone and paired in the A to T and C to G fashion described earlier.

Watson and Crick's research paper, "Molecular Structure of Nucleic Acids: A Structure for Deoxyribose Nucleic Acid," appeared in *Nature* on April 24, 1953. It is only one page long, but it is undoubtedly the most important paper to appear in that prestigious journal since the announcement of nuclear fission 15 years earlier by Meitner and Frisch in an equally short communication. The Watson-Crick paper is written with the customary passionless detachment of contemporary scientific prose. Even the most significant statement in the communication is delivered with typical British understatement: "It has not escaped our notice that the specific pairing we have postulated immediately suggests a possible copying mechanism for the genetic material."

The history of science is full of examples of how a single discovery can release a flood of related research. So it was with the discovery of the structure of DNA. Scientists immediately set out to discover the molecular details of how DNA is replicated, how it encodes genetic information, and how that information is translated into physiological characteristics. **Replication,** the process by which copies of DNA are made, is now well understood, and it is diagramed in Figure 12.8.

12.5 Your Turn

The DNA in each human cell consists of three billion base pairs. Calculate the length of this DNA. Does this length agree with that given in Section 12.1? Why or why not?

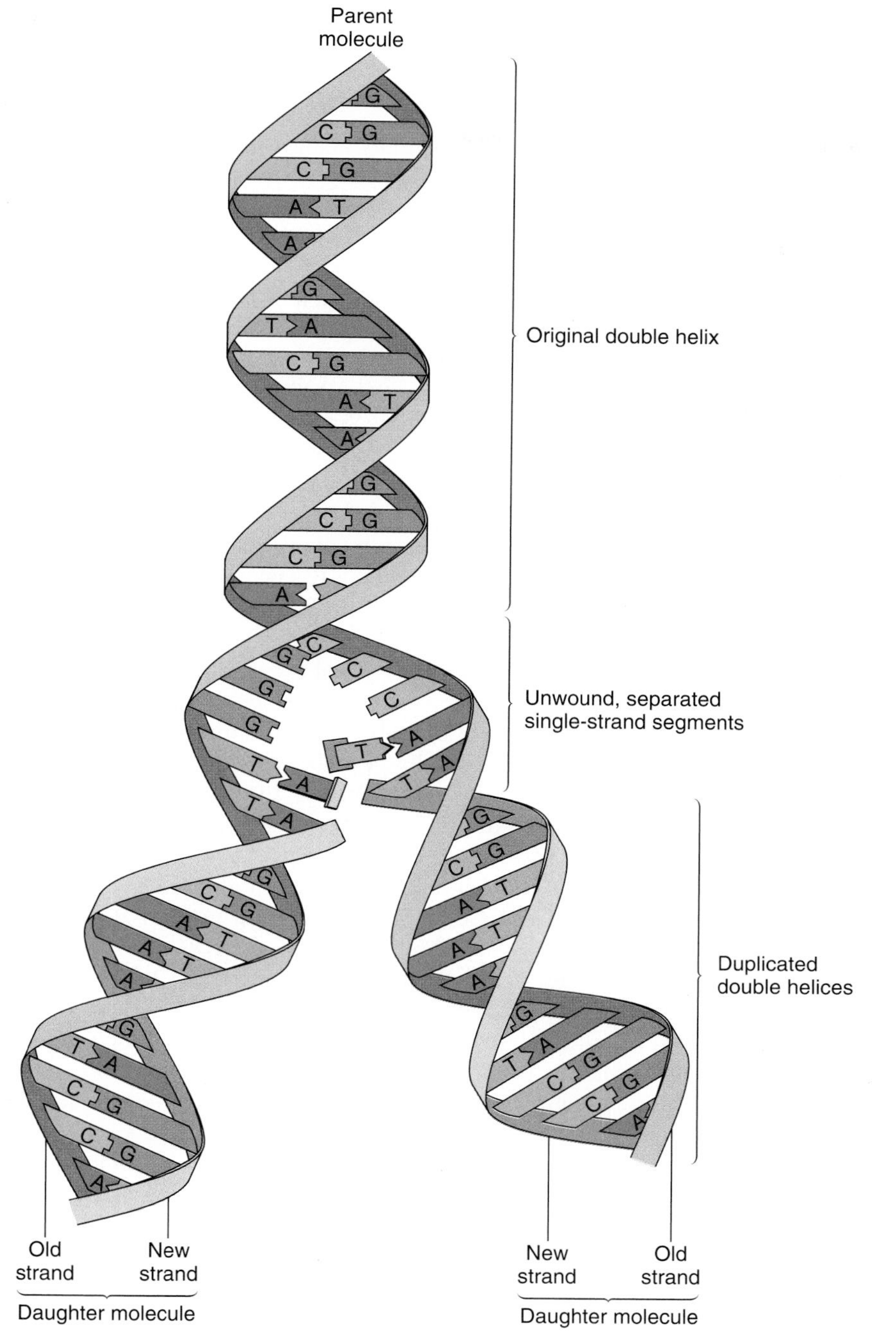

Figure 12.8

Diagram of DNA replication. The original DNA double helix (top portion of figure) partially unwinds and the two complementary portions separate (middle). Each of the strands serves as a template for the synthesis of a complementary strand (bottom). The result is two complete and identical DNA molecules.

12.6 Consider This: Checking All Bases

The structural features of a DNA molecule are more apparent if you can look at a three-dimensional (3-D) representation. At the *Chemistry in Context* web site, you can view and rotate several different molecular structures.

a. How is a 3-D structure of DNA shown at that site similar to the one shown in the text? How is it different?

b. Look carefully at the structures of the four bases that comprise DNA. What are their structural common features? How do they differ?
c. Look again at the large DNA molecule. Can you find the bases? How are they aligned?

Before a cell divides, the double helix partially and rapidly unwinds (i.e., at a rate of about 10,000 turns per minute). This results in a region of separated complementary single strands of DNA, as pictured in the middle portion of Figure 12.8. Individual nucleotides in the cell are selectively hydrogen-bonded to these two single strands that serve as templates: A to T, T to A, C to G, and G to C. Held in these positions, the nucleotides are bonded together (polymerized) by the action of an enzyme. Every minute, about 90,000 nucleotides are added to the growing chain. By this mechanism, each strand of the original DNA generates a complementary copy of itself. The original template strand and its newly synthesized complement coil about each other to form a new double helix, a daughter molecule identical to the first. Similarly, the other separated strand of the original molecule twines around its new partner, another new daughter molecule. Thus, where there was only one double helix, there are now two, represented at the bottom of Figure 12.8. As the nucleus splits and the cell divides into two daughter cells, one complete set of chromosomes is incorporated into each of them. This process is repeated again and again, so that each of the 2×10^{12} nucleated cells in a newborn baby contains all the genetic information first assembled from parental DNA when the sperm combined with the ovum.

In 1968, James Watson published a highly personal account of the research that led to the determination of the DNA structure. This readable book, entitled *The Double Helix*, is recommended to anyone who still doubts that scientists are flesh and blood with feet of clay. In the book, Watson candidly describes the process of scientific discovery: the competition and ambition, the lucky guesses and the blind alleys. His account does not always conform to a textbook definition of the scientific method but then neither does scientific research. Two decades later, Crick followed with *What Mad Pursuit*, his own reminiscences of those heady days in Cambridge.

It is noteworthy that Watson and Crick were not experts in the field of genetics when they began their research on DNA. Moreover, they did few experiments themselves. Instead, they drew on the work of experts such as Erwin Chargaff, Rosalind Franklin and her crystallographer colleague Maurice Wilkins, and the American chemist Linus Pauling. At the time, all these scientists were better known and more highly regarded than Francis Crick and his young American collaborator. But Watson and Crick seem to have brought a fresh point of view to the problem of DNA structure, hence they saw what more experienced and better-informed scientists missed. Some rightly might argue that Rosalind Franklin's very significant contribution of crystallographic data was not sufficiently acknowledged, either in 1953 or in *The Double Helix.* But few would quarrel with the decision to award Watson, Crick, and Wilkins the 1962 Nobel Prize in Physiology or Medicine. Unfortunately, Franklin had died of ovarian cancer by that time, and the Nobel Prize is not awarded posthumously.

12.7 Consider This: Discovering Rosalind Franklin

In 1958, Rosalind Franklin died an untimely death from ovarian cancer at age 37. Thus, she did not live long enough to add her own account to the history that was written (and rewritten) about the discovery of DNA. Because her work was long minimized or ignored, some historians now assert that both DNA and Rosalind Franklin were discovered. To set the record straight, several excellent biographies of Franklin are now available. You can find reviews of these books as well as other accounts of her life on the Web by searching for Rosalind Franklin. What were her contributions to the structure of DNA? Why was her work not given full credit during her time? What questions would you ask her if you could interview her?

12.8 Your Turn

What role did knowledge of Chargaff's rules play in the discovery of the DNA double helix?

12.3 Cracking the Chemical Code

The discovery of the molecular code in which the genetic information is written is arguably history's most amazing example of cryptography. Key to the code is the sequence of bases in the DNA. The three billion base pairs repeated in every human cell provide the blueprint for producing one human being. Although these specifications are carried in DNA, they are expressed in proteins. Proteins are everywhere in the body: in skin, muscle, hair, blood, and the thousands of enzymes that regulate the chemistry of life. It follows that, by directing the synthesis of proteins, DNA can dictate the characteristics of the organism.

Chapter 10 and especially Chapter 11 contain a good deal of information about proteins. They are large molecules formed by the combination of individual **amino acids.** The 20 amino acids that commonly occur in proteins can be represented by this general formula.

• For more information about amino acids and proteins, see Section 11.6.

```
H      H  O
 \     |  ||
  N—C—C—OH
 /     |
H      R
```

The amine or amino group is —NH_2, the acid group is —COOH, and R represents a side chain that is different in each of the 20 amino acids. When the amino acids combine, the —COOH group of one of them reacts with the —NH_2 group of another, forming what is known as a peptide bond and eliminating an H_2O molecule. A **protein** is therefore a long chain of amino acid residues, as these structural units are called once they have been joined together.

The biochemists who set out to decipher the genetic code concentrated on translating base language into amino acid language. They assumed that somehow the order of bases in DNA determines the order of amino acids in a protein. The hypothesis that the code is related to the sequence of base pairs is the only reasonable one. The phosphate and deoxyribose units are identical in all DNA. Therefore, they could not supply the variability in the DNA structure to account for the individuality among species, and variation is essential in genetic material. Only the base pairs provide the opportunity for variability in the structure of DNA.

It was obvious at the outset that the code could not be a simple one-to-one correlation between bases and amino acids. There are only four bases in DNA. If each base corresponded to an individual amino acid, DNA could encode for only four amino acids. But 20 amino acids appear in our proteins. Therefore, the DNA code must consist of at least 20 distinct code "words," each word representing a different amino acid. And the words must be made up of only four letters—A, T, C, and G—or, more accurately, the bases corresponding to those letters.

Some simple statistics can help us determine the minimum length of these code words. To find out how many words of a given length can be made from an alphabet of known size, one raises the number of letters available to a power corresponding to the number of letters per word.

$$\text{Number of words} = (\text{number of letters available})^{\text{number of letters per word}}$$

Thus, using four letters to make two-letter words generates 4^2 or 16 different two-letter words. Similarly, DNA bases read in pairs (akin to two letters per word) could encode for only 16 amino acids. Again, this vocabulary is too limited to provide a unique representation for each of the 20 amino acids. So we repeat the calculation, this time assuming that the code is based on three sequential base pairs or, if you prefer, that we are dealing with three-letter words. Now the number of different triplet-base combinations is 4^3 or $4 \times 4 \times 4 = 64$. This system provides more than enough capacity to do the job.

12.9 Your Turn

Suppose that the DNA code used four sequential base pairs instead of a triplet-base code. How many different four-base sequences would result?

Answer
$4 \times 4 \times 4 \times 4 = 4^4 = 256$ different four-base sequences

Obviously, more than mathematical reasoning was required to prove the molecular basis of genetics. Once again, Francis Crick was a leader in this research. His work clearly established that the genetic code is written in groupings of three DNA bases, called **codons.** And today, thanks to Marshall Nirenberg, Har Gobind Khorana, and others, this triplet-base code has been cracked and specific amino acids have been related to particular codons.

- The markings on the Rosetta Stone helped to decipher Egyptian hieroglyphics.

No Rosetta Stone was available to aid these scientists in their efforts at translation. Instead, they relied on elegant and imaginative experiments that ultimately yielded a genetic dictionary. If you were to use the letters A, T, C, and G in a game of Scrabble, you could generate 64 different three-letter combinations. A few, CAT, TAG, and ACT, for example, make sense. Most are like AGC, TCT, and GGG and are meaningless—at least in English. Nature does far better than that; 61 of the 64 possible triplet codons specify amino acids. Thus, the codon sequence GTA in a DNA molecule signals that a molecule of the amino acid histidine should be incorporated into the protein, AAA codes for phenylalanine, and GGC stands for proline. The three-base sequences that do not correspond to amino acids are signals to start or stop the synthesis of the protein chain. An example of a nine-base DNA segment and how it codes for three amino acids is shown in Figure 12.9.

Because there are more codons than amino acids, redundancy is built into the code. Some amino acids are represented by more than one codon. Leucine, serine, and arginine have six codons each. On the other hand, tryptophan and methionine are each represented by only a single codon. Significantly, the code is identical in all living things. The instructions to make Albert Einstein, bacteria, or trees are written in the same molecular language.

The amount of information carried by your deoxyribonucleic acid is truly phenomenal. The DNA in each of your cell nuclei consists of approximately 1×10^9 (one billion) triplet codons. You have just read that each triplet is at least potentially capable of encoding 1 of the 20 amino acids found in human protein. If each codon could be assigned a letter of the English alphabet, rather than an amino acid, your DNA could encode 1×10^9 letters or about 2×10^8 five-letter words. These words would fill 400,000 pages of 500 words each, or 1,000 volumes of 400 pages each. And you carry that library in two meters of a helical thread, invisible to all but electron microscopists. The miniaturization of this information to the molecular level puts to shame the most sophisticated supercomputers.

Unfortunately, scientific fact interferes a bit with the hyperbole of the previous paragraph. It has been determined that less than 2% of human DNA actually constitutes unique gene sequences. The human genome contains multiple copies of some genes that code for frequently used proteins. Moreover, there are many copies of DNA sequences that are too short to function as genes. For example, there are millions of copies of sequences consisting of only 5–10 base pairs. But the presence of this "junk" DNA in no way detracts from the wonder of molecular genetics or from its challenge. It merely gives scientists something else to study.

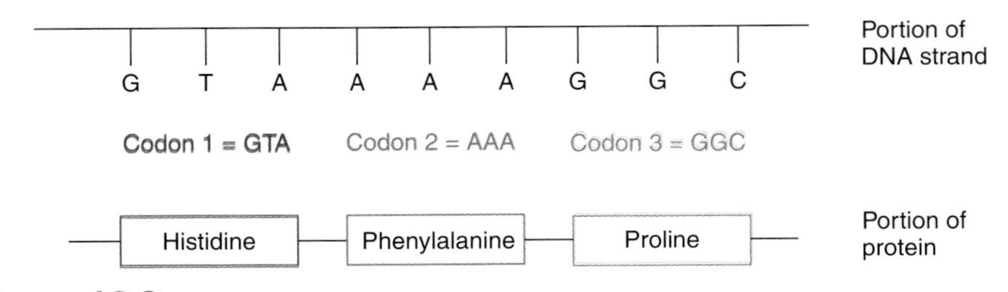

Figure 12.9
A nine-base DNA sequence demonstrating three separate codons.

12.10 Your Turn

Suggest some advantages of a genetic code in which several codons represent the same amino acid.

12.11 Your Turn

There are about 3 billion base pairs in the human genome, but only about 2% of this DNA consists of unique genes. The number of genes is estimated at 30,000. Use this information to calculate the average number of base pairs per gene.

Answer

$$\frac{3 \times 10^9 \text{ base pairs}}{\text{human genome}} \times \frac{2 \text{ unique genes}}{100 \text{ base pairs}} \times \frac{1 \text{ human genome}}{3 \times 10^4 \text{ genes}} \times \frac{1 \text{ base pair}}{1 \text{ unique gene}} = 2 \times 10^3 \; \frac{\text{base pairs}}{\text{gene}}$$

12.4 Protein Structure and Synthesis

The mechanism by which DNA directs protein synthesis is known in great detail—too much detail for this text. For our purposes, it is sufficient to recognize that the transfer of information and matter is extremely complicated. Given this complexity, it is amazing that errors in protein synthesis are very rare. Consider, for example, chymotrypsin. This protein, an enzyme that catalyzes the digestion of other proteins, consists of 243 amino acid residues. Chymotrypsin is just one of the proteins that can be made from 20 different amino acids by using 243 amino acid residues. Statistically, 20^{243} different protein molecules could be formed. Expressed relative to the more familiar base 10, this number corresponds to 1.4×10^{316}, a number larger than the estimated number of atoms in the universe! Each member of this immense group of molecules has its own unique **primary structure,** the identity and sequence of the amino acids present. One and only one primary structure is the biologically correct form of chymotrypsin with the desired enzymatic properties. The fact that the body unfailingly (or almost unfailingly) synthesizes this particular protein out of 1.4×10^{316} possibilities is evidence of a molecular blueprint and a cellular assembly line of almost incomprehensible specificity and accuracy. And, of course, similar considerations apply to each of the proteins in the entire organism.

- All enzymes are proteins, but not all proteins are enzymes.

After the amino acids are strung together in the correct sequence, the resulting protein chain should be able to twist and turn into an infinity of shapes. Surprisingly, it does not. Rather, the protein molecule assumes a characteristic shape that is the result of structural features but can also be influenced by variables such as temperature and pH. Once again, X-ray diffraction provides a means of determining this structure. For chymotrypsin, the result is pictured in Figure 12.10. What looks like a jumble of videotape is the carefully ordered backbone of the protein molecule. The figure shows helical segments and parallel chains that constitute the intermediate level of molecular organization, called the **secondary structure.** The overall shape or conformation of the molecule is termed its **tertiary structure.** Evidence suggests that the three-dimensional conformation of a protein molecule is stabilized by the interaction of various functional groups. Hydrogen bonds are particularly important in stabilizing secondary structural subunits that occur in many proteins.

The catalytic activity of chymotrypsin and any other enzyme is evidence of the reliable regularity of the tertiary structure of the protein. For an enzyme to carry out its chemistry, functional groups on certain amino acid residues must come close enough to form an active site. The **active site** is the region of the enzyme molecule where its catalytic effect occurs. Sometimes, the amino acids involved are adjacent; in other cases, they are widely separated in the protein chain, but close together in the tertiary structure. Figure 12.10 indicates that the active site in chymotrypsin consists of three amino acids that would be far apart if the protein were unwound. These groups help hold the **substrate,** the molecule or molecules whose reaction is catalyzed by the enzyme. In the case of chymotrypsin, the active site of the enzyme catalyzes the breaking of peptide bonds in the substrate, another protein. In some other enzymes, the active site promotes the formation of chemical bonds. In all cases, the orientation of the active site and the

- The mode of action of many enzymes can be explained by the lock-and-key model described in Section 10.6.

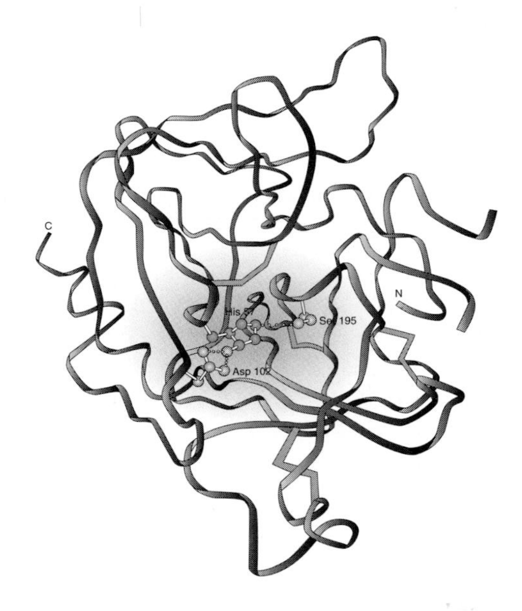

Figure 12.10

Tertiary structure of the enzyme, chymotrypsin. The "ribbon" portion represents the polymerized amino acid chain; the central colorized portion is the active site at which the enzymatic chemistry takes place.

conformation of the rest of the enzyme molecule are of critical importance. The fact that a newly synthesized protein molecule automatically assumes the enzymatically active shape almost suggests that the chain of amino acid residues possesses some sort of molecular memory. In fact, the favored tertiary structure is the most energetically stable conformation.

Sometimes a subtle change in the primary structure of a protein can have a profound effect on its properties. A much-studied example is provided by hemoglobin, the blood protein that transports oxygen, and the condition called *sickle-cell anemia.* When an individual with a genetic tendency toward sickle-cell disease is subjected to conditions that involve high oxygen demand, some red blood cells distort into rigid sickle or crescent shapes (Figure 12.11). Because these cells lose their normal deformability, they cannot pass through tiny openings in the spleen and other organs. Some of the sickled cells are destroyed and anemia results. Other sickled cells can clog organs so badly that the blood supply to them is reduced.

The property of sickling has been traced to a minor change in the amino acid composition of human hemoglobin. There are 574 amino acid residues in a hemoglobin molecule. The only difference between normal hemoglobin and hemoglobin S in persons with the sickle-cell trait is in two of these amino acids. In hemoglobin S, two of the

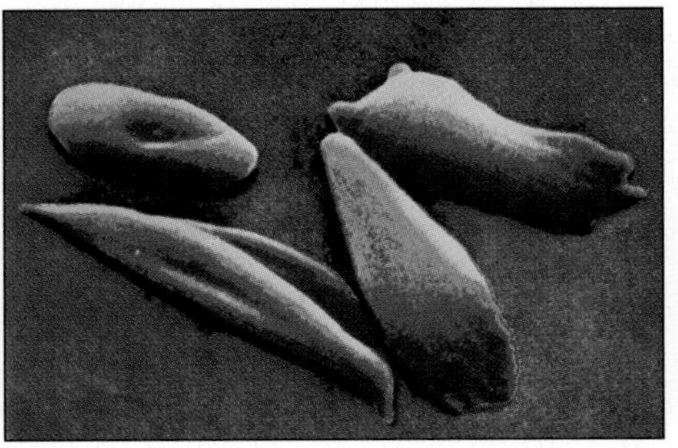

Figure 12.11

Scanning electron micrographs of normal red blood cells (left) and red blood cells showing the effect of sickle cell disease (right).

residues that should be glutamic acid residues are replaced with valine. Apparently this substitution is sufficient to cause the hemoglobin to convert to the abnormal form at low oxygen concentration.

Sickle-cell anemia is hereditary; the error in the amino acid sequence reflects a corresponding error in a DNA codon. Normally, mutations detrimental to a species are eliminated by natural selection. Perhaps sickle-cell trait has survived because it may also convey some benefit. A clue to what the benefit might be comes from studying the carriers of the gene for hemoglobin S. The gene is most common in people native to Africa and other tropical and subtropical regions and in their descendants. The fact that these are also areas with the highest incidence of malaria has led to speculation that an individual whose hemoglobin has a tendency to sickle may be protected against malaria. Specific mechanisms have been proposed to account for this protection. If the hypothesis is correct, it is an interesting example of how a genetic trait that originally had survival advantage can become a detriment in a different environment. Of course, the fact that sickle-cell anemia is a genetic disease at least raises the possibility that genetic engineering may some day eliminate it.

Scientists have not yet synthesized hemoglobin in the laboratory, but they have made simpler proteins. The first successful effort was in 1968, when two groups of scientists, one at Rockefeller University and the other at the pharmaceutical firm of Merck, Sharpe and Dohme, independently prepared bovine ribonuclease A. Ribonuclease is an enzyme that catalyzes the cleavage of ribonucleic acid (RNA). Once the 124 amino acids that make up bovine ribonuclease A were linked in the correct sequence, the resulting molecule possessed the catalytic activity of the naturally produced enzyme. This is additional evidence that the secondary and tertiary structures of the protein are determined by the primary structure.

The laboratory synthesis of bovine ribonuclease A was a great scientific achievement, richly deserving of the Nobel Prize it received. What a cow does in a minute or two required about 18 months of actual work, plus some 50 years of preliminary research. Although this research proved that the method worked, the direct laboratory or industrial synthesis of proteins is seldom carried out. There are easier ways of doing it. Today, we can replace both the cow and the chemist with bacteria, thanks to recombinant DNA technologies. It is to that topic that we now turn.

12.5 Recombinant DNA

Mythology is full of fanciful creatures: the sphinx with the head of a woman and the body of a lion; the griffin, which is half-lion and half-eagle; and the chimera with a lion's head, a goat's body, and a serpent's tail. In 1973, two American scientists, Herbert Boyer and Stanley Cohen, created another unnatural hybrid. They introduced a gene for manufacturing a protein from the African clawed toad into a common bacterium, *E. coli.* Upon replication, the *E. coli* bacteria produced the toad's protein. To be sure, humans have been manipulating the gene pool for thousands of years. We have created mules by crossbreeding horses and donkeys, dogs as diverse as Chihuahuas and Saint Bernards, and fruits and vegetables that never existed in nature. All these were done by selective breeding. Boyer and Cohen created their chemical fantasy in laboratory glassware through the manipulations of genetic engineering.

A statue of a griffin.

To illustrate the technique, consider a real response to a very real need. *Insulin* is a small protein consisting of 51 amino acids. It is produced by the pancreas, and it influences many metabolic processes. Most familiar is its role in reducing the level of glucose in blood by promoting the entry of that sugar into muscle and fat cells. People who suffer from a common type of diabetes have an insufficient supply of insulin, and hence elevated levels of blood sugar. Left untreated, the disease can result in poor blood circulation, especially to the arms and legs, blindness, kidney failure, and early death. But, diabetes can be controlled by diet, exercise, and insulin injections.

Before 1982, all insulin used by diabetics was isolated from the pancreas glands of cows and pigs, collected in slaughterhouses. It turns out that the insulin produced by cattle and hogs is not identical to human insulin. Bovine insulin differs from the human hormone in three out of 51 amino acids; porcine and human insulins differ in only one. These differences are slight, but sufficient to undermine the effectiveness of bovine and porcine insulin in some human diabetics. For many years, there seemed to be no hope of obtaining enough human insulin to meet the need. Although insulin has been synthesized in the laboratory, the process is far too complex for industrial adaptation. However, since 1982 the lowly bacterium, *E. coli*, has been tricked into making human insulin.

This unlikely bit of inter-species cooperation is a consequence of using **recombinant DNA** techniques to introduce the gene for human insulin into this simple organism (Figure 12.12). Bacteria contain rings of DNA called **plasmids.** These rings can be removed and cut by the action of special enzymes. Meanwhile, the gene for human insulin is either prepared synthetically or isolated from human tissue. This human DNA is inserted into the plasmid ring by other enzymes. The result is inter-species recombinant DNA. The modified plasmids (also called **vectors**) are then reintroduced into the bacterial "host." Once inside the cell, the biochemistry of the bacterium takes over (Figure 12.12). Every 20 minutes, the *E. coli* population doubles, and soon there are millions of copies or clones of the "guest" (human) DNA.

Clones are a collection of cells or molecules identical to an original cell or molecule. It is possible to harvest the cloned DNA, but in the insulin example we are interested in a supply of the protein, not its gene. Therefore, the bacteria are allowed to synthesize the proteins encoded in the recombinant DNA. Although the *E. coli* has no use for human insulin, it generates it in sufficient quantities to harvest, purify, and dis-

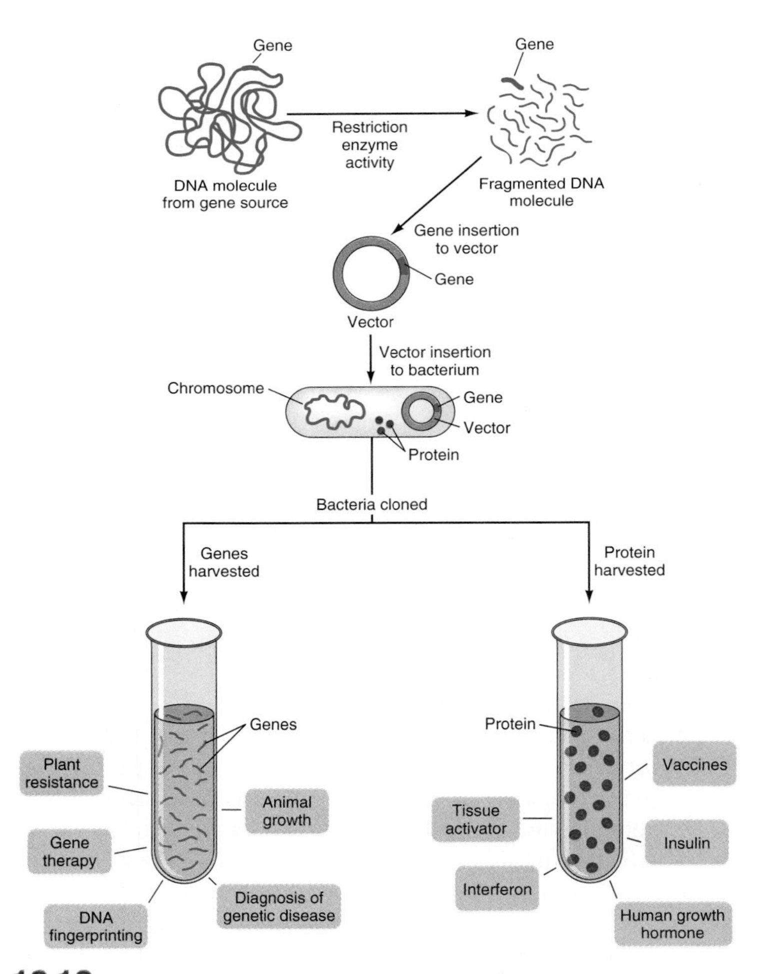

Figure 12.12

A general schematic of genetic engineering.

tribute to diabetics. Today, the cost of bacterially produced insulin is less than that of insulin isolated from animal pancreas. Currently, more than three million Americans use genetically engineered insulin to treat their diabetes.

Figure 12.12 is a representation of the recombinant DNA techniques just described. The actual operations are a good deal more complicated than the figure suggests, and many details have been omitted. A variety of vectors and host organisms have been used in molecular engineering. Other bacterial species are sometimes used instead of *E. coli*, and yeasts and fungi are often employed. DNA with specific properties is fragmented and the appropriate portion introduced into a vector that carries it into a host organism. Cloning of the host, a bacterium in this case, gives rise to two categories of end products: cloned genes (left) and proteins (right) Figure 12.12.

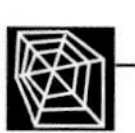

12.12 Consider This: Where Does Insulin Come From?

If a person must take insulin daily, how would that person determine the source of the insulin? Does the source make a difference in how the insulin acts in the body? Use the resources of the Web or consult a pharmacist to answer these questions.

Another success for genetic engineering has been the synthesis of *human growth hormone* (HGH). This protein, produced by the pituitary gland, stimulates body growth by promoting protein synthesis and the use of fat as an energy source. Children with insufficient HGH fail to reach normal size. If the condition is diagnosed early, injections of the hormone over eight to ten years can prevent dwarfism. Formerly, a year's HGH therapy for one person required the pituitary glands from about 80 human cadavers. That source is no longer used, thanks to the production of human growth hormone in bacteria. However, the cost of treatment, even with cloned HGH, can be as high as $20,000 per year.

12.6 Engineering New Drugs and Vaccines

You have just read about two examples of replacement therapy, in which an insufficient natural supply of an essential protein is augmented with a genetically engineered supplement. Similar biochemical methods are also being used to create new drugs or larger supplies of already known drugs. The gene coding for the drug is introduced into a host organism, which then synthesizes the desired product. This is currently one of the most rapidly growing applications of recombinant DNA technology.

Other products of biotechnology appear to be effective against viruses. **Viruses** are simple, infectious, almost living biochemical species. We say "almost living" because viruses do not have the necessary biochemical machinery to carry out metabolism or to reproduce by themselves. A typical virus consists of nucleic acid and protein; many viruses are essentially inert. When it invades a cell, the virus takes charge, forcing the cell to make more viral nucleic acid and protein. In effect, the virus does naturally what genetic engineers accomplish with recombinant DNA techniques. Because they are so simple, viruses are notoriously difficult to combat or defend against. Thus, pneumonia, "strep," or other bacterial infections, though potentially far more dangerous than a common cold, are much easier to treat than a cold, which is caused by a virus. Pneumonia and strep are bacterial infections, which can be destroyed by penicillin or other antibiotics.

Genetically engineered **interferons** may help change all that. Interferons are nature's way of providing protection against viruses. Over 20 distinct naturally occurring interferons have been identified. All of them are proteins, and some also contain carbohydrate portions. The mechanism by which these molecules defend against viral invasion is not fully understood, but it has been exploited. Thus far, genetically engineered interferons show promise against hepatitis, herpes zoster (shingles), a type of multiple sclerosis, and a variety of cancers including some forms of leukemia, malignant melanoma, multiple myeloma, and certain kidney cancers. The use of an interferon nasal spray to control the common cold may still be years away because of the high costs and biochemical complexity associated with cloning these proteins.

For the treatment of many diseases, the ultimate goal is not just the development of a drug to treat it, but the creation of a **vaccine** to prevent contracting the disease. Vaccines work by mobilizing the body's own defense mechanism. The idea is to expose the body to a molecule or organism closely related to the virus or bacterium that causes the disease. The immune system responds to this stimulus by generating **antibodies** against it. These antibodies remain in the body, where they offer protection against subsequent infection by the virus or bacterium. Of course, it is important that the vaccine does not itself cause the disease. Therefore, vaccines are typically made from bacteria or viruses that have been killed or weakened or from fragments or subunits of the virulent invaders. The latter approach is the preferable one, because there is no danger that the disease will be transmitted in the process of vaccination. Fortunately, it is here that genetic engineering is most promising. The DNA encoding for a characteristic but noninfectious part of a virus—for example, its protein coat—can be introduced into plasmids. The bacteria consequently produce this particular protein. The protein is then isolated, concentrated, and used as a vaccine that carries essentially no risk of infection. This technique has been employed to synthesize a vaccine against hepatitis B. Because hepatitis is transmitted by blood, health care professionals are often vaccinated against the disease. If and when a vaccine is developed against HIV, it may be through a similar technology. Anti-viral vaccines now exist for Influenza A and RSV (respiratory syncitial virus, a virus that manifests itself as the common cold in adults). Anti-viral drugs are available for a series of infections including herpes, chicken pox, and forms of meningitis.

12.7 Diagnosis Through DNA

Until very recently, the most sensitive methods of diagnosing disease were based on the detection of enzymes or antibodies in an infected organism. The antibodies are generated by the host organism in response to the invasion. This means that infection is often well established before a positive test can be obtained. Fortunately, genetic engineering has enabled diagnosticians to identify the DNA of the infectious agent, even at early stages and a low concentration.

- DNA probes were used to identify *Helicobacter pylori* as the causitive agent of gastric ulcers and also a likely cause of stomach cancer.
- Radioactivity is discussed in Sections 7.7–7.9.

Such sensitivity is possible only because of the development of two important techniques: DNA probes and the polymerase chain reaction. *DNA probes* are engineered so that they are complementary to some segment of the infecting viral or bacterial DNA (the target). The probes are single-stranded DNA, with from 10 to over 10,000 bases, and they are usually labeled with a radioactive isotope. These radioactive probe molecules are introduced into a sample of biological fluid or cytoplasm that is suspected of containing an infectious agent. The test is carried out at a temperature and pH at which the DNA double helix of the infectious agent separates into single strands. If the probe encounters a strand with a complementary segment of bases, hydrogen bonds form between A and T and C and G, and the probe sticks to the target molecule. Because the probe is radioactive, it can easily be traced using a radiation detector. The radioactivity level in a certain fraction indicates that the infectious DNA is indeed present.

Successful early diagnosis involves detecting the infecting virus or bacterium before the disease is well established. Even the most sensitive DNA probes will not work if the concentration of the invading DNA is too low. Here, the **polymerase chain reaction (PCR)** has proven to be of great utility. This technology makes it possible to start with a single segment of DNA and make millions or even billions of copies of it in a few hours. The PCR process, which won its inventor, Kary Mullis, a Nobel Prize in 1993, is diagrammed in Figure 12.13. A researcher starts with a sample, denoted as a double-stranded DNA at the top of the figure. In step (a), the sample is heated to unwind and separate the two complementary strands, cooled, and combined with short segments of "primer" DNA. These **primers** are synthetic, single-stranded nucleotides that bracket and identify the section of the DNA to be copied. In step (b), the enzyme poly-

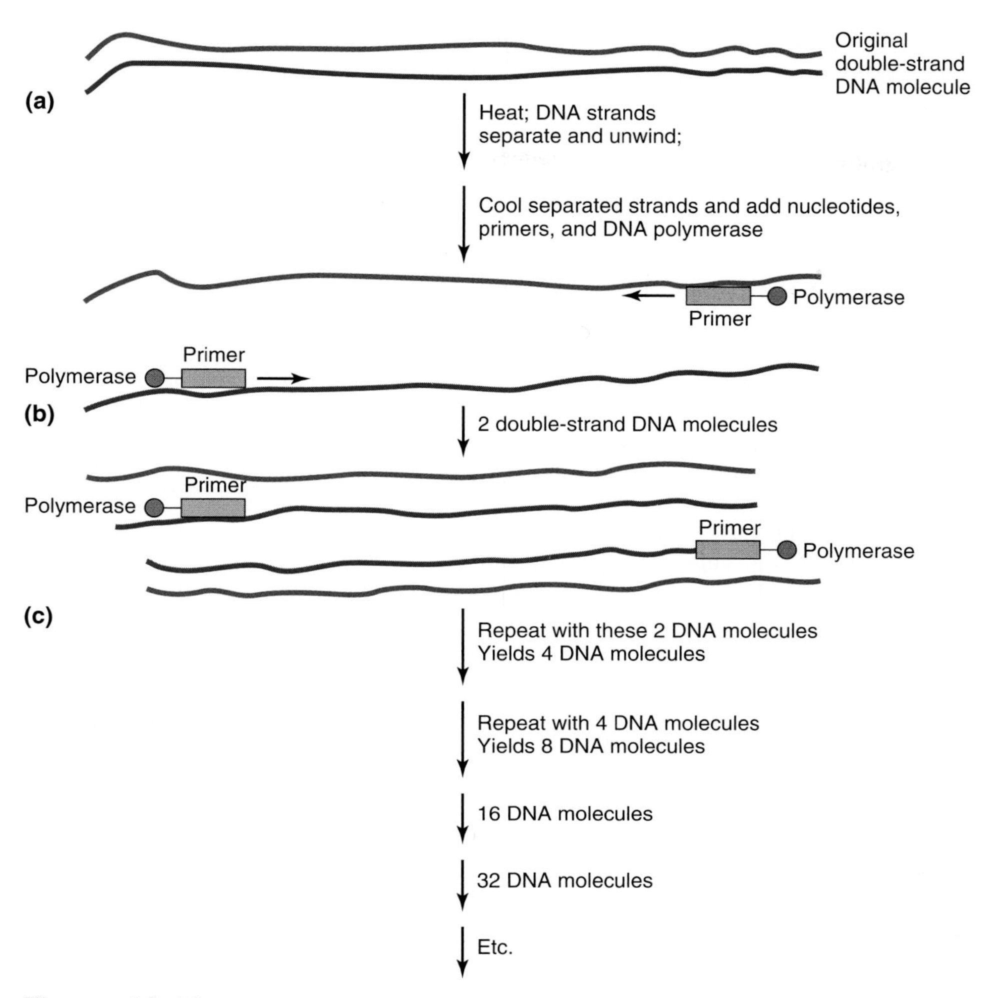

Figure 12.13
Diagram of the polymerase chain reaction.

merase and an ample quantity of the four free nucleotides bind the separated DNA strands. The polymerase enzyme starts at the location of the primer and directs the formation of a new strand using the free nucleotides, thereby creating a complementary copy of the single strand. In this way (step c), a copy is made of each of the two original strands. This sequence of steps yields four strands derived from the original double strand. Repeating the process causes all four strands to be reproduced. The overall yield of DNA molecules is 2^n, where n is the number of times that this process is repeated. One can see that yield of identical copies of the original double strand grows very quickly. This should be obvious by considering that $2^{10} = 1.024 \times 10^3$, $2^{20} = 1.048 \times 10^6$, and $2^{100} = 1.27 \times 10^{30}$ molecules.

The polymerase chain reaction is very rapid; only one to two minutes are needed for a complete cycle. Thus, 100 cycles would require just two to three hours. But, the reaction would run out of starting materials long before then. A little arithmetic (12.13 The Sceptical Chymist) shows that 1.27×10^{30} double-stranded DNA molecules, each of 100 base pairs, would weigh over 139,000 tons!

12.13: The Sceptical Chymist: 140,000 Tons of Base

It is a good idea to check the assertion that 1.27×10^{30} double-stranded DNA molecules of 100 base pairs each would weigh almost 140,000 tons.

Hint: Start by using Avogadro's number to determine how many moles of DNA are represented by this large number of base pairs. Assume an average mass of 300 g per mole of nucleotide.

12.14 Your Turn

A technician starts with a single DNA molecule. How many cycles of PCR must take place to form each of these numbers of DNA molecules?

a. 16 DNA molecules
b. 256 DNA molecules
c. 1.0×10^6 DNA molecules

Answers
a. 4 cycles ($1 \rightarrow 2 \rightarrow 4 \rightarrow 8 \rightarrow 16$) **b.** 8 cycles **c.** 20 cycles

PCR technology has proven indispensable in any procedure where a small sample of DNA must be dramatically amplified to get a sufficiently large supply for subsequent studies. Thus, it is used in developing "DNA fingerprints" in criminal cases, in studying archeological remains, and, of course, in diagnosing disease. In the latter instance, the DNA in specimens thought to contain infecting or defective DNA is multiplied by PCR so that DNA probes can be used.

The early diagnosis of HIV infection is one of the achievements of this new technology. The DNA of human immunodeficiency virus can be detected several weeks before antibodies to HIV build up to the point at which they can be identified. Similarly, recombinant DNA methods have been used to speed up the diagnosis of tuberculosis. DNA probes and PCR have also been used to identify a number of hereditary diseases. Defective genes have been identified for cystic fibrosis, Huntington's disease, some forms of Alzheimer's disease, and amyotrophic lateral sclerosis (ALS), better known as Lou Gehrig's disease, after the great New York Yankee first baseman who died of it in 1942. Scientists identified the altered gene responsible for the disease in 1993. As in sickle-cell anemia, the mutation is slight and subtle. A single amino acid is altered in superoxide dismutase, an enzyme that eliminates free radicals, highly reactive chemical species with unpaired electrons. Free radicals appear to accumulate in Parkinson's disease, Alzheimer's disease, and in normal aging. In the case of ALS, their buildup results in the destruction of motor nerves. ALS gives little early warning. Early diagnosis, based on detection of the altered gene, might permit preventative treatment.

- Free radicals are also involved in stratosphere ozone depletion (Sections 2.11 and 2.13), in addition polymerization (Section 9.3), and in cooking foods (Section 11.2).

The "might" in the previous sentence indicates a major problem in the diagnosis of hereditary diseases and defects. In many cases, the ability of modern science to respond to the effects of a defective gene has not equaled the ability to detect the gene. The tendency to develop genetic diseases is programmed in the DNA, and it may or may not be stimulated by infection. One can well ask what is the advantage of knowing that an individual is the carrier of one of these genes when there is no way to prevent or even to treat its deleterious effects. Is it helpful to know that you have a high probability of acquiring ALS or Alzheimer's disease when there is nothing you can do about it but wait?

To be sure, such knowledge could be useful in case new therapies are developed. Moreover, some carriers of inheritable diseases choose not to have children or to use *in vitro* (literally "in glass") fertilization and genetic screening. A married couple, both of whom were carriers of the gene for cystic fibrosis, recently took the latter approach. Five ova taken from the woman were fertilized with her husband's sperm, and the resulting embryos were analyzed for the cystic fibrosis gene at the eight-cell stage. An embryo with no or only one copy of the defective gene, neither of which would result in the disease, was implanted into the woman's uterus, and a healthy baby was born.

Genetic screening is more frequently used on embryos and fetuses that are further developed. If there is a likelihood of the parents passing on a defective gene, they sometimes request **amniocentesis.** In this procedure a sample of the amniotic fluid is withdrawn from the mother's uterus. This fluid contains fetal cells that are then analyzed for their genetic makeup. By this means, the defective gene for Down's syndrome and other hereditary conditions can be detected. If they are present, the parents face a difficult decision with a significant ethical component: whether to abort the fetus. Such painful choices could be avoided if science were to develop ways of actually changing the DNA of the fetus or even of children or adults. We now consider this prospect.

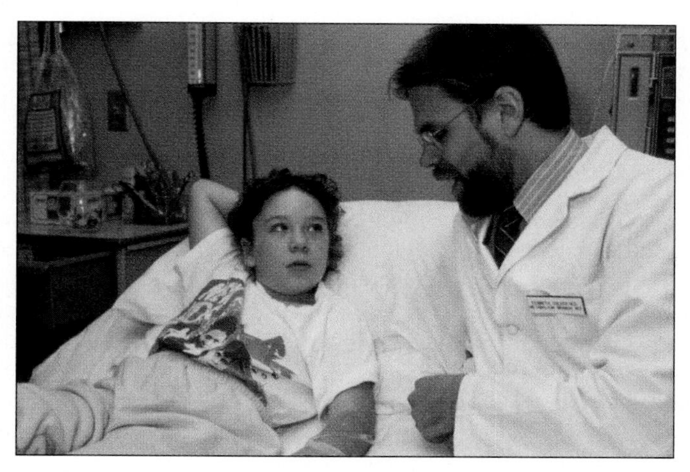

Figure 12.14

Curing disease through genetic engineering. One of two young girls who were the first humans "cured" of a hereditary disorder by transferring into their bodies healthy versions of the gene they lacked. The transfer was successfully carried out in 1990, and the girls remain healthy.

A revolutionary approach to fight or prevent disease involves an attempt to correct the basic problem by altering the genetic makeup of some of a patient's own cells. Thus, the goal of **gene therapy** is to supply cells with normal copies of missing or flawed genes. Medical researchers estimate that about 4000 known genetic disorders in addition to cancer, heart disease, arthritis, and other illnesses seem to be ideal candidates for these treatments. Simply stated, cells are taken from a patient, altered by the introduction of normal genes, and then returned to the patient. If all goes well, the imported genes function normally.

Gene therapy was first applied successfully to a human being was in 1990, when the technique was used to treat a four-year-old girl suffering from *severe combined immunodeficiency disease* (*SCID*), a very serious, and fortunately very rare, condition. Because of a genetic defect, a specific enzyme is not synthesized. The absence of this enzyme leads to the destruction of the white blood cells that protect the body against infection. Children suffering from SCID have essentially no functioning immune system, and the slightest infection can prove fatal. In the past, some children with SCID survived for a few years, but only by living in sterile isolation chambers.

- SCID received previous publicity because of David, the boy in the plastic bubble.

In the procedure used with the first patient, the gene that encodes for the missing enzyme was identified and isolated from other sources. Special viruses were used to introduce copies of this gene into cells that had been removed from the patient. These modified cells were then reintroduced into the girl's body. The new genes have continued to function well, producing the previously absent enzyme. As a result, the concentration of white blood cells has increased significantly and the girl's antibody-producing defense mechanism is working (Figure 12.14).

Although gene therapy holds great promise for humans, it cannot be categorized as a general success to date. Some have characterized the efforts of the last decade as having made virtually no progress. The process is perhaps best characterized by a quote from *The Scientist* (12[10]:4, May 11, 1998): "The gene therapy field resembles a toolbox containing instruments researchers haven't quite mastered, and the number of devices—viral and nonviral vectors—in this toolbox keeps increasing . . . 'People are still trying to figure out what tools to use for what diseases.'"

12.15 Consider This: Gene Therapy

There are promising developments in using gene therapy for treating serious human conditions, but there is not universal success.

a. Why do you think that experimental protocols with gene therapy have shown mixed results? Explain some of the factors that you feel may influence outcomes.

b. Given the mixed results, under what conditions would you consider gene therapy a valuable tool? Explain the reasons for your criteria.

12.8 Genetic Fingerprinting

A woman has been sexually assaulted. She cannot identify her attacker, who was masked. The police have two suspects, but neither man was seen in the neighborhood on the night of the attack. There seems to be little evidence, except for several drops of semen on the victim's clothing and in her vaginal canal. That may be sufficient. Thanks to DNA fingerprinting, it may prove possible to identify the attacker with a probability approaching certainty.

DNA fingerprinting is based on the fact that each of us has his or her own unique DNA. It is not surprising that the really important genes, those that encode for insulin, hemoglobin, chymotrypsin, and most other proteins, are identical in almost all of us. Here, as we have already noted, mutations are rare. We differ primarily in the "junk" DNA that makes up about 98% of the three billion base pairs in each human cell nucleus. Therefore, forensic scientists look to this apparently nonessential DNA when they seek to determine "who dun it."

The authors recognize that jurors typically turn off when expert witnesses go into great detail about the biochemistry of DNA fingerprinting. For that reason we will try to be brief without sacrificing accuracy. In short, the technique is based on the fact that every individual appears to have a unique set of the DNA segments that serves as the spacers or punctuation marks between genes. Consider the semen sample found on the victim's clothing. Even if it is a very tiny spot, it contains more than enough DNA for a reliable genetic fingerprint; 1×10^{-9} g of DNA is sufficient. First the DNA is extracted, then it is multiplied many times over by PCR. These copies are exposed to the action of enzymes that cut the DNA strands before and after the spacer segments just mentioned.

The spacer fragments are then subjected to **electrophoresis,** a method of separating molecules based on their rate of movement in an electric field. In the technique used in DNA fingerprinting, samples are applied to a strip of a polysaccharide gel, and electrophoresis is carried out in this medium. Because the phosphate groups of the DNA are negatively charged, the fragments migrate toward the positive electrode or pole. The speed at which a DNA segment moves depends on the magnitude of its electrical charge and its size or molecular mass. Shorter strands of DNA, consisting of fewer base pairs, move faster than longer strands, which encounter more resistance from the gel.

It is necessary to see and measure how far the DNA fragments have traveled in a fixed period. This is done by using radioactive markers that can be detected because they expose photographic film. The fingerprint thus consists of a film with black smudges or bars, each one corresponding to the distance migrated by a particular segment of DNA. The heaviest and longest segments are closest to the point of application, the lightest and shortest ones are the farthest away.

In most criminal cases, the electrophoresis pattern produced from the evidence collected at the crime scene is compared to the electrophoresis pattern made by DNA obtained from the suspect or suspects. Figure 12.15 is the electrophoretic evidence in the sexual assault case we are investigating. Each spot indicates how far DNA segments of specific size migrated during the electrophoresis experiment. Rows 1, 5, 8, and 9 are reference markers of DNA exhibiting a known range of molecular masses. Spots at the far left represent the longest, heaviest segments of DNA; spots at the far right represent the lightest, shortest segments. Row 3 is DNA from a semen sample found on the victim's clothing, and row 6 is semen DNA obtained by swabbing her vaginal canal shortly after the attack. Not surprisingly, the positions of the spots in these two rows are identical. To avoid possible misidentification, it is also important to include a sample of the woman's own DNA, which gives the pattern in row 7. Now look at rows 2 and 4.

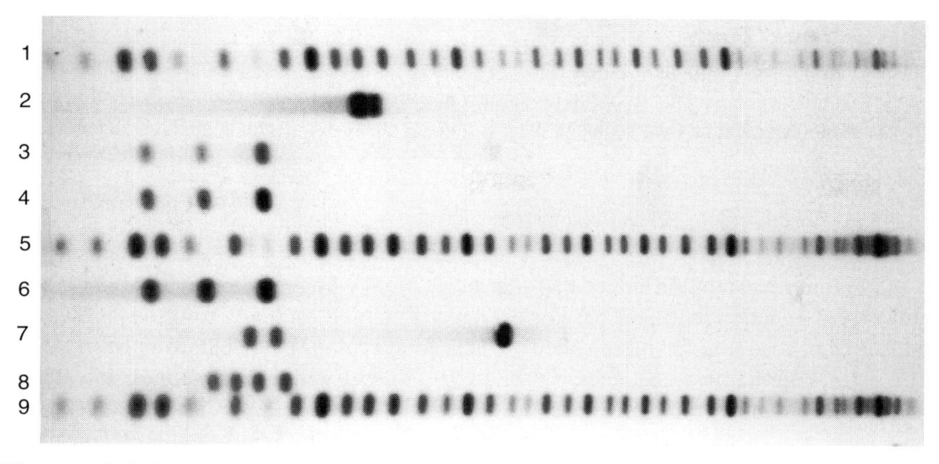

Figure 12.15

DNA fingerprints in a sexual assault case. The various rows represent the electrophoretic migration pattern of DNA segments.
Rows 1, 5, 8, 9: Reference markers of a mixture of DNA segments of known length and mass.
Row 2: DNA from a sample of blood from suspect A.
Row 3: DNA from a semen sample found on the victim's clothing.
Row 4: DNA from a sample of blood from suspect B.
Row 6: DNA obtained by swabbing the victim's vaginal canal.
Row 7: DNA obtained from a sample of the victim's blood.

Row 2 is characteristic of the DNA in a blood sample obtained from suspect A; row 4 is from a blood sample from suspect B. B's genetic fingerprint matches the DNA from the semen samples in rows 3 and 6.

In the case represented by Figure 12.15, the evidence strongly indicates that suspect A is innocent and that suspect B is probably guilty, but caution is required. A matching DNA fingerprint does not *absolutely* prove the guilt of a suspect. It is possible that DNA from two individuals might yield the same electrophoresis patterns, but it is highly unlikely. To increase the odds of accurate identification, comparisons of the sort just described are typically done on DNA from three or more different chromosomes. As the number of determinations increases, so does the improbability of finding any two individuals with identical DNA fingerprints.

To see how this works, suppose that the statistical data base indicates that one person in 100 will exhibit a particular DNA pattern obtained from chromosome number 1. This would mean that there is a 1 in 100 chance that a suspect with that pattern is *not* the source of the sample. But also assume that the frequency of the observed pattern from another chromosome, say, number 8, is 1 in 1000; and that the frequency of the observed pattern from chromosome number 12 is again 1 in 1000. Probabilities are multiplicative, therefore,

$$\text{Total probability} = \frac{1}{100} \times \frac{1}{1000} \times \frac{1}{100} = \frac{1}{100{,}000{,}000}$$

In this example, the odds that any two individuals would have the same genetic fingerprints, are 1 in 100 million. Conversely, if the DNA from a suspect matches a sample from this crime scene, chances are 99,999,999 out of 100,000,000 that the person *was* the source of the original sample.

Such probabilities can be very convincing, especially if the suspect has a motive and can be otherwise placed at the scene of the crime. However, some juries have been skeptical of DNA data. In the O. J. Simpson murder trial, defense attorneys suggested that the blood samples taken from the scene of the crime had been contaminated or planted. Apparently, the doubts raised in the jurors' minds were enough to outweigh the evidence of the DNA fingerprints.

12.16 Your Turn

Suppose that a DNA fingerprint is based on three different chromosomal segments. The frequency of a match is determined to be 1 in 10 based on the first segment, 1 in 100 based on the second segment, and 1 in 1000 based on the third segment. What is the probability of a match in all three tests?

Answer

$\frac{1}{10} \times \frac{1}{100} \times \frac{1}{1000} = 1$ in 10^6 or 1 in a million

Criminal investigation is not the only use of DNA fingerprinting. It is also a powerful tool in genetic identification. For example, it is used routinely to prove or disprove paternity, because it is far more specific than blood typing. DNA can even be extracted from bone, and the base sequence can be used to identify human remains. Such studies were recently carried out on bones exhumed from a mass grave near Yekaterinburg, Russia, the site of the massacre of Czar Nicholas II and his family in 1918. The DNA was compared with samples from living relatives of the last Russian royal family and found to match sufficiently well to conclude that the remains were indeed those of the Romanovs. In 1998, DNA fingerprinting was used to determine the identity of a soldier who was killed during the Vietnam War and buried in the Tomb of the Unknowns at Arlington National Cemetery. The tests resolved the identity of the entombed soldier, which had been narrowed to two possible candidates. The profile of DNA taken from the bones of First Lieutenant Michael Blassie matched sufficiently that of a DNA sample taken from his mother, thus allowing proper identification.

DNA analysis is not merely confined to the living and the recently deceased. Researchers have cloned and investigated DNA samples obtained from a 2400-year-old Egyptian mummy and some even older human remains. Scientists have interpreted the results to gain information about the relationship of ancient peoples, their migration routes, and their diseases.

12.17 Consider This: Lincoln's DNA

Some researchers have speculated that Abraham Lincoln suffered from a genetic condition known as Marfan's syndrome, which causes a person to grow tall and gangly. DNA fingerprints could answer the question. Do you support exhuming Lincoln's body from the Oak Ridge Cemetery in Springfield IL? Give some reasons to explain your decision.

The current DNA age record is held by a bee and a termite that lived about 30 million years ago. Since that time, the insects had been entombed and protected in amber, which is solidified plant resin. In 1992, researchers released the perfectly preserved bodies, extracted their DNA, and subjected it to a number of studies. This research yielded important information about the evolutionary connection of these ancient organisms to modern species. But, could 30-million-year-old DNA yield more? Could it be cloned into a living fossil? That is the premise of *Jurassic Park.* In the science fiction film and the novel, the chief interest is not in the fossilized insects themselves, but in the dinosaur blood they contain. The blood is the source of the DNA that is cloned and introduced into crocodile egg cells, where it replicates until it creates modern copies of long-extinct creatures. "Could this happen in real life?" That question is posed by I. Edward Alcamo in *DNA Technology: The Awesome Skill*, a book that is a useful source of information and illustrations for this chapter. Professor Alcamo's answer may be mildly reassuring for those who would rather not encounter a *Tyrannosaurus rex:* "Possibly. But you would need an entire set of dinosaur chromosomes, and only a minuscule fragment has been obtained up to now. And that's only the first of a thousand problems that must be solved."

- Science fiction books and films to the contrary, dinosaurs and humans never coexisted, as confirmed by fossil records. Dinosaurs were long extinct before the first humans appeared on Earth.

12.18 Consider This: Science Fiction Success

One of the reasons that science fiction is successful is that it starts with a known scientific principle and extends, elaborates, and sometimes embroiders it. Perhaps you can be as successful as Michael Crichton was with Jurassic Park. Start by identifying a scientific principle from this or another chapter, and then writing a one- or two-page outline for a story based on that principle. Be sure to identify the chemical concepts that you plan to include and any pseudoscience that you might employ.

12.9 Mixing Genes: Improving on Nature(?)

One of the more sensational dimensions of genetic engineering has been the creation of higher plants and animals that share the genes of another species, so-called **transgenic organisms.** Inserting foreign DNA becomes progressively more difficult as one moves up the evolutionary ladder from bacteria through plants to animals. Nevertheless, some of the most spectacular successes of recombinant DNA technology have involved modifications of agricultural crops. Earlier researchers resorted to a "shotgun approach" to introduce DNA into plant cells. Millions of microscopic tungsten spheres are coated with the DNA to be inserted into the host, and these projectiles are fired into a group of cells. Some of the DNA finds its way into the plant chromosomes, but it is largely a hit-or-miss proposition. The use of plasmid carriers, similar to those used with bacteria (Section 12.5) is generally more dependable because of the much greater frequency of single-site insertions of the foreign DNA and the resulting greater ease of monitoring the process.

In spite of some rather formidable difficulties, altering the genetic makeup of plants by genetic engineering is faster and more reliable than relying on traditional cross-breeding. Moreover, some of the species that have been genetically combined are so dramatically different that interbreeding is impossible. A list of available, commercial plants and some in development are shown in Table 12.2.

Table 12.2

Transgenic Plant Experiments

Plant	Objective of Change
Transgenic plant	
soybeans corn canola cotton	Resistance to Roundup® and Liberty® herbicides
corn cotton potato	Plant produces a gene from Bt (*Bacillus thuringiensis*), a soil bacterium that releases a delta-endotoxin in insects
papaya	Resistivity to pests and diseases
Plant in development	
corn	Resistance to corn rootworm
tomato	Enhanced nutritional factors (e.g., lycopene, a naturally occurring nutrient related to Vitamin A) Delayed ripening (to enhance flavor) Resistance to bacterial speck disease
rice	Golden rice capable of synthesizing beta-carotene, the precursor of Vitamin A
canola	Improved nutritional value of canola oil with higher Vitamin E content or modified content of fatty acids
sunflower	Disease resistance, anti-pest, herbicide resistance
turfgrass	Herbicide tolerance Disease and insect resistance Reduced growth rates Tolerance to drought, heat, and cold

Among the most promising combinations have been those that confer on the host plant built-in pest resistance. Plants have been engineered to produce their own internal insecticides. For example, DNA from a soil bacterium (Bt or *Bacillus thurnigiensis*) that produces a compound that is toxic to various caterpillars has been inserted into the genome of cotton and corn plants. When cotton bollworms or corn borers attempt to eat the plants, they consume the toxin and die. Insects that do not feed on the genetically altered cotton or corn are unaffected, though they might well be killed with a conventional insecticide spray. Similar advances that reduce agricultural dependence on pesticides and herbicides are likely to be welcomed by some environmentalists. Some herbicides are "broad spectrum" and unselective in their actions, killing many crops as well as weeds. To protect against this, a segment of DNA from *E. coli* inserted into soybeans, corn, cotton, and other crops, makes the plants resistant to certain herbicides used for weed control. Thus, the weeds are killed without damaging the crop. Starlink® is an example of transgenic corn being grown but not for human consumption. Its accidental inclusion into human supplies provided the introduction to the chapter. Recall that the chief safety issue is whether proteins generated as part of the transgenic behavior could cause allergy problems in humans.

During 2000, transgenic plants were grown on approximately 110 million acres worldwide (Table 12.3), an increase of about 11% over the previous year. For example, in 1999, approximately half the U.S. soybean crop and about 25% of the U.S. corn crop was planted in transgenic varieties. Most of those plant varieties are either herbicide tolerant or insect- and pest-resistant. Minor acreage was planted to transgenic potatoes, squash, and papaya.

Someday, genetic engineering may be able to make a significant contribution to better nutrition. Researchers are investigating ways of incorporating the DNA of nitrogen-fixing bacteria into wheat, rice, and corn. These bacteria are present in the root systems of soybeans, alfalfa, and other legumes. Thanks to the bacteria, these plants can absorb N_2 directly from the atmosphere and use it in biochemical reactions. Other plants also need nitrogen, but they can use it only in soluble form from mineral sources in the ground and water. Hence, nitrogen-containing fertilizers are applied to help these plants grow well. But the need for fertilizers would be greatly reduced if all food crops were able to use atmospheric nitrogen directly.

Researchers also have developed sweet potatoes with enhanced protein content by inserting a gene coded for storage of a protein with a high content of essential amino acids. Such an improvement is important in many poorer tropical countries where high-quality protein sources are expensive, but sweet potatoes are a dietary staple food. A

Table 12.3

Millions of Acres of Transgenic Crops Planted in 2000

Country	Area Planted in 2000	Crops Grown
U.S.	74.8	soybeans, corn, cotton, canola
Argentina	24.7	soybeans, corn, cotton
Canada	7.4	soybeans, corn, canola
China	1.2	cotton
South Africa	0.5	corn, cotton
Australia	0.4	cotton
Mexico	[a]	cotton
Spain	[a]	corn
Germany	[a]	corn
France	[a]	corn

[a]Minor area

great benefit of these genetically altered plants is that the newly acquired trait is conserved in the seeds and passed on to the next generation.

The grocery store of tomorrow may also contain other products of rearranged genes. For example, it may prove possible to genetically induce cows to give human milk for newborn babies, lactose-free milk for those suffering from lactose intolerance, naturally iron-enriched milk for anemic persons, casein-rich milk for cheese-makers, and skim milk for weight watchers—not all from the same cow, of course. Bovine growth hormone (BGH) is now produced in bacteria and injected into cattle to increase milk and beef production. The FDA has ruled that this practice presents no hazards to human health because BGH is biologically inactive in humans. However, critics have correctly noted that cows exposed to elevated BGH levels are more prone to infectious disease and consequently are administered higher levels of antibiotics than normal. Some of these drugs can be carried over into the milk.

12.19 Consider This: Where in the World Is BGH?

In 1985, the U.S. FDA approved the use of BGH in cows for boosting milk production. Although regulatory agencies around the world have reached the same conclusion, the debate over whether BGH should be used this way is far from over. Search the Web for "bovine growth hormone" (or its synonyms "bovine somatotropin" or "BST") to check the current situation and report the arguments on both sides. If you can, document the situation in countries other than the United States.

12.20 Consider This: Turning Over a New Leaf

The New Leaf Superior potato has been genetically engineered to produce its own insecticide. This gives the potato the ability to resist attack by potato beetles, a destructive insect that causes significant damage to potato crops. You may have already eaten some of these potatoes without even knowing it, for they do not have to be labeled as a food produced through biotechnology.

a. Identify some of the benefits and risks associated with genetically engineered potatoes.
b. Would you knowingly eat potato chips made from New Leaf potatoes? Explain the reasons for your opinion.

By using a cloned growth hormone, salmon and rainbow trout grow to sufficient size to be sold in supermarkets in one year rather than the two or three years normally required. Pork with only 10–20% of the usual amount of fat, and sheep whose fleece can be pulled off like a sweater during shearing season are also on the genetic engineering drawing boards. Genetic engineers are also exploring the possibilities of using plants to produce edible vaccines or insulin, plastics, and even a naturally grown blend of cotton and polyester. Extracting plastics from plants on a commercial scale would conserve significant quantities of petroleum, the conventional current feedstock for all commercial plastics.

"Molecular pharming" is the term coined to describe the practice of using domesticated animals to produce drugs and other medically significant substances. Transgenic beef and dairy cattle, hogs, and fish have been developed by transferring foreign DNA into the nuclei of eggs' cells or embryos. For example, transgenic pigs grown from pig embryos injected with human hemoglobin genes produce human hemoglobin, which may be used as a blood substitute. Genetic engineering, however, is more than simply a collection of techniques. There is a human face to it among those who have benefited medically from it. Before recombinant DNA, there was no hepatitis C vaccine, insufficient erythropoieten, a protein used to stimulate red blood cell growth in dialysis patients, and a lack of tissue plasminogen activator (TPA) to dissolve blood clots in cardiac patients.

Using transgenic plants and animals to make compounds currently produced mainly in vats and vials is a far-reaching concept from the standpoint of economics and resource management. In the October 23, 1998, issue of *Science*, DuPont board chair Jack Krol commented about this matter: "In the 20th century, chemical companies made most of their products with nonliving systems. In the next century, we will make many of them with living systems." In support of that premise, DuPont has research underway to use microbes and plants to produce a wide range of compounds, from chiral drugs to plastics. Jerry Caulder, an agricultural genetic engineering entrepreneur, is also "bullish" on this approach. In the same issue of *Science*, he asked: "Organisms are the best chemists in the world. Why not use them to produce the chemical feedstocks you want rather than using petroleum?"

Have the potential benefits of genetically modified foods outlined in this section been overstated while potential problems are being "swept under the table" by those who benefit the most economically? Much of the opposition to "Frankenfoods" is based on the concern that "natural" plants could be changed forever and perhaps irreversibly on exposure to pollen from altered crops. Opponents also fear that currently unanticipated health problems could arise in people and animals that consume genetically modified (GM) food. Writing in the May 1999 issue of *Nature*, Cornell researchers reported that Bt-corn, a GM product, is a serious threat to monarch and other butterflies. In laboratory tests, monarchs who were fed milkweed leaves dusted with transgenic corn pollen ate less, grew more slowly, and suffered a higher mortality rate. Nearly half of the larvae who ate leaves dusted with transformed corn pollen died, compared to virtually none who ate leaves dusted with non-transformed corn pollen. Subsequent researchers, however, report that the risk to monarchs is minimal when the leaves are dusted with pollen levels mimicking those actually found in natural settings.

A 1998 report from a research institute in the United Kingdom attacking the safety of genetically modified potatoes is largely responsible for the "Frankenfood frenzy" in Europe. That document reported that GM potatoes have a specific and negative effect on organ development and the immune system in rats. In June 1999, the Royal Society in Britain conducted a thorough, independent review and concluded that the report and the experiments on which it was based were flawed. They cited poor experimental design, possibly exacerbated by the lack of "blind measurements," uncertainty from the small number of samples, the application of inappropriate statistical techniques, possible dietary differences due to nonsystematic dietary enrichment of protein in the rats, and a lack of consistency of findings within and between experiments. Critics, however, remain undaunted and vocal in spite of few reproducible experiments to support any health concerns.

12.10 Cloning Mammals

Although easily fleeced sheep are not yet available, a sheep did make headlines in 1984, an ironically appropriate year, given what was reported. Researchers cloned a sheep by a technique called **nuclear transfer.** Using a very thin, hollow needle, nuclei were removed from the cells of sheep embryos and transferred into unfertilized sheep eggs from which the nuclei had been removed. The inserted nuclei gave the eggs a complete set of genes. The eggs were then implanted into a sheep's uterus, carried full term, and a lamb was born, the first mammal produced by cloning. Soon after, by using nuclear transfer, embryonic cells were cloned into cows, rabbits, goats, and rhesus monkeys.

Dolly and the lamb she gave birth to by conventional means.

In 1997, headlines worldwide heralded the birth of another lamb, Dolly. But Dolly was no ordinary lamb, even for a cloned one. She was the product of the first cloning of a mammal using nuclei from *adult* cells, long thought to be an extremely difficult, if not impossible, process. That barrier fell in July 1996 with the birth of Dolly at the Roslin Institute, near Edinburgh, Scotland. She has a unique pedigree—a mammal who is an exact genetic copy, a clone, of an *adult*, but without a father. Ian Wilmut, the surrogate "father" of Dolly, used nuclei removed from mammary glands of a 6-year-old ewe and transferred them, one into each unfertilized sheep egg from which the nucleus had been removed. These "fertilized" eggs were then implanted into the uterus of a sheep, culminating in the birth of Dolly in July 1996. This process was similar, but not identical, to earlier clonings of mammals (Figure 12.16).

It took 277 such transfers to create Dolly. Until she came along, conventional wisdom had it that adult cells lacked the workable versions of all the genes necessary to create an entire organism. To circumvent this possible difficulty, Wilmut and his coworkers deprived the mammary cells of nutrients for five days before extracting their nuclei, forcing the cells out of their normal growth pattern and into a resting stage. Speculation

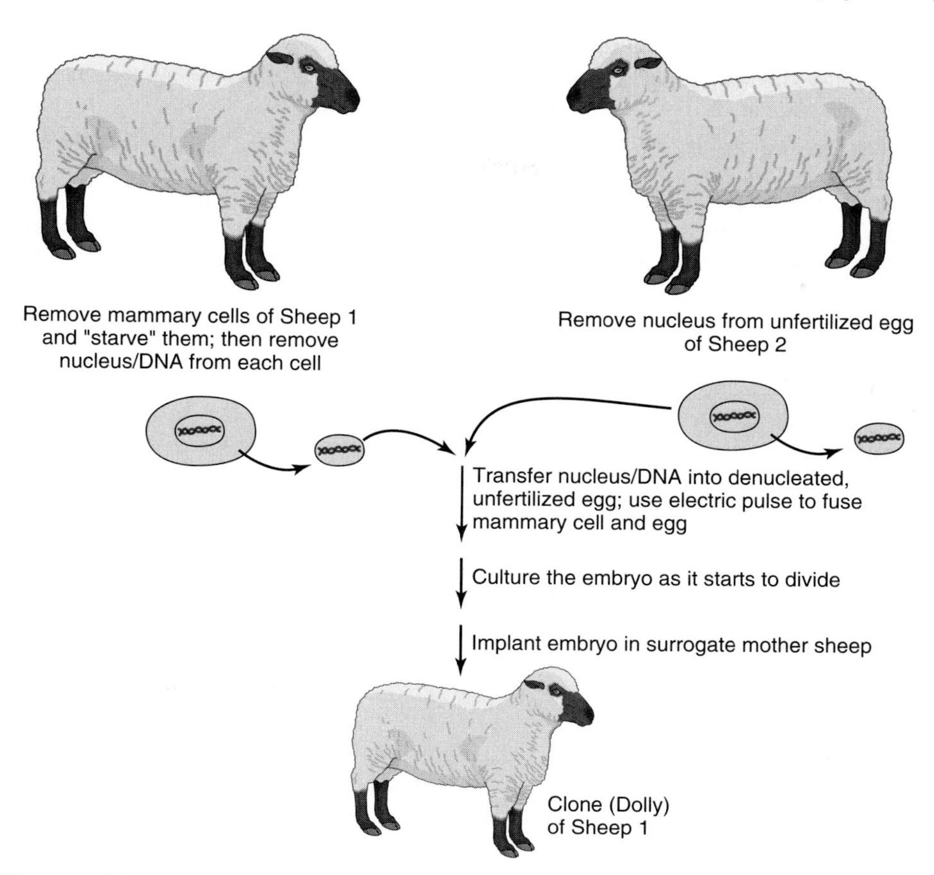

Figure 12.16

The cloning of Dolly from a mature animal. The mammary cells' nuclei were removed using a very thin, hollow needle and then fused, one into each egg, with an electric pulse.

is that this procedure might have increased the likelihood that once implanted into an egg, the chromosomes could be reprogrammed to establish the growth of an entire organism. Since Dolly, other researchers have used nuclear transfer with adult tissue to clone mice and cows, thus demonstrating the applicability of the technique to other mammals.

- Dolly's cells contain the DNA of only one parent, unlike identical twins who share DNA from both parents.

It should be kept in mind that only the nucleus of the adult ewe's donor cell was transferred to the egg. Thus, Dolly shares most, *but not all*, her genes with her nuclear donor. Several dozen genes are outside the nucleus in her cell's mitochondria. These genes came from the recipient egg, not the donor nucleus.

- Mitochondria in cells, but not within cell nuclei, are vital to energy production.

The goal of Wilmut's work, supported by PPL Therapeutics, is to develop a cloning technique to produce animals on a large scale that can be used to create pharmaceuticals in their milk. The target drugs are ones that are difficult and expensive to obtain by conventional methods. PPL Therapeutics has created animals that provide a protein in commercially useful quantities that could possibly be used to treat cystic fibrosis. The protein inhibits the enzyme elastase that breaks down connective tissue and is found in excess in the lungs of patients with the disease. However, improvements will need to be made to increase the yield from the cloning attempts before the overall process becomes a viable source of such molecules.

12.11 The Human Genome Project

The **Human Genome Project** is the effort to map all the genes in the human organism. After more than a decade of research, on June 26, 2000, scientists announced that a rough draft of the project to decode the genetic makeup of humans had been completed.

The goal, to determine the sequence of all three billion base pairs in the entire genome, was completed for the approximately 30,000 genes found on the 46 human chromosomes. The Human Genome Project was supported by the U.S. National Institutes of Health (NIH), the Wellcome Trust (a philanthropic organization based in London) and by Celera Genomics, a private company in Maryland. The completion of the goal was announced in a joint statement by NIH and Celera representatives. Upon the completion of this first phase, this massive enterprise was described as ". . . the most important, wondrous map ever produced by humankind . . ." by then-President Bill Clinton and as ". . . the outstanding achievement not only of our lifetime but perhaps in the history of mankind . . ." by Dr. Michael Dexter of the Wellcome Trust. It is anticipated that it will take several years to discover the traits the genome conveys.

The Human Genome Project began in 1989, with James Watson of DNA fame as its first director. At an estimated $3 billion, the project cost about one dollar per base pair. Instrumental in accomplishing so much so quickly was the high level of cooperation among international teams of researchers, and the competition to be the first to complete the project between publically funded efforts (NIH) and corporate research (Celera).

Often the reason why scientists undertake a research project is not unlike the reason why mountain climbers climb a mountain: "Because it's there." But the Human Genome Project involves more than the spirit of adventure or the thrill of discovery. The more we know about our genetic makeup, the more likely we will be to diagnose and cure disease, understand human development, trace our evolutionary roots, recreate our family tree, and utilize these discoveries for, one hopes, the benefit of our species and our fellow species.

You may well wonder whose DNA has been selected for this unprecedented scrutiny. In fact, most of the DNA being analyzed in the Human Genome Project happens to come from members of over 60 multigenerational French families whose lineage is well documented. But it does not really matter; any one of us could serve as a DNA donor and a representative of *Homo sapiens*. In spite of our apparent differences and our long history of disputes based on those differences, the DNA of all humans is remarkably similar. It differs from individual to individual by about 0.1% of the base sequences. Within that tiny fraction resides our genetic uniqueness. Biology and chemistry provide irrefutable evidence of a lesson we as nations and as individuals have been slow to learn, a lesson stated by former President Clinton: "The most important fact of life on this Earth is our common humanity."

Determining the sequence of the DNA base pairs was difficult and time consuming. The smallest human chromosome contains 50,000,000 base pairs. Given those numbers, researchers had to develop automated base sequencers that were both fast and accurate. The accuracy is very important, because a single missed base will throw off all the subsequent base assignments, just as a skipped buttonhole is transmitted down the length of a shirt. The target error rate is one in 10,000 bases, a 99.99% level of accuracy. Celera Genomics and NIH researchers used different methods. The public release of the data in February 2001 enabled some comparisons between the two groups of data.

Groups of scientists across the world argued for free access to all of the information as it was discovered. Unfortunately, the spirit of cooperation was already strained as the project progressed. Celera Genomics funded its efforts by patenting genes it identified and selling access to some of its data. The first patent application created a controversy that involved both scientists and the broader public. Because NIH funded, supervised, and completed much of the research, NIH applied for some of the first patents. Spokespersons for NIH argued that by securing patents on some of the genes, they would be protecting the public funds invested in the project. Critics responded that because government sources provided most of the funds, the results should belong to the public. Others reasoned that free and open exchange of information is essential in such a complex international project, and that patents would inhibit this communication. James Watson called the idea of patenting genes "sheer lunacy." In 1992, the U.S. Patent Office ruled that gene fragments could not be patented unless they had some known function. The patent problem remains unresolved.

We have already mentioned some of the potential benefits of the Human Genome Project, but the enterprise is not without its critics. There are those who point out that a genetic map ("being caught with your genes down") represents the ultimate invasion of privacy. Information about an individual's genetic makeup might be used by insurance companies to discriminate against those with a hereditary tendency toward certain diseases, by businesses to refuse to hire people who may be genetically at risk, or by ruthless governments to identify the "genetically inferior." Indeed, shortly before the June 2000 human genome completion announcements, a CNN-Time magazine poll reported that 46% of the respondents said they expected harmful results from the endeavor while 40% expected benefits; 41% thought that the project is morally wrong compared to 47% who disagreed with that position. Nevertheless, a majority of respondents, 61%, reported that they would like to know if they were predisposed to developing a genetic disease, compared with 35% who would not want to know.

Research continues to refine the initial data and on producing the genetic map for other species. Comparison of such data across species is likely to generate information about very fundamental and common biochemical functions in living organisms. The even longer task remains—turning the base pair sequences into meaningful and useful information regarding the specific details of their cellular functions.

12.21 Consider This: The Human Genome Project

In 2000, the Human Genome Project completed the sequence for all of the base pairs at least to the level of a "rough draft." What are likely to be some of the problems with the early version? Are there any ways to know if there were mistakes? In 2001, scientists completed the genetic maps of a weed, the thale cress plant, and the bacterium that causes cholera. Are any of these accomplishments useful to the sequence data for humans?

12.12 The New Prometheus(?)

Nature is an indispensable aid and ally in medicinal and biological chemistry. Much of our success has come from understanding and imitating natural processes. For centuries, animal breeders, agricultural researchers, and observant farmers have brought about genetic transformations in animals and plants by selective breeding. Recent applications of chemical methods to biological systems have greatly increased our capacity to effect such changes. Molecular engineering has made it possible to create nucleic acids, proteins, enzymes, hormones, drugs, and other biologically important molecules that do not exist in nature.

The new molecules can be designed to be more efficient catalysts than their naturally occurring counterparts, more effective and less toxic drugs for treating a wide range of diseases, or modified hormones that actually work better than the original. It is not at all fanciful to imagine a whole range of enzymes, engineered to consume environmentally hazardous wastes that are impervious to naturally occurring enzymes. Or creating an enzyme that far surpasses the one that catalyzes photosynthesis, one of the most inefficient of all natural enzymes. As Mark Twain suggested: "Predictions are extremely difficult, especially those about the future." Yet, it is likely that because of genetic engineering, AIDS and at least some forms of cancer may some day become as infrequent as polio or smallpox are now. Even more tantalizing is the possibility of eradicating certain genetic defects. Our growing knowledge of the human genome, coupled with our understanding of the chemistry of genetics, holds the promise of altering our inherited gene pool. Prospects include the elimination of sickle-cell anemia, diabetes, hemophilia, phenylketonuria, and dozens of other debilitating hereditary traits.

The next logical step would seem to be the creation of new organisms. Scientists have already cloned "new and improved" animals and vegetables. Mammals have been cloned from adult cells. In 1999, South Korean researchers claimed to have cloned a four-cell human embryo that was an identical genetic copy of a 30-year-old woman. In 1998, a Ph.D. physicist turned biologist, announced plans to launch a human cloning clinic in Chicago. In 2001, the U.S. House of Representatives held hearings on human cloning and heard from a former university professor about his plans to clone a human in one or two years, and from the leader of a sect that holds genetic engineering as part of its beliefs. The skepticism of the congressional subcommittee matched pronouncements in spring 2001 from England, Italy, and Germany to regulate, control, or ban human cloning activities. Twelve countries currently ban cloning of humans.

12.22 Consider This: The "Frankenfood" Frenzy

The European Union is pushing for stricter controls over the importation of genetically modified foods. Opposition to GM foods was a major theme of American protesters at the 2000 World Trade Organization meeting in Seattle and IMF/World Bank meetings in Washington, DC.

Recent tactics by transgenic food opponents have included a class-action suit against Monsanto, a producer of seeds. The Food and Drug Administration announced a new initiative to engage the public about bioengineered products. Search for and describe articles about recent developments in the "Frakenfood" controversy. Try to establish the bases for concerns by critics, and any evidence refuting them.

Human cloning has been dealt with in fiction such as Huxley's *Brave New World*—using eugenics to create different classes of people, and in films like Woody Allen's *Sleeper*—attempts to clone a dead dictator, and *The Boys From Brazil*—cloning Nazis. How should we view the possibility of cloning human beings, perhaps even "new and improved" ones? The question goes straight to our identity as a species and as individuals. With the possible exception of the South Korean case, cloning has been done only with somatic (non-reproductive) cells; ova and sperm (germ cells) have not been used in cloning or even in cases of gene therapy. This means that although the genetic makeup of the individual may be slightly altered by the gene therapy, her or his reproductive cells (germ cells) are not changed. Any therapeutic effect benefits only the individual and not his or her descendants. One could reason that the best way to treat a genetic disease or disability is to remove the defective gene from the gene pool by intentionally altering the suspect DNA in the sperm or ova. But one could also argue that *Homo sapiens* is not yet sufficiently wise to assume such god-like power. Our species has had some tragic experiences in the past century with political leaders who used less subtle methods of "genetic cleansing."

12.23 Consider This: A Cloning Clinic?

Plans to develop human cloning clinics have been announced in the last several years. What risks and benefits are associated with such clinics? Has such a clinic been completed at this time? Speculate on why or why not.

Cloning through genetic engineering raises not only the possibility for humans to duplicate themselves by unconventional means but also the chilling potential to design a master race or to subjugate or eliminate "defectives" through genetic manipulation. Hence, there is an intentional irony in the title of this final section. Prometheus was the demigod who stole fire and the flame of learning from the gods and brought these incomparable gifts to humanity. "The New Prometheus" is the subtitle of *Frankenstein*, Mary Shelley's classic study of scientific knowledge run amok. Using our ever-growing knowledge of the chemistry of heredity wisely and well will surely be one of the greatest challenges of the 21st century. As Ian Wilmut carefully points out about his discovery: "We are aware that there is potential for misuse, and we have provided information to ethicists and the Human Embryology Authority [of Britain]. We believe that it is important that society decides how to use this technology and make sure it prohibits what it wants to prohibit. It would be desperately sad if people started using this sort of technology with people" (*New York Times*, February 23, 1997, p. 22).

12.24 Consider This: Send in the Clones

The late Isaac Asimov, noted science fiction writer and biochemist, co-authored this verse, called "The Misunderstood Clone."

Oh, give me a clone
Of my own flesh and bone
With its Y chromosome changed to an X.
And when it has grown
Then my own little clone
Will be of the opposite sex.

Evaluate whether the verse is actually describing gene therapy or cloning. Support your answer with an explanation of the two different processes.

12.25 Consider This: Cloning Humans: For Good or Evil?

James D. Watson of DNA fame had this to say about manipulating germ cells to create superpersons, "When they are finally attempted, germ-line genetic manipulations will probably be done to change a death sentence into a life verdict—by creating children who are resistant to a deadly virus, for example, much the same way we can already protect plants from viruses by inserting antiviral DNA segments into their genomes."

From *Time*, "All for the Good," January 11, 1999, Vol. 153, No. 1.

Draft a statement supporting or opposing Watson's position on this issue. Your argument should be grounded in the scientific principles learned in this chapter.

CONCLUSION

The last chapter of this book, like almost all those that preceded it, ends with a dilemma: How can we balance the great potential benefits of modern chemical sciences and technology and the risks that seem inevitably to be part of the Faustian bargain that brought us knowledge? Throughout this text, the authors have occasionally looked, with myopic professorial vision, into the cloudy crystal ball of the future. It is in the nature of science that we cannot confidently predict what new discoveries will be made by tomorrow's chemists. Nor can we know the applications of those discoveries, good or bad. Such uncertainty is one of the delights of our discipline. A chemist must learn to live with ambiguity, indeed, to thrive on it, in the search to better understand the nature of atoms and their intricate combinations, in all their various guises.

But all citizens of this planet must at least develop a tolerance for ambiguity and a willingness to take reasonable risks, especially considering that life itself is a biological, intellectual, and emotional risk. Of course, we all seek to maximize benefits, but we must recognize that individual gain must sometimes be sacrificed for the benefit of society. We live in multiple contexts—the context of our families and friends, our towns and cities, our states, our countries, our special planet. We have responsibilities to all. *You*, the readers of this book, will help create the context of the future. We wish you well.

Chapter Summary

Having studied this chapter, you should be able to:

- Understand the chemical composition of deoxyribonucleic acid (DNA), a polymer of nitrogen-containing bases, deoxyribose, and phosphate groups (12.1)
- Recognize the utility of Chargaff's rules: A = T, G = C (12.1)
- Interpret evidence for the double helical structure of DNA and its base pairing (12.2)
- Understand DNA replication (12.2)
- Describe the genetic code—codons of three bases corresponding to specific amino acids (12.3)
- Relate to the amount of information encoded in the human genome (12.3)
- Discuss the primary, secondary, and tertiary structure of proteins (12.4)

- Describe the molecular basis of sickle-cell anemia (12.4)
- Understand the production of human proteins (insulin and human growth hormone) in bacteria (12.5)
- Recognize the need to establish priorities for the use of limited supplies of drugs produced by recombinant DNA technology (12.5)
- Recognize recombinant DNA techniques—insertion of foreign DNA into bacteria (12.5)
- Relate how new drugs to treat hepatitis, herpes, cancer, heart attack, strokes, AIDS, etc., and new vaccines are developed via genetic engineering (12.6)
- Understand the genetic diagnoses of ALS, Alzheimer's disease, and other conditions (12.7)
- Understand the connection between DNA probes and the polymerase chain reaction and the factors related to appropriate responses to genetically diagnosed hereditary diseases (12.7)
- Describe gene therapy as a treatment for severe combined immunodeficiency disease, malignant melanoma, and other diseases (12.7)
- Discuss DNA fingerprinting for identification and evolutionary and anthropological studies, and the technical and legal issues associated with DNA fingerprinting (12.8)
- Refute the possibility and wisdom of cloning long-extinct animals (12.8)
- Describe the laboratory techniques in DNA fingerprinting (12.8)
- Describe transgenic organisms and their uses: pest-resistant and nitrogen-fixing plants; animals with genes to produce human proteins and other modified biochemicals (12.9)
- Discuss ethical issues associated with transgenic organisms and the fears about "Frankenfood" (12.9)
- Understand mammalian cloning using nuclear transfer (12.10)
- Describe the role of nuclear transfer in the cloning of mammalian cells (12.10)
- Relate to the Human Genome Project: its aims, significance, and ethical implications (12.11)
- Debate issues associated with the prudent and ethical applications of mammalian cloning and genetic engineering (12.12)

Questions

Emphasizing Essentials

1. The letters DNA have been called three of the most important letters of the late 20th century. What do they literally stand for?
2. a. Use Figure 12.1 to determine the percent increase in human life expectancy at birth in the United States between 1940 and 1980 for males and for females.
   b. Are these percentages the same as those calculated in 12.1 Your Turn? Why or why not?
3. Consider the structures in Figure 12.2.
   a. What functional groups are in adenine?
   b. What functional groups are in deoxyribose?
   c. What functional groups are in the phosphate group?
4. Consider the structural formula of deoxyribose given in Figure 12.2.
   a. Why is deoxyribose classified as a monosaccharide?
   b. What is the molecular formula for deoxyribose?
   c. Why isn't deoxyribose an acid in aqueous solution?
5. Equation 12.1 shows the general case for a nitrogen-containing base reacting with water. Use the structural formula of thymine in Figure 12.2 to write an equation showing how thymine could react with water to generate hydroxide ions.
6. a. What three types of units must be present in a nucleotide?
   b. What type of bonding holds these units together in a nucleotide?
7. Table 12.1 lists the base composition of DNA for various species. The four bases are adenine, cytosine, guanine, and thymine. What relationships exist among these bases, no matter what the species?
8. a. What happens experimentally during X-ray diffraction?
   b. The first X-ray diffraction patterns were of simple salts, such as sodium chloride. X-ray diffraction studies of nucleic acids and proteins did not come until much later. Suggest why.
9. Figure 12.6 shows the pairing of nucleotide bases in DNA.
   a. What type of *intramolecular* bonding occurs within each base?
   b. What type of *intermolecular* bonding holds the base pairs together?
10. Identify the base sequence that is complementary to each of these sequences.
    a. ATGGCAT   b. TATCTAG
11. Given that the distance between adjacent bases is 0.34 nm, how many base pairs are present in a chromosome that is 3.0 cm long? 1 m = 10^2 cm; 1 m = 10^9 nm.

12. During cell division, as many as 90,000 nucleotides per minute can be added to the growing DNA chain.

a. The shortest human chromosome contains 50 million bases. What is the minimum time required to form a strand of this chromosome?

b. Determine the length of this chromosome (in cm) that would be formed in one minute if the distance between bases is 0.34 nm. 1 m = 10^2 cm; 1 m = 10^9 nm.

13. Twenty amino acids commonly occur in proteins.

a. What is the *general* structural formula for all of these amino acids?

b. What functional groups are present in all amino acids?

14. The text states that if you were to use the letters A, T, C, and G in a game of Scrabble, you could generate 64 different three-letter combinations. (Your Scrabble opponent would surely challenge some of these combinations!) Nature pairs the four bases A, T, C, and G, and uses the pairs to encode for amino acids. Write down all the possible paired combinations of A, T, C, and G to find the maximum number of amino acids that can be encoded from these four bases. *Hint:* In nature, unlike Scrabble, a letter can be used more than once in forming a pair.

15. What is a codon and what is its role in the genetic code?

16. Only 61 of the possible 64 triplet codons specify certain amino acids. What is the function of the other three codons?

17. Describe what is meant by the primary, secondary, and tertiary structure of a protein. Is one of these more important than the others? Why or why not?

18. Explain how an error in the primary structure of a protein in hemoglobin causes sickle-cell anemia.

19. How can "recombinant DNA" technology be used to overcome a shortage of insulin for use by diabetics?

20. a. What basic molecular building blocks are found in viruses?

b. Why are viruses referred to as ". . . almost living . . ." biochemical species?

c. Why is it not possible to treat a viral infection with antibiotics?

21. The letters DNA have been called three of the most important letters of the late 20th century. A long-shot candidate for this honor might be PCR.

a. What do those letters stand for?

b. What problem did PCR technology successfully address?

c. Name some examples of the success of PCR technology for diagnosis.

22. What is the minimum number of PCR cycles that would be needed to convert two DNA molecules to

a. 5000 DNA molecules?

b. 50,000 molecules?

c. 500,000 molecules?

23. How does electrophoresis separate different molecules?

24. What is meant by the term "molecular pharming"?

25. a. What is the human genome project?

b. Why is it a significant step in understanding the genetic basis of humans?

c. What progress has been made in mapping the human genome?

Concentrating on Concepts

26. What is meant by the term "cloning"?

27. Life expectancy in the United States increased dramatically during the 20th century, as shown in Figure 12.1. Is a similar increase possible during the 21st century? Explain your answer.

28. Compare this representation with Figure 12.4.

```
            Base 1                 Base 2                 Base 3
              |                      |                      |
—Phosphate — Sugar — Phosphate — Sugar — Phosphate — Sugar —
```

Both show a segment of deoxyribonucleic acid. Discuss the strengths and weakness of each representation in conveying similar information.

29. Consider Chargaff's discovery that there are equal percentages of adenine and thymine and of cytosine and guanine in DNA. Was his discovery as important to understanding the nature of DNA as Crick and Watson's discovery of the double helix? Give reasons for your answer.

30. Use Figure 12.6 to help explain why stable base pairing does *not* occur between adenosine and cytosine, thymine and guanine, adenine and guanine, and thymine and cytosine.

31. Errors sometimes occur in the base sequence of a strand of DNA. But, not all of these errors result in the incorporation of an incorrect amino acid in a protein for which the DNA codes. Explain how this happens, and why it is advantageous.

32. Human insulin and human growth hormone have both been made through the use of recombinant DNA technology. Which do you believe is a more significant use of this technology and why? Discuss what factors have influenced your opinion.

33. DNA probes are being used for diagnosis. Describe the principle of using DNA probes to identify *Helicobacter pylori*, the cause of gastric ulcers and possibly stomach cancer.

34. a. The first successful application of gene therapy to a human being was in 1990. What is meant by "gene therapy" and what disease was treated in that landmark case?

b. Has gene therapy been used successfully for treating other diseases as well?

c. Will it work for all diseases?

35. Lou Gehrig's disease is caused by the alteration of a single amino acid in the enzyme superoxide dismutase. How many base pairs are responsible for specifying this amino acid in the gene that codes for the protein? What is the minimum number of base pairs that would have to be changed to produce the disease?

36. Do you favor the patenting of genes? What are the advantages and disadvantages of this approach?

37. How widely available is the New Leaf Potato? If you wanted to plant this in your garden, would you be able to obtain these potato plants? What are the reasons why you would want to obtain this plant?

38. Consider the information in Table 12.3. Use a pictorial representation of this information to convey the differences among the countries in using transgenic crops.

39. Consider the idea of mixing genes as an improvement on nature.

a. What are transgenic organisms?

b. Why is the alteration of the genetic makeup of plants by genetic engineering preferred to traditional crossbreeding methods?

40. Consider some of the successful transgenic plant experiments given in Table 12.2. What generalizations can be drawn between the source of the genes, the transgenic plant, and the objectives of the experiment?

Exploring Extensions

41. Dolly's birth surprised genetic engineering experts and shocked members of the media responsible for reporting the birth and the method. Locate one or two different early reports about the cloning of Dolly. Evaluate these reports for their scientific accuracy and what they reveal about the opinion of experts at that time.

42. Use this information to act as a Sceptical Chymist in checking the correctness of the claim that the DNA in an adult human would stretch from the Earth to the moon and back more than a million times.

Useful Information	
Distance, Earth to moon	3.8×10^5 km
Adult human	1×10^{13} DNA-nucleated cells
Length, one stretched human DNA thread	2 m

43. Consider the structural formula of deoxyribose shown in Figure 12.2. The prefix "deoxy" means without oxygen; the —OH group is replaced by a hydrogen atom. In the specific case of deoxyribose, the —OH group replaced by a hydrogen was bonded to the only carbon in the ring that bonds to two hydrogens in deoxyribose. Use this information to draw the structural formula of ribose side by side with the structural formula of deoxyribose.

44. The text states that the more closely species are related, the more similar the DNA base compositions are in those species. The Sceptical Chymist is having trouble believing this, particularly if it means a close relationship between a fruit fly and a bacterium. Use the information in Table 12.1 to determine whether it supports the generalization about similar base pairs in similar species.

45. Perhaps you have learned some memory aids when taking music lessons (Every Good Boy Does Fine), memorizing the names of the Great Lakes (HOMES), or learning about oxidation and reduction (OIL RIG). One of the authors learned "All-Together, Go-California" as the mnemonic to remember the correct base pairings in DNA.

a. What is the relationship in this mnemonic to DNA base pairing?

b. Design a different memory aid that will help you remember such base pairings.

46. Of the major players in the discovery of DNA's structure, only Rosalind Franklin had a degree in chemistry. What was her background and experience that enabled her to make significant contributions? Did her contributions receive adequate credit and recognition? Write a short report on the results of your findings, giving references to Web information or other citations.

47. Classify the objectives of the experiments in Table 12.2 into different categories and rate the relative importance of each of these categories.

48. Transgenic plants have not been widely accepted in all countries. What are some of the reasons for their rejection in some European markets?

49. Genetic diseases are also called inborn errors of metabolism. You may be familiar with some of these diseases, such as hemophilia, PKU, Tay-Sachs, or sickle-cell anemia. One that does not get much attention is a condition known as Niemann-Pick disease. Find out what inborn metabolic error causes this condition, how many children are born with this disease in the United States each year, what treatments are available, and whether a cure is possible. Write a report to be discussed with your classmates.

50. Remains of a soldier shot down over Vietnam in 1972 and buried in the Tomb of the Unknown in 1984 were identified in 1998 by the Armed Forces DNA Identification Laboratory (see Section 12.8). The confirmatory tests were based on matching mitochondrial DNA (mtDNA) from the soldier's bones with mtDNA from his mother.

a. Why were these tests not done before the remains were placed in the Tomb?

b. Find more information about the specificities of test results when using nuclear DNA rather than mtDNA.

c. As of this writing, scientists at the National Institute of Standards and Technology are preparing an mtDNA standard for forensic laboratories to measure the accuracy of their results. Has that standard been issued?

APPENDIX

1

Measure for Measure

Conversion Factors and Constants

Metric Prefixes

deci (d)	$1/10 = 10^{-1}$	deka (da)	$10 = 10^{1}$
centi (c)	$1/100 = 10^{-2}$	hecto (h)	$100 = 10^{2}$
milli (m)	$1/1000 = 10^{-3}$	kilo (k)	$1000 = 10^{3}$
micro (μ)	$1/10^{6} = 10^{-6}$	mega (M)	$= 10^{6}$
nano (n)	$1/10^{9} = 10^{-9}$	giga (G)	$= 10^{9}$

Length

1 centimeter (cm) = 0.394 inch (in.)
1 meter (m) = 39.4 in. = 3.28 feet (ft) = 1.08 yard (yd)
1 kilometer (km) = 0.621 miles (mi)
1 in. = 2.54 cm = 0.0833 ft
1 ft = 30.5 cm = 0.305 m = 12 in.
1 yd = 91.44 cm = 0.9144 m = 3 ft = 36 in.
1 mi = 1.61 km

Volume

1 cubic centimeter (cm^3) = 1 milliliter (mL)
1 liter (L) = 1000 mL = 1000 cm^3 = 1.057 quarts (qt)
1 qt = 0.946 L
1 gallon (gal) = 4 qt = 3.78 L

Mass

1 gram (g) = 0.0352 ounce (oz) = 0.00220 pound (lb)
1 kilogram (kg) = 1000 g = 2.20 lb
1 metric ton (mt) = 1000 kg = 2200 lb = 1.10 ton (t)
1 lb = 454 g = 0.454 kg
1 ton (t) = 2000 lb = 909 kg = 0.909 mt

Time

1 year (yr) = 365.24 days (d)	1 day = 24 hours (hr or h)
1 hr = 60 minutes (min)	1 min = 60 seconds (s)

Energy

1 joule (J) = 0.239 calorie (cal)
1 cal = 4.184 joule (J)
1 kilocalorie (kcal) = 1 dietary Calorie (Cal)
= 4184 J = 4.184 kilojoule (kJ)
1 kilowatt-hour (kWh) = 3,600,000 J = 3.60×10^{6} J

Constants

Speed of light (c) = 3.00×10^{8} m/s
Planck's constant (h) = 6.63×10^{-34} J·s
Avogadro's number (N_A) = 6.02×10^{23} objects per mole
Atomic mass unit (m) = 1.66×10^{-24} g

APPENDIX

2

The Power of Exponents

Scientific or exponential notation provides a compact and convenient way of writing very large and very small numbers. The idea is to use positive and negative powers of 10. Positive exponents are used to represent large numbers. The exponent, written as a superscript, indicates how many times 10 is multiplied by itself. For example,

$$10^1 = 10$$

$$10^2 = 10 \times 10 = 100$$

$$10^3 = 10 \times 10 \times 10 = 1000$$

Note that the positive exponent is equal to the number of zeros between the 1 and the decimal point. Thus, 10^6 corresponds to 1 followed by six zeros or 1,000,000. This same rule applies to 10^0, which equals 1. One billion, 1,000,000,000, can be written as 10^9.

When 10 is raised to a negative exponent, the number is always less than 1. This is because a negative exponent implies a reciprocal, that is, 1 over 10 raised to the corresponding positive exponent. For example,

$$10^{-1} = 1/10^1 = 1/10 = 0.1$$

$$10^{-2} = 1/10^2 = 1/100 = 0.01$$

$$10^{-3} = 1/10^3 = 1/1000 = 0.001$$

It follows that the larger the negative exponent, the smaller the number. The negative exponent is always one more than the number of zeros between the decimal point and the 1. Thus, 1×10^{-4} is equal to 0.0001. Conversely, 0.000001 in scientific notation is 1×10^{-6}.

Of course, most of the quantities and constants used in chemistry are not simple whole number powers of 10. For example, Avogadro's number is 6.02×10^{23} or 6.02 multiplied by a number equal to 1 followed by 23 zeros. Written out, this corresponds to $6.02 \times$ 100,000,000,000,000,000,000,000 or 602,000,000,000,000,000,000,000. Switching to very small numbers, a wavelength at which carbon dioxide absorbs infrared radiation is 4.257×10^{-6} m. This number is the same as 4.257×0.000001 or 0.000004257.

Your Turn

Express these numbers in scientific notation.

a. 10,000 **b.** 430 **c.** 9876.54
d. 0.000001 **e.** 0.007 **f.** 0.05339

Express these numbers in conventional decimal notation.

a. 1×10^6 **b.** 3.123×10^6 **c.** 25×10^5
d. 1×10^{-5} **e.** 6.023×10^{-7} **f.** 1.723×10^{-16}

APPENDIX

3

Clearing the Logjam

You may have encountered logarithms in mathematics courses but wondered if you would ever use them. In fact, logarithms (or "logs" for short) are extremely useful in many areas of science. The essential idea is that they make it much easier to deal with very large *ranges* of numbers, for example, moving by powers of 10 from 0.0001 to 1,000,000.

It is certainly likely that you have met logarithmic scales without knowing it. The Richter scale for magnitudes of earthquakes is one example. On this scale, an earthquake of magnitude 6 is 10 times more powerful than one of magnitude 5. An earthquake of magnitude 8 would be 100 times more powerful than one of magnitude 6. Another example is the decibel scale. Each increase of 10 units represents a tenfold increase in sound level. Therefore, a normal conversation at 1 meter (dB 60) is 10 times louder than quiet music (dB 50). Loud music (dB 70) and extremely loud music (dB 80) are 10 times and 100 times as loud, respectively, as a normal conversation.

A simple exercise using a pocket calculator can be a good way to learn about logs. You will need a calculator that "does" logs and preferably has a "scientific notation" option. Start by finding the logarithm of 10. Simply enter 10 and press the "log" button. The answer should be 1. Next find the log of 100 and then the log of 1000. Write down the answers. What pattern do you see? (The pattern may be more obvious if you recall that 100 can be written as 10^2 and 1000 is the same as 10^3.) Predict the log of 10,000 and then check it out. Then try the log of 0.1 or 10^{-1} and log of 0.01 (10^{-2}). Predict the log of 0.0001 and check it out.

So far so good, but we have only been considering whole-number powers of 10. It would be helpful to be able to obtain the logarithm of any number. Once again, your handy little electronic wizard comes to the rescue. Try calculating the logs of 20 and 200, then 50 (5×10^1) and 500 (5×10^2). Predict the log of 5×10^3 or 5000. Now for something slightly more tricky: the log of 0.05. Finally, try the log of 2473 and the log of 0.000404. In each of the three cases, does the answer seem to be in the right ballpark? If you see any other interesting relationships, you may want to experiment further.

In Chapter 6, the concept of pH is introduced as a quantitative way to describe the acidity of a substance. A pH value is simply a special case of a logarithmic relationship. It is defined as the negative of the logarithm of the H^+ concentration, expressed as molarity. Expressing this relationship as a mathematical equation, $pH = -\log (M_{H^+})$. The negative sign indicates an inverse relationship; as the H^+ concentration diminishes, the pH increases. Let us apply the equation by using it to calculate the pH of a beverage with a hydrogen ion concentration, M_{H^+}, of 0.000546 mole/liter. We first set up the mathematical equation and substitute the hydrogen ion concentration into it.

$$pH = -\log (M_{H^+}) = -\log (5.46 \times 10^{-4} \text{ M})$$

Next, we take the negative logarithm of the H^+ concentration by entering it into a calculator and pressing the log button, then the "minus" key. This gives 3.26 as the pH of the beverage. Apply the pH relationship to calculate the pH of milk with a hydrogen ion concentration of 2.20×10^{-7} M.

If we can convert hydrogen ion concentration into pH, how do we go in the reverse direction, that is, how to convert pH into a hydrogen ion concentration? Your calculator can do this for you if it has a button labelled "10^x." (Alternatively, it may use two buttons: first "Inv" and then "log.") To demonstrate the procedure, suppose you wish to find the hydrogen ion concentration of human blood with a pH of 7.40. Proceed as follows: Enter 7.40 and hit 10^x (or follow whatever steps are appropriate for your calculator). The display should give the hydrogen ion concentration as 3.98×10^{-8} M. Now apply the same procedure to calculate the H^+ concentration of an acid rain sample with a pH of 3.6.

Test yourself further by determining the H^+ concentration of tomato juice with a pH of 4.8; the pH of milk of magnesia with an H^+ concentration of 3.16×10^{-11} M.

APPENDIX 4

Answers to Your Turn Questions

Not Answered in Text

Chapter 1

1.5 From the table we can see that the average for CO (9 ppm) is approximately 100 times greater than that for O_3 (0.12 ppm), while that for SO_2 (0.030) is 4 times less than that for O_3.

1.8 $\frac{1050\ \mu g}{15\ m^3} = 70\ \mu g/m^3$. It does not exceed the standard 80 $\mu g/m^3$.

1.13 **d.** element **e.** compound **f.** mixture

1.15 **c.** hydrogen, oxygen **d.** carbon, hydrogen, oxygen

1.17 **a.** magnesium bromide **d.** sodium sulfide

1.18 **c.** calcium sulfide **d.** lithium nitride

1.20 **b.** A molecule of the compound represented by the formula SO_2 consists of one atom of the element sulfur combined with two atoms of the element oxygen.
c. A molecule of the compound represented by the formula N_2O_4 consists of two atoms of the element nitrogen combined with four atoms of the element oxygen.
d. A molecule represented by the formula O_3 consists of three atoms of the element oxygen.

1.21 **b.** sulfur dioxide **d.** trioxygen (commonly called ozone)

1.22 **b.** Balanced equation: $N_2 + 2\ O_2 \rightarrow 2\ NO_2$

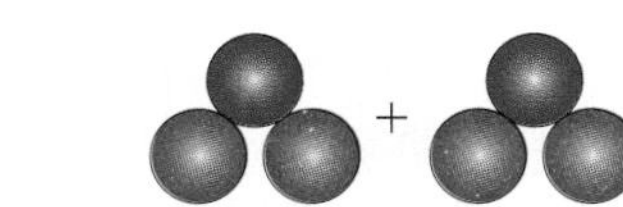

1.25 **b.** $2\ C_4H_{10} + 13\ O_2 \rightarrow 8\ CO_2 + 10\ H_2O$

1.26 In equation 1.6 there are 16 C, 36 H, and 34 O on each side of the equation.

1.31 **a.** Mexico City **b.** Tokyo or London
c. Cairo (lead from leaded gasoline); Mexico City (CO and NO_2)

1.35 $\frac{2 \times 10^{17}\ \text{molecules}}{6 \times 10^{9}\ \text{people}} = \frac{3 \times 10^{7}\ \text{molecules}}{\text{person}} = \frac{30{,}000{,}000\ \text{molecules}}{\text{person}}$

Chapter 2

2.2 **c.** 20 protons, 20 electrons **d.** 24 protons, 24 electrons

2.3 **c.** 8 (Group 8A) **d.** 2 (Group 2A)

2.4 Oxygen, sulfur, selenium, and tellurium are all in Group 6A. Each atom has 6 outer electrons.

2.5 **c.** 29 protons, 29 electrons, 35 neutrons
d. 79 protons, 79 electrons, 119 neutrons

2.6 **b.** There are 7 outer electrons per atom of iodine, for a total of 14 outer electrons in I_2. This is the Lewis structure.

:Ï:Ï: (with lone pairs above and below each I)

2.7 **b.** In dichlorodifluoromethane, CCl_2F_2, the carbon atom has 4 outer electrons, each of the 2 chlorine atoms has 7 outer electrons, and each of the two fluorine atoms has 7 electrons. The total number of outer electrons is 32. This is the Lewis structure.

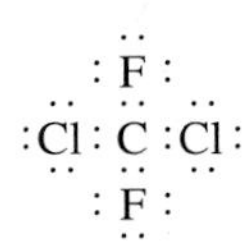

2.8 **b.** In sulfur dioxide, SO_2, the sulfur atom has 6 outer electrons and each of the two oxygen atoms has 6 outer electrons. The total number of electrons is 18. These are the Lewis structures.

:O::S:O: $\longleftrightarrow$:O:S::O:

2.10 **b.** Green light has a shorter wavelength and higher frequency than that of red light.

2.12 **b.** Microwave, infrared, visible, ultraviolet
c. They are different because wavelength and frequency are inversely related; as one increases, the other decreases.

2.22 No, because the concentration of water vapor in the stratosphere is too low to cause the steep decrease in ozone there.

2.28

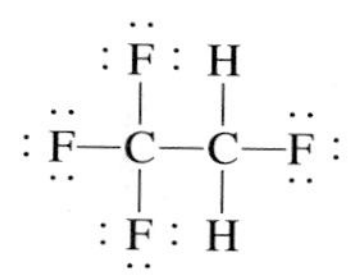

```
     :F: H
      |  |
  :F—C—C—F:
      |  |
     :F: H
```

Chapter 3

3.9 **b.** tetrahedral, F—C(Cl)(F)(Cl), 109.5° **c.** bent, H—S̈—H, <109.5°

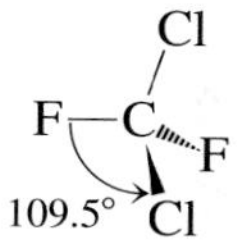

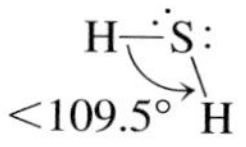

3.10 **a.** bent, :Ö—S̈=O:, 120°

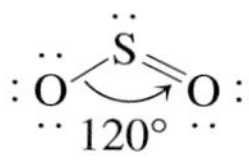

b. triangular planar, :O:‖S, :Ö—S—Ö:, 120°

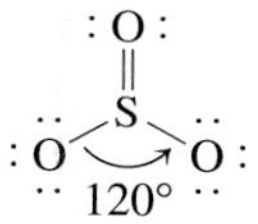

3.14 **a.** deep ocean **b.** 81% **c.** 11%

3.17 **a.** Au-197 has 79 protons, 118 neutrons, and 79 electrons.
b. Au-198 has 79 protons, 119 neutrons, and 79 electrons. Only the number of neutrons has changed.

3.18 Given that there are only two isotopes, the other isotope must have a smaller number of neutrons as the weighted average of the two isotopes is just under 7. The other isotope is Li-6, which makes up 7.5% of all lithium atoms.

3.19 **b.** 1×10^{-10} g **c.** 1×10^{-7} g

3.20 **b.** 30.0 g/mole NO **c.** 137.5 g/mole CCl_3F

3.21 **c.** 0.636 g N/1.00 g N_2O; 63.6% N

3.22 **b.** 71.1 million metric tons S released

3.24 **a.** Yes; 1.8×10^{-4} % = 1.8 ppm **b.** 0.040 ppm
c. percent in common use; trace constituents

Chapter 4

4.1 **c.** 390 kJ **d.** 3.2×10^3 blocks

4.8 The ultraviolet radiation absorbed by O_2, which has greater energy than the radiation absorbed by O_3, has a shorter wavelength. Energy is inversely proportional to wavelength.

4.14 **a.** 18% oxygen in MTBE **b.** 35% oxygen in ethanol

4.27 **b.** 1.8×10^3 lb CO_2 **c.** 31 lb SO_2

Chapter 5

5.6 **a.** soluble **b.** very soluble **c.** insoluble
d. soluble **e.** insoluble
f. partially soluble (pure aspirin)

5.7 **a.** 975 mg/day **b.** 25 500-mL bottles

5.8 **a.** 16 ppb; 1.6×10^{-2} ppm
b. No; the lead concentration is over the standard of 15 ppb

5.9 **a.** 7.5×10^{-1} mole of NaCl in 500 mL of 1.5 M NaCl and 7.5×10^{-2} mole of NaCl in 500 mL of 0.15 M NaCl
b. To calculate molarity, divide moles of solute by liters of solution: 0.5 moles of NaCl in 250 mL of solution is 2 M NaCl; 0.6 moles of NaCl in 200 mL of solution is 3 M NaCl, so the second solution is more concentrated.

5.10 **a.** H—F The electrons are more strongly attracted to the F atom.
b. O—H The electrons are more strongly attracted to the O atom.
c. N—O The electrons are more strongly attracted to the O atom.

5.14 **b.** Mg^{2+}, the Lewis structures are ·Mg· and Mg^{2+}
c. O^{-2}, the Lewis structures are ·Ö· and $[:\ddot{O}:]^{2-}$
d. Al^{3+}, the Lewis structures are ·Al· and Al^{3+}

5.15 **b.** KF **c.** Li_2O **d.** $SrBr_2$

5.16 **c.** sodium hydrogen carbonate or sodium bicarbonate
d. calcium carbonate
e. magnesium phosphate

5.17 **b.** Li_2CO_3 **c.** KNO_3 **d.** $BaSO_4$

5.19 **b.** soluble **c.** insoluble **d.** insoluble

5.20

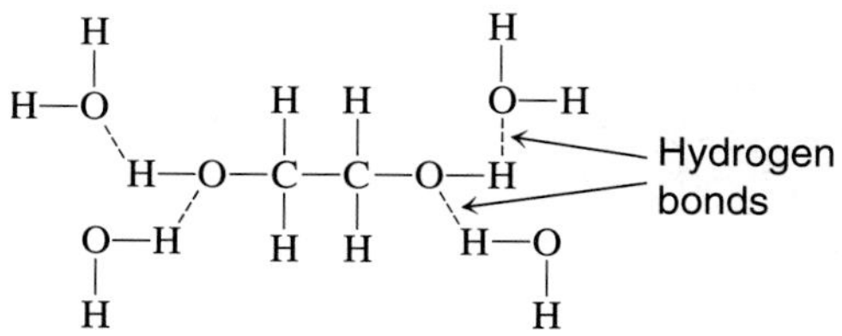

H—O—C(H)(H)—C(H)(H)—O—H with four surrounding water molecules (H—O—H) joined by hydrogen bonds

Hydrogen bonds

5.27 **a.** $Ca(HCO_3)_2$ **b.** $MgSO_4$ **c.** $MgCl_2$

5.30 **a.** 20 ppb = 20 μg/L; higher concentration than 0.003 mg/L = 3 μg/L = 3 ppb
b. 20 ppb > 15 ppb standard; 3 ppb < 15 ppb standard

5.32 **b.** 0.3 ppm
c. Reading is out of the range of the calibration curve; dilute the sample

Chapter 6

6.2 **b.** $HNO_3(aq) \longrightarrow H^+(aq) + NO_3^-(aq)$
c. $H_2SO_4(aq) \longrightarrow H^+(aq) + HSO_4^-(aq)$

6.4 **a.** $KOH(s) \longrightarrow K^+(aq) + OH^-(aq)$
b. $LiOH(s) \longrightarrow Li^+(aq) + OH^-(aq)$
c. $Ca(OH)_2(s) \longrightarrow Ca^{2+}(aq) + 2\ OH^-(aq)$

6.5 **b.** $H_2SO_4(aq) + 2\ NaOH(aq) \longrightarrow Na_2SO_4(aq) + 2\ H_2O(l)$
$2\ H^+(aq) + SO_4^{2-}(aq) + 2\ Na^+(aq) + 2\ OH^-(aq) \longrightarrow 2\ Na^+(aq) + SO_4^{2-}(aq) + 2\ H_2O(l)$
$2\ H^+(aq) + 2\ OH^-(aq) \longrightarrow 2\ H_2O(l)$
$H^+(aq) + OH^-(aq) \longrightarrow H_2O(l)$
c. $2\ H_3PO_4(aq) + 3\ Mg(OH)_2(aq) \longrightarrow Mg_3(PO_4)_2(aq) + 6\ H_2O(l)$
$6\ H^+(aq) + 2\ PO_4^{3-}(aq) + 3\ Mg^{2+}(aq) + 6\ OH^-(aq) \longrightarrow 3\ Mg^{2+}(aq) + 2\ PO_4^{3-}(aq) + 6\ H_2O(l)$
$6\ H^+(aq) + 6\ OH^-(aq) \longrightarrow 6\ H_2O(l)$
$H^+(aq) + OH^-(aq) \longrightarrow H_2O(l)$

6.6 **b.** basic, $M_{H^+} = 1 \times 10^{-8}$
c. basic, $M_{OH^-} = 1 \times 10^{-4}$

6.7 **a.** $OH^-(aq) = K^+(aq) > H^+(aq)$; basic
b. $H^+(aq) = NO_2^-(aq) > OH^-(aq)$; acidic
c. $H^+(aq) > SO_3^{2-}(aq) > OH^-(aq)$; acidic

6.8 **a.** The lake water with pH = 4 is 10 times more acidic and has 10 times more H^+ than the rain water with pH = 5.

6.12 **c.** 3.36×10^4 tons SO_2
d. The SO_2 gets carried by the wind into the surrounding areas. It eventually dissolves in the water droplets of clouds, fog, and rain to form acidic precipitation.

6.13 The NO_x figure shows that emissions from industrial processing and the miscellaneous category did not change, whereas fuel combustion (includes electric utility plants), on-road (motorized vehicles) emissions, and off-road (lawn and garden engines, marine engines) emissions did. The SO_2 figure shows that the largest increases in the 1970s came from fuel consumption and industrial processes such as metal smelting and refining.

6.14 $MgCO_3(s) + 2\ H^+(aq) \longrightarrow Mg^{2+}(aq) + CO_2(g) + H_2O(l)$

6.17 $4\ Fe(s) + 2\ O_2(g) + \cancel{8\ H^+(aq)} \longrightarrow \cancel{4\ Fe^{2+}(aq)} + \cancel{4\ H_2O(l)}$
$\cancel{4\ Fe^{2+}(aq)} + O_2(g) + \cancel{4\ H_2O(l)} \longrightarrow 2\ Fe_2O_3(s) + \cancel{8\ H^+(aq)}$
$4\ Fe(s) + 3\ O_2(g) \longrightarrow 2\ Fe_2O_3(s)$

6.19 **a.** $Ca(OH)_2(aq) + 2\ HCl(aq) \longrightarrow CaCl_2(aq) + 2\ H_2O(l)$
b. $CaO(s) + H_2O(l) \longrightarrow Ca(OH)_2(aq)$

6.20 The troposphere is in the region of the atmosphere in which we live. The stratosphere starts about 10 kilometers up. Ozone is a pollutant in the troposphere but protects us by absorbing UV radiation in the stratosphere. The pollutant gases that contribute to acid deposition are in the troposphere.

Chapter 7

7.4 There are 92 protons in all uranium atoms, including U-234. There are 234 − 92 = 142 neutrons.

7.5 **b.** ${}^{1}_{0}n + {}^{235}_{92}U \longrightarrow [{}^{236}_{92}U] \longrightarrow {}^{137}_{52}Te + {}^{97}_{40}Zr + 2\ {}^{1}_{0}n$

7.6 ${}^{1}_{0}n + {}^{235}_{92}U \longrightarrow [{}^{236}_{92}U] \longrightarrow {}^{143}_{54}Xe + {}^{90}_{38}Sr + 3\ {}^{1}_{0}n$

7.7 1.00×10^{14} J/day; 1.11 g

7.8 The cloud is condensed water vapor. People sometimes call this "steam", although technically this is not correct, as steam is water vapor, which is invisible until it condenses. The cloud does not contain any nuclear fission products.

7.13 ${}^{238}UF_6$ has a molar mass of 352 and ${}^{235}UF_6$ has a molar mass of 349. The difference in mass is less than 1%.

7.16 **b.** Daughter product is uranium-235.
${}^{239}_{94}Pu \longrightarrow {}^{235}_{92}U + {}^{4}_{2}He$
c. Daughter product is xenon-131.
${}^{131}_{53}I \longrightarrow {}^{131}_{54}Xe + {}^{0}_{-1}e$

7.19 Answers will vary with individuals based on their exposure to background radiation.

7.22 The iodine atom is $:\ddot{\underset{..}{I}}\cdot$ with 7 valence electrons.

The iodine molecule, I_2, is $:\ddot{\underset{..}{I}}:\ddot{\underset{..}{I}}:$, a molecule with a single bond between the iodine atoms. The iodide ion has 8 valence electrons, giving it a minus one charge,

$[:\ddot{\underset{..}{I}}:]^-$

7.23 **a.** The radon was produced from uranium, a radioactive atom that decays through several steps to form radon.
c. The radon is continually produced as long as there is uranium in the rocks and soils.

7.29 **a.** U-238, the most abundant isotope.
b. Thorium is found with uranium because uranium decays to produce thorium. See Figure 7.14.

Chapter 8

8.2 **a.** $Zn \rightarrow Zn^{2+} + 2\ e^-$
b. $Cu^{2+} + 2\ e^- \rightarrow Cu$

8.5 **a.** Pb **b.** PbO_2
c. $PbSO_4(s)$ is both oxidized and reduced during recharging.

$2\ PbSO_4(s) + 2\ H_2O(l) \rightarrow Pb(s) + PbO_2(s) + 2\ H_2SO_4(aq)$

8.18 The difference in the energy, (286 kJ − 242 kJ) or 44 kJ, represents the amount of energy needed to change liquid water to gaseous water.

8.19 a. Energy absorbed to break two moles of H—O bonds in water is 2(467 kJ) = 934 kJ; energy released when H_2 and CO formed is (436 + 1073) kJ = 1509 kJ. The difference is 575 kJ.

b. Bond energies apply to elements and compounds in the gaseous state. The equation shows carbon in the solid state. Also, some carbon-to-carbon bonds in C(*s*) must also be broken, leading to a net input of approximately 131 kJ.

8.21 The element gallium, Ga, is in Group 3A, and has 3 outer electrons. The element arsenic, As, is in Group 5A, and has 5 outer electrons. When they bond in a similar array to germanium, as shown in Figure 8.19, each will have a share in a stable octet of electrons.

8.23 a. U.S. Sunrayce 2001 (3700 km) is longer than the Australian Sunrace (2300 km).

b. Ultimate Solar Car Challenge (4000 km) is longer than the U.S. Sunrayce (3700 km).

Chapter 9

9.8

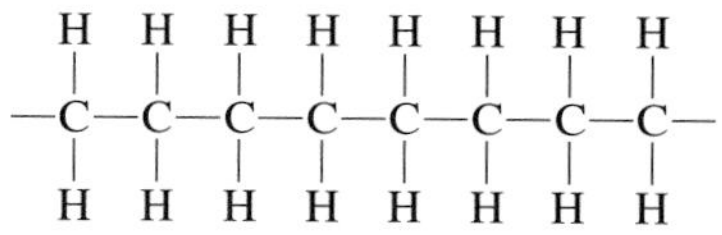

```
 H  H  H  H  H  H  H  H
 |  |  |  |  |  |  |  |
—C—C—C—C—C—C—C—C—
 |  |  |  |  |  |  |  |
 H  H  H  H  H  H  H  H
```

9.10

```
 H  H  H  H  H  H  H  H  H  H  H  H  H  H  H  H
 |  |  |  |  |  |  |  |  |  |  |  |  |  |  |  |
—C—C—C—C—C—C—C—C—C—C—C—C—C—C—C—C—
 |  |  |  |  |  |  |  |  |  |  |  |  |  |  |  |
 H  ⌬  H  ⌬  H  ⌬  H  ⌬  H  ⌬  H  ⌬  H  ⌬  H  ⌬
```

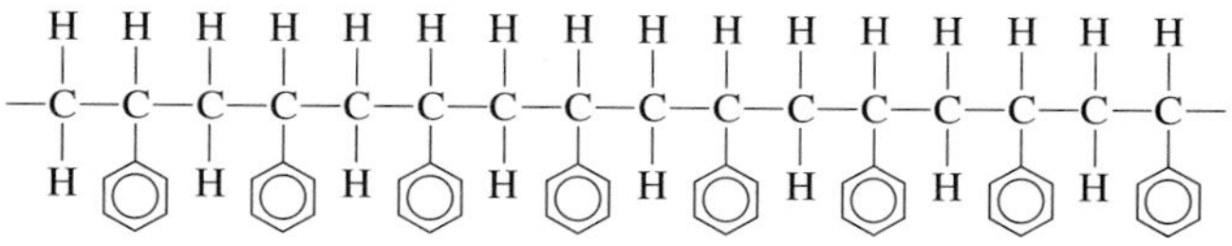

This arrangement is favored over the head-to-head arrangement because it minimizes electrostatic repulsion that would occur between the rings in a head-to-head arrangement.

9.11

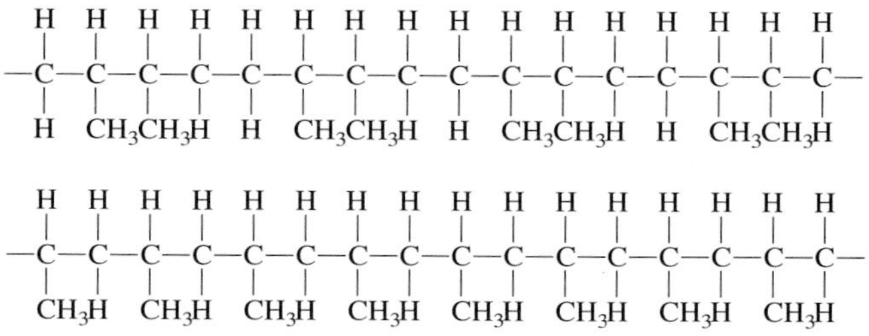

```
 H  H  H  H  H  H  H  H  H  H  H  H  H  H  H  H
 |  |  |  |  |  |  |  |  |  |  |  |  |  |  |  |
—C—C—C—C—C—C—C—C—C—C—C—C—C—C—C—C—
 |  |  |  |  |  |  |  |  |  |  |  |  |  |  |  |
 H  CH3CH3H  H  CH3CH3H  H  CH3CH3H  H  CH3CH3H

 H  H  H  H  H  H  H  H  H  H  H  H  H  H  H  H
 |  |  |  |  |  |  |  |  |  |  |  |  |  |  |  |
—C—C—C—C—C—C—C—C—C—C—C—C—C—C—C—C—
 |  |  |  |  |  |  |  |  |  |  |  |  |  |  |  |
 CH3H  CH3H  CH3H  CH3H  CH3H  CH3H  CH3H  CH3H
```

9.12

a.

```
 F  F  F  F  F  F  F  F  F  F  F  F  F  F  F  F
 |  |  |  |  |  |  |  |  |  |  |  |  |  |  |  |
—C—C—C—C—C—C—C—C—C—C—C—C—C—C—C—C—
 |  |  |  |  |  |  |  |  |  |  |  |  |  |  |  |
 F  F  F  F  F  F  F  F  F  F  F  F  F  F  F  F
```

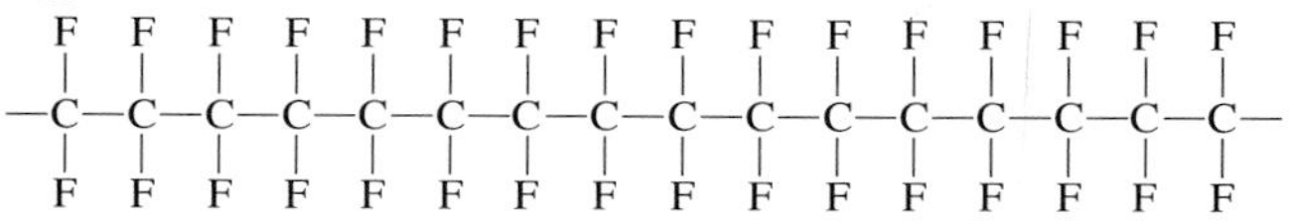

b. Head-to-tail arrangements are not possible because all the groups attached to the carbon chain are the same.

9.19

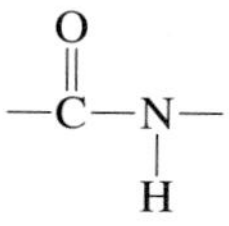

```
   O
   ‖
—C—N—
     |
     H
```

9.20

```
HO—C(=O)—⌬—C(=O)—N(H)—⌬—N(H)—C(=O)—⌬—C(=O)—N(H)—⌬—N(H)—C(=O)—⌬—C(=O)—N(H)—⌬—NH2
```

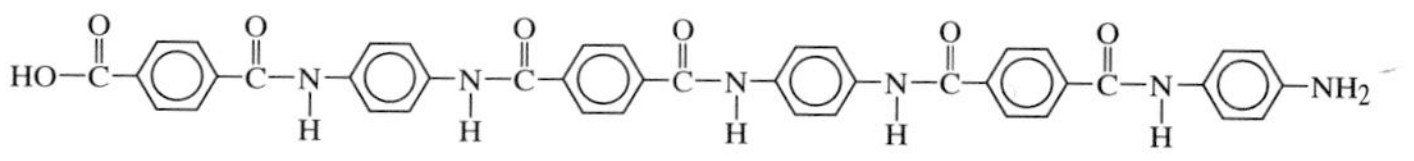

9.25 Propylene is $H_2C{=}CHCH_3$ or C_3H_6.

$$2500\ C_3H_6 + 11{,}250\ O_2 \rightarrow 7500\ CO_2 + 75{,}000\ H_2O$$

Chapter 10

10.3 *Note:* The hydrogen atoms have been omitted to make the linkage of the carbon atoms clear.

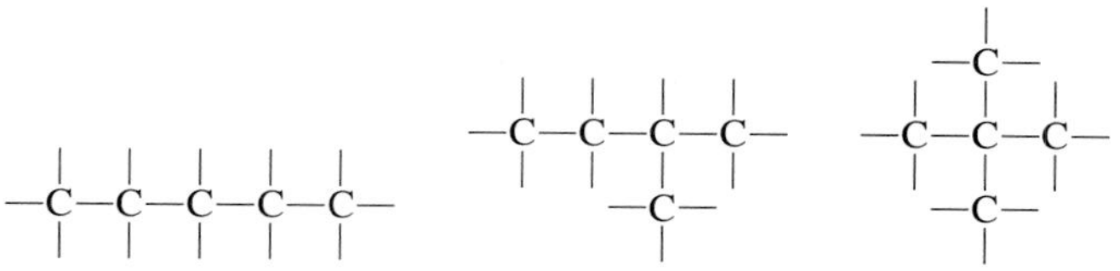

```
                          |  |  |  |            —C—
  |  |  |  |  |         —C—C—C—C—             |  |  |
—C—C—C—C—C—               |  |  |  |         —C—C—C—
  |  |  |  |  |              —C—               |  |  |
                              |                —C—
                                                 |
```

10.4 All three molecules have a benzene ring with functional groups attached that can hydrogen bond with water. Aspirin and ibuprofen also have a carboxylic acid group, —COOH.

10.8 There is only one carbon that has four different groups attached, making it the chiral carbon.

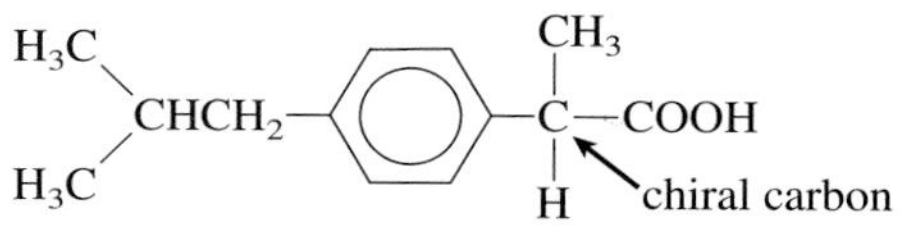

```
H3C                    CH3
   \                    |
    CHCH2—⌬—C—COOH
   /                    |  ↖
H3C                    H   chiral carbon
```

10.9

a.

estradiol	testosterone
—OH group on the D ring	—OH group on the D ring
$-CH_3$ group on C-D ring intersection	$-CH_3$ group on C-D ring intersection

b.

estradiol	progesterone
—OH group on the D ring	C=O group on the D ring
$-CH_3$ group on C-D ring intersection	$-CH_3$ group on C-D ring intersection

c.

cholic acid	cholesterol
no double bonds in any ring	no double bonds in A, C, or D rings
—OH group on the A ring	—OH group on the A ring
$-CH_3$ group on C-D ring intersection	$-CH_3$ group on C-D ring intersection

10.12 Androstenedione and testosterone have the same structural features, except that androstenedione has a C=O on the D ring. In the D ring of testosterone there is a —C—O—H in place of the C=O ring.

Chapter 11

11.3 The number of atoms of an element that are present is proportional to the number of moles of that element present in every 100 g of body weight. Oxygen has a higher atomic mass than hydrogen, so even though it is about 2.5 times more abundant, oxygen has a smaller number of moles present per 100 g of body weight, when compared with the same values for hydrogen.

11.5 Will vary by individual.

11.7 Salivary enzymes break down the complex carbohydrates in unsweetened crackers into simple sugars, which are responsible for the sweet taste.

11.8 Enzymes catalyze breaking bonds during digestion. There must be a close "lock-and-key" fit between the enzyme and the substrate molecule upon which it acts. The molecular architecture of human enzymes apparently is not compatible with the beta linkage between glucose units in cellulose, but the enzymes fit the alpha linkage in starches.

11.10 **a.** Canola oil is 6% saturated (likely capric, lauric, myristic, or palmitic acids), 36% polyunsaturated (linoleic or linolenic acids), and 58% monounsaturated fat (oleic acid).

b. Olive oil is 14% saturated (likely capric, lauric, myristic, or palmitic acids), 9% polyunsaturated (linoleic or linolenic acids), and 77% monounsaturated fat (oleic acid).

c. Lard is 41% saturated (likely stearic acid), 12% polyunsaturated (linoleic or linolenic acids), and 47% monounsaturated fat (oleic acid).

11.14 No, it does not meet these guidelines.

$$16.5 \text{ g saturated fat} \times \frac{9 \text{ Cal}}{1 \text{ g fat}} = 148.5 \text{ Cal from saturated fat.}$$

$$\frac{148.5 \text{ Cal}}{1230 \text{ total Cal}} \times 100 = 12\% \text{ Calories from saturated fat}$$

11.15 At this time, the American Heart Association's recommended values for cholesterol, HDL, and LDL have not changed from those reported in the text. Answers may change with time.

11.18 GlyGlyGly, GlyGlyAla, GlyAlaAla, GlyAlaGly, AlaAlaAla, AlaAlaGly, AlaGlyGly, AlaGlyAla

11.24 **a.** $\frac{130 \text{ Calories}}{2000 \text{ Calories}} \times 100 = 6.5\%$

b. Assuming that the serving size is the same for both the bagel chips and the potato chips, the potato chips are slightly higher in Calories per serving. Remembering that fats release 9 Cal/g when digested, a serving of bagel chips releases 4 g × 9 Cal/g = 36 Cal. The percentage of daily caloric intake from this fat is only 1.8% based on a 2,000-Calorie diet. A serving of potato chips releases 10 g × 9 Cal/g = 90 Cal, which is 4.5% of the caloric intake based on a 2000-Calorie diet. It appears that these bagel chips are slightly lower in fat and Calories per serving than these particular potato chips. Both types of chips are snack food and not meant to be a major component of the diet.

11.25 Different individuals will calculate different answers, resulting in different comparisons.

11.27 Vitamin A has only one —OH that can hydrogen bond with water. The rest of the molecule is composed of nonpolar carbon-carbon and carbon-hydrogen bonds, making this vitamin soluble in nonpolar fats. Vitamin C has several —OH groups, making it a polar molecule that is soluble in water through hydrogen bonding. Both are examples of the general principle that "like dissolves like."

Chapter 12

12.1 **a.** Between 1900 and 1950 for males (from 48 to 66), a 38% increase and for females (from 51 to 72), a 41% increase. Between 1950 and 1998, for males (from 66 to 72), a 9% increase and for females (from 72 to 79), a 10% increase.

b. Factors that have been proposed are that women have better diets and hormonal differences that offer some protection against heart attacks before menopause. Other suggested factors that no longer offer as much difference are the percentage who smoke and the amount of job-related stress.

12.5 $3 \times 10^9 \text{ base pairs} \times \frac{0.34 \text{ nm}}{\text{base pair}} \times \frac{1 \text{ m}}{1 \times 10^9 \text{ nm}} = 1 \text{ m}$

Section 12.1 notes that the length is 2 meters, but this is for the combined length of both DNA strands.

12.6 **a.** The 3-D structure of DNA gives a better representation of the actual arrangement of atoms in this complex molecule, making the text figure more understandable.

b. The four bases all have a six-membered ring containing two nitrogen atoms. Adenine, guanine, and cytosine have a $—NH_2$ group attached to the six-membered ring. The six-membered ring of adenine and of guanine is also connected to a five-membered ring containing two nitrogen atoms.

c. The bases are always paired in the same way. Two hydrogen bonds connect thymine to adenine and cytosine is paired with guanine by three hydrogen bonds.

12.8 Chargaff's Rules pointed to the complementarity of the base pairs, leading to the conclusion that adenine and thymine must occur together, as do guanine and cytosine.

12.10 Redundancy in the code means that a mistake in the amino acid produced by the codon and introduced into a protein may not necessarily result in a mistake in the amino acid introduced into the protein. Multiple codons for certain amino acids may also speed up protein synthesis.

APPENDIX

5

Answers to Selected End-of-Chapter Questions Indicated in Color in the Text

Chapter 1

1. $\frac{1L}{1 \text{ breath}} \times \frac{15 \text{ breaths}}{1 \text{ minute}} \times \frac{60 \text{ minutes}}{1 \text{ hour}} \times \frac{8 \text{ hours}}{1 \text{ working day}} = 7200 \text{ L}$

4. $9000 \text{ ppm} \times \frac{100 \text{ parts per hundred}}{1{,}000{,}000 \text{ ppm}} = 0.9 \text{ parts per hundred or } 0.9\%$

6. $50{,}000 \text{ ppm} \times \frac{100 \text{ parts per hundred}}{1{,}000{,}000 \text{ ppm}} = 5 \text{ parts per hundred or } 5\%$

8. The percentage is calculated by comparing the difference in the two readings to the standard.
$\frac{(0.15 - 0.12)}{0.12} \times 100 = 25\%$ above the standard

10. **a.** 85,000 g
b. 10,000,000 gallons

11. **a.** 7.2×10^7 cigarettes
b. 1.5×10^{4}°C

13. **a.** Group 7A
b. fluorine, chlorine, bromine, iodine, and astatine

16. **a.** potassium oxide
b. aluminum chloride

18. **a.** One atom of the element carbon, two atoms of the element hydrogen, and one atom of the element oxygen are combined to form one molecule of formaldehyde.
b. Two atoms of the element hydrogen are combined with two atoms of the element oxygen to form one molecule of hydrogen peroxide.

19. **a.** $N_2(g) + O_2(g) \rightarrow 2\ NO(g)$

20. **a.**

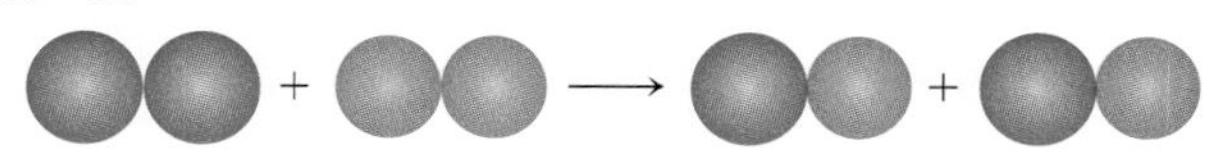

24. **a.** Platinum = Pt, palladium = Pd, rhodium = Rh.
b. All three metals are in Group 8B on the periodic table. Platinum is directly under palladium, and rhodium is just to the left of palladium.
c. These metals are solids at the temperature of the exhaust gases, so they must have relatively high melting points. Also, they do not undergo permanent chemical change in catalyzing the reaction of CO to CO_2 in the exhaust stream.

32. **a.** The order of increasing length is 5.0×10^{-3} m, 1 m, and 3.0×10^2 m.
b. If 1 meter is 1 year, then 3.0×10^2 m is equivalent to 300 years and 5.0×10^{-3} meters is equivalent to 0.0050 years. This is about 1.8 days.

36. Sample **a** represents a compound since two different atoms are joined.
Sample **b** represents a mixture since two different types of atoms are shown.

42. Formaldehyde can be released from cigarette smoke, and from synthetic materials such as foam insulation, and from the adhesives used in dying and gluing carpet pads, carpets, and laminated building materials. The air indoors is often not well circulated, leading to an accumulation of formaldehyde and other pollutants. Efforts to make homes airtight, leading to greater energy efficiency, have led in some cases to making problems of indoor air pollution worse, rather than better.

Chapter 2

2. **a.** Yes, diamond and graphite are two distinct forms of the same element, carbon.
b. No, water and hydrogen peroxide cannot be allotropes because they are not elements, but compounds.
c. Yes, white phosphorus and red phosphorus are allotropes because they are two distinct forms of the same element. Red phosphorus forms when several P_4 combine into a chain structure.

6. **a.** 8 protons, 10 neutrons, 8 electrons
b. 16 protons, 19 neutrons, 16 electrons
c. 92 protons, 146 neutrons, 92 electrons

9. **a.** $\cdot\text{Ca}\cdot$
b. $\cdot\ddot{\text{N}}\cdot$ (with one dot below)
c. $:\ddot{\text{Cl}}\cdot$ (with two dots below)
d. He :

11. **a.** 2(5) = 10 outer electrons :N:::N: and :N≡N:
b. 1 + 4 + 5 = 10 outer electrons H:C:::N: and H—C≡N:
c. 2(5) + 6 = 16 outer electrons :N::N::Ö: and :N=N=Ö:

13. **a.** $5\ \text{cm} \times \frac{1\ \text{m}}{10^2\ \text{cm}} = 5 \times 10^{-2}\ \text{m}$ This is in the microwave region of the spectrum.

14. $c = \nu \cdot \lambda$ and $\nu = \frac{c}{\lambda}$; $c = 3.0 \times 10^8\ \text{m}\cdot\text{s}^{-1}$

a. $\nu = \frac{3.0 \times 10^8\ \text{m}\cdot\text{s}^{-1}}{5 \times 10^{-2}\ \text{m}} = 6 \times 10^9\ \text{s}^{-1}$

15. $E = h\cdot\nu$ and $h = 6.63 \times 10^{-34}\ \text{J}\cdot\text{s}$
a. $E = 6.63 \times 10^{-34}\ \text{J}\cdot\text{s} \times 6 \times 10^9\ \text{s}^{-1} = 4 \times 10^{-24}\ \text{J}$

20. **a.** Cl, 7 outer electrons :C̈l·; NO_2, 17 outer electrons :O=N—Ö:; ClO, 13 outer electrons ·C̈l:Ö:; HO, 7 outer electrons H:Ö·
b. Each of these species has less than a full octet of outer electrons on one atom. Their reactivity is based on trying to attain the full octet of outer electrons for each atom capable of holding an octet.

23. **a.** Methane, CH_4, has 4 + 4(1) = 8 outer electrons.

```
    H
    |
H—C—H
    |
    H
```

Ethane, C_2H_6, has 2(4) + 6(1) = 14 outer electrons.

```
    H   H
    |   |
H—C—C—H
    |   |
    H   H
```

b. Fourteen different CFCs can be formed from methane

28. SO_2 has an identical Lewis structure to ozone. This should not be surprising as sulfur is in the same family as oxygen, so it has the same number of outer electrons. The difference is that the electrons in the sulfur atom in SO_2 are at a further distance from the nucleus than the oxygen atoms in O_3. This difference does not show in the Lewis structures. Here are the resonance structures.
:Ö:Ö::Ö: ⟷ :Ö::Ö:Ö: and :Ö:S::Ö: ⟷ :Ö::S:Ö:

35. UV-C radiation is extremely dangerous, but it is completely absorbed by normal oxygen, O_2, as well as by ozone, O_3, before it can reach the surface of the Earth.

39. **a.** There are 4 + 3(7) + 6 = 31 outer electrons available.

```
     :F:
:F:C:O·
     :F:
```

b. This free radical is quite reactive in the troposphere, so it does not last long enough to reach the stratosphere.

43. O_2, O_3, and N_2 all have an even number of electrons to place in their Lewis structures. N_3 would have 15 electrons. Molecules with odd numbers of electrons are generally more reactive than those in which all electrons are present in pairs.

48. **a.** For CFC-12, add 90 and get 102. The compound contains 1 carbon, no hydrogens, and 2 fluorines. Therefore, there must be 2 chlorines. The formula is CCl_2F_2.
b. There is 1 carbon, no hydrogens, and no fluorine atoms in CCl_4. The code number plus 90 must be 100 and 100 minus 90 gives 10. This is CFC-10.
c. 22 + 90 = 112. This means 1 carbon, 1 hydrogen, and 2 fluorines. This coincides with the formula, which is $CHClF_2$.

Chapter 3

2. **a.**

$$6\ CO_2(g) + 6H_2O(l) \xrightarrow[\text{sunlight}]{\text{chlorophyll}} C_6H_{12}O_6(aq) + 6\ O_2(g)$$

b. The number of atoms of each element on either side is the same. C = 6, O = 18, H = 12.
c. The number of molecules is not the same on either side of the equation. There are 12 on the left, but only 7 on the right. The large molecule glucose has formed on the right-hand side of the equation, using 24 atoms per molecule.

4. **a.** The concentration of CO_2 in the atmosphere at the present time is about 370 ppm; 20,000 years ago, the concentration was about 190 ppm, far lower. However, 120,000 years ago the CO_2 concentration was about 270 ppm, closer to today's levels.
b. The mean temperature at present is somewhat above the 1950–1980 mean temperature of the atmosphere. 20,000 years ago, the mean temperature was lower by about 9°C. However, 120,000 years ago, the mean temperature was lower by only about 1°C.
c. While there appears to be a *correlation* between mean temperature and CO_2 concentration, this figure does not prove *causation* of either factor by the other.

7. These are the two Lewis structures. H—H and H—Ö—H. For H_2 with only two atoms, the atoms can only be arranged in a straight line. For H_2O, even though the Lewis structure has been *shown* as a straight line, this does not mean that the molecule is linear. In fact, the bent structure of water is so well known that the Lewis structure is often written in this manner.

$$\ddot{\text{O}}\text{ bonded to two H atoms (bent): H}\diagup\ddot{\text{O}}\diagdown\text{H}$$

9. **a.** $4 + 2(1) + 2(7) = 20$ outer electrons; The Lewis structure is H—C—H (with :Cl: above and :Cl: below C) and the shape is tetrahedral.

b. $4 + 6 = 10$ outer electrons. The Lewis structure is :C≡O: and the shape is linear. (This could have been predicted from the fact that there are just two atoms.)

c. $1 + 4 + 5 = 10$ outer electrons. The Lewis structure is H—C≡N: The shape is linear.

d. $5 + 3(1) = 8$ outer electrons. The Lewis structure is H—P̈—H (with H below P). The shape is triangular pyramidal, which can also be called trigonal pyramidal.

16. $C_6H_{12}O_6(aq) \xrightarrow{\text{yeast}} 2\ C_2H_5OH(aq) + 2\ CO_2(g)$

19. The weighted average for silver is 107.870 g/mole.
$[107 \times 0.52] + [\mathbf{x} \times 0.48] = 107.870$
$0.48\ x = 107.870 - 55.64$; $\mathbf{x} = 109$;
The other isotope of silver is silver-109.

21. **a.** $2(1) + 16 = 18$ g/mole
b. $12 + 2(19) + 2(35.5) = 121$ g/mole
c. $12 + 16 = 28$ g/mole

26. **a.** $\dfrac{370 \text{ ppm} - 280 \text{ ppm}}{280 \text{ ppm}} \times 100$
= 32% increase in the concentration of CO_2.

b. $\dfrac{1.8 \text{ ppm} - 0.70 \text{ ppm}}{0.7 \text{ ppm}} \times 100$
= 160% increase in the concentration of CH_4

$\dfrac{0.31 \text{ ppm} - 0.28 \text{ ppm}}{0.28 \text{ ppm}} \times 100$
= 11% increase in the concentration of N_2O.
CH_4 has shown the biggest percentage increase.

29. Drilled ocean cores can be analyzed for the number of type of microorganisms present. Another correlating piece of evidence is the changing alignment of the magnetic field in particles in the sediment over time. Another possibility is to analyze the deuterium to hydrogen ratio in ice cores.

33. In BF_3, there are $3 + 3(7) = 24$ outer electrons. This is the Lewis structure. :F̈—B—F̈: (with :F: below B) In NH_3, there are $5 + 3(1) = 8$ outer electrons. This is the Lewis structure. H—N̈—H (with H below N)

Note that, in BF_3, there are just three pairs of electrons around the central boron. To maximize separation of these three mutually repelling electron regions, a fluorine-boron-fluorine bond angle of 120° is predicted. Thus, BF_3 has a triangular planar geometry. In NH_3, however, there are four pairs of electrons—three bonded pairs and one lone pair. This determines the triangular pyramidal geometry. Although the predicted hydrogen-to-nitrogen-to-hydrogen bond angle is 109.5°, the repulsion of the lone pair with the bonding pairs closes the angle somewhat, so that the experimentally measured angle is 107.5°.

37. Forests < fossil fuels <<< oceans. (see Table 3.1).

42. **a.** It does *not* prove that global warming theories are correct. It is a short-range observation consistent with global warming trends that are predicted, but it may not predict large-scale or long-range changes.
b. It does *not* prove that global warming theories are correct. While it is not definite proof, it is important because it provides one more piece of experimental evidence for the predicted effects of global warming on biological species.

46. **a.** 20 electrons are required if three atoms join. This is the general Lewis structure. :Ẍ—Ÿ—Z̈:
b. This molecule will be bent. There are four electron pairs around the central **Y** atom. However, two of the pairs are bonding pairs, and two are lone pairs. Repulsion between the two lone pairs and their repulsion of the bonding pairs is predicted to cause the bond angle to be less than 109.5°.
c. There are new possibilities if double and triple bonds are allowed.

Number of Outer Electrons	Lewis Structure	Shape	Predicted bond angle
18	:Ẍ—Ÿ=Z̈:	bent	less than 120°
16	:Ẍ=Y=Z̈:	linear	180°
16	:Ẍ—Y≡Z:	linear	180°

Chapter 4

2. **a.** Estimating values from the graph, domestic oil production accounted for approximately 60% in 1970, 30% in 1990, and is predicted to be about 18% in 2010.

4. **a.** The temperature is the same, 70°C, in each container. Temperature is commonly measured in °C for scientific work, or in °F for household applications.

b. The heat content of the water is not the same. Heat depends on both the temperature and the mass of the sample. There is twice the heat in the water in Container 1 compared with the water in Container 2 because twice the mass of water is present in Container 1.

7. $$\frac{650{,}000 \text{ kcal}}{1 \text{ day}} \times \frac{365 \text{ days}}{1 \text{ year}} = \frac{2.4 \times 10^8 \text{ kcal}}{1 \text{ year}}$$

This value can be related to each of the energy sources.

a. $$\frac{2.4 \times 10^8 \text{ kcal}}{1 \text{ year}} \times \frac{1 \text{ year}}{65 \text{ barrels}} = \frac{3.7 \times 10^6 \text{ kcal}}{\text{barrel}}$$

10. **a.** $2\ C_2H_6 + 7\ O_2 \rightarrow 4\ CO_2 + 6\ H_2O$

b.

```
      H  H
      |  |                 ..   ..
2  H—C——C—H  +  7 :O==O:
      |  |
      H  H
                   ..      ..             ..
        ——→  4 :O==C==O:  +  6  H—O:
                                          |
                                          H
```

c.

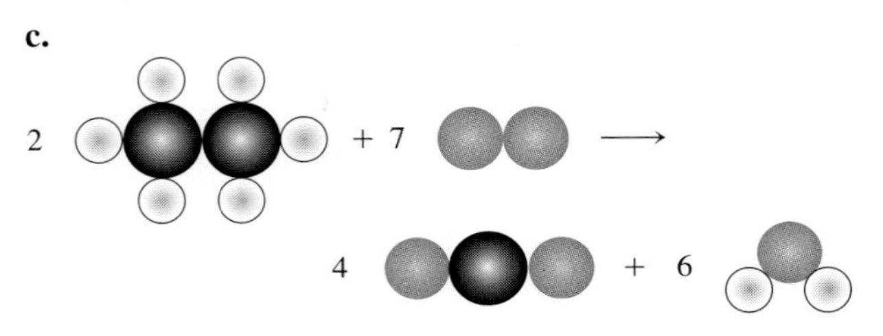

11. $$\frac{52.0 \text{ kJ}}{1 \text{ g } C_2H_6} \times \frac{30.1 \text{ g } C_2H_6}{1 \text{ mol } C_2H_6} = \frac{1560 \text{ kJ}}{\text{mol } C_2H_6}$$

14. **a.** Exothermic; a charcoal briquette releases heat as it burns.

b. Endothermic; liquid water gains the necessary heat for evaporation from your skin, and your skin feels cool.

15. Bonds broken in the reactants

2 carbon-to-oxygen triple bonds	= 2(1073 kJ)	= 2146 kJ
1 oxygen-to-oxygen double bonds	= 1(498 kJ)	= 498 kJ
Total energy *absorbed* in breaking bonds		= 2644 kJ

Bonds formed in the products

4 carbon-to-oxygen double bonds	= 4(803 kJ)	= 3212 kJ
Total energy *released* in forming bonds		= 3212 kJ

Net energy change is (+2644 kJ) + (−3212 kJ) = −568 kJ

Notice that the overall energy change has a negative sign, characteristic of an exothermic reaction.

18. **a.** $2\ C_5H_{12}(g) + 11\ O_2(g) \rightarrow 10\ CO(g) + 12\ H_2O(l)$

$8(356 \text{ kJ}) + 24(416 \text{ kJ}) + 11(498 \text{ kJ}) \rightarrow 10(1070 \text{ kJ}) + 24(467 \text{ kJ})$

The net energy change is −3598 kJ, and the reaction is highly exothermic.

22. There are two different isomers.

```
   H  H  H  H                 H  H  H
   |  |  |  |                 |  |  |
H—C——C——C——C—H    and    H—C——C——C—H
   |  |  |  |                 |  |  |
   H  H  H  H                 H  |  H
                               H—C—H
                                  |
                                  H
```

28. All of these refer to the same idea, which is that energy is not consumed during a chemical reaction. Energy can be transformed, but the total energy is constant during a chemical reaction.

30. **a.** The C—F bond requires 485 kJ/mole, the C—Cl bond requires 327 kJ/mole, and the C—Br bond requires 285 kJ/mole to break the bond. The C—Br bond is the least energetic, and the bromine free radical can interact with ozone even more effectively than the chlorine free radical.

b. This is the Lewis structure for C_2HF_4Cl.

```
       ..   ..
      :F: :F:
 ..    |    |    ..
:F—— C —— C ——Cl:
 ..    |    |    ..
      :F:   H
       ..
```

The C—F bond requires 485 kJ/mole, the C—Cl bond requires 327 kJ/mole, and the C—H bond requires 416 kJ/mole to break the bond. This means that the chlorine free radical forms with the least energy into the system. While this seems contradictory in a replacement for halons, this compound's lifetime in the atmosphere is significantly shorter than the halons, making it less effective in depleting ozone.

39. The entropy values increase with increasing complexity and randomness. Diamond is a highly ordered crystalline structure and so has a low entropy. Methanol is a liquid, and the molecules of methanol have greater complexity and more opportunity for random movement.

Chapter 5

1. **a.** $3 \times 10^9 \text{ gallons now} \times \frac{1 \text{ gallon twenty years ago}}{10 \text{ gallons now}}$
$= 3 \times 10^8$ gallons twenty years ago

b. $(3 \times 10^9$ gas now$) - (3 \times 10^8$ gal twenty years ago$)/3 \times 10^8$ gal twenty years ago $\times 100 = 900\%$ growth

3. The drinking water supply is limited by the amount of available fresh water. Figure 5.4 shows 97.4% is salt-water, which means only 2.6% is fresh. Further restricting the amount of fresh water is the fact that 90% of fresh water is frozen in Antarctica, and that 80% of fresh water in the U.S. is used for irrigating crops and cooling electric power plants. The water available for drinking is only 0.26 L or 260 mL.

$$500 \text{ L total} \times \frac{2.6 \text{ parts fresh}}{100 \text{ parts total}} \times \frac{10 \text{ parts not frozen}}{100 \text{ parts fresh}}$$

$$\times \frac{20 \text{ parts available}}{100 \text{ parts not frozen}} = 0.26 \text{ L available}$$

5. **a.** 55 mg of Ca^{2+} per liter of bottled water is the same as 55 ppm.

$$\frac{55 \text{ mg } Ca^{2+}}{1 \text{ L } H_2O} \times \frac{1 \text{g}}{10^3 \text{ mg}} \times \frac{1 \text{ L}}{10^3 \text{ mL}}$$

$$\times \frac{1 \text{ mL } H_2O}{1 \text{g } H_2O} = \frac{55 \text{g } Ca^{2+}}{10^6 \text{ g } H_2O} \text{ or } 55 \text{ ppm } Ca^{2+}$$

10. Water is a very good solvent for other polar substances and for many ionic substances.

14. The arrow points to a hydrogen bond, an example of an *inter*molecular force, which is a force *between* water molecules not *within* each water molecule.

18. **a.** $250 \text{ g } H_2O \times \frac{1 \text{ cal}}{\text{g}\cdot°\text{C}} \times (100\ °\text{C} - 15°\text{C}) =$ $+2.1 \times 10^4$ cal of heat absorbed

b. $500 \text{ g } H_2O \times \frac{1 \text{ cal}}{\text{g}\cdot°\text{C}} \times (55\ °\text{C} - 95°\text{C}) =$ -2.0×10^4 cal of heat released

21. **a.** Na_2S sodium sulfide

b. Al_2O_3 aluminum oxide

28. These values are not easy to compare because the units are not the same. There are 100 mg of magnesium per tablet, which is usually swallowed with a minimum of water. The concentration of magnesium reaching the body is probably far greater from the multivitamin tablet than it is from this brand of bottled water.

30. **a.** Electronegativity values generally increase from left to right across a period (until group 8A is reached) and from bottom to top within any group. This means that element labeled 2 is predicted to have the greatest electronegativity value.

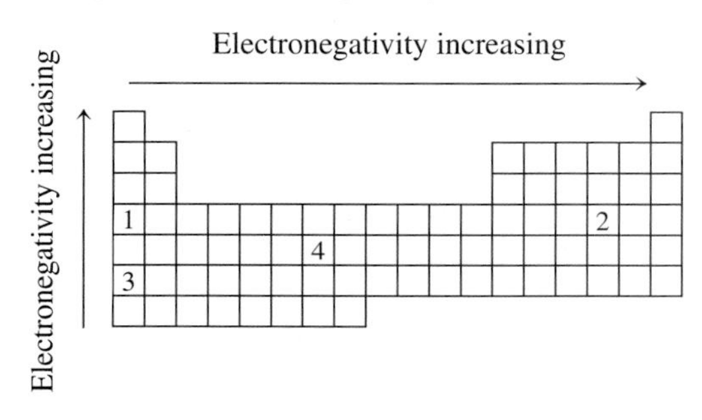

33. **a.** This is the Lewis structure for ethanol.

```
    H   H
    |   |   ..
H — C — C — O — H.
    |   |   ..
    H   H
```

This is a polar molecule, and it will exhibit hydrogen bonding.

b. The cube sinks because, as is the case for most substances, the density of the solid phase is greater than the density of the liquid phase. In solid ethanol, the ethanol molecules are closer together than water molecules are in ice. Therefore, solid ethanol has a greater density than liquid ethanol.

36. With the exception of contaminants that are known carcinogens, MCLG and MCL values are usually very close to being the same.

Chapter 6

1. **a.** Acids have a sour taste, react with carbonates, and impart a characteristic color to acid-base indicators.

b. Acids release hydrogen ions, H^+, in aqueous solution, and impart a characteristic color to acid-base indicators.

3. **a.** $HBr(aq) \longrightarrow H^+(aq) + Br^-(aq)$

b. $H_3C_6H_5O_7(aq) \longrightarrow H^+(aq) + H_2C_6H_5O_7^-(aq)$

4. **a.** Bases taste bitter and have a slippery feel in water.

b. Bases release hydroxide ions, OH^-, in aqueous solution.

6. **a.** $RbOH(s) \longrightarrow Rb^+(aq) + OH^-(aq)$

7. **a.** $KOH(aq) + HNO_3(aq) \longrightarrow KNO_3(aq) + H_2O(l)$

9. **a.** Basic; the concentration of $OH^-(aq)$ is greater than the concentration of $H^+(aq)$.

12. $S(s) + O_2(g) \longrightarrow SO_2(g)$

15. **a.**

$$\frac{14 \text{ g nitrogen}}{[(135 \times 12) + (96 \times 1) + (9 \times 16) + (1 \times 14) + (1 \times 32)]} \times 100 = 0.73\% \text{ N}$$

b. Use equation 6.18: $3 \text{ tons coal} \times \frac{0.73 \text{ ton N}}{100 \text{ ton coal}}$

$$\times \frac{0.5 \text{ ton} \cdot \text{mole } N_2}{1 \text{ ton} \cdot \text{mole N}} \times \frac{2 \text{ ton} \cdot \text{mole NO}}{1 \text{ ton} \cdot \text{mole } N_2} = 0.22 \text{ ton NO}$$

19. The relative importance of fuel combustion and transportation is different for SO_2 and NO_x. The value of SO_2 from fuel combustion is higher because of the burning of sulfur-containing coal. The value of NO_x from transportation is higher because of the large number of automobiles (and jet engines) that produce NO_x.

22. This requires using the molar masses of SO_2 and $CaCO_3$. Note that it is not necessary to change to grams. The ratio of the number of grams per mole is the same as the ratio of the number of kg per kilomole or the number of tons to the number of ton · moles.

$$1.00 \text{ ton } SO_2 \times \frac{1 \text{ ton} \cdot \text{mole } SO_2}{64.1 \text{ tons } SO_2} \times \frac{2 \text{ ton} \cdot \text{mole } CaCO_3}{2 \text{ ton} \cdot \text{mole } SO_2}$$

$$\times \frac{100 \text{ tons } CaCO_3}{1 \text{ ton} \cdot \text{mole } CaCO_3} = 1.56 \text{ tons } CaCO_3$$

Chapter 7

1. E represents energy, m represents mass lost in a nuclear transformation, and c represents the speed of light.

3. **a.** 94 protons
b. 52 neutrons
c. 92 protons and 146 neutrons

5.

Isotope	Number of Protons	Number of Neutrons
C-12	6	6
Ca-40	20	20
Zn-64	30	34
Sn-119	50	69

For the lighter elements with an even number of protons, there is a matching number of neutrons in a stable isotope. From these data, an atomic number of 20 seems to be the limit for this case. For heavier elements with an even number of protons, the number of neutrons must be greater than the number of protons to create a stable nucleus. The lower limit given in the data is atomic number equal to 30 or above. No generalization can be reached for elements with an odd number of protons or for the elements with atomic numbers 11–29.

8. **a.** Because there are only two stable isotopes, there must be a higher percentage of B-11. In fact, there is about 80% of B-11 and only 20% of B-10.

10. **a.** ${}^{2}_{1}H + {}^{2}_{1}H \longrightarrow [{}^{4}_{2}He] \longrightarrow {}^{1}_{0}n + {}^{3}_{2}He$
b. ${}^{238}_{92}U + {}^{14}_{7}N \longrightarrow [{}^{252}_{99}Es] \longrightarrow {}^{247}_{99}Es + 5\ {}^{1}_{0}n$

12. **a.** The sum of the masses of the reactants is 5.02838 g and the sum for the products is 5.00878 g. This means that the mass difference, 0.0196 g, has been transformed to energy following Einstein's equation, $E = mc^2$.
b. $E = mc^2$; $E = 1.76 \times 10^{12}$ J

14. **A** = control rod assembly, **B** = cooling water out of the core, **C** = control rod, **D** = cooling water into the core, **E** = fuel rod

16. There are two advantages to using water rather than graphite as the moderating material in U.S. reactors: (1) Water has a higher heat capacity than graphite, and (2) When water gets hot, it does not burn the way the graphite did in the Chernobyl reactor.

22. **a.** ${}^{131}_{53}I \longrightarrow {}^{131}_{54}Xe + {}^{0}_{-1}e$
b. ${}^{238}_{92}U \longrightarrow {}^{234}_{90}Th + {}^{4}_{2}He$

24. Two half-lives = $(1/2)^2$ = 1/4 or 25% remaining and 75% decayed; four half-lives = $(1/2)^4$ = 1/16 or 6.25% remaining and 93.75% decayed; six half-lives = $(1/2)^6$ = 1/64 or about 1.6% remaining and 98.4% decayed.

25. After seven half-lives, $(1/2)^7$ = 1/128 is remaining, or 0.78 %. This means that over 99% has decayed, which is reasonably close to being "gone" for small samples. However, the radioactivity actually is *not* gone, as 0.78% of the sample still remains. Furthermore, if you start with a large amount of a radioactive substance (say 1000 pounds), after seven half-lives have passed you have just under 1% (close to 10 pounds) left, which is a sizeable amount.

Chapter 8

1. **a.** Oxidation is a process in which an atom, ion, or molecule *loses* one or more electrons.
b. Reduction is a process in which an atom, ion, or molecule *gains* one or more electrons.
c. Electrons must be transferred from the species losing electrons to the species gaining electrons.

3. **a.** Anode is Zn(*s*).
b. $Zn(s) \rightarrow Zn^{2+}(aq) + 2\ e^-$
c. Cathode is Ag(*s*).
d. $2\ Ag^+(aq) + 2\ e^- \longrightarrow 2\ Ag(s)$

5. A galvanic cell is a device that converts the energy released in a chemical reaction into electrical energy. A collection of several galvanic cells wired together constitutes a battery. Many times, the term *battery* is used incorrectly to refer to a galvanic cell. An example of a cell is a "D" alkaline cell (or "battery"). An example of a true battery is a lead-acid automobile battery, in which several cells are connected.

6. **a.** $Li \rightarrow Li^+ + e^-$ half-reaction of oxidation
$I_2 + 2\ e^- \rightarrow 2\ I^-$ half-reaction of reduction
b. $2\ Li + I_2 \rightarrow 2\ LiI$ overall reaction in the cell
c. Oxidation occurs at the anode, so this is the half reaction: $Li \rightarrow Li^+ + e^-$
Reduction occurs at the cathode, so this is the half reaction: $I_2 + 2\ e^- \rightarrow 2\ I^-$

8. In every electrochemical process described in this chapter, energy is produced through electron transfer. Such transfer takes place because of a chemical reaction, such as takes place in galvanic cells and batteries. The transfer may be because light strikes a photovoltaic cell, resulting in the movement of electrons.

9. Yes, both AAA and D cells produce 1.54 V. The electron transfer that takes place is identical in each type of these cells. The larger D cell will generate a larger current, but the same voltage as the smaller AAA cells.

11. In both 1970 and 1980, the major use for mercury was for batteries. By 1990, the major use was in the chlor-alkali process. By 1990, awareness of the dangers of mercury in urban trash had grown. Safer batteries and the need to recycle batteries led to the passage of the Mercury-Containing and Rechargeable Battery Management Act (The Battery Act) in 1996.

13. **a.** The electrolyte provides a medium for transfer of both electrons and ions.
b. KOH paste
c. Sulfuric acid solution

15. This must be the process of *reduction,* for electrons must be *gained* to go from oxygen gas to combined oxygen in water. This is the reduction half reaction.

$$\frac{1}{2}\,O_2(g) + 2\,H^+(aq) + 2\,e^- \rightarrow H_2O(l)$$

19. **a.** This is the balanced equation.
$2\,Mg(s) + SiCl_4(l) \rightarrow 2\,MgCl_2(l) + Si(s)$
Each magnesium atom loses two electrons and there are two magnesium atoms. Silicon in $SiCl_4$ picks up the four electrons.
b. The silicon in $SiCl_4$ gains 4 electrons to form silicon atoms, which is a process of reduction.

20. In current usage, the term "hybrid car" refers to the combination of a gasoline engine together with a nickel-metal hydride battery, an electric motor and an electric generator. Other hybrids using fuel cells will continue to be developed.

23. **a.** To check this equation, $H_2(g) + \frac{1}{2}\,O_2(g) \rightarrow H_2O(l)$ + energy, first consider the Lewis structures for reactants and products.

$$\text{H—H} + \frac{1}{2}\ :\ddot{O}=\ddot{O}: \longrightarrow \text{H}\diagup\ddot{\text{O}}\diagdown\text{H}$$

Energy needed to break bonds:
436 kJ + $\frac{1}{2}$(498 kJ) = +685 kJ
Energy released as new bonds form:
2(467 kJ) = −934 kJ
Overall, according to this calculation, the reaction releases 249 kJ of energy.
b. Average bond energies are based on bonds within molecules in the gaseous state. In the given chemical equation, the H_2O formed is present as a liquid rather than as a gas. Additional energy is released when gaseous water condenses to the liquid state, so the stated value of 286 kJ is greater than the 249 kJ calculated in part a.

24. Note that there are 8 electrons around each silicon atom, but there are 9 electrons around the central atom. Each carbon atom has 4 outer electrons, so the central atom in the figure must have 5 outer electrons. This is consistent with arsenic, which is in group 5A. The additional electron forms an *n*-type silicon semiconductor.

30. Equations 1 and 3 represent redox reactions. Electrons must be transferred to change an element into its combined form. Equation 2 is a neutralization reaction in which ions combine to form a soluble salt as well as covalently bonded water molecules. Electron transfer does not occur in this case.

Chapter 9

1. Plastics are synthetic polymers. This means that plastics are polymers, but not all polymers are called plastics.

6. **a.** 357 lbs/person
b. 155 lbs/person
c. 294% change
d. 230% change

8.

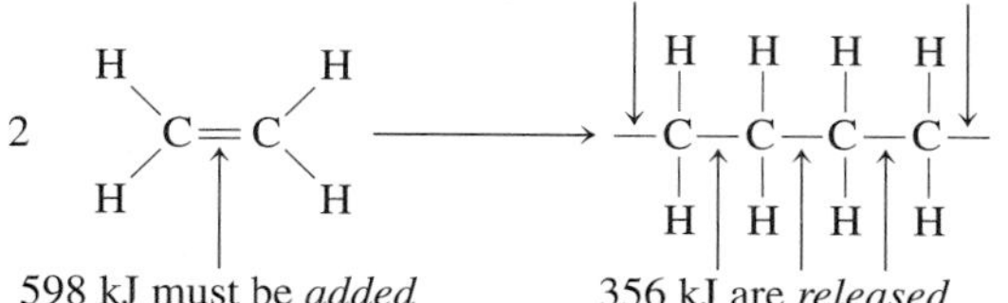

Altogether, note that 1196 kJ of energy were added to the system (2 × 598), and (1068 kJ + 356 kJ) = 1424 kJ were released from the system. This means that 228 kJ of energy are released from the system (−228 kJ); the reaction is exothermic.

10. **a.** 1430 monomer units
b. 2860 carbon atoms

13. There are *three* regions of electrons around each carbon in the monomer, making the geometry around the carbon trigonal planar and the Cl—C—H bond angle 120°. Each carbon in the polymer has *four* regions around the carbon, making the geometry tetrahedral and the bond angle 109.5° in the polymer.

15. The bottle on the left is likely made of flexible, low-density branched polyethylene. The one on the right is likely made of rigid, high-density, linear polyethylene. The structures of LDPE and HDPE, found in Figure 9.5, can be used to help explain this difference in properties at a molecular level. The low-density polyethylene is highly branched, preventing close interactions between the chains and allowing the plastic to be softer and more easily deformed. The high-density polyethylene chains are linear. The chains can more closely approach each other, creating opportunity for interactions. The bottle on the right is less flexible and more rigid than the one on the left.

17. All except #1, PETE, share a common structural feature. They have the same basic structure as the ethylene molecule, but one of the hydrogen atoms has been replaced with a different atom or group of atoms.

20. This is the head-to-head, tail-to-tail arrangement of PVC forming from three monomer units. Note that the carbon containing two hydrogen atoms attached to another carbon containing two hydrogen atoms, and the carbon containing two chlorine atoms attached to another carbon containing two chlorine atoms.

22. The CO_2 most likely replaces CFCs formerly used for this process. Most of the CO_2 used comes from existing commercial and natural sources, so no additional CO_2 is contributed to global warming. CFCs are implicated in the depletion of the ozone layer and can no longer be used for blowing Styrofoam. For several years, pentane, C_5H_{12}, has been used to replace CFCs, but pentane is flammable and CO_2 is not.

24. **a.** This is the Lewis structure.

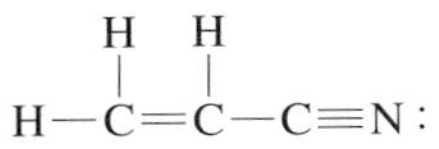

```
    H   H
    |   |
H—C=C—C≡N:
```

b. When Acrilan fibers burn, one of the products is the poisonous gas hydrogen cyanide, HCN.

29. **a.** Yes, each division represents 20 billion pounds.
b. Yes, each division represents a 5-year period.

30. Several other features of polymer chains can influence the properties of the polymer formed. These include

1. the length of the chain (the number of monomer units)
2. the three-dimensional arrangement of the chains
3. the branching of the chain
4. the chemical composition of the monomer units
5. the bonding between chains
6. the orientation of monomer units within the chain

36. The "Big Six" polymers are almost completely nonpolar molecules and therefore do not dissolve in polar water molecules. The generalization, developed in Chapter 5, is that "like dissolves like." Some of the "Big Six" dissolve or soften in hydrocarbons or chlorinated hydrocarbons because these nonpolar solvents interact with the nonpolar polymeric chains.

38. A monomer must have a carbon-to-carbon double bond in its structure. Although carbon-to-carbon double bonds in rings may be present, the double bond used for addition polymerization must be along the chain. That carbon-to-carbon double bond must be accessible to attack by a free radical, resulting in a single carbon-to-carbon bond that is left along the chain. As that reaction repeats itself, the addition polymer grows. An example is the formation of PVC.

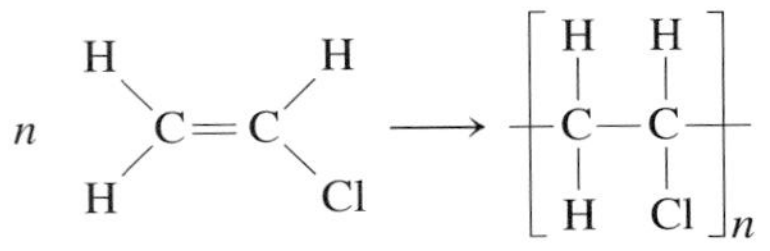

```
   H       H          ┌ H   H ┐
    \     /           │ |   | │
n    C=C      ⟶    ─┼─C—C─┼─
    /     \           │ |   | │
   H       Cl         └ H  Cl ┘n
```

Chapter 10

1. **a.** An antipyretic drug is intended to reduce fever.
b. An analgesic drug is intended to reduce pain.
c. An anti-inflammatory drug is intended to reduce inflammation, which is redness, heat, swelling, and pain caused by irritation, injury, or infection.
d. Yes. Aspirin is an example of a drug with all three properties.

3. The condensed formulas are $CH_3CH_2CH_2CH_2CH_3$ (or $CH_3(CH_2)_3CH_3$), $CH_3CH_2CH(CH_3)CH_3$, and $CH_3C(CH_3)_2CH_3$.

5. Four possible isomers have the formula C_4H_9OH. Here is the structural formula for each isomer.

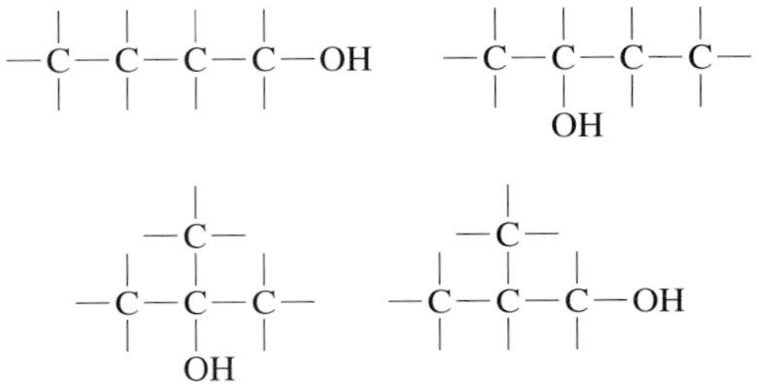

```
 |  |  |  |               |  |  |  |
—C—C—C—C—OH             —C—C—C—C—
 |  |  |  |               |  |  |  |
                             OH

     |                        |
    —C—                      —C—
 |   |   |                |   |   |
—C—C—C—                  —C—C—C—OH
 |   |   |                |   |   |
    OH
```

7. Alcohols, aldehydes and acids all have examples with one carbon atom. The others require more than one carbon atom by the nature of their functional groups.
a. alcohol An example is methanol, CH_3OH.
b. aldehyde An example is formaldehyde, CH_2O.
c. carboxylic acid An example is formic acid, HCO_2H.
d. ester This does not have a possible one-carbon example. The carbonyl group must have a carbon-containing group on either side, one bonded to the carbon and the other to the singly bonded oxygen of the carbonyl group.
e. ether This does not have a possible one-carbon example. The central oxygen atom must have a carbon-containing group on either side.
f. ketone This does not have a possible one-carbon example. There must be a carbon-containing group on either side of the carbon that is double bonded to an oxygen. If one or both positions are occupied by hydrogen atoms, the compound is an aldehyde.

11. **a.** amine group (—NH_2) and carboxylic acid group (—COOH)

b. two amide groups
```
 O  H
 ‖  |
—C—N—
```

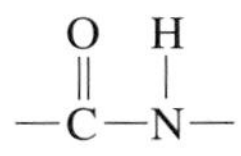

c. Two amide groups $-\overset{O}{\overset{\|}{C}}-\overset{H}{\overset{|}{N}}-$, carboxylic acid group (—COOH); also sulfide (—S—) in ring.

13. There are several nonpolar groups, so it is likely to be very soluble in the lipids that form the cell membrane. Only the amine group is polar, helping the molecule to dissolve in the fluid inside and outside the cells.

17. "Superaspirins" are a new class of medicines that are selective in preferentially blocking the COX-2 enzyme that makes prostaglandins associated with inflammation, pain, and fever. These "superaspirins" have no effect on COX-1 enzymes, which means fewer side effects such as stomach irritation and kidney disfunction.

20. **a.** This cannot exist in chiral forms. Two groups are the same, $-CH_3$.
b. This can exist in chiral forms, as all four groups attached to the central carbon are different.
c. This cannot exist in chiral forms. Two groups are the same, CH_3.

24. **a.** Although technically not an anabolic steroid, androstenedione is converted to testosterone when ingested and testosterone is a banned anabolic steroid.
b. Androstenedione is not an anabolic steroid and therefore does not fall under the policy of being banned.

28. There are just three distinct isomers shown here—1, 2, and 6; others are duplications.
"Isomers" 1 and 5 are different pencil-and-paper representations of the *same* isomer.
"Isomers" 2, 3, and 4 are different pencil-and-paper representations of the *same* isomer.
Isomer 6 is an isomer different from those given for isomers 1 and 5, and 2–4.

34. L-dopa is effective because the molecule fits the receptor site, but the nonsuperimposable mirror image of R-dopa does not. This is the structure of L-dopa, with the chiral carbon atom marked. Note that there are four different groups attached to the starred carbon atom.

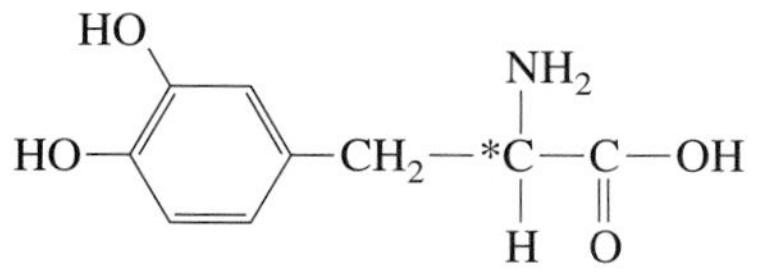

$$(HO)_2C_6H_3-CH_2-{}^{*}\overset{NH_2}{\underset{H}{C}}-\underset{O}{\overset{}{\underset{\|}{C}}}-OH$$

Chapter 11

1. The four fundamental types of materials provided by food are water, energy sources, raw materials, and metabolic regulators.

3. **a.** Among these are the countries of sub-Saharan Africa, many former Soviet Union countries, Afghanistan, and Cuba.
b. The countries of North America—U.S., Canada, and Mexico—grow and/or import sufficient food.

5. There is too much carbohydrate for steak, and too much protein for chocolate chip cookies. The pie chart is likely a representation of peanut butter. (See Figure 11.2 for confirmation.)

10. Although hydrogen atoms are far more abundant than oxygen or carbon atoms, hydrogen has a far smaller mass than either oxygen or carbon.

13. **A.** 6–11 servings of bread, cereal, grain, and pasta
B. 3–5 servings of vegetables
C. 2–4 servings of fruits
D. 2–3 servings of milk, yogurt, and cheese
E. 2–3 servings of meat, poultry, fish, dry beans, eggs, and nuts
F. fats, oils, and sweets used sparingly

16. The structure of glucose is based on a six-member ring of five carbons and one oxygen. The structure of fructose is based on a five-membered ring of four carbons and one oxygen. Glucose has one $-CH_2OH$ side chain, and fructose has two.

20. **a.** Fats and oils are both composed of nonpolar hydrocarbon chains. Edible fats and oils both contain some oxygen. Most fats and oils are triglycerides, which are esters of three fatty acid molecules and one glycerol molecule. Both oils and fats feel greasy and are insoluble in water.
b. Oils tend to contain more highly unsaturated fatty acids and smaller fatty acids than fats. If the triglycerides are solid at room temperature, the material is usually termed a fat. If the triglycerides are liquid at room temperature, the material is usually termed an oil.

27. According to Figure 11.2, peanut butter is 26% protein, which makes it a very good source of protein indeed. However, it is also 52% fat, which is quite high if one needs to limit fat in the diet. It is, however, relatively low in saturated fat if not hydrogenated.

30. It would be unlikely to meet the guidelines. Only 2–3 servings per day are recommended for milk, yogurt, and cheese; 6–11 servings of bread, cereal, grains, and pasta are required, 3–5 servings of vegetables, 2–4 servings of fruit, and 2–3 servings of meat, poultry, fish, dry beans, eggs, and nuts. If 40% came from milk or dairy products, there is not enough room in the diet for the other food groups to be well represented.

Chapter 12

1. DNA stands for **d**eoxyribo**n**ucleic **a**cid.

3. **a.** The base adenine has the amine group, $-NH_2$.
b. The sugar deoxyribose has several —OH groups.
c. There are no functional groups *in* the phosphate, but the phosphate itself is a functional group. Do not mistake the doubly bonded oxygen atom for a ketone, for there is no bond to a carbon.

6. **a.** A nucleotide must contain a base, a deoxyribose molecule, and a phosphate group linked together.
b. Covalent bonding holds the groups together.

8. **a.** A beam of X-rays, which have relatively high energy and relatively short wavelengths, is directed at a target. The X-rays are then diffracted at certain angles, which are related to the distance between atoms.
b. Ions in a salt like sodium chloride have a very regular structure that is easily determined by X-ray studies. Atoms in nucleic acids and proteins do not show the same well-known patterns of crystalline regularity, making the interpretation of the X-ray diffraction pattern more difficult.

13. **a.** This is the general formula for an amino acid, where R represents a side chain that is different in each of the 20 amino acids.

$$\begin{array}{c} \text{H} \\ | \\ \text{H}_2\text{N}-\text{C}-\text{COOH} \\ | \\ \text{R} \end{array}$$

b. The functional groups are —COOH, which is the carboxylic acid group, and the —NH_2 group, the amino group.

15. A codon is a grouping of three DNA bases. Codons are used to signal that a molecule of a certain amino acid should be incorporated into a protein.

18. There is only a minor change in the amino acid composition of human hemoglobin that leads to sickle-cell anemia. In hemoglobin S, two of the residues that should be glutamic acid are replaced with valine. This seemingly innocuous change has rather drastic results for the person with this genetic disease.

21. **a.** PCR stands for **p**olymerase **c**hain **r**eaction.
b. Using DNA for diagnostic probes was not effective if the concentration of the invading DNA is too low. PCR technology made it possible to start with a single segment of DNA and make millions or billions of copies of it in a relatively short time period.
c. Amplification of DNA has made possible early diagnosis of HIV infection, for example, as well as the diagnosis of several genetic diseases.

24. This is the term coined to describe the practice of using domesticated animals to produce drugs and other medically significant substances.

29. The discovery that %A = %T and that %C = %G provided the basis for asking *why* this pattern was observed. Chargaff's contribution was in finding that the bases were paired. Crick and Watson took this information a step further to discover both *how* and *why* they were paired, and the influence the pairing had on the structure of DNA.

42. The statement is false. Calculations show that the DNA in an adult would stretch just about 26,000 times to the moon and back.

$$\frac{2 \text{ m}}{1 \text{ DNA thread}} \times \frac{1 \times 10^{13} \text{ cells}}{1 \text{ adult}} = 2 \times 10^{13} \text{ m of DNA in an adult}$$

$$3.8 \times 10^5 \text{ km} \times \frac{10^3 \text{ m}}{1 \text{ km}} \times 2 = 7.6 \times 10^8 \text{ m, the distance to the moon and back}$$

The ratio of these two distances is

$$\frac{2 \times 10^{13} \text{ m}}{7.6 \times 10^8 \text{ m}} = \frac{26{,}000}{1}$$

Glossary

The numbers indicate the pages where these terms are defined and explained in context.

A

absolute (Kelvin) temperature scale Zero on the Kelvin scale is absolute zero or −273.15°C, the lowest possible temperature *181*

acid a substance that releases hydrogen ions, H^+, usually in aqueous solution *246*

acid anhydride a compound, typically an oxide of a nonmetallic element, that reacts with water to generate an acid *253*

acid deposition the process by which acid is deposited through precipitation, fog, or air-borne particles *251*

acid neutralizing capacity (ANC) the capacity of a lake to resist change in pH when acids are added to it *267*

acid precipitation (*see acid deposition*)

acid rain (*see acid deposition*)

acidic solution an aqueous solution in which the H^+ ion concentration is greater than the OH^- concentration; a solution with a pH less than 7.0 *248*

activation energy the energy necessary to initiate a chemical reaction *163*

active (receptor) site the region of an enzyme molecule where the catalytic activity occurs *406, 491*

addition polymerization a process in which monomeric molecules combine to form a polymer without the elimination of any atoms *367*

aerosol a form of liquid in which the droplets are so small that they stay suspended in the air rather than settling *31*

alkali a term applied to some bases *247*

allotropes two forms of the same element that differ in their molecular or crystal structure and hence in their properties *49*

alpha particle a particle given off during radioactive decay consisting of 2 protons and 2 neutrons; it has a mass of 4 amu and a charge of +2 *300*

amine a basic organic compound with the generic formula RNH_2 *375*

aminFo acid a compound containing a carboxylic acid group (—COOH), a basic amino group ($—NH_2$), and a characteristic R group; amino acids polymerize to form proteins *375, 489*

amino acid residues amino acids that have been incorporated into a peptide chain *454*

amniocentesis the procedure in which a sample of the amniotic fluid is withdrawn from the mother's uterus and tested *498*

ampere a unit of electrical current *331*

anabolic steroid a compound that promotes muscle growth but can have serious negative side effects *417*

androgens male sex hormones *413*

anion a negatively charged ion *213*

anode the electrode at which oxidation occurs *329*

antagonist a molecule that occupies the active site of an enzyme but exhibits no activity *416*

antibody a protective biological agent generated by the body in response to infection *496*

aqueous solutions solutions in which water is the solvent *202*

aquifer a large natural underground reservoir *201*

atmospheric pressure the force with which the atmosphere presses down on a given surface area *13*

atom the smallest unit of an element that can exist as a stable independent entity *19*

atomic mass the mass of an atom expressed relative to a value of exactly 12 for carbon-12 *118, 120*

atomic mass unit (amu) a unit used to express the mass of individual atoms and molecules, equal to 1.66×10^{-24} g *118*

atomic number the number of protons in an atomic nucleus, equal to the number of electrons in an electrically neutral atom *50*

atomic weight (*see atomic mass*)

Avogadro's number the number of objects in one mole, 6.02×10^{23} *120*

B

basal metabolism rate (BMR) the number of Calories necessary to support basic body functions *461*

base a substance that releases hydroxide ions, OH^-, usually in aqueous solution *246*

basic solution an aqueous solution in which the OH^- concentration is greater than the H^+ concentration; a solution with a pH greater than 7.0 *248*

beta particle an electron released during radioactive decay; it has a mass of 1/1838 amu and an electrical charge of −1 *300*

biomass materials produced by biological processes *176*

biotechnology technology based on the manipulation and alteration of biological materials, especially genetic material *496*

bond energy the amount of energy that must be absorbed to break a specific chemical bond, usually expressed in kJ/mole of bonds *158*

breeder reactor a fission reactor that converts U-238 to fissionable Pu-239 while generating energy *299*

C

calibration graph a graph of the absorbances versus the concentrations of several solutions of known concentration *231*

calorie the amount of heat necessary to raise the temperature of exactly 1 gram of water by 1 degree Celsius; 4.184 J *151*

Calorie the amount of heat necessary to raise the temperature of exactly 1 kilogram of water by 1 degree Celsius; 1000 cal; used in nutrition *152, 436*

carbohydrate a compound containing carbon, hydrogen, and oxygen, the latter two in the same 2:1 atom ratio as found in water *443*

carbon cycle the cyclic process by which carbon and its compounds circulate through the animal, vegetable, and mineral kingdoms *115*

carboxylic acid an acidic organic compound with the general formula RCOOH *374, 401*

catalyst a chemical substance that participates in a chemical reaction and influences its speed without undergoing permanent change *81, 370*

cathode the electrode at which reduction occurs *329*

cation a positively charged ion *213*

Chapman cycle the set of four related reactions that represents the natural steady-state formation and destruction of ozone in the stratosphere *71*

Chargaff's rules the generalization that, in DNA from all species, the percent of adenosine equals the percent thymine and the percent guanine equals the percent cytosine *484*

chemical change (*see chemical reaction*)

chemical equation a representation of a chemical reaction using chemical symbols and formulas *23*

chemical formula a representation of the elementary composition of a chemical compound *20*

chemical reaction a process in which substances described as reactants are transformed into different substances called products *23*

chemical symbols one- or two-letter symbols that represent the chemical elements *16*

chiral isomers two forms of a compound, with the same formula and the same number and elementary identity of atoms, whose molecules are nonidentical mirror images of each other; also known as optical isomers *408*

chlorination disinfection of water supplies with chlorine gas, sodium hypochlorite, or calcium hypochlorite *223*

chlorofluorocarbon a compound composed of the elements chlorine, fluorine, and carbon *48, 78*

chromosomes thread-like strands within cell nuclei that are the repository of genetic information in the form of DNA molecules *480*

clone an identical copy of a molecule, cell, or organism *494*

codon a sequence of three nitrogen-containing bases in a DNA molecule that encodes for a specific amino acid during protein synthesis *490*

coenzyme a substance, generally consisting of small molecules, working in conjunction with an enzyme to enhance the enzyme's activity *464*

combustion, burning; the rapid combination of oxygen with a flammable material, accompanied by the evolution of heat energy *23, 154*

complementary bases the DNA base pairs: adenine with thymine and guanine with cytosine *485*

compound a pure substance made up of two or more elements in a fixed, characteristic chemical combination and composition *18*

concentration the ratio of amount of substance (solute) to amount of water (solvent or solution) *205*

condensation polymerization a process in which monomeric molecules combine to form polymers by the elimination of small molecules such as H_2O *373*

conservation of energy, law of (first law of thermodynamics) energy is neither created nor destroyed; the energy of the universe is constant *152*

conservation of matter and mass, law of in a chemical reaction, matter and mass are conserved; the mass of the reactants converted equals the mass of products formed *24*

copolymer a polymer consisting of two or more different monomeric units *373*

covalent bond a chemical bond created when two bonded atoms share electrons *54*

covalent compound a compound consisting of molecules that are in turn made up of covalently bonded atoms; a molecular compound *218*

cracking breaking down of large molecules in petroleum into smaller ones in the gasoline range *172*

current the rate at which electrons flow; measured in amperes *331*

D

daughter the isotope formed by the radioactive decay of the "parent" isotope *301*

density mass per unit volume; usually expressed in grams per cubic centimeter or grams per milliliter *211*

deoxyribonucleic acid (DNA) the compound constituting the genetic material of all living things *481*

desalination any process that removes ions from salty water, such as sea or brackish waters *237*

dipeptide a compound composed of two amino acid units *455*

distillation a purification or separation process in which a solution is heated to the boiling point and the vapors are condensed and collected *169, 237*

DNA fingerprinting the technique of DNA matching that can be used to identify the individual source of a DNA sample *500*

DNA probes relatively short segments of single-stranded DNA used to specifically bind to other DNA *496*

double bond a covalent bond consisting of two pairs of electrons shared between two bonded atoms *57*

double helix description of the molecular structure of deoxyribonucleic acid (DNA) *484*

E

efficiency the fraction of heat energy that is converted to work in a power plant *181*

electrochemical cell (battery) a device that converts the energy released in a spontaneous chemical reaction into electrical energy *329*

electrode an electrical conductor that serves as the site of a chemical reaction in an electrochemical or electrolytic cell *329*

electrolysis the electrical decomposition of a compound into its constituent elements *346*

electrolyte a solution that conducts electricity or a compound that conducts electricity when melted or dissolved in water *213*

electrolytic cell a device in which applied electrical energy is used to bring about a nonspontaneous reaction *329*

electromagnetic spectrum the entire range of radiant energy, including X ray, gamma, ultraviolet, visible, infrared, microwave, and radio wave radiation *60*

electron a subatomic particle with a mass of 1/1838 amu and a charge of −1 unit that is of great importance in atomic structure and chemical reactivity *50*

electronegativity a measure of the attraction of an atom for the electrons that constitute its covalent bond *208*

electrophoresis a method of separating molecules based on their rate of movement in an electric field; the speed at which a molecule travels depends on its size (mass) and electric charge *500*

element a substance that cannot be broken down into simpler stuff by any chemical means *15*

endothermic absorbing heat *157*

energy the capacity to do work *151*

entropy a measure of randomness in position or energy *184*

enzyme a biochemical catalyst; a protein that influences the rate and direction of a chemical reaction *491*

essential amino acid an amino acid that cannot be synthesized by the body and must be supplied in the diet *456*

essential fatty acid a fatty acid that cannot be synthesized by the body and must be supplied in the diet *450*

ester an organic compound with the general formula RCOOR′; formed by the condensation reaction of a carboxylic acid and an alcohol *374*

estrogens female sex hormones *413*

exothermic releasing heat *154*

exposure the amount of a substance encountered *9*

F

fat a triglyceride; a compound made from fatty acids and glycerol *446*

fatty acid an acidic compound with a long hydrocarbon chain; a component of fats and oils *446*

first law of thermodynamics (law of conservation of energy) energy is neither created nor destroyed; the energy of the universe is constant *152*

fission (nuclear) a reaction in which a large atomic nucleus, such as uranium-235, splits when struck by a neutron to form two smaller fragments, and releases neutrons and large quantities of energy *288*

fraction a component separated from bulk crude oil (petroleum) by fractional distillation based on the fraction's boiling point *169*

free radical an unstable chemical species with an unpaired electron *76*

frequency in wave motion, the number of waves passing a fixed point in one second *58*

fuel cell a cell in which a fuel such as hydrogen is allowed to react with oxygen under controlled conditions and the chemical energy is converted to electricity *334*

functional groups groupings of atoms that confer characteristic properties on the molecule and the compound *373, 401*

G

galvanic cell a device that converts the energy released in a spontaneous chemical reaction into electrical energy *329*

gamma rays short-wavelength, high-energy electromagnetic radiation released during radioactive decay *160, 301*

gas chromatography (GC) an analytical method that uses the differential absorption of components in a mixture carried by a gas as they move down a packed column; a detector indicates the emergence of each component from the mixture as it leaves the column *233*

gene therapy the introduction of normal genes into patients lacking them *499*

generic drug a drug that is equivalent to a pioneer drug but not able to be marketed until the patent protection of the pioneer drug runs out (20 years) *424*

genetic engineering the manipulation and alteration of genetic material (DNA) for a wide variety of purposes *495*

global greenhouse effect the return of 84% of the energy radiated from the surface of the Earth *103*

green chemistry designing chemical products and processes that reduce or eliminate the use and/or generation of hazardous substances *30*

greenhouse effect the process by which atmospheric gases such as CO_2, CH_4, and H_2O trap and return a major portion of the heat (infrared radiation) radiated by the Earth *103*

greenhouse factor a number that represents the relative contribution of a compound to global warming *126*

green revolution the development of high-yield grains through genetic modification *470*

groundwater water pumped from wells that have been drilled into aquifers *201*

H

half-life the time required for the level of radioactivity to fall to one-half its initial value *307*

hard water water containing a significant concentration of magnesium or calcium ions *225*

heat the form of energy that flows from a hotter to a colder body *151*

heat of combustion the quantity of heat released when a fuel is burned; variously expressed in cal/g, cal/mole, J/g, or J/mole *156*

heavy metal a member of a rather ill-defined group of metallic elements with high densities and large atomic masses *228*

high-density lipoprotein (HDL) a combination of lipid and protein that transports cholesterol from dead or dying cells back to the liver; the HDL's density depends on the ratio of lipid to protein *452*

high-level nuclear waste (HLW) waste typically from spent fuel taken from commercial nuclear reactors and from nuclear weapons production *309*

hormones substances produced by the body's endocrine glands that can have a wide range of physiological functions, including serving as "chemical messengers" *403*

human genome the totality of human genetic information *507*

Human Genome Project an international effort to map all the genes in the human organism and determine the DNA base sequence *507*

hybrid car an automobile that uses a gasoline engine and an electric motor in combination as an alternative method of propulsion *343*

hydrocarbon a compound of hydrogen and carbon *26, 168*

hydrogen bond a relatively weak electrostatic attraction between a hydrogen atom bearing a net positive charge and lone pair electrons on a nitrogen, oxygen, or fluorine atom bearing a net negative charge; hydrogen bonds exist between some molecules and within some large molecules *209*

hydronium ion the H_3O^+ ion responsible for acidic properties in solution *246*

I

infrared radiation heat radiation; the region of the electromagnetic spectrum adjacent to the red end of the visible spectrum and characterized by wavelengths longer than red light *60*

interferons naturally occurring protein-based molecules that provide protection against viruses *495*

ion an electrically charged atom or a group of covalently bonded atoms; can be positive (cation) or negative (anion) *213*

ion exchange a process in which ions are interchanged, usually between a solution and a solid *228*

ionic bond a chemical bond created by the electrostatic attraction between oppositely charged ions *213*

ionic compound a compound consisting of positively and negatively charged ions *213*

isomers different compounds with the same formula and the same number and kinds of atoms; isomers differ in molecular structure, that is, the way in which the constituent atoms are arranged *172, 398*

isotopes two (or more) forms of the same element whose atoms differ in number of neutrons and therefore in mass number and atomic mass *53*

J

joule a unit of energy corresponding to $kg \cdot m^2/s^2$ *62, 151*

K

Kelvin scale the absolute temperature scale whose zero corresponds to −273.15°C, the lowest possible temperature *181*

kilocalorie, kcal an energy unit equal to 10^3 calories or 1 nutritional Calorie (Cal) *460*

L

lethal dose-50 (LD_{50}) the minimum dose required to kill 50% of test animals *421*

Lewis structure a representation of molecular structure based on the octet rule that uses dots to represent electrons *54*

lipids fats, oils, and related compounds *446*

liter, L a unit of volume; 1 L = 10^3 mL *205*

low-density lipoprotein (LDL) a combination of lipid and protein that transports cholesterol from the liver to peripheral tissues; the LDL's density depends on the ratio of lipid to protein *452*

low-level radioactive waste (LLW) nuclear waste contaminated with relatively small quantities of radioactive materials; other than high-level nuclear waste *314*

M

macrominerals minerals such as calcium, phosphorus, chlorine, potassium, sulfur, sodium, and magnesium, necessary for life. The adult Recommended Daily Allowances for macrominerals is from one to two grams *465*

macromolecule a molecule with large molecular size and a high molar mass; term often applied to polymers *363*

macronutrients the major classes of compounds required for nutrition: carbohydrates, fats, and proteins *438*

malnutrition a condition caused by a diet lacking in the proper mix of nutrients, even though enough Calories are eaten daily *438*

mass a measure of the quantity of matter in a body, often expressed in grams or kilograms, and measured by weighing the object with a balance *19*

mass number the sum of the number of protons and the number of neutrons in any atomic nucleus *53, 289*

maximum contaminant level (MCL) the legal limit of a contaminant, expressed in ppm or ppb *221*

maximum contaminant level goal (MCLG) the level, in ppm or ppb, at which a person weighing 70 kg (154 lb) could drink 2 L of water containing the contaminant every day for 70 years without suffering any ill effects *221*

mesosphere the region of the atmosphere above an altitude of 50 kilometers *13*

microminerals the microminerals are zinc, copper, iron, and fluorine *465*

microwave radiation electromagnetic radiation with relatively long wavelengths, low frequencies, and low-energy photons; stimulates molecular rotations; used in microwave ovens and in radar *114*

millirem (mrem) one thousandth of a rem *305*

minerals in general, inorganic chemical substances; more specifically, the inorganic chemical substances required for healthy nutrition and body function; classified as macrominerals, microminerals, and trace minerals *465*

mixture a physical combination of two or more substances (elements or compounds) present in variable amounts *15*

molar mass the mass of one Avogadro's number (1 mole) of atoms, molecules, or whatever particles are specified; usually expressed in grams *121*

molarity, M the number of moles of solute present in 1 L of solution *206*

mole one Avogadro's number of anything; 6.02×10^{23} atoms, molecules, electrons, etc. *121*

molecular compound a compound consisting of molecules; a covalent compound *218*

molecular mass the mass of a molecule expressed relative to a value of exactly 12 for carbon-12 *121*

molecular pharming the practice of using domesticated animals to produce drugs and other medically significant substances *505*

molecularweight (*see molecular mass*)

molecule a combination of a fixed number of atoms, held together by chemical bonds in a certain geometric arrangement *20*

monomer a small molecule that combines with other monomers to yield a polymer *363*

monounsaturated having one carbon-carbon double bond per molecule; usually applied to fats or fatty acids *447*

N

n-type semiconductor a material that will not normally conduct electricity well, but will do so under certain conditions through the movement of electrons *350*

neutral solution an aqueous solution containing equal concentrations of H^+ and OH^- ions; a solution with a pH of 7.0 *249*

neutralization the chemical reaction of an acid and a base *247*

neutron a subatomic particle with a mass of 1 amu and no electrical charge *50*

nitrogen balance a condition in which the body excretes as much nitrogen as it ingests *456*

nonelectrolyte a substance that does not conduct electricity, either melted or in solution *212*

nonspontaneous a process that will not occur by itself, but only if energy is supplied from some external source *184*

nuclear transfer the use of a very thin, hollow needle to remove a cell's nucleus to transfer it to an unfertilized egg from which the nucleus has been removed *506*

nucleotide the repeating unit of DNA, consisting of a nitrogen-containing base, a deoxyribose sugar, and a phosphate group *482*

nucleus (atomic) the center of an atom *50*

O

octet rule a generalization that, in most stable molecules, all atoms except hydrogen will share in eight outer electrons *55*

oils triglycerides that are liquid at room temperature *446*

optical isomers (*see chiral isomers*)

osmosis the natural tendency for a solvent to move through a membrane from a region of higher solvent concentration to a region of lower solvent concentration *237*

outer electrons the electrons in the outer energy level of an atom; the outer electrons are chiefly responsible for the chemical properties of that particular element *51*

oxidation a process in which an atom, ion, or molecule loses one or more electrons *329*

oxygenated gasoline gasoline blended with oxygen-containing compounds such as MTBE, ethanol, or methanol *174*

P

p-type semiconductor a material that will not normally conduct electricity well, but will do so under certain conditions through the movement of positively charged "holes" *350*

parent the isotope undergoing radioactive decay *301*

parts per million (ppm) a measure of concentration that can be expressed in units of mass or in numbers of atoms, molecules, and/or ions *6*

peptide bond the molecular linkage bonding amino acid residues in proteins and monomers in nylon *375*

periodic table an organization of the elements in order of increasing atomic number and grouped according to similar chemical properties and similar electron arrangements *16*

pH a number, typically between 0 and 14, that indicates the acidity of a solution; also, the negative logarithm of the hydrogen ion concentration when the concentration is expressed as moles of H^+ ion per liter of solution; $pH = -\log(M_{H^+})$ *249, App. 3*

phenyl group a common molecular fragment based on a hexagon of six carbon atoms (the benzene ring); $—C_6H_5$ *369*

photon a "particle" of radiant energy *62*

photosynthesis the process by which green plants use sunlight to power the conversion of carbon dioxide and water into sugars, such as glucose, plus oxygen *100*

photovoltaic cell a device that converts radiant energy into electricity *349*

pioneer drug the first version of a drug to be marketed under a brand name *424*

Planck's constant the proportionality constant relating the energy of a photon to the frequency of radiation; 6.63×10^{-34} J·s *62*

plasmid a ring of bacterial DNA; used as a vector for introducing new genes in recombinant DNA research and technology *494*

polar covalent bond a covalent bond in which the electrons are not equally shared but displaced toward the more electronegative atom *208*

polyamide a polymer formed by the condensation reaction of amino acids (in which case the polyamide is a protein) or of diacids and diamines (in which case the polyamide is nylon) *375*

polyatomic ion a group of covalently bonded atoms bearing a positive or negative electrical charge *215*

polyester a polymer formed by the condensation reaction of diacid and dialcohol monomers *374*

polymer a substance consisting of long macromolecular chains *363*

polymerase chain reaction (PCR) a technique for rapidly making many copies of a DNA segment *496*

polyunsaturated having more than one carbon-carbon double bond per molecule; usually applied to fats and oils *447*

potable water fit for human consumption *223*

potential energy on a molecular scale, the energy related to the positions of atoms and molecules *54*

prescription drugs drugs that could be habit forming, toxic, or unsafe for use except under medical supervision *420*

primary structure the identity and sequence of the amino acid residues present in a protein molecule *491*

primers single-stranded nucleotides that bracket and identify the section of DNA to be copied in the polymerase chain reaction *496*

products the substances formed from reactants as a result of a chemical reaction *23*

prostaglandins a group of hormone-like compounds produced by the body where they cause a variety of responses including fever, swelling, and pain; inhibited by aspirin *403*

proteins polymers made up of various amino acids as the monomeric units; essential components of the body and the diet *453, 489*

proton a subatomic particle with a mass of 1 amu and a charge of +1 *50*

Q

quantized separated into discrete energy levels, as, for example, the electronic energy levels in an atom *62*

R

racemic mixture a mixture consisting of equal amounts of each optical isomer *409*

radiation absorbed dose (rad) the absorption of 0.01 joule of radiant energy per kilogram of tissue *304*

radioactivity the phenomenon in which certain unstable atomic nuclei emit radiation and thereby undergo nuclear transformation *300*

reactants the starting materials in a chemical reaction that are transformed into products during the reaction *23*

receptor site a site on a cell or molecule where a hormone or other biologically active molecule can bind *406*

recombinant DNA DNA that has incorporated into it DNA from another organism *494*

reduction a process in which an atom, ion, or molecule gains one or more electrons *329*

reformulated gasolines (RFG) oxygenated gasolines that contain a lower percentage of certain volatile hydrocarbons *175*

rem (roentgen equivalent mammal) a unit of radiation, rem = number of rads × Q, a quality factor characteristic of the type of radiation *304*

replication the process by which copies of DNA are made *486*

resonance possible structures of a molecule for which more than one Lewis structure can be written, differing by the arrangement of valence electrons but having the same arrangement of atomic nuclei *57*

reverse osmosis a process for water purification in which water is forced through a membrane, and ions and other contaminants are filtered out *237*

risk assessment the process of analyzing and balancing the risks and benefits associated with some particular course of action *9*

S

saturated having only carbon-carbon single bonds; usually applied to fats and oils *447*

scientific notation a system for writing numbers as the product of a number, usually with one digit to the left of the decimal point, and 10 raised to the appropriate power or exponent, for example,

6.02×10^{23} *10*

second law of thermodynamics it is impossible to completely convert heat into work without making some other changes in the universe; heat will not of itself flow from a colder to a hotter body; the entropy of the universe is increasing *184*

secondary structure helices, parallel chains, and other localized structural features in the overall structure of a protein molecule *491*

semiconductors materials that do not normally conduct electricity well but will do so under certain conditions *349*

Sievert, Sv a unit of radiation dosage equal to 100 rem *304*

significant figures the number of numerals that correctly represents the accuracy with which an experimental quantity is known *39*

singlebond a covalent bond consisting of one pair of electrons shared between two bonded atoms *54*

softwater water that contains low concentrations of magnesium, calcium, or iron ions *225*

solute a component that dissolves in a solvent to form a solution *202*

solution a homogenous mixture at the atomic, molecular, and/or ionic level, consisting of a solute (or solutes) dissolved in a solvent *202*

solvent in a solution, the component present in the largest concentration, usually the liquid component in which the solute dissolves *202*

source reduction decreasing the amount of plastic waste generated by reducing the quantity of plastics produced and used *380, 385*

specific heat the quantity of heat energy that must be absorbed to increase the temperature of 1 gram of a substance by 1°Celsius *211*

spectrophotometer an instrument in which light of a desired wavelength is passed through a sample into a detector where the light is converted into an electrical signal *230*

spent fuel material remaining in fuel rods after they have been removed from a nuclear reactor *299, 310*

spontaneous a process that can occur by itself, though it may be necessary to initiate the reaction *184*

steady state a condition in which a dynamic system is in balance so that there is no net change in the concentration of the major participants in the reactions *71*

steroids a class of ubiquitous and diverse organic compounds that contain three six-membered carbon rings and one five-membered ring *410*

storage battery batteries that store electrical energy, such as a lead-acid storage battery *333*

stratosphere the region of the atmosphere between 15 and 50 kilometers above sea level; location of the ozone layer *13*

substrate the species on which an enzyme acts *491*

surface water lakes, rivers, and reservoirs *201*

T

temperature a property that determines the direction of heat flow; when two bodies are in contact, heat always flows from the object at the higher temperature to that at the lower temperature *151*

tertiary structure the overall three-dimensional structure of a protein molecule *491*

tetrahedron a regular figure with four identical sides, each one an equilateral triangle; the four corners of the tetrahedron correspond to the locus of the four electron pairs (bonding and/or nonbonding) around an atom that obeys the octet rule *107*

thermal energy energy characterized by the random motion of molecules *183*

thermoplastic polymers that tend to be flexible and can be melted and shaped *366*

titration the reaction of a reagent with a measured volume of a solution of known concentration *226*

toxicity the intrinsic hazard of a substance *9*

trace minerals dietary minerals generally needed in microgram quantities; iodine, selenium, vanadium, chromium, manganese, cobalt, nickel, molybdenum, and tin *465*

transgenic organism an organism having the genes of more than one species *503*

triglyceride an ester composed of three fatty acids and glycerol; fats and oils are typically triglycerides *446*

triple bond a covalent bond consisting of three pairs of electrons shared between two bonded atoms *57*

troposphere the part of the atmosphere that lies on the surface of the Earth *12*

U

ultraviolet the region of the electromagnetic spectrum adjacent to the violet end of the visible spectrum and characterized by wavelengths shorter than violet light *60*

undernourishment a condition in which the daily intake of food is insufficient to supply the body's energy requirements *438*

unsaturated having at least one carbon-carbon double bond per molecule; often applied to fats and oils *447*

V

vaccine a biological agent that produces or increases immunity to a particular disease *496*

vector a molecular or cellular component, for example a plasmid, used to import foreign DNA into a host cell *494*

virus a biochemical species consisting of nucleic acid and protein that can be replicated in a host cell, where it often causes disease *495*

vitamins organic compounds that serve a variety of functions essential to life, often by promoting metabolic processes *464*

volt a unit of electrical potential *329*

voltage a measure of difference in electrical potential *329*

W

wavelength in wave motion, the distance between successive peaks *58*

weight a measure of the attraction of gravity on an object, proportional to mass *19*

work work is done when movement occurs against a restraining force; equal to the force multiplied by the distance over which the motion occurs *151*

X

X ray electromagnetic radiation with short wavelengths, high frequencies, and high-energy photons; used in medical diagnosis and therapy, and in determining crystal structures; can damage biological tissue *60*

X-ray diffraction a procedure for determining crystal and molecular structure by interpreting the pattern formed when X rays are scattered by the constituent atoms *484*

Z

zeolite a claylike mineral made up of aluminum, silicon, and oxygen; often used as a water softener or catalyst *228*

Credits

Text and Line Art

CHAPTER 2

FIGURE 2.5 From *Chemistry: The Molecular Science*, 1st edition, by J. W. Moore, C. L. Stanitski, P. C. Jurs © 2002. Reprinted with permission of Brooks/Cole, an imprint of the Wadsworth Group, a division of Thomson Learning. Fax (800) 730-2215. **FIGURE 2.6** From *An Introduction to Solar Radiation* by Muhammad Iqbal, copyright 1983, Elsevier Science (USA), reproduced by permission of the publisher. **FIGURE 2.9** Reprinted by permission of John E. Frederick, University of Chicago. **FIGURE 2.11** From *Scientific American*, Vol. 275, No. 1, July 1996. Copyright © Laurie Grace. Reprinted by permission. **FIGURE 2.21** From R. D. Sojkov, *The Changing Ozone Layer*. Copyright © 1995, World Meteorological Association. Reprinted by permission.

CHAPTER 3

CARTOON, p. 106 From the *Wall Street Journal*, Saturday, August 10, 2001. **FIGURE 3.15** Courtesy of Jeffrey A. Draves, University of Central Arkansas. **FIGURE 3.18** Courtesy of Conrad Stanitski, Univ. of Central Arkansas. **FIGURE 3.22** Copyright © 2001 by Myles Allen. Reprinted by permission of Myles Allen, Oxford University.

CHAPTER 4

FIGURE 4.5 From *Chemistry: The Central Science*, 5/E by Brown/LeMay/ Bursten. © 1985. Reprinted by permission of Prentice-Hall, Inc., Upper Saddle River, NJ. **TABLE 4.2** From J.W. Moore and E.A. Moore, *Environmental Chemistry*, p. 94. Academic Press, 1976. Used with permission of the authors. **FIGURE 4.12** From *The Chemical World: Concepts and Applications*, Second edition by Moore, Stanitski, et al., copyright © 1998 Saunders College Publishing, reproduced by permission of the publisher. **FIGURE 4.15** Reprinted with permission from *Chemistry in the Community* (ChemCom), 1988. Copyright © 1988 American Chemical Society. **FIGURE 4.17** From *The Chemical World: Concepts and Applications*, Second edition by Moore, Stanitski, et al., copyright © 1998 Saunders College Publishing, reproduced by permission of the publisher. **FIGURE 4.22** From *Chemistry: Imagination and Implication* by A. Truman Schwartz, copyright © 1973 by Harcourt, Inc., reproduced by permission of the publisher.

CHAPTER 5

FIGURE 5.2 DENNIS THE MENACE ® used by permission of Hank Ketcham Enterprises and © by North America Syndicate. **FIGURE 5.12** From Jacqueline I. Kroschwitz et al,

Chemistry: A First Course. Copyright © 1995 The McGraw-Hill Companies. All Rights Reserved. Reprinted by permission. **FIGURE 5.14** From Jacqueline I. Kroschwitz et al, *Chemistry: A First Course*. Copyright © 1995 The McGraw-Hill Companies. All Rights Reserved. Reprinted by permission. **FIGURE 5.20** Courtesy of Conrad Stanitski, Univ. of Central Arkansas. **FIGURE 5.27** © International Bottled Water Association. Reprinted by permission. **FIGURE 5.28** From World Resources 2000-01: *People and Ecosystems* ISBN 1-56973-443-7, p. 111, as adapted by *Scientific American* (http://www.sciam.com/1197issue/1197scicit5.html). Reprinted by permission of World Resources Institute.

CHAPTER 7

FIGURE 7.5 From Martin S. Silberberg, *Chemistry: The Molecular Nature of Matter and Change*, 2nd Edition. Copyright © 1997. Reproduced with permission of The McGraw-Hill Companies. **FIGURE 7.7** From Northern States Power Nuclear Plant, Monticello, MN. Reprinted by permission. **FIGURE 7.14** From M. Silberberg, *The Molecular Nature of Matters & Change*, 2nd edition. Copyright © 1997 McGraw-Hill Companies. All Rights Reserved. Reprinted by permission. **TABLE 7.4** Reprinted with permission from *Chemistry in the Community* (ChemCom). Copyright © 1988 American Chemical Society. **FIGURE 7.16** From *The Nuclear Waste Primer*. Copyright © 1985 The Lyons Press. Reprinted by permission. **FIGURE 7.21** Reprinted with permission from *Disposition of High-Level Waste and Spent Nuclear Fuel*. Copyright 2001 by the National Academy of Sciences. Courtesy of the National Academy Press, Washington, DC

CHAPTER 8

FIGURE 8.6 From R. Chang, *Chemistry*, 6th edition. Copyright © 1998 The McGraw-Hill Companies. Reproduced with the permission of The McGraw-Hill Companies.

CHAPTER 12

FIGURE 12.1 From *The Enzymes: Peptide Bond Hydrolysis*, Third Edition, Volume 3, edited by Paul Boyer, copyright 1971, Elsevier Science (USA), reproduced by permission of the publisher. **FIGURE 12.2** From Robert H. Tamarind, *Principles of Genetics*, 4th edition. Copyright © 1993 McGraw-Hill Companies. All Rights Reserved. Reproduced with permission of The McGraw-Hill Companies. **FIGURE 12.6** From *Chemistry: Imagination and Implication* by A. Truman Schwartz, copyright © 1973 by Harcourt, Inc., reproduced by permission of the publisher. **FIGURE 12.7** From Sylvia S. Mader, *Biology*, 6th edition. Copyright © 1998 The McGraw-Hill Companies. All Rights Reserved. Reproduced with permission of The McGraw-Hill Companies. **FIGURE 12.8** From Sylvia S. Mader, *Biology*, 6th edition. Copyright © 1998 The McGraw-Hill Companies. All Rights Reserved. Reproduced with permission of The McGraw-Hill Companies. **FIGURE 12.12** From Edward Alcamo, *DNA Technology: The Awesome Skill*. Copyright © 1996 The McGraw-Hill Companies. All Rights Reserved. Reproduced with the permission of The McGraw-Hill Companies. **TABLE 12.2** From Edward Alcamo, *DNA Technology: The Awesome Skill*. Copyright © 1996 The McGraw-Hill Companies. All Rights Reserved. Reproduced with the permission of The McGraw-Hill Companies. **12.24 CONSIDER THIS** From *The Sun Shines Bright* by Isaac Asimov. Copyright © 1981 Nightfall, Inc. Used by permission of Doubleday, a division of Random House, Inc. and Ralph Vicinanca Agency. **INSIDE FRONT AND BACK COVERS** Source: Raymond Chang, *General Chemistry: The Essential Concepts*, Third Edition, Copyright 2003 The McGraw-Hill Companies, New York, NY.

Photographs

CHAPTER 1

OPENER: © STONE/Getty Images; **FIGURE 1.1:** Image provided by ORBIMAGE. @ Orbital Imaging Corporation and processing by NASA Goddard Space Flight Center; **FIGURE 1.2:** © Shelley Gazin/The Image Works; **FIGURE 1.5:** © NASA; **FIGURE 1.9:**

Courtesy IBM, Almaden Research Center; **FIGURE 1.10:** © The McGraw-Hill Companies, Inc./Photo by Bob Coyle; **FIGURE 1.11:** Maryland Department of the Environment; **FIGURE 1.14:** © Courtesy, Corning, Inc.; **FIGURE 1.16:** © The McGraw-Hill Companies, Inc./Photo by Ken Karp; **p. 39:** © Sheila Terry/Photo Researchers; **FIGURE 1.17:** © David M. Grossman/Photo Researchers.

CHAPTER 2

OPENER: National Oceanic and Atmospheric Administration; **FIGURE 2.1 (all):** © The McGraw-Hill Companies, Inc./Photo by Stephen Frisch; **FIGURE 2.3:** © Philip Schermeister/National Geographic Image Collection; **FIGURE 2.13:** Courtesy Blue Lizard Products; **FIGURE 2.19 (all):** © The McGraw-Hill Companies, Inc./Photo by Stephen Frisch; **p. 82:** Courtesy, Dr. Susan Solomon. Photo by Carlye Calvin; **FIGURE 2.22:** National Oceanic and Atmospheric Administration; **FIGURE 2.25:** © Courtesy Pyrocool Technologies, Inc.

CHAPTER 3

OPENER: © John Maier, Jr./The Image Works; **FIGURE 3.1:** NASA; **FIGURE 3.2:** © Maria Stenzel/National Geographic Image Collection; **p. 107:** © David Young-Wolff/PhotoEdit; **p. 115:** Reproduced from Chang, Raymond. (2002) CHEMISTRY, 7th edition. © The McGraw-Hill Companies, Inc.; **FIGURE 3.18:** Courtesy Conrad Stanitski; **FIGURE 3.19:** U.S. Department of Energy; **FIGURE 3.23 (bottom):** © Michele Burgess/Stock Boston; **FIGURE 3.23 (top):** © Ira Kirschenbaum/Stock Boston; **FIGURE 3.27:** © NASA.

CHAPTER 4

OPENER: © Zephyr Picture/ Index Stock Imagery/PictureQuest; **FIGURE 4.1:** © Charles D. Winters; **FIGURE 4.7:** Courtesy DaimlerChrysler Corporation; **FIGURE 4.9:** © Bettmann/Corbis Images; **FIGURE 4.10:** © Charles Winters/Photo Researchers; **FIGURE 4.14:** © Werner H. Muller/Peter Arnold; **FIGURE 4.18:** © Michael Newman/PhotoEdit; **FIGURE 4.19:** © AP/Wide World Photos; **p. 175:** © David Young-Wolff/PhotoEdit; **p. 177:** Courtesy Wilmer Stratton; **FIGURE 4.20:** © Bob Daemmrich/Stock Boston; **FIGURE 4.21:** Courtesy A. Truman Schwartz; **FIGURE 4.26:** © H. Schwarzbach/Peter Arnold; **FIGURE 4.28:** © AP/Toyota/Wide World Photos.

CHAPTER 5

OPENER: © Network Productions/The Image Works; **page 198:** Madison, Wisconsin Water Utility; **FIGURE 5.1:** © David Young-Wolff/PhotoEdit; **p. 199:** © Michael Newman/PhotoEdit; **FIGURE 5.3:** © D. Cavagnaro/Visuals Unlimited; **FIGURE 5.11a–c:** © Tom Pantages; **FIGURE 5.18:** © Jerry Mason/SPL/Photo Researchers; **p. 224:** © R.W. Jones/Corbis Images; **FIGURE 5.19:** Courtesy Betz Corporation; **FIGURE 5.20 (both):** Courtesy Conrad Stanitski; **FIGURE 5.22:** © The McGraw-Hill Companies, Inc./Photo by Bob Coyle; **FIGURE 5.31:** © PUR " Drinking Water Systems," a division of Recovery Engineering, Inc.

CHAPTER 6

OPENER: © Rafael Macia/Photo Researchers; **FIGURE 6.1:** © Ted Spiegel/Corbis Images; **FIGURE 6.2:** © PhotoDisc/Vol. #18; **FIGURES 6.3, 6.4, 6.5:** © The McGraw-Hill Companies, Inc./C.P. Hammond, photographer; **FIGURE 6.10:** © E.R. Degginger/Color-Pic; **p. 257 (top):** © Kennon Cooke/Valan Photos; **p. 257 (bottom):** © Stock Portfolio/ Stock Connection/ PictureQuest; **FIGURE 6.15 (left):** © NYC Parks Photo Archive/Fundamental Photographs; **FIGURE 6.15 (right):** © Kristen Brochmann/Fundamental Photographs; **FIGURE 6.16:** © A.J. Copley/Visuals Unlimited; **FIGURE 6.17 (both):** U.S. National Park Service; **FIGURE 6.18:** © Pittsburgh Post Gazette Archives, 2002. All rights reserved. Reprinted with permission; **p. 269:** © M. Edwards/Peter Arnold; **FIGURES 6.20, 6.21:** Wil Stratton; **p. 276:** © Mark Gibson/Visuals Unlimited; **p. 280:** Courtesy Cathy Middlecamp.

CHAPTER 7

OPENER: Courtesy, Nuclear Energy Institute; **FIGURE 7.1:** © The McGraw-Hill Companies, Inc./C.P. Hammond, photographer; **p. 285:** © Cindy Charles/PhotoEdit/PictureQuest; **FIGURE 7.2:** Courtesy, Nuclear Energy Institute; **FIGURE 7.3:** © Rob Crandall/The Image Works; **FIGURE 7.4:** Photo by Heka, courtesy AIP Emilio Segrè Visual Archives, Physics Today Collection; **FIGURE 7.8:** © Joe Sohm/The Image Works; **FIGURE 7.10:** © Igor Kostin/Corbis-Sygma; **FIGURE 7.11 (both):** EC/IGCE, Roshydromet (Russia)/Minchernobyl (Ukraine)/Belhydromet (Belarus); **FIGURE 7.12:** © Bettmann/Corbis Images; **FIGURE 7.13:** Courtesy USEC, Inc.; **p. 300 (top):** © AP/Wide World Photos; **p. 300 (bottom):** © AIP Emilio Segre Visual Archives/W.F. Meggers Collection; **FIGURE 7.18:** © Southern Illinois University/Photo Researchers; **FIGURE 7.19:** © Vladimir Siomin; **FIGURE 7.20:** © Visuals Unlimited; **FIGURE 7.22:** © Science Source/Photo Researchers; **FIGURES 7.23, 7.24:** © U.S. Dept. of Energy/SPL/Photo Researchers.

CHAPTER 8

OPENER: © PhotoDisc/Vol. #73; **FIGURE 8.1:** © Courtesy, BLACK & DECKER; **FIGURE 8.3:** © Tony Freeman/PhotoEdit; **FIGURES 8.8, 8.9:** Courtesy, Ballard Power Systems; **FIGURE 8.11:** Courtesy of Georgetown University; **FIGURE 8.12:** © Reuters/Wolfgang Rattay/Corbis Images; **FIGURE 8.13:** © AP/Wide World Photos; **p. 342:** © Brett Dewey; **FIGURE 8.14 (both):** © AP/Wide World Photos; **FIGURE 8.17:** Courtesy A. Truman Schwartz; **p. 349:** © Tony Freeman/PhotoEdit; **p. 350:** © Westinghouse/Visuals Unlimited; **FIGURE 8.23:** © Ken Lucas/Visuals Unlimited; **FIGURE 8.24:** © A.J. Copley/Visuals Unlimited; **FIGURE 8.25:** © AP/Wide World Photos; **FIGURE 8.26:** © AP/Tribune-Star/Wide World Photos.

CHAPTER 9

OPENER: © Corbis/Vol. #49; **FIGURE 9.1:** © Hank deVre/Mountain Stock; **FIGURE 9.4a:** © Bill Aron/PhotoEdit; **FIGURES 9.7, 9.8:** Courtesy Dupont; **FIGURE 9.12:** © The Garbage Project, University of Arizona; **FIGURE 9.13a:** © Courtesy of ACS, photo by Mike Ciesielski; **FIGURE 9.13b:** © Gayna Hoffman/Stock Boston; **FIGURE 9.15:** © DaimlerChrysler Corporation; **FIGURE 9.16:** Courtesy, Automotive Design and Composites, Ltd.; **p. 389:** © Martha G. Clarke.

CHAPTER 10

OPENER (top left): © Bushnell/Soifer/Getty Images; **(top right):** © Michael Newman/PhotoEdit; **(bottom):** © Quill/Getty Images; **FIGURE 10.1:** © Terry Wild Studio; **FIGURE 10.3 (both):** © The McGraw-Hill Companies, Inc./Bob Coyle, photographer; **FIGURE 10.6:** © Felicia Martinez/PhotoEdit; **FIGURE 10.9:** Courtesy of The Alexander Fleming Laboratory Museum, St. Mary's Hospital, Paddington, London; **FIGURE 10.13:** Courtesy Stephen J. Cato, Chemical Design; **p. 413:** © PhotoDisc Website; **FIGURE 10.24:** © Michael Newman/PhotoEdit; **p. 428 (both):** © Michael P. Gadomski/Photo Researchers.

CHAPTER 11

OPENER: © Michael Newman/PhotoEdit; **FIGURE 11.8 (left):** © Lund/Custom Medical Stock Photo; **FIGURE 11.8 (right):** © Roseman/Custom Medical Stock Photo; **FIGURE 11.10:** © Len Lessin/Peter Arnold; **FIGURE 11.14:** © Tony Freeman/PhotoEdit.

CHAPTER 12

OPENER: © Corbis/Vol. #123; **p. 478:** © Corbis/Vol. #102; **p. 484:** © Oesper Collection in the History of Chemistry; **FIGURE 12.5:** © Courtesy Biophysics Department, King's College London; **p. 485:** © Bettmann/Corbis Images; **FIGURE 12.11 (both):** © Bill Longcore/Photo Researchers; **p. 493:** © Michael Nicholson/Corbis Images; **FIGURE 12.14:** Courtesy Dr. Ken Culver, photo by John Crawford, National Institutes of Health; **FIGURE 12.15:** Courtesy Lifecodes Corporation; **p. 506:** © APTV via Media/AP/Wide World Photos.

Index

References followed by *t* and *f* refer to tables and figures, respectively.

B

C

F

G

H

I

J

K

L

O

P

S

T

U

V

W

X

Y

Z